The Social Organization of Work

Third Edition

TERESA A. SULLIVAN

Ohio State University

RANDY HODSON

The University of Texas at Austin

WADSWORTH
™
THOMSON LEARNING

Australia • Canada • Mexico • Singapore • Spain
United Kingdom • United States

WADSWORTH

THOMSON LEARNING

Sociology Editor: Lin Marshall
Assistant Editor: Analie Barnett
Editorial Assistant: Reilly O'Neal
Marketing Manager: Matthew Wright
Project Manager, Editorial Production: Jerilyn Emori
Print/Media Buyer: Tandra Jorgensen
Permissions Editor: Joohee Lee

Production Service: Peggy Francomb / Shepherd, Inc.
Copy Editor: Carey Lange
Cover Designer: Hiroko Chastain / Cuttriss
  & Hambleton
Cover Images: PhotoDisc
Text and Cover Printer: Webcom, Ltd.
Compositor: Shepherd, Inc.

Printed in Canada
1 2 3 4 5 6 7 05 04 03 02 01

Library of Congress Cataloging-in-Publication Data
Hodson, Randy.
  The social organization of work/Randy Hodson, Teresa A.
    Sullivan.—2nd ed.
    p. cm.
  Includes bibliographical references and index.
  ISBN 0-534-55278-1 (alk. paper)
  1. Industrial sociology. I. Sullivan, Teresa A., 1949- II. Title.

HD6955 .H58 2001
306.3'6—dc21   2001039120

Wadsworth/Thomson Learning
10 Davis Drive
Belmont, CA 94002-3098
USA

For more information about our products, contact us:
Thomson Learning Academic Resource Center
1-800-423-0563
http://www.wadsworth.com

International Headquarters
Thomson Learning
International Division
290 Harbor Drive, 2nd Floor
Stamford, CT 06902-7477
USA

UK/Europe/Middle East/South Africa
Thomson Learning
Berkshire House
168-173 High Holborn
London WC1V 7AA
United Kingdom

Asia
Thomson Learning
60 Albert Street, #15-01
Albert Complex
Singapore 189969

Canada
Nelson Thomson Learning
1120 Birchmount Road
Toronto, Ontario M1K 5G4
Canada

Brief Contents

I	HISTORY AND METHODS	1
1	The Evolution of Work	3
2	Studying the World of Work	34

II	THE PERSONAL CONTEXT OF WORK	59
3	Work and Family	61
4	Meaningful Work	89
5	Barriers and Disruptions at Work	110
6	Collective Responses to Work	140

III	INDUSTRIES AND TECHNOLOGIES	171
7	Technology and Organization	175
8	From Field, Mine, and Factory	198
9	The High-Technology Workplace	226
10	Services	254

IV	OCCUPATIONS AND PROFESSIONS	277
11	Professions and Professionals	281
12	Executives, Managers, and Administrators	307
13	Clerical and Sales Workers	328
14	Marginal Jobs	352

V	WORK IN THE TWENTY-FIRST CENTURY	375
15	The World of the Large Corporation	377
16	Work in a Global Economy	402
17	The Future of Work	433

Contents

I HISTORY AND METHODS 1

1 THE EVOLUTION OF WORK 3

Changes in the World of Work 4

    The Social Organization of Work 4

    Consequences of Work for Individuals 6

    Consequences of Work for Society 7

    Social Stratification 8

A History of Work 8

    Hunting and Gathering Societies 8

    Early Agricultural Societies 10

    Imperial Societies 13

    Feudal Society 15

    Merchant Capitalism 18

    The Industrial Revolution 20

    The Factory System 21

    Mass Production under Monopoly Capitalism 26

    Postindustrial Society 28

2 STUDYING THE WORLD OF WORK 34

Techniques of Analysis 34

    Ethnographies 35

    Case Studies 37

    Sample Surveys 39

Units of Analysis 40

    The Worker and the Labor Force 40

    Industry 46

Occupation	47
Workplaces	54
Other Units of Analysis	54
Problems in Studying Work	54
Lack of Information	54
Hard-to-Measure Characteristics	55

II THE PERSONAL CONTEXT OF WORK 59

3 WORK AND FAMILY	**61**
The Life-Cycle Perspective	62
Individual Life Cycle	62
The Career	63
The Family Life Cycle	64
Socialization and Work	65
Informal Socialization	65
Formal Socialization	66
Socialization in the Workplace	67
The Working Years	69
Entering the Labor Force	69
Occupational Mobility	72
Retirement	74
Alternative Cycles	76
Integrating Work and Family Life	76
Role Conflict and Role Overload	76
Work Arrangements among Couples	79
The Arrival of Children	82
Homemakers and Home Production as a Career	83
The Income Squeeze	84
The Impact of the Family on Work	85
The "Empty Nest"	85
Proposals for Combining Family and Work	85
Repackaging Jobs	85
Family-Related Fringe Benefits	86

4   MEANINGFUL WORK   **89**

What Is Job Satisfaction?   90

Theories of Alienation   90

Theories of Self-Actualization   92

What Determines Job Satisfaction?   93

Self-Direction   93

Belongingness   95

Technology   96

Organizational Structure and Policies   97

Participation   98

Individual Differences in the Experience of Work   99

Great Expectations   101

Responses to Work   101

Attitudes toward Work   101

Behavioral Responses to Work   103

The Future of Job Satisfaction   107

5   BARRIERS AND DISRUPTIONS AT WORK   **110**

Discrimination in Hiring   111

Equal Rights Legislation   112

Continuing Forms of Hiring Discrimination   114

Discrimination in Pay and Promotions   118

Racial Discrimination   118

Gender Discrimination   119

The Debate over Comparable Worth   120

Sexual Harassment   122

Managing the Diverse Workforce of the 2000s   124

Unemployment   124

Layoffs   125

Coping with Unemployment   127

Hazardous Work and Disability   128

Industrial Accidents   129

Occupational Diseases   130

Regulating Workplace Safety and Health   134

Stressful Jobs   134

Environmental Degradation	135
Living with Disability	136
Safety and Health in the Workplace of the Future	137

6   COLLECTIVE RESPONSES TO WORK   140

Why Do People Need Labor Organizations?	141
Union Membership	142
An Outline of North American Labor History	142
Local Craft Unions	143
Workers' Political Parties	143
Early National Unions	144
General Unions: The Knights and the Wobblies	146
The AFL and Craft Unionism	149
The CIO and Industrial Unionism	150
The Postwar Retrenchment	152
Facing New Challenges	153
Lessons from Labor's History	156
Labor Unions at the Beginning of the Twenty-First Century	156
Current Union Roles	157
Growing and Declining Unions	160
Innovative Union Programs for the 2000s	163

III   INDUSTRIES AND TECHNOLOGIES   171

7   TECHNOLOGY AND ORGANIZATION   175

Defining Technology	176
Operations Technology	176
Materials	176
Knowledge	176
Defining Organization	177
How Does Technology Influence Work?	178
Changing Technologies	178
What Is Skill?	179
Acquiring New Skills	180

How Do Organizations Influence Work? 181

   The Division of Labor and Changing Organizational Structures 181

   Organizational Structure as Labor Control 182

   Rediscovering the Worker 184

The Growth of Bureaucracy 185

   Defining Bureaucracy 185

   Bureaucratic Control 186

   Customizing Bureaucracies 188

   Informal Work Cultures 188

Limitations of Bureaucracy 190

   Top-Heavy Management 190

   The Centralization of Control in the Economy 190

   Reduced Creativity 191

   Corporate Accountability 191

Direct Worker Participation 193

Technological and Organizational Determinism 195

8 FROM FIELD, MINE, AND FACTORY 198

Postindustrial Society? 199

Occupations and Industries 200

Raw Materials: Agriculture, Forestry, and Fishing 200

   Agriculture 200

   Forestry 204

   Fishing 205

Mining 206

Construction 208

Manufacturing 210

   Craft Workers 210

   Machine Operators and Assemblers 211

   Unskilled Labor 213

   Working-Class Culture 215

Three Key Manufacturing Industries 215

   Automobiles 215

   Steel 217

   Textiles 218

Global Competition and the New World Order 220
   The Wrong Policies at the Wrong Time 220
   Unexplored Alternatives 223

9 THE HIGH-TECHNOLOGY WORKPLACE 226

Competing Views of High Technology 227
Microprocessor Technologies and Skill Requirements 228
   The Skill-Upgrading Thesis 229
   The Deskilling Thesis 230
   The Mixed-Effects Position 231
   Training for Changing Skill Requirements 233
Changing Job Content 234
   Engineering 235
   Assembly Jobs 236
   Machine Work 237
   Skilled Maintenance Work 238
   Clerical Work 238
   Middle Management 238
   Technical Workers 239
   Telecommuting 239
Job Displacement and Job Creation 241
   Job Displacement 241
   Job Creation 243
   Increasing Segmentation? 244
   Public Policy and Employment 245
Working with High Technology 246
   Computer Technology and the Meaning of Work 246
   Computer Technology and Organizational Dynamics 246
   Union Responses 250

10 SERVICES 254

What Are Services? 255
   Characteristics of Services 255
   Sources of the Demand for Services 257

The Rise of the Service Society   258
   Sectoral Transformation   258
   Tertiarization   259
Types of Service Industries   261
   Professional Services   261
   Business Services   261
   Producer Services   261
   Distributive Services   262
   Social Services   262
   Personal Services   264
Compensation in Services   265
Service Interaction   266
   Standards   266
   The Role of Employers   267
   The Worker's Perspective   271
The Future of Service Work   273

IV OCCUPATIONS AND PROFESSIONS 277

11 PROFESSIONS AND PROFESSIONALS 281
How Sociologists Recognize Professions   282
   Abstract, Specialized Knowledge   283
   Autonomy   285
   Authority   285
   Altruism   287
Evaluating the Four Hallmarks   288
   How Powerful Are the Professions?   288
   Monopolizing Knowledge   288
   Power within the Professions   289
   Changes in the Professions   290
   Are the Professions Meritocracies?   292
Changing Degrees of Professionalization   295
   Professionalization   295
   Deprofessionalization   297

The Semiprofessions and the Paraprofessions   300

    The Semiprofessions   300

    The Paraprofessions   302

The Future of the Professions   303

**12   EXECUTIVES, MANAGERS,
AND ADMINISTRATORS**   **307**

Types of Management Roles   308

    Executives   308

    Managers   308

    Administrators   309

    Staff and Line Managers   309

Executives, Managers, and Administrators at Work   309

    Demand for Managers   309

    The Self-Employed Worker   310

    Supply of Managers   311

    The Managerial Career   313

Continuities and Discontinuities in Management Roles   315

    Changes in Scale   316

    Changes in Environment   316

    Changes in Specialization   317

    Changes in Technology   319

Tracking Management Performance   320

    The Behavioral Approach   321

    The Organizational Culture Approach   322

The Future of Executives, Managers, and Administrators   324

13   CLERICAL AND SALES WORKERS   **328**

History of Clerical Work   329

    Demand for Clerical Workers   330

    Supply of Clerical Workers   331

Transforming the Clerical Occupations   335

    Office Technology   335

    Work Reorganization   338

The Future of Clerical Workers　　　341

History of Sales Work　　　341

Demand for Sales Workers　　　342

　Product Marketing　　　342

　Type of Firm　　　345

　Knowledge Base　　　347

Supply of Sales Workers　　　348

The Future of Sales Workers　　　349

14　MARGINAL JOBS　　　**352**

What Is a Marginal Job?　　　353

　Illegal or Morally Suspect Occupations　　　353

　Unregulated Work　　　353

　Contingent Work　　　355

　Underemployment　　　356

How Do Jobs Become Marginal?　　　356

　Marginal Occupational Groups　　　357

Employers Who Marginalize Jobs　　　361

　By Industry　　　363

　By Firm　　　364

　By Employment Contract　　　365

Why Are Some Workers Considered Marginal?　　　366

　Geographic Isolation　　　366

　Educational Level　　　368

　Disabling Conditions　　　368

　Job Displacement　　　368

　Age　　　369

　Race and Ethnicity　　　370

　Gender　　　370

　Interacting Characteristics　　　371

Marginal Workers and Social Class　　　371

The Future of Marginal Jobs　　　372

　Dual Labor Markets　　　372

　Internal Labor Markets　　　373

V   WORK IN THE TWENTY-FIRST CENTURY   **375**

15   THE WORLD OF THE LARGE CORPORATION   **377**

The Power of the Large Corporation   378

Public Concerns about Corporate Power   378

Types of Corporate Market Power   380

The Legal Status of Corporations   382

Merger Mania   384

The First Five Merger Waves   384

The Current Megamerger Frenzy   386

Increased Diversification   388

The Effects of Increasing Size and Concentration   390

A Slowdown of Mergers?   392

Intercorporate Linkages   392

Interlocking Directorates   392

The Role of Banks   393

Subcontracting   394

The Small-Firm Sector   396

Satellites, Loyal Opposition, and Free Agents   396

The Birth of New Jobs   398

Economic Revitalization   398

16   WORK IN A GLOBAL ECONOMY   **402**

How Has the Global Economy Developed?   403

Theories of Industrial Development   403

Emergence of the Contemporary World Economy   406

The World Economy Today   408

The Role of Multinational Corporations   408

Slowed Growth in the Industrialized Nations   410

The End of U.S. Economic Dominance   411

Protectionism, Free Trade, and Fair Trade   412

Trading Blocks: Regional Solutions to Lagging Growth   413

*Combined and Uneven Development
in Less Developed Nations*   413

How Do Work Practices Differ around the Globe?   415

    Least Developed Nations   415

    Developing Nations   416

    State-Regulated Capitalism   417

    German Codetermination   418

    Scandinavian Autonomous Work Groups   419

    Macroplanning in Japan   421

    China   424

    The Four Tigers   424

    Eastern Europe and Russia   425

    Competing Organizational Forms   426

    International Labor Solidarity   427

17   THE FUTURE OF WORK   433

Pivotal Work Trends   434

    Computer Technology   434

    An Integrated World Economy   435

    Female and Minority Workers   435

The Face of Work in the Twenty-First Century   436

    The Innovative Sector   436

    The Marginal Sector   444

Achieving a Brighter Future   448

    Increasing Innovation   448

    Reducing Marginal Employment   450

    Expanding Leisure   450

    Expanding Public Goods   451

APPENDIX TABLE 1: EMPLOYED CIVILIANS BY DETAILED OCCUPATION, SEX, RACE, AND HISPANIC ORIGIN, 2000   454

GLOSSARY   460

REFERENCES   468

INDEX   494

Preface for Instructors

For the past half-century few topics have so fascinated social scientists as the study of work. Scholarship has flourished in the sociology and anthropology of work, industrial sociology and psychology, labor economics, organizational studies, economic sociology, gender and work, and labor force demography. Although this growth has generated great interest and discussion among those of us with research interests in the area, we have not always been able to assimilate the new information and insights into the classroom curriculum as quickly as we would like.

One response to the new scholarship has been innovative course offerings. Many departments still offer industrial sociology and occupations and professions, but now departments are also offering courses with titles such as the sociology of work, work and family, women and work, and technology and work. Finding adequate, up-to-date information for these courses often means coordinating a series of monographs and articles and relying heavily on the class lectures and discussion to provide integrative themes.

We have faced these issues in our own courses. Hodson initially taught industrial sociology, and Sullivan taught occupations and professions. As we discussed our classes together, we began to borrow from each other's knowledge and insights. We found it difficult to teach industrial sociology without also providing material on occupations, and difficult to make sense of contemporary changes in occupations without knowledge of industrial structures and their changing dynamics. We discovered vast bodies of scholarship that neither of us had explored but that our students found exciting. We pored over materials on new microprocessor technologies and tried to understand and communicate the changes occurring in the nature of work as a result of these technologies. We lamented the lack of materials to help our students understand the methods we used to conduct our own research on work. Finally, we began to develop a more unified view of the sociology of work, a view represented in this book. This unified view highlights key themes of technology, class, gender, race and ethnicity, and globalization and allows the book to be adapted to courses focusing on occupations and professions, gender and work, or industrial sociology. Although we collaborated on each of the chapters, Hodson had primary responsibility for Chapters 1, 4–9, and 15–17. Sullivan had primary responsibility for Chapters 2, 3, and 10–14.

Students are vitally interested in an analytic approach to work, and with good reason. The social sciences are not merely part of a "liberal arts" education; they are literally "liberating" because they give students a vocabulary and perspective for understanding the world around them. Given the tremendous importance of work in our lives, understanding the work world is both intellectually satisfying and pragmatic.

Although the content of this book deals mainly with substantive issues concerning the world of work, we also hope that the book will be useful in developing students' skills of analysis, reasoning, and argumentation. We have tried to be fair in presenting competing theoretical arguments, but we have also indicated on which side we believe the weight of the evidence lies. You and the students may disagree with us. Our own students often do, and some of our best class sessions are generated from these disagreements. We have tried to identify prejudices and cultural biases that affect perceptions of work and workers. In particular, we have integrated the discussion of women and minorities into every chapter. We have also grappled with the profound changes surrounding the microelectronics revolution and the rapidly changing global economy. We discuss the influence of technology and globalization on economic development and class relations throughout the book. We have tried to be frank about the gaps that exist in social scientists' current knowledge and to point out alternative scenarios for future developments.

We also provide support for more general curricular goals by including frequent boxes that highlight cross-cultural issues and by providing tables and graphs to help students develop the skill of interpreting data. Every chapter ends with a list of key concepts and questions for thought. These materials are useful for student review, for written assignments or homework, and for examinations. We also provide a brief annotated list of additional library, internet, and media sources at the end of every chapter. Students can use these sources for further exploring issues developed in the chapter or for assistance in preparing term papers. Both of us encourage our students to write, and the subject of work lends itself to creative and thoughtful student papers.

Changes to the Third Edition

One of the most significant changes we have made in the Third Edition is to highlight throughout the book five key themes that help to organize the book. These themes are technology, class relations, gender, race and ethnicity, and globalization. We also use boxed and inserted material to further highlight these themes. Boxed and inserted material relevant to these five themes is highlighted in the text with special icons. These icons are first presented and explained in the "Preface for Students."

We have also added more first-hand ethnographic material in which workers speak with their own voices. In addition, we highlight the mid-range conceptual underpinnings of each section through extensive use of paragraph-level headings. We have systematically updated data, concepts, and sources, and we have rigorously edited the manuscript for length and style so that each chapter can be read in one sitting. We also give increased attention to new concerns in family-work linkages and new developments in worker participation programs. Other changes include more emphasis on women's issues and on the expansion of marginal employment, expanded coverage of globalization, and a heightened emphasis on the role of microprocessor technology in transforming work.

Supporting Materials

An *Instructor's Manual* is also available with the third edition and we recommend you write or call the publisher or your local Wadsworth representative to receive a copy. The *Instructor's Manual* includes a multiple-choice test bank, suggested films, role-playing exercises, lecture frameworks, and many other pedagogical suggestions and aids. The material in the *Instructor's Manual* is also available at *http://sociology.wadsworth.com*. In addition, instructors may also find a great deal of material useful for classroom purposes in the websites listed at the end of chapters and throughout the text.

We hope that instructors will be able to use this book in a variety of educational settings and course titles under both the semester and quarter systems. For a course on *occupations* we recommend Part I, which provides a historical overview and discusses research methods for studying the world of work; Part II, which discusses individual and collective adaptations to work; Part IV, which discusses the major occupations; and Chapter 17, which discusses the future of work. For a course on *industrial sociology* we recommend the same starting sequence but the substitution of Part III, which discusses organizations, manufacturing, the microelectronics revolution, and service industries, for Part IV. Chapters 15 and 16, on large corporations, mergers, and the world economy, will also fit well into an industrial sociology course, depending on the number of weeks available. For a course on *women and work,* we recommend Chapters 1–3 on history, methods, and the work-family connection, followed by Chapter 5 on barriers at work. Chapter 10 on service work and Part IV on occupations will also be essential for a course on women and work, as well as Chapters 16 and 17 on the global economy and the future of work. For a semester course on the sociology of work we recommend the entire book, with about one chapter assigned per week along with whatever supplementary readings the instructor chooses. For a quarter-length course on the *sociology of work,* several chapters can be skipped while retaining the core of the book. Depending on the instructor's preferences, omitted chapters might include Chapter 4, on the experience of work, Chapter 6, on unions, Chapter 7, on organization and technology, Chapter 14, on marginal work, or Chapter 15, on large corporations.

We enjoy teaching, and we enjoy becoming better teachers. If you have questions about our text or if you have ideas for improving the text or for using the material in a particular setting,

we would like to hear from you. Our addresses appear at the end of this preface. Our own teaching has been improved by our collaboration, and we are eager to continue the dialogue with others.

Acknowledgments

We would like to acknowledge our debts to the many colleagues who have unfailingly assisted us. We have not always taken their advice, but we have always appreciated it, and the book has been substantially improved by their contributions. We appreciate the careful editorial work that Meera Dash and Charles M. Bonjean devoted to every chapter. We learned first-hand about corporate acquisition and reorganization when Dorsey Press, our original publisher, was acquired by Wadsworth. Paul O'Connell, Serina Beauparlant, Sheryl Fullerton, and Lin Marshall offered us extremely helpful editorial assistance and taught us much about textbook publishing. Many colleagues have shared with us their pedagogical and scholarly expertise by reading and commenting on various chapters. These include Andrew Abbott, Howard Aldrich, Robert Althauser, Ronald Aminzade, James Baron, Vern Baxter, John Bodnar, David Brain, Harley Browning, Phyllis Bubnas, Beverly Burris, Johnny Butler, Catherine Connolly, Daniel Cornfield, Sean Creighton, Tom Daymont, Nancy DiTomaso, Frank Dobbin, Michael Dreiling, Lou Dubose, Sheldon Ekland-Olson, Joe Feagin, Neil Fligstein, Ramona Ford, Eliot Freidson, Omer Galle, Maurice Garnier, Tom Gieryn, Michael Givant, Jennifer Glass, Norval Glenn, Mark Granovetter, Larry Griffin, Ein Haas, Richard Hall, John Hannigan, Heidi Hartmann, Jeff Haydu, Jane Hood, Gregory Hooks, Arne Kalleberg, Robert Kaufman, Jacqueline King, James Kluegel, Judith Langlois, Eric Larson, Laura Lein, Sanford Levinson, Susan Marshall, Garth Massey, Ruth Milkman, Delbert Miller, Joanne Miller, Jeylan Mortimer, Mary Murphree, Jan Mutchler, Janet Near, Annette Nierobisz, Brigid O'Farrell, Toby Parcel, Alan

Ponak, Brian Powell, David Rabban, Sabine Rieble, Pamela Robers, Rob Robinson, Nestor Rodriguez, Patricia Roos, Rachel Rosenfeld, Arthur Sakamoto, Paul Schervish, Carmi Schooler, Peter Seybold, James Shockey, Ken Spenner, Suzanne Staggenborg, Robin Stryker, Joyce Tang, Peggy Thoits, Charles Tolbert II, Linda Waite, Michael Wallace, Sandy Welsh, Christine Williams, James Wood, and Gloria Young. We are also grateful to a number of research assistants, whose help has been invaluable: Dick Adams, Bill Brislen, Robert Dixon, George A. Harper, Jr., Laura Hartman, Robert Parker, and Matthew B. Ploeger. And of course we are grateful to our students, who are both our toughest critics and our greatest supporters. They have provided the essential ingredient that makes teaching, writing, and learning such a satisfying and rewarding experience.

Finally, we would like to acknowledge the creative insights and support of our spouses, Susan Rogers and Douglas Laycock. They provided detailed comments on every chapter and always supplied whatever we lacked at the moment, whether it was conviction, energy, courage, or just appreciation.

Randy Hodson
Department of Sociology
Ohio State University
Columbus, Ohio 43210
www.soc.sbs.ohio-state.edu/rdh

Teresa A. Sullivan
Vice President and Dean of Graduate Studies
University of Texas
Austin, Texas 78712-1111
www.la.utexas.edu/socdept/faculty/sullivan.html

Preface for Students

Most people will work throughout their adult lives. Work will absorb the best part of their days. College students are naturally interested in the world of work, how it is changing, and the implications of those changes for themselves and their families. In our own classes we find that students are very concerned, even worried, about their roles as future workers. We hope that this text will help you explore some of these issues by yourself and with your classmates and instructor.

The intellectual backbone of any course on work concerns the process through which work becomes more and more specialized, the transformation of specialization into stratification and inequality, and the organizational context of work. This skeletal framework informs this text, though it will often be part of the only faintly visible background. You will spend most of your time reading about topics such as the impact of the microelectronics revolution, the rapidly changing roles of women at work, and the constantly evolving world economy. Five themes, in particular, are highlighted through the use of boxed materials identified with special thematic icons. These themes are:

- *Technology*
- *Class relations*
- *Gender*
- *Race and ethnicity*
- *Globalization*

These themes are developed throughout the book from the first to the last chapter.

Part I provides background material for the study of work. Chapter 1 offers an overview of work in past societies and identifies key themes that will be followed throughout the book. Chapter 2 explains how we study work in contemporary society. This chapter will be of value both to those wishing to specialize in the sociology of work and to others interested in understanding research findings based on studies of individuals, groups, and organizations.

Part II, made up of Chapters 3–6, deals with our work roles and how these influence our daily lives. The topics covered here include the life cycle, careers, integrating work and family, finding meaning at work, job problems, such as unemployment, disability, discrimination, and participation in unions and other collective organizations at work.

Part III, made up of Chapters 7–10, deals with the technology and organization of work. The chapters parallel the major economic sectors: agriculture and manufacturing, high-technology industries, and services. Changes in the technology and organization of work give rise to the transformation of occupations discussed in the next section.

Part IV, composed of Chapters 11–14, deals with the occupational roles that we hold and with the unique sets of skills that are needed to perform these roles. The chapters in this section focus on professionals, managers, clerical workers, sales workers, and marginal workers. (Manufacturing and service workers are discussed in Part III.)

Part V, made up of Chapters 15–17, focuses on societal-level consequences of the changing nature of work. The topics covered in this final part include the world economy and the role of huge transregional and transnational corporations in molding the world of tomorrow.

We are pleased that your instructor has adopted our book. Since you have become our student by proxy, we would like to share some of the study tips we give our own students in class. Educational studies show that the more actively you are engaged in reading and reviewing text material, the more likely you are to understand, integrate, and retain ideas. An active reader brings several senses to play in every study session. We recommend that you read and study with a pen or pencil in hand and that you make frequent notes to yourself as you identify and learn new ideas. It also helps to read key passages aloud and to use a tape-recorder or note cards to highlight core ideas for review. Try to study regularly. You will enjoy the material more if you set yourself a regular schedule for studying and reading, giving yourself sufficient time to assimilate the material.

When you begin a study session, preview the chapter to learn about its contents. At the end of every section, quiz yourself about the main points of the section and underline points that you consider to be important. At the end of every chapter, review the key concepts. If you cannot recall the meaning of a concept, return to the text and reread the relevant paragraph or look the concept up in the glossary. All the boldface key concepts are defined in the glossary. Every chapter ends with thought questions. Even if your instructor does not assign them, try to answer them. Some are designed to be easy, and others are hard. Some do not have one correct answer but provide an opportunity to apply the material you have read. We find that students who practice these questions generally write more insightful essay exams and term papers. Read the tables. The information in them is the most current we could find. Data interpretation is an important skill for you to develop, regardless of your occupational destination.

When you review a chapter before an examination, begin with the chapter summary. It is often helpful to develop hypothetical test or essay questions to assist your review and to identify points that you want to bring up in class before the test. Additionally, discussing ideas and concepts with your classmates helps to permanently cement your learning.

The multimedia and websites at the end of each chapter will be helpful if you want to learn more about the topics in the chapter. In addition, at the back of the book there is a list of references detailing the source of every study we have cited; you might want to look some of these up in the library to deepen your knowledge or to help prepare a paper.

We are college professors by occupation, and we find our work very rewarding. We hope that you, too, will find a place in the world of work that is both satisfying and challenging. And we hope that this book will help you become better prepared for that world. After reading this text, let us know your views, either positive or negative. We are very responsive to suggestions from students. Only with feedback from students will we know if our efforts have been successful. Our addresses are listed at the end of the "Preface for Instructors."

PART I

History and Methods

1

The Evolution of Work

There is nothing better for a man, than that he eat
and drink, and tell himself that his labor is good.

ECCLESIASTES 2:24

Every morning at six I drove myself out of bed, did not shave, sometimes
washed, hurried up to the Place d'Italie and fought for a place on the
Metro. By seven I was in the desolation of the cold, filthy kitchen, with
the potato skins and bones and fishtails littered on the floor, and a pile of
plates, stuck together in their grease, waiting from overnight. I could not
start on the plates yet, because the water was cold, and I had to fetch milk
and make coffee, for the others arrived at eight and expected to find coffee
ready. Also, there were always several copper saucepans to clean. Those
copper saucepans are the bane of a *plongeur's* life. They have to be scoured
with sand and bunches of chain, ten minutes to each one, and then
polished on the outside with Brasso. Fortunately, the art of making them
has been lost and they are gradually vanishing from French kitchens.

(ORWELL, 1933:107–108)

These two quotations point out the contradictory nature of work: it is both
a salvation and a curse. Work creates prosperity and meaning in life, but it
also creates poverty and alienation. This chapter will review changes in the

nature of work across time so that you can better understand its possibilities and limitations. This entire book is an effort to sort out the varied experiences of workers in order to make sense of work in modern society.

What is work? Work is the creation of material goods or services, which may be directly consumed by the worker or sold to someone else. Work thus includes not only paid labor but also self-employed labor and unpaid labor, including useful work done in the home. Work provides material and personal benefits, but it can also be a source of frustration and aggravation.

CHANGES IN THE WORLD OF WORK

In this chapter we explore some of the ways in which work has changed as well as the consequences of these changes for individuals and for society. The organization of work has varied greatly over time. The division of labor and the extent of social inequality also vary over time. The changing nature of work thus has important implications for personal satisfaction, for the cohesiveness of society, for relations between men and women, and for our ideas about the meaning of work and its place in social life.

The Social Organization of Work

The social organization of work is the set of relations among people at work. In this section we outline a set of themes that describe the social organization of work and that will be explored throughout the remaining chapters of the book. These themes highlight the organizational and technical aspects of work, the consequences of work for social inequality, the demographics of the workforce, and the meaning of work for individuals.

The Division of Labor   The most fundamental transformation in the nature of work over time has been the increasing **division of labor** (Durkheim, 1966 [1897]). In primitive societies, each member engaged in more or less the full range of work activities. The only differences in work activity were those based on age and gender. In later feudal society, most workers were engaged in agricultural work, but some specialized in a single product so that they became, for example, tailors, cobblers, or bakers. In modern industrial societies, work has become so specialized that each trade is broken down into seemingly innumerable specialties. The meat-packing industry provides a good example of an extremely specialized division of labor: "In the slaughter and meat-packing industry one can specialize as: a large stock scalper, belly shaver, crotch buster, gut snatcher, gut sorter, snout puller, ear cutter, eyelid remover, stomach washer (sometimes called belly bumper), hindleg puller, frontleg toenail puller and oxtail washer" (Wilensky and Lebeaux, 1986:33). Specialization constantly creates new lines of work that require new and different skills; however, the division of labor can also reduce the *range* of skills needed to perform jobs. A much narrower range of skills, for example, is needed to be a "gut snatcher" than to be a butcher offering a full line of services. Greater specialization can thus have both positive and negative consequences for workers as they struggle for dignity and meaning at work.

The modern division of labor also occurs between different regions and different nations—some areas specialize in agriculture, some in different types of manufacturing, and others in service industries such as computer software development, biotechnology research, or banking and investment.

Technology Technological developments have multiplied productivity over time. Starting with only simple hand tools, people advanced through the use of steam- and electric-powered tools to the automation of many aspects of production through assembly-line technology and robotics. Today, computer technologies are again revolutionizing work in what many are calling a second industrial revolution. The workers and nations that successfully harness the power of computer technologies will be the winners in the competitive global economy of the twenty-first century (Burris, 1998).

Inequality Not only does the division of labor and technology change over time but also the ways in which people work together. Work always involves social relations between people. Relationships exist between employer and employee, between colleagues and co-workers, between trading partners, and between suppliers and consumers. These relations are called **social relations of production.** Social relationships at work can be *cooperative and egalitarian,* as in primitive societies, or *hierarchical and unequal,* as in industrial societies. In primitive societies, people decided jointly how to proceed with a given task and shared equally in its results. Cooperative arrangements were grounded in the reality that most skills were held in common. In societies with a more advanced division of labor, such egalitarian arrangements are replaced by more hierarchical ways of organizing work, in which some skills are held to be more important than others and in which some societal members have vastly greater power and wealth than others. The relationship between peasants who till the land and landowners is among the earliest hierarchical organizations of work. The most important contemporary form of hierarchy at work is the relation between owners and workers (Perrucci and Wysong, 1999; Tilly and Tilly, 1998).

Women, Minorities, and Immigrants Who works, and in what capacity, is also key to

understanding the nature of work and its consequences. During most of industrial society, men have been more likely than women to leave home to work in factories and offices. This difference is rapidly eroding today with women making up over 46% of the labor force in the United States and Canada. Minority ethnic populations in many societies have traditionally been segregated into lower-paying occupations and trades. The spread of various forms for protective legislation for minority populations addresses these inequalities, although the attainment of full equality has often been illusive (Wilson, 1997). New immigrants to a country also typically occupy the lowest rungs of the occupational ladder with succeeding generations climbing to greater heights. The accelerated movement of people around the world in the twenty-first century has increased the significance of immigrant populations and workers in many nations and the challenges of assimilating these workers and their families (Jasso et al., 2000).

Bureaucracy The nature of economic enterprises is also crucial to the meaning and experience of work and to the success of enterprises. In the contemporary workplace, hierarchical relations often take the form of **bureaucracy.** A bureaucracy is a hierarchical system with clearly designated offices and responsibilities and a clearly defined chain of responsibility leading to the top position. The behavior of all parties in a bureaucracy, no matter how high up, takes place within the dictates of clearly stated rules. Bureaucracies exist in both modern corporations and in modern governments and are the major way in which work is organized in contemporary society (Crozier, 1964; Gouldner, 1964; Scott, 1998). Postbureaucratic forms of work organization based on greater worker involvement and initiative are, however, emerging today and are expected to have a defining influence on the experience and meaning of work in the twenty-first century.

The Professions The hierarchy of authority in the workplace is further complicated by the growing significance of highly educated professional workers. These workers claim unique rights and privileges based on their possession of specialized knowledge gained through long periods of study. At the beginning of the twentieth century, only 4% of the labor force in the United States was made up of professional workers. At the beginning of the twenty-first century, 20% of the labor force is made up of professional workers, making this group one of the largest occupational categories. Many college students are following courses of study that prepare them to become members of a particular profession.

Meaning and Dignity in Work People's ways of thinking about the role of work in their lives also change over time. In primitive societies people did not experience work as an activity separate from the broader round of daily events. In agricultural societies, work was seen as an inevitable burden, made even heavier by the abuses of greedy landlords, bad weather, and variable market prices. Capitalism saw the emergence of a work ethic that identified work with piety and grace. Many fear that the work ethic grounded on frugality and unquestioning effort has been lost in contemporary society. Perhaps it has. But if so, it has been replaced by a vision of work as a route of upward social mobility (Ospina, 1996). These different visions of the meaning of work become parts of social **ideologies**—systems of ideas that justify the economic and political arrangements of a society as appropriate and desirable. In all settings, however, workers desire autonomy and respect in order to experience dignity in their daily lives at work (Hodson, 2001).

Globalization An ever greater proportion of economic exchange occurs between nations. In addition, large corporations located in industrially advanced nations typically have many branch plants and joint ventures outside their nation of origin. As a result, the world economy is characterized by dense networks of economic links between nations and between transnational corporations. These realities increase world competition, pushing down prices for many commodities; but they also allow corporations to transfer production to areas with lower-priced labor, thus placing the workers of each nation in ever sharper competition with each other and creating downward pressures on wages and health, safety, and environmental protections (Chase-Dunn, Kawano, and Brewer, 2000).

Consequences of Work for Individuals

Individual workers often seem reduced in importance in the large-scale, bureaucratic world of modern organizations. Their individual contributions seem interchangeable or even expendable. This reality can lead to a sense of *alienation* from work, in which people feel detached from their activity, from one another, and, eventually, from their own selves.

Alienation from work is further increased by the reduced importance of the family as the basic unit of economic life. In preindustrial societies, work was a family activity in which all members participated according to their abilities. With the advent of factories, people increasingly worked outside the home. They came to spend their time at work alone or with others with whom they shared little in common.

The modern organization of work also has positive consequences for individuals. Many individuals receive a share of the expanded productivity of modern industry. Industrial societies produce a much wider range of material goods and services than preindustrial societies; these goods include more and better food, and many items that would have been considered luxuries in previous societies or that were completely unavailable such as central heating and television. Improved services include better medical care and higher education. For many people, work experiences also continue to be a primary source of fulfillment and self-realization (Hochschild, 1997).

Consequences of Work for Society

Self-Interest The very nature of society has been fundamentally altered by the changing organization of work. The most significant change is the transformation from rural society, based on deeply felt *bonds of commonality,* to urban society, based on more fragmented, fluid, and changing relationships, often grounded in *self-interest.* Traditional rural societies placed a high value on conformity and on maintaining solidarity in the face of external threats. These values were necessary because of the harshness and vulnerability of peasant life. In industrial societies the relationships between people are based on distinct yet interdependent contributions rather than on commonly shared abilities and positions (Durkheim, 1966 [1897]). Modern societies thus make greater allowances for, and may even encourage, diversity and competition among their members.

Organizational Size The greater *size* of economic organizations has also produced changes in how both work and society are organized. Products that once were manufactured in millions of households and, later, in thousands of small factories are manufactured today for the entire world by a handful of corporations. Giant companies that produce automobiles, tobacco products, petroleum, and computers provide some of the clearest examples. Such corporations have tremendous power over workers, customers, and even nations. The increasing size of organizations further contributes to the loss of individuality and intimacy at the workplace. Direct personal relationships are often replaced in these corporations by bureaucratically administered rules. In this context, the way in which individuals experience work depends to a significant extent on whether their smaller work groups provide opportunities for them to realize positive self-images.

Markets Over time, the role of markets in the organization of work has expanded. In primitive societies, several people got together for the hunt, or several people joined together to dig roots or gather edible plants. All members of the group, perhaps fifteen to twenty in number, then shared equally in the proceeds of these labors. Today, thousands of people labor together at one site to manufacture, for example, saline solution bags for hospitals, while tens of thousands of other employees of the same corporation work in different parts of the world to manufacture other pharmaceutical products. These workers are integrated in a network of markets in which they exchange their earnings from manufacturing saline solution bags for a diversity of goods and services such as housing, food, and entertainment. These workers may never have occasion to make use of the products they themselves manufacture. Modern industrial systems have brought about the organization of the world into a single, interconnected economic unit. Today we live in a truly global economy. As Figure 1.1 points out, these transformations in social arrangements are paralleled by changes in how we think about and understand not only work but other aspects of our lives as well.

Greater organizational size and bureaucracy allow increased productivity because of efficiencies associated with producing a great number of similar or identical things. These sorts of efficiencies are called **economies of scale.** Rationalized planning further increases productivity. Giant economic organizations produce standardized products in quantities unthinkable under previous industrial systems. The resulting power of large corporations over people, the environment, and society often necessitates at least some government regulation of many economic activities and industries. (In Chapters 7 and 15 we also consider some of the limitations on productivity imposed by standardization and large organizational size.)

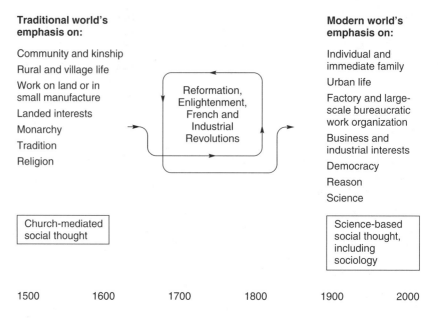

FIGURE 1.1 Traditional Societies versus Modern Societies

SOURCE: Tony J. Watson, 1980, *Sociology, Work and Industry.* London: Routledge and Kegan Paul, p. 8. Reprinted by permission of International Thomson Publishing Services, Ltd.

Traditional world's emphasis on:

Community and kinship

Rural and village life

Work on land or in small manufacture

Landed interests

Monarchy

Tradition

Religion

Reformation, Enlightenment, French and Industrial Revolutions

Church-mediated social thought

Modern world's emphasis on:

Individual and immediate family

Urban life

Factory and large-scale bureaucratic work organization

Business and industrial interests

Democracy

Reason

Science

Science-based social thought, including sociology

1500   1600   1700   1800   1900   2000

Social Stratification

Modern forms of work produce a great abundance of goods and services, but they distribute these goods and services unequally. Members of society receive shares based on their position in the division of labor and in the structure of power. In this way, the organization of work in modern societies influences not only our work lives but also our lives outside of work. Some members of society labor long days and weeks, perhaps even holding down two or more jobs, but receive relatively little for their efforts. Some cannot find work and suffer poverty throughout all or parts of their lives. Many work what has come to be considered a standard forty-hour week and receive a reasonably good living for their efforts depending on their exact location in the division of labor. Some do not have to work at all but have inherited riches unimaginable to the majority of people.

A HISTORY OF WORK

The developmental model on which the following outline is based argues that the *social organization of work* and the *technology* at a given stage

of history determine the nature of society, including its degree of social inequality. These factors also create contradictions and limitations that set the stage for the next level of development (Marx, 1967 [1887]; Lenski, 1966). Social organization, as you have seen, refers to the relations among those involved in work. Technology refers to the tools and skills used in the process of work. An understanding of these broad changes in the nature of work will help you understand the evolution of work and its future possibilities.

Hunting and Gathering Societies

By about 300,000 B.C., the human species, *Homo sapiens,* had evolved to its present form. Humans lived as nomadic hunters and gatherers until about 8,000 B.C. Thus, the hunting and gathering stage included about 97% of the collective life of our species and continues in isolated areas even today. Hunters and gatherers did not perceive "work" as a separate sphere of life. Activities necessary to secure sustenance took place throughout the day and were not clearly distinguished from leisure activities—as people gathered berries, dug edible roots, and hunted game, they did so in a rather leisurely manner, depending on the circumstances of the moment. People did not work hard because there was no point in creating a surplus. Surpluses of food or possessions could not be stored or transported for future usage. Work, leisure, and socializing formed an integrated flow of activities.

The Band   A hunting and gathering band consisted of fifteen to twenty members, depending on how many people the vegetation and animal life would support. The group's hunting and gathering activities eventually depleted the resources in the area immediately around the encampment, and the group was forced to move on. These nomadic movements were cyclical; the group would move through the same areas year after year, following the seasons and the food supply.

Technology was very simple. The most important elements of technology were the various skills that each member of the band learned and used. Without the skills to make and bank fires and to locate and harvest edible plants and animals, the band would have quickly starved. The technology also included hides, lodge poles, and other equipment necessary for survival such as bone needles, stone cutters and scrapers, and wooden spears. Beginning about 35,000 years ago, people began to use the bow and arrow.

Most skills were shared in common so that any single member could do all or most of the tasks required of the group as a whole. However, there were rudimentary forms of the division of labor based on gender and age. Young people tagged along with their elders and performed helping functions such as gathering wood or picking berries. In this way they received the equivalent of modern on-the-job training. Older people who lacked the stamina or mobility for hunting and gathering tended the fire and prepared food or made tools.

The life cycle of people in hunting and gathering societies involved three stages: childhood, adulthood, and old age. Children engaged in helping activities, which also provided anticipatory socialization for their later adult roles. With the onset of puberty, the individual was recognized as a fully functioning member of the group whose opinions must be recognized and given consideration. This transition was typically celebrated through a rite of passage involving an extensive set of ceremonies. The transition to old age was a slower process involving a gradual reduction of activities, though not necessarily status.

The Gender Division of Labor   A division of labor based on gender initially resulted from biological differences between men and women. Few people lived beyond childbearing age and so women spent much of their adult life either pregnant or nursing. These activities restricted their physical mobility, so that they were less often involved in hunting large game. Women specialized in gathering roots, berries, and other edible plants and in hunting small animals such as rabbits and other rodents. There is considerable evidence that women's gathering and small-game hunting were often more productive economic activities than men's hunting of large game—in most hunting and gathering societies, roots, berries, and small game made up a larger part of the diet than did large animals (Lee, 1981).

The social position of women throughout most of human history has been subordinate, to some degree, to that of men (Friedl, 1975). Gender inequality was less extreme in hunting and gathering societies than in many later societies. In hunting and gathering societies, the greater power of men rested on their monopoly over large-game hunting, which provided rare periodic surpluses of meat, and on their central role in the important arenas of contact, trade, and conflict with other groups (Friedl, 1975:135).

Sharing   The early forms of the division of labor based on age and gender did not typically result in any great inequality of material rewards among the members of a society. All members shared equally in the food secured from the environment. This arrangement produced optimal benefits for everyone because of the unpredictability of hunting and gathering activities. If one person or family were successful on a certain day, they could not store or transport the surplus. Through equal sharing, however, all members were assured a share of the bounty of others when their own efforts were unsuccessful. Thus, people ate or went hungry together.

Because food and other possessions could not be accumulated in quantity, hunting and gathering societies did not develop **social classes** based on possession of different amounts of wealth. Instead, equal sharing of resources prevailed and provided a fundamental precondition for the continuation of these societies—in hunting and gathering societies there was more to be gained by sharing than by hoarding.

Few true specialists existed in such societies. Some members may have taken on a leadership role more often than others but even this role was often rotating. Incumbency in this role was based on personality traits such as cleverness or leadership ability and carried with it few if any privileges of position. Similarly, some individuals may have received recognition for their ability to minister to the sick, forecast the weather, or predict the movements of the animals. These individuals sometimes took on the role of a *shaman,* or medicine man.

Motivation to Work In hunting and gathering societies, the motivation to work was straightforward. The band lived a day-to-day existence. If one did not engage in purposive activity on a regular basis, then one either went hungry or relied on others to share a portion of their food. Hunger and social pressure to participate in the group's activity provided daily motivation to work. In an account of the Andaman Islanders, the anthropologist A. Radcliffe Brown describes these norms in the following way: "Should a man shirk this obligation, nothing would be said to him, unless he were a young unmarried man, and he would still be given food by others, but he would find himself occupying a position of inferiority in the

camp, and would entirely lose the esteem of his fellows" (Brown, 1922:187).

Hunting and gathering people thought of work in ways that would seem quite foreign in the modern world. They did not view work as a distinct activity; life as a whole was seen in a sacred context in which the various forces impinging on the group were held in awe and reverence. Groups that were seasonally dependent on a particular plant or animal, for example, might consider this item sacred and worship its spirit to assure its continued presence. Such entities, called *totems,* were seen as representing and protecting the group. The spiritual relation of the group with the totem represented the shared interests, unity, and lineage of the group. The worship of such totems was the earliest form of religious expression. Other daily activities of hunting and gathering included sacred aspects, which were undertaken with appropriate piety and observation of ritual. In this sense primitive people experienced work in a spiritual context.

The differentiation of society into classes did not take place until the group was able to settle in one place and accumulate a surplus of goods. In rare situations hunting and gathering societies settled and flourished in environments that had an abundance of naturally occurring food sources. Such settings were almost totally restricted to fishing communities, such as those in the Pacific Northwest of North America. In general, settled societies awaited the development of agriculture. The nature of economic life in a hunting and gathering society is illustrated in Box 1.1.

Early Agricultural Societies

The major difference between settled and nomadic societies is that settled societies can accumulate a surplus of food and other goods. Accumulation becomes possible because agricultural and pastoral activities are more productive than hunting and gathering and because the group is able to accumulate and store goods over

---

BOX 1.1   Acorn Gathering in Sacramento Valley

The Maidu were one of the principal tribes of the Sacramento Valley and adjacent sierras. Their country followed the eastern banks of the Sacramento River and encompassed the modern city of Sacramento, California. The collection and preparation of acorns for food were among the most important industries of the Maidu, in common with most of the Central California tribes. At the time in the autumn when the acorns are ripe, everyone is busy. The men and the larger boys climb the trees and, by the aid of long poles, beat the branches, knocking off the acorns. The women and the smaller children gather these in burden baskets, and carry them to the village, storing them in granaries or in the large storage baskets in the houses. . . . In addition, eels were speared, split and dried. In preparing them for food, they were usually cut into small pieces, and stewed. Salmon were split, and dried by hanging

them over a pole. When thoroughly dry, the fish was usually pounded till it was reduced to a coarse flour, and kept in baskets. Deer and other meat was cut into strips and dried. Usually this was done in the sun; but occasionally a fire was lighted under the drying meat to hasten the process, and to smoke the product slightly. Except on their hunting trips, the Maidu seem not to have been travellers. They rarely went far from home, even on hunts. The Northeastern Maidu traded with the Achoma'wi Indians, getting chiefly beads, and giving in exchange bows and deer hides. Those in the higher sierra traded for beads, pine nuts, salt, and salmon, giving in exchange arrows, bows, deer hides and several sorts of food.

SOURCE: Excerpted from Roland B. Dixon, 1977, "The Northern Maidu." In *A Reader in Cultural Anthropology*, edited by Carleton S. Coon. Huntington, N.Y.: Krieger, pp. 262–291. Reprinted by permission of the publisher.

---

time. These differences distinguish settled societies from nomadic societies even today.

Agriculture developed independently in several places around the world from 9,000 to 3,000 B.C. These areas include Southeast Asia, the Persian Gulf, and Mesoamerica. The development of agriculture started with the harvesting of wild grains, such as wheat, barley, and corn, and wild tubers and the eventual development of techniques to encourage the growth and yield of these plants. The technologies included the use of the digging stick and, later, the hoe.

The First Surplus   With the development of agriculture and the domestication of animals came tremendous changes in the organization of society. A surplus of food was produced, though at first it was quite small. The work of perhaps 80 to 100 farmers was required to support one nonfarmer. On this scanty basis a new social order came into being. Instead of everyone in society occupying the same role, specialized positions came into being with differentiated activities. These positions included warrior, priest, and,

eventually, other official positions such as scribe and tax collector. The production of everyday goods was still carried out mainly by the agricultural worker. Later, craft positions specializing in the production of religious, civic, and military goods also developed.

The life of the agriculturalist differed little from that of the hunter and gatherer though it perhaps lasted a little longer because of the better protection from famine afforded by greater ability to store food. Children still helped with basic work activities until they were able to take on a fuller role, and the elderly returned to a helping role as dictated by declining strength and stamina. The relative positions of men and women also changed little. "Since men had been hunting, men were the inventors of systematic herding. Since women had been gathering plants, women were the inventors of systematic agriculture" (Deckard, 1979:199). Based on their continuing contributions to the household economy, men and women enjoyed roughly equal access to the goods and services produced by society. With the development of

BOX 1.2   Agriculturalists: The Ghegs of Northern Albania

In the mountainous regions on the eastern shore of the Adriatic Sea the Ghegs lived as agriculturalists until recent times. Over 90% of the Ghegs are farmers with other occupations including charcoal burner, blacksmith, carpenter, saddler, hatter, tailor, shoemaker, butcher, cook, and pastry cook.

Cutting tools are of iron. They are made by blacksmiths in the market towns in which the Ghegs do their trading. With hammer and tongs, a small anvil, and a pair of foot-pump bellows they forge nearly all of the metal objects needed by the mountaineers in their farming, herding, transport, and household carpentry. They make small anvils, hammers, nails, axes, adzes, knives, ploughshares, shovels, hoes, toothed sickles, door hinges, horse shoes, and horse hobbles.

The Gheg at home in his mountains is a jack-of-all-trades. He may not be able to manufacture the special objects listed above, but he can adze out beams for his house, put together a new plough

during the winter months when there is little work out of doors, or build one of his massive chairs. Professional carpenters are rare, although some men show more skill in this than others.

The interiors of most houses are quite bare, with no carpets on the hewn plank floors, and little decoration on the walls. In the older houses a fireplace covers about 10 square feet of floor area, and the smoke finds its way out through the thatch or roof tiles. In so doing, it cures meats hung in the rack above the fire.

The basic cloth is woven at home. The farmer gives his wife the necessary wool from his sheep, both white and black. She cards it with a homemade device, either a flick bow or a nail studded card, and spins it. Her spinning kit consists of a distaff, a spindle, and a basket. Holding the distaff in one hand, she twists the wool with the other, and winds it onto the spindle in the basket. Whenever she has nothing

---

the plow drawn by a team of draft animals, women's relative contribution and position may have declined somewhat.

The orientation of the agriculturalist to work differed from that of the hunter and gatherer. For the farmer, the land took on a sacred status similar to that of animals and plants for hunters and gatherers. A new element in the peasant's orientation to work was a focus on the importance of bountiful harvests. Whereas the hunter and gatherer had no need or use for surplus, the agriculturalist actively sought as large a surplus as possible to ensure survival through the winter. A large harvest was critical because the agriculturalist could not wander in continual search of food, as the nomad had done. Rather, agriculturalists were tied to one spot all year in order to protect their investment of time and resources in planted fields and to wait for and protect the harvest. They thus depended on the land they farmed to produce a surplus that would sustain them throughout the year.

Plunder and Warfare With the accumulation of surplus also came the possibility of plunder by outside groups. This possibility spurred the creation of a warrior class and fostered the inward-looking nature of agricultural society. Some sociologists believe that this increased importance of warfare accounts for women's more subordinate position in agricultural societies. Because men assumed the principal responsibility for warfare, their role in society grew in power and importance (Sanday, 1981:211).

The warlord and priest were powerful roles in the organization of agricultural society. The focus of religion in agricultural society shifted away from personalized demons and spirits toward more abstract and distant deities that appeared to be under the direct influence of these privileged classes.

An Expanded Division of Labor Improvements in agricultural technology gradually allowed more and more people to leave agricultural work.

else to do, or when she is walking along the trail, the Albanian housewife dutifully spins, and her spinning has soon become a reflex action, like knitting or bead-telling.

Farming is a family affair. The man buys the iron implements which he needs in town, and makes the rest of wood. He breaks the soil with a spade, and ploughs it with oxen. As he ploughs each furrow, his wife walks behind him with a basket of seed, sowing it. Later the women will do the weeding, and the whole family comes out to reap. If the crop is wheat, they cut off each head of grain with a sickle and put the heads in a bag; if maize, they pick the ears and braid the husks for drying. They thresh small grains and beans with flails. When the tobacco leaves are ripe, the men cut them carefully and hang them on the verandas or the sides of their houses to cure. The Mountain Ghegs also grow much fruit on trees near their houses. The chief species are

quinces, pears, and apples. Women have charge of milking the cows and making butter, curds, and cheese. Small boys are usually employed as shepherds, and lead their flocks high on the mountains. In the summertime many Gheg families drive their cattle up to the Alpine meadows on the mountain passes, and keep them there weeks at a time while they make butter and cheese. While on these heights they live in small temporary houses.

Animal husbandry furnishes the Ghegs not only with much of their food, but also with a supply of energy, for they use oxen in ploughing, and horses for travel and the transport of goods. In wintertime they keep their cows and horses indoors.

SOURCE: Excerpted from Carleton S. Coon (Ed.), 1977, "The Highland Ghegs." In *A Reader in Cultural Anthropology.* Huntington, N.Y.: Krieger, pp. 347–356. Reprinted by permission of the publisher.

These improvements included terracing and irrigation, the use of animal and human fertilizers, and advances in metallurgy that led to the proliferation of metal tools (Lenski, 1966). Box 1.2 on the Gheg of Albania describes the life of a typical agriculturalist. Agriculture continued to be the dominant form of economic activity in Western Europe until well into the Middle Ages. Throughout this long period few changes occurred in the life of the average person. This period, however, included the births and deaths of what are known as the classical civilizations and the emergence of feudal society.

Imperial Societies

Imperial societies were based on the subjugation of smaller and weaker agricultural societies by larger and more militaristic societies and the extraction of food, goods, and slaves as tribute. Based on the subjugation of these smaller societies and on improvements in agricultural technology,

the classical empires grew to immense size. For instance, historians estimate that the Inca Empire included 4 million people at the time of the Spanish Conquest (Lenski, 1966). Other examples of such civilizations include the Mayans and the Aztecs of Central America, the Azande of East Africa, the Phoenicians and the Egyptians of the Mediterranean, and the Imperial Chinese (Lenski, 1966:149). Imperial societies gave rise to the first large cities. In these cities, several thousand people lived off the agricultural surplus of the surrounding areas. New craft trades emerged in the cities to produce more refined products for the rising tastes of the empires' rulers, officials, and attendants.

Slave and Free Labor Perhaps as much as two-thirds of craft work in the classical empires was done by slave labor (Childe, 1964). Urban workshops were typically quite small, employing fewer than a dozen artisans. Most were smaller, consisting only of an artisan who hired free labor or bought

BOX 1.3   Slave Labor in the Roman Mines

The miner of ancient times was nearly always either a slave or a criminal. This explains why the means used remained almost unchanged for thousands of years. . . . It was considered unnecessary to make the work easier for the slave, whose hard lot inspired no sympathy, although it kept him to the end of his days buried in the gloomy depths of the earth, suffering all sorts of torments and privations. There was mostly a superabundance of slaves. After campaigns there were usually so many that great numbers of them were massacred. So there was no dearth of labour. And so it happened that in almost all the mines of the ancients only the simplest means were adopted.

The tunnels constructed in the rock by these simple means are often of astonishing length. It has been computed and confirmed by observing the marks of wedges that in even relatively soft stone the progress made amounted to about half an inch in twenty-four hours. This low efficiency was compensated to some extent by making the tunnels very low, by working only along the seams of the ore and by avoiding as far as possible the removal of unnecessary stone. Consequently, the galleries and tunnels were so narrow that a slave could squeeze himself through only with great difficulty. In many mines, in particular those of the Egyptians, Greeks and Romans, children were employed, so that as little

stone as possible would have to be removed. Although the slaves must have become weakened by their sojourn in the mines and by the unhealthy posture during work, as well as through sickness—in lead mines particularly through lead-poisoning—they must often have used very heavy tools. Hammers have been found that weighed between 20 and 26 pounds.

At the same time there were no precautions against accidents. The galleries were not propped up and therefore often collapsed, burying workmen beneath them. In ancient mines many skeletons have been found of slaves who had lost their lives in this way while at work. Nor were attempts made to replenish the supply of air or to take other steps for preserving health. When the air in the mines became so hot and foul that breathing was rendered impossible the place was abandoned and an attack was made at some other point. These conditions must have become still more trying wherever, in addition to the mallet and chisel, the only other means of detaching the stone was applied, namely fire. The mineral-bearing stone was heated and water was then poured over it. There was no outlet for the resulting smoke and vapors.

SOURCE: Excerpted from Albert Neuberger, 1977, "The Technical Skills of the Romans." In *A Reader in Cultural Anthropology,* edited by Carleton S. Coon. Huntington, N.Y.: Krieger, pp. 517–518. Reprinted by permission of the publisher.

slave labor to increase the output of pottery, cloth, woodenware, or metal tools. Because of the ready availability of slave labor, few technological advances occurred in craft production during the period of classical civilization. Slaves possessed neither the opportunity nor the motivation to be innovative at work. Such advances would await the development of free trades and free labor in the Middle Ages. Box 1.3 describes the harsh conditions of slaves used as miners during this period.

In Rome, craftworkers formed **guilds** to regulate the standards of their trade and provide religious and social services for their members. Guild membership was restricted almost exclu-

sively to men. During the decline of the Roman Empire and throughout much of the Middle Ages, in an effort to stabilize production, the authorities forbade leaving a guild or refusing to follow in one's father's trade (Tausky, 1984). "The Venetian government, for instance, strictly prohibited the emigration of caulkers, and from a document of 1460 we learn that a caulker who left Venice risked six years in prison and a two-hundred-lire fine if apprehended" (Cipolla, 1980:189). Although such efforts were of questionable success, the guilds thrived during this period and took on an important role in organizing production throughout the Middle Ages.

The End of Classical Civilizations In the centuries following the birth of Christ, Roman civilization declined because of the difficulties of maintaining a worldwide empire and because of direct challenges from Germanic tribes from Northern Europe and Mongolian tribes from Western Asia. The end of the period of classical civilizations is typically dated substantially later, sometimes as late as the fall of Charlemagne's empire in Western Europe in the early 800s. People left the large cities and returned to rural areas during these centuries. However, agricultural work was no longer undertaken by independent cultivators who were members of agricultural societies or by slaves laboring in large holdings. Instead, agriculture was organized around large estates in which local landlords ruled from fortified manors and the cultivators were legally tied to the land.

Feudal Society

In many ways, **feudal society** was simply an extension of agricultural society. The majority of people still tilled the land in the same traditional ways. However, the way in which agricultural surplus was extracted from peasants changed. In simple agricultural society and in imperial societies, peasants had given up a portion of their crops as tax to feed the rulers, priests, and warriors or they were forced to work as slaves. In feudal society, landlords extracted surplus both as a share of the peasants' crops and in the form of forced labor on the land-lords' land. The latter imposition was called *corvee labor* and peasants working under the feudal system were called *serfs*. Forced labor averaged three days per week. The burden was lessened to the extent that the labor could be performed by any member of the peasant's family, but it was increased to the extent that requirements were greatest at planting and harvest when the peasants most needed to tend their own fields. The movement of peasants off the manor was either forbidden or might result in a fine. Fines might be levied, for example, when a daughter married away to a different manor or a son moved to the city.

Extreme Inequality Incremental improvements in technology expanded agricultural productivity during the Middle Ages. These advances included the horseshoe, the padded horse collar, the wheeled plow, and the three-field system of crop rotation. These resulted in a gradual growth of population, but they did little to improve the position of the peasant who on average subsisted on a scant 1,600 calories per day (Polanyi, 1957).

The ruling class absorbed the additional surplus in what was perhaps the most extreme period of inequality in human history. Historians have estimated that between 30% and 70% of serfs' crops were expropriated in the form of taxes or duties by feudal lords or by the Catholic Church. In much of the feudal system as practiced in Western Europe, "when a man died, the lord of the manor could claim his best beast or most valuable movable possession, and the priest could often claim the second best" (Lenski, 1966:269).

Artisans and Guilds What is most distinctive and significant about the Middle Ages from the standpoint of work is the growth of a new class of producers, the free artisans. **Artisans** were typically the sons and daughters of serfs who had escaped the rural servitude of their parents and had moved to a town or "bourg." The citizens of medieval towns had a large degree of self-governance and were free from the obligations that characterized the rural manors. The trades of artisans included baking, weaving, and leather working. Instead of performing these diverse activities as part of their daily round of duties, the artisans specialized in a particular trade, producing superior goods for the other town dwellers who included church officials, soldiers, and merchants engaged in intercity trade.

The practitioners of the various crafts formed guilds for their mutual benefit. These guilds regulated the quality of goods, acceptable hours of work, and even prices. For example:

> Chandlers had to use four pounds of tallow for each quarter pound of wick. Makers of bone handles were forbidden by their guilds to trim their products with silver lest they pass them off for ivory. . . . In the interest of preserving fair competition, work on Sundays and saints' days was banned, thus preventing the impious from gaining an unfair advantage over the pious. Night work was forbidden both in the interest of fair competition and because poor lighting compromised meticulous workmanship. (Kranzberg and Gies, 1986:2–3)

Merchants, too, organized themselves into guilds to regulate and standardize their activities. Prices, hours, and first rights to bid on cargo were among the many regulated practices (Tausky, 1984). Medieval guilds were similar to those of the classical civilizations, but they played a considerably broader role in organizing the economic and political life of the medieval city. In the classic civilizations the king or the emperor and powerful landowners held political power. In the medieval cities, by contrast, the guilds were major actors in both economic and political life.

The training of new artisans was also strictly controlled by the guilds, again with the goal of regulating quality and thus protecting the reputation and status of the guild. An additional goal was limiting the number of people who could practice a trade, in order to ensure work for existing guild members. Apprentices were recruited from the extended families of artisans within the trade, from other artisans' families, and from rural areas. After years of on-the-job training, the apprentice would produce what was judged to be a masterpiece and would then be admitted formally to full guild membership as a master craftsman. Over the course of the Middle Ages, however, opening one's own shop became increasingly difficult as more and more craftsmen competed for limited markets. As a result, guilds became more restrictive in their membership criteria and a new position emerged called the **journeyman.** A journeyman artisan had successfully completed his apprenticeship training but did not own his own shop. Instead, he traveled from artisan to artisan, and sometimes from city to city, looking for work as an artisanal helper.

A final significant change in work that occurred during the Middle Ages was the disappearance of slave labor. Massive public works were constructed during this period, just as during the period of classical civilization. These monuments, cathedrals, and monasteries were financed by surplus extracted primarily from feudal serfs and secondarily from taxes on intercity trade. The labor, however, was not provided by slaves or by serfs. Rather, these structures were constructed by skilled artisanal workers, such as masons, carpenters, blacksmiths, and glaziers. The use of free skilled labor for these construction projects provided an important impetus to the growth of the artisanal classes during the Middle Ages.

The nature of daily life in the Middle Ages depended completely on one's position in society. Male members of the nobility ruled the manor, hunted, ate, and told stories. They were also responsible for defending the manor and its surrounding lands. This duty, however, generally required only their presence and availability; only in the rarest of occasions did it require actual defense of the manor against invaders. Female members of the nobility sewed, ate, and socialized. Serfs bore the task of eking a living out of the soil and worked long days toward that end. Religious holidays, relatively frequent by today's standards, provided welcome breaks from work for the serf.

Urban artisans had a lifestyle that was very distinct from that of the rural areas. They worked long hours with the help of family members and apprentices to produce goods of sufficient quality to be purchased by the nobility and by other town dwellers. Through their guilds they sought to regulate the number of hours of work allowable and to establish what they considered fair prices, so that they were not reduced to cutthroat competition against one another in the marketplace.

A New Vision of Work The advent of free artisanal work brought with it important new perspectives on the meaning of work and on relations between members of society. Artisans strongly supported traditional, highly skilled ways of work. Producing large quantities of goods quickly was not the overarching goal because there were no mass markets; the artisans served a very restricted clientele. The guilds therefore encouraged group solidarity to lessen the danger of being undercut in limited markets by price-cutting or by the sale of shoddy merchandise. As members of a guild, artisans identified their interests with preservation of high-quality standards rather than with getting ahead

of other artisans in their guild. These ideals of equality and solidarity would later help inspire the revolutionary demands of the artisans and the peasants as they sought to overthrow feudal society and install a society based on freely producing craft labor.

Economic Expansion and the End of Feudal Society Many changes led to the passing of feudal society and the transition to modern industrial society. This transition provides a fulcrum for much of the social history written since that time. It was also a central concern for the forerunners of modern sociology—Karl Marx, Max Weber, and Emile Durkheim. The transition from feudal society to industrial society was brought about by an expansion of population, trade, and markets. Between the years 1000 and 1500, more than a thousand new towns sprang up in Europe. Connected by usable roads, these towns provided the basis for *regional specialization,* that is, for a division of labor based on the unique resources of different regions (Kranzberg and Gies, 1986:16). The Crusades also spurred the growth of important new markets for European products, especially woolen goods. The exploration of the New World and the opening of trade routes to Asia were also boons to the European economy. Finally, the huge store of precious metals seized from Indian civilizations in Central and South America allowed a dramatic growth in the money supply and facilitated the expansion of credit and trade.

The period between feudal society and industrial society was one in which increased trade provided the impetus for changes in the organization of work. This intermediate period, called the age of Merchant Capitalism, lasted from the fourteenth century to the advent of the first modern factories in England in the mid-eighteenth century. Figure 1.2 provides a time line that summarizes major historical transformations in the organization of work.

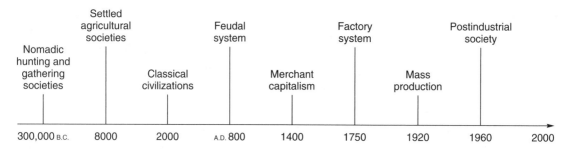

FIGURE 1.2 Time Line of the Organization of Work

Merchant Capitalism

The earliest form of capitalism grew not as a way to organize production but as a way to organize trade. Previously, craftworkers had bought their own raw materials and retailed the finished goods from their own shops. Under **merchant capitalism** the merchant capitalist increasingly took on these networking roles. This change frequently occurred because of the merchant's monopoly over lucrative intercity markets for finished goods or agricultural products. For example, a leather merchant might have had a corner on the purchasing of hides from a cattle-raising region or on the market for leather shoes sold in neighboring areas. Local craftsmen thus had to work through the merchant to participate in these markets.

The Merchant as Labor Contractor The system of production under merchant capitalism was called **putting-out industry,** because the merchant would "put out" the raw materials to be worked up and would later collect the finished products to be sold.

> In England the typical form of cottage or domestic industry was wool and, later, cotton weaving. . . . Merchants brought raw materials to rural cottages and then picked up the woven cloth which they had finished in towns or large villages. By having cloth woven in the countryside, the merchants managed to escape the control of the guilds. (Tilly and Scott, 1978:14)

In essence, craftworkers became subcontractors for merchants and were paid a piece rate for their work. The system also brought many nonguild workers into production because the merchant capitalist would put out simple tasks, such as preparing and softening leather or cleaning and carding wool, to less skilled workers whose labor was cheaper. These workers included seasonally underemployed peasants, recent immigrants from rural areas, widows, young women waiting to marry, and wives of underemployed husbands (Gullickson, 1983). Apprentices and journeymen who could not find employment as artisans because of the encroachment on craft markets by the merchant capitalists were also recruited into the putting-out system. In rural areas this system was called *cottage industry.* In urban areas it was called *sweatshop production* because the work typically took place in the attics of people's homes where it was hot, cramped, and often dirty.

Putting-out arrangements were the earliest form of the system of wage labor typical of industrial production throughout the world today. The system was successful because it undercut the pricing structure of guild regulations. The artisan made a full range of goods in his trade and wanted to be paid accordingly. A tailor, for example, would want to make a fair day's wage for his labor, regardless of whether he worked that day on petticoats, shirts, or jackets. However, some tasks, such as making petticoats, did not require the full range of the artisan's skills and could be done reasonably well by less skilled workers. The merchant

BOX 1.4 Sweatshops—Home Work in Early Capitalism

The three main features of the sweat-shop have been described as insanitary conditions, excessively long hours, and extremely low wages. The shops were generally located in tenement houses. As a rule, one of the rooms of the flat in which the contractor lived was used as a working place. Sometimes work would be carried on all over the place, in the bedroom as well as in the kitchen. Even under the best of conditions, this would have made for living and working in grime and dirt. . . . A cloakmaker used one room for his shop, while the other three rooms were supposed to be used for domestic purposes only,

his family consisting of his wife and seven children. In the room adjoining the shop, used as the kitchen, there was a red-hot stove, two tables, a clothes rack, and several piles of goods. A woman was making bread on a table upon which there was a baby's stocking, scraps of cloth, several old tin cans, and a small pile of unfinished garments. In the next room was an old woman with a diseased face walking the floor with a crying child in her arms.

SOURCE: Excerpted from *America's Working Women*, 1976, edited by Rosalyn Baxandall, Linda Gordon, and Susan Reverby, pp. 101–102.

would put out such work to a seasonally under-employed peasant or to a journeyman tailor unable to set up his own shop, and would pay this worker less than an artisan expected to receive as a living wage. On this basis, the merchant capitalists undercut the artisans' prices and encroached further and further into their markets.

Guild Resistance to Merchant Capitalism The guilds resisted the putting-out system by implementing civic laws regulating the number of journeymen or apprentices that one person could employ. However, the merchant capitalists' control of the intercity markets and their flexibility in putting out work to rural areas afforded them options unavailable to the craftworkers of any given city. As a result, the putting-out system eventually replaced the guild system in the manufacture of many basic commodities, most importantly in textiles. The craft system of work typical of the guilds lingers on to this day in such areas as specialty tools and the manufacture of some luxury items. The legacy of the guilds is also felt in modern unions and professional associations. However, it will never again be the way in which the majority of goods are manufactured in society.

The social relations of work were profoundly transformed by the putting–out system. In place

of free artisans, two classes emerged with distinct and even antagonistic relations: the merchant capitalists and those whom they employed in the putting-out system. The merchant capitalists sought to pay as little as possible for each type of work they put out. Those who worked under this system sought to secure a living wage for their labor, a goal often hampered by the availability of cheaper labor in another city or region. These relations set the stage for the historic conflict between *capital* and *labor.*

The daily lives of artisans were dramatically affected by the advent of merchant capitalism. Wages fell for craftworkers as they were forced to cut their prices to retain a share of markets (Hobsbawm, 1969). They were forced to work longer hours and in general their positions as members of the middle class were gravely threatened. Even the average age of marriage among journeymen increased substantially at this time because of the difficulty of securing a position as a master craftsman who could support a family (Aminzade, 1993). On the other hand, the technical aspects of work tasks were in many ways little affected by the rise of merchant capitalism. Most tasks were still done in traditional ways. Box 1.4 describes the working and living conditions of a family making cloaks in their urban apartment.

Under the guild system, women had worked as helpers in their husbands' crafts and sometimes as members of their own guilds, though generally at lower earnings (Deckard, 1979:208). Under the putting-out system women were often employed directly by merchant capitalists, who used their low wages as leverage to undercut artisanal wages. By working at home, women were able to combine various forms of productive activity including paid work, domestic activity, and care for children. Because paid work was only one part of their productive activity, they were often willing to undertake this work for lower wages than were skilled artisans, who needed to secure their entire livelihood in this way (Tilly and Scott, 1978:3).

Spiritual Grace and Worldly Success Merchant capitalism also witnessed the emergence of new theologies based on the thought of Martin Luther and John Calvin, whose writings gave voice and content to the Protestant Reformation. These theologies suggested a new vision of work sometimes called the **Protestant work ethic** (Weber, 1958 [1904]). This vision identifies successful pursuit of one's occupational calling with spiritual grace. If one prospers through diligent work, this prosperity is seen as evidence that one is among those chosen to go to heaven. The Protestant work ethic was well matched with the emerging worldview of the merchant capitalist, who was engaged in a competitive struggle for success on earth. This ethic identifies worldly success as a sign of spiritual grace and provides both a justification and a motivation for the pursuit of earthly endeavors. The Protestant work ethic also encourages savings and thus lays a foundation for the accumulation of capital to be reinvested in business and commerce. Savings are encouraged by the ethic's call for worldly asceticism to be realized through frugality, austerity and plain living. The ethic also provides a basis for despising those who are less successful because their lack of success implies that they are among those whom God has forsaken. Although few would subscribe completely to this view today,

elements of the early Protestant work ethic are retained in many aspects of modern Western culture (Applebaum, 1998).

The Industrial Revolution

The transition from putting-out industry to industrial capitalism was a violent one. It involved the forcible movement of large numbers of peasants off the land and into factories. In the words of Karl Marx, it is a history "written in letters of fire and blood." The Industrial Revolution took place first in England, partly because the English were involved in an expanding woolen trade with the north European region of Flanders. England's role in the woolen trade set the stage for the forcible removal of peasants from the land and their replacement by grazing sheep. This displaced peasantry, in turn, provided a ready pool of labor for the early factories. The woolen trade also generated capital for investment in new factories and machinery.

Replacing Agriculture with Industry Peasants were forced off the land through "**enclosures,**" in which land previously held in common by the peasants and the landlord and used for grazing livestock, was enclosed with fences (Hobsbawm, 1969). The land was then used for raising sheep. This change caused a dramatic deterioration in the situation of the peasants, who were no longer able to use this land to support their few farm animals. Marx notes the actions undertaken by the Duchess of Sutherland between 1814 and 1820 as an example of this process:

> [The] 15,000 inhabitants, about 3,000 families, were systematically hunted and rooted out. All their villages were destroyed and burnt, all their fields turned to pasturage. British soldiers enforced this eviction, and came to blows with the inhabitants. One old woman was burnt to death in the flames of the hut, which she refused to leave. Thus this fine lady

appropriated 794,000 acres of land that had from time immemorial belonged to the clan. (Marx, 1967 [1887]:729–730)

After the peasants had been forced off the land to make room for sheep, they were further hounded as vagabonds until they entered the early factories, often as forced labor.

The government of England enforced the movement of displaced peasants into the early factories through what Marx called "bloody legislation against vagabondage":

Beggars old and unable to work receive a beggar's license. On the other hand, whipping and imprisonment for sturdy vagabonds. They are to be tied to the cart-tail and whipped until the blood streams from their bodies. . . . For the second arrest for vagabondage the whipping is to be repeated and half the ear sliced off; but for the third relapse the offender is to be executed as a hardened criminal and enemy of the common weal. . . . Thus were agricultural people, first forcibly expropriated from the soil, driven from their homes, turned into vagabonds, and then whipped, branded, tortured by laws grotesquely terrible into the discipline necessary for the wage system. (Marx, 1967 [1887]:734–737)

Vagabondage laws, poor laws, and head taxes were used to force displaced agricultural workers into the early factories. The earliest factory workers were thus frequently the victims of penal sanctions rather than freely hired wage labor (Wilensky and Lebeaux, 1986:29).

The emergence of these factories was the final blow against the feudal guilds and their apprenticeship systems for recruiting and training skilled workers. Indeed, many laws prohibiting the existence of guilds, unions, and other combinations of workers—and even defining union membership as a crime punishable by death—were passed in England in the sixteenth, seventeenth, and eighteenth centuries (Hobs-bawm, 1969). Factories also signaled an end to merchant capitalism, which was based on expanding production by putting out work to more and more home-based workers, and ushered in the next stage of society and economic history—industrial capitalism.

The Factory System

What made the early factories so terrible that people refused to enter them except under the force of law? To understand the position of the early factory workers, it is helpful to know something about how these factories operated. Workers were centralized under one roof in the factory system. Such centralization avoided the costs of transporting partly finished goods from one location to another as required by the putting-out system. It also forced workers to work according to the dictates of the owners rather than according to their own pace and rhythm. The centralization of work also meant that in order to have access to any work at all, a worker had to be willing to work the hours and days demanded by the employer. The result was an expansion in the length of the working day, an increase in the intensity of work, and a decrease in the number of religious and personal holidays allowed.

The massing of many workers in one location also allowed the development of machinery to do repetitious tasks. A seemingly endless variety of tasks could be broken down into their simplest components and mechanized. The widespread introduction of machinery had not been feasible under the putting-out system, because too few repetitions of a task were performed at any one location to warrant the expense of machinery. The social organization of the factory thus encouraged the introduction of machinery.

The tendency toward longer working days was further intensified by the factory owner's need to run the new machinery as many hours per day as possible in order to justify its purchase. The machinery heightened productivity but ironically resulted in lower wages for workers because fewer skills were needed. Increased productivity and decreased costs allowed factory owners to sell their products at prices that drastically undercut artisanal producers and even merchant capitalists. These dynamics resulted in the dominance of factory production over older forms of production.

Textiles and the Industrial Revolution The **Industrial Revolution** in England had begun in the woolen industry. The focus shifted to cotton textiles as the import of cotton from the New World grew and markets for British textiles expanded in Asia and North America. Several inventions also facilitated the rapid growth of the textile industry, including the flying shuttle (1733), the spinning jenny (1767), the water frame (1769), the spinning mule (1779), the power loom (1787), and the cotton gin (1792) (Faunce, 1981:14). These developments generated tremendous growth in British industry in a very brief period. "Output of printed cotton rose from 21 million yards in 1796 to 347 million yards in 1830. Pounds of raw cotton consumed increased from 10.9 million in 1781 to 592 million in 1845" (Faunce, 1981:13). This boom in cotton cloth production also sparked the rise of cotton plantations in the American South and fueled the slave trade as a source of cheap labor for the cotton fields.

The consequences of growth in the textile industry spread rapidly to other British industries. Coal was needed to provide steam for power looms. Steel was needed to build railways to carry the coal. More coal was needed to make coke for smelting iron and steel. Machine tools were needed for building and maintaining textile, mining, and railway machinery. Ship manufacture advanced perhaps more rapidly than any other field because of the dependence of the Industrial Revolution on foreign trade. By the end of the nineteenth century, huge steel-bodied, steam-powered ships would replace the slower, smaller sailing vessels allowing a further expansion in the tonnage of goods shipped around the world.

The Detailed Division of Labor The division of labor into finer and finer activities advanced more rapidly during this period than at any other in history. The English economist Adam Smith described this process for the simple trade of pin manufacture:

> One man draws out the wire; another straightens it; a third cuts it; a fourth points it; a fifth grinds it at the top for receiving the head; to make the head requires two or three distinct operations; to put it on is a peculiar business; to whiten the pin is another; it is even a trade by itself to put them into the paper; and the important business of making a pin is in this manner divided into about 18 distinct operations, which in some manufactories are all performed by distinct hands, though in others the same man will sometimes perform two or three of them. . . . Ten persons, therefore, could make among them upwards of 48,000 pins in a day. (Smith, 1937 [1776]:3)

Under the artisanal system a pin maker might have repeated each of the eighteen distinct operations several times before proceeding with the next step, thus increasing efficiency by not having constantly to take up and put down each task. However, these tasks would never have been assigned to different workers. Social theorists have offered strongly divergent appraisals of the consequences of this detailed division of labor. Smith applauded the system because it lowered the price of pins. Marx condemned it because it reduced the skills and lowered the wages of the workers who made the pins.

In the early factories work was organized under the supervision of foremen. These foremen were more similar to subcontractors than to modern-day supervisors. They hired their assistants, trained them in their tasks, set them to work, supervised them on the job, and paid them out of the piece-rate earnings they received for the goods produced. This system was typical of the textile industry, iron making, gun making, saddle making, coach building, and most other trades until the beginning of the twentieth century (Braverman, 1974).

The position of workers deteriorated rapidly during the early stages of the Industrial Revolution. The earnings of weavers fell to as little as a tenth of what they had been before the introduction of machinery (Hobsbawm, 1969). This resulted partly from the reduced need for skilled workers in the early factories. The early factory workers were more likely to be women, children, indentured laborers, vagabonds forcibly placed in poorhouses, and, in the New World, slaves and indentured servants. **Indentured laborers** were workers under contract to work for a certain amount of time—typically eight to ten years—for a set price or as part of their penalty for being found guilty of a crime such as petty theft or vagabondage. Handicraft weavers simply could not compete with the lower wages and the new and more productive technologies of the factory system.

The rights of those who labored in the early factories were minimal. Slavery and indenturement were common, especially in the New World. The role of indentured laborers in settling the new lands started as early as 1607 with their use by the Virginia Company. Conditions for these workers were terrible. They suffered high levels of mortality and were treated with great cruelty. In response some indentured workers ran away to live with the Indians. The Virginia colony's governor dealt firmly with recaptured laborers: "Some he apointed to be hanged Some burned Some to be broken upon wheles, others to be staked and some to be shott to death" (quoted in Galenson, 1984:4). A quarter of the labor force in the United States in the early 1800s was made up of slaves. From the standpoint of the slave owners, this provided a workable solution to the problem of labor shortages in the New World and, simultaneously, a solution to the problem of retaining and controlling their workers. Wage laborers thought less well of this solution as it set a harsh standard against which their own labor was measured. Slaves were not consulted on the system.

Women and Children In 1838, only 33% of the workers in British textile factories were adult males (Hobsbawm, 1969:68). Women and children thus played a central role in the early Industrial Revolution. They were employed because it was acceptable to pay them much less than men and because they were easier to bully into the harsh discipline of mechanized production. Young women were also considered more expendable to agricultural work and were thus more likely to be available for factory work. Between 1850 and 1950, women made up 30% to 35% of the manufacturing labor force in the industrialized nations of France and England (Tilly and Scott, 1978:70). Given the conditions of early factory work, it would be inaccurate to consider the participation of women at this time as a sign of their emancipation (Kessler-Harris, 1982). The plight of women in the early factories is illustrated in a petition to the state of Massachusetts that the women of Lowell brought against their employers (see Box 1.5).

Government involvement in the economy during the period of early industrialization occurred in the form of military support for the establishment of British colonies and the expansion of trade with these colonies. Indeed, this period of British political and economic history is sometimes called *mercantilism* to indicate the pivotal importance of establishing international markets. Government involvement did not, however, include any but token efforts to establish minimum working or safety standards in British factories.

Inside the factories the round of daily life was extremely monotonous. The strict routine was quite unlike preindustrial rhythms of work, which had been based on seasonal variations, frequent holidays, and a degree of personal discretion in organizing one's daily tasks (Hobsbawm, 1969). The hours were extremely long. Neither before nor after have people worked longer or harder than during early industrial capitalism: "In some huge factories from one fourth to one fifth of the children were cripples or otherwise deformed, or permanently injured by excessive toil, sometimes

BOX 1.5   Complaint by Lowell Factory Women, 1845

The first petitioner who testified was Eliza R. Hemingway. She had worked 2 years and 9 months in the Lowell factories; 2 years in the Middlesex, and 9 months in the Hamilton Corporations. Her employment is weaving—works by the piece. . . . She complained of the hours for labor being too many, and the time for meals too limited. In the summer season, the work is commenced at 5 o'clock, a.m., and continued til 7 o'clock, p.m., with half an hour for breakfast and three quarters of an hour for dinner. During eight months of the year but half an hour is allowed for dinner. The air in the room she considered not to be wholesome. There were 293 small lamps and 61 large lamps lighted in the room in which she worked, when evening work is required. . . . About 130 females, 11 men, and 12 children (between the ages of 11 and 14) work in the room with her. She

thought the children enjoyed about as good health as children generally do. The children work but 9 months out of 12. The other 3 months they must attend school. Thinks that there is no day when there are less than six of the females out of the mill from sickness. Has known as many as thirty. She herself is out quite often on account of sickness. . . . She thought there was a general desire among the females to work but ten hours, regardless of pay. . . . She knew of one girl who last winter went into the mill at half past four o'clock, a.m., and worked till half past 7 o'clock, p.m. She did so to make more money.

SOURCE: Excerpted from *America's Working Women,* edited by Rosalyn Baxandall, Linda Gordon, and Susan Reverby, pp. 49–50. Copyright 1976 by Rosalyn Baxandall, Linda Gordon, and Susan Reverby. Reprinted by permission of Random House, Inc.

by brutal abuse. The younger children seldom lasted out more than three or four years without some illness, often ending in death" (quoted in Faunce, 1981:16). Beginning in the early 1800s, the working day was further extended from twelve hours to as many as sixteen hours by the use of gaslight. As a result, workers labored from before dawn well into the night (Hobsbawm, 1969:60).

Industrial Cities The cities in which this emerging working class lived were no more healthy or pleasant than the factories in which they labored:

> The industrial town of the Midlands and the North West was a cultural wasteland; . . . Dumped into this bleak slough of misery, the immigrant peasant, or even the yeoman or copyholder was soon transformed into a nondescript animal of the mire. It was not that he was paid too little, or even that he labored too long—though either happened often to excess—but that he was now existing under physical

conditions which denied the human shape of life. (Polanyi, 1957:98–99)

Not until the 1890s did sewage systems begin to catch up to the need for them, even in the largest industrial cities (Tausky, 1984:37). Epidemics of cholera and typhoid took an appalling toll on workers in the early industrial cities. Domestic cooking and heating as well as factory production depended on coal, and a thick layer of soot settled over everything in the industrial city. As a result, thousands died from tuberculosis and other respiratory ailments—or had their lives shortened and crippled by these diseases (Hobsbawm, 1969:86).

The introduction of factories further separated work from the home. If people wanted to find gainful work, they had to leave their families and venture out alone. The removal of work to the factory undermined the family's function as the primary unit of economic production. To the extent that women were unable to leave the house in the search for work because of the demands of domestic activity and child rearing, their relative contribution to the household's economy was undermined (Matthei, 1982). Women often supplemented their contribution, however, by taking in boarders, doing laundry or sewing at home, or performing other wage-earning activities compatible with their domestic responsibilities (Tilly and Scott, 1978).

The Skilled Trades Centralized work in factories under close supervision, with machines dictating the pace, robbed workers of the skills and autonomy necessary to take pride in their work. It is no wonder that early factory workers were alienated and resentful (Wilensky and Lebeaux, 1986:35). It would be inaccurate, however, to characterize all workers as having experienced a loss of skill and pride in their work at this time. Older, skilled ways of work prevailed among blacksmiths, shoemakers, stone masons, and the new machinist trades whose growth was spurred by the Industrial Revolution. These workers continued to experience strong feelings of pride and of solidarity with their fellow craftworkers. The skilled trades provided the basis for early attempts at unionization. In North America these efforts produced a variety of unions, including local craft unions, the Knights of Labor, and national craft unions organized under the banner of the American Federation of Labor. We discuss these early attempts at unionization further in Chapter 6. Successful efforts to establish unions in the larger factories did not occur until the 1930s with the emergence of new strategies for organizing workers in the mass-production industries.

Industrial Capitalists The new class of factory owners, called *industrial capitalists,* and those who aspired to become members of this class, developed new ideologies to represent their view of work. One popular ideology among the capitalist class in North America in the latter part of the nineteenth century was Social Darwinism, patterned after Charles Darwin's evolutionary theory of the survival of the fittest (Perrow, 1979:60). The Social Darwinists viewed extreme competition and inequality as not only just but also as being in the best interests of society, because they

would ensure that the most able individuals would rise to the top. Similar lines of reasoning were embodied in the philosophy of *laissez-faire,* which argued that economic systems work best when undisturbed by any government stimulus or regulation. The philosophy of laissez-faire argues that unregulated markets maximize outcomes for individuals and for the community as a whole. The growth of such self-oriented ideologies and the waning of craft ideologies stressing pride in work and solidarity left little middle ground for the development of a shared workplace ethic suitable for reconciling the two opposing classes of industrial capitalism (Hill, 1981:22). Attempts at reconciliation would await the development of monopoly capitalism and postindustrial society.

In the long run industrial capitalism allowed for some improvement in the position of the working class. In the modern world "nearly every indicator of physical well-being corresponds to the extent of industrialization" (Tausky, 1984:30). Increased well-being is based on greater agricultural and manufacturing productivity and on workers' abilities to demand a share of these rewards. The latter, in turn, is based on the continuing need for new skills as technology advances and on the organization of workers into unions and their successful petition for legal rights and safeguards in the workplace.

Early Trade Unions In 1824, trade unions ceased to be illegal in England though every effort to destroy them was still made by employers (Hobsbawm, 1969:122). By 1870, other important legal changes were occurring in England, including the passage of child labor laws and a law limiting the working day to ten hours. By the 1880s, all children under the age of ten were required to be full-time students. Factory inspections also increased to provide at least some enforcement for these laws. These changes occurred about fifty years later in North America, which experienced a slower process of industrialization and a later emergence of unions. In aggregate, these changes produced an increase in life expectancy for workers and improved standards of living.

By the early years of the twentieth century, industrial capitalism was displaced by monopoly capitalism (Burawoy, 1985). This change occurred because of the greatly expanded size of companies. Large companies could utilize more efficient types of mechanization and more powerful marketing techniques than smaller companies and thus gradually came to replace them in more and more industrial lines. Power became centralized in many industries in an *oligopoly* of a few large companies. Other industries became dominated, at least regionally, by a single company called a *monopoly*. Today, railroads and power companies are good examples of complete or almost complete regional monopolies.

Along with greater oligopoly and monopoly power came an increased involvement of government in the productive process. Government involvement grew for a number of reasons. For one, there was a heightened need to establish secure markets for the greatly expanded productivity of the large companies. Industrial nations struggled for a share of world markets, and governments took on an important role in securing such markets through diplomatic and military action (Hobsbawm, 1969:131). The First and Second World Wars were to a significant extent a result of such global economic competition. In addition, competing demands on the state by monopoly capitalists, small capitalists, and the growing working class encouraged greater government involvement in the economy to stabilize class relations and to mediate between these competing interests. This intervention was frequently very uneven with powerful groups successfully using the government to support their own interests.

Mass Production under Monopoly Capitalism

The end of the nineteenth century and the beginning of the twentieth century was a time of great changes in the organization of enterprises. Increased size and centralization were the order of the day. "In 1851 there were fifty telegraph companies; in 1866 Western Union operated fifty thousand miles of line and had bought up all but a few remaining telegraph companies. . . . In the 1890s Carnegie's steel factories controlled two-thirds of steel production in the United States" (Tausky, 1984:46–47). American Tobacco Company controlled over 90% of cigarette sales by 1890. The Westinghouse Electric Company and General Electric Company controlled electrical equipment. The F. W. Woolworth Company held a huge share of sales in general merchandise stores. Sears, Roebuck, and Company and Montgomery Ward and Company dwarfed their competitors in the mail-order business. Several dozen early car manufacturers were overshadowed and forced out of business by the wildly successful Ford Motor Company. E. I. DuPont de Nemours and Company dominated chemical production, and the Standard Oil companies, owned by the Rockefellers, dominated petrochemicals. These successes were partly based on efficiencies associated with economies of scale. They were also based on the ability to develop and deploy new technologies and the ability to use economic and political power to undercut competitors. Smaller companies had a difficult time competing against these corporate giants. They were forced to operate in areas yet unconquered by the large companies, or to become suppliers of parts and services to the giant corporations (Averitt, 1968).

Along with greater size and concentration of companies came increased bureaucracy. Bureaucracy entails the systematic use of rules and the creation of formalized job positions with clearly delineated duties. (See Chapter 7 for further discussion of bureaucracy.) Standardized procedures had to be developed as the internal workings of companies became more complex and there was less and less reliance on various forms of subcontracting to foremen. Foremen no longer hired their work crews, for example. Instead, a central personnel office did the hiring. For workers, such bureaucratic procedures can have both positive and negative consequences. Standardized practices protect workers from some of the worst abuses and favoritism of foremen, but they also tend to further depersonalize the work environment.

Mass Production The immense size of economic organizations at the beginning of the twentieth century brought about the possibility of new forms of technology and new means of organizing work and controlling workers. The most notable of these was the **assembly line.** The full-fledged use of the assembly line occurred first in the automobile industry, but it built directly on existing forms of mechanized production being used in other industries at the time:

> An unskilled operator snapped engine blocks, for example, onto specially designed tables and watched a machine mill them automatically and accurately. Made this way, parts such as cylinder heads and engine blocks could be fitted together without the need for hand scraping of surfaces during assembly.
>
> Once hand fitting had been eliminated and the specialized machines had been arranged in the sequence of manufacturing operations, the next and revolutionary steps were to reduce the transit time of work pieces from machine to machine and to systematize their assembly. . . . The assembly line may have been inspired by the disassembly lines in Chicago slaughter houses, which circulated carcasses from butcher to butcher, or by Ford's own gravity slides and conveyors. . . . The first crude moving line cut the time needed for final assembly from just under 12.5 to about 5.8 man-hours. (Sabel, 1982:33)

With assembly-line production, job skills become highly specific to the technology and procedures used in a given plant. Such jobs are considered *semiskilled,* because they require a specific skill but one that can be learned in a relatively short time,

often one to two weeks. A smaller portion of workers are required to have broader-based skills, such as those of the machinist, electrician, or tool and die maker.

The Assembly Line The assembly line sets a rapid pace for workers and keeps them at required tasks much more closely than even the harshest foreman (Thompson, 1983:144). Assembly lines and other forms of advanced mechanization are organized under the principles of *scientific management.* Scientific management is identified with the work of Frederick Taylor, an American industrial engineer, who believed that all thinking should be removed from the realm of the worker. Instead, the worker was to execute diligently a set of motions engineered to ensure the most efficient performance of a given task. This so-called scientific plan came from first observing how workers did the task, then progressively redesigning the task to increase efficiency. The theory of scientific management was highly compatible with the new assembly-line system of production. In combination, however, these new forces sparked fierce resistance from workers, who felt that the new production systems treated them like automatons rather than human beings.

The daily life of production workers under monopoly capitalism was somewhat better than it had been in early industrial capitalism. The working day was progressively shortened because of heightened productivity and because of pressure from working-class trade union and political activity. However, the monotony of work, if anything, was increased. In North America, the emergence of monopoly capitalism coincided with a period of heavy immigration of workers from Western Europe, in the 1870s, and from Eastern Europe, in the 1910s. These immigrants, both male and female, provided labor for many of the expanding mass-production industries. Ethnic groups often moved through the factories in successive generations, as one group moved upward toward craft and white-collar jobs, another group replaced them on the factory floor.

The shared situation of large numbers of semiskilled workers in the mass-production industries facilitated solidarity and the identification of common grievances among these workers (Hill, 1981:217). Workers in such industries as automobiles and steel organized into industrial unions to promote their collective welfare. In North America there was a dramatic growth in union organizing during the Great Depression of the 1930s. Demands for higher wages were not the primary motivation for these organizing drives. Rather, the workers resented the dehumanizing practices related to scientific management and the speed-ups imposed on the already rapidly paced assembly lines. We discuss these union organizing drives in greater detail in Chapter 6.

A More Complex Class Structure The class structure of monopoly capitalism was significantly different from that of early industrial capitalism. Society remained strongly divided between owners and workers, who were themselves divided between skilled craftworkers and less skilled workers. However, there was a rebirth of the middle class as a result of a heightened need for clerical, managerial, and professional workers in the giant new companies and in the growing public sector. Society also experienced an expanded demand for professional services, such as education and health care, as a result of a rising standard of living.

Ideas about the meaning of work were also changing among capitalists, and especially among the new class of hired managers. The old ideology of "survival of the fittest" provided little guidance on how to placate angry workers or encourage their productivity. As a result, new ideologies focusing on persuasion and cooperation emerged. One ideology, associated with Elton Mayo and Chester Barnard, early industrial sociologists, came to be known as the *human relations* approach to industrial management (Perrow, 1979:68). Workers were seen as tractable and as desiring recognition and personal treatment. Such a school of thought reflected the reality that

workers had been robbed of such individual recognition and personal treatment by the advent of industrial capitalism and the subsequent development of mass-production systems.

Postindustrial Society

The transition from mass production to the present stage of industrial society, sometimes called **postindustrial society,** resulted from the immense productivity of mass-production systems. This transition began with two decades of stable economic growth in the world following World War II. By the 1960s, a smaller and smaller percentage of the labor force was needed to manufacture goods. Similarly, farm productivity continued to rise. As a result, new employment growth has taken place mainly in clerical, service, and professional work. Some of these jobs are rewarding and exciting; others are as monotonous as factory work and frequently pay worse. In combination, their growth signals the advent of postindustrial society.

Also associated with the development of postindustrial society is an increasingly international division of labor, which has reduced the significance of local markets and economies. The economic fate of every nation is highly dependent on its position in the world economy. Thus, the nature and rewards of work are not solely determined by social and economic relations with others in one's work group or employing organization. Rather, the nature of one's work is importantly determined by its position in the global division of labor.

The heightened international division of labor has intensified inequalities between nations. Some nations appear trapped in the role of agricultural production or mineral extraction; others are becoming centers of low-wage assembly work reminiscent of early industrial society; still others are characterized by the growth of service and professional work. The postindustrial world includes all of these different types of work and societies. The international division of labor and the relations between the industrially

---

BOX 1.6 Electronics Assembly in Malaysia

The large-scale entry of labor-intensive industries such as garment manufacturing, food processing, but especially electronics assembly greatly increased the absorption of rural Malay women into the industrial labor force. . . . They represented a fairly well-educated labor pool, which was often overqualified for the mass semiskilled factory occupations. About 50 percent of these workers had at least lower secondary education, and many had aspired to become typists, secretaries, trainee nurses, or teachers. . . .

Maximum product output was extracted from these rural women by the factories. Quickly exhausted operators were replaced by the next crop of school leavers. By keeping the wages low, the factories motivated the operators to work overtime on a regular basis, to take on more unpleasant tasks (which exposed them to fumes and acids), or to work at an increased pace in order to earn special cash allowances. Freshly recruited workers were routinely assigned to the production processes that required continual use of microscopes. Thus most workers, by the end of a couple of years, suffered from eye strain and deterioration of their eyesight. . . .

The industrial firms not only exploited the workers in this manner, but they also attempted to limit their employment to the early stage of adult life, a strategy that ensured fresh labor capable of sustained intensive work at low wages. In one factory new workers were employed on six month contracts so that they could be released or rehired at the same low wage rates. Government legislation for the protection of pregnant female workers has had the unintended effect of reinforcing factory policy to discourage married women from applying, although employed workers who got married would stay on. Married workers were given advice on family planning and provided with free contraceptives by the factory clinic.

The rapid exhaustion of the operators also resulted in most of them leaving on their own accord after three to four years of factory employment, although an increasing number remained working, even after marriage. Operators leaving the factories have not acquired any skills which would equip them for any but the same dead-end jobs.

SOURCE: Aihwa Ong, 1983, "Global Industries and Malay Peasants in Peninsular Malaysia." In *Women, Men, and the International Division of Labor,* edited by June Nash and Maria Patricia Fernandez-Kelly, pp. 429–431. Copyright 1983 State University of New York Press. Reprinted by permission of the publisher.

---

developed and the industrially developing world are key to understanding the contemporary world of work. The growth of manufacturing in the Third World is directly linked to the decline of manufacturing in the industrially developed nations and the growth of the service sector in these societies, with all its positive and negative consequences. The report on Malaysian electronics workers in Box 1.6 illustrates how the conditions of the early factories are being reproduced today in manufacturing establishments in the Third World, even those producing high-technology products for global markets. The conditions of such workers and the nature of the modern global economy are discussed further in Chapter 16.

Service Industries A growing proportion of workers in the industrially advanced nations are employed in such service industries as transportation, wholesale and retail trade, finance, insurance, real estate, professional and business services, public administration, entertainment, and health. In 1940, less than half of the U.S. labor force was employed in service industries (Bell, 1976). Today the proportion exceeds 80%.

What is work like in this postindustrial society? Work in such a society includes only a small number of people engaged in agriculture. This number is slowly decreasing toward 2% of the labor force in the United States (Carey and Franklin, 1991:46). A larger proportion of people, about 20%, are engaged in manufacturing. The remainder are in

BOX 1.7   Engineering the Future

Developing new product lines is the glamorous work. This is seen as the essence of creative engineering, what engineering is all about. It is high-pressure work: crunches, slips, and other forms of organized hysteria accompany the pressure to be creative, to produce, to be smart. In development, engineers typically work on projects. . . . Within development, engineers sort themselves out by the type of work they do and their perceived skill. Engineering is a highly competitive arena in which formal statuses are supplemented by informal ratings. Informally, engineers are categorized by their skill. There are

the "brilliant" and the "geniuses," their status sometimes debated ("the only way he made the list of 100 brightest scientists is if he mailed coupons from the back of cereal boxes") and sometimes acknowledged ("Peter is brilliant. There is no question about that; he is a crackerjack engineer"); and there are journeymen (and the occasional journey women), who might be "solid citizens—no rah rah."

SOURCE: Excerpted from Gideon Kunda, 1992, *Engineering Culture*. Philadelphia: Temple University Press, pp 39–40.

---

service industries, but these jobs vary greatly. Some are low-skill jobs in food service and health care, or repetitive and poorly paid clerical and data-entry jobs. These jobs are predominantly filled by women and minorities. Other jobs entail highly skilled professional work. Examples include computer programming and systems analysis. The overall effect of this variability is an increase in the *diversity* of employment situations rather than a clear improvement or deterioration in the quality of available work. In addition, the pace at which new skills are needed has accelerated dramatically as a consequence of rapid technological changes based on the widespread use of microprocessors and computers in the workplace. This results in an additional set of problems of adjustment and retraining for workers.

The number of highly skilled professional workers has grown in postindustrial society. These workers hold a privileged position in the division of labor based on their possession of knowledge and expertise not widely available without rigorous, extended study and preparation. As a result of their expertise, they command relatively high wages and a certain degree of autonomy in decision making (Abbott, 1988). Box 1.7 describes the work environment of engineers in a high-technology company.

The proportion of skilled and semiskilled manual workers in the labor force has remained relatively stable in postindustrial society. A new class of low-paid service workers, however, has come into being. Their wages and conditions are considerably worse than those of skilled craftworkers and those of most semiskilled workers. Their conditions more closely resemble those of the declining segment of unskilled labor. In many ways their actual jobs are also similar to those of unskilled laborers. Service workers move hamburgers around the fast-food outlet or bedpans around the hospital, just as laborers move materials around the factory floor or construction site.

The class structure of postindustrial society is the most diverse of any society to date. It includes a capitalist class, a managerial class, a large professional class, a large manual class, and a large service class. The disparate situations of these classes include both opulence and continuing poverty. It was once believed that postindustrial society would bring about an end to poverty (Bell, 1976). Contemporary social scientists now believe that such an expectation was grossly exaggerated (Wilson, 1997).

Work Motivation What motivates people to work in postindustrial society? Some observers assert that the work ethic has died in recent years and that apathy rather than motivation typifies the workforce (Pascarella, 1984). This vision only represents a partial truth, however. People whose jobs pay no more than a subminimal level have little reason to work enthusiastically; but even so, they often strive for dignity by taking pride in their work. Alternatively, for those with advanced education, more desirable job opportunities are available. These workers are motivated by the high level of rewards they can expect, by the pride associated with advanced training in a professional specialty, and by the expectation of their profession and their employer that they will be committed to their work (Hochschild, 1997).

Being committed to one's profession or to one's organization provides a meaningful orientation to work for a significant portion of the labor force in postindustrial society. This focus is distinct from earlier visions, which exhorted workers to succeed as a sign of God's grace or to work in solidarity with other members of their class to improve their collective position. This new vision includes a strong theme of individual attainment but also retains the idea of commitment to broader goals.

A major change in work life that distinguishes postindustrial society is related to the idea of commitment. Work takes on an overriding importance in people's lives, tending to overshadow family and community attachments that prevailed in previous periods. Many factors amplify this tendency. Greater demands for geographic mobility preempt family attachments and curtail friendships of long duration. The fear of unemployment intensifies competition for available positions. Retirement, access to health care, and social status are all attached to one's job. In postindustrial society, work has become a "master status," determining a person's overall position in society and his or her sense of dignity and identity.

Women's Liberation? The position of women has significantly improved in postindustrial society. Women have entered jobs that were once exclusively male preserves, including those of airplane pilot, firefighter, and heavy-equipment operator. The improved position of women is partly the result of a reduction in the demands on women to perform homemaking duties. These roles have declined in significance because of delayed marriage, reduced birth rates, labor-saving technologies, and the substitution of paid services in such areas as child care for services that women previously performed at home. Women have also secured better jobs because of their high rate of college graduation in combination with the increased importance of higher education in postindustrial society. In spite of these gains, however, women's earnings for full-time work still equal only about 75% of men's. We explore some of the reasons for the limited nature of these improvements in Chapter 3 on work and the family.

An Increasingly Diverse Workforce The movement of people around the world in the search for greater economic opportunities is growing rapidly in the twenty-first century. As a result, the economies of both developed and developing countries are characterized by increasing racial and ethnic diversity. This diversity represents both a challenge and an opportunity. The *challenge* is to train and integrate the new workforce while minimizing discrimination and resentment from other workers. The skills and motivations of these new workers, however, also represent an important *opportunity* for heightened economic growth and prosperity for all. Chapter 5 describes some of the disruptions and barriers that workers sometimes face, with a special focus on the challenges of female, minority, and immigrant workers.

The Future Work is here to stay. The future, however, will in all likelihood bring vast changes in its nature. The international division of labor among nations specializing in agricultural and extractive products, manufacturing products, and services will generate continuing changes, tensions, and conflicts in the international organization of production (Chase-Dunn et al., 2000). Automation will increase productivity but reduce the number of workers needed in many industries. Automation will also lead to the creation of new products and new jobs in other industries. The future seems destined to bring greater participation for workers in determining the direction of their enterprises; but the nature and degree of this participation remains uncertain. The future may bring greater or lesser safety, security, satisfaction, and dignity for workers. It may bring greater equality for women, or it may not. Which of these possibilities will be realized will be determined by the actions of all of us as productive members of society.

SUMMARY

The nature of work has changed dramatically over time. The division of labor has grown and the level of social inequality has increased and then declined. These changes have given rise to, and have sometimes been a result of, the development of new technologies. The position of women declined in agricultural and industrial society relative to hunting and gathering society, but has improved again in postindustrial society.

Drastic changes have also occurred in the family, which has lost many of the functions it once had as a center of economic production. Market forces have entered almost all areas of social life and exert a profound influence on the way we live in modern society. Most recently, this means that our lives as producers and citizens are being profoundly influenced by our position in the international division of labor. Finally, the increasing division of labor and the growth of large, complex organizations has encouraged the expansion of complex governmental structures charged with the function of ensuring the stability and success of a highly complex economic system.

Changes in the nature of work have brought affluence for many, but they have also produced alienation and poverty for many others. Changes in the organization of work have been a pivotal force, moving the world from a past dominated by localism and tradition to a modern society dominated by rapid change and an interconnected world economy.

KEY CONCEPTS

division of labor	guilds	enclosure movement
social relations of production	feudal society	Industrial Revolution
bureaucracy	artisans	indentured labor
ideology	journeyman	assembly line
economies of scale	merchant capitalism	postindustrial society
social stratification	putting-out industry	service industries
social classes	Protestant work ethic	

QUESTIONS FOR THOUGHT

1. What are some of the ways in which the nature of your work has influenced your life? How have your parents' work roles influenced your life?

2. What changes in people's lives resulted from the development of agriculture? What are the most important consequences of these changes?

3. How did the wealthy extract surplus from those who produced food in feudal society? How is this different from how the wealthy get rich today?

4. Describe two ideologies that have been used to define the meaning of work at different periods in history. What different aspects of work do they highlight?

5. What are the key characteristics of postindustrial society, and why has it come into being? What do you think will be the most important developments in work between now and 2050 when today's college students will begin to reach retirement age?

MULTIMEDIA RESOURCES

Print

Carleton S. Coon, ed. 1977. *A Reader in Cultural Anthropology.* Huntington, N.Y.: Krieger. Contains a wealth of fascinating examples of work and life in primitive societies.

E. J. Hobsbawm. 1969. *Industry and Empire.* Middlesex, England: Penguin. The most accessible and well-written history of the Industrial Revolution in England.

Gerhard Lenski. 1966. *Power and Privilege.* New York: McGraw-Hill. Considered by many to be the best sociological account of the progressive development of more and more complex societies.

Alice Kessler-Harris. 1982. *Out to Work: A History of Wage-Earning Women in America.* New York: Oxford University Press. A detailed history of the experiences of working women in North America.

Karl Marx. 1967 [1887]. *Capital, Volume 1.* New York: International Publishers. The original critical interpretation of the Industrial Revolution and the rise of capitalism.

Adam Smith. 1937 [1776]. *The Wealth of Nations.* New York: Random House. The original positive interpretation of the expanding division of labor and its benefits for society.

Juliet B. Schor. 1992. *The Overworked American.* New York: Basic Books. A compelling analysis of why workers in advanced economies continue to work such long hours.

Internet

American Sociological Association (ASA). *www.asanet.org* The official website of the professional association for sociologists in the United States provides the latest news, research, and events in American sociology.

Canadian Sociology and Anthropology Association (CSAA). *www.arts.ubc.ca/csaa/* Provides the latest news, research, and events in Canadian sociology.

Organizations, Occupation and Work—section of the ASA. *www.campus.northpark.edu/sociology/oow* This ASA section focuses specifically on issues concerning work.

Jobs and Employment Practices. *www.workindex.com* Provides human resources information on hiring, compensation, benefits, employment law, training, and retirement. Information is valuable for employees, employers, and human resources professionals.

Economic History. *www.eh.net* Provides information and services for students and researchers interested in economic history.

RECOMMENDED FILM

The Good Earth. 1937. A profound and moving depiction of the challenges of peasant life in pre–Second World War China. Based on a novel of the same title by Pearl Buck.

2

Studying the World of Work

Tinker, tailor, soldier, sailor,
rich man, poor man, beggar man, thief,
doctor, lawyer, merchant, chief.

This counting rhyme reminds us that children are aware of only a few occupations, usually through exposure to their parents' work and to workers whom they have met in education, medicine, and other social services. But as you saw in Chapter 1, modern industrial economies are specialized. Even adults may know relatively little about the world of work beyond their own jobs. Because of evolving technologies, new jobs are rapidly being created while old jobs become obsolete.

This chapter examines how sociologists study the world of work, especially when that world is expanding and changing. The chapter will explain some concepts and techniques used to gather information about work and workers. Finally, it will look at problems that researchers encounter when studying work. Anthropologists, economists, psychologists, and other social scientists use some of the same methods to study work. The techniques and tools discussed may also help students or workers who are changing jobs find useful information about the labor market.

TECHNIQUES OF ANALYSIS

Work is so varied and important an aspect of human life that sociologists need many methods to study it. Any social science method should be both **valid** and **reliable.** A valid method yields accurate information about the phenomenon being studied. A reliable method produces the same results if it is used repeatedly or if a different investigator uses it. The strong emphasis on valid and reliable methods is one important distinction between social scientific studies of work and journalistic accounts in newspapers or magazines.

This section presents three major techniques that sociologists use to study work and workers: ethnographies, case studies, and sample surveys. These techniques do not exhaust the available methods. For example, some sociologists may study work or workers using experiments (Valian, 1999) or historical approaches (Bernstein, 1997), but the three methods described here are among the most important.

Ethnographies

One way to learn what workers actually do on the job and how they interact with their fellow workers is through an **ethnography.** An ethnography is a careful analysis of a work situation written by a knowledgeable observer, usually after six months to a year of observation. The observer seeks not only to explain the work from the worker's perspective, but also to describe and explain larger patterns that may be invisible to individual workers. This narrative account of work is familiar to us because it superficially resembles friends' or relatives' accounts of life on the job. It is different, however, because the trained observer is sensitive to subtle features of the job and interactions among the workers. Their observations are also more detached. Each ethnography helps social scientists understand a work role or work settings. Evidence cumulated from many ethnographies provides an even stronger basis for conclusions (Hodson, 2001).

There are several types of ethnography. In **participant observation** the observer actually becomes a worker for a period of time. This technique gives the observer an intimate, everyday familiarity with the job content and the actual interactions among the workers. Sociologist Everett C. Hughes and his students at the Univer-

sity of Chicago in the 1950s and 1960s popularized participant observation with provocative studies of medical students, janitors, and taxi dancers, among other occupations. Box 2.1 provides a participant observer's account of work at a fast-food restaurant. Such studies are valuable for the rich detail they provide about working and about interactions among workers and between workers and supervisors.

Both the validity and reliability of participant observation involve some limitations. Participant observers can typically study only a limited range of jobs. It is unlikely that the sociologist observer would have the skills or access necessary to participate, even for a short time, in highly technical jobs or in top-level management positions. In addition, the participant observer may inadvertently choose an atypical work site or join a work group that is atypical. Different observers of the same work situation might also interpret aspects of the job quite differently because of their different backgrounds, predispositions, or experiences.

In **nonparticipant observation,** a trained observer does not actually become a part of the work group. One famous example of nonparticipant observation is the study of the Bank Wiring Room in the Western Electric Company plant in Hawthorne, Illinois (Roethlisberger and Dickson, 1939:379–408). Fourteen men worked in this room, wiring, soldering, and inspecting electrical boards. An overt nonparticipant observer sat with them for a number of days, watching their work and interactions. Initially, the observer noticed how the workers joked with and teased one another, or occasionally helped one another with their work.

The observer also noticed that the group's productivity was basically constant, despite company efforts to increase it. The observer eventually learned that the small work group had developed an informal *norm* defining an appropriate level of productivity. A **norm** is a rule that a group develops for thinking, feeling, or behaving. Laws are examples of formal norms, but most norms are informal. A worker who could not

BOX 2.1   An Ethnography of Hamburgers

Ester Reiter's ethnography on the fast-food industry was a participant observation at a Burger King in Toronto. Fast food is popular in Canada, but the industry is dominated by multinational companies headquartered in the United States. The Toronto Burger King outlet where she worked opened in 1979, and by 1980 it was the highest-volume Burger King in Canada.

Fast-food establishments standardize their products in part by standardizing the way in which workers do their jobs. One objective of Reiter's participant observation was to learn the impact of this standardization on the workers. Making hamburgers, she found, was one of the most enjoyable jobs.

The store had two conveyors that could broil up to 835 patties per hour. Near the meat conveyor were two bun chains that toasted the buns and dropped them into a chute near the cooked patties, a process that took about thirty seconds.

A worker keeps the freezer near the broiler filled by hauling boxes stored in the walk-in freezer located on the other side of the kitchen. During busy times, one worker keeps the chains of the conveyor belt broiler filled with meat and buns, while another worker stands at the other end. The worker at the "steamer" end of the belt uses tongs to pick up the cooked patties as they fall off the belt, and places them on the "heel" of a bun (the bottom half). The bun is then "crowned" with the top half and the ungarnished hamburger is placed in a steamer, where, according to Burger King policy, it can remain for up to ten minutes. Jobs at the broiler-steamer are often assigned to new workers as they can be quickly learned.

The burger board, where the large and small hamburgers are assembled, is made of stainless steel and can be worked from both sides. When the store is busy, the larger hamburgers, called "whoppers," are produced on one side, and the smaller hamburgers on the other. The trays of garnish are placed in a long well or trough located in the centre of the board: cheese slices, pickles, onions, mayonnaise, lettuce, and tomatoes. The ketchup and mustard, stored in plastic bottles, are located near the steamer. One or more workers take the ungarnished hamburgers out of the steamer, add the condiments, place the burger in a box or wrapper, and reheat it in the microwave oven located just above the counter. It is then placed in a chute. . . .

First the whopper cartons are placed printed side down on the table, and the patty removed from the steamer. The bottom half, or the bun heel, is placed in the carton, and the pickle slices spread evenly over the meat or cheese. Overlapping the pickles is forbidden. Then the ketchup is applied by spreading it evenly in a spiral circular motion over the pickles, starting near the outside edge. The onions (1/2 oz.) are distributed over the ketchup. Mayonnaise is applied to the top of the bun in one single stroke and 3/4 oz. of shredded lettuce placed on the mayonnaise, holding the bun top over the lettuce pan. Then the two slices of tomato are placed side by side on top of the lettuce. If the tomatoes are unusually small, the manager will decide whether or not three tomato slices should be used that day.

SOURCE: Excerpt from Ester Reiter, 1991, *Making Fast Food: From the Frying Pan and into the Fryer.* Montreal: McGill-Queen's University Press, pp. 99–100

reach the normatively defined productivity level would be helped by others, but a worker who produced too much would be teased, called a "speed king," or eventually be subjected to "binging" (a thump on the upper arm). The observer reasoned that the workers, concerned about job security as many workers were during the Great Depression, feared that increased productivity would become an excuse for laying off workers.

Nonparticipant observation is useful to sociologists who cannot study a job as participant observers. It would be difficult, for example, to be admitted to medical school to do a participant observation of medical students. Instead, one group of researchers "accompanied students on rounds with attending physicians, watched them examine patients on the wards and in the clinics, and sat in on discussion groups and oral exams. We had meals with the students and took night calls with them" (Becker et al., 1961:26).

Nonparticipant observation also has disadvantages. The nonparticipant observer may have a more difficult time than the participant observer in winning the confidence of the workers being observed, and workers who appear to be acting naturally may nevertheless be quite conscious of the observer. Workers may also change their behavior to please the observer, a phenomenon similar to *experimental bias*. An example of experimental bias may have occurred in early research at the Hawthorne plant. The researchers had designed experiments to test the relationship between levels of lighting and worker efficiency. Workers maintained their productivity even under conditions of very low light (Roethlisberger and Dickson, 1939:14–18; Schwartzman, 1993). One interpretation is that the factory workers were trying to please the experimenters, regardless of the lighting conditions. The phrase "Hawthorne effect" came to refer to an experiment in which participants try to do what the experimenters want. Experimental bias is difficult to detect, and there is dispute over whether it really occurred at the Hawthorne plant (Jones, 1992; Gillespie, 1991). In principle, however, experimental bias threatens both the validity and the reliability of the observations.

Regardless of whether the researcher is a participant or a nonparticipant in the research, an ethnography may be either overt or covert. If the observation is overt, the other workers may know the observer's true "cover story," which is often that the observer is writing a book. The workers may initially feel uneasy or suspicious at the presence of an overt observer, but many researchers find that after a few days their presence is no longer noted.

In covert studies, the workers do not know that their fellow worker is an observer. Alterna-tively, the covert observer may be disguised as a customer, an inspector, or some other stranger with a right to be in the workplace. Covert observers may avoid experimental bias, but this enhancement of the validity of the study is counterbalanced by the ethical issues raised by subterfuge and pretense. Covert observers may not be able to ask clarifying questions, and so their interpretations may be superficial or incorrect. For these reasons researchers sometimes suspect the validity and reliability of data collected by covert observation.

Case Studies

Ethnographic studies are usually limited to fairly small work groups during a specific period. The reader of an ethnography typically learns the point of view of one group of actors in a workplace, typically a group of workers. A **case study** attempts to bring in several perspectives to understand a workplace issue, perhaps in addition to the views of workers, supervisors, customers, suppliers, union leaders, and others. Thus a case study is usually larger in scope, uses more types of data, and is usually conducted over a longer period of time. A case study typically examines a work site using combinations of personal interviews, analyses of written documents, and observations. Both official documents and personal records of workers may be consulted. Case studies frequently analyze entire companies or large divisions within companies. The findings and conclusions emerge from all the materials and people that the researcher consults.

Ethnographies typically present a work group at a particular point in time—the time frame during which the observer was there. By using written documents, the case study can provide information about the history of a work site and how existing arrangements came about (Feagin, Orum, and Sjoberg, 1991). Case studies may illustrate how an organization solves a problem or they may identify new problems faced by workers. Case studies are often used to examine the effects of recent job changes. For example, a case

BOX 2.2   "Praise-Addiction" among Secretaries at Indsco:
An Example of a Case Study

Rosabeth Moss Kanter served as an outside consultant for several years to the company she calls the Industrial Supply Corporation (Indsco). During this period she collected materials and developed a network among employees. Convinced that a case study of a large corporation was needed, she began to analyze many sources of information, including group discussions, conversations, and documents. She also used participant observations of meetings, and she analyzed data from employee surveys (Kanter, 1977:293–298). She could then check each source of information against the information available from other sources. Specific incidents were reported to illustrate the more general principles that she developed.

She uses the term *praise-addiction* to describe a condition she observed and heard others remark on in describing secretaries. Kanter's identification of praise-addiction is one sort of finding that can result from a case study.

> The emotional-symbolic nature of rewards in the secretarial job; the concern of some bosses to keep secretaries content through "love" and flattery; and the continual flow of praise and thanks exchanged for compliance with a continual flow of orders—all of these elements of the position tended to make some secretaries addicted to praise. Praise-addiction was reinforced by the insulation of most secretaries from responsibility or criticism; their power was only reflected, the skills they most exercised were minimal, and authority and discretion were retained by bosses. Thus, many years in a secretarial job, especially as private secretary to an executive, tended to make secretaries incapable of functioning without

their dose of praise. And it tended to make some wish to avoid situations where they would have to take steps that would result in criticism rather than appreciation. Their principal work orientation involved trying to please and being praised in return.

One older executive secretary with long tenure at Indsco was a victim of praise-addiction. Though happy as a secretary, and well-respected for competence, she accepted a promotion to an exempt staff job because she thought she should try it. After a year and a half, it was clear to her and to those around her that she could not take the pressures of the new job. Her nervousness resulted in an ulcer, and she asked to return to the secretarial function. In the exempt job she had supervisory responsibilities and had to make decisions for people—sometimes unpleasant ones, such as terminations. Her manager thought she spent much too long making such decisions, "moaning" afterwards even if she knew she had made the right decision. But she felt herself to be in an intolerable position. She had a feeling she was not appreciated. No one said "thank you" for her work in the new job. As the manager put it, "She was used to lots of goodies from her boss—'Hey, that's a good job.' Here we have to be of service to managers as well as subordinates. The managers feel we're one of them, so they don't go out of their way to thank us. And subordinates don't thank managers. So she was missing something she had been used to."

SOURCE: Excerpt from Rosabeth Moss Kanter, *Men and Women of the Corporation.* Copyright © 1977 by Rosabeth Moss Kanter. Reprinted by permission of Basic Books, Inc., Publishers.

study might examine a work site before and after the introduction of computerized workstations. Because different management teams implement innovations in different ways, a researcher might develop case studies to compare the effects of the innovation in different work settings.

Sociologist Rosabeth Moss Kanter presents a case study of gender roles in a large company in

her book *Men and Women of the Corporation* (1977). Box 2.2 is a brief selection from her work that illustrates the types of conclusions that can be drawn from a case study. Because case studies use several kinds of information, the researcher can search for agreement and disagreement among the various sources. This cross-checking tends to improve both the validity and the reliability of the evidence. A good case study nearly always requires the cooperation of the employer. It is unlikely that the researcher will receive access to written records in any other way. Some companies are so eager to have the research conducted that they will commission and pay for it. Even these companies, however, often insist that published research refer to the company using a disguised name. For example, Kanter refers to the company she studied as "Indsco."

Some workplaces, however, do not welcome research. They may place certain documents off-limits to the researcher, or allow access only if their documents are not quoted. Case studies are especially threatening to companies that are in fiercely competitive economic situations, are closely regulated by the government, have a record of hostile labor relations, or are suspected of wrongdoing by citizen's watch groups, environmentalists, or others (Cornfield and Sullivan, 1983). These, of course, may be the very companies that are of greatest interest to the sociologist.

Sample Surveys

The **sample survey** is widely used to study many social phenomena, including work. A survey is conducted by asking a uniform set of questions of a systematic **sample** of people. The people who answer the survey are called respondents. They are selected according to the principles of a branch of mathematical statistics known as sampling theory, so that they will be representative of the **population,** the larger group from which they were selected. The sample may be selected to represent all workers in the United States, all employers, the workers in a particular workplace, or any other population of interest.

The set of questions, or questionnaire, may include questions of fact ("How long have you worked at your present job?") and questions of opinion ("How satisfied would you say you are with your current job—very satisfied, somewhat satisfied, or not at all satisfied?"). Researchers administer questionnaires in three basic ways: (1) personal interviews are conducted face-to-face

by a trained interviewer; (2) telephone interviews take place over the respondent's home telephone, again with a trained interviewer asking the questions; and (3) self-administered questionnaires are handed, mailed, or e-mailed to respondents who answer the questions at a convenient time.

A cross-sectional survey is administered once to a sample of respondents. The same questionnaire might be administered again to a different sample of respondents. The repeated use of cross sections is useful for detecting trends in job satisfaction, work commitment, and so on. In a longitudinal or panel study, the researchers return several times to survey the same sample of respondents. Longitudinal studies are useful for such things as studying job changes among a group of workers.

Sample surveys are extensively used in many countries to study work. Every month, the U.S. Bureau of the Census and Bureau of Labor Statistics conduct the Current Population Survey (CPS) (*http://www.bls.census.gov/cps/*). This survey of approximately 60,000 households asks a variety of questions about whether members over the age of sixteen are looking for work or have jobs. Those with jobs are asked additional questions about hours of work, type of work, and earnings. The Bureau of the Census also conducts the quarterly Survey of Income and Program Participation (SIPP), which asks detailed questions concerning earnings and such fringe benefits as pensions and medical insurance (*http://www.census.gov/pub/dusd/MAB/ sipp~1.html*). The National Longitudinal Surveys (NLS), supported by the Department of Labor, interview the same samples of workers several times over a period of years to examine changes in employment, earnings, and work-related attitudes (*http://www.bls.gov/nlshome.htm*).

Other surveys are conducted occasionally. The U.S. Department of Labor funded two Quality of Employment Surveys (QES) in 1977 and 1983. These surveys asked detailed questions about working conditions and worker satisfaction Various **establishment surveys** sample employers to ask questions concerning the characteristics of their companies and employees. The National Organizations Survey is an example of such a study. Each of these samples is designed to represent the entire United States.

Sample surveys may also be designed to represent certain groups of workers or certain regions or states. Professional associations or unions survey their members on workplace issues. Trade associations survey employers or owners who are members. Many businesses survey their customers to evaluate customer satisfaction. Innovations based on the use of laptop computers, e-mail, and web-based surveys are extending the range of contemporary survey practices.

Compared with ethnographies or case studies, surveys have the advantage of being more easily generalized to the population they were designed to represent. Sampling theory allows the researcher to estimate by how much a survey will vary from the "true" answer—the answer that would have resulted from interviewing the entire population. By directly questioning workers, a survey can measure subjective indicators such as job satisfaction. Changes in facts and attitudes can be traced and studied if the same question is asked in repeated surveys.

Potential disadvantages also arise in the use of survey methods. One problem is selection bias in which only certain types of people respond to a survey. Respondents may mistrust the interviewer or fear what use might be made of their replies. Some respondents refuse to cooperate at all with surveys. If their refusals cluster within an important population subgroup, the resulting sample is no longer representative of the population. For example, if rich people refuse to answer questions about their income, estimates of overall income will be too low.

Another common problem is **response error,** which results when a respondent misunderstands a question or intentionally gives an untrue answer. Response error may happen if the questionnaire contains ambiguous or double-barreled questions. A double-barreled question includes more than one issue, so that the answer cannot be clearly interpreted: "Have your hours of work or your working conditions recently changed?" A "yes" answer might mean that either hours of work or working conditions or both had changed. Selection bias and response error may also result if the questions pry into areas that respondents consider sensitive or confidential. Response error is difficult to detect, and it threatens the validity and reliability of the information gathered.

The collection and analysis of survey information forms a specialized area within the social sciences. Sociologists, political scientists, and economists make use of survey data. Box 2.3 contains examples of the questions asked every month in the CPS. The answers are used by sociologists who study work, and by economists, marketing analysts, and government planners.

UNITS OF ANALYSIS

Ethnographies, case studies, and sample surveys are examples of how sociologists study work; specific units of analysis are what and whom they study. The unit of analysis may be individual workers or groups of workers. Or the unit of analysis may not be people at all—it may be a group or organization. For example, the sociologist may study unions, businesses, factories, or corporate networks. One important unit of analysis is the **labor force,** a collective term for all the workers within a country. (Outside the United States an equivalent term is the *economically active population.*)

The Worker and the Labor Force

The most straightforward unit of analysis is the individual worker. Workers can be analyzed in terms of their background, or **demographic characteristics.** These include **ascribed** characteristics such as gender, race, or age. Although the worker has no control over ascribed characteristics, employers and co-workers may react strongly to them. Demographic characteristics also include such **achieved** characteristics as educational background, work experience, and skills, over which the worker does have some control.

The U.S. Bureau of Labor Statistics reports the size and composition of the labor force every month by using information from the CPS. Anyone is eligible to be counted in the labor force who is aged sixteen or older and who is not institutionalized (for example, in a prison or a residential hospital). Members of the labor force can be either employed or unemployed.

According to the government's definition of employment, **employed** people in the labor force are those who in the week preceding the survey (1) worked at least one hour for pay or profit, (2) worked at least fifteen hours without pay in a family business, or (3) were temporarily not working because of illness, vacation, or similar reasons. The **unemployed** are not merely those without jobs; rather, they are people who are not employed but who actively sought work during the four weeks preceding the survey and were currently available to take a suitable job. In addition, people are counted as unemployed if they do not meet the criteria for being employed and are temporarily laid off or are waiting to report to a new job in the near future (U.S. Department of Labor, Bureau of Labor Statistics [BLS], 1992b). An eligible person who does not fall into either of these categories is termed *NILF* (not in the labor force). Most of the NILF people in the United States are students without jobs, retirees, disabled people, or people who are keeping house.

Using these concepts, the Bureau of Labor Statistics publishes every month two rates to describe the status of the labor force. The first rate, the civilian **labor force participation rate,** is the number of persons in the labor force divided by the number of persons eligible to be in it, multiplied by 100 to convert to a percentage. This can be expressed as:

$$\text{LFPR} = \left(\frac{\text{labor force}}{\text{all noninstitutionalized persons aged 16 +}} \right) \times 100$$

The labor force participation rate indicates what proportion of the eligible population is economically active. In September 2000, the U.S. participation rate was 66.9%. Trends in the rates for certain groups, such as women, teenagers, and the elderly, indicate their levels of incorporation into the economy. Nearly every industrialized country has experienced a phenomenal increase in women's labor force participation rates since World War II. In September 2000, the rate for U.S. women aged twenty years or more was 60.6%, compared with 76.5% for men (BLS, 2000a:Table A-3).

The **unemployment rate** is the number of unemployed people divided by the number of people in the labor force, multiplied by 100. This may be expressed as:

$$\text{UR} = \left(\frac{\text{unemployed}}{\text{labor force}} \right) \times 100$$

Box 2.4 presents recent data on labor market indicators and it includes important information on how to read statistical tables. These data indicate overall economic activity as well as the differing labor market experiences of workers from various demographic characteristics.

A rise in the unemployment rate often indicates that the business cycle is about to enter a downturn; conversely, a decline often indicates economic improvement. The unemployment rate is high in economically depressed areas and lower in prosperous ones, so unemployment rates indicate local labor market conditions. Historically, the unemployment rate for black workers has been at least twice the rate for whites.

As useful as labor force statistics are, the definitions used by the government trouble some observers (National Commission on Employment and Unemployment Statistics, 1979). The definition of the labor force parallels in some ways the measurement of the gross national product (GNP). The gross national product is the value of all the goods and services produced for the market during a year. People who produce goods or services for sale in the market are

BOX 2.3   Questionnaires by Laptop Computer

A good survey has two essential ingredients, a representative sample and a good questionnaire. The representative sample ensures that the people questioned adequately represent the population from which they were drawn. A good questionnaire requires careful attention to the wording of questions and to the order in which they are asked.

Figure A presents examples of questions asked each month in the Current Population Survey (CPS) of a representative sample of U.S. households (Polivka and Rothgeb, 1993). This survey is conducted by personal interviews with answers entered into a laptop computer. The interviewer asks every respondent the same question in the

same way, but the computer is programmed to help the interviewer ask the questions efficiently.

These data are used to derive labor force information about the entire population. To reduce response error, the interviewer first defines "last week" by mentioning specific dates and learns if there is a household business. Then the interviewer asks question 20, "LAST WEEK, did you do ANY work for (either) pay (or profit)?" Depending on the answer the respondent gives, the computer will follow a skip pattern to prompt the interviewer to ask some questions but not others. For example, if the respondent answers "No" to question 20, the computer provides questions that inquire about retirement or

FIGURE A

Q19. LABOR	I am going to ask a few questions about work-related activities LAST WEEK. By last week I mean the week beginning on Sunday, January 12, and ending on Saturday, January 18.			
NOTE: Q19A. BUS	This item is asked only once, after demographics for household have been asked. Does anyone in this household have a business or a farm?			
	Yes	0 → Whose business	__ __ (line no.)	BUSL1
	No	0 or farm is it?	__ __ (line no.)	BUSL2
[blind]	Don't know	0	__ __ (line no.)	BUSL3
[blind]	Refused	0	__ __ (line no.)	BUSL4

At Work

	(IF Q19A is "yes" then parentheticals should be filled.)		
Q20. WK	LAST WEEK, did you do ANY work for (either) pay (or profit)?		
	Yes	0	(Skip to Q20C)
	No	0	(Go to Q20-CK)
	Retired	0	(Go to Q20-CK)
	Disabled	0	(Go to Q20-CK)
	Unable to work	0	(Go to Q20-CK)
[blind]	Don't know	0	Skip to Q20B-a)
[blind]	Refused	0	(Skip to Q20B-a)

disability status (not shown), and then prompts, "LAST WEEK (in addition to the business), did you have a job either full or part time? Include any job from which you were temporarily absent" (not shown). If the respondent answers "Yes" to question 20, the questionnaire will skip to question 20C (not shown), which is, "LAST WEEK, did you have more than one job (or business), including part-time, evening, or weekend work?"

The ordering of the questions is designed to identify first the employed, then those who have a job but are not at work, and then the people who are actively seeking work, or unemployed. The question wording is designed carefully so that it can be understood by respondents and can

produce reliable results (Bowie, Cahoon, and Martin, 1993). Changes in questionnaires are pretested on people similar to the respondents to identify questions that are ambiguous or confusing.

The computer can also be programmed to tabulate the results. A survey analyst examines the responses and publishes the results. The Bureau of Labor Statistics uses the CPS data to estimate the numbers of employed and unemployed people, the size of the labor force and the number of people not in the labor force, the labor force participation rate, and the unemployment rate (Bregger and Dippo, 1993). Box 2.4 shows examples of the resulting data.

Q20-CK. BUSCK1	CHECK ITEM		
	Q19A is "Yes"	(Ask Q20-1)	
	Q19A is "No" "D" or "R"	(Skip to Q20-CK2)	

Q20-1. BUS1	LAST WEEK, did you do any unpaid work in the family business or farm?		
	Yes	0	(Ask Q20-2)
	No	0	(Skip to Q20-CK2)
[blind]	Don't know	0	(Skip to Q20-CK2)
[blind]	Refused	0	(Skip to Q20-CK2)

Q20-2. BUS2	(If Q19A is "yes" and Q19A line number EQ person number, then plug Q20-2 "yes" and skip to Q20E-A.)		
	Do you receive any payments or profits from the business?		
	Yes	0	(Skip to Q20E-A)
	No	0	(Go to Q20E-A)
[blind]	Don't know	0	(Go to Q20E-A)
[blind]	Refused	0	(Go to Q20E-A)

SOURCE: BLS, CPS Questionnaire Evaluation Work Group, 1993, "Composite Questionnaire for CATI/CAPI Overlap (CCO) Test."

BOX 2.4   How to Read a Table

Sociologists frequently present their data in tables, which condense a great deal of information within a small space. Because of their concentrated information, however, tables can be difficult to read and understand. We will be presenting many tables in this text, and this box is designed to provide a method of reading tables to glean the maximum information.

Table A presents some information about the U.S. labor force. The table has also been marked to indicate the principal parts of a table. Reading the table in the order of the numbered parts will convey efficiently and accurately the information in the table. The parts are defined on the facing page.

① **Table A**

Major Indicators of Labor Market Activity, Seasonally Adjusted

② (Numbers in thousands)

③ Category	Monthly Data[a]	
	1999	**2000**
Labor Force Status	**Sept.**	**Sept.**
Civilian labor force	139,475	140,639 ④
Employment	133,650	135,161
Unemployment	5,825	5,477
Not in labor force	68,790	69,522
Unemployment Rates		
All workers	4.2	3.9
Adult men	3.4	3.2
Adult women	3.7	3.5 ⑦
Teenagers	14.6	12.8
White	3.6	3.5
Black	8.3	7.0
Hispanic origin	6.6	5.6

⑤ [a]Data in 2000 reflect revised population controls used in the household survey.

⑥ SOURCE: BLS. Employment and Earnings 47,10 (October 2000), Table A.1, p. 3.

Parts of a Table:
1. Headline
2. Headnote
3. Stub
4. Column headings
5. Footnote
6. Source note
7. Entries

included in the labor force. The labor force definition excludes many people who perform useful services outside the market economy. For example, homemakers and volunteers perform needed services but not for pay or profit. If there were no homemakers or volunteer workers, then families, churches, and hospitals would have to hire workers to perform those duties or leave them undone. Although the newly hired workers would be in the labor force, homemakers and volunteer workers doing the same work are not in the labor force.

The measurement of unemployment is also controversial. **Discouraged workers** are avail-

1. The *headline* tells the reader which data are presented in the table, for which groups. The headline in Table A indicates that the table contains major indicators of labor market activity. Sometimes a headline specifies the time and place in which the data were collected. This headline notes that the data are *seasonally adjusted;* that is, seasonal idiosyncrasies that might affect the labor force—for example, bad weather in the winter months—have been statistically removed from the data.

2. The *headnote* is a parenthetical expression that contains information important in interpreting the table. The headnote in Table A indicates that the numbers in the table represent thousands of persons. Thus, the reader knows that in September 2000 there were over 140 *million* persons in the labor force, not 140 *thousand!*

3. The *stub* is the left-hand column of a table. The categories indicate which data appear in the horizontal rows of material in the table. Reading down the stub, the reader sees that information will be presented first on labor force status and then on unemployment rates. Each of these categories is further subdivided to present a total figure and then data for each of several subcategories.

4. The *column headings* indicate which data are given in the vertical columns of material in the table. In Table A, the two column headings indicate that the data refer to September 1999 and 2000.

5. A *footnote* contains information that is important for interpreting some, but not all, of the entries of the table. Not every table has a footnote. In Table A, the footnote indicates that the population number used for sampling changed in 2000.

6. The *source note* is important because it tells the reader where the data were obtained. The reader can refer to the original source for additional information or to check the accuracy of the data.

7. The *entries* of the table should be read last. The reader draws conclusions by carefully reading the entries and comparing them across rows and down columns. We have already noted, for example, that the unemployment rate for blacks has consistently been at least twice that of whites. The data in Table A indicate that in September 2000 the unemployment rates for whites and blacks were 3.5% and 7.0%, respectively. Thus, the unemployment rate for blacks was exactly twice the rate for whites.

What other conclusions can you draw from the data in Table A?

able for work and have searched for work during the preceding year, but have stopped seeking because they believe no work is available. They are not classified among the unemployed but are considered to be NILF. The Bureau of Labor Statistics estimates that there were 250,000 discouraged workers in the United States in September 2000 (BLS, 2000a). By excluding the discouraged workers, the measured unemployment rate is arguably too *low*. Others argue that many unemployed people conduct only halfhearted searches for work, perhaps because they are required to look for work to continue unemployment compensation benefits. In the view of such critics, the measured unemployment rate is too *high,* for it includes people who are not really interested in working.

One might wonder why data on unemployment insurance benefits are not used to estimate unemployment. The principal reason is that not every worker is covered by unemployment compensation. In 1997 the number of people receiving unemployment benefits amounted to only about 34.5% of the number of unemployed people (U.S. Census, 1999:Table 626). Some types of work, such as agricultural labor, are not covered. Unemployment insurance laws require that a worker to have been employed in a covered occupation for a specified length of time, and some workers have not worked long enough to qualify. For example, young people seeking their first jobs are unemployed, but they are not eligible for unemployment compensation. Other workers continue to be unemployed after exhausting their unemployment compensation benefits. Consequently, unemployment benefit records seriously underestimate total unemployment.

Industry

Industry provides another unit of analysis. **Industry** refers to a branch of economic activity that is devoted to the production of a particular good or service. The good or service may be quite specific; thus, we might speak of the fast-food industry, but it, in turn, can be considered part of the restaurant industry or part of the even larger personal service industry.

Knowing a worker's industry is important for several reasons. First, conditions of economic competition tend to be quite specific to industries. Some industries experience heavy pressure from foreign competition, for example, while others do not. Some industries are closely regulated by the government, and others are unregulated. Second, the nature of production varies by industry. An industry with an electronically automated production process differs substantially from an industry that still requires large inputs of hand labor. The industrial process determines

which occupations are needed in the industry and what the working conditions within the industry are—specifically, what hazards workers may face on the job, what skills are needed for employment, how much training is needed, and so on.

Finally, workers experience economic consequences from their industries (Sullivan, 1990). Declining industries are often less productive and provide sporadic, lower-paid work. Growing industries are more likely to be productive and to provide better wages, benefits, promotion opportunities, and job security. The earnings in industries vary a good deal, from $19.61 an hour in manufacturing motor vehicles to $8.50 in child day-care services (BLS, 2000a).

Individual respondents in sample surveys are asked what product or service is produced at their place of work. Their answers are used to develop classification codes to assign each respondent to an industry. Industrial codes enable researchers to contrast the characteristics of workers in different industries. Businesses are classified by products or services they produce. The government classifies industries using the Standard Industrial Classification (SIC) codes. SIC codes contain up to six digits, with each consecutive digit specifying more precisely what good or product is produced. Thus, the two-digit code 23 tells us that a business manufactures apparel, 232 specifies men's and boys' furnishings, and 2325 is men's and boys' trousers and slacks.

A company's sales, profits, or production can be expressed as a proportion of all the sales, profits, or production within the SIC code. The larger a proportion attributable to a single company, the more dominant or important that company is within its industry. The importance of a single company may be measured by the extent to which the company dominates its four-digit or three-digit SIC code. An industry is concentrated when a small number of firms with the same SIC code account for a major proportion of the production, sales, or profits for the industry. The four-firm concentration ratio is the proportion of all production, sales, or receipts accounted for by the largest four firms. For example, in 1987 the food and kindred products industry, SIC code 20, was a $330 billion industry, with a four-firm concentration ratio of 11 for the value of shipments. This means that the top four firms accounted for 11% of the value of all shipments. But within this large industry, the four-firm ratios for specific industries varied a great deal: 32 for meat packing (code 2011), 44 for flour and other grain products (code 2041), 87 for sugar cane refining (code 2062), and 96 for chewing gum (code 2067) (Census, 1992a:Table 4, pp. 6-4 and 6-5).

Table 2.1 presents information on the employment changes projected to 2005 for rapidly growing and declining industries. The SIC codes are also provided for reference.

Industry differs importantly from occupation. Industry identifies what a worker helps to produce, but occupation identifies the kind of work a worker does. This distinction is sometimes difficult to understand. Some occupations are found in every industry. Nearly every industry, for example, requires clerical workers, maintenance workers, and managers. Other occupations are heavily concentrated within a single industry— for example, nurses within the health-care industry or lawyers within the legal services industry. Even these examples have their exceptions, however; a few nurses work in factories, camps, or schools, and many lawyers work as house counsel for firms in manufacturing or service industries. A few occupations work only in a single industry; an example would be taxi drivers in the transportation industry. As a rule, however, we must consider both occupation and industry for a full understanding of working life.

Occupation

An **occupation** is a cluster of job-related activities constituting a single economic role that is usually directed toward making a living. Occupation refers to the type of work someone does. Some workers have several occupations because they have more than one job or are able to do more than one kind of work. Occupations are

also a unit of analysis. Chapters 11 to 14 will look at specific occupational groups.

White-Collar–Blue-Collar Division Perhaps the simplest occupational classification is the white-collar–blue-collar division. This classification is simple but also increasingly outdated and misleading. Blue-collar workers—mostly factory and craft workers—once did only manual labor. White-collar workers—office workers and most professionals—had clean working conditions that made it possible for them to wear white shirts. Traditionally they earned more than blue-collar workers, but today a factory or craft worker may earn more than a clerical or sales worker.

The white-collar–blue-collar classification is less useful today for several other reasons. First, there are now many service workers, some of whose work resembles blue-collar jobs and some of whose work is more like white-collar jobs. For example, the cook in a fast-food restaurant and the elite chief of police in a large city are both service workers. The fast-food cook may experience factorylike conditions reminiscent of blue-collar work. The police chief has job training and autonomy on a par with many white-collar management jobs.

Second, some jobs that are now classified as either blue collar or white collar may seem misclassified when the actual work conditions are considered. Technicians, for example, are considered white-collar workers. Many of them are highly educated, like the white-collar workers, but they spend most of their day working with machinery, just as blue-collar workers do. Some factory operatives, on the other hand, work in industrial laboratories that are not just clean, but sterile. Their day-to-day job responsibilities may look very much like those of the technicians, but they are classified as blue-collar workers.

Finally, the white-collar–blue-collar distinction ignores the so-called pink-collar workers. These workers labor in occupations, such as nurse, secretary, or child-care worker, that were traditionally filled by women. Pink-collar jobs are usually characterized by very low pay, although

Table 2.1   Employment Change in Selected Industries, 1992–2005
(Numbers in thousands)

Standard Industrial Classification	Industry Description	Wage and Salary Employment			
		Level		Change	Annual Rate of Change
		1992	2005	1992–2005	1992–2005
	Fastest Growing				
836	Residential care	535	1,335	800	7.3
737	Computer and data processing services	831	1,626	795	5.3
807, 8, 9	Health services, n.e.c.	833	1,577	744	5.0
835	Child day-care services	449	777	328	4.3
732; 7331, 8; 7383, 9	Business services, n.e.c.	903	1,543	640	4.2
874	Management and public relations	655	1,110	455	4.1
832, 9	Individual and miscellaneous social services	703	1,162	459	3.9
472	Passenger transportation arrangement	183	300	117	3.9
735	Miscellaneous equipment rental and leasing	205	325	120	3.6
872, 89	Accounting, auditing, and services, n.e.c	553	876	323	3.6
736	Personnel supply services	1,649	2,581	933	3.5
61, 7	Nondepository; holding and investment offices	615	949	334	3.4
833	Job training and related services	271	418	147	3.4
7334, 5, 6; 7384	Photocopying, commercial art, photofinishing	190	291	102	3.4
494, 5, 6, 7, pt. 493	Water and sanitation including combined services	197	299	102	3.3
801, 2, 3, 4	Offices of health practitioners	2,387	3,617	1,229	3.2
3728, 3769	Aircraft and missile parts and equipment, n.e.c.	170	255	85	3.2
805	Nursing and personal care facilities	1,543	2,306	763	3.1
752, 3, 4	Automobile parking, repair, and services	719	1,071	352	3.1
823–9	Libraries, vocational, and other schools	208	310	102	3.1
	Most Rapidly Declining				
313, 4	Footwear, except rubber and plastic	69	39	−30	−4.4
3483, 3489	Ammunition and ordnance, except small arms	46	27	−19	−4.0
3731	Ship building and repairing	124	77	−46	−3.6
311, 5, 6, 7, 9	Luggage, handbags, and leather products, n.e.c.	51	32	−18	−3.4
386	Photographic equipment and supplies	95	62	−33	−3.2
3571, 2, 5, 7	Computer equipment	353	237	−117	−3.0
231–8	Apparel	807	556	−251	−2.8
341	Metal cans and shipping containers	45	31	−14	−2.8
3761	Guided missiles and space vehicles	105	73	−33	−2.8
3578, 9	Office and accounting machines	38	27	−12	−2.7
	Federal electric utilities	28	19	−8	−2.7
3466, 9	Stampings, except automotive	83	59	−24	−2.6
12	Coal mining	126	90	−36	−2.5
88	Private households	1,116	802	−314	−2.5
3661	Telephone and telegraph apparatus	108	81	−28	−2.3
362	Electrical industrial apparatus	158	119	−38	−2.1
3482, 3484	Small arms and small arms ammunition	20	15	−5	−2.1
21	Tobacco manufacturers	49	37	−12	−2.1
365	Household audio and video equipment	82	63	−20	−2.1
291	Petroleum refining	120	92	−28	−2.0

n.e.c. = not elsewhere classified.

SOURCE: James C. Franklin, 1993, "Industry Output and Employment." *Monthly Labor Review* 116,11 (November):41–57.

Table 2.2 Occupational Groupings Based on the 1970 and 1980 Census Classification Systems

1970	1980–present
White-Collar Workers	Managerial and Professional Specialty
Professional and technical workers	Executive, administrative, and managerial
Managers and administrators, except farm	Professional specialty
Sales workers	
Clerical workers	
Blue-Collar Workers	Technical, Sales, and Administrative Support
Craft and kindred workers	Technicians and related support
Operatives, except transport	Sales occupations
Transport equipment operatives	Administrative support, including clerical
Nonfarm laborers	
Service Workers	Service Occupations
Private household workers	Private household
Other service workers	Protective service
	Service, except private household and protective
Farm Workers	Precision Production, Craft, and Repair
Farmers and farm managers	
Farm laborers and supervisors	Operators, Fabricators, and Laborers
	Machine operators, assemblers, and inspectors
	Transportation and material moving occupations
	Handlers, equipment cleaners, helpers, and laborers
	Farming, Forestry, and Fishing

SOURCE: Gloria Peterson Green, Khoan tan Dinh, John A. Priebe, and Ronald R. Tucker, 1983, "Revisions in the Current Population Survey Beginning in January 1983." *Employment and Earnings* 30, 2. Washington, D.C.: U.S. Department of Labor, p. 10.

some, such as nursing, require highly specialized skills. Nurses are classified with professional workers, but their pay may not be as high as that of other upper-level white-collar workers. Such jobs are difficult to classify as either white collar or blue collar.

Major Occupational Groups A more useful system categorizes workers into the **major occupational groups.** These broad categories of occupations are used by government agencies to report job data. Occupations are listed in a rough order of skill and prestige, from the most prestigious to the least (Census, 1943). Each group contains a number of more specific occupational codes. For example, the "administrative

support" category includes secretaries, stenographers, bookkeepers, office machine operators, and other related occupations.

Table 2.2 compares the major occupational groups used in the 1970 and 1980 censuses by the U.S. Bureau of the Census. Between these two decennial censuses, the occupational codes were substantially revised. In 1970, the classification distinguished among skilled manual workers (craft and kindred workers), semiskilled manual workers (operatives and transport operatives), and unskilled manual workers (laborers). Among agricultural workers, the farmers and farm managers, many of whom are highly skilled, were classified separately from farm laborers. Even in 1970, however, the service workers were too heterogeneous to fit easily into the hierarchy of skill or prestige. Later in this section we will return to the issue of occupational prestige.

In 1980 and for all subsequent government studies, the occupational groups were modified, partly to reflect the increased significance of the technical and precision production occupations and partly to correspond more closely to the International Standard Classification of Occupations (ISCO). ISCO is a method of coding occupations that was developed at the International Labour Office, an affiliate agency of the United Nations. As you can see, however, the 1980 classifications are a mixture of industry and occupation; for example, "farming, forestry, and fishing" is listed as a separate occupational group, although it is composed of three industries. For this reason, much of the social science research on occupations still uses the 1970 groups.

The most detailed occupational classification is the *Dictionary of Occupational Titles,* also called the DOT. The fourth edition of the DOT was published by the U.S. Department of Labor in 1991 (*http://www.oalj.dol.gov/libdot.htm*). It contains over 20,000 occupational titles. Each title includes a detailed description of what people in the occupation do, including actions they perform; machines, tools, or equipment they use; materials they use; products they make or services they render; and instructions they must follow or judgments they must make. The descriptions are based on over 75,000 observations of workers and work sites (Miller et al., 1980; Cain and Treiman, 1981).

The DOT has more than an exhaustive listing of occupations, however; it also contains coded information on the skills needed to hold the occupation. A companion volume, *Selected Characteristics of Occupations Defined in the Dictionary of Occupational Titles,* provides additional information concerning physical demands of the work, hazards the worker may face, and how much training is required to learn the job (U.S. Department of Labor, Employment and Training Administration, 1993a). Job counselors use this information to help place workers in suitable jobs; for example, they may seek to place injured workers in jobs requiring similar skills but less strenuous physical demands. Box 2.5 explains how information is condensed into the brief DOT listings. (U.S. Department of Labor, Employment and Training Administration, 1993b). The U.S. Department of Labor is developing a new tool called O★Net that will perform similar functions (*http://www.doleta.gov/programs/onet/glance.asp*).

Detailed occupational information is valuable for analyses of workplaces and worker skills. Occupation, however, has many social ramifications beyond its instrumental and economic consequences. Sociologists often consider occupation as a proxy for one's position in the social class structure. People of similar occupation, besides having similar incomes and work experiences, often pursue similar patterns of leisure and consumption, share distinctive lifestyles, and are perceived in similar fashion by other members of the society (Trice, 1993). It is for this latter characteristic, how occupations are perceived by others, that sociologists have developed measures of **occupational prestige.**

As we mentioned earlier, the major occupational groups are roughly ordered by skill and prestige, but there are more precise ways to measure prestige. One of the earliest was developed using survey techniques, and it is often called the NORC scale because the National Opinion Research Center carried out the research (North and Hatt, 1947; Hodge, Siegel, and Rossi, 1964). A large number of survey respondents were asked to rate occupations in terms of how much standing members of that occupation would have in the community. The ratings were combined and transformed into a ranking of the occupations on a 100-point scale. Supreme Court justice received the highest ranking (89), and shoeshiner the lowest (27). Similar prestige scales have been developed in many countries, and the findings in one country tend to approximate closely those in other countries (Treiman, 1977).

Occupational prestige, education, and income tend to be closely related, but there are exceptions. For example, members of the clergy may receive relatively low incomes despite their

Box 2.5   Using the Dictionary of Occupational Titles

The first item in a DOT occupational definition is the nine-digit occupational code. In the two examples given in Figure B, the codes are 141.081–010 for motion picture cartoonist and 611.482–010 for forging-press operator. In the DOT classification, each set of three digits has a specific meaning, and together the nine digits uniquely identify an occupation (U.S. Department of Labor, Employment and Training Administration, 1977).

FIGURE B

Two Examples of DOT Occupational Definitions

141.081-010 CARTOONIST, MOTION PICTURES (motion pic.; radio & tv broad.) animated-cartoon artist; animator.
Draws animated cartoons for use in motion pictures or television: Renders series of sequential drawings of characters or other subject material which when photographed and projected at specific speed becomes animated. May label each section with designated colors when colors are used. May create and prepare sketches and model drawings of characters. May prepare successive drawings to portray wind, rain, fire, and similar effects and be designated CARTOONIST, SPECIAL EFFECTS (motion pic.; radio & tv broad.). May develop color patterns and moods and paint background layouts to dramatize action for animated cartoon scenes and be designated CARTOON-BACKGROUND ARTIST (motion pic.; radio & tv broad.).

611.482-010 FORGING-PRESS OPERATOR (forging) I die-machine operator; forging-machine operator.
Sets up and operates closed-die power press to produce metal forgings, following work order specifications, using measuring instruments and handtools: Alines and bolts specified dies on ram and anvil of press, using rule, square, shims, feelers, and handtools. Turns knobs to set pressure and depth of ram stroke. Pulls workpiece from furnace, when color of workpiece indicates forging temperature, and positions workpiece on lower die, using tongs. Depresses treadle or pulls lever to lower ram that compresses metal to shape of die impressions. May move workpiece through series of dies for progressively finer detail.

The first three digits specify a particular occupational group. There are nine broad categories represented by the first digit. The 1 in the first example identifies the cartoonist with the professional, technical, and managerial occupations. The 6 in the second example identifies the forging-press operator as being in the machine trades occupations. The second digit refers to a division within the broad category, and the third digit defines the occupational group within the division. The cartoonist belongs to group 141, which is made up of designers and artists. The forging-press operator belongs to group 611, which is forging-press occupations.

The second three digits indicate the extent to which the worker is expected to function in relationship to data (digit 4), people (digit 5), and things (digit 6). Worker functions requiring more complex responsibility and judgment are assigned *lower* numbers.

Data (Fourth Digit)

0 Synthesizing
1 Coordinating
2 Analyzing
3 Compiling
4 Computing
5 Copying
6 Comparing

People (Fifth Digit)

0 Mentoring
1 Negotiating
2 Instructing
3 Supervising
4 Diverting
5 Persuading
6 Speaking-signaling
7 Serving
8 Taking instructions–helping

Things (Sixth Digit)

0 Setting up
1 Precision working
2 Operating-controlling
3 Driving-operating
4 Manipulating
5 Tending
6 Feeding-offbearing
7 Handling

Continued

Box 2.5   Continued

The codes for our two examples indicate that the forging-press operator must be able to compute data (data score) and operate and control machinery (things score). Working with things is the highest skill required. For the cartoonist, the highest skill is synthesizing data (data score); in working with things, the cartoonist needs to do precision work. In terms of dealing with people, taking instructions is the only functional requirement for either occupation.

The last three digits of the DOT code indicate the alphabetical order of titles within the six-digit code groups. They serve to differentiate an occupation from all others with the same six first digits. No two occupations will have the same nine digits.

After the DOT code the main title of the occupation is given, with the relevant industry or industries indicated in parentheses. Any titles that follow the industry list are alternate titles sometimes used for the same job. In both examples here, alternate job titles are in use. The remainder of the description tells what the occupation requires. In the case of the cartoonist two variations of the basic occupation are also included.

Using the DOT code as a reference, the reader can use the *Selected Characteristics of Occupations Defined in the Dictionary of Occupational Titles* (U.S. Department of Labor, Employment and Training Administration, 1981) to learn even more about the occupation (see Figure C). The physical demands of our two occupations are quite different. The cartoonist is rated as sedentary (S), with some need for reaching and handling and for visual acuity

(codes 4 and 6). The forging-press operator, by contrast, has a job rated heavy (H) in physical demands. This is defined as a job that requires lifting a maximum of 100 pounds with frequent lifting or carrying of objects that weigh up to 50 pounds. The environmental-conditions code indicates that both example occupations are conducted primarily indoors (code I). Display designer, another occupation on the list, has the code B, indicating that both indoor and outdoor work may be required. The additional codes 3 and 5 for the forging-press operator indicate that this job also involves exposure to extremes of heat, temperature changes, and noise and vibration.

The column headed M indicates mathematical development needed for the job, and in this case the two occupations require only the ability to add, subtract, multiply, and divide (code 2). The language (L) requirement for the forging-press operator, also a code 2, implies a need for a passive reading vocabulary of 5,000 to 6,000 words and the ability to speak clearly and distinctly. The cartoonist, with the higher rating of 4, needs to be able to read journals, manuals, and dictionaries, among other things; to be able to write basic business letters and reports; and to speak well enough to participate in a panel discussion or to speak without preparation on a variety of topics.

Finally, the column headed SVP, for "specific vocational preparation," indicates that the forging-press operator can learn the job in three to six months (code 5). The cartoonist, however, would need from two to four years (code 7) to learn the job.

---

extensive education and considerable prestige within the community. Nevertheless, occupation, prestige, and income are so closely related that we can often predict the general status of an occupation by knowing the average education and earnings of its members. Such predictions are called **socioeconomic status (SES) scores.** Data on occupation, education, and income collected from censuses or in periodic surveys are combined using a technique called **multiple regression**

FIGURE C Selected Characteristics of Some DOT Occupations

DOT Code	DOT Title and Industry Designation	Physical Demands	Environmental Conditions	M	L	SVP
141.031-010	ART DIRECTOR (profess. & kin.)	S 4 5 6	I	3	5	8
141.061-010	CARTOONIST (print. & pub.)	S 4 6	I	2	4	7
141.061-014	FASHION ARTIST (ret. tr.)	S 4 6	I	2	4	7
141.061-018	GRAPHIC DESIGNER (profess. & kin.)	S 4 5 6	I	3	4	7
141.061-022	ILLUSTRATOR (profess. & kin.)	S 4 6	I	2	4	7
141.061-026	ILLUSTRATOR, MEDICAL AND SCIENTIFIC (profess. & kin.)	S 4 6	I	4	5	7
141.061-030	ILLUSTRATOR, SET (motion pic.; radio & tv broad.)	S 4 6	I	3	4	8
141.067-010	CREATIVE DIRECTOR (profess. & kin.)	S 5 6	I	3	5	8
141.081-010	CARTOONIST, MOTION PICTURES (motion pic.; radio & tv broad.)	S 4 6	I	2	4	7
141.081-014	COMMERCIAL DESIGNER (profess. & kin.)	S 4 6	I	3	4	7
142.031-010	ART DIRECTOR (motion pic.; radio & tv broad.)	S 5 6	I	3	5	8
142.051-010	DISPLAY DESIGNER (profess. & kin.)	S 4 6	B	3	4	7
142.061-014	CLOTH DESIGNER (profess. & kin.)	S 4 5 6	I	3	4	7
142.061-022	FURNITURE DESIGNER (furn.)	S 4 5 6	I	4	4	7
142.061-026	INDUSTRIAL DESIGNER (profess. & kin.)	S 4 5 6	I	4	4	7
142.061-042	SET DECORATOR (motion pic.; radio & tv broad.)	S 6	I	2	4	8
142.061-046	SET DESIGNER (motion pic.; radio & tv broad.)	S 4 6	I	3	4	8
610.462-010	DROP-HAMMER OPERATOR (forging)	H 3 4 5 6	1 2 3 5 6 7	2	2	6
611.462-010	UPSETTER (forging)	H 4 6	1 3 5 6	2	2	5
611.482-010	FORGING-PRESS OPERATOR (forging) I	H 4 6	1 3 5	2	2	5
612.682-010	BUCKSHOT-SWAGE OPERATOR (ammunition)	H 3 4	1 5 6	2	2	4
612.682-014	FORGING-ROLL OPERATOR (forging)	H 4 6	1 3 5 6	2	2	4
614.382-010	WIRE DRAWER (wire)	H 4 6	1 5 6	2	1	4
614.482-010	DRAW-BENCH OPERATOR (iron & steel; nonfer. metal alloys)	H 4 6	1 5	1	2	4
615.482-010	ANGLE SHEAR OPERATOR (any ind.)	H 4 6	1 5 6 7	2	2	3
615.482-014	DUPLICATOR-PUNCH OPERATOR (any ind.)	H 3 4 6	1 5 6	3	3	5

analysis to develop the SES scores (Duncan, 1961; Stevens and Cho, 1985).

Occupation is an important concept for sociologists, for studying both work life and life off the job. For this reason there are several ways to study occupations, ranging from simple dichotomies (blue collar–white collar) to the complex DOT scores, prestige scales, and SES indicators. Industry and occupation intersect in a specific job, and jobs are found in specific workplaces.

Workplaces

Many workers go to work each day for enormous corporations. Some of them may not even know from day to day just which corporate entity is their employer because of reorganizations, acquisitions, or mergers (see Chapter 15), but workers have a good idea of approximately how many people work at their particular work site. Sociologists are interested in both the local work site and its position within the larger organizational context, often using the workplace as a unit of analysis.

The place to which one reports for work can be called an **establishment.** Some employing organizations have only one establishment; others may have many. For most workers the establishment is important, because it is where they perform their daily tasks and interact with other workers. Even for workers whose jobs require travel, the establishment serves as a base of operations. The establishment can be distinguished from the **firm,** which is the employing organization. Firms may be organized as corporations, partnerships, professional practices, or sole proprietorships.

Firms may be organized into complicated forms. A number of firms may be bought or controlled by a **parent company.** In this case some workers may be employed directly by the parent company and others by the **subsidiary** companies it owns. Burger King, the company described in Box 2.1, was a subsidiary of Pillsbury. "Families" of companies are owned by a single parent company and make related products. Pillsbury's traditional products—flour, baking mixes, and other convenience foods—have a "family" tie to the fast-food industry. If the firms are all in different industries, the parent company may be called a **conglomerate.** If the establishments or firms are located in different countries, as Burger King is, the parent company is called a **multinational company.**

Firms are often linked to one another through complex networks of suppliers and customers, subcontracts, credit lines and other financing agreements, and "tie-ins" of one product line with another. An additional source of links is **interlocking directorates,** which occur when a director of one corporation is also a director or officer of another corporation.

Sociologists collect information on workplaces directly from workers or employers, and information on many firms is available from annual reports, government regulatory agencies, and other sources. Social scientists analyze data on firms to provide information on conditions that affect workers. In later chapters we will discuss the effects of work organizations on workers and their jobs.

Other Units of Analysis

Sociologists also analyze other social units that affect work. Examples include unions and professional associations, which represent groups of workers. **Trade associations,** organizations of firms within the same industry, are significant in understanding the economic conditions and technological considerations affecting an industry. Government agencies, especially those with local, state, or federal regulatory power, are also important units for sociologists to study.

PROBLEMS IN STUDYING WORK

Work is a complex human phenomenon with far-ranging effects. Although researchers continue to refine their techniques and expand their studies, there remain many aspects of work life about which we have little information. This section examines some of the problems researchers encounter in studying work.

Lack of Information

Even with the many sources of information we have already discussed, there are many gaps in sociologists' knowledge and understanding of the world of work. Partly because of the definition of the labor force used by most of the world's governments, we are just beginning to examine nonmarket work such as homemakers (Oakley, 1975), volunteer workers (Daniels, 1988), neighbors who exchange labor, and so forth. In addition, we have very little useful information on the production of illegal goods and services such as prostitution, gambling, illegal drugs, or weapons, to mention just a few.

Other goods and services are legal but the employment they generate is hidden to avoid taxation. For example, some workers are paid cash to avoid income tax withholding and the payment of employer's and employee's Social Security tax. Illegal aliens are sometimes employed in this fashion. Smuggling goods or bartering goods and services also constitute unmeasured economic activity and employment. These uninspected aspects of employment, which are not captured in official labor force statistics, will be discussed in greater detail in Chapter 14.

Hard-to-Measure Characteristics

Some characteristics of work are important but difficult to measure. Social scientists are very interested in issues such as job commitment and underemployment, but there is little agreement about how they should be measured (Sullivan, 1978; Hodson, 1991). One reason such characteristics are hard to measure is that they have both objective and subjective elements. For example, workers who are subjectively bored with their jobs may consider themselves uncommitted to their jobs and underemployed. A researcher might reach a different conclusion for these same workers by looking at indicators such as rates of absenteeism (for job commitment) or hours and wages (for underemployment).

Even productivity, a concept that is relatively easy to measure in manufacturing industries, is difficult to measure in service industries. Is a service worker more productive because more customers have been served, or because fewer customers have been served but have greater feelings of satisfaction about the service they received? Developing methods to measure and study such characteristics is an important frontier for research on the sociology of work.

SUMMARY

Sociology and the other social sciences seek to develop valid and reliable information on the world of work. In advanced industrial societies, work is very complex and heterogeneous, and the study of work involves subjective elements. For these reasons, sociologists have devised different ways to examine the world of work. Although each method has its disadvantages, each also illuminates certain aspects of work situations.

Three important ways to examine the world of work are ethnographies, case studies, and sample surveys. Using different methods, sociologists study individual workers or collective groups of workers such as the labor force. Labor force studies include the study of the demographic characteristics of workers and their rates of labor force participation and unemployment. Social scientists also study occupations, industries, firms and other workplace units, unions, and government regulatory agencies. Although some aspects of work are not yet being adequately studied, the existing methods have yielded important and substantial findings about the complex modern world of work. In subsequent chapters we will present some of these results.

KEY CONCEPTS

validity

reliability

ethnography

participant observation

nonparticipant observation

case study

sample survey

sample

population

establishment surveys

response error

labor force

demographic characteristics

ascribed and achieved
   characteristics

employed

unemployed

labor force participation rate

unemployment rate

discouraged workers

industry

occupation

major occupational groups

occupational prestige

socioeconomic status (SES) scores

multiple regression analysis

establishment

firm

parent company

subsidiary

conglomerate

multinational company

interlocking directorates

trade associations

QUESTIONS FOR THOUGHT

1. According to the definitions used by the U.S. Bureau of Labor Statistics, what was your labor force status last month? Try to locate the most recent unemployment rate for your state or locality. (Note: Check the federal statistical website as provided for regional data.)

2. Give an example of how workers in the same occupation, but in different industries, might have different working conditions.

3. As the children's counting rhyme at the beginning of the chapter illustrates, most people know about relatively few types of work. How can a student or worker learn about more jobs?

4. A researcher wants to discover what happens to workers who are laid off when their factory closes. Compare and contrast the advantages of using an ethnography, a case study, or a sample survey.

5. "To understand work, you need only understand the workers." Do you agree with this statement? Why or why not?

6. Ponder the economic future of the United States, Canada, or Mexico. What sources of data might be useful for planning our collective future and achieving as positive a future as possible?

MULTIMEDIA RESOURCES

The U.S. Bureau of the Census annually publishes the *Statistical Abstract of the United States*. It contains information on labor force, unemployment, industries, occupational distributions, and earnings.

Other good compilations of data may be found in two monthly publications of the U.S. Bureau of Labor Statistics. *Monthly Labor Review* and *Employment and Earnings* provide detailed information on the numbers of persons employed in various occupations, unemployment rates, and rates of pay. Both of these sources include information from establishments as well as from surveys of individual workers.

The U.S. Department of Labor also publishes *The Dictionary of Occupational Titles, Occupational Outlook Quarterly,* and *The Occupational Outlook Handbook.* The DOT includes information about occupations as described in this chapter. The occupational outlook publications contain efforts to project the numbers of workers who will be needed in various occupations and to specify what qualifications will be required of them.

For international information, consult the *International Labour Review* or the *Yearbook* produced by the International Labour Office.

International Labour Office. *http://www.ilocis.org/ iloencyc.html* The ILO, an agency of the United Nations, maintains an encyclopedia of work available at this site.

U.S. federal statistics. *http://www.fedstats.gov/* Over seventy agencies collect some data. This is an umbrella website that describes the standards for statistics and provides links to many of the agencies.

U.S. Bureau of Labor Statistics. *http://www.bls.gov/* The Bureau of Labor Statistics collects data on the labor force, employment, unemployment, and many other topics. These programs are discussed at *http://www.bls.gov/proghome.htm,* and the Current Population Survey is described in detail at *http://www.bls.gov/cpshome.htm.*

U.S. occupational data. For details on specific occupations, including the needed education, a description of the job, and employment projections, see *http://www.bls.gov/ocohome.htm.*

RECOMMENDED FILM

Erin Brockovich (2000, Universal Studios). Directed by Steven Soderberg and starring Julia Roberts, this film is loosely based on a real story about a divorced woman who uses her research skills to bring a large company to account. Rated R for language.

❖❖

PART II

The Personal Context of Work

For small children, work is nearly indistinguishable from play. For adults, however, there is a sharper distinction between work and leisure, with work often separated by time and place from family and leisure activities. As you saw in Part I, the characteristic types and places of work have changed throughout history, and social scientists have devised many ways to identify and analyze work.

Many people view work as their most important activity. Their earnings provide food, clothing, and other necessities, but work fulfills other needs as well. In this section we will examine how work affects the worker's life and family. Work can be an avenue for expressing creativity, perfecting knowledge and skills, and interacting with others. But every job also has its negative aspects, and many jobs are unremitting drudgery. Unsatisfying jobs and unemployment are major sources of tension and stress. Although there are many types of work within the economies of advanced industrial societies, all workers face certain common problems.

Table A shows how median income varies with gender, race, and family status. Married-couple families have the highest median incomes, whereas families supported by women have the lowest median incomes. Moreover, striking differences exist among racial/ethnic groups, even when controlling for marital status. Black families supported by women have a median annual income that is more than $10,000 a year less than the income of white families supported by women.

Such data raise questions about the context of work. Because marital status shows such a consistent association with labor force activity, analysts must explore the issue of *how* families affect work. The data also indicate gender and race differences in work. Why do these patterns persist despite laws to eliminate discrimination? Analysts cannot tell from Table A alone whether employers are discriminating or whether women supporting households might have lower skill levels, less job experience, or greater geographic immobility than men. Moreover, many women may prefer shorter workweeks because of family responsibilities. An analyst must explore such factors in greater detail before drawing a conclusion.

Social scientists begin to answer such questions by studying workers. The four chapters of this section introduce important problems that affect workers. In each chapter we will show how recent work trends alleviate some problems, but also create new problems for workers. Chapter 3 provides an overview of the trends and problems that accompany fitting work into the life of the worker and the worker's family. Chapter 4 examines the personal significance of job satisfaction and dissatisfaction. Chapter 5 discusses a wide range of common disruptions of work, including layoffs, industrial accidents, occupational disease, stress, and employment discrimination. Chapter 6 explores the efforts of workers to organize themselves to solve work-related problems.

Table A   Median Income for Families, by Race and Ethnicity, 1999

	Total Families	White Families	Black Families	Hispanic Families[a]
Median Annual Income[b]				
All families	$49,940	51,912	33,805	33,077
Married-couple families	$56,827	57,242	50,758	37,583
Families supported by women	$26,164	29,629	19,133	20,765
Number of Families[c]				
All families	72,025	60,251	8,664	7,561
Married-couple families	65,311	48,790	4,244	5,133
Families supported by women	12,687	8,360	3,814	1,759

[a] Hispanics may be of any race.

[b] Reported in dollars.

[c] Reported in thousands.

SOURCE: U.S. Census, 2000, "Money Income in the United States," *Current Population Reports,* pp. 60–209 (September), Table 8, p. xi, Table 1, pp. 1–5.

3

Work and Family

> All the world's a stage,
> And all the men and women merely players.
> They have their exits and their entrances,
> And one man in his time plays many parts,
> His acts being seven ages.

Shakespeare, writing in *As You Like It* (Act II, Scene 7), poetically describes the life cycle, or stages through which most humans pass. He enumerated the relevant stages in Elizabethan England from the infant and the school child all the way to "second childishness and mere oblivion." Today, although the stages have changed somewhat, we still can identify a distinctive life cycle both for individuals and for families. As both individuals and family members, we associate some stages with work and others with preparation for work or with leisure.

One important transformation of human work has been its restriction to certain times, places, and ages. For most of human history, "work" was what adults did with most of their time. Work shifts, careers, and retirement are relatively recent concepts that describe the packaging of work into hours, days, and years.

Commuting came about as work was isolated into places separate from the home. Even people who work at home often designate part of their home as an office. As work became temporally and spatially separate, work time came to be separate from family time (Hochschild, 1997).

This transformation had many consequences. Instead of describing the activity of most adults, work came to mean employment, the activity undertaken for pay or profit. People who worked unpaid at home, especially homemakers, were no longer considered workers. As you saw in Chapter 2, the government does not count homemakers as members of the labor force. For people who were employed, integrating work into their lives and the lives of their families became important issues. This chapter will examine these issues within the context of the individual and family life cycles. This chapter describes challenges and problems that workers face at different stages in their life cycles, from their preparation for work to their retirement. We will also examine problems of integrating the family life cycle with work, including such issues as child care, that dual-earner couples may face. This chapter also discusses some possible solutions to these problems.

THE LIFE-CYCLE PERSPECTIVE

The **life cycle** is the ordering of roles from infancy to death. Many tasks are age related; that is, they are considered appropriate to different stages of the life cycle. Rules called norms help us understand what is appropriate at each stage. **Norms** are social rules for thinking, feeling, or behaving, and the existence of norms creates expectations for individuals. It is normative in North America for middle-aged men to be steadily employed; it is the expected behavior. Norms are somewhat flexible depending on mediating circumstances. For example, the expectation of steady employment would be relaxed if a man became disabled or if he and his co-workers experienced a layoff.

Norms help define the roles that people play. Sociologists understand a **role** as a set of behaviors associated with a particular position in society. All of us play such family roles as daughter, sister, parent, and spouse. Many of us also have work roles within the economy such as employer or employee. The work role may be further differentiated by occupation (cook, barber, machinist, secretary), by hierarchical position (apprentice, supervisor, manager, head nurse), and by permanence of employment (part time, temporary, tenured). Most of us play multiple roles outside our work and family roles. We are citizens, consumers, and taxpayers; in addition, we may be licensed drivers, church or club members, and tennis players.

Individual Life Cycle

In modern industrial societies, as in Shakespeare's society, it is normative that people acquire some roles sequentially. We expect that children will be in school; that adolescents will stay in school, go to work, or join the military; that young adults will seek employment; and that at some point elderly persons will retire. These stages of the **individual life cycle** are of great importance. Psychologist Erik H. Erikson described the challenges that people face during their life cycle. The resolution of each challenge shapes the personality, which in turn affects how the person faces the next challenge the life cycle brings. Box 3.1 describes how Erikson's analysis of the life cycle applies to work-related challenges.

When applied to work, the normative ordering of roles is sometimes called the *sequential life plan* (Best and Stern, 1977). In the sequential life plan, education comes first as a preparation for work; then come the working years, and finally the leisure years of retirement. The normative ordering is flexible, however. For example, a teenager may work part time while still in school. The sequential life plan has traditionally allowed women some variations to permit time for child rearing or for remaining a full-time homemaker. One challenge in modern work is the search for greater flexibility in the sequential life plan, a topic to which we will return.

BOX 3.1   The Life Cycle, Work, and Personality Development

Erik H. Erikson related the stages of life to the development of personality (Erikson, 1963:247–274). As the life cycle progresses, he argued, everyone encounters important challenges that can be resolved in one of two ways. One resolution leads to good emotional health, but the other resolution leads to poor emotional health and impedes the successful resolution of later challenges. Each challenge, he contended, is related to an important area of social life.

The first two challenges occur in early childhood and have little to do with work. The first challenge, "basic trust versus basic mistrust," refers to infants' eventual belief that others will provide for their needs, especially food, warmth, and comfort. Organized religion is the area of social life related to this stage. The child learning toilet training faces the second challenge, "autonomy versus shame and doubt." Law and order is the social area related to this stage.

The next challenges deal more explicitly with work. The third challenge, "initiative versus guilt," is related to the economy ethos. Erikson wrote of the child at this stage: "He is eager and able to make things cooperatively, to combine with other children for the purpose of constructing and planning. . . . He looks for opportunities where work-identification seems to promise a field of initiative" (1963:258).

The fourth stage, "industry versus inferiority," is also related to eventual work behaviors, and Erikson linked it to the technological ethos of the society. The child learns to win recognition by producing things, developing skills, and performing assigned tasks. Children at this stage learn about the tools and skills of their society, and a sense of inadequacy or inferiority will arise if they despair of learning to work successfully.

The fifth stage, "identity versus role confusion," usually occurs during adolescence.

Young people at this stage become preoccupied with their peers' view of them and with the issue of how to connect the skills and tools they have learned with the existing occupational types. Erikson saw explicit work roles becoming important at this time: "In most instances . . . it is the inability to settle on an occupational identity which disturbs individual young people" (1963:262). The social areas related to this stage are what Erikson termed *ideology* and *aristocracy*, implying that young people at this age are concerned about whether the best people rule. A sociologist might describe the same issues as part of *social stratification*.

The sixth stage, "intimacy versus isolation," challenges the young adult to develop close relationships or, alternatively, to remain isolated and alone. The social area to which this stage corresponds is that of sexual selection, cooperation, and competition. In the seventh stage, "generativity versus stagnation," mature adults discover their need to be needed. Generativity refers to the nurturing and guiding of the next generation, either through one's children or through one's work. Generativity includes, but is broader than, productivity or creativity. This challenge is related to child rearing, but also to work.

Erikson referred to the eighth stage as "ego integrity versus despair." This challenge can be met only by the person who has successfully met the previous seven challenges, who "has taken care of things and people and has adapted . . . to the triumphs and disappointments adherent to being, the originator of others or the generator of products and ideas" (1963:268). At this stage the older adult assesses both personal relationships and contributions to work. In the integrity that comes from integrating one's life experiences, the human loses fear of death, the final stage of life.

The Career

Sociologists distinguish the individual life cycle from the career. The life cycle refers to a variety of events that occur within one's life, including work events, family-related events, and other age-related events. When sociologists speak of **career,** however, they refer specifically to the sequence of events within a person's work history. Sometimes people use the word *career* to distinguish a good job from a bad job. "I don't want just a job," they might say. "I want a career." From the sociological perspective, *every worker has a career.*

Just as we speak of the typical, or normative, life cycle, while admitting that there are many exceptions in the lives of real individuals, we also speak of the typical career. The typical career of a physician, for example, may include medical school, residency, the opening of a private practice, and eventual retirement. Sociologists term this an *orderly career,* in which job changes follow predictable patterns. Many people prefer to have orderly careers, with steady advancement in responsibility and compensation.

As a particular physician experiences a career, however, it may have idiosyncratic or disorderly aspects, just as an individual's life cycle may not be normatively ordered. For example, the physician might become a hospital administrator or incur a disabling illness. This physician still has a career, but it may not follow the orderly career pattern. More recently, sociologists have identified the "boundaryless career," one that may be pursued with multiple employers and that builds on the experiences and learning in a variety of work contexts (Baker and Aldrich, 1996).

The nature of a career appears to affect the worker's personality. Work that remains challenging and interesting stimulates workers and keeps them mentally flexible and optimistic. Some important elements of the challenge appear to be having a variety of things to do, some choice about when and how to perform tasks, and some complexity or conceptual difficulty to the tasks. The tasks do not need to be physically demanding or hard to do, but they do need to present new opportunities for learning and problem solving. Autonomy, or the ability to make some decisions about work performance and timing, is perhaps the most significant job characteristic in promoting healthy functioning of the worker (Langfred, 2000).

Many workers hold jobs with few opportunities for an orderly career. Virtually every job, even one that is normally challenging and interesting, has its aspects of drudgery and routine. Some jobs, however, offer little else, and workers receive no

opportunity to learn or show initiative. Workers in such jobs become more cautious, conservative, and less flexible. One series of studies found that jobs with low requirements for skill and complexity tended to dull the workers' outlook (Kohn and Schooler, 1983; Kohn et al., 1990). On the other hand, careful design of a job may keep workers interested and committed (Casey, 1996).

Such studies suggest a relationship between someone's career and personality changes. Sociologists have paid particular attention to bureaucratic careers, because more jobs are now located in bureaucratic, hierarchical organizations. Bureaucratic organizations will be discussed in more detail in Chapter 7. For now, it is enough to note that in a bureaucracy, each worker tends to have specific, specialized duties, a detailed set of rules to follow, and a supervisor. Communication in the entire organization is expected to go "through channels," that is, from supervisors to subordinates through several layers of supervision. Most bureaucracies have job ladders, so that workers understand the possible promotions and can plan an orderly career in the organization.

Many young bureaucrats begin their jobs with enthusiasm and excitement. Some retain this enthusiasm and are promoted several times. Inevitably, however, there are fewer promotion opportunities than eligible workers, and so some workers eventually feel stuck in their jobs. The "bureaucratic personality" they may develop can lead them to strictly enforce all rules and to avoid any creativity in applying them (Merton, 1968). To an observer they may seem to be going through the motions of their jobs, and their goal changes from getting ahead to playing it safe. In times of downsizing, not even this tactic may be sufficient to ensure job security (Tivendell and Bourbonnais, 2000). In Chapter 4 we will discuss some other worker reactions to blocked opportunities.

The Family Life Cycle

Besides the life cycle of each individual, there is also a **family life cycle,** which describes the stages of formation, growth, and dissolution of the nuclear family. A nuclear family, which consists of a married couple or of one or more parents and children, is the most common family structure. Important events in the family life cycle include marriage, birth of the first child, birth of the last child, departure of the last child from the home (the "empty nest"), retirement from work, and the death of one spouse. Divorce and remarriage are additional stages that occur in the life cycle of many families (Hill, 1986; Spanier and Glick, 1980). Although there have been changes in the timing of these events—for example, the age at marriage has risen in recent decades—studies in such diverse countries as Japan, India, Sweden, Belgium, and Canada show remarkable consistency in the phases of the family life cycle (Sussman, Steinmetz, and Peterson, 1999; Hill, 1986).

Three important points need to be made about the family life cycle. First, because of variations in family membership, the model does not describe all families. Childless couples omit the stages involving child rearing, and single-parent families may miss the stages involving a spouse. Some families experience additional stages as the result of the temporary or permanent addition of family members, such as a grandparent who moves in or an adult child who returns home.

Second, the timing of the family life cycle is highly variable. A divorced parent who remarries and subsequently has children with the second spouse may be "recycling" through earlier stages of the family cycle (Cherlin et al., 1991). The death or the divorce of a spouse may occur at any time after the marriage, regardless of the family "stage" with regard to children.

Third, different social and economic problems accompany various stages of the cycle. Getting married, becoming a parent, or facing the empty nest may all engender crises of adjustment. Each of them is also likely to be accompanied by changes in the financial situation of the family.

An important issue for sociologists of work is the joint sequencing of events in the individual life cycle with those in the family life cycle. Many of these issues are already familiar to college stu-

dents. Should you get married before completing your schooling? Should you be secure in your job before deciding to have children? How will having children affect a parent's career? If both the husband and wife are employed, how will the family react if one of them receives a job transfer to a distant city? In the remainder of this chapter we will discuss the interaction of the individual's work roles and family roles, the problems people experience in integrating them, and some of the consequences of these opportunities and difficulties for workers.

SOCIALIZATION AND WORK

Sociologists use the term **socialization** to refer to the process of learning norms, roles, and skills. The "teachers" in this learning process are called *socializing agents.* Education, training, workplace orientations, and similar arrangements are *formal socialization.* In formal socialization the typical socializing agents are educators at all levels as well as training personnel in the armed services, corporations, and elsewhere. Perhaps more important, however, is the *informal socialization* that goes on with parents, peers, or the mass media as the socializing agents. Sociologists of work study both kinds of socialization as a process that prepares children for work and helps workers learn their jobs or adjust to changes in their jobs.

Informal Socialization

Children are first socialized at home, and parents are among the most important socializing agents. Parents teach children language and basic living skills, which are essential to the socialization that will occur later in school. Indirectly, parents also model what it means to be a worker, how one goes about working, and the importance of work relative to other activities. Very young children play games of pretending to work, and adults reinforce the significance of work by asking children, "What do you want to be when you grow up?" Little girls may still sometimes receive the message that paid work is not as important for them:

> From the smallest age you don't picture yourself in a job. And if you talk to someone like my husband, he pictured himself in jobs. They were different jobs at different times but that's how you picture yourself. You know, this is what my life is supposed to be like—carry my little briefcase into work. (Dinnerstein, 1992:23)

Different parents teach their children different lessons about work. Sociologists have studied socialization in families of different social classes. Parents with manual jobs tend to have somewhat different child-rearing practices than parents whose jobs involve office work or intellectual work (Kohn et al., 1990; Sears, Maccoby, and Levin, 1957). Parents' own job experience, according to this view, indicates that some behaviors and personality traits are more important than others in being successful, and these are the behaviors the parents emphasize with children (Gecas, 1979; Kohn, 1976). For example, being independent and showing initiative may be very important in professional jobs, but being obedient and taking orders may be more important in manual work.

This type of socialization is most useful when children will take up the same work as their parents did. When this happens, we say that there has been *occupational inheritance.* Occupational inheritance takes place through the inheritance of capital, as when a child inherits a parent's business, but it can also occur because the parent has prepared the child with the same skills and knowledge the parent has. Many children in the advanced industrial societies, however, will occupy jobs in new or rapidly changing industries and require skills their parents never had. For this reason, formal socialization is increasingly important.

Formal Socialization

In school, students' teachers and peers constitute the socializing agents. Schools provide training in literacy, calculation, communication, and other basic skills useful in any job. Even in elementary schools the formal curriculum includes instruction in occupational skills such as computer programming or keyboarding. At advanced levels entire courses of study are devoted to explicit occupational training. In high schools vocational agriculture, business training, and home economics are examples. In college some degrees qualify students for specific occupations, such as engineering or accounting, while others, such as the liberal arts, provide general expertise in problem solving and communication.

It would be a mistake, however, to view schools as preparation for work solely in terms of the curriculum. Sociologists have studied other ways in which schooling prepares the young person for life as a worker. For the young child the school represents the first separation of "workplace" from home (Dreeben, 1968). At home the child is accepted more or less unconditionally; failings in terms of one's skill or attitude are often overlooked or offset with compensating virtues. The long-lasting relationships within the family are stressed and nurtured. In school, by contrast, the child is evaluated for performance. Being charming does not compensate for inadequacies in math. The child learns to relate to others in terms of their positions: Although the person who is the teacher may change, every teacher will expect similar behavior. In the same way, workers are evaluated principally for their job performance and relate to co-workers and supervisors primarily in terms of positions within the firm. Thus, the school takes children from a setting in which they are valued as individuals and prepares them for a setting in which they will be valued for productivity and competence. This transition may be very gradual in kindergarten and first grade, where the teachers may view themselves as surrogate parents as well as educators. But the process is far advanced by the time the student enters college. Relationships with professors, at least on many college campuses, are likely to be remote and formal, resembling relationships with supervisors in later jobs.

The *hidden curriculum* of the school conveys many other lessons. Students learn the importance of punctuality, orderliness, and learning and following rules. For example, students learn to raise their hands and wait for recognition before speaking, a behavior that persists in adult gatherings. Students learn to identify and respond appropriately to people with authority, and they come to accept as legitimate the right of authorities to assign tasks and to define how to complete them.

Schools reflect the economy and the society that supports them. Thus, students will learn an ideology, or way of explaining the existing economic and political patterns. They will also be exposed to any pattern of discrimination that characterizes their society. One aspect of socialization is the "steering" of children toward some jobs and not others based on their race, gender, or social class (Bowles and Gintis, 1976). Box 3.2 gives an example of such steering.

Socialization in the Workplace

Socialization for work takes place in other settings besides the school. The workplace is an important socialization site. Employers or co-workers can teach many jobs to new workers. Dorinne Kondo describes learning a job in a Japanese bakery:

> The primary morning task was to wrap individual slices of cake with cellophane. Though seemingly simple, this task was hard to do without disturbing the frosting decorations, difficult to accomplish in one smooth, deft movement. My first day on the job, I remember the division chief, Akita-san, coming out from his room after I'd been wrapping for a half hour or so. "Relax your shoulders!" he said, with a concerned look on his face. "Don't worry—you'll get the hang of it before long." But it took a month before he could say to me, "Looks like you've gotten used to it. You're quicker, and you don't seem to get as tired as you did before." (Kondo, 1990:287)

Employers, unions, or government agencies may offer the appropriate training. Unions working jointly with management often organize apprenticeship programs that provide the trainee with half pay during a two- to four-year apprenticeship.

One important difference among jobs is the extent to which the employer is willing to invest the time and cost to provide training. Some companies provide lengthy programs with on-site classroom instruction. For example, banks often offer training programs so that new employees will understand how the federal government regulates the banking system and how bank security is provided. A recent study indicates that 63% of young workers (aged twenty-four and under) received training provided by their employers. On average, employees received 13.4 hours of formal training and another 31 hours of informal training. Women, college graduates, professional and technical workers, and workers with five to ten years employment in their current job were most likely to have received employer-provided training (U.S. Census, 1999:439).

Many employers assume that the worker will know most of what is important before being hired. The employer may rely on a credential from a school or union to certify that the new worker has the requisite background. One observer has argued that the increase in women's labor force participation represents increased *demand* for jobs such as schoolteachers, nurses, social workers, and secretaries—jobs that were stereotypically held by women and for which workers received most of their training before applying for work (Oppenheimer, 1970). The employer's training responsibility is limited to a basic orientation concerning the specific rules and procedures at the work site. As a result, the employer's costs are lower.

Socialization does not end when schooling ends, nor does it end even when formal on-the-job training ends. Co-workers and supervisors continually socialize workers. New electrician apprentices, for example, carry many tools of varying quality and spend time learning specialized

BOX 3.2   Occupational Steering

Multicultural and multiracial societies often develop a "cultural division of labor" in which some racial and ethnic groups typically perform some work and others perform different types. In the Southern United States during the Jim Crow era or in South Africa under apartheid, laws or quasi-legal arrangements officially sanctioned job segregation. In other circumstances adults maintain job segregation unofficially by steering young people into jobs thought "appropriate" to their racial and ethnic categories. Malcolm X (born Malcolm Little) was an American black leader during the early 1960s. In his influential *Autobiography* (Little, 1965:36), he describes an example of such "steering" that occurred while he was a high school student in Lansing, Michigan:

> [Mr. Ostrowski, my English teacher] probably meant well in what he happened to advise me that day. I doubt that he meant any harm. It was just in his nature as an American white man. I was one of his top students, one of the school's top students—but all he could see for me was the kind of future "in your place" that almost all white people see for black people.
>
> He told me, "Malcolm, you ought to be thinking about a career. Have you been giving it thought?"
>
> The truth is, I hadn't. I never have figured out why I told him, "Well, yes, sir, I've been thinking I'd like to be a lawyer." Lansing certainly had no Negro lawyers—or doctors either—in those days, to hold up an image I might have aspired to. All I really knew for certain was that a lawyer didn't wash dishes, as I was doing.
>
> Mr. Ostrowski looked surprised, I remember, and leaned back in his chair and clasped his hands behind his head. He kind of half-smiled and said, "Malcolm, one of life's first needs is for us to be realistic. Don't misunderstand me, now. We all here like you, you know that. But you've got to be realistic about being a nigger. A lawyer—that's no realistic goal for a nigger. You need to think about something you *can* be. You're good with your hands—making things. Everybody admires your carpentry shop work.

Why don't you plan on carpentry? People like you as a person—you'd get all kinds of work."

Because this event made Malcolm X aware of how his race seemed to be more important than his achievements, he referred to it as the "first major turning point" in his life.

Occupational steering may also be based on gender or class in addition to race or ethnicity. Girls are sometimes encouraged to play with "appropriate" toys or to take music or art courses instead of science and mathematics. Eventually, girls may be steered into a fairly narrow range of occupations that are thought suitable for women. Occupational steering may be directed toward working-class students because middle-class teachers and counselors may hold stereotyped ideas of their abilities and interests. For example, students from working- or lower-class backgrounds may be encouraged to take vocational-education programs instead of college-preparatory programs, even if their academic abilities would predict success in college.

There are several ways to counter occupational steering. One important method is to provide *role models,* people whose occupational roles become an example for others. Recall that Malcolm X mentioned that there were no black lawyers or doctors "to hold up an image I might have aspired to." As more minority group members and women enter nontraditional occupations, students and younger workers who share their characteristics may also be able to pattern their behavior using their example. Affirmative action, a program that seeks to place women and minorities in jobs in which they are underrepresented, may provide additional role models. (Affirmative action is discussed in more detail in Chapter 5.) Other ways to avoid occupational steering include training teachers and counselors more effectively, using the mass media to provide career information, and making students and parents more aware of the possibility of steering.

SOURCE: Excerpt and quotations from Malcolm X, with Alex Haley, *The Autobiography of Malcolm X*, p. 36. Copyright © 1965 by Alex Haley and Betty Shabazz. Reprinted by permission of Random House, Inc.

terms. They eventually learn from the master electricians to carry only a few well-worn, high-quality tools, and they become completely familiar with the jargon and procedures of the craft (Riemer, 1977). Such socialization prepares workers for eventual promotion or job changes, and it is the way in which they learn new roles: to be a supervisor, to be on strike, to be unemployed, and to retire (Trice, 1993:112–137).

THE WORKING YEARS

School is the bridge between home and the more impersonal world of work. Young adults face several important and related life tasks. Not everyone will perform each of these tasks, but most will. These tasks are leaving the parents' home; deciding whether to continue schooling; deciding whether to seek permanent, full-time employment; and deciding whether to start a new family. For men the normative order of the last three events is to finish school, find the job, and then get married. There are other, nonnormative orderings of these decisions. For example, one may get married, then finish school, then find a job. One study of young men finds that the highest occupational rewards go to those who follow the normative order; those who follow other orderings usually achieve lower positions (Hogan, 1981). Women may follow the same path as men, but it would also be considered normative for women to make the decision to begin a family instead of continuing school or working.

Entering the Labor Force

The choice of an occupation is never made in a vacuum. It is influenced by a variety of factors, including personality and social context. Long before their formal schooling ends, young people acquire a self-image and attitudes toward work involvement and occupations. These personality characteristics play an important part in the jobs that they will eventually take and in the success they will experience (Mortimer, Lorence, and

Kumka, 1986). Resources for new workers are described at the School-to-Work website (*http://www.stw.ed.gov/*).

The mix of available occupations affects the incorporation of new workers into the labor force. In Chapter 1 we mentioned changes in occupational distribution, a topic we will address in greater detail in Part IV. In general there has been a long-term shift from agricultural work to manufacturing work to service work. This process entailed a parallel shift from a labor force that does predominantly manual work to one that does bureaucratic and service tasks. Relatively new occupations tend to attract fairly young workers who have just completed the requisite training. Thus, the computer field has relatively young workers in it.

Occupations vary in their *barriers to entry*. It is relatively easy to become a janitor, in terms of skill, because the duties are thought to be within the capability of most workers. To be a physician, on the other hand, requires long years of specialized training, testing, and licensing, and many fewer people are thought to have the ability to master these skills. It is always possible to change occupations, but young workers who choose an occupation with low barriers to entry may eventually regret not having acquired the qualifications for entry into a more desirable occupation. We will discuss in later chapters some of the mechanisms for creating barriers to entry, including craft unions, licensing, and professionalization.

Regardless of the occupation chosen, every young worker enters the labor force in a specific job. Even within the same occupation, jobs differ a great deal. A nurse, for example, may find that a job in a hospital is very different from working in a physician's office or a nursing home. An entry-level job is a relatively low-level one with minimal job requirements. There are different types of entry-level jobs; one distinction we can make is between so-called dead-end jobs and entry-port jobs.

Dead-End Jobs A **dead-end job** requires relatively little skill, often has a high turnover, and rarely if ever leads to promotions, higher pay, or more responsibility. Such jobs often pay only the minimum wage and characteristically provide little security and few fringe benefits. **Fringe benefits** refer to the nonwage compensation and perquisites ("perks") of workers. Some common fringe benefits include pension coverage, health and other forms of insurance, sick leave, and paid vacation. These benefits may cost the employer 25% or more of the salary or wages. An example of a perquisite would be an employee discount on purchases of the company's product. Many dead-end jobs are part time or seasonal, or they may be subject to reduced hours or layoffs when business conditions are adverse. Because of the relative unattractiveness of the dead-end job, the employer expects little preparation from the workers, and provides little on-the-job training, as the following account shows:

> Between four and eight women took three different colors (and lengths) of wire and inserted them into a small plastic block an inch square and a quarter of an inch deep. This was by far the hardest and most tedious job. Each lead had a square terminal on the end which had to be pushed into a square channel in the plastic block until it locked. . . . It took a certain amount of force and some finesse as well. If you held the lead too far back you bent the terminal. If you held the lead too close you banged your fingers. You could always tell who was new on the job by their bandaged fingers. (Juravich, 1985:47)

From the employer's viewpoint one advantage of the dead-end job is that "anyone can do it." Assuming that labor is plentiful, it will be relatively easy to recruit someone to do the job, because few skills are needed. Recruiting and interviewing prospective workers takes little time. The employer can fire difficult or undisciplined workers and hire new workers. If business is good, the employer can easily create more such jobs. When business lags, the employer can lay off workers or reduce their work schedules.

A dead-end job also may be the only possible arrangement for even a well-meaning employer. For a small employer, or for one whose work flow is seasonal or erratic, such as a subcontractor who does "overflow work" for a larger firm, there may be no reasonable alternative to creating a dead-end job. Such an employer might prefer better trained or more ambitious workers, but those workers will be attracted to larger or better capitalized firms with better working conditions. Many private householders create what are, in effect, dead-end jobs for housekeepers, gardeners, and child-care workers, simply because there is no realistic way for them to create better jobs.

Dead-end jobs have some disadvantages for the employer, too. Replacing workers is easy if labor is plentiful, but difficult and costly when labor is scarce. Furthermore, because the best workers prefer better jobs, the pool of available workers is likely to have relatively little experience and training. High turnover can disrupt production. In service industries an uncommitted worker can irritate clients and customers and cause them to take their business elsewhere.

It is easy to see the disadvantages of the dead-end job for the workers, yet these jobs do serve some positive functions for young workers. Part-time or summer jobs provide the first working experience for many young people. Although the job may offer no formal training, working itself is new to them, and so they learn some general lessons about this new world. The employer on a young person's first job is often willing to provide references to colleges, credit bureaus, or other employers. Part-time jobs are attractive to students and young parents who cannot work full time. For middle-class college students, these jobs often serve as a bridge between their schooling and eventual permanent jobs.

Dead-end jobs are detrimental to the worker who is never able to find better employment. Such a worker may change jobs frequently, looking for better pay or working conditions, only to find that the job is once again a dead end. In this case the worker may never successfully bridge the gap between dead-end jobs and jobs with better prospects. In time, the history of *job hopping,* rather than indicating ambition and a desire for self-improvement, may come to be interpreted as the inability to hold a job or as more generalized instability. Thus, characteristics that originally described the *job* are used instead to label the *worker.* In fact, workers who hold only dead-end jobs learn few new skills and have few incentives to improve their work habits. (Related topics will be discussed in Chapter 14, on marginal jobs.)

Knowing about dead-end jobs helps to explain why the nonnormative orderings of life events may lead to lower job rewards. Students who drop out or who have a new family to support may need the income from a job right away, but their lack of education may limit them to dead-end jobs. Although to new workers such jobs may seem to pay well, there will be few annual pay raises and few opportunities to progress to better paid, higher skill jobs.

Entry-Port Jobs In contrast with the dead-end job, some entry-level jobs are called **entry ports** because they offer the worker the possibility for training, greater responsibility, improved pay and fringe benefits, and promotion. Entry ports are usually filled through formalized recruitment procedures. Many of the jobs have well-established criteria for hiring. Recruiting and screening are done carefully to ensure that the best available workers are hired. The employer may pay particular attention to education, including degrees earned, the quality of school, and any additional certificates or licenses held. Employers also pay attention to previous work experience, including part-time jobs, summer jobs, and volunteer work.

The conditions of work also differ. Entry ports usually require at least an orientation period (if not a training program) to introduce the new worker to company policies and regulations. There is likely to be a well-defined job description. Many entry ports have a probationary period, during which a new worker is carefully watched and evaluated. It is usually easy to dismiss a probationary worker, but afterward the workers are likely to be protected by a job security arrangement and also by formal procedures that prescribe how they may be disciplined. The entry port usually pays more than the minimum wage and provides some fringe benefits. Commonly, there is provision for regular reviews of performance and salary.

Sometimes entry ports offer low initial pay, although above minimum wage. One reason for the relatively low pay, according to economists, is that the workers are subsidizing their own on-the-job training, but their eventual promotions within the firm will compensate them for their low starting salaries. The most distinguishing characteristic of an entry port is that it makes the worker eligible to compete for advancement within the firm. The positions to which one can be promoted are the rungs of what is called a *job ladder.* Job ladders vary among employers in both their height and the intensity of competition for the next rung. Some ladders are very short. A worker hired into an entry port may anticipate only one or two promotions until the ladder ends. In many companies, for example, it is possible to be promoted to supervisory positions, but managerial positions require a different, higher entry port for candidates with a different set of credentials. In other companies "working your way to the top" may be the company policy, and the job ladder extends, at least in principle, all the way to the top.

A number of factors determine the competitiveness of promotions. One factor is business conditions: A company that is expanding can promote its workers, whereas one that is considering layoffs is unlikely to offer promotions. Another factor is the relative *steepness* of the company hierarchy—that is, whether there are many or few possible levels for promotion. A third factor is the *span of control* for those higher on the job ladder. They may each supervise only a few workers, in which case more of them will be needed. If, on the other hand, each supervisor has responsibility for a large number of workers, then relatively few supervisors will be needed. Many such factors are invisible or uninteresting to the beginning worker, at least initially, but they can have an effect on how long the worker is willing to remain in the firm.

Most entry-level workers who want permanent jobs would prefer an entry port to a dead-end job. Entry ports generate some advantages for employers by attracting higher quality workers and inducing them to stay with the firm. But employers must also invest more training and compensation in their workers in entry-port jobs.

Regardless of other issues, including the company's financial position, the preferences, beliefs, and strategies of managers affect the characteristics of entry-level jobs. Some companies create entry ports for employees who work closely with the production process and whose growing expertise will be needed by the company in future years. In the same company, other types of work, such as routine maintenance or low-level clerical work, may be packaged as dead-end jobs with little prospect for promotion. Even if dead-end jobs have high turnover, the company can still produce its major good or service.

Small firms create about two-thirds of the new jobs in the United States, most of these in the service industry (Bednarzik, 2000). Between 1995 and 1996, about 915,000 more new jobs were created from the opening of new establishments as opposed to the jobs lost through the closing of establishments. New jobs, however, are not necessarily good jobs. Sociologists are concerned that the majority of new jobs will be dead-end jobs, because small firms have few resources for job ladders. Moreover, small firms are most vulnerable to market forces and so are very likely to create jobs that are short term, seasonal, temporary, or part time.

There is a possibility that employers may feel pressure to create more attractive entry-level jobs because the birth rates of the industrialized countries have been low since the 1970s. Thus, there will be relatively slow growth in the number of young workers entering the labor market in the early 2000s. Employers may compete for young workers with better wages and working conditions. Moreover, the industrialized countries,

which have recently relied on immigrant workers to fill dead-end jobs, have tightened their immigration laws. In 1986, for example, the United States enacted civil and criminal penalties against employers of illegal aliens.

Occupational Mobility

In modern industrial societies there is a great deal of **occupational mobility,** or movement among jobs. There are two kinds of occupational mobility. The first kind, intergenerational mobility, refers to the child's pursuit of an occupation different from that of the parents. Intragenerational mobility, or career mobility, occurs when a worker moves from one employer to another, from one position to another, or even from one occupation to another. About one in ten employed workers changes occupations every year (U.S. Department of Labor, Bureau of Labor Statistics [BLS], 1992a).

Intergenerational mobility occurs for a number of reasons. Some have to do with the individual abilities and ambition of the child. During the late 1800s, Horatio Alger wrote stories about poor children who became multimillionaires because of their ambition and hard work. But much occupational mobility is *structural*—that is, it results from large-scale changes in the economy, in technology, and in the mix of goods and services produced (Blau and Duncan, 1967; Hauser and Featherman, 1977).

Opportunity Structure A good example of intergenerational mobility emerges from the increased productivity of farms beginning in the late 1800s. All the children of farmers were no longer needed to produce crops and livestock. If only one or two adult children remained on the farm and used the new agricultural equipment, they could produce as much as ten workers did in their parents' generation. Simultaneously, in the cities the demand for factory labor grew. Farmers' children flocked to the cities to work in factories and foundries. By the time *their* children and grandchildren were ready to work, opportunities were opening in entirely new fields of communication, information technology, and so on. The *opportunity structure*—that is, the kinds of jobs that are available—sets the limits for occupational mobility. When many prestigious jobs are being created, more young workers can attain them. Most of the intergenerational occupational mobility in the United States in recent years may be attributed to shifts in the opportunity structure (Hauser and Featherman, 1977).

Vertical Mobility Occupational mobility (both intergenerational and intragenerational) is often conceptualized in terms of vertical distance. *Upward mobility* refers to movement into more prestigious, better paid, or more responsible positions. Upward intergenerational mobility occurs when an adult daughter or son enters an occupation of higher standing than that of the parents. Upward intragenerational mobility occurs when a worker is promoted to a more important or responsible job.

The United States has more upward intergenerational mobility than some other industrial countries and a little less than others. A study released in 1985 indicated that 41% of a sample of U.S. men had experienced upward mobility, compared with 44% of French and Austrian men (Haller et al., 1985:589). American men's mobility is greater than that of men in Italy (Pisati, 1997). Men and women in the United States have similar intergenerational mobility (Li and Singelmann, 1998). Canada also has relatively high mobility (Wanner and Hayes, 1996). One reason for this upward mobility is the continuing shift in the opportunity structure, so that there are more jobs calling for high levels of skill.

North Americans value upward occupational mobility for their children. Many parents hope that their children will "do better than I have done and have it easier than I have had it." Nevertheless, the desire for upward mobility for their children may cause a dilemma for the parents. They can socialize their children only for the types of jobs they have known. In preindustrial societies a son or daughter could learn virtually every needed skill

from a parent or relative, and occupational inheritance was commonplace. In industrial societies, parents must work together with schools to provide formal training for their children. With rapidly changing technologies, many parents cannot even imagine the skills and occupations that might be available for their children. Many parents now find that they do not really understand the jobs held by their adult sons and daughters.

Downward mobility refers to movement into less prestigious, less well paid, or less responsible positions. Downward intergenerational mobility occurs when an adult son or daughter enters an occupation of lower standing than that of the parents. Downward intragenerational mobility may occur as a result of a disabling condition, formal demotion, layoff, or the "bumping" of high-level workers to lower-level jobs instead of laying them off. Firm downsizing during the 1980s led to substantial amounts of downward mobility.

> When David Patterson's boss left frantic messages with the secretary, asking him to stay late one Friday afternoon, his stomach began to flutter. Only the previous week David had pored over the company's financial statements. Things weren't looking too good, but it never occurred to him that the crisis would reach his level. He was, after all, the director of an entire division, a position he had been promoted to only two years before. But when David saw the pained look on the boss's face, he knew his head had found its way to the chopping block. He was given four weeks of severance pay. . . . (Newman, 1988:1–2)

Career Shifts The average American changes jobs about three times during a career, so intragenerational or career mobility is frequent. The reasons for changes vary through the life cycle. Young workers, in particular, often move from the entry-level job to another job within three years as they adjust to the labor force. Many of these changes represent lateral, or *horizontal,* movement within the opportunity structure.

In their middle working years, workers often feel a restlessness that leads them to find a new job, perhaps in a different occupation or industry. For others, an orderly career assumes a change in responsibilities and duties:

> Electrical and chemical engineers, operations researchers, accountants, clergy, and college teachers all face probabilities of more than 50% of leaving by age 40 the census categories in which they began their careers. (Abbott, 1989:280)

For some workers this change represents the fulfillment of a lifelong dream to own their own business or to enter a more exciting field.

Taking a new job is not always voluntary. Sometimes it is motivated by an unexpected layoff or dismissal. Middle-aged workers may be laid off as technologically obsolete because they have not learned to use newer technology. Their layoff also saves the employer more labor costs, because middle-aged workers are usually paid more than younger workers. Finally, companies that are eager to become "lean and mean" may lay off middle levels of supervision and management that are filled by middle-aged workers.

Middle-aged workers tend to have a harder time finding new jobs. Many job openings are at the entry level, but middle-aged workers' experience and skills qualify them for more responsible positions. The middle-level or advanced jobs for which they might be qualified are often not advertised, because they are filled from within the company. Further, many middle-aged workers are unwilling to suffer the loss of income that a lower-paid entry-level job would require. In these circumstances moving to a new occupation or industry may be the only viable option. Some workers return to school or formal training to start over again.

> [Sandra] worked for five years in a variety of low-level jobs in semiprofessional engineering work. At the end of that time she was working side-by-side with a man who had an engineering degree. "Of course he was getting two to three times more than I was. I realized we were doing the same thing. And I was thinking, 'If this yo-yo can do it, I am sure I can.' So I said, 'I am going back to school.'" But she was worried that at thirty-five she was too old to be successful. . . . She told her advisor at the university, "Well, I'm thirty-five." He reassured her, "Oh, that is nothing. I went back when I was forty-two." (Dinnerstein, 1992:190–191)

Retirement

The labor force, as measured in the United States, has a lower age limit of sixteen but no upper age limit. In the 1980s, mandatory retirement at a given age was outlawed for most private-sector jobs in the United States. A few occupations are exceptions because the physical aging process can be demonstrably linked to poorer job performance. Airline pilots, for example, still have a mandatory retirement age because the gradual deterioration of vision, hearing, and reflexes in older pilots might increase the safety risks of air travel. Some workers will take advantage of this legal change to stay at work as long as possible. In particular, observers expect professional workers to remain in the labor force for a relatively long time, while manual workers in harsh environments or doing heavy physical labor are likely to retire earlier.

One reason to continue working is financially based. Many workers have small or inadequate pension plans, and only 42% of workers are covered by private pension plans (U.S. Census, 1999:395). Legal reforms during the 1970s and 1980s required "vesting" of a worker's pension plan after ten years. Vesting means that the worker is assured of benefits at retirement. Vesting rates range from 33% of workers under the age of thirty to 55% of workers aged fifty to fifty-nine (Census, 1993b). These reforms came too late to help some older workers, and workers who changed jobs often, worked in industries without pension coverage, or worked part time.

Although most workers are covered by Social Security benefits, these payments are relatively small, especially when compared with the income the worker received on the job. About 38% of current retirees rely on Social Security to pay living expenses; and 76% of workers surveyed in 2000 said that they were saving for their retirement, compared with only 61% in 1993 (Gallop-Goodman, 2000).

Women are especially likely to be hard pressed at retirement because they are more likely to have worked intermittently or part time and to have received lower wages. A widow who earned less than her husband is eligible for Social Security benefits based on her husband's earnings. About 92% of wives born in 1930–1934 and 82% of wives born in 1955–1959 will receive widow's benefits if their husbands die, because the benefit will be greater than benefits based on their own earnings (Iams, 1993). A widow may also be entitled to a private pension if her late spouse was covered, but frequently the widow's coverage is a fraction of what the couple would have received in retirement. A divorced woman may be entitled to nothing from the estate or pension coverage of her former husband. Such financial concerns may encourage a worker to stay on the job as long as possible.

Many workers, however, prefer to retire early. Between 1970 and 1998 the labor force participation of men over age fifty-five decreased, whereas that of women in the same age group increased (U.S. Census, 2000:411). Early retirement is frequently linked to illness or disability. In other cases, older workers who are laid off may report themselves retired because they believe that no work is available. Significant numbers of male and female workers arrange a transition period of partial retirement between full-time work and complete retirement.

In the United States laws require employers not to discriminate against workers older than forty and not to make retirement mandatory on the basis of age alone. They may, however, still encourage retirement at or around the customary age of sixty-five. Retirements provide turnover to allow the hiring of younger workers. In addition, the younger workers are more likely to have acquired the latest technological skills. Older workers are usually better paid, especially if wages increase with seniority. Thus, the retirement of an older worker may provide the salary to hire two younger workers, or the opportunity to save on payroll expenses. Finally, although data do not indicate that older workers are any less productive than younger workers, employers may believe them to be less productive and thus encourage them to "move on." There is nevertheless concern that the retirement of large numbers of baby boomers starting in 2010 will create staffing shortages in skilled jobs (Dohm, 2000).

There is also an interesting countertrend in the hiring of elderly people. Several firms in the fast-food industry have begun to substitute older part-time workers for younger workers, who are becoming less available. Their television advertisements often make a point of showing older workers serving food. Employers report that elderly workers make good employees because they are mature and have good work habits (Pereira, 2000).

Historically, preparation for retirement was solely the worker's responsibility. The company might provide the customary dinner and gift of a gold watch to mark the transition to the new status:

> An elaborate ceremony at the time of retirement serves . . . as a rite of separation from both the occupation of firefighting and one's engine company. Its most prominent feature is a verbal "roast" during which the audience and various speakers confront the retiree with many of his misdeeds and indiscretions and generally review his overall performance as a firefighter. . . . These "roasts" are intermingled, however, with compliments and expressions of respect and appreciation. At the end of the dinner, the former firefighter appears publicly in front of his fellow firefighters as a retiree. He is forced into isolation and made to realize the inevitable and irreversible change. (Trice, 1993:123)

Not surprisingly, many retired workers today find that the long-awaited leisure is not what they had anticipated. "Disengagement," or relinquishing their work roles, is difficult. For those with strong work commitments, it is disorienting to have no workplace to report to every morning, and they lose their strong role identity as breadwinners. Others find that because of failing health or financial difficulties, the activities they had planned for their retirement years are no longer possible. Inflation may gradually eat away at carefully saved reserves, so that the dreams of extensive travel or endless days of golf are never fulfilled.

Some companies encourage their retirees to consider seasonal or part-time work with the company, and they invite retirees to company social events. Retirees are often eligible to continue participation in the company credit union, athletic teams, and other activities. During the 1980s, some larger employers began retirement-planning seminars for their older workers. These seminars are not yet widespread, but if successful, they could help the workers think realistically about retirement income, activities, and continuing work alternatives.

Alternative Cycles

In the sequential life plan, human work is organized sequentially into a period of education, a period of work, and then a period of leisure (retirement). Our review of these periods, however, indicates problems with the sequential life plan in at least two areas, education and leisure.

- *Education.* With rapidly changing technology, the skills even of young workers may become obsolete. There is a need for workers in many fields to update their education periodically through continuing study, sabbaticals, and other means. It is probably cheaper in the long run for employers to make such opportunities available rather than constantly hiring new workers (and losing the expertise of the older workers). These educational needs cannot be met so long as

education is relegated solely to the adolescent and early adult years.

- *Leisure.* Postponing leisure to the retirement days means that many workers will never enjoy it, because of either failing health or low income. Meanwhile, most workers have relatively short vacations that often do not provide sufficient time to refresh themselves for the return to work. With the average age at retirement likely to increase, longer vacations and sabbaticals for workers become increasingly attractive options

The "cyclical life plan" would intersperse periods of education, work, and leisure throughout the life cycle (Best and Stern, 1977). Such a plan might include more internships and work experiences for students, education and leisure sabbaticals for workers, and longer vacations throughout the life span. Possible advantages of the cyclical life plan include reduced boredom, especially in the middle years of work, and enhanced job productivity. The disadvantages of the cyclical life plan include its cost to employers, both to fund the sabbaticals and to provide replacement workers for the absent workers. Workers may also miss important workplace developments during a sabbatical.

INTEGRATING WORK AND FAMILY LIFE

Most workers have families, and so besides the issues they face in their careers as they progress through the individual life cycle, issues also arise from the family life cycle. As with the individual life cycle, the family life cycle presents significant challenges that change with each stage.

Role Conflict and Role Overload

The young adult years are associated with a number of life-cycle tasks, including the completion of schooling, entry into the labor force, leaving

Table 3.1   Single-Parent Households with Own Children under Age 18 by Race

	Thousands of Households					
	1950	1960	1970	1980	1990	1998
White						
Single parent	1,200	1,638	2,382	3,760	4,579	8,077
Mother-Child	971	1,394	2,058	3,166	3,608	6,328
Father-Child	229	244	324	594	971	1,749
Two parents	16,990	21,625	22,268	20,997	19,777	22,237
Black						
Single parent	331	552	989	1,727	2,127	3,493
Mother-Child	285	497	901	1,568	1,897	3,211
Father-Child	46	55	88	159	230	282
Two parents	1,326	1,845	1,951	1,950	1,780	2,111

SOURCE: U.S. Bureau of the Census (1999), table 76; Spain and Bianchi, 1996, table 2.5. Used by permission of Russell Sage Foundation.

home, and starting a new home. These tasks entail giving up some roles, such as student; reducing the salience of others, such as son or daughter; and learning new roles, such as worker, spouse, and parent. These role shifts may pile up within just a few months, or they may stretch out over a period of years. For example, since the 1970s the age at marriage has been rising, especially among the college educated, and there is a longer interval between marriage and the birth of the first child. Other people, particularly those who begin work and marry just after high school, may acquire these new roles more quickly.

Multiple life roles can be a source of satisfaction, whereas people with only one life role are at higher risk for depression. Women who play more than one role report greater satisfaction, and married people are more likely than single people to report that they like their jobs (Bersoff and Crosby, 1984). Taking on the many new roles of young adulthood, however, may create **role overload,** a tension caused by trying to do too much at once. Even worse, the roles may actually be in conflict. **Role conflict** occurs when someone occupies two roles with contradictory expectations of what one should be doing at a certain time. For example, a worker may be expected both to be at work on time and to be

caring for her sick infant. Job-related travel, changing work shifts, child care, and unexpected emergencies either at home or on the job are important sources of role conflict. The resolution of role conflict requires setting priorities within both the workplace and the home.

For women, some role conflicts arise from *structural ambiguities,* in which institutionalized and agreed-upon arrangements for integrating work and family are not yet available (Gerson, 1985:123–124). Men have avoided these ambiguities to some extent because of the widespread assumption that husbands would take principal responsibility for work and that wives and mothers would take principal responsibility for children. The employment of both spouses is often necessary for financial reasons, but the family frequently assumes that the woman's commitment to the home will be undiminished (Hochschild, 1989). Many employed women experience role conflict from the need to maintain both their work tasks and their home tasks.

Another problematic situation for integrating work with family is the one-parent family. Table 3.1 shows the increase in the number of single-parent families between 1950 and 1998. In about 32% of the households with children, there is only one adult present, usually the mother

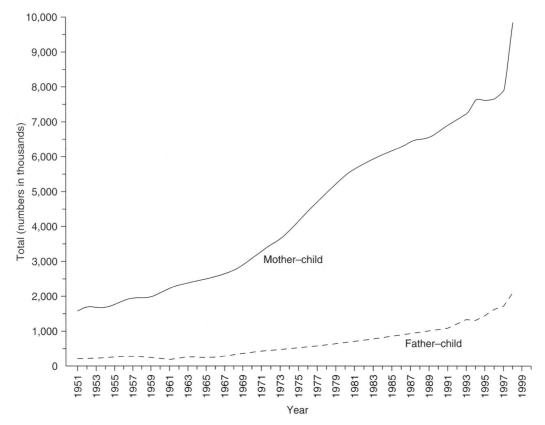

FIGURE 3.1   U.S. Growth in the Number of Mother–Child and Father–Child Families, 1950–1998

NOTE: Numbers shown are three-year moving averages.

SOURCES: Suzanne M. Bianchi, 1994, "The Changing Demographic and Socioeconomic Characteristics of Single-Parent Families." *Marriage and Family Review* 20(1/2):71–97. "Household and Family Characteristics: March," Series P-20, various years, available at *http://www.census.gov.*

(Census, 1999:Table 76; see Figure 3.1). No division of labor between spouses is possible; the single parent must provide both income and child care. Almost half of the persons in female-headed households (49%) are below the poverty level (Census, 1999:Table 766). Of the 11.6 million mothers who were custodial parents in 1990, only 7.1 million had been awarded child support and 1.8 million did not receive any of the support they had been awarded (Census, 1999:Table 637). Low-paying jobs also contribute to poverty in the single-parent home.

The care of home and children remains important to most young adults, but it often holds second priority to job and career. In manual work, especially when the job is covered by a collective-bargaining agreement, overtime and evening work are usually carefully monitored and compensated, although they may be involuntary. In salaried professional and technical jobs as well, workers feel that they have too little control over their time. Full-time workers in 1998 averaged 43.1 hours of work per week (Census, 1999:Table 664). Many workers also bring home bulging briefcases of work for evenings and weekends (Schor, 1991). Superiors expect that workers will routinely make themselves available for additional work whenever a project is nearing completion or a deadline is approaching.

Occupations vary in the extent to which workers draw boundaries between their home space and home time and their work space and work time. Sociologist Christena E. Nippert-Eng (1996) found that workers differed in the extent to which they sharply differentiate work space and work time from their home space and home time. Ed, a machinist whom she interviewed, draws a sharp distinction between work and home. His wife has never visited his work space, and he does not talk about personal things with his co-workers. Nippert-Eng identifies this behavior as "segmentation." John, a scientist, works in the same lab with Ed, but John is married to his scientific collaborator, works at home, stays long hours at the lab, and talks about both work and home in both places. This behavior is called "integration." Table 3.2 shows the criteria that Nippert-Eng used in exploring the continuum from segmentation to integration. Even segmenters, however, may spend longer hours at work if they find home conditions boring or stressful (Hochschild, 1997b).

Work Arrangements among Couples

The setting of priorities between work and home is complicated by the commitments of larger numbers of couples to dual careers and dual earnings. Women's labor force participation has risen in many countries, not just the United States. Box 3.3 reviews this rise. Although many working women are not married, the large number of wives who work for pay represents a major change within recent decades.

For couples in which both spouses work, issues of *time* and *location* are closely related to the issues of priority. The time issues can include the possibility that one or both of the spouses is expected to work overtime, at nights, or on weekends or holidays. The couple may work different days of the week or different shifts during the day. Such a schedule leaves little time for them to spend together in leisure, in home tasks, or in parenting. When one or both workers is also bringing work home, the home becomes an extension of the workplace, and even the time spent there is not available for sharing. Sharing the housework also presents a *time* problem for the working couple.

The proximity of the home to work and the availability of transportation are issues involving *location*. A limited budget for commuting may force one spouse to look for work near home or use less convenient methods of transportation. If there are children, one parent may also consider working near the children's school or caregiver. Many families develop complex logistics for delivering adults to work and children to school, and for completing the necessary errands during lunch hours or after work.

An even more serious location problem is posed by the commuter marriage. Commuter marriages arise when suitable jobs in the same geographic area are unavailable to both spouses. This term is somewhat misleading, for everyone who works outside the home travels a certain distance to work. But in the commuter marriage the distances are so far that returning home every night is not possible. Thus, the commuter marriage maximizes the spatial separation between workplace and home. As an alternative to the commuter marriage, one partner may leave the labor force, accept a job in a different field, or become a part-time worker. These alternatives are costly in terms of both income and the career continuity of the affected spouse.

In some corporations as well as in the military and certain government agencies, continued progress on the job requires occasional transfers. More than half of all moves in the United States are believed to be work related. Promotions, new job responsibilities, or even just job retention are sometimes attached to geographic moves. Job transfers have been linked to depression in wives of corporate managers, and they can be stressful to young children or adolescents (Stroh, 1999; Feldman and Bolino, 1998). Transfers can be especially difficult for the couple in which both partners work. A few corporations have begun offering relocation counseling and referrals for the spouses of their transferred employees.

Table 3.2   The Boundary Work of "Home" and "Work" along the Integration-Segmentation Continuum

Integration	Segmentation
Calendars	
Pocket calendar	Two wall calendars, one at home, one at workplace; no overlap in contents
Keys	
Home and work keys on one ring	Home and work keys on two rings; no overlap in contents
Clothes and Appearance	
One all-purpose home and work wardrobe; changing in morning and evening insignificant	Distinct "uniforms" for home and work; changing in morning and evening crucial
Many work- and home-related items in purse/wallet	Few work items in purse/wallet
Eating and Drinking	
Same foodstuffs and drinks consumed in same (un)routinized ways at home and work, throughout day and week	Different foodstuffs and drinks consumed in distinctly different, (un)routinized ways at home and work throughout day and week
Money	
Same monies used for personal and work expenses, incurred at home and workplace	No overlap in accounts or uses of personal and work monies, places where they are spent, or their respective bills, receipts, and IRS forms
Multipurpose bills, receipts, and tax forms	
Talk	
Cross-realm talk within and about both realms	No talk about work at home; no talk about home at work
Same style of talk used in both realms	Realm-specific talk styles
People and Their Representations	
Addresses and phone numbers for all acquaintances kept in one book	Addresses and phone numbers for work and home acquaintances kept on separate lists in separate places
Photographs of coworkers at home; photos of family kept at workplace	Photos of coworkers kept in workplace; photos of family kept at home
Coworkers come to house to socialize with family; family comes to workplace to socialize/work with coworkers	Coworkers socialize together without families, in workplace during workday; family does not come to workplace
Reading	
"Work"- and "home"-related material read and stored anywhere, anytime	"Work" material read and stored only at workplace, during work time; "personal" material read only during "personal" time, away from workspace
Breaks	
No distinction between worktime and personal time during day or year	Distinct pockets of personal time during workday when no wage labor is done; distinct annual vacations when no wage labor is done
Commutes	
"Two-way bridges"; no transformative function	"One-way bridges"; crucial for achieving transformations between realm-specific selves
Phone Calls	
Frequent, random cross-realm calls; intrarealm calls include cross-realm subject matter	No cross-realm calls; intrarealm calls include only realm-specific subject matter

SOURCE: Nippert-Eng, 1996, Table 1, pp. 149–151. Used by permission of the University of Chicago Press.

BOX 3.3 Working Women around the World

The level of female labor force participation is highly dependent upon cultural attitudes. In the traditional Muslim culture that predominates in Western Asia and North Africa, women are discouraged or prohibited from leaving the safety and sanctity of their homes to work for others. In Egypt, less than 10% of women are engaged in labor force activities according to official statistics. In sub-Saharan Africa and in some Asian countries, women traditionally are involved in agriculture and market activities, and labor force participation is much higher. In Latin America, female activity has risen sharply since 1950. About 20% of Latin American women of working age were in the labor force at the beginning of the 1990s, compared with 33% in South Asia and sub-Saharan Africa (Figure A).

Politics also plays a role. In many socialist countries, such as China and Vietnam, women are expected to contribute equally with men. Therefore, female labor force rates are high.

Some of the regional differences in female labor force participation reflect variations in labor statistics. Where culture frowns upon women being involved in economic activity, people underreport it in labor surveys and censuses. This exclusion occurs most often in agricultural societies in which the lines between household chores and economic activity may be blurred. Women may consider raising chickens or vegetables part of their housework, but if they sell or trade the proceeds of their work, their labor is "economic activity" under most definitions. Agricultural work is often seasonal, which can affect statistical reporting. A 1976 labor survey taken during Indonesia's busy harvesting season estimated the female agricultural labor force at 14 million, while one conducted the same year, but during the slack period, estimated it at 10 million. Cultural attitudes, along with misunderstandings or disagreements about what constitutes economic activity, cause an undercount of female labor force participation in developing countries. A careful labor survey in Egypt, for example, estimated that women made up 42% of agricultural workers in 1960, not the 4% reported in the census.

The shift toward more productive industrial and service sectors is decreasing, not increasing, the female share of the labor force in some developing countries. Despite the gains in female education and lower fertility, which were associated with increased female employment in industrial countries, women comprise only about 24% of the non-agricultural work force in developing countries. Between 1950 and 1985, the female share of the labor force has remained at 34% to 35%.

As the agricultural share of an economy shrinks, the share of women in the labor force may also decline. Further education of women increases their chances of employment in the non-agricultural sectors of the economy only if there are sufficient employment opportunities for both sexes.

SOURCE: David E. Bloom and Adi Brender, 1993, "Labor and the Emerging World Economy." *Population Bulletin* 48,2 (October):8–9.

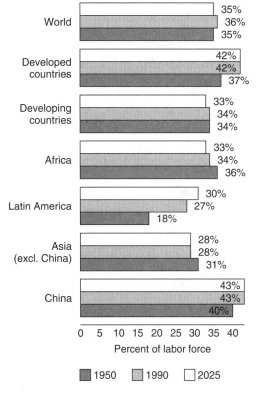

FIGURE A Female Share of the Labor Force in Major World Regions, 1950, 1990, and 2025

SOURCE: International Labour Office (ILO), 1986, *Economically Active Populations.* Geneva: ILO.

The Arrival of Children

All of the advanced industrial countries have experienced a long-term decline in birth rates and in the average number of children borne by each woman. This trend was briefly reversed during the baby boom following World War II, but today it is firmly reestablished. In 1998, 51% of all family households had no children under the age of eighteen, up from 48.7% in 1992. The percentage of households with three or more children has declined from 15.3% in 1940 to 10% in 1998 (Census, 1999:62,65; Census, 1993e:59–60; Helmick and Zimmerman, 1984:403). Moreover, roughly 10% of all U.S. couples report that they intend to remain childless.

Nevertheless, most couples will become parents and will have children at home for part of their married life. Moreover, many adults will be single parents for at least part of the time that their children are minors. This means that the issue of combining child care with work will still arise for most adults, but it is being handled differently today than it was in the past.

Besides having fewer children, many women are also waiting longer to bear their first child. An even more dramatic change has occurred in the way in which women synchronize their childbearing and work decisions. It was once common for women to leave the labor force at the time their first child was born. In more recent years this practice has nearly disappeared, at least statistically. Figure 3.2 graphs the changes in women's labor force participation rates, by age, from 1950 to 1998. (For a definition of the labor force participation rate, see Chapter 2.) Each successive year's rates are higher, indicating that labor force participation was generally increasing, but the increase shows an interesting variation by age. As the graph shows, women in the prime childbearing years did not participate in the labor force at the same rate in 1950 and 1960 as they did in 1970. Instead, there was a decline between the ages of twenty and twenty-four and twenty-five and thirty-four, followed by a rise after the age of thirty-five. This M-shaped

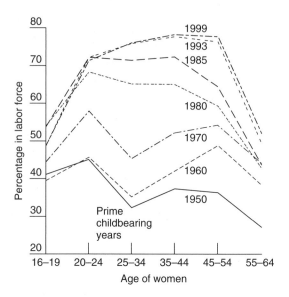

FIGURE 3.2 Labor Force Participation for Women, by Age, 1950–1999

SOURCES: Martin O'Connell and David E. Bloom, 1987, "Juggling Jobs and Babies: America's Child Care Challenge." *Population Trends and Public Policy* 12 (February). Reprinted by permission of the Population Reference Bureau. 1993 data from BLS (1994a). 1999 data from U.S. Department of Labor (2000).

pattern of women's labor force participation can be seen most clearly in the figure by looking at the graph for 1970. By 1980, there was hardly any decline at all in labor force participation in the childbearing years, and by 1985 labor force participation leveled off during those years. Today the M-shape has disappeared.

More continuous employment has probably helped the careers of many women and has provided more continuous income for their families. It has also challenged the long-held view that mothers should be the primary providers of daytime care for their young children. In 1998, 64% of wives with children under age six were working, and 62% of mothers with babies one year of age or younger were working (Census, 1999:417). Both proportions have increased dramatically since 1970 (O'Connell and Bloom, 1987). A lack of affordable child care has prevented many more mothers from seeking work (Cattan, 1991).

A couple with children has many options for combining child care with work. One option is for a parent to provide all the child care. One parent (usually the mother) can leave the labor force, or one parent can substitute part-time for full-time work. When one parent is away, the other parent worker cares for the children. About 2.3 million children under the age of six have mothers who are not in the labor force and are cared for only by their parents (U.S. Census, 1999:402). Many women choose part-time work in jobs that accommodate more flexible schedules (Glass and Camarigg, 1992). Another 200,000 children under the age of six have mothers who work part time and receive all of their care from parents. Another possibility is for the husband and the wife to work separate shifts, so that one parent is always home with the children (*www.shiftworker.com,* Presser, 1988). About 23 million workers work night shifts, and one in five workers work outside the traditional 9-to-5 schedule. Some parents turn to self-employment or to paid work done at home to provide more flexible working conditions. About 9% of preschool children are cared for by their mother while she works at home or at another location (Census, 1994c).

If the mother is a full-time worker, it is more likely that preschool children will be cared for by a different relative (33%), in another home (32%), or in an organized child-care facility or school (39%). (These numbers do not add to 100% because some children receive multiple forms of care. [U.S. Census 1999:402]). School is the primary source of child care for children aged six to thirteen. Many of these children have after-school care of some sort. Older children are much more likely than younger children to care for themselves after school.

On-site child care at the work site or nearby has become an important alternative for parents. Many on-site centers permit parents to eat lunch with their children or to spend their breaks playing with them. Cameras linked to secure internet sites permit parents to check on their children during work hours. Some employers provide child-care vouchers to off-site day-care centers, reserve sites in day-care centers for children of their employees, or offer subsidies for child care as a fringe benefit.

Homemakers and Home Production as a Career

Q. On the whole, do you prefer raising your children to working outside the home?

A. Oh, yes. I never plan to go back. I'm too spoiled now. I'm my own boss. I have independence; I have control; I have freedom, as much freedom as anyone is going to have in our society. No job can offer me those things. (Gerson, 1985:129)

Substantial numbers of adults, most of them women, do not work for pay but remain engaged in home production. **Home production** refers to the nonmarket production of goods and services, usually for the family but occasionally on a volunteer basis for schools, churches, or other groups. What is traditionally known as "housework" is only one aspect of home production, which also includes household budgeting, shopping, care of dependents, and other tasks that go beyond cleaning, cooking, and laundry. If the cooking, sewing, chauffeuring, child care, and so on were monetarily compensated, they would account for billions of dollars each year. Workers engaged in home production without pay are called *homemakers.* In Chapter 9 we will also discuss high-technology workers who work for pay at home; these workers and others who work for pay, but at home, are called *home workers* to distinguish them from homemakers. Home workers are counted in the labor force; homemakers are not.

Home production offers workers the opportunity to schedule their own work and set their own priorities. Many parents highly value staying at home with their children. On the other hand, home production also has disadvantages, many of which are financial. The homemaker is not covered by pensions, insurance, or Social Security, and is economically dependent on another person (Vanek, 1988). Almost all homemakers are women. About ten years ago observers began to draw attention to the plight of the "displaced homemaker," who when divorced lost her "work" as well as her financial support. Several European countries make provisions for covering homemakers with some benefits, and for compelling families to provide homemakers an annual vacation.

Perhaps the greatest problem for homemakers is the social devaluation of their work (Oakley, 1975). Housework is not considered a "real" job, because many of the tasks can be postponed, and the tasks are thought to be low-skilled and repetitive. Despite evidence to the contrary (Schor, 1991), many people believe that "labor-saving" devices such as vacuum cleaners and automatic washing machines do all the housework. Some people consider work outside the home for pay to be more prestigious, perhaps even more "adult." A homemaker who is referred to as "just a housewife" feels devalued because the work has been devalued. Social evaluations of the homemakers appear to be a major factor in their relative happiness; when they feel valued, they are happier (Ferree, 1984).

The Income Squeeze

In the middle stages of their work career, many workers are well established on the job. Compared with the earlier period of role transitions, this stage of work life may be less eventful. Many middle-aged workers find that this is their most productive period, with steady increments in pay and responsibility. Others find that even good jobs, with job ladders and job security, have become boring, that the realistic opportunities for promotion are infrequent, or that economic and organizational changes have made their jobs insecure.

With both young and middle-aged workers, however, the family is likely to rely more on job-related income than it did in earlier years. Although young couples may still receive some financial support from their families, by the middle stages of the career there are few outside sources of income, and indeed, elderly parents may begin to need financial support. Not only is the income from the job crucial, but also the fringe benefits take on greater significance. With

children to care for, the job-related benefits of health insurance, disability insurance, and life insurance become much more important. Also, middle-aged workers become concerned about their pensions. Staying with the same company ensures continuity of fringe benefits, and this may have the effect of "tying" the worker to the company. This is especially true if some fringe benefits or perquisites are pegged to seniority. One reason that layoffs are so disruptive to middle-aged workers is the loss of fringe-benefit coverage for their families (Sullivan, Warren, and Westbrook, 2000).

As children become older, the expense of caring for them also increases. Teenagers eat more than younger children; in addition, providing clothing, transportation, educational expenses, and leisure expenditures for older children is more expensive than it is for babies and preschoolers. College tuition is an even greater expense. This means that many families at this stage need an *increase* in income. Occupations vary in the extent to which middle-aged workers can expect their incomes to increase. In general, manual occupations are likely to level off in earnings while the workers are relatively young. Although there may be cost-of-living raises thereafter, it is unlikely that there will be any further spurts in income. In many white-collar occupations, by contrast, merit raises and seniority increases continue well into middle age. For the white-collar worker the increases in income may not slow or stop until after the children have already left home and the expenses have begun to decline. Manual workers may experience an **income squeeze,** which means that just as the expenses of their families increase, the increments in their annual income stop (Oppenheimer, 1974).

Workers find several ways to deal with the income squeeze. Many borrow money to finance new purchases or to send children to college. Some workers "moonlight"; that is, they take on a second job, often part time or seasonal, to supplement their income. In 1998, 7.9 million workers had second jobs, and over two-thirds of them were between twenty-five and fifty-five

(U.S. Census 1999:421). A third alternative is to increase the family labor supply. Homemakers may enter or reenter the labor force to supplement the family's income. In addition, the teenage members of the family may also start to work, usually part time, to help cover some of their own expenses or to contribute to the family's expenses.

The Impact of the Family on Work

At every stage, the family may affect the workers' performance on the job. For men a stable family life is usually considered an asset to life on the job. Indeed, the family may be more or less incorporated into the job through the expectation that the wife will be available to entertain business guests or to take on volunteer activities that will benefit the company (Pavalko and Elder, 1993). This is sometimes called the two-person career (Mortimer and London, 1984:25–26). Some companies, especially those with an image of their employees as one big happy family, expect workers' families to attend holiday parties and company picnics.

Employers may view women's families, however, as a potential distraction from work. In the middle stages of the work career, the employer is less likely to worry about time lost to care for sick children or to find child care. Nevertheless, there may be continuing concern that the female worker's attachment to her family will be detrimental to her work. Moreover, the company usually does not expect the husband to be available for business entertainment or social events. One researcher, summarizing the views of managers she had interviewed, writes: "Thus, while men symbolically brought two people to their jobs, women were seen as perhaps bringing less than one full worker" (Kanter, 1977:107).

The "Empty Nest"

By the time of retirement, a worker's children are usually forming their own families. This may ease the financial burden on the older worker, although today an increasing number of retirees still have *their* parents to care for. Aside from the financial problems associated with retirement, other important family problems arise at this stage of the life cycle. The retiree may feel bored and unwanted, stripped of the productive role that is so important in modern life. Lip service is paid to the significance of volunteer work and home production, but many retirees find little prestige is actually attached to these pursuits, and they miss the activity of their former work environment.

The retiree (usually male) may disrupt the nonworking spouse's pattern of daily activities in his (or her) effort to be useful. In a dual-career family the problem may be that one spouse has retired while the other is still working. Even though the married couple has been looking forward to being together again, they may find that the reality is that they get on each other's nerves. As the proportion of the elderly in the population grows, the issue of retirement and its alternatives will surely take on greater significance.

PROPOSALS FOR COMBINING FAMILY AND WORK

A number of changes in the timing of work have been suggested to help improve the fit between work and family. Many of these changes would also have other benefits in terms of increasing the worker's sense of control and productivity.

Repackaging Jobs

One suggestion that is already in use is **flextime.** Flextime allows workers to set their own hours, within some limits. For example, employers may insist that all employees be on the premises between 10 A.M. and 2 P.M., but workers may design their own schedules for when they will arrive and when they will leave, arranging to work the full work week with flexible hours. Flextime is popular with both workers and employers (Bohen and Viveros-Long, 1981). It allows the parents of small children to help get them off to school in the morning or to meet them in the afternoon. If both parents have flextime, it is much easier to program child care. Flextime also allows workers who are fresher in the morning—or perhaps late in the evening—to work during their most productive hours. When a number of employers in a city use flextime, traffic congestion during the rush hour may be relieved. About 15% of all workers are on flextime, and it is most prevalent in business and repair services, entertainment and recreation, finance, insurance, real estate, and public administration (BLS, 1992c).

A second change is **block scheduling,** or putting together the traditional forty-hour work week in nontraditional ways. There are several long-standing examples of this scheduling that require the worker to reside briefly at the workplace. For example, many offshore oil crews work one week on duty and the next week off duty. Firefighters often have a "one-day-on-one-day-off" rule. For the twenty-four hours that they are on duty, they sleep and eat in the fire station. Other versions permit the worker to live at home but schedule longer hours for fewer days. Some hospitals now have a forty-hour weekend plan for nurses. The nurse works a full forty hours in one weekend and then has the remainder of the week free for home obligations. Under the 4/10 program, the workers work ten-hour days for four days a week. One study shows that male workers on this schedule spend more time with their children and with traditional male household chores, such as mowing the lawn (Maklan, 1977a, 1977b). For employers the principal problem with these plans is that workers' efficiency may decay with the long hours. The principal advantage is in extending the usual business hours.

A third change is **work sharing.** In one version, a full-time job is partitioned into two half-time jobs. These two jobs may be shared by a husband and wife or by two unrelated workers. Especially if it involves a husband and wife, work sharing may allow parents to care for their children themselves. For unrelated workers who, for

whatever reason, cannot work full time, work sharing makes a new part-time job available. The principal disadvantage for the workers is reduced income; in addition, many part-time jobs carry no fringe benefits. Moreover, part-time workers may be excluded from the company's job ladder, if there is one. The company saves on fringe benefits, although there are additional bookkeeping costs. There may also be an advantage in keeping workers who might otherwise go elsewhere. It is important for the two part-time workers to communicate with each other about the job; otherwise, the employer will completely lose the continuity that was provided by having only one worker do the tasks.

Family-Related Fringe Benefits

Family-related work benefits can be an important mechanism for helping workers coordinate their families and their jobs (Ferber and O'Farrell, 1991). Maternity, paternity, and family leaves may help ease the transition into parenthood. Canada provides parenting and maternity benefits through unemployment insurance. Most advanced industrial countries have national legislation guaranteeing the right to employment leave and the protection of the mother's job (Ferber and O'Farrell, 1991:161; Kamerman, 1986:60).

The United States lagged behind these other countries. In 1991, for example, maternity leave benefits were available to only about half of young women workers (BLS, 1993b). In 1993, Congress enacted the U.S. Family and Medical Leave Act of 1993. This law requires employers with fifty or more employees to provide eligible employees up to twelve weeks of *unpaid* leave for their own serious illness, the birth or adoption of a child, or the care of a seriously ill child, spouse, or parent. To be eligible, an employee must have been employed at least one year and have worked at least 1,250 hours within the previous twelve months. For most families, however, the problem of coordinating work and family is not limited to the actual birth of the babies; rather, it is a long-run problem that needs longer-term solutions.

Many large employers are considering a *cafeteria approach* to benefits that would give workers the ability to choose among different benefits that best serve their family needs. In this the worker is given a dollar amount of benefits and asked to choose among the available alternatives. Besides traditional insurance benefits, some employers offer child-care subsidies, college tuition assistance, or elder-care assistance.

SUMMARY

The transition from adolescence to adult life is affected both by the personality of the young workers and by the opportunity structure that confronts them. The opportunity structure changes through occupational shifts and through the efforts of employers to create entry ports as opposed to dead-end jobs for young workers. Entry ports and career ladders are characteristics of jobs that are more likely to be well compensated, have job security, and have fringe benefits. Such jobs, especially when they offer the worker autonomy in daily tasks, are associated with better mental functioning in middle-aged workers as well as greater financial security.

Socialization for work and work itself affect both the individual life cycle and the family life cycle. The separation of work from home leads to issues of both time and space in reconciling one's job or career with one's family. Although different work-related challenges must be met at different stages of life, the issues of resolving work and family are always present. There is also a reciprocal, but weaker, effect of family life on work life. Men and women often experience these effects differently because of the continuing assumption that mothers have the primary responsibility for child care.

Many recent changes have affected the ease of integration between family and work. Factors that tend to worsen the conflict are the increased employment of working mothers, the increased number of single-parent homes, the number of jobs that require long hours or travel, difficulty in finding child care, difficulty in finding two jobs in the same geographic area, and the "income squeeze." Factors that help to reduce the conflict include novel methods of repackaging jobs into different hours or days, and the development of family-related fringe benefits.

KEY CONCEPTS

life cycle	socialization	role conflict
norms	dead-end job	home production
role	fringe benefits	income squeeze
individual life cycle	entry-port job	flextime
career	occupational mobility	block scheduling
family life cycle	role overload	work sharing

QUESTIONS FOR THOUGHT

1. What are the ways in which parents affect their children's work behavior? How does this influence change at different points in the family life cycle and at different points in the children's life cycles? How do children affect their parents' work behavior?

2. Identify the principal problems that workers face in synchronizing their careers with the family life cycle. How are the problems different for men and women workers?

3. What role, if any, do you believe employers should play in helping workers accommodate family and career? What role, if any, should the government play?

4. Take a proposal for change that was mentioned in this chapter, such as flextime. Suppose that all employers adopted this change. Trace the likely effects, both positive and negative, for the labor force and for the economy.

5. Why is housework devalued? What, if anything, could be done to change the devaluing of housework?

MULTIMEDIA RESOURCES

Print

Tracy Bachrach Ehlers. 2000. *Silent Looms: Women and Production in a Guatamalan Town.* University of Texas Press, revised edition. A study of how economic development policies have changed the lives of women weavers in a Guatamalan region.

Arlie Hochschild and Anne Machung. 1997. *The Second Shift: Working Parents and the Revolution at Home.* Avon Books, reproduced edition. An excellent account of women who work their "second shift" when they care for their families.

Leslie Perlow. 1997. *Finding Time: How Corporations, Individuals, and Families Can Benefit from New Work Practices.* Cornell University Press. A collection of pragmatic suggestions for improving the fit between home and work.

Websites

http://www.ssa.gov The official website of the Social Security Administration. Information on disability support, Supplementary Security Income, Medicare, and retirement planning.

http://www.workfamily.com An information clearinghouse for work-life professionals. Contains a great deal of information on child care and a list of best practices.

http://www.bc.edu/bc_org/avp/csom/cwf/center/overview.html The Boston College Center for Work and Family website, with information on workplace flexibility and other current topics.

http://www.stw.ed.gov/ The school-to-work website. Contains information on how different countries structure their transition from school to work; also features best practices in K–12 teaching and learning.

RECOMMENDED FILM

Mrs. Doubtfire. (1993). Directed by Chris Columbus. Robin Williams plays a voiceover actor who cannot keep a steady job; he and his career-oriented wife eventually divorce. Dressed as "Mrs. Doubtfire," he comes back to do housework and child care. A funny but provocative portrayal of work, divorce, housework, and child care. Rated PG-13.

4

Meaningful Work

You try to fill up your time with trying to think about things: what you're going to do on the weekend or about your family. You have to use your imagination. If you don't have a very good one and you bore easily, you're in trouble. Just to fill in time, I write real bad poetry or letters to myself and to other people and never mail them. The letters are fantasies, sort of rambling, how I feel, how depressed I am.

. . . I always dream I'm alone and things are quiet. I call it the land of no-phone, where there isn't any machine telling me where I have to be every minute.

The machine dictates. This crummy little machine with buttons on it—you've got to be there to answer it. You can walk away from it and pretend you don't hear it, but it pulls you. You know you're not doing anything, not doing a hell of a lot for anyone. Your job doesn't mean anything. Because *you're* just a little machine. A monkey could do what I do. It's really unfair to ask someone to do that.

. . . I'll be home and the telephone will ring and I get nervous. It reminds me of the telephone at work. It becomes like Pavlov's bell. (Laughs.) It makes the dogs salivate. It makes me nervous. The machine invades me all day. I'd go home and it's still there.

I changed my opinion of receptionists because now I'm one. It wasn't the dumb broad at the front desk who took telephone messages. She had to be something else because I thought I was something else. I was fine until there was a press party. We were having a fairly intelligent conversation. Then they asked me what I did. When I told them, they turned around to find other people with name tags. I wasn't worth bothering with. I wasn't being rejected because of what I had said or the way I talked, but simply because of my function. After that, I tried to make up other names for what I did— communications control, servomechanism. . . .

(TERKEL, 1974:57–60)

Work sometimes falls short of providing workers with self-esteem and a positive identity. Work can cause workers to be alienated or separated from themselves, from the products of their labor, from others, and from an active sense of purpose in the world. A good job provides the material necessities of existence, but it also contributes to workers' self-esteem, identity, and sense of order, which are important keys to finding meaning and satisfaction in life through work.

In this chapter we discuss the major theories of alienation and self-actualization at work. We also discuss attitudinal and behavioral responses to work and explore the future of alienation, satisfaction, and meaning at work.

WHAT IS JOB SATISFACTION?

Job satisfaction is the summary evaluation that people make of their work. People's levels of job satisfaction are the result of their job tasks, the characteristics of the organization in which they work, and individual differences in needs and values. The character and meaning of work in modern society have provided a major focus for social research and social philosophy (Gamst, 1995). This section attempts to make sense of these theories and their principal conclusions. People can be either satisfied or dissatisfied with their work—they can find their work either highly meaningful or utterly meaningless. These different experiences give rise to different theories that focus either on alienation or self-actualization.

Alienation occurs when work provides inadequately for the human needs for identity and meaning. Work is alienating if one does it only because of economic necessity, not for its intrinsic pleasures. **Self-actualization** occurs when work contributes to the fulfillment of these broader human needs. A self-actualizing job provides for material needs, but if one's material needs were met in some other way, one would want to continue the work anyway, for its own rewards.

Theories of Alienation

Modern understandings of alienation owe a large debt to the early work of Karl Marx. He observed that as the world of material goods increased in value, the value placed on individuals seemed to diminish (Marx, 1959 [1844]). Early industrial capitalism brought into being not only unprecedented productivity, but also some of the most wretched living and working conditions in human history (see Chapter 1). Marx believed that these conditions resulted from the denial of workers' rights to control their work activity and the products of their labor. Thus, he described work under capitalism as "wage slavery."

Marx's Four Aspects of Alienation Workers are alienated in four ways under industrial capitalism, Marx argued. First, they are alienated from the *products* of their labor. They no longer determine what is to be made, nor how to dispose of it. In primitive societies, workers had a direct relationship to the products of their labor. They owned the products and these products became an important part of their physical world. In industrial societies, workers no longer have such a direct relationship to their products. Work on these products becomes a *means to an end;* that is, rather than being an end in itself, work is a means to acquire money to buy the material necessities of life. Because workers are robbed of a meaningful relationship to the products of their labor, they come to relate to the products they produce as alien objects rather than as useful manifestations of their ideas, skills, abilities, and efforts.

Second, workers are alienated from the *process* of work. Someone else controls the pacing, patterns, timing, tools, and techniques of their work. Because workers no longer control their moment-to-moment activity, work becomes less meaningful to them. When workers are emotionally separated from their activity on the job, they become alienated from themselves and their identity as human beings. Such identity must be secured outside the workplace through leisure or family pursuits. Recall the receptionist in the chapter-opening example. She experienced her work as being directed by the telephone; part of her self-image was that she was "just a little machine." She daydreamed about the weekends and about her family while she was on the job. Nonalienating work, in contrast, is virtually indistinguishable from leisure. The worker experiences work and hobbies with the same enthusiastic absorption.

In Marx's third aspect of alienation, workers are denied the ability to be creative. He believed that the unique, defining characteristic of humans is the ability to be creative. The capacity for self-directed *creative activity* to meet changing needs is what distinguishes human beings from animals. If workers cannot express their *species being* (their creativity), they are reduced to the status of animals or machines. Alienated labor is noncreative and is only a means to secure the material condi-

tions of existence. Extreme specialization makes each worker's contribution to the final product obscure. To be self-actualized, human beings have to create freely like artists rather than like bees, which labor under the drives of instincts.

Fourth, alienated labor is an isolated endeavor, not part of a collectively organized effort to meet a group need. As a result, workers are alienated from *others* as well as from themselves (Marx, 1959 [1844]). Human beings are social animals by nature and their work always involves others, either directly or indirectly. When labor is alienated, control of these social relations is removed from the worker, who interacts with others on the job only as directed by some higher authority. Alienated workers are not part of an integrated team engaged in collectively determining the nature and goals of their activity. As a result, workers are isolated from others in the realm of purposive activity. Again, the receptionist, who spoke only to disembodied voices all day, is a good example of an isolated worker.

Alienation not only isolates the worker while at work but also affects the nature of the worker's involvement in the broader society. Those who control the process and products of labor are able to control the direction in which society develops. The alienation of workers, which starts at the workplace, thus extends to society as a whole, and workers find no place, no thing, and no experience that they can truly call their own. "[The worker] spends his time doing things in which he is not interested, with people in whom he is not interested, producing things in which he is not interested" (Fromm, 1968:38). Thus, according to Marx, industrial capitalism denies meaningful participation for the vast majority of people in society.

When Marx wrote about work in 1844, the physical conditions of labor were much harsher than they are today. His writings remain the touchstone for studies of alienated labor, however, because harsh working conditions have not disappeared and because alienation persists even in more physically pleasant work settings. For example, a telephone receptionist's work area might be comfortable and well lit, but she may still be alienated because of the unceasing flow of superficial interactions. In some ways Marx's writings are even more relevant today because of increased mechanization, standardization, and bureaucratization.

Subjective Definitions of Alienation Contemporary researchers have extended Marx's theory by analyzing the subjective experience of alienation. The American sociologist Melvin Seeman (1959) hypothesized that workers experience alienation in terms of powerlessness, self-estrangement, meaninglessness, isolation, and normlessness.

Seeman's components of the subjective experience of alienation correspond roughly to Marx's components of alienating conditions. *Powerlessness* is the expectation that one cannot control the events in one's life. This idea corresponds roughly to Marx's concept of alienation from one's products. *Self-estrangement* refers to the individual's lack of rewarding and engaging activities. It roughly parallels Marx's concept of alienation from the process of work. Individuals who act only in response to material rewards are estranged from their immediate activity and therefore from an important element of their identity. *Meaninglessness* arises when workers feel that they cannot adequately predict the future and when their efforts seem to have few worthwhile results. Meaninglessness corresponds roughly to Marx's concept of alienation from human creativity. *Isolation* arises from a disparity between the goals, values, and expectations of an individual and those of the rest of society. It closely parallels Marx's notion of alienation from others. Seeman added the concept of *normlessness* (French, *anomie*) from the work of the French sociologist Emile Durkheim. Normlessness indicates a state in which either the appropriate standards of behavior are unknown or there is insufficient reason to abide by them. The concept extends the analysis of the subjective experience of alienation outside the workplace and into broader society. In this way it parallels Marx's own effort to link workplace alienation to alienation from broader society.

Theories of Self-Actualization

Maslow's Hierarchy of Needs Contemporary theorists have broadened the analysis of job satisfaction by including a consideration of the self-actualizing effects of work as well as alienating ones. Much of this work was motivated by the American psychologist Abraham Maslow's theory of *hierarchical need satisfaction* (1954). It argues that human beings have (1) physiological needs for such things as food and sex; (2) safety needs for a secure physical and emotional environment; (3) belongingness needs for acceptance and friendship; (4) esteem needs for recognition, attention, and appreciation; and (5) self-actualization needs for developing to one's fullest potential. These needs are arranged hierarchically; that is, individuals are concerned with their physiological and safety needs first, and only after these lower-order needs are met do they turn their attention to higher-order needs.

In Maslow's theory the most alienating jobs are those that do not provide for minimum levels of material sustenance and security. Less alienating jobs may satisfy these needs but still fail to provide a feeling of self-worth or self-esteem. The best jobs are those that not only provide material necessities, security, belongingness, and esteem, but also help workers develop to their highest potential.

Maslow's Followers Maslow's theory has been applied to the world of work by several contemporary social theorists including Frederick Herzberg, Chris Argyris, and Douglas McGregor. Herzberg (1966) observes that workers consider different issues when determining whether they are satisfied or dissatisfied. Satisfied workers stress professional growth, achievement, recognition, responsibility, and advancement. Dissatisfied workers stress negative factors such as managerial incompetence, close supervision, low wages, and poor working conditions. Such differences in emphasis led Herzberg to develop a *two-factor theory of job satisfaction*. Workers are dissatisfied if they have to work in unpleasant physical or social settings. Herzberg referred to these characteristics as "hygiene factors." Such negative conditions have to be removed or neutralized to avoid alienation; however, the removal of these conditions is not sufficient for satisfaction. To be satisfied, workers must have their needs for personal development met. Herzberg refers to these factors as "motivators." He includes such things as autonomy, challenge, recognition, and opportunity for developing new skills.

Argyris (1973) developed a similar theory of job satisfaction based on a development from passivity to activity and from dependence to independence. He believes that frustration of the natural development from passivity to activity and from dependence to independence is a major source of organizational problems (see Pfeffer, 1998).

McGregor (1967) offers another version of this argument with his *two theories of management.* Theory X assumes the need for various forms of external control of workers. It is based on a carrot-and-stick approach to work in which workers are motivated only by rewards and punishments. Theory Y calls for greater recognition and respect for workers in order to achieve higher levels of motivation and productivity. These theories of management are discussed further in Chapter 7.

The models of job satisfaction based on theories of hierarchical needs have some important limitations. These theories rest on unproven assumptions about human nature. It has not been empirically demonstrated that so-called lower-level needs have to be met before higher-level needs can be addressed. The research that has attempted to evaluate these assumptions has provided only limited support for their relevance or accuracy (Fein, 1976). It may well be that human beings attempt to meet whatever of their needs can be met in any given situation, with little inherent ranking into higher-level and lower-level needs. Nevertheless, it is clearly useful to consider a range of needs and motivations if we are to understand the complex underpinnings of meaningful work.

Marx's theory about objective conditions that produce alienation, Seeman's theory about the subjective experience of alienation, and Maslow's and his followers' theories about self-actualization have provided the basis for a vast body of research on job satisfaction. Contemporary researchers retain these early theorists' insights into the all-embracing nature of the experience of work, but they have extended their analysis by looking at the detailed characteristics of jobs and personal expectations. It is to these analyses that we now turn.

WHAT DETERMINES JOB SATISFACTION?

In this section we discuss some of the characteristics of work and of workers that lead to alienation or self-actualization. Factors that determine the degree of satisfaction include the nature of job tasks, technology, organizational characteristics, workers' participation in decision making, individual differences, and prior expectations. A general model of job satisfaction and dissatisfaction is presented in Figure 4.1.

Self-Direction

The nature of one's daily tasks on the job is the most important determinant of self-actualization at work. Work that has autonomy, complexity, and diversity can be self-actualizing. **Job autonomy** is the extent to which workers control their own work and relations with others at work, including both co-workers and supervisors. The *complexity* of the task is equally important. Men and women whose job responsibilities are more complex and allow for greater self-direction are less psychologically distressed and more intellectually flexible, and they come to value self-direction more for both themselves and others (Kohn, 1990). *Diversity* of tasks is also a significant factor—repetitive work, lacking in variety, done at a forced pace, is among the most brutally alienating forms of work.

FIGURE 4.1 Causes and Consequences of Alienation or Self-Actualization

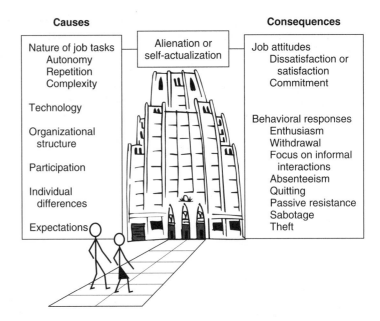

Causes

Nature of job tasks
  Autonomy
  Repetition
  Complexity

Technology

Organizational
  structure

Participation

Individual
  differences

Expectations

Alienation or
self-actualization

Consequences

Job attitudes
  Dissatisfaction or
    satisfaction
  Commitment

Behavioral responses
  Enthusiasm
  Withdrawal
  Focus on informal
    interactions
  Absenteeism
  Quitting
  Passive resistance
  Sabotage
  Theft

The assembly line is a classic example of work organized in an alienating way. An assembly line in a cosmetics factory reveals what work is like under these conditions:

> All along the belt women in blue smocks, sitting on high stools, pick up each mascara tube as it goes past. They insert brushes, tighten the brushes, tamp on labels, encase the tubes in plastic and then cardboard for the drugstore display.
>
> At the Brush-On Peel-Off Mask line, a filler picks an empty bottle off the belt with her right hand, presses a pedal with her foot, fills the bottle with a bloop of blue goop, changes hands, and puts the filled bottle back on the line with her left hand as she picks up another empty bottle with her right hand. The bottles go past at thirty-three a minute. (Garson, 1975:56)

Because the jobs on the line lack complexity and diversity, they provide few social or psychological benefits. The work of these women is truly alienated, allowing for no human creativity and for little sense of unique personal identity or social connectedness.

Clerical work, as well as blue-collar assembly-line work, can also be repetitive and alienating. Data-entry clerks often encounter dull, rapidly paced job tasks:

> Take a payroll for example. You have a form with an employee's name, address, salary, tax info. . . . Aida and I are very good, high performers—bored, but still high performers. So we thought racing might be our special thing. Then a girl, Janet, told us she did the same thing. I guess it was the only kind of entertainment you could have. Like I said, your hands were occupied, your eyes were occupied, you couldn't move your body, couldn't talk. You only had numbers on the sheets and the sounds of the other machines. (Garson, 1975:153–155)

Automation has made some clerical work just as routinized as the most repetitious manual work. Because numeric data can be standardized and regimented more easily than many physical products, some clerical work may be in the process of becoming even *more* alienating than many manual jobs. In Chapter 12 we discuss contemporary changes in the nature of clerical work.

BOX 4.1   Friendship and Solidarity in Female and Male Work Groups

While working as a participant observer in an electronics assembly plant with a predominantly *female labor force,* an ethnographer reports being overwhelmed by the generosity of the poorly paid women with whom she works:

> [After returning from a two-week sick leave without pay,] I was talking to Anna when she stuffed a £10 note in my trouser pocket so quickly I wasn't even really sure what it was. She was giving it to me because I would be short, having lost two weeks' wages. . . . I was quite overwhelmed by her generosity; the gift was completely genuine and she really didn't want the money back. The whole attitude toward money and seeing that others had got enough was so different from my previous job where although we earned much more, people remembered who owed whom a cup of coffee. All the women were very generous, sharing out sweets and crisps and whatever they bought for themselves. (Cavendish 1982:62)

Group solidarity is also evidenced in an ethnography of an underground mine with a predominantly *male labor force.* The ethnographer reports the following episode in which a lead worker and his men gather at the head of a mineshaft to search for co-workers trapped by a fire:

> Suddenly Jimmie Isom picks up a mask from the jeep. "Put one on me, Dan," Jimmie says. Dan stares at his friend, with the deep-etched lines from his heart attack. Dan usually works Jimmie on the outside crew these days, afraid of working him inside. Now Jimmie is volunteering to go into the smoke. Dan doesn't know how to turn him down. (Vecsey 1974:190)

SOURCES: (1) Excerpted from Ruth Cavendish, *Women on the Line,* p. 62. London: Routledge and Kegan Paul. Copyright 1982. (2) Excerpted from George Vecsey, *One Sunset a Week: The Story of a Coal Miner,* p. 190. New York: E.P. Dutton. Copyright 1974.

By contrast, pride in work based on significant self-direction is also a common workplace experience. The work of engineers and scientists who have great latitude in self-directing their work provides an example. An ethnography describing the development of the polymerase chain reaction (PCR), which profoundly transformed the human potential to identify and reproduce segments of genetic code, reports on the extreme pride that the chemists and biologists who were engaged in the project felt in their work. The development of PCR has made possible not only cloning but also a vast array of genetic interventions in the areas of medicine, agriculture, biology, and related fields. A young biochemist recalls:

> Probably I worked harder in that time in my life than I ever had in terms of hours. I was very interested in the job. It was really fun to learn how to synthesize DNA. . . . It was the heyday of biotechnology. There were all kinds of bold ideas floating around

all the time . . . and there was absolutely no constraint on the imagination. . . . The company was really fun. . . . [We] were right in the middle of something that was a red-hot kind of an area. (Rabinow, 1996:90)

Belongingness

Self-actualization is also influenced by the extent to which meaningful interaction is possible on a job. Most people prefer to work as a member of a group rather than in isolation: "Workers prefer jobs that permit interaction, are more likely to quit jobs that prevent peer interaction, and cite congenial peer relationships as among the major characteristics of good jobs" (Kahn, 1972:189). Peer support and solidarity are essential building blocks for a meaningful experience of work in many settings (Fantasia, 1988). Box 4.1 describes the important role of positive co-worker relations in both female and male work groups.

Technology

Because the nature of job tasks is such an important cause of job satisfaction, a great deal of research has focused on technology as both a cause and a potential solution to the problem of alienation. The technology one uses depends primarily on one's occupation. Accordingly, occupational differences in job satisfaction have frequently been used as indicators of job quality. When workers are asked how satisfied they are with their job as a whole, farmers indicate the greatest satisfaction, with 67% reporting that they are very satisfied. By contrast, only 40% of clerical workers and machine operatives report that they are very satisfied with their jobs (see Table 4.1). Noise, dirt, and dangerous working conditions are also important sources of dissatisfaction for workers. When workers are asked what problems they experience at work, safety and health are their most common areas of concern (Rothman, 1998).

Blauner's Theory of Technology and Alienation Robert Blauner, in his famous book about industrial society, *Alienation and Freedom* (1964), explored the possibility that technology would bring an end to alienation. Blauner based his analysis on case studies of four manufacturing industries at various stages of technological development: printing, textiles, automobiles, and chemicals.

The printing industry was organized along traditional craft lines. Job tasks were complex and rewarding. Printers were very involved in their work and experienced a good deal of self-actualization. Workers in the textile industry were more alienated than printers. The work was organized around machine tending, with workers being responsible for highly repetitive tasks such as keeping threads attached to moving spindles. However, the social setting of textile mills in small Southern communities seemed to ameliorate their alienation by providing other avenues of satisfaction and involvement. The repetitive assembly-line technology of the automobile

Table 4.1   Job Satisfaction across Occupations $

	Very Satisfied	Fairly Satisfied	Dissatisfied
Professional	58%	31%	11%
Managerial	62	30	8
Sales	46	44	10
Clerical	40	52	9
Craft	42	51	7
Operators	40	41	19
Unskilled	44	39	17
Service	58	31	11
Farm	67	27	7
Total	52	37	11

SOURCE: National Opinion Research Center, *General Social Survey, 1998* (public use tape).

industry produced even greater alienation among workers because of the increasing size of organizations, the heightened intensity of assembly-line work, and the location of the new factories in regions removed from workers' origins.

By contrast, workers in the chemical industry were less alienated. Blauner argued that there was less alienation because the automated **continuous-process technology** of the chemical industry allowed a less repetitive arrangement of work, greater worker autonomy, and greater worker responsibility for the care and maintenance of expensive automated equipment. Based on these observations, Blauner proposed an **inverted U-curve of technology and alienation** (see Figure 4.2). He believed that alienation increases as machine pacing and assembly-line technologies replace the craft organization of work, but that it decreases again with more advanced continuous-process technologies. In this view self-actualization at work can be achieved either in craft settings or in settings that use advanced technology, but not in mass-production manufacturing settings.

Modern Technological Developments Developments since Blauner's analysis have neither confirmed nor convincingly refuted his conclusions

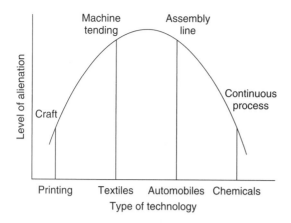

FIGURE 4.2 Blauner's Inverted U-Curve of Technology and Alienation

SOURCE: Adapted from Robert Blauner, 1964, *Alienation and Freedom.* Copyright © 1964 The University of Chicago. Reprinted by permission of The University of Chicago Press.

that advanced technology reduces alienation. Subsequent research has suggested "that workers accept technological change as normal and that they like their machines" (Form and McMillen, 1983:175). Other researchers have found that although advanced technologies may reduce alienation for some workers, technology undermines the skills of other workers and increases their sense of alienation (Shaiken, 1984; Spenner, 1985). Increased alienation has been especially characteristic of clerical workers in data-entry jobs (Braverman, 1974). In recent decades the printing industry, which Blauner selected to represent the craft organization of work, has experienced a technological revolution due to widespread use of electronic typesetting. This revolution eliminated the jobs of many skilled craft workers and replaced them with somewhat less skilled data-entry and programming jobs. We discuss controversies about the role of advanced technology in determining the nature of work further in Chapter 9.

The most reasonable conclusion that we can draw from these findings is that technology has no single, unidirectional effect on alienation. Rather, alienation can increase or decrease, depending on the nature of the technology and the nature of the jobs the technology creates, eliminates, or changes. For example, clerical work can be *routinized* by computerized systems that narrow a worker's range of activities, or it can be *humanized* by selecting technologies that expand workers' capabilities. Too often, the emphasis on control and monitoring of workers results in the design and selection of technologies that restrict self-actualization at work (Noble, 1997). It is up to us as members of society to design and promote technologies that improve working conditions. The possibilities are there, but the outcome is by no means assured.

Organizational Structure and Policies

Job tasks and technology are important determinants of alienation, but workers are also strongly influenced by the characteristics and policies of the organizations in which they work. The size of an organization and its pay and promotion policies are particularly important in this regard.

Pay Being paid a living wage for one's work is a necessary condition for self-actualization. Workers consistently rank pay as a crucial characteristic of a good job (Freeman and Rogers, 1999). High wages may not be sufficient to compensate for an alienating job, but the provision of wages adequate to meet basic needs is a fundamental requirement before a job can be experienced as rewarding and meaningful.

Size Workers prefer small companies to large corporations. They would also rather work in small departments or groups within companies than in larger units. Large size produces feelings of powerlessness and isolation, because workers have difficulty in identifying the overall purpose and direction of their organization and in feeling that they are a significant part of that purpose. Workers report greater satisfaction across a variety of dimensions in small, locally owned companies than in regionally based companies or in the nation's largest corporations (Hodson and Sullivan, 1985).

Promotion Policies Workers can also feel alienated by a lack of opportunity for promotion. Because almost all complex organizations are organized as a hierarchical pyramid, the ratio of high-level jobs to beginning jobs is generally low and frequent promotions can occur only for a small number of workers. Workers who have a low likelihood of promotion often come to feel disenchanted with their jobs. If there are few prospects for promotion, workers often lose hope of advancing to new and more challenging jobs. For a job to be fulfilling, it is necessary to have not only rewarding tasks but also a meaningful career trajectory. This problem is even more acute when people define their self-worth in terms of their advancement at work, as is so common today (Opsina, 1996).

Dignity and Respect An essential foundation for good work is to be treated with dignity and respect. Abusive bosses who yell at workers, fire them without just cause, and mismanage the workplace demoralize their workforces. Conversely, bosses who respect workers' rights and treat them with dignity find that workers are capable of great enthusiasm and loyalty (Hodson, 2001; Pfeffer, 1998). One of the most important contributors to organizational productivity—and to a meaningful experience of work—is having supervisors who are competent, dependable, and trustworthy (Lorenz, 1992).

Organizations *can* make work more fulfilling. Such a process would entail a reduction in the division of labor so that jobs become more diverse and interesting, rather than the opposite. Technologies can be selected that are compatible with a reduced division of labor. Organizations can also provide job rotation and other forms of continuous learning. Managerial competence can be trained and a respectful work-life experience can be assured. All of these changes would encourage greater involvement on the part of workers and would contribute to their self-actualization and to organizational productivity.

Participation

Organizations can facilitate self-actualization through their style of supervision and through allowing workers to participate in making decisions. When workers have an opportunity to participate in the day-to-day decisions influencing their work lives, they experience greater self-actualization. Alienation increases when they are closely supervised and blocked from participation through centralized systems of authority and decision making (Wolf, 1995). An example of the positive effects of participation on job satisfaction in an automobile factory is presented in Box 4.2.

Worker participation is a multifaceted phenomenon that is difficult to nail down and define. It can include an informal openness on the part of supervisors to the complaints and ideas of workers. At the other end of the continuum, it also includes formal participation in determining the overall direction of an organization, as among the members of a worker-owned cooperative (Russell and Rus, 1991).

Union Membership Union membership provides an important form of participation for many workers. Elected union representatives negotiate with employers to determine the rules governing working conditions. In North America this is the most widespread and significant form of participation available to workers. Unions also promote "due process" in the workplace through the *grievance system*. Under the grievance system, a worker who files a formal complaint can receive judicial review from an arbitrator jointly agreed on by the employer and the union. The provision of due process for workers is one of the most important benefits of union membership and provides an important sense of fairness at the workplace (Leicht, 1989). We discuss the role of unions in the workplace further in Chapter 6.

Producer Cooperatives Involvement in an organization can include any combination of formal, informal, and representative participation.

BOX 4.2 Participation and Job Satisfaction at Volvo

Throughout the company, we have a hierarchy of works councils, with representatives from both management and the employees. Some of these councils have been required by collective agreement. Others, like the Corporate Works Council, have been created on a voluntary basis to meet our own needs for consultation.

Bedeviled by an inherently noisy and dirty process, body-shop workers and managers got together and chose a working group to assess the various problems, suggest some solutions, and figure out the costs of the alternatives. . . .

The results were promising. The architects suggested ways to cut the noise from jigs and grinding machines. They also proposed a fundamental color scheme with red, orange, blue, and green. These suggestions were put together in a special exhibition in the shop, and employees came to see and decide for themselves. Many had further suggestions and related problems to offer. The response was quite positive. It took several years to implement all the proposals, but today the body shop is one of the brightest spots in the corporation. . . .

In the upholstery shop job rotation was the opening wedge. It started for very practical reasons. Employees complained of sore muscles from doing the same operation over and over. They discovered that if they paired off and traded jobs every day or so, they were able to use different sets of muscles. From that ergonomic, down-to-earth beginning have grown most of the other changes that focus on the quality of work life more generally. . . .

Most of the employees felt they understood the overall process a little better, and had more tolerance for their colleagues when they learned that the other jobs were not as easy as they thought. Improved contact with fellow workers sometimes led one person spontaneously to help another. The physical pains (and associated absenteeism) dwindled. . . .

As these changes took place, rather gradually, there was a concomitant increase in team spirit and individual commitment. Employee turnover and absenteeism changed dramatically, and the upholstery quality improved because team members understood the entire process and felt much more responsible for the product of their work.

SOURCE: Excerpted from Pehr G. Gyllenhammar, *People at Work*, pp. 81–91. Reading, MA: Addison-Wesley. Copyright 1977. Reprinted by permission of the publisher.

The more forms of participation that are included, the more likely alienation is to be reduced (Rothschild and Ollilainen, 1999). Direct participation reaches a peak in collectively owned cooperatives, such as those in the plywood industry in the Pacific Northwest (Greenberg, 1980). In these cooperatives the workers are responsible for the governance of the enterprise. They hold regular shareholder meetings and informal group meetings on the mill floor. As a result, only a fourth as many supervisors are needed in cooperative mills as in privately owned mills. In addition, the supervisors primarily coordinate the flow of materials rather than scrutinizing the work of their fellow workers. Spontaneous cooperation and informal job rotation are the rule in the producer cooperatives, providing workers with active participation not only in the overall direction of their enterprise but also in the day-to-day content of their jobs. The range of possible forms of participation open to workers today is a primary focus in Chapter 17.

Individual Differences in the Experience of Work

Workers' feelings about their jobs depend not only on the nature of the work but also on the background, values, and needs that they bring to the job. Whether workers are satisfied with a job thus depends to some extent on what they hope to get out of it. Gender, race, and age are important differences among individuals that influence the needs and values they bring to the workplace. It is thus reasonable to ask if workers of different genders, races, and ages have different experiences of work.

Gender Female workers tend to have lower-paying jobs, more repetitive work, fewer chances for advancement, and fewer opportunities to exercise high levels of skill than male workers. Despite these differences, research consistently indicates that women are about as satisfied as men with their jobs (Tolbert and Moen, 1998). What accounts for this paradox? The explanations that have been offered include the idea that women are socialized to be more acquiescent, and therefore simply do not verbalize their complaints. The explanation that has received the most support, however, is that women evaluate their jobs on a different basis than men—that is, relative to those of other women, not relative to those of men. Thus, they do not feel particularly deprived (Crosby, 1982; Hodson, 1989). The use of other women as a **comparison group** is learned in childhood and applied later in school and on the job. The use of different comparison groups by men and women for evaluating their jobs is further reinforced by occupational segregation into different lines of work. Many women work in occupations that are strongly female dominated (see Chapter 5). This occupational segregation increases women's contact and feelings of a shared work situation with other women while decreasing their opportunities to make comparisons with men.

Differences have also been reported in the relative importance that men and women place on various aspects of work. Women have been found to be more interested in complex tasks, interesting work, and congenial co-workers (Miller, 1980). Men have been found to be more interested in authority, advancement, and freedom from close supervision (Ross and Wright, 1998). These reported differences, however, have not been broadly replicated across different research projects. The only finding that has been consistently replicated is that women are just as satisfied with their work as men, in spite of the frequently more alienating nature of their work. Given the increased participation of women in jobs that were once the exclusive domain of men, however, in the future women may increasingly come to evaluate their jobs relative to those of men.

Race Members of racial and ethnic minorities are also more likely to be stuck in poorer paying, less skilled, and less rewarding jobs (Wilson, 1997). Unlike women, however, they voice strong discontent with this situation. For instance, the negative job satisfaction of black workers relative to that of white workers is among the strongest contrasts in job attitudes found between different social groups (Mueller et al., 1999). Two differences between blacks and women account for their different reactions to inferior jobs. First, blacks, unlike women, compare themselves with dominant groups when they evaluate their jobs. Second, most blacks believe strongly that they should be treated equally with whites. This ideology of racial equality gives voice and legitimacy to their anger at being stuck in inferior jobs (Form and Hanson, 1985).

Age Young workers are often more unhappy with their jobs than older workers (Glenn and Weaver, 1985). Part of the reason for this difference is that younger workers are more educated and thus expect more from their jobs. Another part of the explanation is that older workers have either been able to find work adequate to their needs or have made downward adjustments in their aspirations.

Tenure Length of time on the job also makes a difference in how workers feel about their work. New workers are more interested in the importance of their job than are workers with greater tenure, but they are not as concerned about having a high degree of autonomy. Once workers have been on the job for several years, their attitudes and desires begin to change, and they become more interested in autonomy and variety and less interested in the centrality of their task within the organization (Hall, 1994).

Great Expectations

The expectations that workers bring to their jobs arise from their prior socialization. As you saw in Chapter 3, socialization begins early in childhood with one's parents and continues throughout life. Workers are more or less satisfied with what their jobs offer depending on the weight that their prior socialization leads them to give to specific aspects of work, such as autonomy, congenial co-workers, fringe benefits, or the chance to develop new skills. Early occupational experiences can also create values and dispositions that influence later occupational choices and satisfaction with those choices (Kohn, 1990). Satisfaction is thus determined not only by the characteristics of the job, but also by the **job-worker fit,** that is, the fit between a worker's values and the job characteristics. Prior values have been shown to be one of the most important determinants of overall satisfaction with one's job (Kalleberg et al., 1996).

One's level of education influences job satisfaction through creating expectations about the rewards that the workplace should offer. Thus, holding job quality constant, more educated workers are less satisfied! Why does this occur? In college, students have a great deal of autonomy to decide their curriculum and pattern their study habits. Later, when they enter the full-time world of work, many college graduates find themselves in highly routinized white-collar jobs, which have few if any of these characteristics. Instead, their jobs may be boring, lack challenge, and use few of their skills (Burris, 1993). After graduating from college, students may find a mismatch between what they had come to expect in college and the jobs that are actually available in the world of work.

RESPONSES TO WORK

In this section we analyze the nature of workers' attitudes toward work and their behaviors at the workplace. We start again with the general concepts of alienation and self-actualization and then move to more specific attitudes and behaviors.

Attitudes toward Work

Sociologists have developed a variety of measures for appraising workers' attitudes toward their work. These include measures of intrinsic and extrinsic job satisfaction and commitment to the job.

Job Satisfaction Social scientists use two approaches to measure job satisfaction. In one approach they ask respondents how satisfied they are with their jobs or with specific aspects of their jobs. Common phrasings of such questions are "How satisfied are you with your job as a whole?" or "How satisfied are you with your chance to use your abilities on your job?" An alternative approach is to ask whether the respondents would stay in their current jobs if other opportunities were available. Common phrasings include "If you had it to do over again, would you go into the same line of work?" and "Would you continue to work if you didn't have to?" These two approaches produce markedly different proportions of reportedly satisfied workers. About 75% of workers report being "fairly" or "very" satisfied with their jobs when asked directly about their level of satisfaction. Only about 40% of workers, however, would continue in the same line of work if other options were available (Hamilton and Wright, 1986).

What accounts for this discrepancy? For one thing, phrasing a question in terms of satisfaction with a current job limits respondents' options; that is, it forces them to consider the current constraints on their options. In this situation, most respondents give cautiously positive evaluations of their jobs. When questions offer workers broader possibilities by asking if they would enter the same job again or would prefer not to work at all, they voice greater reservations about their current jobs. In reality, such options do not exist for many people because of limits imposed by such factors as fixed expenses, limited training, family responsibilities, and regional ties. Given the opportunity, however, most people would forsake their current job for something better. "For most workers it is a choice between no

---

BOX 4.3   San Francisco Scavengers

With all the problems and discontents in the work of the garbage collector, there are plenty of scavengers who will answer positively when you ask them, "What's good about this job?" . . .

"I guess you have to go into the psychology of it," Lenny said. The "psychology of it" had a great many different aspects. One of these was what might be called the variety of the work, as unlikely as that may seem to the uninitiated. Ron told me: "Thing about this job is that you do different things." If he had taken the somewhat comparable job of pick-up man on a city street-cleaning crew, Ron said, he would "go crazy. That would be monotonous." The tasks on his route were varied enough to break the monotony of the work's routine—operating the blade, solving problems with customers, blanket work and can work, driving the truck, and so forth. But there was also a different kind of variety, the unexpected in the human events of the work day—like the burglar on that day with Freddie and his crew. When I met Freddie's young son working with his father, he told me that he had refused a nine-to-five job that paid more: "Staying in one place all day? Naw!" . . .

Another part of "the psychology of it" was being able to set one's own pace, Freddie had said. "You are your own boss." A crew could work as fast or as leisurely as they wished. As I observed, they usually worked as swiftly as humanly possible—if only to go home early or, as Lenny told me from his own experience, to be in a bar waiting for another crew to come in and be able to say, "What's been keeping *you*?" . . .

Being outdoors is a big advantage that even workers in less clement climates emphasize. To be indoors all the time is a drag.

SOURCE: Excerpted from Stewart E. Perry, *San Francisco Scavengers,* pp. 110–117. Berkeley, CA: University of California. Copyright 1978. Reprinted by permission of the author.

---

work [and work] burdened with negative qualities. In the circumstances, the individual has no difficulty with the choice; he chooses work, pronounces himself moderately satisfied, and tells us more only if the questions become more searching" (Kahn, 1972:179). What this means is that people can be satisfied and dissatisfied at the same time. Most people try to make the best of their situation at work and report being moderately satisfied with it. Verbalizing a mildly positive attitude about work is also socially more acceptable and psychologically safer than admitting that one is dissatisfied.

People also have a tremendous capacity to gain at least some satisfaction and meaning from even the most tedious jobs. It is possible to build lives, careers, and even communities around work that is, in fact, quite alienating:

> Workers have a number of ways of dealing with monotony. . . . One of the older women on the floor had a routine she followed religiously. Every day at morning coffeebreak she went to the corner store and bought a newspaper. She brought it to her table and then went to the bathroom for a paper towel that she spread on her table. She then proceeded to eat half of her sandwich, no more, no less, every working day. There were numerous other examples of women "setting up" their meager possessions—radios, cigarettes, and coffee cup—in similar fashion. (Juravich, 1985:56)

Even garbage collectors can take pride in their work and realize a significant degree of self-actualization from it (Perry, 1978). Are workers satisfied or not? It depends on what we mean by satisfaction and how we ask the question. Box 4.3 describes the work of garbage collectors in a San Francisco cooperative and how they find meaning in work that many would avoid.

Intrinsic and Extrinsic Satisfaction Researchers sometimes divide job satisfaction into intrinsic and extrinsic factors. **Intrinsic rewards** are realized on the job and include such things as the freedom to plan one's own work, the chance to use one's abilities, the absence of close supervision, and positive relations with co-workers.

Extrinsic rewards are realized off the job and include such things as pay, fringe benefits, and job security.

Regardless of the measure or aspect of job satisfaction analyzed, studies show a moderate decline in job satisfaction in North America since the 1950s, when social scientists first began to conduct research on job satisfaction (Hamilton and Wright, 1986). What accounts for declining job satisfaction? It does not appear that a declining quality of jobs is responsible. Changes *have* occurred in the content of many jobs, but these changes have been gradual and offsetting (Spenner, 1985). A more likely interpretation relates to greater dissatisfaction among young workers. Over time, with increasing prosperity and educational attainment, expectations have risen. As a result, young people tend to voice greater dissatisfaction with their jobs today than in the past.

Commitment   Above and beyond being satisfied with their jobs, workers can also be more or less committed to their jobs. **Commitment** develops when workers perceive that their own needs will be met through continued employment in the job and when they perceive that the goals and values of the occupation or the employer are compatible with their own goals and values. Commitment implies a willingness both to retain long-term membership in the group and to give one's full energy and abilities to the group's ongoing tasks (Lincoln and Kalleberg, 1990).

Other reasons for commitment exist besides a compatibility of goals and values. Lengthy association with a profession or an organization increases the number and importance of benefits that are dependent on continued group membership. For example, pension plans, medical benefits, and friendship networks may all depend on retaining one's current job. As a result, people are often reluctant to leave a job that they have held for a long time, in spite of dissatisfactions with it, because their lives have come to depend on it in many different ways. Japanese corporations have intentionally heightened these "side-bets" by directly tying such peripheral aspects of workers' lives as housing and access to vacation retreats to continued employment with the company. This strategy has resulted in increased commitment among Japanese workers, but it has not resulted in greater satisfaction among Japanese workers than among American workers (Lincoln and Kalleberg, 1996).

Evidence suggests that, as with job satisfaction, commitment to employing organizations may be declining. A college-educated worker doing routine clerical work in a large corporation reports the following sentiments: "People don't care about the company. They could care less if [it] burned down tomorrow" (Burris, 1983:144). This reduced commitment occurs at a time when increased commitment and extra effort are being expected of employees in order to increase productivity. We explore the consequences of this dilemma further in the concluding chapter of this book.

Ideologies about Work   Workers may also subscribe to more elaborate belief systems concerning work and its place in their lives. Many manual workers hold a **working-class ideology** that stresses solidarity against the more powerful capitalist and managerial classes. This orientation appears most commonly in traditional manual occupations such as shipbuilding and coal mining. Alternatively, manual workers may subscribe to a more **instrumental orientation** in which the primary purpose of work is to provide money for family and leisure needs (Goldthorpe et al., 1969). Similarly, white-collar workers may subscribe either to a bureaucratic orientation, in which workers' duties to the organization are of utmost importance, or to a more instrumental orientation focusing on consumption. These ideologies have great significance not only for workers' attitudes but also for their day-to-day behaviors. Such behaviors are the focus of the next section.

Behavioral Responses to Work

Workers' attitudes toward their jobs, in combination with the options available to them, strongly influence how they will respond to the workplace. Workers, including both those who are quite enthusiastic about their jobs and those who are disgruntled, make behavioral adjustments at work to maximize their benefits and minimize their costs. These adjustments range from leaving the job to various forms of accommodation and resistance. A common response to alienating work is passive resistance through making work into a game (Burawoy, 1991), restricting one's output (Roy, 1952), or focusing on aspects of work life that are tangential to the main productive activity (Collinson, 1992). For instance, workers often adjust to alienating situations by focusing on interactions with their peers. Managers label such behavioral responses as "poor performance." However, such behaviors do not necessarily result from incompetence or laziness; rather, they may be straightforward responses to having to work at tedious, repetitive, and alienating tasks. These responses are difficult to predict from workers' levels of job satisfaction or commitment. Workers who are very committed to their work may be the ones most likely to resist alienating conditions. Those who are less committed may simply quit or grudgingly suffer in silence (Mortimer and Finch, 1996).

Pride and Enthusiasm Positive responses to work are likely to occur when the job is adequately rewarding in terms of both extrinsic and intrinsic satisfaction and when the worker is committed to the profession or the organization because of shared goals and values. In this situation, employees may work with exceptional enthusiasm and are often willing to make great personal sacrifices (Walsh and Tseng, 1998). Sacrifices may include working overtime without pay, even forsaking leisure and family activities (Hochschild, 1997). These responses are not necessarily altruistic. Rather, such behaviors may be self-actualizing—being engaged in genuinely nonalienating work helps people realize their identity, goals, and values. People are also more creative under these conditions, and their productivity is likely to be immeasurably higher than that of workers who consciously withhold their best efforts because of alienation (Pfeffer, 1998).

Pride in work is exemplified in an ethnography of ironworkers. The worker responsible for maintaining the crane used to lift the heavy steel girders evidences great pride in his daily chores:

> Most oilers are nearly invisible, fueling and lubricating their rigs before the day begins for the rest of us, vanishing to God knows where during the bulk of the day, reappearing at 4:00 to preside over putting the rig to bed. Beane, however, was not of that stripe. He fussed over the crane like a stage mother, constantly wiping away puddles of oil or grease, touching up scratches with fresh paint, agonizing loudly whenever a load banged into the stick. (Cherry, 1974:166)

Jobs that provide self-actualization and encourage enthusiasm can also have positive consequences for workers outside the workplace. For example, self-actualizing work can result in greater community involvement and in improved relations with family members. Conversely, alienation arising at the workplace can color a worker's whole life. Alienated workers are more likely to abuse alcohol and drugs and to have troubled interpersonal relations outside the workplace, including divorce, spouse and child abuse, and social isolation (Rothschild and Russell, 1986).

Absenteeism Absenteeism is a highly visible response to alienating work. On any given day, between 3% and 4% of the workforce will not show up for work (Leigh, 1986). Each worker is absent between seven and twelve days a year and only a relatively small part of this absenteeism is due to illness. In some industries the daily absenteeism rate runs as high as 10% to 20%.

Quitting Quitting is also a common response to an unrewarding job. Turnover rates average about 15% per year for the labor force as a whole, and for some industries they are much higher. For example, nearly half the nation's secretaries leave their jobs within two years, representing a 25% annual turnover rate (Vreeland, 1985). In the hotel industry, and in other service industries that employ large numbers of part-time workers, turnover is estimated to be even higher. Quitting occurs most frequently where job rewards are lowest. If wages are high and workers have greater resources, such as scarce skills, they are more likely to form a union. In this way they seek to improve their jobs through collective action rather than to leave through quitting (Freeman and Rogers, 1999).

Resistance and Sabotage Alienated workers who do show up for work are often on the lookout for ways to vent their frustrations. When workers have no opportunity for redress of their grievances in consensual ways, the behavioral responses of sabotage and theft become increasingly attractive options.

The word *sabotage* originated in the 1400s in the Netherlands, where workers would throw their sabots (wooden shoes) into the wooden gears of the textile looms to break the cogs. The purpose of this activity was to force concessions from their employers. The Luddite movement of the early 1800s in England practiced organized sabotage as a bargaining tactic with employers to forestall new weaving technologies that were causing widespread displacement of workers (Randall, 1991). Thorstein Veblen, one of the founding figures in American sociology, coined the phrase "conscious withdrawal of efficiency" to describe the phenomenon of sabotage (1921). In the modern industrial world, sabotage can take a variety of forms. Lower-level administrative workers may intentionally slow a bureaucracy by scrupulously sticking to regulations, by intentionally destroying or misplacing files, or by refusing to fill out required paperwork (Sprouse, 1992). Factory workers may intentionally break machinery or use it in ways that hasten its breakdown. For example, a welder in an automobile factory reports the following events:

> But if they make us work without any time off then we wreck a gun and take a few minutes while it is being repaired. This happens all through the shop. Many times the guns could be easily repaired. A worker sees his gun going bad. He has no interest in

saving it so he'll let it go completely wrong and burn clear up before calling the repairman. Many times we know what is wrong and if we feel good we repair it ourselves. The workers put things in their guns or break them on purpose. A worker was fired for this not long ago. Every time he got mad he would take his knife and cut the rubber hose. He would put something on it to make it look as if it had burst. This happened twice every day. The company got the foreman to hide and watch what was happening. (Denby, 1978:139–140)

Workers in some situations routinely practice sabotage. One of the authors worked at a sawmill where workers frequently tore metal casings off machinery and threw them down the exit chute that took scrap wood out of the mill. The metal casings broke the teeth on a chipper that shredded the scrap wood, resulting in brief unscheduled breaks. Box 4.4 describes how a domestic worker resists what she perceives to be excessive demands by her employers.

Theft Theft is also a common response to unaddressed grievances at the workplace. The number of employees engaged in some sort of workplace theft has been estimated at 28% in manufacturing, 33% in the hospital industry, and 35% in retail (Geis et al., 1995). Common forms of theft include taking merchandise and tools, getting paid for more hours than were worked, purposely undercharging a friend or coconspirator, and being reimbursed for more money than was spent on business expenses. Theft is especially common among young workers and is more common among men than women.

In some service jobs, theft is an expected part of the employment arrangement between worker and employer. The sociologist Elliot Liebow describes this situation for a dish-washing job held by one of the subjects in his study of street corner society:

> Tonk's employer explained why he was paying Tonk $35 for a 55–60 hour

---

BOX 4.4   Refusing to Work on One's Knees

"I didn't do everything those folks told me to do. Some I did and some I didn't. They would tell me to get on my knees and scrub the floor and I didn't do it. I didn't mess up my knees. I told one lady, 'My knees aren't made for scrubbing. My knees are made to bend and walk on.' " . . .

Ms. Ryder has no hint of servility or obsequiousness about her; on the contrary, she has a self-possessed quality that could be, one would imagine, threatening to some employers. Ms. Ryder enjoyed telling me stories of her resistance to exploitation. Her self-respect was sometimes expressed in overt feistiness: "I remember one Sunday morning, this woman told me to scrub her kitchen floor on my hands and knees. I got mad at her and said to her, 'You sit right down there and wait until I scrub it.' So I got a whole lot of

ammonia and clorox and the stuff with the twin kids on the box. And I just poured it over the floor. And then half wiped it up. And you know what it looked like when I got finished! She just looked at that floor. That floor looked bad for two or three days. I wouldn't wash it 'cause I told her I'd already scrubbed it. That floor was so bad I didn't even like to walk on it 'cause it was muddy and sticky. . . . She had wanted me to get on my knees and scrub it. And I wasn't *thinking* about getting on my knees and scrubbing it. And, after that, I could just mop it up and it would look nice. No, my knees weren't made for walking all over the floor!"

SOURCE: Excerpted from Judith Rollins, *Between Women: Domestics and Their Employers*, pp. 142–143. Philadelphia: Temple University Press. Copyright 1985.

---

workweek. These men will all steal, he said. Although he keeps close watch on Tonk, he estimates that Tonk steals from $35 to $40 a week. What he steals, when added to his regular earnings, brings his take-home pay to $70 to $75 per week. The employer said he did not mind this because Tonk is worth that much to the business. But if he were to pay Tonk outright the full value of his labor, Tonk would still be stealing $35–$40 per week and this, he said, the business simply would not support.

This wage arrangement, with stealing built-in, was satisfactory to both parties, with each one independently expressing his satisfaction. Such a wage-theft system, however, is not as balanced and equitable as it appears. Since the wage level rests on the premise that the employee will steal the unpaid value of his labor, the man who does not steal on the job is penalized. And furthermore, even if he does not steal, no one would believe him. (Liebow, 1967:37–38)

Workplace theft has been estimated at about 1% of the gross national product (Geis et al., 1995).

In spite of the magnitude of this figure, most employers prefer to look the other way, discounting theft as inventory or operating losses. For example, only a small minority of supervisors indicate that they would report a suspected pilferage to their boss if it were the first incident for the employee involved.

The English urban anthropologist Gerald Mars (1982) has developed a typology of workplace theft. The typology is based on closeness of supervision and on the presence or absence of a strong work group. Workers in closely supervised jobs with a weak work group ("donkey jobs"), such as cashiers, are likely to attempt to retake control of their jobs through stealing time and through creative accounting practices. Workers in closely supervised jobs with strong work groups ("wolfpack jobs") are more likely to engage in elaborate and systematic pilferage. In longshoring, such practices are called "lightening the cargo." Workers in weakly supervised jobs with strong group structures ("vulture jobs"), such as cab drivers and delivery workers, are often successful in organizing stable systems of pilferage, misaccounting of funds, or both. Box 4.5 gives Mars's account of pilferage in a "vulture job" involving

BOX 4.5 The Pastry Scam

Over a long period of time "mistakes" can actually be built into a round or sales area to become an unalterable feature of the landscape. The newcomer or trainee has a choice: he can accept the fiddle [scam] and by so doing implicitly recognize the whole landscape of fiddling; or he can reject it, and by so doing draw attention to an individual act of dishonesty.

> There's one woman . . . who runs the canteen at the tech. college. She orders trays of pies. Every day she orders three, every day she sells three, and for years she's always thought there were sixty pies to a tray when there should be seventy-two. This fiddle was passed on to me when I took the job on and the bloke who showed me had it passed on to him when he started. It must have been going on for years because that woman had been there thirteen years when I took her on and that was two years ago. The fiddle is possible because she'll always sign for three trays with no mention on the voucher of the number of pies.

A salesman who could not accept this handed-down fiddle might well alert the customer to the fact that she had consistently been given thirty-six pies less than she should have been given every day for thirteen years. In doing so he might clear his own conscience though he would also, at the every least, lose his company a valuable customer. But it says much for the careful acclimatisation of newcomers, at least to this company, that the subterfuge should have lasted for so long without disruption.

SOURCE: Gerald Mars, *Cheats at Work*. London: Unwin Hyman, pp. 121–122. Copyright 1982 by Gerald Mars. Reprinted by permission of the publisher.

---

pastry delivery. Finally, workers in weakly supervised jobs with a weak work group ("hawk jobs"), such as professionals and managers, are likely to abuse expense accounts and to spend company time and equipment on private endeavors. An example would be a stockbroker who uses privileged "insider" information to build a personal investment portfolio. Similarly, a professor might use money from a research grant to buy a computer for a private consulting business. The more privileged of these workers may be motivated less by alienation from work than by simple greed.

Employees can be quite inventive in devising systems of pilferage. One of the authors worked for several years at a truck stop ("donkey job") where he observed an accounting scam involving several employees. The scam involved misreporting the license plate numbers of out-of-state trucks to the cashier in order to secure an "in-state" discount. The employees pocketed the money from the discount, and neither the cashier nor the out-of-state truck driver was the wiser. The scam lasted for about six months before it was discovered. The originator of the scam, who had been doubling his take-home pay, was fired but the other practitioners were neither identified nor disciplined.

THE FUTURE OF JOB SATISFACTION

Marx argued that alienation reaches its zenith under capitalism and would be reduced under communism because workers' basic needs would be met and the class division between capitalists (who control the work process) and the working class (who are controlled) would be eliminated. However, alienation did not appear to have been eliminated in socialist societies such as those in Eastern Europe (Burawoy, 1985). Marx correctly perceived that alienation is a fundamental characteristic of capitalist society; however, the replacement of the capitalist class by state directors did not eliminate alienation. Extreme division of labor, hierarchical forms of organization, centralized control, bureaucracy, and pressures toward economic growth typify socialist societies as well as capitalist ones and produce alienation from work regardless of the underlying economic system (Haraszti, 1978).

The elimination of alienation entails more than supplementing private interest with public interest in formulating economic policy at top levels. It must also involve the redistribution of control and decision making on as broad a base as possible. Without providing substantial opportunity for participation at the workplace, industrial economies of whatever economic and ideological orientation will be unable to guarantee their citizens meaningful, nonalienating work.

In North America, the recent past has witnessed decreased satisfaction with work. It is possible that work may become even more alienating in the future. Indeed, as people become more and more educated, alienation will tend to increase unless the quality of jobs also improves. In addition, low-level jobs in the service sector are expected to increase and these are some of the least rewarding jobs. On the positive side, professional jobs are also expected to increase and these jobs are typically more rewarding and satisfying.

A prediction about the future of job satisfaction is difficult to make under these circumstances. However, there are hopeful signs that we are learning how to create jobs that produce greater self-actualization while increasing or at least maintaining productivity. Industrial societies are evidencing increased interest in redesigning jobs to increase worker participation, involvement, and commitment. We discuss such strategies for organizational redesign in Chapter 17.

SUMMARY

Alienation at work arises from repetitive working conditions, lack of autonomy, large bureaucratic organizations, and blocked opportunity. Job attitudes depend not only on alienating conditions but also on individual differences among workers. Prior expectations based on educational attainment and comparison groups constitute important criteria that individuals use to evaluate the quality of their jobs.

Workers' attitudes toward their jobs are based on complex sets of evaluations, involving both intrinsic and extrinsic rewards of the job. There is substantial evidence that these attitudes can vary independently, so that workers can be simultaneously satisfied and dissatisfied with their jobs. Workers respond to their job conditions in a variety of ways. Most typically these responses involve greater or lesser enthusiasm for work, but they can also involve absenteeism, turnover, and even sabotage and theft in order to redress felt grievances.

Contrary to Marx's prediction, alienating working conditions are not a temporary situation that will automatically pass as new economic formations emerge. Levels of job satisfaction in the future are difficult to predict. Expectations will continue to rise because of increased education. Whether the quality of jobs will improve to a commensurate degree is an unanswered question.

KEY CONCEPTS

job satisfaction

alienation

self-actualization

job autonomy

continuous-process technology

inverted U-curve of technology
    and alienation

comparison group

job–worker fit

intrinsic rewards

extrinsic rewards

commitment

working-class ideology

instrumental orientation

sabotage

QUESTIONS FOR THOUGHT

1. What creative behavioral strategies have people devised to cope with their jobs at places where you have worked?

2. How is Marx's theory of alienation useful for understanding the rewards and frustrations of work in modern society? What are the theory's limitations?

3. What job factors are needed for self-actualization? What jobs have you held that have been most rewarding?

4. What are the major causes of alienation? Which of these have been most relevant in the jobs you have held?

5. Why are women about as satisfied with their jobs as men in spite of lower average pay and job autonomy?

6. Outline a future scenario resulting in greater self-actualization for workers. Outline one resulting in greater alienation.

MULTIMEDIA RESOURCES

Print

Melissa Everett. 1995. *Making a Living While Making a Difference: A Guide to Creating Careers with a Conscience*. New York: Bantam. A great guide to meaningful alternative careers.

Robert Denby. 1978. *Indignant Heart*. Boston: South End Press. The autobiography of a black youth who moves from the South to Detroit and finds work, alienation, and prejudice in the automobile industry.

Barbara Garson. 1988. *The Electronic Sweatshop*. London: Penguin. A lively account of how work in clerical and professional jobs is being transformed by new computer-based technologies.

David Halle. 1984. *America's Working Man*. Chicago: University of Chicago Press. An overview of work and community life in a chemical company town.

Tom Juravich. 1985. *Chaos on the Shop Floor*. Philadelphia: Temple University. A theoretically informed and well-written account of the author's experiences working in a wiring harness factory.

Melvin L. Kohn. 1990. *Social Structure and Self-Direction*. Cambridge, Mass.: Blackwell. The leading scholarly account of the effects of working conditions on intelligence, creativity, open-mindedness, and other aspects of cognitive functioning.

Gerald Mars. 1982. *Cheats at Work*. London: Unwin. A fascinating account of the variety of scams practiced at work.

Websites

Alternatives to Traditional Employment. *www.essential.org/goodworks* A clearinghouse of employment opportunities for "jobs that make a difference." Includes contacts, salaries, openings, and application materials.

Social Science Information Gateway. *http://sosig.ac.uk* An information clearinghouse for practitioners and researchers in the social sciences, business, and law in the United Kingdom.

How to Fire Your Boss. *www.geocities.com/CapitolHill/Senate/3671/DirectAction.htm* Strategies for taking control of your work life.

Institute of Labor and Industrial Relations, University of Michigan. *www.ilir.umich.edu* A central site for current high-quality research on a wide range of workplace issues.

RECOMMENDED FILM

Mr. Holland's Opus. 1994. Richard Dryfus is an underemployed musician who turns to teaching high school music to support his family and finds his true calling.

5

Barriers
and Disruptions
at Work

The sociological imagination enables us to grasp history and biography and the relations between the two within society. . . . Perhaps the most fruitful distinction with which the sociological imagination works is that between "the personal troubles of milieu" and "the public issues of social structure." This distinction is an essential tool of the sociological imagination and a feature of all classic work in social science.

Troubles occur within the character of the individual and within the range of his immediate relations with others; they have to do with his self and with those limited areas of social life of which he is directly and personally aware. . . . *Issues* have to do with matters that transcend these local environments of the individual and the range of his inner life. They have to do with the organization of many such milieux into institutions of an historical society as a whole, with the ways in which various milieux overlap and interpenetrate to form the larger structure of social and historical life. . . .

In these terms consider unemployment. When in a city of 100,000, only one man is unemployed, this is his personal trouble, and for its relief we properly look to the character of the man, his skills, and his immediate opportunities. But when in a nation of 50 million employees, 15 million men are unemployed, that is an issue, and we may not hope to find its solution within the range of opportunities open to any one individual. . . . To be aware of the idea of social structure and to use it with sensibility is to be capable of tracing such linkages among a great variety of milieux. To be able to do that is to possess the sociological imagination.

(C. WRIGHT MILLS, *THE SOCIOLOGICAL IMAGINATION,* 1959:6–11)

Discrimination, unemployment, and occupational hazards are personal troubles for individuals. When such experiences occur for many people, however, they become social problems. In this chapter we explore these social problems and the personal troubles they create.

DISCRIMINATION IN HIRING

I have a dream that one day this nation will rise up, live out the true meaning of its creed: we hold these truths to be self-evident, that all men are created equal. (Martin Luther King, August 27, 1963)

Throughout most of the history of the United States, African-Americans, Hispanics, and other minorities have been restricted to low-paying agricultural, factory, and service jobs (Wilson, 1997). Similarly, women have been largely restricted to clerical and service work and selected professions such as nursing and teaching. This **cultural division of labor** channels minorities and majorities and men and women toward different occupational roles. (See Chapter 3 on occupational steering.) For women, the cultural division of labor is further reinforced by the assumption that women will carry the main responsibility for child care and housework (Parcel, 1999).

For women in particular this cultural division of labor is deeply ingrained in social, family, and workplace relations. One important manifestation of the cultural division of labor for men and women is the gender-typing of occupations into jobs thought most appropriate for men, such as construction and engineering, or most appropriate for women, such as secretarial work, nursing, and primary-school teaching. Gender-typing of employment has thus channeled women into a

restricted set of occupations that are relatively poorly paid, at least in part because they have been predominately filled by women (Reskin and Padavic, 1994). The gender-typing of occupations can also result in statistical discrimination against women who are trying to enter male-typed jobs. Women in general, as a group, may be assumed not to possess the interests and abilities to do jobs characteristically filled by men, simply because these jobs have historically been filled by men.

In recent years women and members of racial and ethnic minorities have made important gains in employment in the United States and Canada. Legal actions brought by female and minority employees against companies that have discriminated in hiring and promotions have been one important tool in promoting these changes. Multimillion-dollar settlements were secured against American Telephone and Telegraph Company in the 1970s, General Electric in the 1980s, and Ford Motor Company in the 1990s to resolve histories of past discrimination. These settlements typically include restitution to the individual parties damaged by the discrimination as well as a commitment that a certain proportion of new hires will be women or minorities as dictated by the specifics of each situation (Powell, 1999).

These highly visible settlements between large companies and female and minority workers have helped reduce employment discrimination. Unfortunately, employers can still discriminate in many ways that are hard to eradicate. The following section explores the gains made by women and minorities as well as the continuing problems they face in gaining equal access to good jobs. In a subsequent section we discuss progress and shortfalls in the struggle for equal pay and promotions once a female or minority worker has been hired.

Equal Rights Legislation

The legal basis for the elimination of employment discrimination is contained in Title VII of the U.S. Civil Rights Act of 1964:

> . . . It shall be an unlawful employment practice for an employer (1) to fail or refuse to hire or to discharge any individual or otherwise to discriminate against any individual with respect to his compensation, terms, conditions, or privileges of employment, because of such an individual's race, color, religion, sex, or national origin; or (2) to limit, segregate, or classify his employees or applicants for employment in any way which would deprive or tend to deprive any individual of employment opportunities or otherwise adversely affect his status as an employee, because of such individual's race, color, religion, sex, or national origin.

This law made it illegal for employers to discriminate against women and minority workers in hiring and promotions and thus outlawed the deliberate **occupational segregation** of female and minority workers into restricted sets of job positions. The act also established the Equal Employment Opportunity Commission (EEOC, *www.eeoc.gov*) to enforce the law. Although written primarily with minorities in mind, the law has had far-reaching implications for female workers as well. It has also fueled a general expansion of due process in the workplace (Edelman, 1990).

Height and Weight Requirements From the standpoint of women, one of the most noteworthy legal changes resulting from this legislation has been the removal of formal height and weight requirements as employment criteria if the requirements are irrelevant to performing the job. A minimum height requirement of 5'2" would exclude about one-third of women in the United States while excluding only about 1% of men. A minimum weight requirement of 120 pounds would exclude about 22% of women but only about 2% of men (Wallace, 1982:6). Employment standards that include such provisions have been successfully challenged as women increasingly enter such nontraditional jobs as prison guard, firefighter, and police officer.

Affirmative Action It is also the policy of the U.S. government to seek to remedy the consequences of past discriminatory employment practices. Various executive orders calling for **affirmative action** are an attempt to compensate for past discrimination through hiring goals, preferential consideration among otherwise equal candidates, or active recruitment of women or minority workers. The most important government statement supporting affirmative action is Executive Order 11246, issued by President Lyndon B. Johnson in 1967. It mandates the Federal Office of Contract Compliance to issue government purchasing contracts only to organizations that are making efforts to remedy the effects of past discrimination. This mandate has had far-reaching implications, because many companies sell at least some of their products to the federal government. Thus, defense contractors, utility companies, computer and electronics manufacturers, and many other businesses have been forced to develop affirmative action hiring plans. Such plans can give preference to qualified minorities or women if the purpose of the plan is to erase "a manifest imbalance in traditionally segregated job categories."

Affirmative action plans can be instituted voluntarily by an employer or jointly by an employer and a union. Alternatively, a plan can come about as a result of a discrimination lawsuit. The majority of affirmative action plans have been adopted in large firms with predominately white-collar

BOX 5.1   Racially Divided Union Locals on the Gulf Coast

There were two longshoremen's locals in Corpus Christi—both International Longshoremen's Association, to be sure—one composed of white men and the other of black men. The white local was ILA Local No. 1224 and the black was 1225. "White and colored" was the expression used to describe them. The work was done by solid gangs of white and black. Each local had its own hiring place and conducted its own business. There was an arrangement where the one local would work the forward hatches of one ship, then the aft (or "after") hatches of the next, and so on. The usual freighter of the day was a five-hatch vessel with three hatches forward of the smokestack. If a ship

having four or six hatches was to be worked, the hatches were split two and two or three and three.

The two locals would hold a joint meeting now and then, but there was never any arrangement for scheduled meetings of both memberships. This segregation lasted until the year 1983, when a federal court decreed that black and white locals having the same work jurisdiction in the same port should merge into one local union.

SOURCE: Gilbert Mers, 1988, *Working the Waterfront: The Ups and Downs of a Rebel Longshoreman.* Austin: University of Texas Press, p. 11.

labor forces. Such plans have the potential to redistribute some desirable jobs to previously excluded female and minority workers. Many of the plans, however, have been extremely modest in conception and have called for only the most minimal adjustments necessitated by law (Reskin, 1998). Box 5.1, for example, describes the existence of racially separate trade unions in Gulf Coast ports in the United States as late as the 1980s. In addition, the lack of such plans in smaller firms, where the most blatant forms of discrimination exist, means that minority workers continue to face significant limits on employment in these sectors. It is thus a mistake to assume that the enactment of equal rights legislation automatically eliminated discrimination.

Continuing segregation is evidenced in the different occupational profiles of men and women and whites and minorities. Appendix Table 1 shows the representation of women, African-Americans, and Hispanics across detailed occupations. African-Americans are underrepresented in managerial, professional, and craft occupations. They are overrepresented in service, machine operative, and laboring jobs. Hispanics are also underrepresented in managerial and professional occupations. They are overrepresented

in laboring and agricultural jobs. Women are underrepresented in skilled craft jobs, though not in professional jobs because of the large number of nurses, teachers, and other *female-typed* professions that have long been identified as appropriate for women. Women are overrepresented in clerical and service work.

Affirmative action has had positive consequences for some female and minority workers and has helped break down sexual and racial hiring barriers (Tomaskovic-Devey, 1993). These gains have become apparent as women and minorities have secured jobs in a variety of settings. In addition, equal rights legislation has encouraged the creation of decentralized state, county, municipal, and organizational affirmative action programs. The integration of women into many professional and managerial occupations has thus significantly increased, although the representation of women in the ranks of chief executive officers (CEOs) of major corporations is still very low (Powell, 1999).

Affirmative action programs have also had a positive impact on the economy as a whole through allowing talented individuals to serve in positions from which they would previously have been excluded. Such individuals bring new skills to their positions and valuable new insights based on their unique social backgrounds. Increased workforce diversity thus has the potential to encourage the economic growth that benefits all workers (DiTomaso and Hooijberg, 1996).

An additional, and largely unintended, consequence of affirmative action has been that hiring and promotion decisions are made with increasing openness and rigor. Job openings are more widely advertised than in the past and these openings are also posted for current employees to consider. Personnel officers use more explicit hiring criteria than in the past and they frequently employ detailed rating systems for ranking qualified applicants. Such hiring systems displace previous systems that included a greater role for cronyism, favoritism, and nepotism. Rigorous hiring and promotion systems that favor merit over other characteristics are thus advantageous for all individuals, whether they are minority members, majority members, men, or women (Dobbin et al., 1993).

Affirmative action has not, however, been a cure-all for the problems faced by minorities and women in the United States. For instance, companies have put a lot of effort into hiring African-Americans for highly visible positions. As a result, the wages of black college graduates have risen faster than those of white college graduates (Mishel et al., 1999). Meanwhile, the wages of black high school graduates have fallen even further behind those of their white counterparts. Affirmative action has thus helped create a black middle class, but it has done little to help the large black underclass (Wilson, 1997). These problems have been aggravated by the concentration of African-Americans and Hispanics in regions and urban areas with high unemployment rates (DeFreitas, 1993).

A White Backlash? Affirmative action has sometimes sparked resistance based on the belief that increased opportunities for minorities and women must come at the expense of opportunities for whites and men. Thirty years of affirmative action have clearly not eradicated the inequalities resulting from three hundred years of legal restrictions on blacks in America or thousands of years of gender-based differentiation. From the vantage point of a white male job applicant facing a racially or gender-based hiring goal, however, such considerations may seem secondary to their own needs. The increasing employment challenges faced by all Americans in recent decades have also tended to make whites less sympathetic to the problems of minority workers. Consequently, support for affirmative action has fallen among whites, and even among some blacks (Sears, Sidanius, and Bobo, 2000). Some whites assume that every employed black they encounter owes his or her job to federal pressure rather than to personal qualifications and efforts (Kluegel and Bobo, 1993). Such assumptions can be very unfair to minority workers who have struggled and sacrificed to secure their education or training. Affirmative action policies are also increasingly being challenged by referendum and in court. As a result of these legal challenges, such plans increasingly focus on efforts to *actively recruit* qualified minority or female candidates, rather than on quotas or hiring goals.

The likelihood of further progress toward greater equality of employment opportunity remains uncertain. Such progress depends to a significant extent on the political climate of the next decades. Box 5.2 depicts a factory setting in the United Kingdom in which cooperation between the races has become the norm.

Continuing Forms of Hiring Discrimination

Many aspects of discrimination in hiring are subtle and difficult to eliminate through legislation. One such form of discrimination arises as a result of informal, word-of-mouth methods of recruitment. For example, access to apprenticeship programs in the skilled trades often occurs through personal contacts among union members and their immediate friends and families (Fernandez et al., 2000). Such continuing forms of segregation help maintain **dual labor markets,** in which women and minorities are sorted into

---

BOX 5.2   Cooperation in a Racially Diverse Workforce

There were eight Irish women on my line, six West Indians and me. Most of them had worked at Universal Mechanical and Electrical Components for years and knew each other well. It was known as a friendly line, despite some major conflicts—the West Indians said it was friendly because there were so many of them. They addressed each other as "ladies" and referred to themselves collectively as "girls" . . . The Irish women were all under thirty, except for the two who were deaf and dumb—they were in their early forties. The black women were older and most had grown-up children.

The cooperation between the women on our line made it more efficient. If someone in front had forgotten to put a small clip or peg on an electrical product, we would shout to them to send one along the line and attach it ourselves, or take it up to them to put right. What we were supposed to do, however, was to put out the part as a reject, marking down the fault both on it and on a sheet of paper. Then the reject operator would collect it, mend it, and return it to where it had left the line, also marking down the fault twice. Of course, all that took much longer. . . .

In the wages and personnel offices, practically all the clerical workers were English, including some young black women. It didn't seem to matter what color the clerical workers were so long as they'd gone to school here and reached whatever standard of spoken and written "Queen's English" the personnel office required.

SOURCE: Ruth Cavendish, 1982, *Women on the Line*. London: Routledge & Kegan Paul, p. 18.

---

lower-paying jobs and white males have greater access to higher-paying and more secure jobs (Nelson and Bridges, 1999).

Subjective Screening Criteria Subjective screening criteria also play an important role in eliminating job applicants with personal characteristics that differ from those of employers or current employees. Such criteria are rarely spelled out, leaving ample room for evaluators' subjective impressions of "intelligence," "appearance," "self-confidence," and "vigor." These impressions are easily biased by the evaluators' prior stereotyped notions about women or minorities. Indeed, standards may even arise over time that are shaped to match the characteristics of majority group candidates and to exclude women and minorities. Such standards may include hair and clothing styles or aspects of personal demeanor.

Professional jobs in previously male-dominated fields pose special problems of access to women. The role performance expected in these jobs often includes behavioral styles characteristically associated with men. Women in these jobs face a dilemma. They can either conform to the expected role behavior of the job and appear unfeminine (which may make their male colleagues, as well as themselves, uncomfortable) or they can follow the expected role behavior of their gender (in which case they may be acting out of character for their occupational role). This dilemma creates a "Catch-22" situation in which women have limited options for achieving acceptance and recognition. For example, the American social scientist Barbara Deckard conducted a survey of male lawyers and found about half saying that women lawyers were "tough and masculine." The other half characterized women lawyers as too "weak and feminine," and said that many go to law school only to "catch a man" (Deckard, 1983:128).

Statistical Discrimination Women and minority workers may also suffer from **statistical discrimination.** This sort of discrimination arises when an employer bases decisions on the *average qualifications* of a group or entire category of people, rather than on an individual's qualifications. Thus, an employer may choose a white high school graduate over an Hispanic graduate

---

BOX 5.3   Barriers to Women Attaining Top Management Positions

Pressures for total dedication sometimes serve to exclude women from employment as managers. Women have been assumed not to have the dedication of men to their work, or they have been seen to have conflicting loyalties, competing pulls from their other relationships. Successful women executives have often put off marriage until rather late so that they could devote their time during the important ladder-climbing years to a single-minded pursuit of their careers.

Concerns expressed by men in management about the suitability of women for managerial roles reflect these themes. Questions about turnover, absenteeism, and ambition are frequently raised in personnel meetings at Indsco. The issue behind them often has to do with marriage.

The question of marriage is experienced by some women in professional, managerial, or sales ladders at Indsco as full of contradictory injunctions. Sometimes they got the message that being single was an advantage, sometimes that it was just the opposite. Two single women, one of them forty, in quite different functions, were told by their managers that they could not be given important jobs because they were likely to get married and leave. One male manager said to a female subordinate that he would wait about five years before promoting a competent woman to

see if she "falls into marriage." On the other hand, they were also told in other circumstances that married women cannot be given important jobs because of their family responsibilities: their children, if they are working mothers; their unborn children and the danger they will leave with pregnancy, if currently childless. One woman asked her manager for a promotion, to which he replied, "You're probably going to get pregnant." So she pointed out to him that he told her that eight years ago, and she hadn't. A divorced woman similarly discussed promotion with her manager and was asked "How long do you want the job? Do you think you'll get married again?" One working mother who had heard that "married women are absent more," had to prove that she had taken only one day off in eleven years at Indsco.

A male manager in the distribution function who supervised many women confirmed the women's reports. He said that he never even considered asking a married woman to do anything that involved travel, even if this was in the interests of her career development, and therefore he could not see how he could recommend a woman for promotion into management.

SOURCE: Rosabeth Moss Kanter, 1977, *Men and Women of the Corporation.* New York: Basic Books, pp. 66–67.

---

because of a belief that white high school graduates *on average* are better qualified. A highly qualified Hispanic job candidate may thus suffer discrimination based on stereotypes about Hispanic job candidates as a group. Statistical discrimination need entail no prejudice on the part of the employer, only a willingness to attribute a group's average characteristics to every member of that group. When past discrimination has been pervasive, as in the case of African-Americans and Hispanics, many group characteristics (such as education, training, and work histories) will on average favor the majority group. The very real disparity between group's average characteristics makes statistical discrimination particularly hard

to eradicate. Statistical discrimination is especially likely in situations where employers do not take the trouble to carefully assess applicants' qualifications on a case-by-case basis. Box 5.3 presents an excerpt from a sociological study of barriers to women attaining top management positions due to statistical discrimination.

Gender-Typing of Jobs Female workers face additional hurdles, based on the **gender-typing** of jobs, that make occupational segregation into lower-paying and less secure jobs particularly acute for them (Browne, 1999). Occupations involving helping and serving others, such as nursing and waitressing, have been identified with the female role because women have historically been responsible for these activities within the family. An almost exclusively female labor force characterizes some occupations. Such occupations include secretaries (99%), receptionists (97%), dental hygienists (99%), and child-care workers (97%) (Census, 2000). Unfortunately, such occupations are often relatively poorly paid, and their incumbents are treated as subordinate—a subordination that parallels the traditionally subordinate role of wives in relation to husbands. In other female-typed occupations, such as waitress or receptionist, qualifications of attractiveness and pleasantness may be more important than acquired skills and competencies. Not only are such physical traits fleeting, but they do not accumulate over time as a package of skills, experiences, and abilities necessary for promotion to more responsible, and more highly paying, positions (Williams, 1995).

By contrast, women compose only a very small percentage of the skilled trades: carpenters (1.0%), brick masons (0.3%), electricians (1.5%), machinists (3.7%), and crane operators (0.4%). On the other hand, they fill 25% of lower-paid assembly positions. Table 5.1 shows the overrepresentation of women in the lower ranks of federal civil service jobs and their relative absence from the top ranks.

Female professionals face special problems of gender-typing. One problem results from the channeling of women into a relatively narrow range of professions: nursing, teaching, library work, social work, and other "helping" professions. As a result there is substantial overcrowding in these professions, which leads to lowered wages (Blau et al., 1998). The fact that many female professionals marry male professionals and managers also creates the possibility that geographic moves will be made that facilitate the husband's career rather than that of the wife. Such decisions may be rational to the extent that the husband's job pays more than the wife's. The result, however, may be a cumulative erosion of the wife's career trajectory, which makes future

Table 5.1 Federal Employment of Women, by Civil Service Grade

Grade Groups and Salary Range	Percentage of Women in Group
Senior levels (individually set)	11.1
Grades 13–15 ($52,176–$94,287)	19.0
Grades 9–12 ($30,257–$57,043)	34.7
Grades 5–8 ($19,969–$35,610)	51.4
Grades 1–4 ($12,960–$23,203)	72.6

SOURCE: U.S. Department of Commerce, Bureau of the Census, 2000, *Statistical Abstract of the United States, 1999,* Washington, D.C.: U.S. Government Printing Office.

decisions even more likely to be weighted toward the husband's career.

Gender differences in occupational placement are not well explained by differing educational qualifications. Male and female workers both have an average of 12.7 years of education. Women exceed men in two-year and four-year degrees, but men are more likely to complete advanced degrees—a difference, however, that is rapidly being eroded (see Table 5.2). Women have obviously prepared themselves educationally for the world of work, both in the past and increasingly in the present, but they still experience restricted employment opportunities. Why does gender-based occupational segregation persist in the face of continuing pressures for change?

One reason that gender-based occupational segregation persists is the continuing discrimination against women and minorities described in this section. Women face additional obstacles, however, that arise outside the workplace. The first obstacle is childhood socialization that encourages women to aspire only to certain types of careers. The second is marriage, which results in additional constraints on women's career options, constraints imposed both by husbands and by wives themselves. Even harder to overcome are the constraints arising from bearing and raising children and the assumption that women will take primary responsibility for these activities.

Table 5.2   Women's Educational Attainment (Percentage of college degrees)

Associate	60.5
Bachelor's	55.1
Master's	55.9
MBA	37.5
Law	43.5
Medicine	40.9
Doctoral	39.9

SOURCE: Diana Furchtgott-Roth and Christine Stolba, 1999, *Women's Figures: An Illustrated Guide to the Economic Progress of Women in America,* Washington, D.C.: American Enterprise Institute.

Recent years have also seen increased attention given to the rights of homosexuals in the workplace. Employment discrimination against people on the basis of sexual orientation was not prohibited by the original Civil Rights Act of 1964. Instead, homosexuals are protected by a hodgepodge of federal, state, and local laws that leave many loopholes. Gays and lesbians also have had difficulty getting their partners covered under health benefits and retirement packages that are available to married heterosexual couples. Laws protecting homosexuals and recognizing homosexual marriages vary between states, creating many ambiguities and difficulties for gays and for gay couples in the workplace. In recent years, however, many large corporations have started to extend health coverage to "significant others," regardless of sexual orientation or marital status (Raeburn and Taylor, 1998).

DISCRIMINATION IN PAY AND PROMOTIONS

Equal access to promotions and pay is also an issue for female and minority workers. Many people would like to believe that this is a problem of the past in the United States. Although some progress has been made in opening up access to previously closed occupations, the ratios of black-to-white earnings and female-to-male earnings remain very unequal, even in recent decades. The median family income for blacks remains only about 61% of that for whites; and full-time female workers make only about 75% of the earnings of full-time male workers. In this section we explore the reasons why discrimination in pay and promotions continues to be such a large problem.

Racial Discrimination

Racial discrimination in pay and promotions occurs in a number of ways. Decisions about promotions are sometimes made informally, behind closed doors by an executive group composed predominantly or exclusively of majority whites. Subjective recommendations by immediate superiors play an important role in such decisions, and there are rarely explicit, written standards for such evaluations (Fernandez and Davis, 1999). Because executives feel more comfortable trusting people with whom they have much in common, they are more likely to decide in favor of white candidates over minority candidates. Decisions made in this way rely on stereotypes of minorities and fail to give proper weight to more objective criteria. Such selection becomes more likely as a person moves higher in the authority hierarchy. As a result, minority gains in the professions, management, and the skilled trades have been particularly slow (Baldi and McBrier, 1997).

Overt expressions of racism have declined in the United States (Kluegel and Bobo, 1993). More subtle everyday expressions of racism, however, are still common. These include efforts to marginalize minorities, to identify minorities as the cause of social problems, and to reject complaints by minorities about prejudice and discrimination as invalid (Essed, 1991). Equal treatment for minorities often occurs only when those within an organization actively demand such treatment (Baron, Mittman, and Newman, 1991).

Prior discriminatory practices may also mean that minority workers have less seniority than majority group workers. As a result, even when employers use a relatively objective criterion such as seniority for allocating promotions and layoffs, minority workers will still often be at a disadvantage (Barnett et al., 2000). Thus, the courts have sometimes mandated the use of separate seniority queues for female or minority workers to allocate promotions or layoffs.

Tokenism An additional problem for female and minority workers is that they are highly visible representatives of their group when they enter new occupational fields (Kanter, 1977). In such situations of **tokenism** a worker is under a spotlight and may experience exaggerated pressure to overachieve. There may also be open hostility from majority workers who feel their position threatened by the incursion of "lower-status" workers (Williams, 1987).

As a result of these continuing problems, the position of African-Americans in the United States has improved only moderately, if at all, in recent decades. Black family income grew 1.3% per year in the closing decades of the twentieth century, while white family income grew 1.1% per year, thus reducing the relative income gap. However, because black income was much lower initially, the absolute gap in family income actually widened during this period (Massey, 1996). Currently, black families earn about 61% as much as white families and Hispanic families earn about 60% as much as white families. Table 5.3 shows recent trends in black and white family incomes.

Gender Discrimination

The poverty rate for women is about twice that for men (U.S. Department of Commerce, Bureau of the Census, 2000). What factors account for such a large difference, and how are these similar to or different from the factors involved in racial inequality? The phenomena of statistical discrimination, subjective hiring criteria, and restrictions on training and apprenticeship programs affect women in much the same manner as they affect minorities. Since so many women work in clerical occupations, the consequences of these forms

Table 5.3 Racial Inequality in Family Income

	Median Family Income[a]		Black Income as a Percentage of White
	Blacks	**Whites**	
1965	3,886	7,251	54
1970	6,279	10,236	61
1975	9,476	16,070	59
1980	12,674	21,904	58
1985	16,786	29,152	58
1990	21,423	36,915	58
1991	21,548	37,783	57
1992	21,103	38,670	55
1993	21,542	39,300	55
1994	24,698	40,844	60
1995	25,970	42,646	61
1996	26,552	44,756	59
1997	28,602	46,754	61

[a]Family income is reported in constant (1997) dollars.

SOURCE: U.S. Department of Commerce, Bureau of the Census, 2000, *Statistical Abstract of the United States, 1999.* Washington, D.C.: U.S. Government Printing Office

of discrimination against women are sometimes referred to as a "glass ceiling"—an image that depicts clerical workers being shut out of opportunities to move into male-dominated managerial and professional positions in large modern corporations (Maume, 1999). Several unique barriers also confront female workers. These barriers result from conflicts between women's roles in the family and in the workplace.

Home Duties Many women are expected, and expect themselves, to take principal responsibility for home duties such as cooking, cleaning, shopping, arranging social engagements, and, most importantly, child care. These duties take a tremendous amount of time and energy, leaving less time available for career commitments (Parcel, 1999). Even if a woman does not plan to shoulder these burdens, those in charge of pay and promotions may fear that she will eventually "get married and quit" or "have children and be less willing to give her full commitment to the company"

(Benokraitis and Feagin, 1986). Similar stereo-types do not hinder the promotion of men on the basis of possible marriage or family plans. Indeed, marriage plans by a man may be viewed as exert-ing a favorable influence on his "stability and commitment" (Kanter, 1977:676).

Pregnancy Leave Many women also experi-ence difficulties in arranging their work duties around pregnancy. In 1993, President Clinton signed the Pregnancy Leave Act requiring that employers allow female workers to take a six-week unpaid pregnancy leave. Sometimes women can continue to receive pay during this period under sick leave policies. Most sick leave policies, however, do not cover the six-week minimum recommended convalescence for a normal delivery. As a result, many women risk substantial loss of pay when they give birth—a time when extra bills are sure to accumulate (Waldfogel, 1997). Table 5.4 shows the cumula-tive effects of the special challenges faced by women on their relative earnings.

The Debate over Comparable Worth

The Equal Pay Act of 1964 made it legally possi-ble to challenge situations in which a minority worker is paid less for doing the *same* job. It has been much more difficult to resolve the issue of wages that are *artificially depressed* in an occupation because the work has typically been done by female or minority workers. This issue is partic-ularly important for women because of the extreme occupational segregation they have experienced.

Comparable Rates versus Market Rates The idea that jobs of equal value or contribution should be paid equally is called comparable worth. **Comparable worth discrimination** occurs when companies base their pay on existing market rates for jobs, and such market rates build in discrepancies in pay between jobs that have typically been filled by men and those that have typically been filled by women. These discrepan-

Table 5.4 Comparison of Median Earnings of Year-Round Full-Time Workers, by Sex

	Median Earnings[a]		Women's Earnings as a Percentage of Men's
	Women	Men	
1955	$ 2,719	$ 4,252	64
1960	3,293	5,417	61
1965	3,823	6,375	60
1970	5,323	8,966	59
1975	7,504	12,758	59
1980	11,197	18,612	60
1985	16,252	24,999	65
1990	20,586	29,172	71
1991	21,376	31,092	70
1992	22,167	31,012	70
1993	22,276	31,173	70
1994	22,388	31,334	70
1995	22,497	31,496	71
1996	23,710	32,144	74
1997	24,973	33,674	74

[a]Median earnings is reported in current (real) dollars.

SOURCES: U.S. Department of Commerce, Bureau of the Census, 1991, *Money Income of Households, Families, and Persons in the U.S., 1990*, Washington, D.C.: U.S. Government Printing Office. Also, U.S. Department of Commerce, Bureau of the Census, 2000, *Statistical Abstract of the United States, 1999*. Washington, D.C.: U.S. Government Printing Office.

cies may have little to do with the skill required in the job. The concept of comparable worth is not really new or revolutionary. Job evaluations that establish the comparable worth of different positions are widely used by large enterprises in both the public and private sectors. "Job evalua-tion consists of a formal set of procedures for hierarchically ordering jobs on the basis of their relative skill, effort, responsibility, and working conditions for the purpose of establishing relative pay rates" (Hartmann, Roos, and Treiman, 1985:5). The criteria used may include the edu-cation or intelligence required, on-the-job train-ing, verbal aptitude, numerical aptitude, com-plexity of the task, perceptual skills, dexterity, motor coordination, physical strength, and such social skills as speaking persuasively, supervising,

Table 5.5   Job Evaluation Points and Gender Differences in Pay
(Selected results of the Washington State comparable worth study)

		Average Annual Salary, 1983–1984	
Job Title	Job Evaluation Points	Male Dominated	Female Dominated
Warehouse worker	97	$17,030	
Delivery truck driver	97	19,367	
Laundry worker	105		$12,276
Telephone operator	118		11,770
Data-entry operator	125		13,051
Intermediate clerk typist	129		12,161
Civil engineering technician	133	18,796	
Library technician	152		13,963
Licensed practical nurse	173		14,069
Auto mechanic	175	22,236	
Maintenance carpenter	197	22,870	
Secretary	197		14,857
Chemist	277	25,625	
Civil engineer	287	25,115	
Senior computer systems analyst	324	24,019	
Registered nurse	348		20,954
Librarian	353		21,969

SOURCE: Helen Remick, 1984, "Comparable Worth and Wages: Economic Equity for Women." Manoa, Hawaii: Industrial Relations Center, University of Hawaii at Manoa, Table 3. Reprinted by permission of the Industrial Relations Center, University of Hawaii at Manoa.

instructing, negotiating, and mentoring. About 30% of the difference in pay between men and women would be eliminated if such skills were consistently used in determining pay rates for men and women (England, 1992).

The legal basis for challenging comparable worth discrimination has been debated in a series of court cases dating back to the 1970s and continuing into the 2000s. In 1983, the proponents of comparable worth achieved a major victory when a U.S. federal district judge found "overwhelming" evidence that "intentional and pervasive" discrimination had caused jobs held mostly by women to be paid an average of 20% less than jobs held by men in Washington State public sector employment (Walsh, 1985a). The total settlement for the 15,000 state employees involved approached $1 billion, including back pay and benefits. This and subsequent cases have been

important in forcing employers to reexamine their pay scales and eliminate blatant differences between predominantly male and female jobs. Some of the results from the job evaluation study that established the existence of comparable worth discrimination in the Washington State case are presented in Table 5.5.

Legal debates on the implementation of comparable worth standards continue to this day and involve two primary issues. First, the criteria used to establish job comparability can be endlessly debated. Second, employers can contend that they are merely offering the *prevailing wage* for a particular job. The latter argument, which places the responsibility for inequality outside any particular employer, has sometimes been accepted as part of a successful defense in comparable worth cases; however, it fails to answer the question of how market rates originally got set (England,

1992). Proponents of comparable worth argue that these market rates embody past discrimination. They argue that comparable worth discrimination should be determined by discrepancies between the measured value or skill of an occupation and its pay, not by disparities between a specific employer's pay scale and prevailing market rates for the work.

The future of comparable worth is uncertain. The courts have not established a clear precedent for ruling in favor of it. In addition, there is a movement toward demanding evidence of *intent* to discriminate, which makes defenses based on the concept of prevailing market wages increasingly acceptable. The U.S. Congress seems unlikely to pass additional legislation on this topic, especially given the failure of the Equal Rights Amendment. In their role as employers, local governments have for the most part taken a wait-and-see approach, because it is both easier and less costly than addressing the problem directly. On the other hand, unions that represent substantial numbers of women in public employment can be expected to continue to be mainstays in the campaign for comparable worth. Lawsuits based on comparable worth discrimination will continue. These lawsuits are at least partly responsible for the improvement in women's relative wages evidenced in recent years (see Table 5.4) (Guthrie and Roth, 1999).

Sexual Harassment

A final problem confronting women in the workplace is their treatment as sexual objects. This problem can range from embarrassing jokes and banter to overt propositions and demands for sex. The EEOC defines **sexual harassment** as repeated, unwelcome behavior with a sexual content when submission to such behavior is explicitly or implicitly a condition for the person's hiring or for other employment decisions, or when such behavior creates a hostile, intimidating, or offensive working environment.

How common is sexual harassment in the workplace? Studies of sexual harassment report that, across their careers, as many as a fifth of women have experienced a severe form of sexual harassment, such as sexual blackmail (demanding sex in order to get a job or a promotion). As many as half have experienced moderate forms, such as sexual propositions, verbal innuendoes, or degrading remarks based on gender (Fletcher, 1999). The case of a young file clerk provides a typical example of such "moderate" forms of harassment. Her supervisor regularly asked her to come into his office "to tell [her] about the intimate details of his marriage and to ask what [she] thought about different sexual positions" (Benokraitis and Feagin, 1986). Box 5.4 describes the sexual harassment of rookie policewomen by their male co-workers.

Based on her study of an underground coal mine, Yount (1991) developed a typology of three strategies that female coal miners used to confront sexual harassment. "Ladies" sought to cast their co-workers as gentlemen and socially withdrew when they confronted offensive behavior. "Flirts" interacted with men in ways considered to be seductive. This style encouraged come-ons from men and sometimes resulted in more severe harassment if the men perceived that the women were using this strategy to gain preferential treatment. "Tomboys" emphasized their occupational roles as miners and engaged in joking relationships with their male co-workers. Tomboys experienced a great deal of sexual "razzing" but this was often intended to be friendly and inclusionary. Tomboys, who resisted being placed in traditional female roles, seemed to be the most successful at minimizing harassment while being accepted as capable workers. In contrast, the adaptations of Flirt and Lady, which relied on traditional female role behaviors, provoked either active harassment or reduced respect.

Women in low-status occupations who are in regular contact with men in higher-status occupations (such as secretaries in relation to managers) are the most likely targets for sexual harassment. Women in these relatively powerless positions also have the fewest options in responding. They are less likely than women in higher-status jobs to

BOX 5.4   Sexual Harassment of Policewomen

Despite her eventual acceptance as an officer, Ann is not "one of the boys"; her incorporation in the informal social networks of the department, which is vital to success, did not happen as readily and remains more tenuous than Tim's. Unlike Tim, Ann could not join the men's sexual joking or banter because she was often the object of their remarks. . . .

Ann's social life does not revolve around police officers. She maintains a protective social and emotional distance between herself and other officers. . . . Another policewoman observed that one way to behave in a "professional" manner and maintain respect is to avoid being on a first-name basis with other officers. However, this makes the policewoman "different" since the men are on an informal first-name basis with each other. Policemen I had never met addressed me as "sweetheart" and "dear." By the end of one tour of duty one officer was calling me "hon" and others called me "Suzie" although I had introduced myself as Susan. Several of the men got quite upset when, in response to their question as to whether I was Miss. or Mrs. Martin, I replied that "I'm Ms. Martin." They were unable to relate to me comfortably until they knew whether I was "taken" or "available," assuming that unmarried meant available. . . .

Male officers frequently told me "that uniform does nothing for you," although it obviously conveyed an authority on the street that I lacked in civilian clothes. What was meant, of course, was that it was not sexually alluring and that I would be more attractive if I were not wearing a man's uniform. . . . Others commented on my appearance out of uniform; when I wore a dress for a court appearance, I received a stream of compliments. I wore a skirt to the station one summer day; while waiting for the officer who I was to interview, he and a sergeant discussed my legs. Their remarks made me self-conscious; I had become an object to be discussed rather than a professional person there to do a job. . . .

One female officer reported that when signing out, several officers looked at her chest and remarked, "How long do you think she can float?" as others stood chuckling. She noted, "I know this is sexism and I try to put them in their place, but if I get snippy they'll get me for insubordination." Other women are less concerned with reprisals. One was greeted as a newcomer to the district by the comment: "Officer, I don't mean any harm but I just want you to know that you have the biggest breasts I've ever seen on a policewoman." She replied:

> Officer, I don't mean any harm but I just want you to know that you don't comment on my anatomy in any shape, form, or manner unless I comment on yours. . . . When you see me coming just act like you don't see anything. Don't speak to me and I won't speak to you.

respond assertively and more likely to attempt to placate the person who is harassing them. Similarly, women who are perceived as less powerful are more likely than other women to be harassed (Wilson, 2000). Single and divorced women are more likely to experience harassment than married women. Younger women and minority women are also likely to be targeted (Fletcher, 1999).

Often there are shades of gray between normal, acceptable warmth and sexual harassment. When does an arm around the shoulder cross the line between a show of affection and support and a come-on? When does an off-color joke go beyond harmless banter to become offensive and derogatory? Often the difference lies in a power differential between the parties involved. Greater awareness of the seriousness of sexual harassment and its ramifications for creating a hostile and unproductive work environment may make it possible to enjoy a friendly and supportive workplace in which sexual pressure and innuendo do not make some workers uncomfortable.

Managing the Diverse Workforce of the 2000s

The workforce of the United States in the twenty-first century is more diverse than at any point in history. This diversity represents not only a challenge but also a potential asset. Diverse people bring different contributions to the workplace. In addition, our long history of attempting to absorb and treat fairly people of different cultural backgrounds can become a social foundation for dealing successfully with people of other nations and cultures in an increasingly interconnected world economy. Capitalizing on this potential, however, will not necessarily be easy or automatic and will require sustained effort on the part of all of us (Fernandez and Davis, 1999; Pelled et al., 1999).

UNEMPLOYMENT

Unemployment poses a significant threat for majority workers as well as minority workers. The unemployment rate in the United States has stayed at relatively high levels during much of the period since the end of World War II. It ranged between 3% and 6.8% through the 1950s and 1960s. During the 1970s it varied between 5% and 8.5%, and in the 1980s and 1990s it reached levels close to 10%. The consistently high unemployment rates of the 1980s and 1990s contributed to growing problems of homelessness in the United States. The early years of the 2000s have seen lower rates, but unemployment rates in some occupations and regions, and among some demographic and age groups, remain high. The possibility of high unemployment rates is a chronic concern for policy makers and for the many citizens who are most vulnerable to unemployment. World unemployment rates also remain high, especially in less economically developed countries. Table 5.6 reports trends in unemployment rates since 1950.

As you learned in Chapter 2, people who are actively looking for work but unable to find it are

Table 5.6   Changes in U.S. Unemployment Rates,[a] 1950–2000

Year	All Civilian Workers	White	Black	Hispanic Origin
1950	5.3	4.9	8.9	—
1955	4.4	3.9	7.1	—
1960	5.5	5.0	9.1	—
1965	4.5	4.1	7.4	—
1970	4.9	4.5	8.2	—
1975	8.5	7.8	14.2	12.2
1980	7.1	6.3	14.3	10.1
1985	7.2	6.2	15.1	10.5
1990	5.5	4.7	11.3	8.0
1991	6.7	6.0	12.4	9.9
1992	7.5	6.7	13.9	11.1
1993	6.9	6.2	12.8	10.2
1994	6.1	5.5	11.3	9.0
1995	5.6	4.9	10.4	9.3
1996	5.4	4.7	10.5	8.9
1997	4.9	4.2	10.0	7.7
1998	4.5	3.9	8.9	7.2

[a]Unemployment rates are calculated as a percentage of the civilian labor force.

SOURCE: U.S. Department of Labor, Bureau of Labor Statistics, 1989, *Handbook of Labor Statistics*. Washington, D.C.: U.S. Government Printing Office; and U.S. Department of Commerce, Bureau of the Census, 2000, *Statistical Abstract of the U.S., 1999*. Washington, D.C.: U.S. Government Printing Office.

counted as unemployed in government statistics. In 2000, the unemployment rate in the United States was 4.4%; thus, at any given time in 2000 about 6 million people were actively, but unsuccessfully, seeking work. Some unemployment results from unavoidable delays between jobs, such as when a worker moves to a new location. Such **frictional unemployment** is said to account for an unemployment rate of about 3%. Frictional unemployment could be reduced by programs designed to match job openings more quickly with qualified candidates. During World War II the unemployment rate fell to 1.6%, perhaps a more realistic minimum level of frictional unemployment. The greatest share of unemployment, however, stems either from a chronic gap between the number of jobs available and the number of people seeking work or from short-term downturns in the economy. The chronic gap between the number of jobs the economy provides and the number of people seeking work is called **structural unemployment** (BLS, 1998). This gap between available work and people needing work widens in times of economic downturn. This latter component of unemployment is called **cyclical unemployment.**

Published unemployment rates understate the true number of people who would work if they were able to find employment. Many people give up actively seeking employment because of discouragement and an absence of opportunities. These potential workers, sometimes called **discouraged workers,** desire work and are available to start work immediately, but are not actively looking for work because they are in school, ill, disabled, keeping house, or are convinced they cannot get a job. Adding these discouraged workers to those actively seeking employment would increase the unemployment rate by about half.

Occupational Differences in Unemployment Unemployment is distributed unequally across occupational groups with the least skilled workers facing the highest rates of unemployment. In 1998, when the overall unemployment rate stood at 4.5%, the rate for professionals and managers was 1.8%; for clerical workers, 3.7%; for craft workers, 4.2%; for service workers, 6.7%; and for machine operatives, 6.7%. The rate reached a high of 14.2% for construction laborers, who face high levels of seasonal unemployment (Census, 2000).

Differences between Social Groups Unemployment is also unequally distributed across social groups. The rate for African-Americans and Hispanics is about twice that for whites. Youth unemployment poses a special problem in industrial societies. The unemployment rate among those sixteen to nineteen years old is two to three times as high as the overall rate. Unem-

ployment among young people has had devastating consequences: "The incidence of unemployment among youth and the difficulties which youth face in entering the world of work continue at levels which are unacceptable. The consequences, in terms of antisocial behavior, alienation, lost output, and reduced social mobility, impose both short- and long-term consequences that cannot be ignored" (National Commission for Employment Policy, 1979). Unemployment among minority youths reached the astonishing level of 30% to 40% in the 1990s.

High levels of unemployment have plagued most of the industrialized nations of Western Europe, including Italy (12.3% unemployment in 1998), France (11.8%), and Germany (7.5%) (Census, 2000).

Layoffs

Unemployment can occur for many reasons. These include quitting a job, being fired, becoming disabled, being laid off, and moving to be with a spouse or loved one. **Layoffs** are a particularly important cause of unemployment because they affect so many people at one time and in one place. Older workers who are laid off may have an especially difficult time finding new employment. Layoffs have become more common as bankruptcies have increased among small companies and as large companies have downsized and become more mobile both within the United States and overseas (Sullivan, Warren, and Westbrook, 2000). Production plants are routinely closed in high-wage areas and reopened in areas where companies can pay lower wages (Bluestone and Harrison, 2000).

Layoffs vary in their duration. Short layoffs are more typical for skilled workers employed in large firms that operate in key industries; longer layoffs, including "indefinite layoffs," are more typical for workers in less skilled occupations employed in smaller, more peripheral firms (Grunberg et al., 2000). The term *indefinite layoff* is frequently a euphemism for permanent termination, and even if a recall is possible at some

---

BOX 5.5   Layoffs as Standard Operating Procedure

One day Carroll called a meeting of everyone in the cafeteria. He informed us that one of the companies we supplied had reduced its order for the next two months and that he would have to lay off about half the women. They would be laid off approximately along lines of seniority but if anybody wanted to, they could take a voluntary layoff. I was surprised how well most of them took it. One of the older women later explained to me that this sort of thing happened two or three times a year. Apparently they had grown accustomed to the instability of working for a subcontractor with a varying workload.

A few of the younger women did not take it so lightly. Most had not been working long enough to collect unemployment. I saw one throw a handful of leads into her machine and curse at Carroll. "I was just starting to get my bills paid, and this no good son-of-a-bitch lays me off with a one-day notice." Many could not afford to wait for a call-back, and took work elsewhere.

SOURCE: Tom Juravich, 1985, *Chaos on the Shop Floor: A Worker's View of Quality, Productivity, and Management.* Philadelphia: Temple University Press, p. 109.

---

future date, few of the laid-off workers can afford to wait indefinitely. Box 5.5 describes possible reactions of workers to layoffs.

Social Consequences of Layoffs Layoffs can have devastating consequences for workers. The loss of income and fringe benefits such as health insurance often results in a rapid deterioration of lifestyle, including potential loss of home, delayed education plans, reduced medical care, and reductions in food and recreation budgets. Speaking about his choice of police work as a career, a young African-American worker reports: "I like the work and I like the job security. When I see others laid off and jobless I feel I made the right choice" (Martin, 1980:70). Workers often feel helpless and worthless in the face of lengthy unemployment. In combination these factors can have severe negative health consequences. Based on a study of how unemployment affects health, the American economists Barry Bluestone and Bennett Harrison (1982:65) projected that a 1% increase in the unemployment rate over a six-year period results in 37,000 total deaths (including 920 suicides and 650 homicides), 4,000 state mental hospital admissions, and 3,300 state prison admissions.

Layoffs can have devastating consequences for workers: Very little has gone right for

Harold Johnson, 38, since he was laid off by GM. . . . He has not held another job for more than a few months since.

After briefly working at a seasonal job at a boatyard, Johnson worked at a small factory in rural Michigan that was on strike, but he was laid off when the strike ended. Later, he briefly worked for a small recreational vehicle manufacturer, but the firm quickly went bankrupt. He went to a trade school to become a furnace repairman, but could not find any work in the field because he lacked experience. . . .

In the meantime, his wife has divorced him, and he's had to sell his house to avoid losing it to the bank and he has developed high blood pressure.

"I'm out there looking, but it has sure been discouraging" says Johnson with quiet understatement. (Risen, 1984:I19)

In the 2000s, layoffs in the United States have been particularly severe in certain industries, such as aerospace, which has experienced increased competition with the European confederation of companies manufacturing Airbus.

Careers, marriages, families, and communities are all vulnerable to the devastating effects of layoffs. Higher unemployment rates produce increases in family violence, family breakups, and divorce. Community services such as education, parks and recreation services, police protection, and street repair all typically deteriorate in the face of large-scale layoffs.

Replacement Jobs? Some jobs lost due to lay-offs reappear in other parts of the country as plants close in the Northeast and Midwest and reopen in the South and Southwest. Some workers regain employment by following these jobs, almost always at significant personal costs of severed ties with their families, friends, and communities. Unfortunately, many laid-off workers do not have the qualifications for the new jobs. As workers move out of the unionized Northeast and Midwest and into the predominantly nonunionized South and Southwest, they may also find their wages dramatically reduced. In 2000, annual earnings in Michigan averaged $33,761, whereas in Mississippi annual earnings averaged only $23,772 (Census, 2000). Workers may also find social services in the new industrial boomtowns, as well as the overall quality of life, to be lacking.

Other jobs lost through layoffs reappear overseas when American companies move their operations to other countries. The volume of manufacturing activity exported overseas has been staggering. The value of goods and services produced overseas by American multinational corporations is now larger than the gross domestic product of every country in the world except the United States itself (Barnet and Cavanagh, 1994). This export of jobs overseas to lower-wage areas is one of the reasons why unemployment rates in the industrialized nations have risen and stayed at high levels. We discuss the nature of the world economy and its implications for workers in Chapter 16.

Coping with Unemployment

When workers become unemployed, they have a rather limited set of resources to ease their way. Perhaps the most important of these are financial and emotional help from their families (Kenyon, 2000). Some workers also receive **unemployment compensation** from the state. However, only about 78% of workers are covered by unemployment insurance, and for those who are covered, payments expire after a set period of time (generally twenty-six weeks). Unemployment insurance laws vary from state to state, but those not covered typically include farm workers, personal service workers (such as in-house child-care workers), part-time workers, workers who have been employed for fewer than a specified number of weeks, and workers who quit their jobs voluntarily or were fired for cause. As a result, at any given time only about one of four unemployed workers receives unemployment compensation. Some union contracts also provide *supplemental unemployment benefits* that extend the amount or duration of unemployment insurance benefits. If workers are permanently laid off, they may also receive **severance pay,** a lump sum payment to help pay bills until they can find other work. Finally, workers may also be able to receive some form of welfare (public aid). However, this relief provides minimal benefits, is generally unavailable until the worker and his or her family are completely destitute and have sold off nearly all of their possessions, is time limited, and is often contingent on a willingness to accept work at subminimum wages and conditions.

Legislative Controls Laws that help reduce unemployment are almost nonexistent in the United States. A partial exception is provided by the administrative procedures of state unemployment insurance systems. The procedures mandate that a company's unemployment insurance premiums reflect the number of workers that it has laid off in the past. High premiums thus discourage companies from casually laying off workers.

Some states have also enacted more specific legislative measures to discourage layoffs. Several states have enacted plant-closing legislation that limits the ability of companies to close plants without providing substantial advance warning,

BOX 5.6   The World of the High-Rise Steel Worker

The pile-driving crew was sheet piling a sewer trench 25 feet deep. Steel I-beams were used to brace the pilings and were placed at the top of the sheets across the trench. The men had to go back and forth across the I-beams in order to get from one side of the trench to another. On the day of the accident, the hammer that drove the sheets into the ground (and eventually into rock 25 feet below) was hung up on the crane boom and the operator could not release it. It was dangerous, for if the weight dropped out of control, it could pull the crane over onto the men below. When the foreman saw the situation, he started running across one of the I-beams, slipped, and went down. Directly below him was a

foundation for a manhole with reinforcing rods sticking up in the air.

The foreman luckily missed the steel rods, where he would have been impaled. Instead, he landed hard on a steel mat of crisscrossing reinforcing rods, broke six ribs, and punctured his lung. The problem was how to get him out of the trench. When the fire department ambulance arrived, he was wrapped in blankets and strapped into a stretcher. Then the men in the crew rigged a cable and hook to the stretcher and lifted him out with the crane. He never returned to the job.

SOURCE: Herbert A. Applebaum, 1981, *Royal Blue: The Culture of Construction Workers.* New York: Holt, Rinehart and Winston, p. 79.

compensating workers, and sometimes even compensating communities for their investments in the company's success. There has even been precedent for awarding claims to workers for unjust layoffs where such plant-closing legislation does not exist. In California, 537 laid-off workers at an Atari home computer manufacturing plant were awarded four weeks' pay and legal fees on the basis of a little-cited provision in the state's labor code requiring advance notice of termination of employment (Walsh, 1986:7). Atari had repeatedly assured workers that their jobs were secure, but it suddenly closed the plant and moved its computer assembly operations to Hong Kong. The workers were told at mid-morning on a Tuesday that they could pick up their final checks that Friday at a local high school and then were made to leave by the back door (Lewin, 1984:G6). California Superior Court Judge Peter Stone ruled that Atari's repeated assurances to workers constituted a contractual obligation to provide advance notice of termination.

It is very likely that high levels of unemployment will continue to plague much of the world economy for the foreseeable future. We explore the reasons for this further in Chapter 8.

HAZARDOUS WORK AND DISABILITY

Occupational hazards and diseases are such commonplace phenomena that we often take them for granted. Many people's careers, however, are disrupted by job-related accidents and illnesses. In 1998, 28 of every 1,000 workers suffered injuries and illnesses serious enough to permanently or temporarily disable them or even kill them. Over an average work life of forty years, each worker can thus expect to experience at least one serious work-related accident or illness. Each decade about 70,000 workers die as a result of work-related injuries. This means that an average of about 16 workers die each day from work-related injuries and more than 17,000 are injured each day. Although the annual injury and illness rate has declined moderately in recent decades, the number of lost workdays per year has been relatively stable at about 480 lost workdays per 1,000 workers (Census, 2000). This leaves a mixed answer to the question of whether the workplace is becoming safer or more dangerous. Box 5.6 reports an incident at a construction site involving an unexpected fall.

Industrial Accidents

One reason for the uncertainty regarding workplace safety is that there are serious reporting problems with occupational injuries. Many companies require injured employees to come to work and then assign the workers to light or even trivial duties. In this way they disguise the injury and lower the rate of reported injuries and therefore the costs of workers' compensation insurance. Responding to a company claim of 458 consecutive days without a lost-time accident, Bob Walters, a union safety committeeperson at a chemical plant in Pasadena, Texas, reports:

> When I first went to work out there I got a skin irritation from benzene. My hands swelled up. Oh, they were huge! Well, they brought me back out there and set me at the desk in front of the telephone and asked me to answer the phone . . . to keep from listing a lost-time accident. . . . They'll bring the people back in on stretchers if they have to. (Berman, 1986:126)

Researchers estimate that more accurate reporting might double the number of recorded accidents.

Differential Risks across Industries Safety and health risks are not distributed equally across industries. The incidence of injuries ranges from 135 cases per 1,000 workers per year in automobile and truck manufacturing to 22 cases in finance, insurance, and real estate (Census, 2000).

Many accidents in the construction industry occur because of unsafe practices. For instance, a friend of one of the authors was severely burned in an accident as a result of inadequately shielded power lines at an electric power plant construction site. He was hospitalized for five months and experienced excruciating pain. He eventually lost both legs to gangrene because of the severity of his burns.

Chronic Stress Injuries Other workplace injuries are **chronic stress injuries,** which result from such things as using improper equipment, assuming unnatural postures, or performing repetitious tasks with too few breaks and too little variation. Back injuries are the most common of these, accounting for between a quarter and a third of all reported industrial injuries. Tendonitis, a chronic tendon inflammation in the arms and hands, is a common injury among assembly workers. Carpal tunnel syndrome, an inflammation of the nerves in the wrist, is a particularly painful and debilitating injury—one that has experienced a dramatic rise due to increased computer usage.

Noise is another common source of injury. If a noise is loud and sudden, the eardrum can be punctured, resulting in a significant loss of hearing. This damage occurs somewhere around the threshold of pain at about 140 decibels. Even a noise level of 80 decibels, a level well within legal limits and the actual conditions at many workplaces, can cause traumatized nerve endings and temporary hearing loss. Continued exposure can make the loss permanent. Noise also increases stress and contributes to circulation problems, high blood pressure, and heart problems. Figure 5.1 displays average noise levels in offices and their magnitude relative to other common sources of noise.

Other injuries result from the use of vibrating tools such as jackhammers and cutting, drilling, and assembly tools. Excessive vibration can damage the bones, joints, and soft tissues. Prolonged exposure to heat and cold or to various forms of low-level radiation common in many different work settings can also injure bodily organs such as the eyes, liver, and kidneys.

Workers in less industrialized nations frequently labor under even more dangerous conditions. Manufacturing and mining operations in these countries are notorious for excessive heat, noise, and fumes, and lack of adequate ventilation, clean drinking water, and latrines (Lee, 1998). Unfortunately, similar conditions still exist in some work settings in the United States. Dangerous conditions (and illegal work practices) have been routinely reported in recently reemerging sweatshops that manufacture clothing

Decibels

	Decibels	
Threshold of pain	140	
		Hearing damage
Rock band	130	
Jet aircraft at 500 feet	120	
Motorcycle	110	
		Mental stress
Food blender	100	
Loud street noise	90	OSHA limit for 8-hour day
Tabulating machines Noisy office	80	Task interference
Accounting office	70	
Average radio	60	
Average conversation	50	Invasion of privacy
	40	
Private office Quiet conversation	30	
	20	
Whisper	10	

FIGURE 5.1 Noise Levels at the Office

SOURCE: Joel Makower, 1981, *Office Hazards: How Your Job Can Make You Sick.* Washington, D.C.: Tilden Press, p. 135. Used by permission of the publisher.

in major cities such as Los Angeles and New York (Waldinger, 1996). Examples of unsafe practices are also common in the agricultural sector, which has been systematically excluded from occupational safety and health regulations and from many other laws protecting workers' rights. We discuss the plight of agricultural workers further in Chapter 8.

Occupational Diseases

Even more serious than the underreporting of industrial injuries is the underreporting of **occupational diseases** (Rosner and Markowitz, 1991). This underreporting results from their frequently delayed onset and from the fact that they may have multiple contributing causes. For example, exposure to carcinogenic chemicals may not lead to the development of cancer until twenty years later. By then, it may be difficult to separate the effects of on-the-job exposure from later health problems or from the effects of smoking or diet. By some estimates as many as nine of ten occupational diseases go unreported as such (Steinman and Epstein, 1995).

New Chemicals One reason why occupational diseases have come to play such a significant role in the total injury and illness picture is that over 5,000 new chemicals are introduced into the workplace each year. There are advisory standards for only about 500 chemicals and regulations by the Occupational Safety and Health Administration (OSHA) for only a handful (*www.osha.gov*). Awareness of occupational diseases as a significant health problem, however, predates modern scientific knowledge of the effects of chemicals on health. People used the expression "mad as a hatter" long before scientists discovered the effects of mercury, which was used in hat felting, on the central nervous system.

Industrial chemicals can cause immediate acute reactions such as burns, strong allergic reactions, seizures, and loss of consciousness, as well as chronic reactions that may take years to emerge. Chemicals enter the body mainly through the lungs, where a large surface of tissue is exposed to the air. Some highly solvent chemicals, however, such as benzene, can be absorbed through the skin. Other heavier and less soluble toxins, such as lead, mercury, and arsenic, enter the body only through the mouth and digestive tract through food, dust, or other contaminants.

Once a pollutant enters the body, a number of damaging effects can occur, depending on the nature of the chemical. These effects include severe scarring of the lungs; breakdown of the red blood cells, resulting in a reduction in the blood's oxygen-carrying capacity and possible damage to any or all of the major organs; accumulation of chemicals in the kidneys, liver, or central nervous system with resulting damage to these organs; and, perhaps the most feared possibility, cancer in

Table 5.7 Cancer-Causing Chemicals in the Workplace

Chemical	Target Organs	Other Chronic Health Effects	Workers at Risk	Excess Risk Ratio for Cancer	Number of Workers Exposed Annually	Latency Period for Cancer (Years)
Asbestos	Lungs, gastrointestinal tract	Asbestosis, anorexia, weight loss	Miners; millers; textile, insulation, and shipyard workers	1.5-2X	4,122,000	4–40
Benzene	Bone marrow (leukemia)	Central nervous system and gastrointestinal effects, blood abnormalities (anemia, leukopenia)	Explosives, benzene, and rubber-cement workers; distillers; dye users; printers; shoemakers	2-3X	3,900,000	6–14
Chromium	Nasal cavity and sinuses, lung, larynx	Dermatitis, skin ulceration, nasal system ulceration, bronchitis, pneumonia, inflammation of the larynx and liver	Producers, processors, and users of Cr-acetylene; aniline workers; bleachers; glass, pottery, and linoleum workers; battery makers	3-4X	1,675,000	5–15
Arsenic	Skin, lung, liver, lymphatic system	Gastrointestinal disturbances, hyperpigmentation, peripheral neuropathy, hemolytic anemia, dermatitis, bronchitis, nasal system ulceration	Miners; smelters; insecticide makers and sprayers; chemical workers; oil refiners; vintners	3-8X	1,500,000	10+
Nickel	Nasal cavity and sinuses, lung	Dermatitis	Nickel smelters, mixers, and roasters; electrolysis workers	5-10X (lung) 100X (nasal sinuses)	1,425,000	3–30

SOURCES: Devra Lee Davis, Kenneth Bridbord, and Marvin Schneiderman, 1981, "Estimating Cancer Causes," pp. 285–316 in R. Peto and M. Schneiderman (editors), Banbury Report #9, *Quantification of Occupational Cancer*. Cold Spring Harbor, N.Y.: Cold Spring Harbor Laboratory, Table 1, page 287. Reprinted by permission of Cold Spring Harbor Laboratory and the authors.

parts of the body where the chemical accumulates. The fetus is particularly vulnerable to chemical assaults. Levels of chemical contamination that would pose low risks for adults may pose significant risks for the fetus. As a result, pregnant women have sometimes been excluded from working with certain chemicals, leading to debates about whether to clean up the workplace for all workers or to exclude pregnant women from dangerous processes. Table 5.7 lists some of the most common cancer-causing chemicals used in the workplace and the workers exposed to these chemicals.

Asbestos Asbestos provides a well-known example of chemical contamination. Asbestos is a virtually indestructible inorganic fiber that was once considered a "miracle mineral" because it has so many possible uses. It can be spun and woven for making fire-resistant textiles, used as a strengthening agent in concrete and car brakes, and used as a strong, lightweight, and fire-retardant component in building materials and insulation. As early as 1918, however, life insurance companies in the United States and Canada stopped insuring asbestos workers because of a severe lung-scarring disease called asbestosis. Nothing was done to regulate the use of asbestos, however, until the 1970s when asbestos was first recognized as a carcinogen. In 1972, the standard for the presence of asbestos in the air at the workplace was set at 2 fibers per cubic centimeter of air (Brodeur, 1985). In 1980, the government proposed changing the standard to .1 fiber per cubic centimeter, but the revised standard was never enacted due to resistance from the asbestos industry.

Asbestos continues to be used in a variety of products; however, its use has dropped significantly in recent years as health concerns, and the death toll, have risen. Asbestos production in the United States fell from 875,000 metric tons in 1973 to 9,000 metric tons by 1984. Current fears center on the past use of asbestos as insulation in schools, public buildings, and homes. In a survey of just ten cities, the Environmental Protection Agency found that 733,000 public and private buildings contained asbestos in one form or another. Leaving the asbestos in place as the packaging and binding materials deteriorate guarantees future problems. However, disturbing the asbestos by removing it from walls and ceilings can dramatically increase the current health risks. It is estimated that as many as 100,000 workers died as a result of inhaling asbestos while installing insulation. Many more will die as a result of removing it (Dahl, 1986).

Hazards in High-Technology Workplace
The high-technology workplace of the future has its own health hazards. Workers in high-technology settings are exposed to a variety of chemicals. The California Department of Labor Statistics reports that electronics workers suffer from job-related illnesses at more than four times the rate of other manufacturing workers (Martinez and Ramo, 1980:12). Electronics workers are routinely exposed to arsenic and arsine gas used in the manufacture of microchips, among other hazards. These dangers are compounded by the fact that workers confront so many different chemicals. Because of chronic exposure to toxic chemicals, workers in the electronics industry sometimes develop a condition called

chemical sensitization. This condition is illustrated in the following example:

> The rash can break out at any time: when she walks down the aisle in the grocery store that holds the laundry detergents and soaps; when she puts on a pair of pants fastened with a metal clasp. . . .
>
> During the twelve years that she worked as an assembler for Varian Associates, a Palo Alto, California, manufacturer of scientific instruments, she was regularly exposed to some twenty-five different chemicals ranging from organic solvents like xylene and freon to acids, epoxies, and metal fumes. . . . She is so thoroughly "sensitized" that she can no longer work in the electronics industry. (Howard, 1985:139)

Multiple chemical exposure is also thought to be at least partly responsible for what is sometimes called *mass psychogenic illness.* Workers in electronics firms sometimes simultaneously experience fainting, exhaustion, and other symptoms of distress. Pressure, heat, and long hours are important causes of the outbreak and spread of such phenomena. Some observers have argued that the rapid spread of these outbreaks indicates that they result from "hysteria" among the electronics workers, most of whom are women. More cautious interpretations look at the compound effects of the many chemicals present in electronics factories, effects that are difficult to discover when each chemical is tested separately.

Despite the potential severity of safety and health problems in the electronics industry, semiconductor companies spend less than 1% of their revenues on safety and health programs, in contrast to the 2.5% to 3% spent by firms in other industries (Eisenscher, 1984). This low level of spending may well contribute to the underrecognized, and very likely underreported, safety and health problems in the industry.

VDT Hazards   Other high-technology hazards are associated with the use of computer video display terminals (VDTs). Potential problems from the use of VDTs include eyestrain, chronic

---

BOX 5.7   Proposed Standards for VDT Use

9to5 Bill of Rights for the Safe Use of VDTs

9to5 recommends the following protections while research to determine the causes of problem *pregnancies* is conducted and long-term solutions are developed:

- Workers should have the right to transfer away from VDT work to other work within the company during the course of pregnancy, without loss of pay, seniority, or benefits.
- VDT equipment should be made safe for all workers by the manufacturer through inexpensive metal shielding. Machines already in use should be retrofitted with shields in order to eliminate any possible radiation emissions.

For the safety of *all workers,* including pregnant workers, 9to5 recommends the following standards:

1. All workers should have a rest break of 15 minutes for every two hours of VDT work, or 15 minutes for every hour of intense VDT work.
2. Jobs should be structured so that operators are not required to use a terminal for more than 50% of their working day.
3. All VDT equipment should have detached keyboards, non-glare screen surfaces, brightness and contrast controls, tilt adjustability, characters which are large and sharp enough to read easily, and absence of visible flicker.
4. Furniture, lighting, and work environments should be designed for the comfort and safety of the operator.
5. Employers should provide regular eye examinations for all VDT workers, which should include tests recommended by the American Optometric Association.
6. Stress induced features of automated jobs such as machine pacing or computer monitoring should be eliminated.
7. All VDTs should be regularly cleaned and serviced in compliance with the manufacturer's recommendations.
8. Further research into all potential hazards of VDTs should be conducted without delays.

Some employers have developed guidelines for their own VDT use. But legislation is the most effective way to ensure safe working conditions for *all* VDT operators in *all* workplaces.

SOURCE: Reprinted by permission of 9to5, National Association of Working Women, 614 Superior Ave., NW, Cleveland, OH 44113, (216) 566-9308.

---

muscle fatigue, and radiation exposure, leading to disturbed sleep, headaches, dizziness, indigestion, high blood pressure, heart disease, peptic ulcers, and nervous disorders (Morgensen, 1996). A study of data-entry operators indicated that 60% develop painful pressure points at tendons, joints, and muscles, whereas fewer than 10% of traditional office workers develop such problems. "Self-reports of 'almost daily' pains in the neck, shoulders, right arm and right hand show high incidence in the data–entry group, and low in the control group of traditional office workers" (Fossum, 1983:92–93).

There have also been concerns about the effects of VDTs on pregnant women because of potential low-level radiation emission. These concerns have forced VDT manufacturers to build in shields to limit radiation exposure. Additional safeguards include new screen designs that limit glare and increased attention to office ergonomics—the study of the implications of body posture at work for health and injuries. Conditions have thus improved but long hours spent at terminals for increasing numbers of workers, starting at the very earliest ages of childhood, suggest the possibility of even greater problems in the future. Box 5.7 presents information from a handbill circulated by the National Association of Working Women proposing health and safety standards for workers using VDTs.

"Sick Buildings" The Environmental Protection Agency has estimated that about one-sixth of commercial buildings in the United States pose significant health hazards to those who work in them due to **indoor air pollution.** The problems arise from the construction of energy-efficient but airtight buildings. Dust, vapors, chemicals, molds, and bacteria are the main contaminants and lack of adequate external ventilation is the underlying cause. Important indoor air pollutants include fumes from photocopying fluids and gases and particulates from insulation. The problems show up as coughs, sinus problems, burning eyes, fatigue, and headaches. Asthma, reduced lung capacity, acute and chronic allergies, and even cancer can follow (Williams, 1993). More and more people will be working in offices in the future, but this trend may not be as comforting in terms of health concerns as one might at first think. Research has suggested that there is potentially more damage to human health from indoor air pollution than from outdoor pollution (O'Reilly, 1998).

Regulating Workplace Safety and Health

Many safety and health problems could be reduced or eliminated by more effective regulation of industrial chemicals and processes. The principal agency mandated to regulate workplace safety and health in the United States is the Occupational Safety and Health Administration (OSHA), created in 1970. Currently, any chemical can be used in the workplace until it has been proven to be dangerous. Proof can take years of laboratory testing, and it may take even longer to observe the chemical's influence on illness rates among those who have been exposed to it. The companies that use the chemicals also routinely challenge evidence that they are hazardous. Only large workplaces are regularly inspected and the vast majority of workplaces are never visited at all (Weil, 1991).

Right-to-Know Legislation In the United States, workers do not even have the right to know the names of the chemicals with which they are working, nor are companies under any legal obligation to share information about the potential health hazards posed by these chemicals. A common attitude seems to reflect that of a company doctor for Johns-Manville Asbestos Company, who recommended that a worker suffering from asbestosis "should not be told of his

condition so that he can live and work in peace and the company can benefit by his many years of experience" (Brodeur, 1985:45). Workers in some states and cities have pushed for **"right-to-know" legislation,** which would require that chemicals be clearly labeled and that information be made available on their health hazards. In other states, workers have attempted to raise workers' compensation benefits to pressure companies to adhere to better safety standards.

Most other industrialized nations have more rigorous safety standards than the United States. In Sweden, for example, "safety stewards" have the right to shut down production areas that they deem unsafe until the problem has been corrected. Industry, government, and trade union groups in Sweden have taken an aggressive stand against workplace hazards and have acted to anticipate and prevent hazards rather than dealing with them after they have already taken a toll on workers (Green and Baker, 1991). On the other hand, many less industrialized nations have only minimal occupational safety and health standards. These low standards, unfortunately, are often an added attraction to manufacturing companies moving from the industrialized nations.

Stressful Jobs

Job stress affects us all at some point, but for some workers it becomes overpowering. The sources of job stress include physical characteristics of work, such as heat, noise, and cold. Other stressors include time pressures; abusive or incompetent bosses; job changes such as layoffs, demotions, or promotions; excessive responsibility; ambiguity in role demands; and even chronic boredom. Many of these stresses are aggravated by overwork and understaffing, which have become increasingly common in many workplaces as responses to heightened competitive pressures. Dual-career families face additional stresses from workloads that allow little time for family or leisure. Between 1980 and 2000, the average time worked by middle-income families increased from 3,200 hours to 3,600 hours, nearly the equivalent of two full-time jobs (Mishel et al., 1999).

The consequences of job stress are equally diverse. These include anxiety, depression, drug and alcohol abuse, suicide, and antisocial behavior. For perhaps 1% to 2% of the working population, severe mental disorders such as chronic depression result from job-related stress (Neff, 1985:251). It has been estimated that alcoholism, absenteeism, and medical costs related to job stress cost $75 billion to $90 billion each year. There is even evidence that occupational stress can lead to the onset of schizophrenia and severe psychotic breaks from reality (Link, Dohren-wend, and Skodol, 1986). In many cases a "toxic work environment" in which authoritarian managers exert too much control and undercut the individual's sense of accomplishment and dignity are behind the most severe cases of job stress. Such workplaces are capable of creating tremendous stresses that lead even to outbreaks of violence in which workers return to kill their bosses and co-workers (Baxter, 1994).

Stress affects both office and factory workers. The work of telephone operators, who handle as many as 24,000 calls each day, exemplifies many aspects of job stress. Workers generally have to ask permission to go to the bathroom. Sometimes they must wait as long as an hour before being allowed to leave their work station. It is also important not to dally in the bathroom, because management times the bathroom visits. A long-distance operator reports the following effects of job stress: "People just get hyper. You see them; they're shaking. They can't deal with getting to work on time. They get into a fit of crying or something. One woman fainted. It's a pressure cooker in there" (Howard, 1985:85). Telephone operators also suffer from the stress of shift work. Working a night or evening shift frequently results in sleep problems, related nervous disorders, and digestive and gastric problems. Current research suggests that the body does *not* adapt to shift work over time. Other researchers argue that even more significant are the disruption of family and social life associated with shift work (Presser, 2000). Table 5.8 lists some major sources of stress reported by office workers.

Even among occupations that are relatively privileged, stress can be a major hazard. Doctors

Table 5.8 Sources of Office Stress

Source	Rate
Lack of promotions or raises	51.7%
Low pay	49.0
Monotonous, repetitive work	40.0
No input into decision making	35.1
Heavy workload or overtime	31.5
Supervision problems	30.6
Unclear job descriptions	30.2
Unsupportive boss	28.1
Inability or reluctance to express frustration or anger	22.8
Production quotas	22.4
Difficulty juggling home and family responsibilities	12.8
Inadequate breaks	12.6
Sexual harassment	5.6

SOURCE: Joel Makower, 1981, *Office Hazards: How Your Job Can Make You Sick.* Washington, D.C.: Tilden Press, p. 125. Used by permission of the author.

experience a great deal of stress because of the life-and-death decisions they make. As a result, they have a higher than average rate of suicide (Werth, 1999).

Environmental Degradation

Dangerous industrial practices influence everyone, regardless of occupation, through their impact on the environment. Because there is a substantial delay between the introduction of pollutants into the environment and the appearance of their consequences, we do not yet know how serious the effects of current environmental contamination may be. Acid rain, for example, has the potential to devastate much of the remaining forested area in the Northeastern United States and Eastern Canada and, with it, much of the wildlife and fish in the area. Acid rain is precipitation polluted by sulfuric acid. It forms when water vapor in the air reacts with pollutants containing sulfur, such as smoke from power plants that use fossil fuels. Acid rain and other forms of industrial pollution even threaten Lake Superior, one of the largest bodies of freshwater in the world.

Air contamination resulting from industrial production is also widespread. Residents of many large cities suffer chronic irritation and increased rates of respiratory disease because of air pollution, with Houston, Denver, Phoenix, and Los Angeles being among the worst U.S. cities (Godish, 1997). Large cities in industrializing nations often have even worse air pollution because of lack of regulations with Mexico City, Bangkok, Thailand, and Lagos, Nigeria, being among the worst (Census, 2000). Agricultural workers living near the fields where they work routinely report health problems from the improper application of pesticides through aerial spraying. Each year, the Texas Department of Agriculture, for example, receives almost 800 health complaints arising from pesticide use.

Love Canal Residents of Love Canal, New York, an industrial community adjacent to Niagara Falls, gained national notoriety in the early 1980s when toxic chemicals oozed into their yards and homes from metal containers buried in a nearby abandoned canal. The Hooker Chemical Company had buried the containers there decades earlier but inadequate containers and the passage of time allowed the chemicals to make their way back into the environment. The resulting health problems include deafness, respiratory difficulties, convulsions, cancer, stunted growth in children, and unusually high rates of pregnancy problems and birth defects (Levine, 1982). Today, large sections of the town are abandoned and boarded up as testament to the continuing potency of these chemicals.

Bhopal The widespread use of toxic chemicals also sets the stage for large-scale accidents. On December 2, 1984, toxic methyl isocyanate gas escaped from a pesticide plant owned by the Union Carbide Corporation in Bhopal, India, resulting in the deaths of as many as 3,000 people and serious injury to 300,000 more (Shrivastava, 1987). The following year, the Union Carbide plant at Institute, West Virginia, leaked the same toxic gas into the environment. Fewer injuries resulted because of better warning procedures and a lower population density.

Silkwood and Kerr-McGee The Kerr-McGee nuclear fuel plant near Gore, Oklahoma, gained notoriety in the 1970s because of the death of one of its workers, Karen Silkwood. Her life later served as the focus of a dramatic movie on the dangers of radiation contamination. Silkwood, who had experienced plutonium contamination at the plant, was killed in an automobile accident while traveling to bring documents on the contamination to a lawyer handling her case against the company. Radiation leaks at the Kerr-McGee plant have continued to pose health hazards to workers and residents alike. A local pharmacist, William Young, commented on conditions in the area: "You want me to send you a one-eyed catfish? . . . You don't even have to cook it" (Bernstein and Blitt, 1986:25).

Chernobyl A 1986 explosion and fire at a nuclear reactor at Chernobyl, in the former Soviet Union, offers a disturbing vision of the possibilities of large-scale industrial contamination. Not only were hundreds of workers injured and scores killed from the radiation that leaked at the site, but a great deal of radiation also escaped into the atmosphere, injuring unknown numbers of residents and eventually drifting as far away as Northern and Western Europe. According to organizational sociologist Diane Vaughan (1996), chronic accidents of this sort are nearly inevitable given the complicated technological and organizational structure of modern production systems. We can thus expect to see such accidents repeated as part of the predictable future of industrial society (Perrow, 1984).

Living with Disability

Workers who are disabled or who become disabled on the job generally face a difficult process of economic, social, and psychological adjustment (Vickers, 2000). In the United States, there are over 14 million people who are handicapped, over 40% of whom are considered employable. Unfortunately, the unemployment rate of the handicapped is about twice that of others. These disabled workers include both those injured on the job and others who were injured outside of work or who were born with disabilities. The Americans with Disabilities Act of 1990 provided important protections for these individuals and allowed many of them to return to work or to start paid employment for the first time. However, as with the Civil Rights Act of 1964, these protections provide only a legal starting point, not a realized solution (Perlin, 2000).

Training Learning to live with a disability and to return to a productive life requires tremendous effort by the injured or impaired person as well as support from friends, family, and professional helpers. Before injured workers can resume a productive life, they may need physical therapy, vocational training, specialized job placement services, counseling, and other help to overcome physical or mental limitations (Perlin, 2000). Such help can be expensive and difficult to find. Adding to these problems, less than 60% of workers in the United States are covered by group health plans at work and less than 40% are covered by disability plans (Census, 2000). Handicapped workers also experience hiring and pay discrimination from some employers. It has been estimated that approximately half of the 40% reduction in wages experienced by handicapped workers is due to discrimination rather than reduced capacity (Johnson and Lambrinos, 1985).

Social Support One of the most important resources for workers who have been injured or who experience overpowering stress on the job is the social support they receive from friends, family, and co-workers. Full recovery from serious health problems is much more likely if workers feel that they are "cared for and loved" and "esteemed and valued" and that they belong to a "network of communication and mutual obligation" (House, 1981:16).

Safety and Health in the Workplace of the Future

The increasing use of highly toxic chemicals in manufacturing has led many researchers to pessimistic expectations about safety and health in the workplace of the future (Fox, 1991). These problems, however, *can* be addressed through the development and use of substitute chemicals. Alternatively, containing dangerous processes in closed systems can reduce risk. Young people today have increased awareness of health and environmental issues. This increased awareness has produced a greater focus on preventive measures to ensure workplace safety and health. These developments have also been supported by a growing awareness of the connection between employee health and organizational productivity.

SUMMARY

Female and minority workers still encounter the possibility of discriminatory employment practices in the twenty-first century. This discrimination occurs both at hiring as well as later when pay raises are determined and promotions are considered. These individual troubles are also social problems. Continuing discrimination is hard to eradicate because of the difficulty of enforcing antidiscrimination legislation in the workplace, because it is often subtle, and because of the lingering effects of past discrimination in creating current barriers to employment. There are some reasons for optimism, however, about further reductions in discrimination. The increas-

ing labor force experience of women and the increasing presence of women and minorities in nontraditional jobs are helping erode employment barriers. The experience, income, and skills gained in these contexts are important preconditions for further gains in the future.

In this chapter we have also examined some of the negative consequences that can occur for all workers when their jobs are disrupted by unemployment, illness, or injury. Widespread unemployment can have devastating consequences both for individuals and for the communities in which they live. These negative consequences are an important part of contemporary economic, social, and personal reality for millions of Americans. Occupational accidents and illnesses can also have traumatic, even fatal, consequences for individuals.

As workers struggle to overcome the disruptions, barriers, and stresses encountered in their careers, they are often aided in their efforts by others who have experienced similar burdens.

Social support networks of friends, co-workers, and family members are important resources as workers struggle to cope with the uncertainties of the world of work. Such uncertainties and difficulties also lead workers to organize to seek redress for their shared complaints. Chapter 6 focuses on the efforts of workers to collectively improve their situation in the workplace.

KEY CONCEPTS

cultural division of labor

occupational segregation

affirmative action

dual labor markets

statistical discrimination

gender-typing

tokenism

comparable worth discrimination

sexual harassment

frictional unemployment

structural unemployment

cyclical unemployment

discouraged workers

layoffs

unemployment compensation

severance pay

chronic stress injuries

occupational disease

chemical sensitization

indoor air pollution

right-to-know legislation

job stress

environmental degradation

QUESTIONS FOR THOUGHT

1. Identify the key mechanisms through which discrimination in hiring, pay, and promotions continues to affect female and minority workers. Which of these do you think are most important?

2. How much effect do you think the continuing struggle over comparable worth will have for gender differences in earnings, and why?

3. Have you ever experienced, or do you know of anyone who has experienced, sexual harassment? What changes would be required to effectively reduce sexual harassment at work?

4. What sorts of government policies might be effective in preventing layoffs or lessening the negative effects of layoffs?

5. What two health hazards do you believe are most important in the modern workplace? Propose a policy to address these hazards.

MULTIMEDIA RESOURCES

Print

Douglas S. Massey and Nancy A. Denton. 1993. *American Apartheid.* Cambridge, Mass.: Harvard University Press. A classic study on the pervasive effects of racial inequality for African-Americans.

Philomena Essed. 1991. *Understanding Everyday Racism.* Newbury Park, Calif.: Sage. An insightful book about the subtle forms of racism encountered by minorities in everyday interactions at the workplace.

Work and Occupations, Special Issue on "Gendered Work and Workplaces." 1998, May, Vol. 25, No. 2. Cutting-edge research on the pervasive role of gender in the workplace.

Gary N. Powell (editor). 1999. *Handbook of Gender and Work.* Thousand Oaks, Calif.: Sage. The best compilation available on the relations between gender and work.

Gareth M. Green and Frank Baker. 1991. *Work, Health, and Productivity.* New York: Oxford University Press. A well documented report on the long-term productivity benefits of ensuring a safer workplace.

Barbara Reskin. 1998. *The Realities of Affirmative Action in Employment.* Washington, D.C.: American Sociological Association. A review of the real, but limited, gains of African-Americans resulting from affirmative action programs.

Jack Santiago. 1989. *Miles of Smiles, Years of Struggle: Stories of Black Pullman Porters.* Urbana: University of Illinois Press. A compelling history of the black Pullman porters and their union leader, A. Philip Randolph.

Websites

Statistical Abstract of the United States. *www.census.gov/ statab/www* Latest U.S. statistics on almost everything imaginable.

Statistics on Canada. *www.statcan.ca* Latest Canadian statistics.

U.S. Department of Labor Women's Bureau. *www.dol.gov/dol/wb* The main government agency concerned with women's employment issues.

Sex Discrimination in Australia. *www.hreoc.gov.au/sex_discrimination/workplace* Australian government website on sexual discrimination in the workplace.

National Association for the Advancement of Colored People (NAACP). *www.naacp.org* The oldest-standing organization promoting the interests of African-Americans.

Institute of Latin American Studies at the University of Texas. *lanic.utexas.edu/ilas* Latest research and scholarship on all aspects of Latin America.

Latino/a Affairs. *www.latinoweb.com* Dedicated to Latino and Latina cultural, employment, and lifestyle issues.

The Urban Institute. *www.urban.org* One of America's most respected organizations dedicated to social justice and progressive social policy.

Layoff Report. *www.hrlive.com/local-bin/layoff.cgi* Weekly reports on layoffs around the United States—a fascinating site.

U.S. Center for Disease Control. *www.cdc.gov* The main U.S. government agency dedicated to the study and prevention of human diseases of all kinds.

RECOMMENDED FILM

Roger and Me (1990). Michael Moore's biting comic documentary of the role of General Motors in the decline of Flint, Michigan.

6

Collective Responses to Work

When the Union's inspiration through the workers' blood shall run,
There can be no power greater anywhere beneath the sun.
Yet what force on earth is weaker than the feeble strength of one?
But the Union makes us strong.

Solidarity forever!
Solidarity forever!
Solidarity forever!
For the Union makes us strong.

Is there aught we hold in common with the greedy parasite
Who would lash us into serfdom and would crush us with his might?
Is there anything left to us but to organize and fight?
For the Union makes us strong. [*chorus*]

It is we who plowed the prairies; build the cities where they trade;
Dug the mines and built the workshops; endless miles of railroad laid.
Now we stand outcast and starving, 'midst the wonders we have made;
But the Union makes us strong. [*chorus*]

All the world that's owned by idle drones is ours and ours alone.
We have laid the wide foundations; built it skyward stone by stone.
It is ours, not to slave in, but to master and to own,
While the Union makes us strong. [*chorus*]

They have taken untold millions that they never toiled to earn,
But without our brain and muscle not a single wheel can turn.
We can break their haughty power; gain our freedom when we learn
That the Union makes us strong. [*chorus*]

In our hands is placed a power greater than their hoarded gold;
Greater than the might of armies, magnified a thousand-fold.
We can bring to birth a new world from the ashes of the old.
For the Union makes us strong. [*chorus*]

("SOLIDARITY FOREVER," LYRICS BY RALPH CHAPMIN, FROM *SONGS OF THE WORKERS,* 34TH ED., PP. 4–5. REPRINTED BY PERMISSION OF THE INDUSTRIAL WORKERS OF THE WORLD, 3435 N. SHEFFIELD, SUITE 202, CHICAGO, IL 60657.)

"Solidarity Forever," sung to the tune of the "Battle Hymn of the Republic," is one of the widely recognized union inspirational songs. It was first popularized by the Industrial Workers of the World, a North American union active at the beginning of the twentieth century. Its verses evoke images of workers' collective power to run the industrial world in which they toil.

People who work together often come to see that their personal troubles are not unique, but are shared by others in their workplace. Such common interests provide the basis for collective actions, ranging from bargaining over wages, hours, and working conditions to organizing workers' cooperatives to overthrowing the existing political and economic system through mass insurrection.

This chapter explores the reasons behind workers' collective actions. We provide a brief history of working-class collective activity. We also examine the roles served by labor unions and professional associations at the beginning of the twenty-first century.

WHY DO PEOPLE NEED LABOR ORGANIZATIONS?

Workers organize themselves into unions to bargain collectively with employers over specific grievances in the workplace. These grievances vary greatly. They include inadequate wages and benefits, work that is too rapidly paced, unfair retention and promotion practices, and exclusion from decision making. When a problem first arises, workers often blame themselves for their frustrations. They discuss the problems only with a few trusted co-workers, if at all. As you learned in Chapter 4, the responses at this stage are typically individualistic and may include apathy, withdrawal, and quitting. In the second stage, the informal work group becomes a medium for airing complaints in collective terms. Solutions, however, tend to remain largely personal. Alternatively, workers derive solutions that rely on family, community, or ethnic group ties. Sometimes a specific event stimulates the move from this second stage to organized collective action in the workplace. The transition frequently requires leadership from one or more outspoken workers. Collective action will occur, however, only if the workers are committed enough to their jobs to forgo the option of leaving and if a significant core of workers sufficiently overcomes fears of management reprisal to start organizing their co-workers (Shostak, 1999).

Table 6.1   Who Belongs to Unions?   **$**

Criterion	Percentage Represented
Total	16%
Industry	
Manufacturing	17
Mining	13
Construction	13
Transportation	27
Trade	6
Service industries	7
Government	42
Occupation	
Professional	15
Service occupations	14
Clerical and technical	10
Manual	23
Farm	5
Age	
16–24 years of age	6
25–64 years of age	18
Gender	
Men	17
Women	13
Race	
White	15
Hispanic	13
Black	22
Hours	
Full time	18
Part time	8

SOURCE: U.S. Department of Commerce, Bureau of the Census, 1999, *Statistical Abstract of the United States, 2000*. Washington, D.C.: U.S. Government Printing Office.

Union Membership

What kinds of workers belong to unions? As Table 6.1 shows, membership varies by industry, occupation, age, gender, and full-time versus part-time status. Workers in transportation, government, manufacturing and mining, and construction are more likely to be union members than workers in trade and services. Manual workers are more likely to be union members than are professional, service, clerical, or farm workers. Because of greater commitment to their jobs, workers over 25 years of age are more likely to be members than those under 25. Men are more likely to be members than are women. Black workers are more likely to be members than are whites or Hispanics. Full-time workers are more likely to be members than are part-time workers.

Most local unions are part of a larger national union. Table 6.2 lists the twenty-five largest national unions in the United States. The largest is a professional association, the National Education Association (NEA) with just over 2 million members. Professional associations such as the NEA take on some union roles, such as lobbying in the professions' interests, but do not necessarily bargain directly with employers over wages and benefits. The next largest union is the International Brotherhood of Teamsters (*www.teamster.org*), a union with almost 1.3 million members. As the table reveals, workers have organized into labor unions across a wide range of industries and occupations. Unions in the United States are not confined to blue-collar jobs in the manufacturing sector. In recent years the white-collar, professional, and service sectors have been among the most important growth areas for unions.

Sociologists study unions because through these important organizations workers give voice to their grievances and demands. Unions have thus had an impact on the nature of work in modern society far beyond their own membership.

AN OUTLINE OF NORTH AMERICAN LABOR HISTORY

To provide a background for understanding how unions have shaped work in advanced industrial societies, we briefly overview the history of workers' collective movements, focusing on the United States and Canada.

Table 6.2 The Twenty-Five Largest Labor Unions in the United States

Labor Organization	Members (in thousands)
National Education Association[a]	2,001
Teamsters	1,271
State, County, and Municipal (AFSCME)	1,236
Service Workers (SEIU)	1,081
Food and Commercial Workers	989
Automobile Workers (UAW)	766
American Federation of Teachers	694
Electrical Workers (IBEW)	655
Communication Workers (CWA)	504
Steelworkers	499
Machinists and Aerospace Workers (IAM)	431
Carpenters	324
Laborers	298
Operating Engineers	294
Postal Workers	279
Paperworkers International Union	226
Hotel and Restaurant Employees	225
Plumbing and Pipe Fitting	220
Letter Carriers	210
American Federation of Government Employees	170
Fire Fighters	156
Electronic, Electrical and Technical	128
Transit Workers	98
Bakery, Confectionery and Tobacco Workers	95
Mine Workers (UMW)	81

[a]The NEA is not an affiliated member of the AFL-CIO.

SOURCE: Court Gifford (editor), 1999, *Directory of U.S. Labor Organizations*. Washington, D.C.: Bureau of National Affairs.

Local Craft Unions

Skilled workers formed the earliest unions in the United States. These **craft unions** first developed shortly after the Revolutionary War and included associations of shoemakers (Philadelphia, 1792), carpenters (Boston, 1793), and painters (New York, 1794) (Marshall and Briggs, 1989). These associations had much in common with the medieval craft guilds of which they were direct descendants (see Chapter 1). A local craft union drew its membership from only one trade and from only the local area where direct contact between members was possible. The primary goal of these unions was to provide members with higher wages and a measure of economic security based on such benefits as accident and sickness relief and aid to the widows and orphaned children of deceased members. In addition, these unions performed important social functions such as providing meeting rooms for literacy classes and social gatherings.

The Problem of Solidarity Group **solidarity** based on common interests is essential for all early unions. Solidarity includes mutual defense and support in times of crisis or challenge. Common interests among members of a single craft in a local area, however, did not provide an adequate basis for collective action on issues of general importance that cross regions or craft lines. In a world increasingly dominated by regional and even world economies, such general issues took on increasing importance. The burning issues in the 1800s included extending the rights of free speech and influencing government trade and fiscal policies. The transition from local craft unions to workers' organizations based on broader interests provided the motivation for much of the ensuing history of labor in the United States and Canada.

Workers' Political Parties

Many of the early efforts by workers to organize around broader class interests took the form of political parties. By the 1820s workers had established such parties in over a dozen states. For a brief time these parties held the balance of power between the major political parties. The interests that they represented, however, were often divided along regional and craft lines, and a unified program with broad appeal among workers was not to be forthcoming (Lipset and Marks, 2000).

There are many reasons for the failure of working-class political parties in the United

States. These parties did not, as in Europe, have to lead the working class in a struggle to overthrow feudal society and secure political democracy. Also important was the diffusion of interests resulting from "the spreading of the population across the continent, and persistent infusions of new immigrants from abroad" (Dulles and Dubofsky, 1984:50). The parties also failed because the U. S. political system makes it difficult for third parties to succeed. In this winner-take-all system, which mandates a general election for president, third-party views often go unrepresented. In parliamentary systems, as in England and Canada, the legislative members elect the prime minister. The process of negotiation that goes into this vote allows greater room for third-party influence. A final important factor was violent repression of working-class movements by the government throughout the eighteenth and nineteenth centuries.

These forces combined to keep American workers, and the movements that they built, weak and internally divided. Workers in the United States were unsuccessful in organizing as a unified political party to pursue their collective interests. Instead, they went forward with a more fragmented posture, continually struggling to reconcile the competing interests of different occupations and ethnic groups with the more general interests of the working class as a whole.

Early National Unions

The earliest and most significant national union seeking to coordinate the interests of workers from different trades and regions was the National Labor Union, founded in 1866 by Bill Sylvis of the Iron Molders Union. The principal goal of the NLU was the establishment of an eight-hour workday, which it helped win for federal government workers and for workers in six states. This victory, unfortunately, was largely symbolic. The eight-hour day was impossible to enforce because most workplaces were small and widely dispersed and the government had only a limited ability and willingness to intervene. The withdrawal of the NLU candidate in the presidential election of 1872 resulted in the collapse of the union, and provides an example of the difficulties unions experienced in identifying common mobilizing issues for workers as a whole. New technological developments resulting in larger and more centralized workplaces would, however, assist workers in their efforts to organize.

A Central Role for Railroads The railroad industry, the largest and most concentrated in the United States during the nineteenth century, gave birth to the first large national unions, with memberships in the tens of thousands. During the 1870s and 1880s these unions played a significant role in the tumultuous origins of the modern American economy. This role was facilitated by the increased size of the railroad companies and by the opportunities for contact and communication among workers that they provided. Widespread railroad strikes over wages, hours, and conditions were staged across the country in 1877. In Baltimore, the government called out two army regiments. These soldiers opened fire on the strikers, killing ten strikers and wounding more than twenty. In Pittsburgh when officials called out the local militia, the militiamen joined the strikers and marched to the rail yards, which they helped burn down after a pitched battle with hired company guards (Brecher, 1972). Such strikes took on an increasingly general nature in the 1880s as more and more workers joined in the strike activity.

Strikes involving several trades simultaneously are called **general strikes.** When civic participation occurs in the form of demonstrations, marches, and insurrections, these general strikes become **mass strikes.** Mass strikes are a revolt by the general population against some major aspect of the social order. General strikes leading to mass strikes were crucial events in the nation's history in the 1880s and again in the 1930s.

May Day, 1886 One of the most significant mass strikes in U. S. history was the May Day strike of 1886. It was organized by a consortium of unions headed by the Eight-Hour League under the leadership of Albert Parsons. The organizers planned a nationwide mass strike for May 1, 1886, to demand the eight-hour day in order to humanize the conditions of labor and alleviate unemployment. Over 190,000 workers went on strike, and 340,000 workers and citizens paraded in more than a dozen cities. The strike centered in Chicago because of the city's significance as a hub of rail transportation. Over 80,000 workers struck in Chicago alone (Brecher, 1972).

Near the close of the day's events, the police fired into a crowd of pickets who were attacking strikebreakers as they left the McCormick Harvester plant, one of the few businesses in Chicago that had attempted to remain open. Four pickets were killed. Three days later, people crowded into Haymarket Square to hear speakers protest the killings at the McCormick plant. As the speeches ended and the crowd began to thin, a column of police arrived and ordered the crowd to disperse. A bomb was thrown among the police, killing seven, and injuring sixty-seven others. The police opened fire on the crowd, killing four, and wounding fifty or more. Subsequently, the police went on a monthlong rampage, breaking into union offices, burning pamphlets, smashing printing presses, arresting union leaders, and seeking the deportation of union members who were recent immigrants (Dulles and Dubofsky, 1984). Eight men who had given speeches at Haymarket Square were arrested and tried for murder. Four of them, including Parsons, were subsequently hanged. The American union movement suffered severely in the aftermath of the Haymarket affair. The accusation that unionists were "foreign-born, bomb-throwing anarchists" was used to legitimate widespread repression against labor unions. The difficulty of recruiting members also increased. To this day, workers commemorate these events in May Day marches and demonstrations around the world. The United States is one of the few nations that does not commemorate these events and instead celebrates Labor Day at the end of the summer.

The Pullman Strike The late 1800s continued as a period of widespread labor unrest. In 1894, a strike in Chicago against the Pullman Company rekindled the flames of mass strike and civic protest. The company manufactured passenger cars for the busy railroads of the time. The Pullman strike was brought about by a long series of accumulated grievances against the company and was precipitated by wage cuts. To work for Pullman, workers were forced to live in the company-owned "Pullman Town," where rents were twice as high as in neighboring areas, water and gas were quadruple their price elsewhere, and the newspaper editor and church preacher were hired by Pullman (Litwack, 1962). Given the vastly superior resources of the company, a strike by these workers would have had little chance for success. However, the strike was supported by the newly formed American Railway Union under the leadership of Eugene Debs. The ARU agreed to sidetrack any trains that included Pullman cars. Such sympathy strikes in which one union supports another by refusing to handle struck goods are called **secondary boycotts.** Secondary boycotts are illegal under today's labor laws.

The Pullman strike was eventually broken when a federal judge found the ARU guilty of interference with the mail. The company had transferred the mail to Pullman cars to encourage federal action against the strikers. The hiring of 5,000 armed men who were deputized for the occasion by the state of Illinois but paid by the Pullman Company, the deployment of 6,000 U.S. troops and 3,000 Chicago "temporary" police against the strikers, and the fatal shooting of more than thirty men and women provided the final blows against the strike (Boyer and Morais, 1955). Debs was subsequently convicted in federal court for "restriction of trade" as the first case tried under the Sherman Antitrust Act.

The May Day and Pullman strikes were important battles in the struggle for free trade unions, as well as in the struggle for many other rights and benefits that we take for granted today. The 1980s witnessed a similar wave of general and mass strikes in Poland over such issues as the price of basic commodities, the right to form independent unions, the right to strike, wages, maternity leave, and the five-day workweek. Similar to the situation in the United States in earlier historic periods, strikes started with work stoppages in the Gdansk shipyards but rapidly evolved into mass strikes that involved participants from virtually every sector of society.

General Unions: The Knights and the Wobblies

In the late 1800s and the early 1900s, two **general unions** emerged in the United States that attempted to enlist workers from all walks of life: **Knights of Labor** and the **Industrial Workers of the World.** Although neither survived as an active union, both had a strong influence on the direction of the American labor movement.

The Knights of Labor   The Knights emerged as a national organization in 1878 under the leadership of Terence Powderly. The Knights enlisted workers of all kinds, including craft workers, unskilled laborers, farmers, small-business owners, immigrants, and women. The only occupations that were barred were saloon keepers, professional gamblers, lawyers, and bankers. The goal of the Knights was to improve the position of the "direct producers" through education and social reform. The hiring of workers to read aloud to other workers on the job in order to advance their literacy and education is illustrated in the history of Florida cigar workers, as reported in Box 6.1.

The Knights' program called for government sanctions against the monopolies and encouraged the growth of producers' cooperatives. The union grew rapidly in the 1880s in response to a series of depressions that both impoverished workers and thinned the ranks of craft unions. It had grown to a membership of 50,000 by 1883, only five years after its emergence. The depression of 1884 and 1885 sent large numbers of

unemployed trade unionists into the open ranks of the Knights, and by 1888 membership had jumped to 700,000. In Canada, at their height, the Knights had 252 locals in eighty-three different cities, with their greatest strength in the industrial towns of Toronto, Hamilton, and Ottawa (Palmer, 1983).

The Knights officially opposed strikes, favoring political and social reform over work stoppages. They believed that shop-floor actions tended to address only the specific needs of each craft and that political and social reform could better serve the general interests of the working class. However, much of their membership was taken from the struggling trade union movement and favored direct action on the shop floor as well as longer-range social reform strategies. Because of these divergent membership goals, the Knights became unwillingly embroiled in a spectacular but unsuccessful strike against the Southwest Railroad System in Texas and Louisiana in 1886. The negative political reaction from this defeat, as well as from the Haymarket affair that same year, fell heavily on the Knights as the most visible national labor organization of the times and was an important cause of their demise. The Knights' failure, however, must ultimately be traced to their inability to devise a program behind which a broad-based working-class coalition could stand. An agenda of political and social reform without a strong activist trade union movement on the shop floor did not provide a sufficient organizational basis for sustained working-class collective action.

In spite of their rapid decline, the Knights of Labor left many important legacies. Their vision of worker-owned cooperatives remains one of their lasting contributions. Equally important is the legacy of including women in their locals. Previously, women had largely been excluded from the labor movement. The Knights organized several locals among laundresses and seamstresses, as well as in other predominantly female trades. Both in ideology and in practice they did much to advance the efforts of women to improve their working conditions.

BOX 6.1 Fighting for Literacy, Education and Workers' Rights

During the first third of this century, lectors—people who read to cigar-factory workers—came to Ybor City, Florida, from the cigar factories of Havana and Key West. They were educated men who had great acting abilities and a vast capacity to entertain and educate. . . .

They were paid 25 cents per week by each worker. As many as four hundred workers contributed, so the lector was among the highest-paid employees in the cigar industry. . . .

Lectors usually read the *Tampa Tribune* in the opening hours, which meant that they had to be up early to translate the news to Spanish. Many liked to read from two different types of books. The first literary reading would be from the

classics: Cervantes, Hugo, Shakespeare, or Moliere. The second would be a popular dime novel. The workers voted on what popular novel was to be read. . . .

Over the years, the cigar makers were transformed into the best-educated work force in the world. Since the lectors were their teachers, the workers looked to them for leadership. In time, the lectors also read from political tracts, which were often of a socialist nature, and argued for workers' rights. They supported unionization, better working hours, higher wages, medical benefits, and pension funds.

SOURCES: Excerpted from Ferdie Pacheco, 1997, *Pacheco's Art of Ybor City.* Gainesville, Fl.: University of Florida Press, p. 10.

The Industrial Workers of the World The Industrial Workers of the World (IWW, *iww.org*), also called the Wobblies, was founded in 1905 and included unskilled factory workers, miners, lumberjacks, dock workers, and even cowboys. The IWW organized workers not only in the industrial East but also in the mining towns of the Rocky Mountain West and in the lumber towns and ports of the Pacific Northwest. It called for an overthrow of the capitalist class and its replacement by committees of workers in each enterprise. The IWW believed that "an injury to one is an injury to all" and held that it was the responsibility of all workers, no matter what their occupation, to put down their work and assist their fellow workers whenever and wherever they were in conflict with the capitalist class. Many of the goals of the IWW are included in its preamble, which is reprinted in Box 6.2. Because of its goal of establishing a general union for all workers in order to radically transform society, the IWW was often in competition with the craft unions of the time, which sought more limited goals.

A textile strike in Lawrence, Massachusetts, set up one of the IWW's most significant victo-

ries and marked the height of its power. The strike was foreshadowed by increased national concern with working conditions in the textile and apparel industries. On March 25, 1911, 146 women burned to death or jumped to their deaths in the infamous Triangle fire (*www.ilr. cornell.edu/trianglefire*). This tragedy resulted from the Triangle Company's refusal to provide safety measures of any sort. The workers were trapped in the burning building because the company had locked the doors on each floor of the tall apparel factory from the *outside* to prevent workers from shirking or stealing.

In January 1913, wage cuts and poor working conditions in Lawrence precipitated a strike by more than 20,000 textile workers, many of them women and children. Police brutality against the strikers and extensive press coverage of the strike ironically transformed a desperate struggle among the poorest of workers into one of the most significant victories in American labor history. As a result of the strike not only were wages raised and conditions improved in the textile industry as a whole, but important legislation was also enacted that restricted the exploitation of child and female labor (Marshall and Briggs, 1989).

BOX 6.2   Rallying Cry of the IWW

Preamble of the Industrial Workers of the World

The working class and the employing class have nothing in common. There can be no peace so long as hunger and want are found among millions of working people and the few, who make up the employing class, have all the good things in life.

Between these two classes a struggle must go on until the workers of the world organize as a class, take possession of the earth and the machinery of production, and abolish the wage system.

We find that the centering of the management of industries into fewer and fewer hands makes the trade unions unable to cope with the ever growing power of the employing class. The trade unions foster a state of affairs which allows one set of workers to be pitted against another set of workers in the same industry, thereby helping defeat one another in wage wars. Moreover, the trade unions aid the employing class to mislead the workers into the belief that the working class have interests in common with their employers.

These conditions can be changed and the interests of the working class upheld only by an organization formed in such a way that all its members in any one industry, or in all industries if necessary, cease work whenever a strike or lockout is on in any department thereof, thus making an injury to one an injury to all.

Instead of the conservative motto, "A fair day's wage for a fair day's work," we must inscribe on our banner the revolutionary watchword, "Abolition of the wage system."

It is the historic mission of the working class to do away with capitalism. The army of production must be organized, not only for the everyday struggle with capitalists, but also to carry on production when capitalism shall have been overthrown. By organizing industrially we are forming the structure of the new society within the shell of the old.

SOURCE: *Songs of the Workers,* Thirty-Fourth Edition. Copyright © 1980. Reprinted by permission of Industrial Workers of the World, 3435 Sheffield, Chicago, IL 60657.

Equally famous in IWW history was a massacre in the mining town of Ludlow, Colorado. Workers striking against the Colorado Fuel and Iron Company, owned by John D. Rockefeller, had been evicted from their company housing and were living with their families in tents. On Easter night of 1914, while the men were at a meeting, company gunmen set fire to the tents. Thirteen women and children died in the fire or were shot to death as they were running out of the tents. Five other strikers were shot to death as they tried to help the women and children escape (Boyer and Morais, 1955).

Similarly violent confrontations also occurred in Canada during the early decades of the twentieth century. The most famous of these was the Winnipeg general strike of 1919. Workers throughout Canada were agitating for change in the face of harsh working conditions, low wages, and deteriorating living conditions resulting from

soaring wartime inflation. The situation became a crisis when local employers in Winnipeg refused to recognize and bargain with the building and metal trades workers. The Winnipeg Trades and Labour Council called a general strike in support of union recognition for the trades workers. In response, local employers hired armed vigilantes and convinced the government to use the Royal Northwest Mounted Police to charge through lines of picketers. The general strike was broken, strikers were arrested and jailed, and the workers' demands were left unmet (Krahn and Lowe, 1998).

Such episodes of violent repression, including the long-term imprisonment or deportation of more than a hundred IWW leaders on sedition charges, brought about the eventual decline of the union. In their wake, however, the Wobblies left the labor movement with many cultural heros, such as leaders Bill Haywood and Mother

Jones and songwriter Joe Hill. They also created a lasting image of a distinctly American version of radical mass unionism.

Strong resistance from employers limited the success of these early labor unions. Companies were reluctant to give workers a share in either the profits or the decisions. Employers' strategies of resistance included using *blacklists* to identify and refuse employment to union sympathizers. Employers also used *yellow-dog contracts* in which workers had to sign a promise that they would not join a union or engage in any collective action. (The unions argued that anyone who would sign such an agreement was a "cowardly yellow dog.") Companies also employed spies, *scabs* (replacements for striking workers), professional strikebreakers, and armed guards to undermine unions and intimidate workers. The well-known Pinkerton detective agency first came into national prominence through providing such services.

Employers also enlisted the government to intervene against workers. In the early 1800s unions in the United States were considered conspiracies for the purpose of limiting owners' free use of their property. Unions were forced to exist as secret societies. By the mid-1800s the conspiracy doctrine had been relaxed, and unions were no longer seen as inherently a conspiracy against the rights of property. Instead, unions were legal unless their objectives or the means used to secure these objectives were *conspiratorial* (of possible injury to others). In plain language, this meant that unions were legal, but strikes were still illegal.

As the conspiracy doctrine dwindled in importance, however, unions faced yet another obstacle. This was the **injunction** (court order) to cease and desist a specific action, such as a strike (Marshall and Briggs, 1989). Injunctions were used against strikes throughout the nineteenth century and into the twentieth century. Violating an injunction put strikers in contempt of court. It could also evoke the entire arsenal of state power, including the use of fines against unions, the imprisonment of union leaders, the

deployment of the police and the military, and, ultimately, the use of direct violence against workers. Following the passage of the Sherman Antitrust Act in 1890, workers could also be found guilty of monopolistic practices because of their attempt to bargain wages through collectively withholding their labor. In 1914, the Clayton Act explicitly exempted labor unions from prosecution under the Sherman Antitrust Act. In 1932, the Norris-LaGuardia Act sharply restricted the use of injunctions in labor disputes. But well into the twentieth century government interference in favor of owners represented a major barrier that workers had to overcome in their struggle for improved working conditions.

The AFL and Craft Unionism

Samuel Gompers, the president of the Cigar Makers' Union, was the first head of the **American Federation of Labor,** founded in 1886. The AFL is the oldest existing labor organization in North America and helped set a successful and enduring pattern for what labor organizations and industrial relations look like in the United States and Canada to this day.

Gompers's vision was based on an effort to reconcile the differing needs of each craft or trade. The AFL was organized as a decentralized federation of unions; that is, each member union (the carpenters, the cigar makers, the boilermakers, and so on) had autonomy over its own affairs. The AFL served as an umbrella organization responsible for coordinating and supporting these efforts and pursuing the general interests of the working class at the regional and national levels.

Strikes and Collective Bargaining Gompers emphasized collective bargaining by each trade on specific issues rather than general and mass strikes over broader issues. In his vision, strikes by each trade were the crucial tool in the struggle for better working conditions:

A strike on the part of workmen is to close production and compel better terms and more rights to be acceded to the producers. The economic results of strikes to workers have been advantageous. Without strikes their rights would not have been considered. It is not that workmen or organized labor desires the strike, but it will tenaciously hold to the right to strike. We recognize that peaceful industry is necessary to civilized life, but the right to strike and the preparation to strike is the greatest preventive to strikes. If the workmen were to make up their minds tomorrow that they would under no circumstances strike, the employers would do all the striking for them in the way of lesser wages and longer hours of labor. (U.S. Congress, 1901:606)

With workers in each trade firmly in charge of their own affairs, the AFL could pursue the goals of political and social change through its activities as a pressure group.

By 1890, only four years after its founding, the AFL was the largest labor organization in the United States. It used strict business principles to organize the collection of dues, the raising of strike funds, and the creation of old-age, sickness, and burial funds for its members. From 1890 to 1914, the AFL pursued policies that increased average weekly wages for unionized workers in manufacturing from $17.57 to $23.98 and reduced average hours from 54.4 to 48.9 per week. The AFL achieved these gains by institutionalizing conflict through the use of collective bargaining and abandoning the more volatile policies of mass unionism and social upheaval. By 1914 the AFL had a "membership of 2,021,000 workers for whom it had won higher wages, shorter hours, and increased security" (Boyer and Morais, 1955:181).

Many of the early craft unions organized under the umbrella of the AFL excluded women and minority workers in an effort to keep wages from being underbid by these cheaper sources of labor. Other AFL-affiliated unions, however, actively organized female and minority workers. Between 1910 and 1930 the International Ladies'

Garment Workers Union (ILGWU), composed largely of female workers, was one of the most rapidly growing unions in the United States (Milkman, 1985). Similarly, in 1933, the all-black Brotherhood of Sleeping Car Porters was given an international charter by the AFL to organize railroad porters. On August 25, 1937, the Porters signed a contract with the Pullman Company. This was the first labor contract ever negotiated between an African-American union and a major U.S. corporation (Foner, 1982).

The CIO and Industrial Unionism

The AFL served as the key organizing and coordinating umbrella for trade unions until the Great Depression of the 1930s. By the mid-1930s unemployment stood at over 25%, and AFL membership was hard hit. The federation was also racked with internal dissent. Some members believed that the AFL should include only skilled craftworkers. Others believed that it should attempt to organize the large numbers of semi-skilled workers in the new mass-production industries of automobiles, rubber, steel, and glass. In 1935 John L. Lewis, head of the United Mine Workers and chair of the AFL's Committee for Industrial Organization, withdrew eight unions loyal to his vision of industrial unionism and formed the **Congress of Industrial Organizations** (CIO). In Lewis's vision of **industrial unionism** all workers in an industry, regardless of their particular craft or their level of skill, would be in the union.

Sit-down Strikes in Mass Production Wages had fallen dramatically in the mass-production industries due to the depression. Employers had also sped up the pace of work on the new assembly lines and workers in these industries were eager to be unionized. Withholding their labor in a strike, however, would have been ineffective because, unlike skilled craftworkers, they were easily replaceable. In response, mass-production workers developed a new form of collective action, the **sit-down strike.** In a sit-down strike, workers stayed in their places but stopped working. Sit-down strikes were staged as protests against speed-ups or against specific abuses by foremen, such as unfair firings or disciplinary actions. They also were used to pressure a company to recognize and bargain with the union. They were generally of short duration, lasting from a few minutes to a few hours. But even short sit-down strikes could effectively disrupt production, especially in the highly coordinated mass-production industries. Most importantly, they prevented companies from replacing workers with strikebreakers in order to continue production.

The Flint Sit-down Strike Sit-down strikes spread in the 1930s, becoming both longer and more frequent. Lewis and the CIO capitalized on these strikes to organize workers into the organization. The most famous sit-down strike was staged against the General Motors Corporation in Flint, Michigan, in January and February of 1937. The sit-down started at Fisher Body Buildings I and II. Police attacked the striking workers by breaking windows and firing tear gas shells inside. The workers doused the shells and held their ground against three police assaults by using fire hoses that had been installed as safety devices. The police gunfire wounded fourteen unarmed workers, but the workers refused to leave the plant. The next day the militia massed outside the plant. Meanwhile, workers at the Fleetwood plant had also gone on strike, and the governor of Michigan refused to escalate the volatile situation any further. A siege lasting forty-four days ensued. The siege included the famous "Battle of the Running Bulls," in which the police (bulls) attacked the workers' wives, who were trying to smuggle food into the plant. The police were forced to retreat in front of the workers' fire hoses (Brecher, 1972).

General Motors eventually capitulated to the CIO. Within a three-week period in early 1937 the CIO organized not only the automobile industry but also the steel industry. This brought into being two of the largest and most important unions in American history, the United Auto Workers (UAW, *www.uaw.org*) and the United Steelworkers of America (USWA, *www.uswa.org*). Sit-down strikes became the craze, and even spread to such occupations as soda-fountain clerks, waiters and waitresses, stenographers, dry-goods clerks, and teachers (Boyer and Morais, 1955).

The Fisher Body Plant in Flint produced car bodies for fifty more years after the famous sit-down strike until December 1987, when it was finally closed. Although the sit-down movement gradually faded, it left a lasting mark on American history through the unions it helped organize and the legacy it left of direct action. This legacy helped set the groundwork for similar sit-down tactics in the civil rights and antiwar movements of the 1960s and 1970s, the nuclear-freeze movement of the 1980s, and the environmental movement of the 2000s.

Because of high unemployment, the Great Depression may at first seem an unlikely period to have witnessed rapid gains in union membership. However, the social and economic situation in the Depression was so desperate that there was a perceived need for a radical transformation of the social order. In this context many people considered labor unions an important part of the solution to the problems of unregulated capitalism that had given rise to the Depression.

Legislative Gains Important federal legislation passed during the 1930s solidified workers' hard-won gains into lasting structures. Legislation gave concrete form to workers' demands and also sparked additional union growth by granting legitimacy to the labor movement. The most significant of these acts was the **National Labor Relations Act** of 1935 (NLRA, *www.nlrb.gov*), commonly known as the Wagner Act. The NLRA calls for secret-ballot elections in which workers can choose whether to be represented by a union. If the union wins the election, it becomes the sole bargaining agent for the workers, and the owners must bargain in good faith with the union. The NLRA does not spell out wages and conditions. Rather, unions and companies are required only to meet, bargain in good faith, and put their final agreement in writing. The NLRA thus endorses the process of collective bargaining but leaves the results of the bargaining completely open to the parties involved. It also provides certain safeguards against unfair labor practices. The most important of these is the stipulation that workers cannot be fired for trying to organize a union at their workplace. The NLRA is the legal basis for modern American unionism.

Canadian labor law, though similar in content, came into being in a more piecemeal fashion over a longer period of time. The traditions of English and European labor law influenced Canada more strongly than the United States; this influence helped establish the legal basis of unionism in Canada at an earlier date. As a result, a greater diversity of union organizations developed in Canada than in the United States, including a powerful presence of Catholic labor unions in French-speaking Quebec (Gunderson and Ponak, 2000).

The Postwar Retrenchment

The organizing efforts of the AFL and the CIO, operating under the new protection of the NLRA, resulted in a fivefold growth of union membership between 1933 and 1945. By 1945, the combined membership of the AFL and the CIO numbered 15 million (Brody, 1980). The period immediately following the war, however, saw serious setbacks for the union movement.

The Taft-Hartley Amendments   In 1947, Congress passed the **Taft-Hartley amendments** to the NLRA. These amendments contained setbacks for labor in three major areas. First, they outlawed secondary boycotts and sympathy strikes, which had enabled workers in one union to strike or otherwise pressure their employer not to do business with another company because of its labor practices. This exclusion took away an important tactical weapon, weak-

ening union's bargaining power and eliminating an opportunity for cooperation between workers in different trades.

Second, the Taft-Hartley amendments allowed states to outlaw contracts requiring union membership as a condition of employment, commonly called *union shop contracts.* The NLRA requires unions to share the wage-and-benefits package they win with all workers in their unit, including the full procedure for settling grievances. As a result, most union workers believe that once a union has been elected by majority vote, all workers should be required to be members of the union and should pay dues. Union workers see nonunion workers as getting a free ride on benefits secured by the union. Twenty-one states, however, have enacted so called right-to-work laws, which prohibit contracts from including such compulsory membership clauses. In these states, workers do not have to join a union at their workplace, even if one has been elected by majority vote to represent the workers. Such workplaces are sometimes called "open shops" in contrast to "union shops."

Third, the Taft-Hartley amendments allow the president of the United States to force striking workers back to work in cases of national emergency. Unions feared that this would reintroduce the extensive use of injunctions against strikes and labeled the Taft-Hartley amendments the "slave labor act." The national emergency clause, however, has been invoked only about once a year and in the final analysis has represented a much less significant setback to labor than the exclusion of secondary boycotts and sympathy strikes and the granting of states' rights to prohibit union shops.

McCarthyism and Right-Wing Attacks   The late 1940s and early 1950s saw the growth of anticommunism, McCarthyism, and right-wing attacks on many progressive elements of American society, including labor unions. To confront these and other attacks on unions, the labor movement closed ranks and carried out a series of internal purges. In 1949 and 1950 the CIO expelled eleven unions that had substantial communist membership. In 1955, the AFL and the CIO, under the leadership of George Meany and Walter Reuther, merged to form the AFL-CIO (*www.aflcio.org*). The merger resulted from concerns about a leveling off of membership growth and also from a narrowing of ideological differences between the two organizations. With the continued growth of mass-production industries, both the AFL and the CIO unions had begun to move toward a hybrid form of unionism that included both semiskilled workers and skilled workers in the same unions.

In 1957, the AFL-CIO expelled the graft-ridden Teamsters, led by Jimmy Hoffa, in an effort to clean up labor's image. In 1959, Congress passed its final major piece of legislation dealing with unions, the Landrum-Griffin Act. The act regulates the internal practices of unions, including the election of officers, and encourages political democracy within unions. Illegal use of dues and retirement funds still sometimes occurs and sullies the reputation of a union, but it is increasingly uncommon. In the 1990s, after a series of reforms, the International Brotherhood of Teamsters was readmitted to the AFL-CIO.

Facing New Challenges

Labor unions have faced many challenges since their legalization. They have confronted racism in their own ranks and largely overcome it. And they have expanded their organizing efforts into the growing public sector and into the expanding professional occupations (Mort, 2000).

Racial Equality Well into the period after World War II, some labor unions, especially those in the building trades, kept out minorities or restricted them to jobs in specific neighborhoods or to lower-paying positions within the trade. Practices of this type occurred as late as the 1960s among such unions as the Plumbers, Electricians, and Sheet Metal Workers. However, these practices were greatly reduced by the mid-1960s due to pressure from the government, the public, and

other integrated unions in which African-Americans and other minorities made up a substantial share of membership. By the early 1960s, the AFL-CIO had became a strong supporter of the civil rights movement, and it played an influential role in getting the equal employment opportunity section, Title VII, included in the landmark Civil Rights Act of 1964 (Marshall and Briggs, 1989). Today, blacks are more likely to be union members than are whites (DeFreitas, 1993). Blacks and other minorities have also secured leadership positions in many racially integrated unions (Cornfield, 1989).

During the 1960s the labor movement and the civil rights movement often worked together to improve the position of black people. For example, on February 12, 1968, black sanitation workers went on strike in Memphis, Tennessee. The strike had the active support of the local African-American community, the AFL-CIO, and civil rights organizations such as the Southern Christian Leadership Conference. The Reverend Martin Luther King, Jr., went to Memphis three times, speaking at rallies and leading marches in support of the strikers. He was assassinated in Memphis on April 4, 1968, during his third visit in support of the strike (Foner, 1982).

The union practice of promoting seniority as the basis for allocating layoffs, however, has sometimes had inadvertent negative effects on minority employment. Seniority-based protection against layoffs has tended to favor whites over more recently hired women and minorities. However, court settlements in favor of racial quotas in layoff procedures, and the aggressive organizing of new female and minority workers, have tended to reduce these negative consequences (Rees, 1989). Asian-Americans are the only minority group significantly less represented in the labor movement today than majority whites. Active organizing drives among garment workers, hotel and restaurant workers, and medical workers are attempting to increase Asian participation in the labor movement (Chen, 1993).

Women in Unions Starting in the 1970s, the AFL-CIO began to be strongly influenced by the women's movement. The AFL-CIO became a major sponsor of the Equal Rights Amendment, and a group of female trade unionists formed the Coalition of Labor Union Women (CLUW) (Green, 1980). The CLUW brought women's issues increased attention in the union movement. These issues have included the elimination of the restriction of women from certain occupations because of circumstantial factors such as height or weight, the need for pregnancy leave, and proposals for comprehensive child-care programs (Shostak, 1999). The increased visibility of women in the union movement has met with resistance from some men; however, many others support it. As early as 1979, for instance, 1400 members of the International Woodworkers of America, most of whom were men, went on a successful strike over the unfair firing of a female worker at a plywood plant. The worker's firing had resulted from her efforts to file a discrimination suit against the company because she was restricted from preferred jobs and shifts for which she was qualified (Kauffman, 1979). Such examples of solidarity between male and female workers have an important role in easing tensions as women expand into previously male occupational domains. As a result of aggressive organizing in heavily female occupations and industries, female membership in unions has grown significantly at a time when male membership is still declining. Unions have been particularly successful in organizing women in the fast-growing telecommunications industry. Women's wages in unionized jobs are higher than in nonunion jobs and the disparity between men and women's wages is significantly less. Box 6.3 presents some of the issues highlighted in an organizing drive among largely female clerical workers at Indiana University by the Communications Workers of America (CWA, *www.cwa-union.org*).

Public-Sector Unions In recent decades the public sector has been the fastest-growing area of union organizing. In 1960, fewer than a third of federal employees belonged to labor organizations; by 2000, almost two-thirds were union members. In this same period union membership among state, county, and local employees increased from 1 million to 5 million, bringing the unionization rate among eligible government employees above 40%, higher than any private-sector industry. A majority of states now utilize final and binding arbitration for some or all state employees. This mechanism allows collective bargaining in the public sector without the use of strikes that might disrupt important public services. Most states now have either statutory or tacit recognition of state workers' rights to join unions and negotiate their conditions of employment (Marshall and Briggs, 1989).

Professional Workers Professional workers have also begun to unionize to improve their bargaining position relative to the large bureaucratic organizations in which they are increasingly employed. In the past, professional organizations focused their activities on membership training, the defense of members' legal rights, and legislative lobbying. In recent years many such organizations have become increasingly eager to represent their members in collective bargaining as well. This organizational transformation has resulted in large leaps in union membership (Marshall and Briggs, 1989). For instance, the combined membership of the National Education Association and the American Federation of Teachers is over 2.7 million, making teachers the largest group of organized workers in the United States.

The largest strike ever in the United States among white-collar workers occurred in 2000. Over 23,000 engineers and technicians at Boeing Aircraft went on strike over such classic trade union issues as pay, benefits, and health insurance. A favorable settlement was reached after thirty-seven days off the job.

Even such high-status professions as university professors and medical doctors have begun to organize unions. In California, faculty in many of the state colleges are organized by the American

---

BOX 6.3 Organizing Clerical Workers at Indiana University

IU/CWA Clerical and Technical Organizing Committee

What are some of our concerns?

SALARY EQUITY: Are you being compensated fairly for all the hard work you do? Did reclassification raise your grade but not your salary? We believe CWA can make a difference. With a union we can address the issue of salary equity—no full-time clerical or technical should qualify for state or federal assistance.

JOB SECURITY: You should not have to work fearing that the next day your job could be eliminated. Job security is part of CWA's strength. Putting in many years of hard work should count for something—we will work together to address the issue of job security.

HEALTH INSURANCE: Everyone knows health costs are going up, but is it always necessary that employees absorb the rising costs? Who can afford a higher deductible? Why don't we have vision care? This is another area where CWA can draw upon their past experience for a fresh approach to the problem.

PENSIONS: Everyone should be able to live comfortably when they retire. Who wants to do with less? Administrators and faculty live

comfortably after retirement, why shouldn't we? CWA will help us to address the issue of pensions and work towards a happier retirement for everyone.

CHILD & ELDER CARE: Why don't IU employees have affordable child care? Presently facilities are expensive and available space is limited. The majority of the clerical/technical workforce at IU is female—working women need good child care. Elder care is also becoming an increasing responsibility for many—together with CWA we can work for improved conditions in both child & elder care.

PARKING: Did you ever wonder why you have to pay to park when going into work and why you can't always find a place? You are not alone—employees need adequate parking.

We are the backbone of the university and we deserve respect. We work hard for IU and for our pay. We believe that together with CWA we can change things for the better. All clericals and technicals need to join together and let the university know our concerns so we may work toward bettering our jobs, our lives and our future.

SOURCE: Communications Workers of America.

Federation of Teachers (AFT, *www.aft.org*). Among college and university faculty nation-wide, 170,000 of 400,000 full-time and 300,000 part-time faculty are organized into unions, and labor activists expect additional organizing gains in the future (Shostak, 1999). Many locals of the American Association of University Professionals (AAUP, *www.aaup.org*) have felt competition from the more aggressive AFT and have started to engage in collective bargaining in addition to more traditional lobbying and professional development activities.

Experiencing similar gains, the Union of American Physicians and Dentists has grown in recent decades to over 50,000 members, partly in response to increasing pressures on doctors who work in large health-care organizations (Budry,

1997). The increasing centralization of the U.S. health-care industry in large, for-profit organizations is expected to shift the allegiance of doctors from the American Medical Association to organizations practicing more traditional union strategies. The emergence of these unions evidences professionals' willingness to overcome prejudices against joining unions because of perceived status differences between themselves and other workers.

Farm Workers Farm workers have also been actively organizing in recent decades. California became the first state to include farm workers under labor laws with the passage of the Agricultural Labor Relations Act in 1975. Since then, other workers in heavily agricultural states such as

Florida and Texas have also been actively organizing, though they have met stiff resistance from entrenched agricultural interests in state legislatures. Nevertheless, their efforts have had some success. In the 1980s, for example, the United Farm Workers (*www.ufw.org*) was successful in getting farm workers in Texas covered for the first time under the state unemployment compensation system.

In spite of growth in some unions, membership in the labor movement as a whole has declined. In 1945, more than 35% of employees were in unions; by 2000, only 15.4% of employees were union members. The decline has resulted from two major factors. First, layoffs in manufacturing resulting from technological displacement and increased international competition have decreased the number of unionized workers in the manufacturing sector. Second, management has been increasingly resistant to unions. Union gains in the service and clerical sectors, where workplaces are often small, have been insufficient to keep pace with losses in other sectors, principally manufacturing and mining.

Lessons from Labor's History

We have seen that the labor movement in the United States is not a monolithic organization. Rather, it continues to have deep internal schisms as organizations with different strategies for articulating and realizing workers' goals contend for power (Form, 1985). The movement thus embraces a variety of often divergent viewpoints and needs, ranging from those of doctors and skilled craftworkers to clerical workers and janitors.

Labor unions in the United States have often appeared to *react* to circumstances rather than to take a leadership role in shaping events in the economy. In many ways this is not surprising. Unions have suffered severe repression at various points in American history. The organizations that have survived have generally been conservative and cautious in their behavior. Conservative leadership, however, may be more of a liability than an asset in times of rapid technological change and increased foreign competition such as the present. American labor unions will need to take innovative stands if they are to survive and prosper. History suggests that unions *will* adapt successfully to change. In their 200-year history in the United States, unions have repeatedly had to adapt to changing circumstances. The emergence of the CIO and industrial unionism in the 1930s provides one of the clearest examples of such creative adjustments to altered circumstances. Unions were not created just to bedevil the current generation of managers. They are a key part of American social, economic, and political history.

LABOR UNIONS AT THE BEGINNING OF THE TWENTY-FIRST CENTURY

The struggles of the labor movement have produced many gains for working people in the United States and Canada. These gains include higher wages, shorter hours, safer working conditions, and increased bilateral governance of the workplace by negotiated rules rather than management fiat. The legacy of the union movement, however, is more than just these specific gains. It is also a vision and a program of how these gains can be protected and extended. That vision includes the central role of collective bargaining, the maintenance of effective grievance procedures, job security, and the targeted use of political lobbying.

How do labor unions fulfill their many current roles? Which unions are declining, which ones are growing, and why? How are unions confronting rapid technological change and increased international competition? What innovative programs are they developing to confront the challenges of the twenty-first century? In this section we explore the answers to these questions.

Current Union Roles

Labor unions fulfil their current roles through three major sets of activities: collective bargaining, organizing strikes, and political lobbying.

Collective Bargaining Negotiating wages and conditions and enforcing the contract thus agreed on are the major roles of North American labor unions. After a union wins a certification election, it tries to negotiate a contract with the employer, who is required only to bargain with the union in good faith. There are about 150,000 labor contracts covering the 17 million unionized workers in the United States. Over half of these workers are covered by the 2,000 largest contracts. Labor contracts, and the precedents they set for workplace relations, represent significant extensions of the property rights of workers over their jobs and over their conditions of employment.

How high are union wages? Higher than those of nonunion workers but lower than popular belief frequently portrays them. The hourly wages for union workers in a packing plant in Dallas, Texas, for example, range from $12.26 for machine operators to $9.33 for janitors (see Table 6.3). For workers employed 40 hours per week, fifty weeks per year (a total of 2,000 hours), these jobs paid $18,600 to $24,500 annually. How much higher are these wages than the wages of nonunion workers? Such comparisons are difficult to make, because union workers are typically more highly skilled and are employed in higher paying industries and larger organizations than nonunion workers. In 1998, the average weekly pay of nonunion workers was $499. The average weekly pay of union workers was about a third higher, at $659 (Census, 2000). If the skills of workers, the sizes of workplaces, and similar differences are statistically controlled, then union workers on average earn about 10% to 15% more than nonunion workers (Wallace, Leicht, and Raffalovich, 1999).

There is also a relatively compressed difference in pay between the lowest and the highest paid worker in a typical union contract relative to

Table 6.3   A Typical Union Wage Scale $

Job Title	Hourly Wage
Extruder Operator	$12.26
Apprentice Extruder	9.79
Ink and Film Handler	9.79
Press Operator	12.73
Apprentice Press Operator	9.79
Plate Maker	12.26
Plate Mounter	12.26
Plate Prep	9.79
Bay Machine Operator	12.26
Helper	9.79
Packer	9.79
Lab Technician	10.05
Shipping and Receiving	9.79
Warehouseman	9.79
Driver	9.79
Lift Driver	9.79
Maintenance Mechanic	12.54
Janitor	9.33

SOURCE: Contract between Graphic Communication Union and Princeton Packing, Inc., Dallas, Texas, 1993. Reprinted with permission.

nonunion establishments. A reduction in such differences is important for maintaining a community of interests among workers. It is also a central building block in the union goal of bringing up the wages of the lowest paid employees as a protection for all workers.

Fringe benefits such as health insurance, retirement, and paid holidays comprise an increasing share of total payments to workers. Unions have played an important role in extending these benefits from salaried employees to all workers. Union employees receive 24% of their total compensation in fringe benefits, as compared with 18% for nonunion workers (Freeman and Medoff, 1984). This difference further widens the gap in total compensation between union and nonunion employees.

Collective bargaining agreements are open contracts that can include almost anything to which both parties agree. Besides wages and benefits, contracts also typically include overtime and shift premiums, promotion and layoff procedures, provisions for due process in discharge cases, and a formal procedure for handling workers' grievances. In addition, elected union officials, such as the president or vice president of the local, are often given a certain amount of *release time* from their jobs to attend to union business, such as the handling of grievances. In small locals or in locals with weak contracts, such a clause may be absent or may allow only one afternoon a week, even for the union president. In larger shops, release time is also bargained for a number of **shop stewards,** whose role is to inform workers of their contractual rights and to handle their grievances about violations of these rights.

Provisions concerning work rules are less common. Managers generally regard the setting of work rules as their inalienable right. However, some work rules may be negotiated, especially in hazardous work situations. A union contract for workers at a electrical power plant, for example, may specify that two workers (not one) be sent to work on remote power stations and that both must be qualified electricians. The NLRA explicitly forbids *featherbedding* clauses, which mandate unnecessary workers.

Most union contracts include provisions for due process in discharge cases. The procedures usually include a statement of *"just cause"* and may even permit a third-party review of the discharge on union request. **Seniority** clauses are also common. Seniority may be used to allocate protection from layoffs, first choice of shifts or job transfers, and access to apprenticeship training programs. Seniority clauses are typically strongest in relation to layoffs; they are often weakest in relation to promotions. Managers generally favor promotion criteria based on some measure of merit.

A crucial section in a union contract concerns **grievance procedures**—the set of procedures for handling of workers' complaints about violations of their rights under the contract. The shop steward or other appropriate union official has the right and duty to represent workers when they believe their rights have been violated. If a worker has been docked pay for what he or she believes should have been a legitimate paid sick day, for example, the shop steward will try to resolve the issue with the worker's supervisor. If the issue cannot be resolved at this level, the shop steward will advance the grievance, in writing, to the next highest organizational level, perhaps to the plant manager and the union president. If they cannot reach an agreement, most grievance clauses mandate final and binding arbitration by a third party mutually agreed on by union and management. Most arbitrators are members of the American Arbitration Association. They typically act as arbitrators on a part-time basis; many are also employed as lawyers or university professors.

Providing a fair and effective system for handling workers' grievances is often a union's most important contribution to a workplace. It represents many workers' primary experience of their union. Individually, workers are relatively powerless in relation to managers. When disputes arise, this powerlessness can produce bitterness and resentment. The grievance procedure allows workers the right to an impartial hearing in the event of a dispute that cannot be settled by direct negotiation.

On a broader scale, the establishment of due process in the resolution of grievances is one of the union movement's greatest contributions to expanding workers' rights in the workplace. Many observers argue that due process is in the interests of managers, as well as workers. Access to due process helps channel personal disputes into more peaceful mechanisms of conflict resolution. The successful resolution of such conflicts helps clear the air and prevents the emergence of lingering resentments.

Strikes Strikes and the threat of strikes are the most important mechanisms through which unions win benefits (Wellman, 1995). However, strikes themselves are quite infrequent and are not always successful (Green, 1990). In 1998, there were forty strikes in the United States that involved over a thousand workers. A total of

387,000 workers were involved in these strikes. Lost time due to strikes represented 0.02% of the total time worked by the labor force (Census, 2000). The peak of strike activity in the postwar period occurred in 1971, when 2.5 million workers were involved. Throughout the postwar period an average of less than 0.10% of work time has been lost due to strikes. This amounts to less than one day lost per thousand days worked by the labor force, a tiny fraction of the amount of time lost to the common cold or to industrial accidents.

In recent years workers have been increasingly reluctant to strike because of fears that management will permanently move their jobs elsewhere. This fear has forced unions to look for alternatives to strikes such as political lobbying and worker ownership. (See the section below on "New Organizing and Bargaining Strategies.")

The benefits won by union members directly influence the wages and working conditions of nonunionized workers. Union gains in one plant or industry generally bring up the prevailing wage in related plants and industries. In addition, many large nonunion companies keep their wages and benefits comparable with those in union plants to help undermine union organizing efforts at their plants. The workers in such nonunion plants are indirect beneficiaries of the struggles and sacrifices of union workers, though they pay no dues and have never had to risk their jobs in a strike. Even the benefit packages of professional and supervisory workers in unionized plants are strongly influenced by the benefits won by union workers (Edwards, 1993; Leicht, 1989b).

Lobbying Labor unions also play an important role in the political arena. Unions exert political power by registering voters, especially low-income and working-class voters, and by encouraging them to vote. They also allocate noncash union resources, including staff time and volunteer efforts, to the political campaigns of candidates who support pro-labor legislation. Finally, they lobby as an interest group for legis-

lation favorable to workers. Because unions directly represent less than 20% of the labor force, they have had limited success in securing legislation directly favorable to organized labor, such as liberalization of picketing laws or elimination of states' rights to outlaw union shop contracts. They have enjoyed greater success in areas where they have acted as part of a coalition with other groups, areas such as public education, antipoverty legislation, civil rights, voting rights, health insurance, public housing, and occupational safety and health. Unions have thus played an important role in the enduring New Deal coalition begun by President Franklin D. Roosevelt that has been a core part of American political and economic terrain for almost three-quarters of a century.

Unmet Membership Concerns Unions have also had many failures in the postwar United States. They have not always been able to address some of the pressing goals and concerns of their members. They have had limited effectiveness, for example, in negotiating working conditions and work rules. Management has retained control over determining work rules in almost all settings. Although unions have been successful in establishing seniority as the basis for layoffs they have had little success in combatting layoffs. Unions have also had limited success in pressuring the government and world economic organizations such as the World Trade Organization and the International Monetary Fund to include workers' rights as a prerequisite for opening North American markets to international goods. Unions argue that global standards for workers' rights should include prohibition of child labor, the right to organize collectively, and basic health and safety standards (Dreiling, 2000). They argue that such standards are needed to provide a basis for "fair trade" between workers and companies in different nations.

Unions have also been criticized for a lack of **internal democracy.** Often there is only one slate of candidates for union offices, especially at the local level. Union locals are often run by a

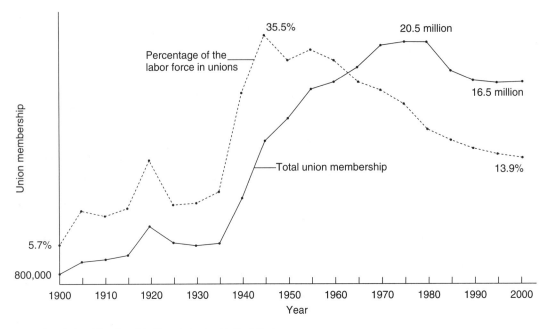

FIGURE 6.1 Union Membership Trends in the United States

SOURCES: Larry J. Griffin, Philip J. O'Connell, and Holly J. McCammon, 1989, "National Variation in the Context of Struggle," *Canadian Review of Sociology and Anthropology* 26,1 (February): pp. 37–68; Census, 2000, p. 453. Reprinted by permission of Larry J. Griffin.

clique of elected union officials who may become less attuned to workers' needs over time. Although some unions have vital internal political processes, the union movement as a whole continues to be typified by a lack of organized internal debate.

Growing and Declining Unions

Union membership in the United States grew rapidly during the Great Depression and steadily, though at a slower pace, during the 1950s, 1960s, and into the 1970s. Figure 6.1 displays trends since 1900 in union membership and in percentage of the labor force unionized. Membership reached a peak in 1978 with 22,757,000 members enrolled and perhaps 10% more covered by collective bargaining agreements but not enrolled as union members. This figure represented 22.2% of the total labor force and 26.2% of nonagricultural employees (Mar-

shall and Briggs, 1989). Union membership as a *proportion* of nonagricultural employment reached its peak during World War II at 35.5%. During the 1970s and 1980s, union membership declined to less than 20% of the labor force. In the 1990s these declines ended and union membership in the United States actually increased slightly. By 2000, the AFL-CIO was growing at a rate of almost 250,000 workers per year, enough to reverse the long-term decline in the percentage of the labor force that is unionized. The UAW, the IBEW, and the Service Workers International Union are among the fastest growing unions. The rate of union membership in Canada, by contrast, is over double that in the United States and was stable or grew throughout the entire second half of the twentieth century. In part these differences reflect a less hostile legal environment for unions in Canada (Rankin, 1990).

What accounts for the decline of union membership in the 1980s and 1990s in the United States and its return to a more stable position in the early 2000s? Ironically, one factor that has limited new union membership is their very success in securing benefits for workers through collective bargaining, social welfare legislation, and spillover effects on nonunionized jobs. It is easy for workers to forget how hard it was to win these benefits. It is relatively more difficult to remain active in the union movement in order to secure and extend these benefits. The erosion of job benefits and job security in the 2000s has been one important factor promoting a resurgence of union membership.

Industrial Shifts Shifts in the composition of industries and occupations have tended to undermine traditional union strongholds. Employment has declined in industries and occupations in which unions became strong in the 1930s, such as mining and steel; and it has increased in white-collar occupations and service industries in which unions have historically been weak. The union movement has had to organize aggressively in areas of employment growth just to maintain membership at current levels. These efforts have been only partially successful. Unionization among white-collar workers increased from 12% in 1970 to 15% by the mid-1990s. Similarly, the proportion of female workers who are members of unions has increased from 10% to 12%. In the most recent decade, well over half of new union members were women.

International Competition The decline of union membership in the 1990s was also caused by additional factors acting to depress manufacturing employment and discourage unionization. High levels of unemployment in manufacturing have resulted at least partially from increased international competition and the movement of American factories and corporations overseas in the search for cheaper labor, lower taxes, and less stringent environmental and safety and health regulations. For example,

from 1970 to 2000, the Machinists and Aerospace Workers Union declined from 754,000 to 431,000. Unionized steelworkers declined from a peak of 1,062,000 in 1975 to just under 500,000 by 2000 (Gifford, 1999).

Based on the declining competitiveness of many American manufactured goods and services in the world economy, many corporations have sought either contractual concessions from labor unions or their total elimination. In the 1990s, many automobile plants in the United States and Canada have closed as market share has been lost to Japanese and European automakers and as U.S. corporations have transferred their production facilities overseas (Rinehart et al., 1997). Demands for wage concessions reached a peak in the early 1980s when unemployment climbed above 9% and have declined since then. Nevertheless, chronically high unemployment has produced a new bargaining environment that encourages a much more aggressive stance on the part of management. Worse, direct plant closings and downsizings have often replaced demands for wage concessions (Milkman, 1997).

Increased Company Resistance More direct attacks on unions have also increased:

> From 1960 to 1980 the number of charges of all employer unfair labor practices rose fourfold; the number of charges involving a firing for union activity rose threefold; and the number of workers awarded back pay or ordered reinstated to their jobs rose fivefold. . . . Managerial opposition to unionism, and illegal campaign tactics in particular, are a major, if not the major, determinant of NLRB election results. . . . The likelihood that an outspoken worker, exercising his or her legal rights under the Taft-Hartley Act, gets fired for union activity is, by these data, extraordinarily high. (Freeman and Medoff, 1984:232–233)

Paid spies and sophisticated antiunion consultants are increasingly used by companies to ferret out and eliminate pro-union workers. It has been estimated that 75% of employers who experience union organizing drives hire antiunion consultants to orchestrate their efforts to avoid unionization (Levitt, 1993). In the 1980s, employers spent over $100 million annually in such efforts (AFL-CIO, 1985).

Why do employers increasingly attack their workers' unions? Are unions responsible for the decline in the United States' competitive position in the world market? The available evidence suggests that union workers are *more* productive than nonunion workers. Their greater productivity results from better training programs, lower turnover, greater tenure, increased commitment to the job, and more professional management in unionized firms (Gunderson and Ponak, 2000). Greater productivity in unionized firms results from management having been forced to develop high productivity routes to competitiveness since low-wage routes to competitiveness are resisted by unions (Michels and Voos, 1992). Unionization does tend to lower the profit rate of firms, because union workers expect higher pay for their more skilled labor. On average, unions reduce the profit rate of companies by about 18% (Freeman and Medoff, 1984:183). The resulting paradox between increased productivity and decreased profitability can create pressure on management to resist unions. Eliminating unions, however, can undermine productivity over the long run.

Growth Areas Even in the face of industrial shifts, international competition, and increased company resistance, many unions have experience growth. Among the fastest-growing unions is the American Federation of State, County, and Municipal Employees (AFSCME, *www.afscme.org*). Similarly, the American Federation of Teachers (AFT, *www.aft.org*) more than quadrupled its membership in the last three decades, moving from 165,000 to 694,000. By 2000, two of the three largest unions in the AFL-CIO, each with a membership of over a million, were AFSCME and the United Food and Commercial Workers, which represent white-collar workers and service workers, respectively.

Workers in large public institutions such as hospitals and schools have been organizing aggressively in recent years. Some commentators argue that public-sector unions kept the labor movement afloat in the troubled 1980s and 1990s. In the public sector, 42% of workers are unionized, whereas in the private sector, only 15% are members. Part of the reason is that public-sector employers are not free to manipulate and intimidate their workers to the same extent as private-sector employers, because government officials must ultimately answer to the public through elections.

Union membership has also grown among clerical workers with the insurance industry, one of the largest employers of clerical workers, being one of the chief organizing targets (Costello, 1991). Clerical workers' motivation to unionize comes partly from the increasing size and anonymity of the organizations in which they work. Another stimulus has come from the wave of office automation, which has increased clerical workers' concerns about job security.

Service workers, including hospital and hotel workers and janitors, are also a growth area. The expansion of unionization into the service sector represents a revival of union effort to organize and serve the poorest segments of society. A nationwide group called "Justice for Janitors" has made widespread organizing gains among janitors. In some large cities, such as Washington, D.C., and Los Angeles, 40% to 90% of janitors are organized. The group uses 1960s-style sit-downs and demonstrations to pressure building owners into negotiating. In 2000, a large janitors' strike in San Diego succeeded in making lasting gains in wages and health benefits for janitors in the city.

International Comparisons Unionization in the United States has declined, and the United States currently has the lowest rate of unionization of all the industrialized Western nations. Slower workforce growth, more gradual indus-

Table 6.4   International Comparisons in Union Membership

	Percent of Total Civilian Wage and Salary Employees								
Year	U.S.	Canada	Australia	Japan	Denmark	Germany	Italy	Sweden	United Kingdom
1955	33	31	64	36	59	44	57	62	46
1960	32	30	61	33	63	40	34	62	45
1965	28	28	46	36	63	38	33	68	45
1970	27	31	43	35	64	37	43	75	50
1975	22	34	48	35	72	39	56	83	53
1980	22	35	47	31	86	40	62	88	56
1985	17	36	47	29	92	40	61	95	51
1990	16	36	43	25	88	39	65	95	46
1995	14	37	35	24	80	29	44	91	33

SOURCES: Clara Chang and Constance Sorrentino, 1991, "Union membership statistics in 12 countries," *Monthly Labor Review* (December), p. 48; *Human Development Report,* New York: Oxford University Press, 1998.

trial restructuring, a tighter labor market, and more efficient organizing strategies all suggest the stabilization or reversal of this trend (McDonald, 1992). Perhaps most importantly, there is a large reserve of workers in the United States who want to be unionized but who do not have a union in their workplace. Among workers who do not have a union, fully 33% say they would vote in favor of union representation (Freeman and Rogers, 1999). Table 6.4 displays unionization rates in some of the major industrialized nations.

Innovative Union Programs for the 2000s

Union strategies for the turn of the century include five broad agendas: (1) national legislative goals, (2) promoting safety and health, (3) programs for making the manufacturing sector more dynamic and competitive so that North American manufacturing jobs can be saved, (4) programs for addressing the needs of the growing segment of low-paid workers, and (5) programs for improving organized labor's image with the public and with its own members.

The National Legislative Agenda Labor unions are seeking reforms in labor law to speed the process of holding union certification elections, extend labor rights to additional groups of

workers (such as farmers and public-sector employees), liberalize picketing laws, eliminate the right of states to outlaw union shops, restrict employers' rights to hire permanent replacement workers, and reinstate the use of sympathy strikes and secondary boycotts. Chief among these goals are those designed to enforce employer noninterference with workers' rights to decide if they want to be represented by a union. Unions have also successfully backed plant-closing legislation in more than twenty states. These laws typically require advance notice of plant closings but may also contain provisions requiring repayment of tax abatements or other incentives given the company to locate in the state.

Union leaders believe that progress is possible on some of these issues, but that sweeping advances are unlikely. Consequently, the union movement puts much of its effort into backing legislation that favors the working class and the working poor more generally, rather than labor unions specifically. First and foremost, this means supporting laws that promote full employment through tax incentives, public works programs, and whatever other means are available. An important part of this package is seeking restrictions on imported goods from nations that employ convict or child labor or otherwise violate basic worker rights. Labor unions also support progressive legislation in education, civil rights, health, housing, and welfare reform.

Promoting Safety and Health Worker safety and health have gained increasing attention as union priorities. This concern results from both increased use of hazardous chemicals at the workplace and increased public awareness of health issues. Unions strongly support right-to-know legislation, which would require the labeling of chemicals at the workplace. They also support more stringent limits on industrial chemicals proven to be hazardous and increased information and education for workers on workplace hazards. Unions have also taken a leading role in developing programs to limit the hazards of video display terminals in the workplace and associated joint problems from prolonged sitting and repetitive wrist movements (Shostak, 1999).

A Broader Role in the Manufacturing Sector In the manufacturing sector, union goals focus on increasing the participation of workers in managing work and setting organizational goals (Appelbaum et al., 2000). The UAW has been a leader in this regard. Current UAW contracts with the major automakers call for extensive training programs jointly managed by the union and the company. These programs help workers learn the skills needed to participate in devising strategies for the most efficient use of new technologies. Innovative programs are also taking place in the GM–Toyota joint venture in Fremont, California. The participation program there includes daily meetings of senior union representatives and plant managers about the operation of the plant, as well as an active role for thousands of teams of workers and supervisors trying to increase efficiency at every level of operation. Similar participation programs also operate at Ford plants where UAW workers have originated over 1,150 proposals for changes in design and production methods. At Ford the participation program is coupled with a profit-sharing plan that results in a bonus of up to $1200 annually for each worker (Schlossberg and Fetter,

1986). The UAW–Ford contract also stipulates that Ford cannot lay off workers if the reduction in force results from transferring employment to overseas subsidiaries or suppliers.

Union goals in the manufacturing sector also include improving the quality of work life, as well as product quality. A successful program was implemented in a joint UAW–Harman Industries worker participation experiment in Bolivar, Tennessee. The plan started with thirty shop-floor committees of workers who suggested changes at the plant, which makes car rear-view mirrors. These suggestions ranged from painting the walls to redesigning the assembly line to starting a "compensatory leave time" arrangement for accumulating overtime credits (in place of overtime pay). Changes also included opening a community child care center, a credit union, and a school open to workers, their families, and residents in the community. The success of this plan rested on using the suggestions of workers, rather than those of managers or consultants, about what would improve the working environment (Zwerdling, 1980).

Unions have a strong interest in programs of worker participation, but they have reservations as well. These reservations stem primarily from a concern for job security. Many large corporations are engaged in antiunion campaigns at the top levels of strategic planning (Appelbaum and Batt, 1994). Such plans call for moving plants to regions of the country less hospitable to unions or even building new facilities overseas to avoid unions. In this context, unions may find that local plant management is a willing partner in participation programs that help fulfill productivity objectives but that top management is simultaneously undercutting American jobs and wages by moving facilities to lower-wage areas and even overseas (Parker and Slaughter, 1994). Unions are thus sometimes suspicious that participation programs may be a short-run strategy to get more work out of workers prior to layoffs. Quality circles and other workplace participation programs have also been used to encourage workers to spy on each other and report union

Mike Parker, 1985, *Inside the Circle: A Union Guide to QWL*. Boston: South End Press, p. 56. Used by permission of the artist, Mike Konopacki.

organizers and sympathizers to management (Grenier, 1988). Unions are suspicious because quality circles can be a carrot with a stick attached. On the other hand, increased productivity and increased quality are essential for the long-term survival of high-paying manufacturing jobs in the United States.

Because of these reservations about the integrity and range of management-sponsored participation programs, unions have also sought to increase the rights of workers over investment decisions. Such rights may be manifest in contractually negotiated job-security clauses. Investment rights can also be realized through various forms of worker ownership. In 1988, for example, 400 union workers bought out their failing shipbuilding company in Seattle. The company emerged from two years of bankruptcy proceedings with workers in control of 73% of the common stock and six of ten seats on the board of directors. The president of the union, Donald Liddle, left to become the chief executive of the company. According to Liddle: "This is the wave of the future and probably the only way to remain competitive. The employees will share in the good and the bad. We stand together not just as united workers, but as partners and owners" (Egan, 1988).

Organizing Low-Wage Workers The growth of low-wage industries and occupations has increasingly moved the United States toward becoming a two-tier society. The upper tier consists of reasonably well-paid jobs, and the other, growing tier consists of marginal employment. The lower tier of jobs provides a major locus of union organizing activities at the beginning of the twenty-first century. Poor wages and virtually nonexistent fringe benefits in these lower-tier jobs have resulted in unions having more success in organizing workers in small firms than in large firms in recent decades (Bronfenbrenner, 1998). Unions promote the theme of social justice for these workers, who are excluded from the mainstream of American society.

Unions have made important advances in organizing low-wage workers in the health-care industry. New members include hospital employees as well as workers in residential homes for the elderly, the mentally handicapped, and the physically disabled. Unions have also made gains in the restaurant and hotel industries. For example, the Service Employees International Union recently organized 400 mostly African-American and Hispanic workers at the Hyatt Regency Hotel in New Orleans (Waddoups, 1999). Such efforts help push back the organizing frontiers of labor among female and minority workers, among poorly paid service workers, and into the historically hard-to-organize South. The fast-food industry is also being targeted by union organizers as a ready pool of poorly paid and poorly treated workers ripe for organizing (Shostak, 1999). Box 6.4 reports on a union movement among topless dancers, a seemingly unlikely occupation to organize, but one with serious demands and objectives similar to those of other workers.

New Organizing and Bargaining Strategies A union movement that was conservative and cautious in the 1980s and 1990s has had to become bolder and more innovative in the 2000s. In this process, unions have developed two new organizing and bargaining strategies:

BOX 6.4   Organizing Topless Dancers

Topless dancers at one of [San Diego's] oldest and best-known bars voted . . . to become California's first unionized nude nightclub.

The dancers—joined by bouncers, bartenders and disc jockeys—are angry about pay and working conditions. . . . Dancers earn $4.35 an hour, plus $100 or more in tips per 6–8 hour shift. But they say club policy forces them to:

- Pay the club $5 an hour for each hour worked.
- Give 15% of their tips to bartenders and disc jockeys, who say they give 40% of all tips to the club.
- Buy costume accessories from the club at inflated prices. Dancers claim cheap garters are sold for $3.50 and last only a night. They say

they buy stockings at $10.50 a pair, which last two or three nights.

Mary, 23, a college student and dancer [at the bar], who won't give her name for fear of being fired, says some nights dancers don't make anything for their work, but owe the club. . . .

Dancers also charge sexual harassment from management, safety violations and lack of job security. . . . Organizer Robert Fisher of Hotel and Restaurant Employees Union Local 30 [believes the workers will now win better conditions].

SOURCE: Ross, Bob. 1993. "Bottom Line for Topless Dancers." *USA Today,* June 4, p. 2. Copyright © 1993, USA TODAY. Reprinted with permission.

---

the corporate campaign and the inside game. The **corporate campaign** expands union activity outside the workplace by targeting financial backers of the company, consumers, and the public. For example, a union might target the principal banks that lend money to a company and distribute literature linking them with wage cuts, unsafe work practices, layoffs, shutdowns, or other problems the workers are experiencing with their employer. Other unions sometimes support such actions by closing their accounts with the lenders, including large retirement funds. Public boycotts of the lenders can also be encouraged. Such negative publicity can have strong effects and lenders will sometimes pressure companies to come to terms with their workers (Peterson, Lee, and Finnegan, 1992).

The **inside game** focuses on informal shop-floor activities. Workers may be encouraged to "work to rule"—that is, to abide to the letter to standard operating procedures. Such exactness in following procedures can bring production to a virtual standstill. Workers can also be encouraged to stop work and to collectively discuss grievances with supervisors, especially grievances about safety issues. Since the inside game often

operates in a gray area between appropriate procedures and sanctionable offenses, shows of solidarity such as wearing union buttons or a similar color or style of clothing on a given day are often used to bolster worker courage and enthusiasm. The strength of the inside game is that it allows workers to stay on the job and draw pay while actively pressuring the company.

Improving the Image of Unions   In an effort to revitalize the union movement, unions are attempting to improve their image with the public and with their own members. The prosperity of the 1950s and 1960s led many people to believe that unions were no longer necessary. The economic stagnation of the last decades of the twentieth century reawakened concerns about job security and wages. However, this awareness came about at a time when union membership was declining because of layoffs and employers were showing increased aggression against unions. In these attacks management blamed unions for the country's economic problems. To counteract these attacks, unions have begun to give greater attention to their image by addressing issues of general interest more vocally rather than issues that affect just one group of workers. As you have seen, these issues include safety and health, plant closings, trade policies, and education.

Unions are also beginning to mount a more active counterattack to defend and bolster their public image. Current union policy priorities of safety and health and child care are selected partly with public appeal in mind. The AFL-CIO also initiated a multimedia project called "Union, YES!" that has probably had its greatest visibility through car bumper stickers. Even the Boy Scouts have been targeted and the AFL-CIO has been successful in getting a labor badge included in the merit badge system of the Boy Scouts. Similarly, the California Federation of Teachers developed a set of lesson plans for teaching about the role of trade unions in resolving workplace conflicts. The lessons involve a role-play exercise in which students work in a fictional company, "Yummy Pizza," and encounter various management restrictions, such as rules against talking and limited restroom breaks. Such strategies are not without effect. Recent national polls suggest that a strong majority of Americans approve of unions and that unions' approval ratings have been improving since bottoming out in the early 1980s (Freeman and Rogers, 1999). Unions are also taking on increasingly active roles in promoting such community service programs as United Way.

Unions are also attempting to increase their level of internal democracy. A study of the International Typographical Union revealed a strong and dynamic role of internal political parties in unions as long ago as the 1950s (Lipset et al., 1956). In the Teamsters, a union often under suspicion for connections between top officials and organized crime, an internal group called the Teamsters for a Democratic Union (TDU, *www.tdu.org*) has grown in importance. The TDU has over 9,000 members in thirty-five chapters, publishes its own newspaper, runs an opposition slate of candidates at the Teamsters' convention, and holds approximately thirty local elected offices (Friedman, 1982). In the UAW a group called the New Directions Movement has hotly contested many UAW policies and has added to internal debate on the role of participation programs in the future of the automobile industry. Organizing campaigns relying on greater member participation are setting the groundwork for greater internal democracy in new and emerging unions as well.

SUMMARY

Workers organize themselves into unions to seek redress for grievances at work. On some issues there will always be a divergence between workers and managers. Modern labor unions emerged as the consequence of a long struggle by workers to improve their situation. In the process of this struggle many alternative forms of organization were explored, including workers' parties, producers' cooperatives, and mass unions. In the United States and Canada the most lasting and significant form of worker organization has been the trade union, which focuses on collective bargaining over wages and conditions. Today, trade unions organize skilled craftworkers and less skilled workers in mass-production industries, as well as increasing numbers of government, service, and professional employees. Unions have been able to coordinate the potentially divergent interests of these groups of workers by combining into national federations, the AFL-CIO in the United States and the Canadian Labour Congress. An important key to the success of unions in the United States and Canada is the simultaneous achievement of trade union autonomy within these federated structures and the development of a united front of labor that can speak with one voice on important national issues.

The primary legacies of the labor movement are heightened job security, the provision of a grievance system, and collective bilateral bargaining with managers over the conditions of work. These rights are secured and protected through the use of strikes. Although strikes are infrequent, they play a vital role in making the collective bargaining system work.

Union membership in the United States declined in the 1980s and 1990s because of employment shifts out of manufacturing and into white-collar and service jobs; movement of jobs overseas or into lower wage, less unionized areas; and increased management attacks on unions. Union priorities for the future include continued efforts to promote full employment, increased participation of workers in setting organizational goals in the manufacturing sector, increased attention to organizing lower-tier jobs in the service sector, and efforts to improve the image of labor as an essential component of advanced industrial society.

In some ways the decline of membership in labor unions in the 1980s and 1990s is reminiscent of a similar decline in the 1920s. On the other hand, "the labor movement today numbers over 23,000,000 members, compared to 3,000,000 in 1929, and today's unions include millions of women, nonwhites, and professional employees absent from the labor movement of the 1920s" (Dulles and Dubofsky, 1984:400). It is impossible to predict whether the next decade will witness a boom in union membership comparable to the 1930s. Further dramatic declines, however, seem unlikely. The bulk of membership losses in the manufacturing sector have probably already occurred, and organizing drives in service and white-collar jobs have been increasingly successful. How successfully unions will respond to the challenges and opportunities of the future remains an open question.

KEY CONCEPTS

craft union	injunction	collective bargaining
solidarity	American Federation of Labor	shop steward
general strike	Congress of Industrial Organizations	seniority
mass strike	industrial union	grievance procedure
secondary boycott	sit-down strike	internal democracy
general union	National Labor Relations Act	corporate campaign
Knights of Labor	Taft-Hartley Amendments	inside game
Industrial Workers of the World		

QUESTIONS FOR THOUGHT

1. Describe some of the different ways in which workers have organized themselves and the reasons for the successes and failures of different types of organization.

2. Why was the Great Depression the pivotal turning point for labor unions in North America?

3. What are the major roles of labor unions in the workplace today? Which of these roles are being pursued successfully and which less successfully?

4. What problems are labor unions facing today? Which of these challenges do you think unions will be able to meet? Which will continue to be problems?

5. Describe current areas of union growth and the reasons that these are growth areas.

6. What strategies are labor unions developing to confront the changing economic and political situation of the 2000s? Do you think these strategies will be successful in helping to revitalize the union movement?

MULTIMEDIA RESOURCES

Print

Richard O. Boyer and Herbert M. Morais. 1955. *Labor's Untold Story.* New York: United Electrical, Radio and Machine Workers of America. Labor history told from the standpoint of the ordinary worker.

John Steinbeck. 1939. *In Dubious Battle.* New York: Modern Library. A compelling fictional account of the rise and defeat of an early unionization effort among California's agricultural workers.

Foster Dulles and Melvyn Dubofsky. 1984. *Labor in America,* 4th ed. Arlington Heights, Ill.: Harlan Davidson. Perhaps the best comprehensive history of the American labor movement.

Richard B. Freeman and James L. Medoff. 1984. *What Do Unions Do?* New York: Basic. The best source book available on the role of labor unions in the contemporary economy.

Richard B. Freeman and Joel Rogers. 1999. *What Workers Want.* Ithaca, N.Y.: Industrial and Labor Relations Press. Reports a comprehensive new survey on the desires of workers for unions and for other forms of participation in the workplace.

Arthur B. Shostak. 1999. *CyberUnion: Empowering Labor through Computer Technology.* Armonk, N.Y.: M.E. Sharpe. An exciting study of the use of new technologies to facilitate the organizing and bargaining activities of labor unions by the senior American authority on trade unions.

Daniel Zwerdling. 1980. *Workplace Democracy.* New York: Harper and Row. Sixteen case studies of workplaces in trouble, attempted solutions, and the successes and failures of those solutions. Several cases involve various forms of worker ownership.

Websites

Labor on Line. *www.laboronline.org* Labor news, resources, and links.

Labor Notes. www.labornotes.org A quarter-/century-old independent workers' news magazine operating from Detroit.

Canadian labour news. *www.geocities.com/CapitolHill/5202/canada.html*

Canadian Auto Workers. *www.caw.ca*

The Steward. *www.thesteward.net/canada* Canadian site dedicated to defending and enlarging human, civil, and worker's rights.

Walter Reuther Labor Archives. *www.reuther.wayne.edu* The most important archives of labor information and history in North America.

Progressive causes network. *www.igc.org*

Cyber Picket Line. *www.cf.ac.uk/socsi/union* Provides information about strikes around the world.

RECOMMENDED FILM

Norma Rae (1979), Sally Fields and Beau Bridges. The true-life story of a poor female textile worker who helps organize a southern mill.

PART III

❖❖

Industries and Technologies

In the first two parts of this book you learned about the history of work, how people enter a particular career, the problems and satisfactions of work, and how people respond individually and collectively to work. In this section we examine in greater detail the nature of work in specific industrial settings.

Companies that produce the same product or service are collectively called an *industry*. Each enterprise in an industry is likely to have a similar, though not necessarily identical, technology and organizational structure. Industries are thus bundles of products or services, with attendant technologies and organizational structures. The industry one works in has great consequences for the nature of one's work; different industries require different skills and have different implications for the stability and quality of employment. In this section we discuss how these differences influence the nature of work.

Why do employment levels in different industries change across time? Many employment changes result from changing technologies that allow more goods and services to be produced with less labor. The most dramatic example of such technologically driven changes in employment is provided by agriculture. In 1790, nearly 90% of the labor force was required to produce food for Americans (Chandler, 1977). By 2000, only about 2% of the labor force was required to produce an even greater quality and variety of food (see Table A).

Table A   Percentage Distribution of American Workers by Industry, 1870–2006

	1870	1880	1890	1900	1910	1920	1930	1940	1950	1960	1970	1980	1990	2006[a]
Agriculture	50.8	50.6	43.2	38.1	32.1	27.6	22.9	19.2	12.7	7.0	3.7	3.4	2.7	2.2
Mining	1.6	1.8	2.0	2.6	2.9	3.0	2.5	2.1	1.7	1.1	0.8	1.0	0.6	0.4
Construction	5.9	4.8	6.1	5.8	6.4	5.3	6.5	4.7	6.2	6.2	5.8	6.2	6.5	6.1
Manufacturing	17.6	18.4	20.2	22.1	22.8	26.4	24.4	23.9	26.3	28.3	25.9	22.1	18.0	16.8
Transportation and utilities	5.0	5.0	6.5	7.3	8.8	10.2	7.6	7.0	7.9	7.1	6.8	6.6	6.9	6.8
Trade	6.5	7.1	8.4	9.6	9.3	9.8	12.6	14.5	15.8	16.1	16.9	20.3	20.6	21.8
Finance and real estate	—	—	—	—	1.4	1.9	3.0	3.4	3.5	4.3	5.4	6.0	6.8	6.5
Domestic service	7.4	6.3	6.5	6.1	6.0	4.1	6.5	5.3	3.2	3.1	1.7	1.3	0.9	0.8
Other personal services	2.0	2.1	2.7	3.4	4.2	4.0	4.7	8.7	8.9	8.2	8.3	8.0	10.8	11.0
Education	1.5	1.9	2.2	2.3	2.5	2.8	3.1	3.5	3.8	5.4	8.6	7.7	7.3	7.7
Other professional services	1.1	1.1	1.5	1.7	2.1	2.6	3.5	4.4	5.4	8.0	10.5	12.0	14.2	15.3
Government	0.8	0.8	0.8	1.0	1.5	2.2	2.8	3.3	4.5	5.2	5.6	5.4	4.8	4.7

[a]Projection based on assumption of moderate economic growth.

SOURCES: U.S. Department of Commerce, Bureau of the Census, *Statistical Abstract of the United States: 1999;* and Bureau of the Census, 1975, *Historical Statistics of the United States,* Washington, D.C.: U.S. Government Printing Office.

Employment changes also result from older goods being replaced by new goods and services. For example, few blacksmiths are employed to shoe horses in a modern economy. Instead, a great many mechanics are employed to maintain and repair automobile engines.

Some industries have grown dramatically in their share of employment. Educational services have grown as a result of the greatly expanded need and desire for literacy and higher education. Professional services, which include medical, engineering, accounting, and legal services, have grown rapidly. Similarly, government (public administration) expanded through the 1960s as a result both of the growth of government-provided services and of the need for more workers to administer an increasingly complex state and economy. Recent growth has occurred only at the state, county, and municipal levels with the federal government actually decreasing in size in recent decades.

Retail and wholesale trade has become the largest employer in the economy. This sector has grown because of the increasing availability of consumer goods and the need to sell these goods. Personal services have also grown over time. Personal services include such industries as hotels, eating and drinking places, day care for children, repair, laundry, and entertainment. Housewives and domestic servants working at home provided many of these services in the past. Thus, the growth of personal service parallels a decline in domestic service and the movement of women out of home work and into the paid labor force. The combined growth of domestic and personal service from about 9% to 12% of the labor force indicates a moderate increase in the demand for personal services coupled with a significant transformation in the way personal services are delivered.

Other industries have experienced more moderate changes. Transportation and utilities reached a peak level of employment in 1920 but have today declined to the level they occupied in the 1890s. This pattern results from a steadily increasing demand for transportation, communication, and utility services, coupled with the development of significant labor-saving technologies in these industries in the 1930s and beyond. These advances include motorized transportation, more efficient electric generation and distribution systems, and the development of highly efficient telecommunications systems. Construction has also maintained a relatively stable percentage of employment. This stability results from moderate growth in the demand for physical structures such as houses, buildings, and roads, and a relatively slow pace of technological innovation in the construction industry.

Manufacturing today employs about 17% of the labor force—a level similar to that in 1870. However, this stability at the endpoints of our time line masks the rise and subsequent decline of manufacturing employment in the United States. In 1960, manufacturing employed over 28% of the labor force. This peak percentage represents the dominance of American manufactured goods in the post–World War II world economy. The subsequent decline in manufacturing jobs is due to increased international competition and technological innovations that have reduced labor requirements in manufacturing.

These employment changes and the concepts necessary to understand them are the subject of the four chapters in Part III of this book. Chapter 7 explores the roles of technology and organization in determining the nature of work. Chapter 8 examines the nature of work in the extractive and manufacturing sectors. Chapter 9 explores current changes in the nature of work resulting from the rapid spread of microchip technology. Chapter 10 examines the nature of work in the expanding service sector.

7

Technology and Organization

The American labor force—indeed, any industrial labor force—
is characterized by (1) the separation of the workplace from the
household, (2) a distinction between the worker as a person and the
position he occupies, (3) widespread employment in large-scale
organizations with both bureaucratic and professional forms of authority,
(4) individual accountability for the performance of tasks judged
according to standards of competence, and (5) by the affiliation of
individuals to organizations through contractual agreements.

(DREEBEN, 1968:114–115)

In industrial societies most work takes place inside large, complex organizations that use highly specific production technologies. The above quotation identifies some of the key characteristics of work in such organizations. No longer do most people work alongside family and community members to produce goods for local consumption. Instead, they enter into contractual employment agreements and work according to the dictates of bureaucratic and professional standards to produce goods and services for mass markets.

Technology and organization are, in essence, two sides of the same coin. That coin, called the *social relations of production,* comprises all the material and nonmaterial means and techniques used to produce goods and services. Thus, social relations of production include the tools and machines used, the skills needed, the formal and informal group structure utilized, and the structure of the larger organization and its relation to other organizations in society. Although technology and organization are thus inseparable, it will be useful to discuss them separately for purposes of presentation. It should always be remembered, however, that they are intricately related.

In this chapter we discuss the ideas of technology and organization and examine how they jointly determine the nature of work. We also examine the nature and limitations of the modern bureaucratic organization of work.

DEFINING TECHNOLOGY

Technology is the application of knowledge and skills for the achievement of practical purposes. It includes both physical apparatus, such as tools and machines, and the knowledge required to build and use them and to solve problems in their application to the production of goods and services. Technology has three components or aspects: operations technology, materials, and knowledge (Bijker, Hughes, and Pinch, 1987).

Operations Technology

The people and machinery that produce a good or service, along with the set of rules and procedures that pattern their use, are called *operations technology*. For example, the operations technology of college education includes professors scheduled to teach specific subjects at certain times, classrooms, libraries, and living and recreational accommodations for students. The operations technology of a chicken-processing factory includes a receiving dock for crates of live chickens, an assembly line where the chickens are slaughtered and dressed, and a packaging area where the dressed chickens are prepared for shipment.

Materials

The *materials* to be used in producing a good or service are also part of the technology of its production. In a chicken-processing plant, for example, live chickens are the essential material. However, scalding water, plastic wrap, waste bins, refrigeration packs for the dressed chickens, and shipping crates also are required as raw materials. These materials are previously manufactured subcomponents used in the production of processed and packaged chickens. Automobile assembly provides another example. Engine blocks, chassis, radiators, transmissions, windshields, and other components are the materials. These materials and their characteristics are part of the essential technology of the manufacture of automobiles. These examples also illustrate how the technology in any one industry (for example, chicken processing or automobile assembly) depends on the technology used in other parts of the economy to produce the needed materials and subcomponents.

Knowledge

The third aspect of technology is *knowledge*. Knowledge is required to operate the various machines, deal with the exceptional cases that are always characteristic of production, and coordinate production activities. Production always involves variability and uncertainty. Unpredictable aspects of production include access to materials, variable quality in these materials, wear and breakdown of tools and machinery, and variations in weather and staffing. Production workers must have a thorough knowledge of the technology and materials in order to anticipate and accommodate the unexpected. For example, the chicken snatchers in a chicken-processing plant must have knowledge of chicken behavior and movement in order to seize and hang their prey on a moving assembly line. Variability in chicken behavior makes these jobs difficult or impossible to automate.

Defining Organization

Organizational structure is the established pattern of relationships among the various parts of an organization and among the various employees in the organization. Organizational structure is not visible in the same sense as the structure of a material object, such as a bridge or archway, but its consequences are just as real. Sociologists use the related but more general concept of **social structure** to describe such diverse phenomena as the family and the state. Families have structures made up of relations between different members of the family. For instance, the family structure of a single-parent family with two teenage girls consists of the relationship between each daughter and the parent and the relationship between the two sisters. Different families have different structures, just as do different governments and different organizations. These structures specify the patterns of obligations and responsibilities that the incumbents of different roles have in relation to one another. These structural relationships also influence the sympathies, affections, and animosities that different members of the group are likely to experience toward one another. Different organizations, such as government agencies, economic organizations, religious organizations, and political parties, have different aims; what they share in common is having identifiable structures for the attainment of these aims. In formal organizations, such as large corporations, specific job positions and the hierarchical and lateral relations among these positions comprise these structures.

According to the organizational behaviorists, French, Kast, and Rosenzweig (1985), the structure of economic organizations consists of the following parts:

1. the pattern of formal relationships and duties (the organizational chart plus job descriptions)

2. how the various activities or tasks are assigned to different departments and people in the organization (differentiation)

3. how these separate activities or tasks are coordinated (integration)

4. the power, status, and hierarchical relationships within the organization (authority structure)

5. the planned and formalized policies and procedures that guide and control the activities and relationships of people in the organization (administrative structure).

Researchers have developed several concepts to describe variations in these structural characteristics. Mary Zey (1998), an organizational sociologist, argues that the most important variations in organizational structure include (1) *role specialization:* are roles in the organization defined in terms of a narrow range of activities as opposed to a broad agenda of responsibilities? (2) *standardization:* do procedures follow specific rules as opposed to being developed on an ad hoc basis for each situation? (3) *centralization* of decision making: are key decisions made by one or a few persons at the top of the organization or dispersed throughout the organization? (4) *autonomy* of the organization: is the organization free to make its own decisions and establish its own agenda as opposed to reporting to a larger governing body, such as a parent company? and (5) *ratio of supervisors to production workers:* does the organization have many managers as opposed to having the majority of its members involved in the production of its product or service?

As you can see, there is clearly substantial overlap between the concepts of technology and organization. For example, are the division of labor and the assignment of workers to different tasks part of the organizational structure, or are they part of the technology of production? Obviously they are both. When examining the nature of work in upcoming chapters we avoid entering into debates about whether some aspect of the social relations of production is a characteristic of technology or of organizational structure. But you should be aware that technology and organization are closely related.

HOW DOES TECHNOLOGY INFLUENCE WORK?

In this section we present a developmental history of technology. We also discuss the concept of skill and how skills are changing because of changing technologies.

Changing Technologies

Technology has developed in distinctive stages, with each stage retaining aspects of the earlier stages and adding to them. Many technologies were discussed in the more general history of work presented in Chapter 1. Here we organize the presentation in terms of distinct technological stages.

Simple Tool Technology The earliest technologies were used by primitive humans to wrest material sustenance from the environment. These technologies included hand tools, made from stone and bone, and woven baskets. Early technologies also included the knowledge needed to make and use these tools and materials. People had to know where to find stones that could be flaked or ground to a sharp edge, how to fashion these stones into usable tools, where to find and how to work with clay or reeds to make bowls or baskets, how to locate edible plants, and how to locate and kill small game. We refer to this level of technology as simple tool technology.

Craft Technology The second level of technological development entails the use of more advanced tools, which require greater effort to manufacture. For instance, the production of ovens for firing clay pots requires a technology to build the ovens in such a way that desired temperatures can be reached and maintained. Similarly, mining and rudimentary smelting of copper or iron ores requires more advanced technology to mine and transport ore and to build kilns. The metals produced can then be used to make a variety of tools and final products. This level of technological development is associated with a division of labor into different crafts, such as ironworking, glass blowing, and boot making, and is accordingly called **craft technology.** The skills needed for craft technology are typically acquired through apprenticeship.

Mass-Production Technology The third level of technological development is that required for the mass production of commodities. **Mass-production technology** depends on the availability of a variety of specialized goods and services and on the integration of these into a mechanized production operation (Noble, 1997). Because tools and products are more specialized under mass-production technology, there is a great need for specialized knowledge. People acquire such knowledge through apprenticeships or college training. Additional knowledge is also required to coordinate diverse production operations. The term "Fordism" is sometimes used as a short-hand expression for mass-production technology, because of the important role of the Ford Motor Company in developing assembly-line technologies in the early years of the twentieth century.

Microchip Technology The latest stage of technological development employs numerical control processes and electronic microprocessors. These technological innovations allow the automation of many production processes and the almost instantaneous collection and tabulation of data about production. Numeric, electronic, and computer-driven technologies have had important consequences for the nature of the skills and knowledge needed in production. This new stage of advanced **microchip technology** will be our focus in Chapter 9.

Even before the widespread use of microprocessors, American sociologist Daniel Bell (1976) argued in an influential book, *The Coming of Post-Industrial Society,* that new knowledge-based technologies would transform the nature of society. Bell maintained that knowledge would become the central resource in society. He predicted that the economy would become *knowledge-intensive* rather than *capital-intensive*. He also forecast that the economy would shift from the production of goods to the production of services and that women would have more job opportunities. Some of these predictions have come true, and others have not. For instance, women have gained new job opportunities; however, these opportunities have come less in the new knowledge-intensive industries, such as engineering, than in more labor-intensive industries, such as personal services. The expanded opportunities of women have thus occurred in those industries that have been *least* influenced by technological change (England, 1992).

In the workplace of the late twentieth century, workers produce more goods, as well as a greater variety of goods, than was ever before possible. The development of new technologies has brought with it increased productivity and the possibility of shorter work hours, increased safety, greater employment security, and material abundance for all members of society. In following sections we will see how organizational structures have both facilitated technologically based increases in productivity and shaped the distribution of rewards in such a way that many people in society are excluded from the benefits of increased productivity.

What Is Skill?

Changing technologies have important consequences for the types of skills needed by workers. Sometimes these skills are more complex than previous skills; sometimes they are simpler. Skills may be based on *mental, manual,* or *interpersonal* aspects of work, or on a combination of these. Social scientists generally measure the skill level of a job in terms of **complexity, diversity,** and **autonomy.** *Complexity* refers to "the level, scope, and integration of mental, interpersonal, and manipulative tasks in a job" (Kalleberg and Berg, 1987:175). *Diversity* refers to the number of different tasks and responsibilities required by a job. *Autonomy* implies self-direction and the potential for creative improvisation (Attewell, 1990).

The amount of training time required to qualify for a job is often taken as an indicator of the level of skill required. Many tacit skills essential for a job, however, can only be learned by long experience on the job. Tacit skills are those bits of information and knowledge that are not easily expressed as formal knowledge but are nevertheless essential for doing work correctly and efficiently. Tacit skills vary with each job but often involve assembling, considering, and weighing a range of considerations in determining how to proceed on a given task. Such tacit skills combine with formal knowledge to define the working knowledge required on a job. Thus, formal training provides only some of the actual skills required on a job (Vallas, 1990). An overreliance on formal training as opposed to on-the-job experience as an indicator of skill tends to exaggerate the skills of those with formal credentials over those with experiential knowledge (Harper, 1987). Box 7.1 describes some of the tacit skills acquired over time and used in the daily work of a machinist.

Many researchers argue that skills have declined because of increased automation (Braverman, 1974). This phenomenon is called **deskilling.** Evidence based on case studies of specific industries, such as printing, often supports such concerns (Wallace and Kalleberg, 1982). Studies based on broader samples of occupations, however, generally find that skills have been relatively stable in the period since World War II. That is, technological changes have had offsetting effects, with some jobs becoming more skilled and others becoming less skilled, so that

BOX 7.1   The Working Knowledge of a Machinist

When you were tempering something you get it to what they call a cherry red. One piece of steel you might need to get to a cherry red, maybe another one a little redder. You cool it in certain ways as you go along. It draws the temper into the steel. Makes it harder. But if you cool it too quickly it gets tempered so hard it's just like glass—you can break it.

They have what they call flame temper, an oil temper, or a water temper. Like if you sharpen a pick—you hammer the point out on a pick and then you want to temper it so it won't burr over when you hit a stone—that's a cherry temper. But if you temper it *too* hard and you hit a stone, it'll pop the end right off. You dip it in the water slow.

And it'll turn a bluish color as the temper works out into it. And your coal temper—a temper out of a coal forge—is a lot better than your gas temper. See, they use gas forges now. Or I can temper with a torch, but you've got to be very careful with it. When you're using gas you're only heating one side at a time. When you're using coal you're poking the metal right into the hot ashes. It heats more evenly, all the way through and around. Where with your gas you don't get that. And you only heat one side with the torch, and it's not as good.

SOURCE: Excerpted from Douglas Harper, 1987, *Working Knowledge.* Chicago: University of Chicago Press, pp. 32–33.

---

the average level of skill demanded has remained fairly stable (Leigh and Gifford, 1999). (See Chapter 9 for a more complete consideration of skill upgrading and deskilling in the context of high-technology work.)

Acquiring New Skills

Workers enter the labor market with a certain level of education and experience. Between 1940 and 2000, the average level of formal education of adult members of the labor force (those twenty-five years and older) increased dramatically from 8.6 years to 13.2 years. Because of changing skill requirements, workers often need additional training after completing high school to qualify for a job. This training is acquired in a variety of ways. Of the workers who needed more than high school training for their current job, 52.1% reported that they had attended a training school or college. In addition, 50.1% said that they acquired needed skills through informal on-the-job training, either at a previous job or on their current job. More than 17.5% said they had learned the necessary skills through a formal company training program. Other significant

sources of skill training included friends and relatives or activities not related to work, such as hobbies or avocations (5.9%), armed forces training (3.5%), and correspondence courses (1.4%) (U.S. Department of Labor, Bureau of Labor Statistics [BLS], 1999). These percentages total more than 100% because some workers report that they make use of more than one source.

Community Colleges and Vocational Training The sources that workers use for acquiring job skills vary by occupation. Enrollments in four-year colleges are steadily increasing. Teachers, engineers, social workers, accountants, and many other professions earn their degrees through four-year colleges. Advanced degrees, such as those for doctors and lawyers, build further on these four-year degrees as a base.

Community college and vocational school enrollments, however, have grown even faster than and now exceed enrollments in four-year colleges. Two-year associate degrees cover a range of fields including medical, dental and biological technicians, computer and electronics technicians, aircraft and powerplant maintenance, computer programming, business management, and hotel management. Vocational training credentials are offered in a range of fields including hairdressing, upholstery, automobile mechanics, refrigeration technology, and truck driving. Thus, additional formal schooling, as provided both by colleges and by vocational schools, is widely used across white-collar, blue-collar, and service occupations.

Apprenticeship Programs Formal **apprenticeship** training programs are more likely to be used by skilled craftworkers. Training programs in the skilled construction trades must adhere to national standards and involve a combination of classroom instruction and supervised on-the-job training that typically lasts from two to six years. Skilled factory workers, police officers, and telephone installers and repairers are also likely to learn their skills in company-sponsored training programs. Many of these programs are operated as joint union–management apprenticeship programs. However, there are no national standards for these programs and completion of such programs may not produce credentials that are universally acknowledged.

On-the-Job Training Both formal and informal **on-the-job training** is also widely used as a source of skill acquisition and upgrading. In the skilled trades, workers often attend company-sponsored training classes as well as honing their skills through experiential knowledge and interaction with peers on the job. Managers and supervisors also learn needed skills through both formal training classes and informal on-the-job learning.

Occupations whose incumbents receive substantial training in the armed forces include aircraft engine mechanics and electricians and electronics technicians. Correspondence courses and paid workshops are used by repairers of electronics and industrial equipment, and by workers in securities, financial sales, and service occupations. Occupations that rely significantly on friends or experiences outside of work for their training include dressmakers, musicians, farmers, carpenters, and painters.

Table 7.1 Workers Using Computers on the Job

Type of Application	1993	1997
Inventory control	45.0%	66.4%
Word processing	44.4	57.0
Communications	38.7	47.0
Analysis/spreadsheets	36.1	40.9
Databases	34.1	34.5
Desktop publishing	22.3	26.1
Sales and telemarketing	16.2	22.1

SOURCE: U.S. Department of Commerce, Bureau of the Census, 2000, *Statistical Abstract of the United States, 1999,* Washington, D.C.: U.S. Government Printing Office, Table 696.

Training in computer skills has become increasingly widespread across a wide range of occupations. Computers are used for inventory control, communication and correspondence, and word processing. The rapid increase in computer use across a variety of tasks is illustrated in Table 7.1.

HOW DO ORGANIZATIONS INFLUENCE WORK?

The skills required at work are determined by the way tasks are divided, by organizational structure, and by the technology used. In this section we discuss the increasing division of labor and the ways in which the social organization of work allows some people to control the work of others.

The Division of Labor and Changing Organizational Structures

The division of labor was initially discussed in Chapter 1. Here we examine its implications for organizational structure. The earliest form of the division of labor is that between men and women and between child and adult. This stage is called the *division of labor by gender and age.* Men specialized in hunting, and women specialized in gathering. Children tagged along as adults went about their activities helping as needed and as their abilities allowed. This activity helped them acquire adult skills. At this stage, organizational structure was identical with family and group structure.

The Social Division of Labor   The social division of labor—the division into different crafts or trades—was typical of work in feudal society. The specialization of workers in crafts results in more efficient production and in better-quality goods. At this stage of development the main organizational structures of work were feudal relations in the rural areas and guild structures in the towns.

The Manufacturing Division of Labor   Most work in industrial society is organized in terms of the manufacturing division of labor. In this stage the different activities involved in each craft are separated. Instead of making soles for a single pair of shoes, then making tops, and then stitching each set together, a shoemaker might make a number of soles and a number of tops and then stitch them together. This breaking down of a task into parts is called the *analysis of labor.* The analysis of labor produces efficiencies in handling materials and the rhythm of work. The manufacturing division of labor may also involve the assignment of the different parts of the job to different workers. In combination, the division and reassignment of tasks allows more efficient mass-production techniques such as those based on the assembly line. The price of finished goods may be further lowered because some workers are paid less for doing parts of the task that require less skill. Starting with this stage, work is increasingly carried out in large, complex organizations.

The manufacturing division of labor also creates the preconditions for a further heightening of productivity through mechanization. Once tasks have been separated into parts, it is possible to develop machinery to do some of the work that was previously done by hand. In 1776, Adam Smith argued that the division of labor produces efficiencies because of the increased dexterity of the worker as he or she specializes in one task, because of the time saved "in passing from one

sort of work to another," and because of the introduction of machinery. In contemporary views, the introduction of machinery is seen as the most significant of these developments. The other two aspects that Smith mentions, increased manual dexterity and reduced time lost in transferring from one aspect of work to another, are frequently negated by the increased alienation of workers and by their lost enthusiasm—factors he did not fully recognize. Such factors are especially important today with more sophisticated production systems that require great worker involvement and commitment.

Organizational Structure as Labor Control

The division of labor produces not only specialized positions but also a vertical differentiation within organizations based on power. The workers whose tasks are finely subdivided suffer a loss of skill, a loss of power, and, finally, a loss of wages. Meanwhile, the power and income of those who organize the labor of others increase. Some social scientists even argue that the *purpose* of assigning detailed tasks to different workers is to lessen their skills and thus lower their wages (Braverman, 1974). The analysis of craftwork into parts and the development of machinery to help with some of the tasks clearly increase productivity. However, some researchers argue the assignment of detailed tasks to different workers is motivated by a drive to deskill labor in order to cheapen its price (Stone, 1974). Thus, the organization of labor also becomes a way in which workers are controlled and manipulated.

Direct Personal Control   To understand how this takes place, the American economist Richard Edwards (1979) categorized the control of the labor process into stages. The first stage is that of **direct personal control.** The owner shows the workers how to do the work and indicates the appropriate pace. The owner also evaluates the workers and rewards or punishes them according to their performance. Early in the Industrial Revolution, punishments included verbal and physical coercion and the threat of firing. Today punishments are mainly restricted to firings, and a greater number and diversity of positive rewards are used, such as promotions and raises. Direct personal control works reasonably well for employers. It typically gets the work out on schedule and according to plan. Employers still use it in small enterprises today. However, the increasing prevalence of large enterprises undermined the ability of employers to use direct personal control. As a result, it is not how most work is organized in industrially advanced economies.

Foreman's Control With the growth of large enterprises, direct personal control is replaced by **foreman's control.** Under this system, the owner hires foremen who take on the duties of recruiting and supervising workers. This system of labor control prevailed in steel manufacturing, railway construction, textiles, and many other industries until about 1900. Foreman's control still exists in agriculture and in construction, but it was largely abandoned in most industries because it proved inadequate for organizing work under mass-production systems that needed more coordination and more standardized procedures. Each foreman had tremendous latitude in organizing tasks. In mass-production industries, large numbers of interchangeable components are needed. Therefore, it is essential to use identical techniques throughout the operations and to coordinate activities in order to ensure the quality and consistency of the final product.

Scientific Management New systems of control were developed in an effort to implement more standardized procedures in mass-production industries. The American industrial engineer Frederick Taylor developed a system that he called **scientific management.** Taylor believed that there was "one best way" to do every task. This way could be discovered by first carefully observing how the workers did the task and then devising a more efficient way to do it. For instance, he experimented with different ways of holding cutting tools on metal lathes in order to produce machine parts as quickly as possible and with different sizes of shovels for more efficiently loading pig iron.

Taylor was also concerned with the problem of workers who resisted working as fast as possible; he saw this problem as a major impediment to efficiency. Taylor observed that many workers engage in **soldiering;** that is, they intentionally work well below their capacity. He concluded:

> The greatest part of systematic soldiering . . . is done by the men with the deliberate object of keeping their employers ignorant of how fast work can be done.
>
> So universal is soldiering for this purpose, that hardly a competent workman can be found in a large establishment . . . who does not devote a considerable part of his time to studying just how slowly he can work and still convince his employer that he is going at a good pace. (Taylor, 1911:32–33)

Taylor's solution was to fire skilled workers and hire less skilled workers to replace them. The new workers could then be trained to do the work Taylor's way. They would receive a higher wage than they had received for less skilled work but would be paid less than the skilled workers whom they replaced. Box 7.2 presents Taylor's outline of the principles of scientific management.

Technical Control Like scientific management, **technical control** was developed to standardize work procedures in the mass-production industries. Under technical control the worker is controlled and paced by the machinery. The classic example of this is the assembly line. The activity and pace of work are directly controlled by the assembly line, which delivers materials to the worker. In many ways the technical control of work is the most rigid form yet developed.

Worker Resistance Scientific management and the technical control of work appear to offer great efficiencies because they specify precisely how the work is to be executed and how quickly it is to be done. However, they also have serious drawbacks and limitations. Chief among these is that they make limited use of the workers' skills. Production rarely occurs exactly as planned, and machinery

---

BOX 7.2   The Gospel of Work According to Taylor

Frederick Taylor is considered the father of scientific management. The following passages express his outline for organizing work:

> The managers assume . . . the burden of gathering together all of the traditional knowledge which in the past has been possessed by the workmen and then of classifying, tabulating, and reducing this knowledge to rules, laws, and formulae. . . .
>
> All possible brain work should be removed from the shop and centered in the planning or laying-out department. . . .

The work of every workman is fully planned out by the management at least one day in advance, and each man receives in most cases complete written instructions, describing in detail the task which he is to accomplish, as well as the means to be used in doing the work. . . . This task specifies not only what is to be done, but how it is to be done and the exact time allowed for doing it. . . . Scientific management consists very largely in preparing for and carrying out these tasks.

SOURCE: Frederick W. Taylor, 1911, *The Principles of Scientific Management.* New York: Harper and Row, pp. 39, 63, 98–99.

---

and parts often fail. When the workers have been robbed of the skills necessary to understand their work and their tools, or have never been given the opportunity to learn these skills, they are not in a position to handle unexpected situations nor to facilitate production. Scientific management and technical control also rob workers of their enthusiasm for work. When workers have been alienated from their work in this way, they make a science of finding ways to allow production to lag, or at least they take little initiative to increase productivity. Scientific management and technical control, because they treat the worker like a machine or an animal, deny the humanity of workers. These systems of organizing work thus do not reap the full benefit of workers' enthusiastic participation and may generate many unanticipated consequences as well. In particular, workers may organize to resist scientific management and technical control. The greatest growth in the American labor movement, for instance, occurred immediately after the advent of these techniques.

Rediscovering the Worker

In 1927, as mentioned in Chapter 2, a series of studies began at the Western Electric Company's Hawthorne plant near Chicago under the direc-

tion of Elton Mayo (Roethlisberger, 1939). Although these studies were motivated by a desire to increase productivity, they yielded many surprising findings that gave birth to what has come to be called the **human relations school** of industrial relations. Initially, experiments were done with various combinations of lighting levels and break schedules in an effort to find the optimum conditions for maximum productivity. The researchers observed that productivity increased in both the experimental group and the control group whenever *any* change in procedures occurred. In an effort to understand this surprising result, additional studies were undertaken in a relay assembly test room and a bank wiring room.

In the relay assembly test room, the experiments continued for two years. The women in this group were placed in a separate room for observation. In addition, they were given regular physical examinations at which cake and ice cream were served. Two women were also removed from this group during the two-year period for "talking too much" and for having generally "bad attitudes" toward the experiment. Again, as in the case with the lighting experiments, productivity tended to increase no matter what the experimental manipulation. This kind of result became known as the **Hawthorne**

effect: added social attention, regardless of its content, increases productivity. The discovery of the Hawthorne effect was taken as a refutation of the vision of "economic man" proposed by scientific management. The economic man could be manipulated by wage rates and incentive plans. In the place of economic man, the human relations theorists proposed a vision of "social man." This theory argued that workers needed a positive social context within which to achieve maximum productivity (Grint, 1991).

In the bank wiring room, the researchers used a different strategy. Here, instead of using experimental manipulation, they kept detailed observations of workers' behavior in order to understand the causes of increased productivity. They observed that there was a prevailing norm among workers for how many switches should be wired in a day. Workers who exceeded that norm would be ridiculed and ostracized by their fellow workers. Workers who exceeded these informal quotas were given such caustic nicknames as "Shrimp," "Rate Breaker," "Slave," or "Speed King." This finding further reinforced the idea that understanding the social aspects of work was of central importance for increasing productivity.

In more recent reflections on the Hawthorne studies, the observations in the bank wiring room have also been interpreted as indicating that workers are capable of resisting productivity drives that they define as exploitive (Jones, 1992). Similarly, productivity increases in the relay assembly test room have been reinterpreted in light of the firing of the two recalcitrant women during the test period. It appears that management coercion and worker resistance do not disappear just because greater attention is given to meeting workers' social needs.

THE GROWTH OF BUREAUCRACY

At the dawn of the twenty-first century, most work takes place inside large, complex enterprises. Modern enterprises are complex in their level of horizontal, vertical, and spatial differentiation. *Horizontal differentiation* refers to the division of labor into component tasks. Thus, in automobile manufacturing some workers are responsible for welding the body of the car and others for installing the windshield. *Vertical differentiation* refers to the creation of multiple levels of hierarchy in complex organizations. In a university, for instance, professors report to departmental heads, who report to deans of colleges, who report to the president of the university, who reports to either a system-wide president or a board of regents. *Spatial differentiation* refers to the geographic dispersion of different aspects of production or different product lines among different plants, sometimes located in different regions, or even in different countries.

Because modern organizations are so large and complex, they have had to develop bureaucratic structures to coordinate their activities. In this section we define bureaucracy, discuss variations in the basic bureaucratic form, and talk about the informal workplace cultures that often emerge within formal bureaucracies.

Defining Bureaucracy

The German sociologist Max Weber initiated the systematic study of **bureaucracy** as a form of social organization at the beginning of the twentieth century. He enumerated six characteristics that distinguish bureaucratic organizations from other forms of social organization:

1. There is the principle of fixed and official jurisdictional areas. . . .

2. The principles of office hierarchy and of levels of graded authority mean a firmly ordered system of super- and subordination in which there is a supervision of the lower offices by the higher ones. . . .

3. The management of the modern office is based upon written documents ("the files"). . . .

4. Office management, at least all specialized office management—and such management is distinctly modern—usually presupposes thorough and expert training. . . .

5. When the office is fully developed, official activity demands the full working capacity of the official. . . .

6. The management of the office follows general rules, which are more or less stable, more or less exhaustive, and which can be learned. (Weber, 1946:196–198)

Although other forms of social organization may share some of these characteristics, such as hierarchy, only organizations that have all or most of these characteristics are appropriately characterized as *bureaucratic*. Weber believed that increasing bureaucracy was the fate of modern society, because bureaucracies are so efficient in achieving their stated objectives. He was not completely favorable toward this future, however, because he also believed that bureaucracies stifled human initiative and creativity. He referred to bureaucracy as the "iron cage of the future."

Bureaucratic Efficiencies Bureaucracies are indeed more efficient than, for example, the administrative structures of feudal society. Feudal administrative structures, such as those used for collecting taxes, relied on bonds of loyalty and obligation. Graft, corruption, and inefficiencies of all kinds were commonplace. Bureaucracy, however, is not necessarily the final word in efficiency. As you will see, bureaucracies generate their own sorts of inefficiencies and may eventually be replaced or modified to allow for greater initiative and innovation by those who work within them. Box 7.3 presents complaints by workers in an automobile factory about the bureaucratic approach of their employer.

Staff and Line Management Contemporary bureaucracies may also involve recent innovations beyond Weber's model. A relatively early innovation was the development of specialized staff positions. In addition to the arrangement of authority in a hierarchy, many modern bureaucracies also include staff positions that are outside the linear chain of command. Positions on the linear chain of command are called **line positions.** Ancillary support positions are called **staff positions.** Staff positions are filled by specialized workers trained in some specific profession, such as safety and health, law, accounting, personnel relations, or other important functions that support the main activity of the organization. These staff personnel report directly to someone in a line position at a given level of the organizational hierarchy. However, they have no direct relationship to those higher up in the hierarchy or to those in subordinate positions. In effect, they are supplementary experts needed at specific levels of the organization, but they are not included in the formal chain of command. Promotion opportunities may be less common for staff workers than for line workers, because they have less clearly defined job ladders.

Matrix Organization A second bureaucratic innovation is the **matrix organization** of authority. Under this system each worker must report to two different supervisors. For instance, suppose that an aircraft manufacturing plant has five different divisions: research and development, engineering, manufacturing, finance and accounting, and marketing. In addition, the plant is working on three major planes, or three different projects. Workers in the plant report both to the supervisor of their functional division and to the supervisor of their project. For example, engineers report to an engineering supervisor and to the manager of their particular aircraft project. Matrix organizations are popular when there is a need to share technical information and coordinate activities between different projects but also a need to allow each project to develop independently (French, Kast, and Rosenzweig, 1985:367).

Bureaucratic Control

We discussed direct personal control, foreman's control, scientific management, and technical control as techniques for controlling labor and getting the maximum amount of work out of employees.

BOX 7.3   Complaints against Bureaucratic Procedures in an Automobile Factory

The following complaints by workers about bureaucratic procedures were taken from interviews conducted by two industrial sociologists at an automobile factory:

It's a big outfit; they're strict. They can't deal with an individual.

The company tries to do some things, but it wants to get the cars out. They are first, and the men are second.

The men on the line have too much work for the time allowed. The company just cares for production, not the men at all.

The company is against the worker. It changes the lunch hour or cuts it off so that a line can catch up. The line goes so fast that guys can't keep up, especially new guys. This results in inferior work, so the company gets behind. Overtime and short lunch hours result.

The worst is the pressure. It's like on a dog-sled. As soon as the whistle blows, they yell "Mush," and away you go producing cars. The company should at least give us a five-minute break. Or the pace could be slower.

I'm left with the impression that the company doesn't think so much of the individual. If it did, they wouldn't have a production line like this one.

The place is run like the Army. They should think more about the men than the product they put out. They could do a lot more.

You're just a number to them. They number the stock, and they number you.

Nobody visits other departments in this plant. The understanding could be better, happier and much easier. Here a man is just so much horsepower. If he's no good, they just kick him out. You're just a cog in the wheel.

As long as they get the cars out, they don't give a damn for the man.

It's a big concern. They are out to make money, and they don't care how they do it. They don't care how the men feel; they only care about money.

The company just thinks of the men as robots. If they get the cars out, they don't care what happens to the men. The bigger the company, the less they do for the men. The engineers never talk to us except on business.

SOURCE: Excerpted from Charles R. Walker and Robert H. Guest, 1952, *The Man on the Assembly Line*. Cambridge, Massachusetts: Harvard University Press, pp. 137–138.

Bureaucracy can be seen as the most recent version of efforts to control labor (Adler and Borys, 1996; Edwards, 1979). In a bureaucracy, procedures are no longer determined by how the boss or foreman says the job should be done or dictated by the machinery or by an industrial engineer. Instead, organizational rules spell out the procedures. These procedures may include the specific techniques to be used for a given task and the criteria used for evaluating and promoting workers.

Internal Labor Markets Bureaucracies rely heavily on rewards and inducements to control workers. For instance, they fill most positions internally from within the organization. Internal recruitment helps create the expectation of advancement (Ospina, 1996). This expectation provides an important source of motivation for employees. Internal recruitment also reduces the costs of training new employees in company procedures and company-specific skills. The job ladders created in this way are called **internal labor markets** and are an important way in which bureaucracies motivate and control workers. Workers enter the organization only at certain jobs that serve as entry ports. Access to higher jobs is mainly through the job ladders starting at these entry ports. For instance, there is usually no way to become a full partner in a law firm without first having been a junior associate. Internal labor markets are an important mechanism for cultivating and retaining experienced workers within a firm. Internal labor markets help to increase the value and effectiveness of **human**

capital within an organization and thus can be an important component of organizational effectiveness (Nahapiet and Ghoshal, 1998).

The spread of such internal labor markets in the latter half of the twentieth century increased average job tenure among workers. Nearly 40% of men aged thirty or older have jobs that they would hold for twenty or more years. Similarly, though women are more likely than men to change jobs, 15% of women thirty years or older can expect to hold their current job for twenty or more years (Osterman, 1999). As we discuss in Chapter 14 on marginality and in Chapter 15 on large corporations, long-term commitments between employers and their employees, however, appear to be weakening and strong internal labor markets may be increasingly replaced by more short-term and contingent employment contracts (Newman, 1999).

Customizing Bureaucracies

Although almost all large, complex organizations are organized and controlled bureaucratically, there are important variations within the bureaucratic form. These variations occur partly because different organizations have different goals. For example, the goals of a department of natural resources may be to preserve these resources and provide recreational facilities for the public. The goal of a manufacturing establishment may be to produce the maximum amount of goods at the minimum price. A goal of an electronics research company may be to foster innovation among its professional staff. The successful attainment of these goals may require different organizational structures. For instance, a bank or a manufacturing company may function best with a strong vertical hierarchy. However, a research, educational, advertising, or architectural firm requiring high levels of creativity may function better with a more dispersed system of authority and greater individual or group autonomy (Perrow, 1986).

Contingency Theory Sociologists also point out that organizational structure depends on the environment that the organization faces. This theory of organizational structure is called **contingency theory** (Itzkowitz, 1996). If the market for the product of a company is highly variable, the organization will fare better if it has a relatively loose organizational structure. If the market is highly predictable, more rigid structures are viable and may increase efficiency. If the environment is so harsh that the survival of the organization is in question, centralized control is generally the most effective form of organization. Armies are good examples of organizations that face very harsh environments and that have highly centralized control. Technologies generate additional contingencies: they place demands and constraints on organizational structure. When standardized production systems are used, highly rigid systems of authority are possible. When there is great uncertainty in materials or in the available technologies or when production systems are highly complex, more flexible systems of authority and organization may be more efficient.

Informal Work Cultures

Although bureaucracies are known for the painstaking detail with which they spell out procedures, most of them also develop **informal work cultures.** Informal cultures are based on social relationships that emerge among the people who work in an organization. For instance, two workers in an engineering firm may become friends and help each other with their work. A third worker in their section may be excluded from this relationship. These informal cultures have a tremendous influence on how the bureaucracy actually operates, in contrast to how it operates "on paper." Informal work groups may facilitate the attainment of stated organizational goals, or they may develop their own goals. Sociologists sometimes refer to informal cultures as "negotiated orders" in recognition of their emergence from the ongoing informal negotiations among different members of the organization (Fine, 1984; Miller, 1991). "The negotiated order on any given day could be conceived of as the sum total of the organization's rules and poli-

BOX 7.4 "Making Out" on the Shop Floor

Sociologist Michael Burawoy coined the phrase *making out* to describe the process through which workers devise ways to reach desired rates of production without exhausting themselves in the process. He describes the process of making out in the following excerpt:

> After the first piece has been OK'd, the operator engages in a battle with the clock and the machine. Unless the task is a familiar one—in which case the answer is known, within limits—the question is: Can I make out? It may be necessary to figure some angles, some short cuts, to speed up the machine, make a special tool, etc. In these undertakings there is always an element of risk—for example, the possibility of turning out scrap or of breaking tools. If it becomes apparent that making out is impossible or quite unlikely, operators slacken off and take it easy. Since they are guaranteed their base earnings, there is little point in wearing themselves out unless they can make more than base earnings—that is, more than 100 percent. That is what Roy refers to as goldbricking. The other form of "output restriction" to which he refers—quota restriction—entails putting a ceiling on how much an operator may turn in—that is, on how much he may record on the production card. In 1945 the ceiling was $10.00 a day or $1.25 an hour, though this did vary somewhat between machines. In 1975 the ceiling was defined as 140 percent of all operations on all machines. It was presumed that turning out more than 140 percent led to "price cuts" (rate increases). . . .

> In 1975 quota restriction was not necessarily a form of restriction of *output*, because operators *regularly* turned *out* more than 140 percent, but turned *in* only 140 percent, keeping the remainder as a "kitty" for those operations on which they could not make out. Indeed, operators would "bust their ass" for entire shifts, when they had a gravy job, so as to build up a kitty for the following day(s). Experienced operators on the more sophisticated machines could easily build up a kitty of a week's work. There was always come discrepancy, therefore, between what was registered in the books as completed and what was actually completed on the shop floor. Shop management was more concerned with the latter and let the books take care of themselves. Both the 140 percent ceiling and the practice of banking (keeping a kitty) were recognized and accepted by everyone on the shop floor, even if they didn't meet with the approval of higher management.

SOURCE: Excerpt from Michael Burawoy, 1979, *Manufacturing Consent,* Chicago: University of Chicago Press, pp. 57–58. Copyright © 1979 by the University of Chicago. Reprinted by permission of the publisher.

cies, along with whatever agreements, understandings, pacts, contracts, and other working arrangements currently obtained. These include agreements at every level of organization, of every clique and coalition, and include covert as well as overt agreements" (Strauss, 1978:5–6). Workers refer to such negotiated orders simply as "office politics."

Making Out Within the context of formal rules and procedures, workers attempt to make an autonomous life for themselves. Michael Burawoy (1979) calls this process "making out."

Making out refers to workers' attempts to meet the requirements of the rules without completely exhausting and alienating themselves in the process. Often, making out entails devising ways to make a game out of meeting workplace quotas and standards. In this way, workers come to terms with the organizations in which they work. The nature of work is thus not determined solely by organizational rules; rather, it is determined by a process of negotiation and sometimes conflict between the various actors in the workplace. Box 7.4 presents a description of the process of making out in a machine tool factory.

It is frequently argued that women and minorities receive fairer treatment in large bureaucracies than they do in smaller organizations. This argument is based on the observation that formal rules make discrimination more difficult. Because of the importance of informal cultures in large organizations, however, women and minorities may still be excluded from the "insider" situations and "old boy" networks in which important coalitions emerge and in which key decisions are made (Kanter, 1983).

LIMITATIONS OF BUREAUCRACY

Bureaucracies may be efficient, but they also have many drawbacks and limitations. These problems are discussed in this section. In the next section we discuss ways of dealing with these problems.

Top-Heavy Management

Perhaps the most important consequence of the rise of bureaucratic organizations has been a significant increase in the proportion of employees engaged in administrative work. In contemporary society, people's lives seem dominated by bureaucrats and red tape. This rise in the number of administrative workers has also reduced the proportion of workers directly engaged in production. Between 1900 and 2000, the proportion of nonproduction employment rose from less than 10% to more than 20% (Census, 2000). Although bureaucracies may be highly productive, they take on extra administrative baggage. Administrative costs may take up too large a share of the budget, even overshadowing the direct costs of production. This top-heavy nature of modern bureaucracies has become a focus of increased concern in the 2000s as U.S. firms attempt to regain their competitive position in the world economy. In recent years, U.S. firms have increasingly sought to reduce middle management and create "lean" managerial structures with fewer intermediate levels and fewer managers at each level (Delbridge, 1998).

The Centralization of Control in the Economy

Not only does bureaucracy centralize control in a large managerial structure, but it also contributes to the **centralization of control** among companies. As firms have grown larger, a greater share of economic activity has come to be controlled by the top managers of a few large firms. In the 1920s and 1930s, this process occurred through the growth of *multidivisional firms* (Chandler, 1977). In multidivisional firms the central office coordinates the activities of a number of operating divisions or product lines through allocating funds, personnel, and other resources. For example, the General Electric Company (*www.ge.com*) operates seven major divisions: technical research, services and materials, power plant systems, industrial products, consumer products, personal finance, and aircraft engines.

Conglomerates In the 2000s, many large firms have come to control even larger shares of the economy by acquiring companies in unrelated product lines. Such *mergers and acquisitions* expanded the control of large companies into product lines that in many cases are outside their core expertise. In 2000, the 500 largest firms in the United States controlled 82% of corporate assets. For example, the telecommunications giant Times-Warner (*www.timewarner.com*) owns America Online, Cable News Network (CNN), Netscape, Warner Brothers, and CompuServe. These 500 corporations represented fewer than 0.10% of the firms in the United States. Control of such huge market shares by a few firms can lead to higher profits for these firms, but also to higher prices for consumers and to reduced employment (Scherer, 1999). (See also the discussions of concentration ratios in Chapter 2 and corporate merges in Chapter 15.)

A Ruling Elite The centralized control of the economy by a few large firms also translates into the existence of a powerful upper class in the United States. Control of corporations rests on the ownership of large blocks of stock. The richest 1% of the U.S. population owns over 50% of corporate stock. The richest 1% also owns over 28% of the *total* wealth in the country. This class constitutes a virtually closed social group with tremendous power. Its members attend exclusive preparatory schools, colleges, clubs, and resorts, and intermarry only with other members of the ruling class (Domhoff, 1998). There are, of course, exceptional self-made tycoons, whose lives parallel the famous rags-to-riches novels by Horatio Alger. In reality, however, the heads of the largest corporations are almost exclusively born into their class positions (Braun, 1997).

Reduced Creativity

The centralization of control in large organizations can also have negative consequences for the employees of the organization and for the organization itself (Burris, 1993). These drawbacks may limit the continued expansion of bureaucratic control. Centralization of control increases alienation among members of the organization, because they have little say in the decisions that influence their lives. Decisions may also be slow or faulty, because it is difficult for a few people to receive and process all the relevant information essential for good decision making.

Bureaucratic Rigidity Innovation and creativity may also be lessened by overconformity to bureaucratic standards. Such bureaucratic rigidity is often a reaction to excessive control by top management:

> An analysis of instances of extreme rigidity in hierarchical organizations reveals that they are usually associated with fear of superiors. . . . Bureaucratic superiors cannot generally censure a subordinate for following official regulations exactly, regardless of how inefficient or ridiculous such action may be in a particular case. . . . Feelings of dependency on superiors and anxiety over their reactions

engender ritualistic tendencies. (Blau and Meyer, 1971:104)

Terkel reports the following account of the damaging effects of bureaucratic rigidity:

> I'll run into one administrator and try to institute a change and then I'll go to someone else and connive to get the change. Gradually your effectiveness wears down. Pretty soon you no longer identify as the bright guy with the ideas. You become the fly in the ointment. You're criticized by your supervisors and subordinates. Not in a direct manner. Indirectly, by being ignored. They say I'm unrealistic. . . .
>
> My suggestions go through administrative channels. Ninety percent of it is filtered out by my immediate superior. I have been less than successful in terms of getting things I believe need to be done. It took me six months to convince my boss to make one obvious administrative change. It took her two days to deny that she had ever opposed the change. (Terkel, 1974:448–449)

In contrast, the conditions that have been found to promote innovation and change include the decentralization of power, low levels of formalization, equity of rewards, low emphasis on volume, low emphasis on cost cutting, and high levels of job satisfaction (Hall, 1996). In brief, excessive hierarchy and bureaucracy may interfere with productivity rather than promote it. At some point, excessive rationality becomes irrational (Ritzer, 2000).

Corporate Accountability

A final problem that we confront in the modern economy is that of establishing **corporate accountability** for the huge organizations that have come to produce most of the goods and services in the world. Because of their size and power, such organizations are often capable of operating beyond the law. Common illegalities include price fixing, manipulation of stock

BOX 7.5 The Firestone Tire Scandal

In the summer of 2000, Firestone announced a recall of several of its tire models for the popular Ford Explorer sport utility vehicle. As the news unfolded over the summer, the public and Congress became outraged to learn that over 100 people had already died because of the tires, with several hundred more seriously injured and the death toll steadily rising.

It was supposed to be a mission of mercy. Victor Rodriguez piled the family into his Ford Explorer over Labor Day weekend to visit a sick aunt at a Laredo, Texas, hospital. But as Rodriguez cruised down Interstate 35, he was startled by a thump and looked back to see the tread shredding off a Firestone Wilderness AT tire on his Explorer. The 53-year-old father was unable to control the vehicle, which flipped, ejecting five of its passengers. Among them: his 10-year-old son Mark Anthony, who died instantly. . . .

Mark Anthony Rodriguez had just become the latest victim of a rapidly widening safety crisis that has driven fear into the hearts of motorists who viewed the sport utility vehicle as the ultimate family car. Even as the Rodriguez family planned Mark Anthony's funeral, top executives from Ford Motor Co. and Bridgestone/Firestone were summoned before congressional panels. . . .

Documents show that Firestone was chronicling a pattern of tire failures for the last three years, while Ford was also getting an early warning about safety problems from its own warranty data. And an internal Ford memo shows suspicions between the business partners about the safety record of the tires.

Firestone executive vice president Gary Crigger said that the data were compiled to assess only the company's financial liability. They were not shared, he said, with the company's safety engineers. . . . For their part, lawmakers have rejected the companies' claims of ignorance about the lethal problem. . . .

The families of the victims are not in a forgiving mood. To them, no apology or excuse will ever make up for the fact that the companies could have warned consumers earlier. "If they would have told people, maybe they wouldn't have made as much money, but it would have saved dozens of lives," says Sara Romero, 12, who survived an Explorer rollover in Florida last December that killed her 37-year-old mother. "My mom didn't have to die. They could have told us."

SOURCE: Newsweek, September 11, 2000, www.msnbc.com/news/457861.

---

prices, illegal environmental pollution, false and misleading advertising, bribery, tax evasion, political payoffs, and the production and sale of unsafe products (Punch, 1996). Box 7.5 describes the Firestone tire scandal as an example of corporate misconduct at its worst.

Externalizing Costs The ability of large enterprises to avoid part of the costs of their economic activity through unlawfully dumping dangerous chemical waste into the environment, instituting dangerous work practices, and producing unsafe products has become a major concern in the 2000s. Individuals and communities absorb the costs of these activities in terms of lost health, destroyed communities, and degraded environments. For example, on March 24, 1989, the largest oil spill in North American history occurred when the tanker *Exxon Valdez* ran aground in Prince William Sound, Alaska. Although Exxon was sued for over $5 billion by the State of Alaska and by other individuals affected by the spill, full payment for the economic costs of the disaster is unlikely ever to be recovered (Picou and Gill, 1993). Smaller scale, repeated abuses, which are unlikely to be recognized or publicized, may cumulatively be even more damaging.

The Price Tag Given the magnitude of the problems resulting from corporate irresponsibility, many social analysts argue that the issue of corporate accountability should receive greater attention (Ermann and Lundman, 1996). The Senate Judiciary Subcommittee on Antitrust and Monopoly estimates that corporate crime costs the public between $174 billion and $231 billion annually. In contrast, "the yearly losses from street crimes [are estimated] at about $4 billion—less than 5% of the estimated losses from corporate crime" (Coleman, 1985:6). Additional casualties due to corporate neglect include 14,000 deaths a year due from industrial accidents and 30,000 deaths a year from unsafe consumer products. By contrast, about 20,000 people are murdered each year in the United States.

Whistle-Blowing One potential way for employees to fight back against corporate malfeasance is by exposing corporate actions that are clearly unlawful. Such activities are called "whistle-blowing" (Vaughan, 1999). Federal laws protect employees from retaliation by the company in situations where employees expose illegal company activities that involve violation of federal laws such as the Clean Air Act or other environmental laws. The reporting of company fraud against the government is also protected. Such laws, however, are difficult to enforce and the negative consequences for employees can be devastating to their lives and careers (Miceli et al., 1999). Corporations have tremendous power and financial resources—they have deep pockets when it comes to hiring lawyers and pursuing lawsuits against employees who they claim have unfairly damaged their reputation. A company can also sometimes succeed in shifting the blame for the misconduct to the worker who reports the problem. Whistle-blowing theoretically has great potential for limiting corporate misconduct, but employees' fears concerning retaliation greatly limit its practical impact (Rothschild and Miethe, 1999). Whistle-blowing occurs only in extreme cases and even then only if someone has the courage to withstand the resulting hailstorm of accusations and counterattacks.

DIRECT WORKER PARTICIPATION

Early theories of industrial relations, such as those of Frederick Taylor, assumed that workers needed to be coerced (by threat of firing) or bribed (by promises of pay raises) into working harder. As noted in Chapter 4, these ideas are sometimes referred to as **Theory X** of organizational motivation. You saw how these theories were replaced by human relations theories, which assumed that workers would be more productive if they received more humane consideration and attention. This latter view is sometimes referred to as **Theory Y.** It still portrays the worker as a passive object to be manipulated by management, though in this vision socially oriented techniques of manipulation replace economically oriented techniques.

Efficiency through Participation In recent years, inefficiencies and rigidities associated with bureaucratic management structures have called both of these theoretical traditions into question. The new theories that are emerging to replace Theories X and Y include a more active role for workers and are sometimes referred to collectively as **Theory Z** (Ouchi, 1981). These theories view productivity as embedded in workers, in their skills, and in their attitudes rather than in specific procedures. In Europe these theories have taken form in the widespread use of autonomous work groups. Under the group organization of work, teams of workers have extensive control over decision making about their day-to-day operations and activities (Ezzamel and Willmott, 1998). Increased **worker participation** is also typical of enterprises that are owned by workers. Box 7.6 describes the high level of participation and cooperation in collectively owned plywood companies in the Pacific Northwest of the United States.

Lifetime Employment In Japan such theories of productivity have been implemented through **lifetime employment.** Guaranteeing employment for workers ensures their job security and,

---

BOX 7.6 Direct Participation in Worker-Owned Plywood Companies

The plywood producer cooperatives are among the most fully developed and enduring democratic industrial enterprises in the United States. . . . The Pacific Northwest plywood cooperatives, most of which have been in existence for twenty-five to thirty years, not only have persisted through the good and bad times characteristic of that industry but have remained enterprises in which those who work also decide policy. . . .

Although the worker-shareholders may choose to delegate many day-to-day concerns and responsibilities to the general manager and the elected board of directors, in a formal sense they are responsible for the total governance and direction of the enterprise. Worker-shareholders may decide to fire the manager and hire a new one, alter hourly wage rates, build a new plant, or reach any other decision about the enterprise that seems appropriate to them. . . . In membership meetings, each shareholder is entitled to only a single vote even if, as in a few exceptional cases, the member owns more than one share. . . .

Central to life in the cooperatives is the sense that the worker-shareholders are in charge, that they run the enterprise, are responsible for what goes on in it, and have the opportunity, within certain boundaries, to make of their environment what they will.

If things get too bad, the stockholders can just say, "Wait a minute. . . we are going to

change this." I think that's great because there's a lot of companies that take advantage of the workers, and there's nothing that can be done about it. . . .

Ownership and participation in the co-ops also fosters an extremely strong sense of collective responsibility and mutuality.

You just find it's kind of a big family attitude. . . . Here you get in and do anything to help. Everybody pitches in and helps. The people stick together, that's the reason we've gone so far and production is so high, cuz everybody works together. . . .

Workers in conventional plants are willing to put in a hard day's work on their assigned tasks, but they are not likely to move beyond those boundaries and act in ways that will enhance the productivity of the entire process. In the cooperatives, the job boundaries are less rigid and more fluid.

If the people grading off the end of the dryer do not use reasonable prudence and they start mixing the grades too much, I get hold of somebody and I say, now look, this came over to me as face stock and it wouldn't even make decent back. What the hell's going on here? That wouldn't happen in a regular mill.

SOURCE: Edward S. Greenberg, 1986, *Workplace Democracy: The Political Effects of Participation.* Ithaca, N.Y.: Cornell University Press, pp. 28–43.

---

thus, their trust and enthusiasm. It also retains their presence in the organization as a repository of skills and knowledge. This benefit is amplified by having workers rotate through a variety of jobs during their careers, thus building their knowledge base. This knowledge is invaluable for coordinating activities among different parts of the organization and is a more efficient mechanism for integrating production than more cumbersome bureaucratic structures that may undermine enthusiasm and commitment (Cole, 1989). Greater reliance on workers' knowledge and decision–making abilities

also reduces the need for first-line supervisors and for middle-level managers, thus producing additional significant cost savings (Pfeffer, 1998).

Although participatory systems of worker involvement and commitment have become identified in the public mind with the "Japanese system," in many ways Japanese industrial relations provide a very incomplete model for worker participation. The Japanese system of lifetime employment is available largely for male workers in large firms in the manufacturing sector—less than a third of the labor force. Women and workers in secondary parts of the economy do not have much security. Even employees who are supposedly covered by lifetime employment may, at mid-career, find themselves transferred to "client organizations" of their company, such as suppliers or wholesalers, as their high-paying jobs are taken over by younger and more aggressive workers. In Chapter 16 we discuss other aspects of the Japanese model of industrial relations. In Chapter 17 we discuss in more depth the possibility of increased worker participation in the workplace of the twenty-first century.

TECHNOLOGICAL AND ORGANIZATIONAL DETERMINISM

Is the future of work determined by technological imperatives? Probably not. Certain technologies cannot be used, or cannot be used effectively, except in moderately large enterprises. For instance, coal-powered electrical generation plants must reach a certain size to be economical. Thus, this kind of technology demands a certain organizational structure—in this case, a minimum size—to be efficient. However, the minimum size required for the effective use of most technologies is quite modest, and, in most cases it is well under that of today's large corporations (Blau et al., 1976).

The reason that key parts of the economy are dominated by large firms is that these firms have the power to do so. There is no technological reason that this has to be the case. In recent decades, international competition and technological change have reintroduced a high degree of uncertainty into the environment of many organizations. In this situation, large bureaucratic organizations may be at a disadvantage. Diverse organizational forms are competing effectively in the world markets of the twenty-first century and this has introduced new opportunities and challenges for economic organizations and those who work within them.

SUMMARY

As technology has advanced from simple hand tools to sophisticated microchip technology, the social organization of production has become more complex. The earliest division of labor, based on gender and age, has been replaced by a detailed division of labor in which tasks are subdivided into many minute processes assigned to different workers. This detailed division of labor and the subsequent hierarchical and bureaucratic organization of the workplace create their own limits because of the inefficiencies introduced by top-heavy managerial structures, lack of organizational flexibility, and lost worker initiative. An additional problem is establishing accountability in a world dominated by large corporations. Increased worker participation has possibilities for ameliorating some of the problems associated with hierarchical and bureaucratic control.

KEY CONCEPTS

technology	microchip technology	apprenticeship
organizational structure	complexity	on-the-job training
social structure	diversity	direct personal control
craft technology	autonomy	foreman's control
mass-production technology	deskilling	scientific management

soldiering

technical control

worker resistance

human relations school

Hawthorne effect

bureaucracy

staff and line positions

matrix organization

internal labor markets

human capital

contingency theory

informal work culture

making out

centralization of control

corporate accountability

externalization of costs

Theory X

Theory Y

Theory Z

worker participation

lifetime employment

technological determinism

QUESTIONS FOR THOUGHT

1. What are the major trends in employment in different industries, and why have these occurred?

2. Define technology; define organizational structure. In what ways do these concepts overlap and in what ways are they distinct? Apply these concepts to an industry in which you have worked. What aspects of the job were reflections of the technology and what aspects were reflections of the organizational structure?

3. How do sociologists define and measure skill? Select one job each that would commonly be considered highly skilled, moderately skilled, and unskilled. Apply the definition to skill to these jobs. In what ways does the definition of skill fit or fail to fit each job?

4. What are the major types of labor control in the workplace? Which of these would you most/least like to work under?

5. Define bureaucracy. What are some of the limitations of bureaucracies? Do you think bureaucracy will be the ultimate form of social organization? Why or why not? What other forms of social organization might replace bureaucracy?

6. Describe some informal work cultures in which you have been involved that encouraged behaviors other than those organizationally prescribed.

7. How might increased corporate accountability be achieved?

8. What other useful government or private human resource internet pages can you find in addition to those listed in the Multimedia Resources section?

MULTIMEDIA RESOURCES

Print

Harry Braverman. 1974. *Labor and Monopoly Capital.* New York: Monthly Review. A very readable history of the development of management from the standpoint of workers. Critiques the efforts of managers to deskill workers.

M. David Ermann and Richard J. Lundman. 2000. *Corporate and Governmental Deviance,* 6th edition.

New York: Oxford University Press. The best current survey of white-collar and corporate crime.

Richard C. Edwards. 1979. *Contested Terrain.* New York: Basic Books. A useful framework for understanding worker–management conflict as these have varied across history and across different settings in the contemporary economy.

Rosabeth Moss Kanter. 1977. *Men and Women of the Corporation.* New York: Basic Books. An analysis of the lives of men and women as they accommodate to their work roles in large, complex organizations.

George Ritzer. 2000. *The McDonaldization of Society.* Thousand Oaks, Calif.: Pine Forge. A compelling discussion of the overrationalized nature of modern society.

Websites

London School of Economics, Department of Industrial Relations. *www.lse.ac.uk/depts/industrial* One of the most renown centers of learning on economic and social issues.

Tavistock Institute for the Study of Human Relations. *www.tavinstitute.org* Origin of the human relations school of industrial relations.

Canada Government Human Resources Homepage. *www.hrdc-drhc.gc.ca* Excellent information on all aspects of employment, training, compensation, and work.

Canadian Workplace Research Network. *www.cwrn-rcrmt.org/eng* Highlights research on workplace change and innovation that fosters efficient and equitable workplace practices.

Workplace Today. *www.workplace.ca* Canadian human resource management site.

Mexico Government Human Resources Homepage. *www.stps.gob.mx*

Economic Sociology and Organizational Studies at Pennsylvania University. *pesos.wharton.upenn.edu* Reports latest research on issues related to work.

The *Economist* news magazine. *www.economist.com* The premier news magazine dedicated to international economic issues.

Cost/Benefit Analysis. *www.qpr-tools.com* A commercial site providing activity-based costing (cost/benefit analysis) for corporations. Offers a look inside high-tech human resource management practices.

RECOMMENDED FILM

The Trial (1996, rerelease), directed by Orson Wells. Gripping adaptation of Franz Kafka's novel—a frightening story of a man accused of an unnamed crime about which he can find no information.

8

From Field, Mine, and Factory

Looking around the shop, first day on the job, I sorted things out. . . . Metal chips were everywhere on the pitted cement floor, scattered in piles formed as they flew off machines. Squinting, intent milling machine operators turned feed cranks with one hand and brushed on cutting oil with the other, pushing metal stock against spinning milling cutters. Lathe workers watched closely while long, spindly chips curled away from their whirling workpieces.

In this little neighborhood non-union shop we were making parts for the military, thousands of them; many were subcontracted from larger corporate defense contractors. The casual, mom-and-pop character of the place didn't seem to fit the deadly nature of its products. . . .

No one in the 30-man shop wore safety glasses. Open buckets of naphtha, a highly flammable cleaning fluid, stood around the floor. Parts were being sloshed in them to get the oil off. . . .

My actual job assignment was far less interesting than the surrounding scene. I was to countersink a particular pair of holes in rotor mounts, aluminum holders for electrical parts in missiles. This meant using a drill press to put a rounded bevel on the top edge of each hole, so that a tap, or screw-thread cutter, could enter it properly. The foreman led me to a new, lightweight drill press made in Taiwan. He showed me how deep to make the countersink, gave me a couple of plug gauges to test its size, and walked off. I countersunk about a hundred pieces, or 200 holes. Then I moved on to tapping the holes. After that I "deburred" the part by filing off the sharp edges. Time crawled. There were two other guys on drills beside me, a young Haitian and another guy my age who I later found was from Barbados. They said very little to me during my first days.

(TULIN, 1984:1–2)

This passage recounts the first day on the job for a young worker in manufacturing. In this chapter we look at the concerns of manufacturing workers and workers in related industries.

In Chapter 7 we discussed the general concepts of *technology* and *organization*. More concretely, economic activity occurs in three principal sectors: extractive, manufacturing, and services. In **extractive industries,** such as agriculture, forestry, fishing, and mining, a product is removed from the environment. The process of extraction may use sophisticated equipment, but the final product retains much the same form as it has in the natural environment. The raw products are further processed in **manufacturing industries** into more usable forms. For instance, vegetables are canned or frozen. Iron ore and coal are used to make steel, which is fabricated into such things as girders and cables. In this way products are manufactured that may be quite far removed from their origins as raw materials. Cars, refrigerators, airplanes, and computers are examples of such products.

Service industries are even more diverse than manufacturing industries. They include financial and accounting services, medicine, entertainment and recreation, education, government administration, and social welfare services. Services are distinct from goods in that they cannot, in general, be stockpiled, stored, or transported. Many services, such as restaurant meals, must be produced at specific times of the day and in specific places. Goods are transferable from buyer to buyer, so that it is possible to have long chains of intermediaries involved in their production. Services, on the other hand, tend to be nontransferable and must be delivered by the producer directly to the final consumer. Because of these factors the production of services tends to be more geographically dis-

persed than the production of goods. Partly as a result, automation has had less of an impact on the production of services, although that may be changing. Service work is the focus of Chapter 10.

POSTINDUSTRIAL SOCIETY?

A hundred years ago, even in what are today industrialized nations, most people were involved in agriculture. Over time, because of increasing agricultural productivity, fewer workers were required to meet society's agricultural needs. The result was a period of employment growth in manufacturing. Later, increasing productivity in manufacturing allowed more workers to be engaged in the delivery of services. The often-used phrase **postindustrial society,** however, is a misnomer for today's society. The current economy, with a large share of the labor force in services, has not gone *beyond* being an industrial society. Rather, our economy is based on a highly productive industrial base in which it is possible to produce more goods with fewer workers. Because the need for manufactured goods can be expanded to a greater extent than the need for agricultural products, the economy is unlikely ever to reach a state in which only a tiny share of employment is in manufacturing.

American society is better characterized as an **advanced industrial society.** This label implies a small but highly productive extractive sector, a larger and also highly productive manufacturing sector, and a growing labor-intensive service sector. This service sector can be expected to continue to grow because the demand for services, such as education, health, and recreation, is even more insatiable than the demand for goods.

OCCUPATIONS AND INDUSTRIES

Many different occupations may be involved in the production of a specific good or service. Some of these occupations are common to many different industries. For instance, all industries employ clerical workers and managers. Because these occupations are not unique to any one industry, we consider them in separate chapters in Part III of this book. In this chapter and in Chapters 9 and 10 we focus on the jobs of those employees who are direct producers of specific goods or services. Thus, in manufacturing industries we focus on blue-collar workers. In service industries we focus on the nonprofessional workers directly engaged in providing services, such as fry-cooks, waitresses, orderlies, janitors, garbage collectors, and taxi drivers.

RAW MATERIALS: AGRICULTURE, FORESTRY, AND FISHING

Agriculture, forestry, and fishing are the oldest extractive industries, and they still utilize the most traditional technologies and organizations of work. Many aspects of production still rely heavily on time-honored techniques.

Agriculture

We used to work early, about four o'clock in the morning. We'd pick the harvest until about six. Then we'd run home and get into our supposedly clean clothes and run all the way to school because we'd be late. By the time we got to school, we'd be all tuckered out. Around maybe eleven o'clock, we'd be dozing off. Our teachers would send notes to the house telling Mom that we were

inattentive. The only thing I'd make fairly good grades on was spelling. I couldn't do anything else. Many times we never did our homework, because we were out in the fields. The teachers couldn't understand that. I would get whacked there also. . . .

The hardest work would be thinning and hoeing with a short-handled hoe. The fields would be about a half a mile long. You would be bending and stooping all day. Sometimes you would have hard ground and by the time you got home, your hands would be full of calluses. And you'd have a backache. Sometimes I wouldn't have dinner or anything. I'd just go home and fall asleep and wake up just in time to go out to the fields again.

If people could see—in the winter, ice on the fields. We'd be on our knees all day long. We'd build fires and warm up real fast and go back onto the ice. We'd be picking watermelons in 105 degrees all day long. When people have melons or cucumber or carrots or lettuce, they don't know how they got on their table and the consequences to the people who picked it. If I had enough money, I would take bus loads of people out to the fields and into the labor camps. Then they'd know how that fine salad got on their table. (Terkel, 1974:34,38 [Roberto Acuna, farm worker])

Rising Productivity Agriculture still involves much hand labor, as described eloquently by Roberto Acuna. Agricultural mechanization and productivity in general, however, has increased dramatically over the last century. This was particularly true during the latter half of the twentieth century because of the expanded use of machinery, fertilizers, and improved varieties of grains and other crops. Because of this increased productivity, fewer people are required to produce agricultural goods. From the standpoint of increased productivity, this is an important step forward into industrial society. From the standpoint of the individual farmer, technology has lightened many physical burdens. But it has also contributed to the chronic treadmill of falling farm prices. As farm productivity has increased, the prices of farm products have fallen. Fewer and fewer family farmers are able to make a living from working the land because they have to compete with large, heavily mechanized corporate farms. In response to falling prices, family farmers may take on second jobs in neighboring towns to continue farming, or they may be forced to mortgage and eventually sell their farms and abandon farming altogether (Lasley et al., 1995).

For many small farmers the problem may simply be having too small a farm to support themselves and their families. Black-owned farms in the South, for example, traditionally have had higher yields per acre than commercial farms as a whole, but many are so small that they cannot produce adequate income to sustain a family. As a result, the amount of farmland owned by blacks in the South fell from 15 million acres in the early 1900s to 5 million acres by the 1970s (McGee and Boone, 1979).

Large corporate farms contribute to the displacement of family farms by further driving down farm prices based on economies of scale. Some of the largest corporate farms have developed huge factorylike production facilities for hogs and chickens. These farms drive down prices for pork and poultry but in so doing undercut small farmers who had been able to sustain themselves with these activities. In addition, in the Carolinas and other parts of the country, large corporate farms have been associated with significant environmental problems resulting from the huge tonnage of animal waste concentrated in single locations.

It is somewhat ironic that at the same time small farms are declining in the largest capitalist nations, they are making a comeback in the formally socialist countries of Eastern Europe. In Hungary, for example, which had a highly collectivized agricultural system of large cooperatives, small, privately owned farms now produce nearly half the fresh vegetables and fruits and farm animals (O'Relley, 1986). In the United States some farmers are attempting to maintain their economic viability by producing specialty products that do not compete directly with those produced on larger farms. Such products include organically grown produce and meats, domestically raised fish, and products for local "farmer's markets."

Rising Costs and Falling Prices Farmers have also been burdened with rising fuel prices and high interest rates, which increase the costs of production while prices for farm commodities continue to fall. In 1998, for example, the average cost of growing a bushel of corn was $2.42. The average price received for a bushel of corn was $1.95, *24% less* than the average cost of its production (Census, 2000:687). Obviously, only farmers with below-average production costs can survive for long in such a market. What occurs in this context is a *shaking-out* process in which smaller or less efficient farms can no longer produce a living income. This process can be very traumatic for the farmers involved:

At dawn, minutes after his wife woke up to discover a note describing his intentions, the Cordell police found Darrel Evans in a trash dumpster near the center of town.

The death August 7 of the 54-year-old grain and cotton grower from Cordell, a town on the high plains 90 miles west of [Oklahoma City], was the latest in a rash of suicides among Oklahoma's farm families—about 100 deaths in less than two years. . . .

Mental health professionals in the Midwest and South say that the long march of hardship on the farm continues for tens of thousands of growers and their families. And for some, the frustration and dread of seeing the work of generations undone in a season or two is too much to live with. . . .

"We find that the loss of a farm or the impending failure is worse for many farmers than the death of a loved one," said

Dr. Glenn K. Wallace, a psychologist and regional program manager for the Oklahoma Department of Mental Health. "The feeling of guilt that a family has when the sheriff's sale sign goes up on the place a great grandfather homesteaded is more than they can bear." (Schneider, 1987:1)

In 1975 there were 2.8 million farms in the United States. By 2000 almost a third of these had folded. Ownership of the remaining farmland is extremely concentrated. The U.S. Department of Agriculture reports that over 75% of the nation's food supply is produced by just 50,000 farms (Census, 2000).

Technological Advances Given their financial constraints, technological advances in agriculture may come as a curse to small farmers. They may not be able to afford the expensive new technology and thus may fall even further behind larger producers. For instance, the use of growth hormones and antibiotics in animal feed has greatly increased yields for farmers who can afford these purchases, but it has also lowered prices for beef and other meats, putting further pressure on small producers. The development of expensive new genetically engineered seeds has further heightened these pressures.

Technological advances in agriculture may also be a mixed blessing for Third World nations struggling to feed their growing populations. In the 1980s, U.S. hogs were imported into Haiti under a joint U.S.–Haitian program, largely in response to a fear that African swine fever among the Haitian hogs would spread to the United States. However, the lifestyle of the U.S. pigs may have been too rich for desperately poor Haiti: "Haitian pigs roamed freely in the countryside and lived off orange peels, mango seeds and garbage. Their American replacements are supposed to live on a concrete floor, eat imported food and be given expensive vaccines. Even worse, unlike the now-defunct Haitian pig, which had long legs and could be led to the market by a string, the American pig refuses to walk" (Oppenheimer, 1986).

Adding to the difficulties faced by farmers is instability in agricultural prices. Prices for farm products are highly dependent on how bountiful the harvest has been. Thus, in good years the farmer has a lot to sell, but prices tend to be low. Annual variations in weather patterns, insect infestations, and diseases all produce dramatic swings in farm productivity. In addition, the *fixed costs* of farming are very high, averaging more than 50% of total costs (Weiss, 1971:38). Fixed costs are those costs that farmers must pay regardless of whether they sell their crops or at what price. They include such things as depreciation on buildings and equipment, property taxes, and interest on land and equipment loans. Fixed costs are a large expense and the farmer must have at least some income to pay them. Thus, it is often rational for the farmer to continue to produce even when the prices received do not cover the full costs of production. The money earned at least helps pay the fixed costs of production. The residual is made up by taking on additional debt, which then demands a greater interest payment the next year. The cycle finally comes to an end when the farmer can no longer take out any more credit or cannot survive on earned income. At this point, the farmer must sell the family farm, pay the proceeds to creditors, and look for nonagricultural employment.

Federal Price Supports Federal farm policy in the United States attempts to slow the decline of the small farmer by supporting prices for selected farm products and providing credit to farmers. Such policies establish a floor for farm commodity prices, thus ensuring the farmer a certain minimum price for products. Unfortunately, such price supports and credit arrangements have benefited mainly large farmers who have the acreage to take advantage of the programs.

Farm prices have also been reduced by increased world competition in agricultural commodities. For example, rapid expansion of Third World countries into sugar production in the 1960s resulted in a fall in sugar prices in the 1970s. This decline almost eliminated sugar production in Louisiana and Hawaii (sugar cane), and Colorado (sugar beets), and it had even more severe consequences for the economies of countries, such as Cuba and the Philippines, that depended heavily on cash income from sugar production.

Farm price supports in the United States were at least partially successful in slowing the decline of family-owned farms until competition in world agricultural markets increased in the 1980s and 1990s. In this new context, high protected prices have made American farm products the choice of last resort on world markets. U.S. farm price supports have thus indirectly encouraged the expansion of agricultural production in Third World nations and further undermined the North American share of markets. The obvious solution of cutting farm subsidies would result in the accelerated demise of the American farmer, a solution that would have profoundly negative social and political costs and perhaps even negative long-run economic costs (Roscigno and Crowley, 2000).

In response to the threat of foreclosure and rising poverty rates in rural areas, the members of farm families have increasingly taken jobs away from home. Often this means that the husband looks for paid employment off the farm. Such arrangements may not be very lucrative, however, because traditionally male jobs in rural areas tend to pay low wages. Farm women are often able to secure higher status employment off the farm (such as clerical work) than are farm men (Lasley et al., 1995). High unemployment in rural areas unfortunately often cuts off even this "safety valve" for rural distress. There are often long lines of unemployed workers waiting for any job openings that become available. The children of farm families, many of whom would prefer to remain in farming, often face disheartening odds. Agriculture in the United States, in spite of high and rising productivity, thus operates in a state of chronic crisis.

Farm Laborers In the United States an additional 3 million people work as farm laborers at some time during the year. These workers are almost as numerous as farmers and unpaid family members who work on farms. Over half a million farm laborers work full time (250 or more days per year). Another half million are long-term seasonal workers who are employed at least two to three months per year in farm work (Census, 2000). The remainder—short-term seasonal workers—are primarily students and housewives or nonfarm workers who take occasional second jobs in agriculture. Full-time farm laborers earn an average of about $14,000 annually. Long-term seasonal workers earn about $8,000 annually; and short-term seasonal workers average less than $3,000 annually.

In contrast to the decline in the number of small farmers, the employment of farm labor is stable and likely to remain so. The employment of farm labor is concentrated on the largest farms, which are growing in size and number. Fewer than 2% of farms hire more than a third of all farm labor. About a third of American farms, however, employ at least some hired labor during the year. Farm work is regionally concentrated in California, Texas, and Florida. As is also the case for farmers and their families, most seasonal farm laborers do not depend on agriculture as their sole source of income.

Farm workers have an intense interest in securing the advantages available to other workers, such as regular employment, health insurance, accident insurance, retirement benefits, and union representation (Dunn, 1985). Farm workers, however, have traditionally been excluded from laws allowing workers to organize into unions and engage in collective bargaining. Cesar Chavez championed the cause of the farm worker from the 1960s until his death at age 66 in 1993. He organized boycotts, hunger strikes, union organizing drives, workers' cooperatives, and marches to highlight the plight of the farm worker. His efforts culminated in the emergence of the United Farm Workers (UFW, *www.ufw.org*), which continues his efforts into the 2000s. Chavez's innovative approach to organizing farm workers is described in Box 8.1.

BOX 8.1   Cesar Chavez and the United Farm Workers

To organize farm workers [Chavez] had to tackle four problems at once. He needed to organize them, to get the leverage so that organizing meant something, to unite bitterly divided regions and nationalities and to simultaneously sustain all these long enough for growers to pay attention. Chavez organized [farm workers] in fraternal service organizations. He got the leverage through boycotts fueled by feverish enthusiasm for liberating America's poorest workers. He united races, nationalities, religions and languages in the fields and cities across the nation by speaking and living a simple life dedicated to justice. And he sustained all of it by his personal example.

"You can't organize on money," he used to tell us. "There isn't enough money to organize now. There never was, and there never will be. Once you depend on money, you're finished." He lived on the UFW stipend of room, board, and $5 a week. . . .

He invented a type of organizing that synthesized the tactics of the civil rights and anti-war movements, cooperatives, labor unions, community organizing and religious ministries. Everyone in the union was uncomfortable with at least one aspect of this medley. But he got us all to live with each other because the union needed every one of these tactics and constituencies. No where else then or since could you find nuns working beside immigrant laborers and college students, a routine combination in the UFW. He didn't care whether he had to march 1,000 miles or set up a shrine outside a vineyard—if it worked, we were going to do it. . . . "There's a moment," he would tell us, "when the growers, politicians, corporations and sheriff's deputies believe we want it more, stronger, and longer than they want to keep it from us. That's the moment we win."

SOURCE: Excerpted from John Gardner, "Community Organizing: Seeds of Justice," *In These Times,* May 17, 1993, pp. 18–19. Reprinted by permission.

To reduce the problems of unstable employment and low earnings faced by farm workers, the U.S. Department of Agriculture has recommended educational programs for employers, labor contractors, workers, and their respective organizations. Such programs would seek to improve the skills of workers and to encourage modern labor-management practices by employers. Such practices would include the provision of benefits such as health care and retirement to provide a more stable employment situation. Increased skills for workers and improved labor-management practices would increase the productivity of farm labor and allow a general increase in farm wages. The Department of Agriculture also recommends further development and equitable enforcement of labor-related laws in the agricultural sector (Wells, 1996).

Forestry

A real artist minimizes the heavy labour for himself by dropping the trees within inches of where he wants them. And that is no mean trick, any novice of the game will find. By dropping the tree mid-way across his skid-pile, the real artist can cut into lengths, strip the branches and pile neatly with little more than a twist of the wrists or the leverage allowed by a handy pike pole. (Radforth, 1986:251)

Forestry employed 91,000 workers in 2000, about 4% of the total employed in agriculture, forestry, and fishing and less than 0.1% of the total civilian labor force (Census, 2000). Forestry is like agriculture in that it requires a high degree of skill. These skills, however, often go unrecognized and unrewarded because they are acquired informally during childhood and adolescence in rural areas.

Skilled work remains the norm in the logging industry. Mechanization in the harvesting of pulpwood, however, has dramatically changed the nature of the skills that are needed in this part of the industry. Pulpwood production is the most rapidly growing part of the forestry industry. Pulpwood, made from small trees not suitable for lumber, is used in paper and paper products. Workers use a machine called a wood harvester to cut 150 to 180 small trees per hour (about 3 trees per minute). The large machine dwarfs its single operator, who "by controlling numerous levers, gadgets, and joy sticks, guides the harvester through the forest, using the machine's huge hydraulic shears to fell the trees, which are then automatically processed into logs and stored at the rear of the vehicle" (Radforth, 1986:261).

Not only do such machines displace workers' traditional skills, but they also greatly increase accident rates. Power equipment was first introduced into the Canadian pulpwood industry in the 1950s. Since then disabling accidents have risen steadily. In Ontario, the center of Canada's pulpwood industry, logging has a higher fatality rate than any other industry.

Forestry shares with other extractive industries extreme vulnerability to price fluctuations. In British Columbia and in the Pacific Northwest of the United States, the lumber industry periodically collapses because of downturns in the housing industry. The economy of this lush region has been described as a "vulnerable hinterland economy based on a rich but limited resource" (Marchak, 1983). When a region is tied to a single raw product, such as lumber or pulpwood, it is highly vulnerable to economic downturns and price fluctuations, no matter how rich its natural resources.

Fishing

The accommodations aboard the [tuna] boats are very comfortable. Most are air-conditioned throughout the main deckhouse and bridge. All have hot showers and nice, even fancy, head (toilet) facilities. The galleys are modern and well-equipped. Many boats have a crew lounge with a card table, color television, and stereo system. . . . Each bunk is large—approximately 4' · 7'—and has a reading and a night light, extra space for books, and

a curtain which closes it off from the rest of the cabin. . . .

Being at sea is dangerous. Three boats in the fleet—big boats—have sunk in the last year and a half. . . . Working around heavy equipment, especially heavy running rigging, is also dangerous. Working with a seine net the size of those on the tunaboats differs from fishing methods employed in other fisheries. . . . The seine net on the tunaboats is primarily simply a way of putting up a "corral" around the tuna, and then slowly tightening the boundaries until the fish are swimming in the net right next to the boat. Only at the very end of the process are the fish in any way touching or suspended by the net itself, and the skippers are very careful to keep that suspension time to a minimum. The net is simply too large (three-quarters of a mile long, 350 feet deep, weighing by itself seven tons) to be controllable after it has been set, and the fish weight contained in a good set (50 to 100 tons) is too great for one to depend on even the strongest of equipment to hold it in the kind of weather in which the boats often have to set. This lack of control and consequent dependence on precise timing make the entire process dangerous for all those involved. (Orbach, 1977:19–20, 28–29)

The Ocean's Limitless Bounty? The year 2000 catch of fish from the world's oceans was over 120 million tons (Census, 2000). As a source of protein for human beings, fish is twice as important as poultry and more than half as important as cattle, pigs, sheep, horses, and goats combined. Between 1950 and 2000, the catch expanded at about 5% annually, significantly faster than the 2.5% annual growth in cereals and the 2% growth in meat.

The largest exporter of fish in the world is Canada, with a yearly catch of 2 million tons, most of which is exported. The United States has a yearly catch of about 7 million tons, with most of this being consumed domestically. Japan both imports the largest amount of fish and has the largest catch in the world (12 million tons, or about 15% of the world total).

At one time ocean fishing was expected to provide relief for the starving millions in the world. In the last decade, however, the world catch has grown at only 1% annually. The reason for this reduced rate of growth is mainly overfishing of coastal waters, which forces fishing boats to move farther offshore. In addition many fisheries have been systematically overfished. Deep-sea fishing today can involve thousands of baited hooks on lines stretching 80 miles across the ocean. Trawlers line up abreast to sweep life from swaths of ocean with nets large enough to cover a small city. In North America, both East Coast cod and haddock fisheries and West Coast salmon fisheries have declined dramatically as a result of overfishing. In coastal waters these problems are sometimes compounded by pollution. Human-generated environmental pollution can greatly reduce or render unfit for consumption harvests of shellfish and other species dependent on coastal habitats.

Overfishing and coastal pollution, combined with rising fuel prices, have limited the expansion of the catch (Apostle and Barrett, 1992). Nevertheless, the U.N. Food and Agricultural Organization estimates the fishing potential of the world's oceans at up to 455 million tons annually (about four times the current figure). Whether this large a catch will ever be economical to harvest is difficult to foretell. In practical terms, ocean fishing may have already peaked and be on the decline unless significant conservation measures are implemented—measures that would have to involve multinational agreements that are difficult to negotiate.

Aquaculture A bright spot in the world fishing outlook lies in the growth of **aquaculture,** also known as *fish farming.* Under this technology fish are raised in ponds or in *fish ranches* (large floating cages in the sea). In 2000, 9 million tons of fish were raised in this way, and the potential for expansion is considerable. Nearly 5 million tons were raised in China alone with only a limited

investment of capital and technology. With improved technology even the existing fish ponds in China are projected to have a potential yield of 40 million tons a year, almost half the current annual catch from the sea.

Technology and Organization in Fishing

The technology and organization of fishing vary from small shore boats with crews of two to three, which are at sea for a day; to larger boats with crews of nine to eighteen, which are gone for up to ten days; to the largest trawlers, which may have crews of twenty or more and be gone from port for up to three weeks or more. Traditionally, fishing has been organized around independently owned boats, each staffed by a crew that shares in the catch. This organization has been preferred over wage-labor systems because of the need for cooperation and teamwork to secure the catch and to face the dangers and challenges of the sea.

In recent years the increased capital investment required in fishing has made it more difficult for individuals to purchase the necessary boats and equipment. Absentee ownership of boats and extension of wage-labor relationships into fishing may erode the cooperative basis of the work and the working conditions of fishermen. Risks are minimized when those in the boat are free to arrive at a consensual decision about when to fish and when to return to port. When this decision is made by an owner back on shore or according to a set of bureaucratic rules, lost opportunities and dangerous situations are more likely to occur (Doeringer, Moss, and Terkla, 1986). In addition, "unlike large corporations, fishing families are much less likely to withdraw their investments from fishing and enter other sectors of the economy during less profitable periods" (Norr and Norr, 1978:169).

MINING

"First time I got hurt, I got covered by a rib roll (a collapsing wall) and got three vertebrae busted and a busted pelvis. I got a

25 percent disability for that, but I came back to work anyway.

"[Eight years later] I slipped in some grease and broke my neck and got my back messed up. It was my fault. I was in too much of a rush.

"Oh, yeah. I've also got first-stage rock dust—silicosis. I got paid a flat $3,100, but my lungs are so bad I can't hardly get around."

His buddies half-listen to Raymond's story. Most of them can match it, fracture by fracture, wheeze by wheeze. . . .

Over the years the men build up friendships with each other that are not equaled in many marriages or families— although most of them would be embarrassed and would deny it if the subject were verbalized. Men call each other 'buddy' with an open affection that seems quite foreign to a visitor from the cold, impersonal urban world. (Vecsey, 1974:19, 124)

Down in the Mine Although mining is dangerous and demanding work, it generates a sense of purpose and a collective identity. This identity is built on the twin pillars of shared group responsibility for the work below ground and geographically isolated work communities above ground. Miners' work ethic prescribes that the work be done according to certain standards of safety and efficiency. This ethic also stresses competition between work crews in the effort to produce the most tonnage. Teams of miners are responsible for organizing their own activities below ground and resent external control (Douglass and Krieger, 1983). The miners' sense of group solidarity is intensified by the many dangers they face. These hazards include inadequate ventilation, poor illumination, accumulation of dust, dangerous gases, use of explosives, poorly supported roofs, unsafe tunnels, flooding, and working with high-voltage electrical equipment (Corn, 1996).

Occupational Solidarity Above ground, this collective identity is reinforced by the frequently isolated location of mining communities. In mining regions, a distinct culture often emerges that provides its members with a shared identity. Sociologists refer to such subcultures as **occupational communities.** They are typical not only of miners but also of lumber workers, longshoring workers, printers, and other trades that work in isolation from the mainstream of society. Occupational communities may even be said to distinguish some professionals, such as doctors, who live their lives in relative social isolation if not in geographic isolation.

In the United States miners have often been a powerful force in national politics. For instance, coal miners had a key role in promoting a more even-handed government approach to trade unions. In the coal strike of 1902 and 1903, President Theodore Roosevelt threatened to nationalize the industry (eliminating the owners' profits) and have the army mine the coal (eliminating the workers' jobs) unless the owners and workers bargained collectively. This episode is frequently cited as the first instance of the federal government taking a neutral stand between capital and labor. Previously, the government had sided with capital by jailing strike leaders and supporters and using troops to protect hired strikebreakers and company property.

American miners, however, have never played as pivotal a role in national politics as British miners have. Their relative weakness has resulted partly from their geographic isolation in Appalachia (the center of the American coal-mining industry) and in the West (the center of much of the metal-mining industry). These chronically depressed, semirural regions have provided a weak springboard for miners' demands. In times of economic downturn, miners have often reverted to rural and agricultural pursuits such as gardening, hunting, and producing eggs and butter for local markets. Even in the best of times American mine workers have kept strong ties to the rural economy. The availability of these alternative pursuits, even if they are only marginally productive, has been an important factor limiting the organizational effectiveness of miners in demanding redress for the problems of their industry (Gaventa, 1980).

Strip Mining   In recent years the coal-mining industry has shifted from Appalachia westward toward the Rocky Mountains. In the West, coal seams are thicker and closer to the surface. As a result deep mining technologies have been replaced by more capital-intensive strip-mining technologies, long used in the West for mining metals such as copper and aluminum. As with agriculture and other extractive industries, improvements in technology have reduced employment in mining even as output has increased.

CONSTRUCTION

The difficulty is not in running a crane. Anyone can run it. But making it do what it is supposed to do, that's the big thing. It only comes with experience. Some people learn it quicker and there's some people can never learn it. (Laughs.) What we do you can never learn out of a book. You could never learn to run a hoist or a tower crane by reading. It's experience and common sense. . . .

This is a boom crane. It goes anywhere from 80 feet to 240 feet. You're setting iron. Maybe you're picking fifty, sixty ton and maybe you have ironworkers up there 100, 110 feet. You have to be real careful that you don't bump one of these persons, where they would be apt to fall off. . . .

There's a certain amount of pride—I don't care how little you did. You drive down the road and say, "I worked on this road." If there's a bridge, you say, "I worked on this bridge." Or you drive by a building and you say, "I worked on this building." Maybe it don't mean anything to anybody else, but there's a certain pride knowing you did your bit. (Terkel, 1974:49–50, 54 [Hub Dillard, crane operator])

Pride in Skilled Work   Like miners, construction workers take great pride in their work. They often have a high level of skill and exercise a good deal of autonomy on the job. Construction is difficult to supervise, because it is spatially dispersed and because much of it is skilled craft work. Often, only the construction worker knows how best to do a given job. Because of these factors, construction workers are not closely supervised. Rather, their work is inspected, either by their employer or by a government building code inspector, *after* it has been completed. Their skill and the resulting autonomy give construction workers a sense of power and pride in work (Applebaum, 1998).

Many construction workers belong to craft unions and the wages and conditions for other workers in construction are strongly influenced by these workers' contracts. Union construction workers are generally more qualified than nonunion workers. "Unions have taken an active role in cooperating with contractors and the government to develop apprenticeship programs that produce craftsmen who are highly skilled in all aspects of their trade, rather than a narrowly defined set of tasks" (Allen, 1984:253). It has been estimated that union craft workers are between 7% and 11% more productive than their nonunion counterparts, even after adjusting for their higher wages (Freeman, 1994). Craft unions also run hiring halls where contractors can secure large numbers of skilled workers on relatively short notice, thus reducing the costs of recruiting, screening, and supervising workers.

Nonunion Workers   Even though they are on average less productive, nonunion workers do an increasing share of construction work. The reason is that, in many settings, price competition is more important than quality. In depressed housing markets, price competition becomes an increasingly important factor. Union construction workers have responded by making wage concessions and stressing their ability to produce high-quality work on schedule, factors that still have to figure prominently in the decision making of construction contractors.

BOX 8.2   Gaining Acceptance as a Female Carpenter

For a woman to survive in the trades, . . . you have to be tactful, not be hostile, not alienate people. You really have to learn professional survival skills, because men's masculinity is threatened by you being there. Society recognizes construction workers as being very macho and virile. When a woman comes along who's five foot three and a hundred and twenty pounds and can get in there and do their type of work, it's a blow to their ego, a real shock. So the men are threatened by it.

The men show that in different ways. If a foreman comes up to a group, he'll look at everyone except you. He'll delegate jobs to everyone but you, and then you have to go up and ask him. It's sort of uncomfortable things like that. Or the carpenters will pair up and you're the one left out. Or men you're working with just won't talk to you. I used to not push in those situations, but lately I've felt more self-confident, and sometimes I play games with those people. Like, for example, today on the job site there is another fourth-year apprentice who is sort of the foreman's pet. He just came over and picked up some wood and started doing some form work on top of the footing another carpenter and I had built. And I yelled at him, "What's going on?" Not in a hostile way, just curious. We'd finished that footing, and I wanted to know what changes had been ordered. He wouldn't answer, so I got a bit

flustered and asked again. He still wouldn't answer. I kept probing, and he kept not answering. Finally I said, "I'm talking to you. I want a response." I was just very aggressive, which I'd never been before. But I've gotten to the point where I'm tired of not getting recognition. Finally he answered me and explained there was a change in plans, not any mistake. I could see he was a little taken aback and I felt so good. I had won something. What was more interesting was the reaction of the other man. Right in the middle of all this he looked up at me and smiled.

The crew I'm working with now—it's never been better. I'm accepted. They kid me like one of the guys. They pay me compliments. They treat me like an equal. It has been a real breakthrough. I've had other crews that have been really nice, but I know enough now so that I can talk business as well as pleasure. Not only am I compatible, but I feel they recognize me as a fairly good carpenter. I haven't had that recognition before. Like if a carpenter I've been working with is talking with another man about something we built, he'll mention my name, say, "Elaine and I were working on this. . . ." Even a week ago . . . I wouldn't have gotten that recognition. So they include me, acknowledge me.

SOURCE: Jean Reith Schroedel, *Alone in a Crowd: Women in the Trades Tell Their Story*, pp. 38–39. Copyright 1985 by Temple University. Reprinted by permission of Temple University Press.

Residential construction has been sluggish across much of the nation during the early 2000s. In addition, excess manufacturing plants and office buildings were built in the 1980s and 1990s, and business construction has been relatively stagnant as a result. Finally, because of strapped government revenues, expansion and repairs of streets, highways, and bridges has often been delayed. All these factors have contributed to a slowdown in construction activity. Because construction work is highly regional, the situation of construction workers can vary dramatically between areas. Due to population shifts, construction activity has been higher in the West and South than in the Northeast and Midwest. In spite of these short-term

problems, the long-term employment outlook in construction is relatively stable. Technological innovations in construction have been largely incremental and, given the nature of construction work, few major employment displacing innovations are expected in the future.

In spite of the current slowdown in construction activity, women have increasingly gained entrance into this previously all-male field. As new entrants into the construction trades, women have had to confront resistance from some of their male co-workers. Box 8.2 reports on the frustration that can be caused by such resistance. It also illustrates that resistance can, in some cases, be overcome.

MANUFACTURING

Manufacturing industries produce a tremendous variety of goods in industrial societies; thus, they employ a wide range of occupations. Manufacturing workers can be broadly classified as craft workers (skilled workers), machine operators and assemblers (semiskilled workers), and laborers (unskilled workers). In this section we discuss these three categories of manufacturing work and look at current developments in four key manufacturing industries. Manufacturing also employs professional, managerial, and clerical workers. We discuss those occupations in Chapters 11, 12, and 13.

Craft Workers

The U.S. Bureau of the Census classified over 12 million workers as **skilled craft workers** in 2000. About 4 million craft workers are employed as precision production operators. These occupations range from tool and die makers to power plant operators. An additional 4 million craft workers are mechanics and repairers, including vehicle and equipment mechanics, telephone line installers and repairers, heating, air conditioning, and refrigeration mechanics, and heavy equipment mechanics (Census, 2000). The remaining 4 million workers in the skilled trades are employed in construction.

Craft Apprenticeships Skilled workers typically learn their trade through an apprenticeship program. At any given time about 250,000 workers are enrolled in apprenticeship programs in the United States. Most of these programs are jointly administered by a company and a craft union. Apprentices receive about half pay during the years they are in the program. Depending on the trade, workers spend 100 to 800 hours a year in classroom education for two to four years. The rest of their time is spent on the job working under the supervision of more senior members of their craft.

Machinists, for example, are required to complete about 570 hours of classroom training and about 8,000 hours of on-the-job training during a four-year program (Wright, 2000; *www.iamaw.org*). Machinists manufacture specialized parts for the machines that mass produce other products. (The workers who operate these machines are classified as "machine operators" and are discussed below.) At the end of their apprenticeship, machinists are expected to be able to work with a variety of metals and machinery, including complex electronic and microprocessor-driven machining tools. Besides knowing how to cut, drill, and shape metal, they must also know how to program their machines and how to make precise measurements, sometimes down to a millionth of an inch, so that the parts they produce will work perfectly.

The skills learned through these programs are the basis of the craft worker's pride and autonomy. Craft workers resent direct supervision and strongly resist unnecessary instructions from management about how to do their work. Because of their skills and training, they believe that only they have the expertise to decide how best to do a given job (Orr, 1996).

As a result of possessing skills that are difficult and time consuming to acquire, craft workers generally have more security against layoffs than do semiskilled machine operators. In addition, automation has affected craft workers less than it has affected less skilled workers. The tasks required of a skilled craft worker are too diverse to be easily automated. The printing industry, however, provides an exception to this pattern. In the past this industry employed a large number of skilled craft workers to set type for newspapers and magazines. With the development of electronic and computer-assisted printing and layout, many of these operations have been eliminated or are now performed automatically by computers. The extent to which other crafts are being affected by computer automation is explored in Chapter 9.

Exclusionary Practices In the past women and minorities were strongly discriminated against in the skilled trades. Exclusionary practices by employers and efforts by craft unions to protect the jobs of their predominantly white, male members were both to blame. In recent years most of these practices have been greatly reduced. As you saw in Chapter 5, however, such practices are deeply ingrained in established prejudices and procedures and are difficult to eradicate completely. In 2000, 14% of white male and Hispanic male workers were in the skilled trades. However, only 9% of black workers and only 2% of women held jobs in the skilled trades (Census, 2000). The strong occupational segregation of women out of the skilled trades continues because of different socialization of men and women, continuing discrimination, and a lack of informal job contacts in the skilled trades for female workers.

Machine Operators and Assemblers

I start the automobile, the first welds. From there it goes to another line, where the floor's put on, the roof, the trunk hood, the doors. Then it's put on a frame. There is hundreds of lines.

The welding gun's got a square handle, with a button on the top for high voltage and a button on the button for low. The first is to clamp the metal together. The second is to fuse it.

The gun hangs from a ceiling, over tables that ride on a track. It travels in a circle, oblong, like an egg. You stand on a cement platform, maybe six inches from the ground.

I stand in one spot, about two- or three-feet area, all night. The only time a person stops is when the line stops. We do about thirty-two jobs per car, per unit. Forty-eight units an hour, eight hours a day. Thirty-two times forty-eight times eight. Figure it out. That's how many times I push that button.

The noise, oh it's tremendous. You open your mouth and you're liable to get a mouthful of sparks. (Shows his arms.) That's a burn, these are burns. You don't compete against the noise. You go to yell and at the same time you're straining to maneuver the gun to where you have to weld. . . .

I don't like the pressure, the intimidation. How would you like to go up to someone and say, "I would like to go to the bathroom?" If the foreman doesn't like you, he'll make you hold it, just ignore you. Should I leave this job to go to the bathroom I risk being fired. The line moves all the time. (Terkel, 1974:221–222 [Phil Stallings, spot welder])

In 2000, 14 million workers were employed as machine operators and assemblers in the United States. Most of these workers are considered **semiskilled.** The Bureau of Labor Statistics defines semiskilled work as requiring *less than two weeks of training.* The period required to become effective at many of these jobs, however, may be much longer.

The largest group of machine operators and assemblers, numbering over 7 million, operate stationary machines. The largest subgroup of these work in textiles and apparel. Machine operators and assemblers also include punching and stamping machine operators, lathe operators, molding and casting machine operators, assembly-line welders, printing machine operators, and laundry and dry cleaning machine operators. The second largest group, just over 5 million, operate transportation equipment. These occupations include 3 million truck drivers, as well as bus drivers, forklift drivers, and other mobile equipment operators. The final group, just under 2 million, are assemblers, testers, and graders. These workers do hand assembly and sorting that does not require the regular use of machinery (Census, 2000.)

Repetitive Work Many semiskilled jobs involve work on mechanically paced lines. Such work is extremely repetitive. It requires a high degree of surface mental attention without corresponding mental absorption (Rankin, 1990). Repetitious work, especially if it is time pressured, can be stressful and it can lead to mental distress and breakdown. Other semiskilled jobs involve sitting at a bench and assembling a subunit. Bench workers take parts from one pile, assemble them, and

---

BOX 8.3 Bored to a Stupor $

Based on over four years of full-time employment and observation as a production worker in the beer bottling industry, sociologist Clark Molstad arrived at the following understandings of why workers sometimes prefer boring jobs in preference to ones with slightly greater diversity.

> When [working on the bottling line] I experienced strong feelings of mental regression. My fantasies became progressively more childlike, until I was actually holding imaginary conversations with the beer cans in my hands. I worried about their dents or streaked labels as though they were animate objects. I wondered where they would be shipped and how they would like it when they got there. I wondered how a can felt as it was being crushed and shredded. It was only with some difficulty and effort that I could muster my consciousness to return to normal after hours of this work. Yet interruptions were troublesome because they required that I return to an adult mode of thought and take up the burden of conscious life in the brewery. . . .

> For that reason, I resisted interruptions and resented problems that forced me to return my attention to the work and the brewery. At times I even regretted going to the lunchroom on breaks because it required focusing my attention on the here and now. I preferred to stay in my fantasies. . . .

> The preference for boring jobs stems not from any inherent attraction but from the lack of control that turns more stimulating jobs into hassles. Jobs in which the workers' control does not match their responsibility are too stressful and will be avoided when possible. In these situations the consequences of having responsibility are mostly negative. These workers get no praise or promotions for jobs well done. The most that they can hope for is to be left alone when the job goes well. Being "left alone" becomes a goal, as it is the only form of autonomy or freedom available.

SOURCE: Clark Molstad, 1986, "Choosing and Coping with Boring Work," *Urban Life* 15,2 (July):221–222, 226–227.

---

place them in another pile. Bench work has the advantage of not being machine paced, but it is still highly repetitive, and workers have to meet production quotas. Box 8.3 describes the experience of doing boring work in a bottling factory and explores why some workers may prefer repetitious and boring jobs on the line to related jobs with slightly more diversity but closer supervision.

Speed versus Quality Workers in semiskilled occupations are also often stressed out because the assembly line or the production quota pushes them so fast that they cannot do quality work or take pride in their work. Workers in an automobile plant registered the following complaints in this regard:

> The cars come too fast for quality. It's quantity instead of quality. I'm doing the best I can, but could do a neater job slower.

> On an assembly line you just do it once; if it's wrong, you have no time to fix it. I get no satisfaction from my work. All I do is think about all the things that went through wrong that should have been fixed. My old job was nothing like this.

> I try to do quality work, but I'm too rushed. This keeps me from getting pleasure from the work. They say "haste makes waste," and they're getting plenty of both.

> I'd rather do less work and do it right. How can you get quality when they don't give you time? The "quality" signs they have mean nothing. (Walker and Guest, 1952:79)

Jobs that demand sloppy work are demoralizing—maintaining pride in one's work is difficult or impossible in such situations. Assembly-line work in some industries has changed since these complaints were registered. As we will see later in this chapter, as well as in Chapters 9 and 17, technological and organizational changes have eliminated some assembly jobs while placing greater demands on other jobs. Assembly workers today are often expected to participate more actively in quality control. However, it is unclear if these jobs are any less stressful today than in the past (Fink, 1998).

Working Ahead In response to job pressures, semiskilled workers devise a variety of creative ways to give themselves at least limited control over the pace of their work. One common technique is to *work ahead*. This may mean either working *up the line* on the assembly line, by moving ahead to work on parts before they arrive at one's station to secure a brief break later, or "building a bank" of parts, by pushing oneself in spurts in order to slack off later. Another strategy is *doubling up,* in which one worker temporarily takes on two jobs while a second worker takes a break. Managers generally oppose such practices, but workers contend that they make fewer errors by working in intense pushes and then slacking off. Workers argue that a continuous grinding pace creates greater boredom, poorer concentration, and more errors (Graham, 1995).

The jobs of machine operators and assemblers can be quite different in core and peripheral parts of the economy. In the *core* economy, made up of large manufacturing companies with a significant share of their market, wages and benefits are generally good and provide at least partial compensation for the repetitive nature of the work. In *peripheral* parts of the economy, dominated by small and financially vulnerable firms, wages may be at or near the legal minimum, and benefit packages may be nonexistent. In these situations the problems of repetitive, alienating work are compounded by problems arising from poverty-level income. Female and minority workers tend to be overrepresented in these latter jobs.

Unskilled Labor

Poultry must be handled, initially, as live birds. Workers (almost invariably male because of the weight handled) must snatch live birds from cages unloaded from tractor trailer trucks and hang them, upside down, on shackles attached to moving conveyor lines. The "hanging" job may even involve 30–40 pound turkeys. The "hangers" are subjected to wing battering by the dirty, squawking birds who not infrequently urinate and/or defecate on the workers handling them. As the flopping, noisy birds move down the line, they undergo an electric shock intended to relax all muscles for a thorough bleeding after the throat is cut. This step also results in additional excretory discharges from the birds. All five senses of the workers are assaulted. One "hanger" who was interviewed revealed, on weekends, he took six to eight showers trying to rid himself of the stench.

After being shocked, the birds are slaughtered by having their throats cut, either by hand with a knife or by a machine with a worker standing by to kill birds where the machine fails to do so. . . .

Poultry processing of necessity involves the commodious use of water, and thus there is an inevitable high degree of dampness and frequency of standing water constituent to this industrial process. The water used may have to be scalding, to remove feathers, or near-freezing, in order to cool the carcasses of the birds. . . . Thus, employees may have to work in damp, cold, or hot rooms, literally standing in water (and sometimes blood, such as in the "killing room"), handle blood, gore, offal, and visceral organs and materials. (Bryant and Perkins, 1986:158)

In 2000, about 5 million workers in the United States were employed as laborers, materials handlers, equipment cleaners, and helpers—jobs all classified as **unskilled** (Census, 2000). These jobs include stock handlers, garbage collectors, hand packagers, and machine feeders and off-bearers. Some unskilled work, such as garbage collection, is relatively autonomous and is not closely supervised. A greater share of unskilled work is closely supervised and involves cleaning, loading, unloading, or preparing for some more complicated step in manufacturing. The loading of chickens onto the "disassembly line" described above is a good example of such work.

Laboring jobs are often closely supervised under the supposition that workers will not do the tasks right unless forced to do so. Whatever the merit of such considerations, close monitoring serves as an additional negative factor in much unskilled work and acts to undermine whatever autonomous motivation workers have.

"No Experience Needed" Most laboring jobs require little if any training. In effect, people can be hired "off the street" to do the work. No training is required beyond a brief explanation of the job and those skills picked up while performing the task. Much of the work is physically demanding and is done under harsh physical conditions. The work of female laborers is no exception to this rule. In describing her job as a "craw puller" (remover of the throat pouch from the dead chicken) on the poultry processing line described above, one woman felt that men could not physically do the job: "Take a man, he won't stick to some of those harder jobs like craw pullin'. A man will stick his finger in there and gets a sore finger and walks the floor all night long—he will quit that job, he won't stick to it" (Bryant and Perkins, 1986:162).

Women and Minorities in Unskilled Jobs Blacks and Hispanics are *over*represented in unskilled work because of a historical pattern of discrimination in better paying jobs. Women are *under*represented in unskilled work largely because many of them have sought employment opportunities in other fields, such as clerical work, as preferred options. Blacks make up 11% of the labor force but 16% of unskilled workers. Hispanics make up 10% of the labor force, but 18% of unskilled workers. Conversely, women make up 46% of the labor force but only 20% of unskilled workers. Blacks are well represented in all types of unskilled labor but most strongly as garbage collectors and as machine feeders and offbearers, where they make up 38% and 23% of workers, respectively. Women are most underrepresented as construction laborers (2%). They are more likely to have jobs as machine feeders and offbearers, where they make up one-third of workers, or as hand packers, where they make up two-thirds of workers (Census, 2000).

Career Mobility The occupational outlook for people in laboring jobs is quite restricted. Of the male workers at the poultry processing plant discussed above, about 25% said they had no particular career aspirations, and about 25% said they wanted to be mechanics. Of the women, 37% said they had no particular occupational aspirations, but those who did have aspirations had somewhat higher ones than their male counterparts. Many aspired, or had aspired, to professional, semiprofessional, or white-collar occupations, such as nursing, teaching, or secretarial work (Bryant and Perkins, 1986:160). Reduced aspirations are a common response to the problems that plague unskilled workers. Among these are the limited job opportunities in local areas, as well as family and other local and regional ties that make workers reluctant to leave home in the search for better work elsewhere. There are only limited opportunities for on-the-job advancement in unskilled work. From the viewpoint of the unskilled worker, dreams of occupational advancement are difficult to fulfill. The work provides little extra money to save for continuing one's education, and about the only skills the worker can demonstrate or perfect on the job are speed and endurance.

Working-Class Culture

The industrial working class has its own sub-culture based on the unique nature of its work. This subculture includes many elements taken directly from the working-class experience on the job, such as pride in doing quality work, fear of economic insecurity, and cooperation with co-workers (Fantasia, 1988). The subculture also contains elements of pride in knowing how industrial plants and procedures really work, as opposed to how they are supposed to work on paper. Workers often guard this knowledge as their own private treasure. This is especially true when they have little incentive to share it with management because they are given no role in decision making (Rothschild and Ollilainen, 1999).

Working-class culture has been described in a variety of sometimes contradictory ways (Form, 1985). Some researchers argue that workers are passive and accepting on the job and find their rewards outside of work through family and consumption (Goldthorpe et al., 1969). Others see workers as chronically dissatisfied but unrebellious. Those in the Marxist tradition see workers as potentially revolutionary but stymied by the greater economic, political, and ideological resources of capitalism (Wright, 2000). Still others see the working class as caught up in a microcosm of informal workplace subcultures and local loyalties (Thomas, 1990).

Working-Class Diversity All of these descriptions represent only partial realities. The working-class experience is simply too broad and diverse to be aptly summarized in a single formula. The working class does not constitute a unitary subculture. Rather, there are distinct segments within it. A minimum of at least three segments is required to capture adequately the working-class experience. First, there is a reasonably well-paid segment of skilled workers employed in large firms with good benefit packages. Second, there is a semiskilled segment employed in more marginal establishments where wages are lower and benefit packages less comprehensive. Finally, there is a marginal segment of unskilled workers who are able to secure only irregular or unstable work at the lowest wages. The working class is further subdivided along race, ethnic, and gender lines.

Although the working class has occasionally been able to organize itself into a unified body, as in the trade union movement or in working-class political parties in Europe, such combinations are fraught with internal divisions as different segments of the working class struggle for different goals. Groups representing the working class as a whole are important in advancing its interests. But the working class is large and has diverse interests. In addition, the capitalist class, which has opposing interests on many issues, is better organized and better funded than the working class. As a result, working-class alliances have tended to be fleeting and partial.

THREE KEY MANUFACTURING INDUSTRIES

Having discussed the major categories of production jobs in manufacturing, we now take a closer look at some of the major manufacturing industries and trends in these industries. We briefly consider the automobile industry, the steel industry, and the textile industry, all of which have faced increased world competition in recent decades.

Automobiles

Automobile manufacturing has been *the* key manufacturing industry since the middle of the twentieth century. Worldwide, over 20 million workers—roughly equivalent to the entire population of Australia—are involved. Organizationally, the production of automobile components and subassemblies is based on one of the most globally dispersed networks of any industry.

The world production of automobiles peaked temporarily in 1978 at 32 million annually, faltered in the 1980s, and resumed growing at a reduced rate in the 1990s, reaching a peak of over 50 million by the early 2000s. Why has this pattern of irregular growth occurred? One reason is that the market for cars in the industrialized nations has become saturated. The market for cars in the less industrialized world is growing, but only slowly, because relatively fewer people in these nations can afford an automobile. These factors are compounded by rising oil prices, which have greatly increased the costs of owning and operating an automobile.

Increased World Competition The problems of a stagnant world market have been amplified for American automobile manufacturers by increased competition from Japan and Western Europe and also from countries in Eastern Europe and Latin America that have begun to build cars to supply their own markets. In 1960 the United States produced over half the automobiles in the world; by 2000 its share had fallen to less than a quarter. An additional reason for this decline is that American consumers have increasingly turned to smaller, better engineered, and more fuel-efficient Japanese cars. Japan exports almost 4 million more cars than it imports. The United States imports almost 3 million more cars than it exports, and a large share of these imports come from Japan. In addition, 35% of the production of General Motors cars and 62% of the production of Ford cars take place outside the United States (primarily but not exclusively in the area of subassembly manufacturing). Thus, many so-called "American cars" contain a substantial proportion, or even a majority, of parts made or assembled overseas (Milkman, 1997).

Why has the United States lost out so dramatically in the market for smaller cars? One reason is that American management was slow to respond to this new market, preferring to specialize in higher-priced cars, light trucks, and sports utility vehicles, with greater profit margins. Unfortunately, this decision has spelled disaster for American automobile workers. Another reason is that American manufacturing technology and work practices have failed to respond to the Japanese challenge. Even when Americans have tried to build small cars, they have been less efficient in doing so than their Japanese counterparts. The American automobile industry has long had the highest rates of layoffs, turnovers, and absenteeism of any major industry. Such organizational practices and worker–management relations have provided an insufficient base from which to compete with Japanese automakers, who employ group production techniques and extensive robotics technology.

Starting in the 1980s and 1990s, Japanese automakers have increasingly opened assembly plants in North America to increase their access to the American market. Two of the largest of these are the joint General Motors–Toyota plant at Fremont, California, and the Nissan plant at Smyrna, Tennessee (Besser, 1996). Other plants are located in Ohio, Kentucky, Indiana, and Michigan. American automobile manufacturers have learned innovative production strategies from these plants. The contrast with work systems in traditional plants is readily apparent. At the California plant, for example, workers have the right to stop the assembly line if problems arise, something strictly forbidden in traditional American manufacturing facilities.

Workers at the Japanese transplant facilities report mixed reactions to the new work systems. Although they like many of the new forms of worker involvement, they also report high levels of pressure, stress, and resulting injuries (Graham, 1995). Canadian workers similarly report resentment about erosions of union power and worker solidarity that are sometimes experienced in Japanese automobile assembly plants (Rinehart et al., 1997). At a Japanese plant in Indiana, workers are organized into "teams" but often balk at cheering for the company, Japanese style, each morning (Graham, 1995).

The challenge of world competition has nevertheless promoted long needed changes in the American automobile industry. A new contract between Buick and the UAW, for instance, stipulates a "pay-for-knowledge" plan in which workers are compensated according to the number of jobs for which they are qualified, rather than solely on the basis of their current job. Chapter 17 considers in greater detail some of the job redesign programs that are gaining increasing acceptance in American industry. Many of these programs have emerged out of the struggles of the American automobile industry in the tumultuous 1990s.

Steel

The steel industry has met with hard times in the United States and Canada. Once again, declining demand in the industrialized countries is partly to blame. A great deal of steel is used to build industrial plants and facilities, but because of excess manufacturing capacity in much of the industrialized world, there is little demand for steel for constructing new plants. Delayed and deferred reconstruction of bridges and other large structures in these countries has further reduced the demand for steel. The demand for steel continues to rise in the industrializing nations, such as Korea and Taiwan. Because of the costs of transportation, however, steel is generally produced near the area where it is to be used. Steel has not been heavily involved in international trade. In addition, its production is strongly *vertically integrated*—that is, closely linked to the mining of iron ore and the production of coke. The production of these raw components, the production of steel, and the production of finished steel products often occur at nearby sites. These factors make it difficult to subdivide the process of making steel and retain selective activities in older locations.

Outdated Equipment The problem of a diminishing market for American steel is compounded by the lack of investment by North American steel companies in new technology. With few competitors after World War II, American steel companies used their profits to buy companies in unrelated industries rather than to upgrade and expand their capacity for manufacturing steel. USX (*www.usx.com*), formerly United States Steel and once the largest steel company in the world, now makes only 11% of its operating profits in the steel industry. It has systematically shifted its operations to other areas, including oil exploration, chemicals, and real estate. As a result, the American steel industry operates with outdated technology and, frequently, with dilapidated equipment. Such "vintage capital" and technology perform poorly in competition with the new capital and technologies used in the German and Japanese steel industries which have been rebuilt from the bottom up since World War II. In addition, just as in automobiles, the Japanese and Europeans have outpaced North American manufacturers in new technological advances. As late as 1970, most American steel was still being manufactured with open-hearth furnaces. By that time 80% of Japanese steel was being made with more advanced basic oxygen process technology (Shorrock, 1983). Although American firms have since moved into this technology, they have been unable to regain the competitive advantage against the Japanese, who also benefit from greater use of automation at all stages of production.

Employment in the American steel industry declined from 450,000 in the 1970s to only about 200,000 by 2000. This reduction in employment has entailed both permanent layoffs and wholesale plant closings. By 2000, the largest steel-producer in the world was Nippon Steel, headquartered in Tokyo. This company produces two and a half times as much steel as its nearest U.S. competitor. The leading steel-producing nations and their share of world production are listed in Table 8.1.

Table 8.1   World's Leading Steel Producers 🌐

Country	Share of World Production
Japan	22.0%
United States	15.2
Germany	7.8
Italy	7.7
India	4.7
France	4.3
South Korea	4.1
United Kingdom	3.3
Brazil	3.1
Spain	2.8
Taiwan	2.7
Mexico	2.7
Canada	1.8
Indonesia	1.7

SOURCE: *International Yearbook of Industrial Statistics,* 2000, United Nations Industrial Development Organization, Vienna, p. 54.

Specialty Steel Products One bright spot in the American steel industry is the recent emergence of smaller companies called **mini-mills,** which use electric furnaces to turn scrap steel into basic steel products, such as concrete reinforcing bars and light construction products. These mills are both productive and profitable and are capturing a significant share of the domestic market.

Advances have also occurred in the employment of women in the steel industry in the 2000s. Women have made these strides even in the face of declining employment in the industry as a whole. The women who have found work in the steel industry report less sexual harassment than women working in such traditional female occupations as clerical work (Deaux and Ullman, 1983). The reason is that women in steel plants are working with men as peers. The majority of female clerical workers are supervised by male managers, and subordination in work roles encourages sexual harassment. On the other hand, 20% of female workers in the steel industry are employed as janitors, whereas only 2% of male workers are so employed. Thus, even though the gender barrier has been broken in the steel industry as a whole, gender-based occupational segregation within the industry remains a problem.

Textiles

The initial task carried out [in the men's lightweight suit production room] was finish cutting. A finish cutter trimmed excess cloth from the cuttings for a jacket's body and sleeves. The cuttings were produced in bulk in a room apart from the workshop's main area. Although finish cutting was considered tailor's work, it required less skill than the other tasks the tailors performed. An apprentice was often assigned to finish cutting before he was given work as a trouser maker. Nevertheless, senior men liked the work. Even though they stood up and were thus more visible than most of the other workers in the room, they said that there was less pressure on them and that they found it less tiring than the other jobs. A cutter could work ahead and then take a break. . . .

The trimmed cuttings, tied in packets, went next to women who did preliminary stitching by machine before the next step, which was called shoulder basting. Basting was done by men, who stitched together the pieces for the sleeves and the bodies of the jackets. Special skill was needed to make the sleeves hang in proper alignment. This operation determined a garment's finished appearance. The basters sat on small stools, bent over their work, which they held on their knees. The low table in front of them held bundles of jackets, needles, and thread.

After basting, the garments were taken to the center of the room for an intermediate pressing, which facilitated subsequent operations. This operation was done by women. Then seamstresses undertook the final machine stitching, which consisted of sewing collars, lapels, buttonholes, and hems. (Savage and Lombard, 1986:167–168)

Textile and apparel manufacturing is a huge worldwide industry. It employs 25 million workers as well as additional millions of unregistered workers laboring in sweatshops and in home production. Worldwide production of textiles increased fourfold from 1950 to 2000 (Census, 2000). During this time the nature of fabrics also changed, with synthetics growing to almost half the total production. Synthetics include fibers derived from wood products, such as rayon and acetate, and petrochemical based fibers, such as polyester, acrylics, and nylon.

The United States is the largest producer of synthetic fibers in the world, with a 25% share of production. This share, however, has dropped from 32% in 1970. Much of the growth in textile employment has occurred in Third World nations, such as Indonesia, Bangladesh, and Sri Lanka, because of cheaper labor costs in these nations, as well as in industrializing nations, such as Taiwan, Korea, and China. Additional employment shares have been captured by technologically advanced Russian and European competitors.

Mill Workers Only a small portion of textile and apparel manufacturing requires skilled labor. The unskilled nature of much of the work has facilitated the spread of the textile and apparel industries to less industrialized countries. Only about 12% of textile workers in the United States are skilled craft workers. Unskilled textile workers have typically been viewed as acquiescing to the demands of their work with little resistance:

> The textile worker's powerlessness is expressed in constant work pressure, an inability to control the pace and rhythm of his work activity, a lack of choice of work techniques, and the absence of free physical movement. In his lack of freedom and control, the textile hand's situation is virtually the polar opposite of that of the free craftsman. (Blauner, 1964:66)

Waves of unionization among textile workers occurred in the United States in the 1930s but turned out to be short lived (Roscigno and Danaher, 2001). Today, textiles remain one of the least unionized American manufacturing industries, partly because much of the industry remains in the South. In this region white workers often receive preference in jobs, treatment, and pay over black workers. As a result, white workers perceive that their self-interest lies in the maintenance of the existing system. The black workers, who are short-changed in this system, have historically been powerless to demand changes, especially in the face of resistance from white workers (Leiter, 1986).

As a result of a declining share of the world market, workers in the American textile and garment industry have experienced layoffs and permanent job losses in recent years. These employment losses result from the importation of less costly foreign-made apparel and from the export of jobs by American textile and apparel companies to areas with cheaper labor, such as border plants in Mexico and production facilities in Pacific Rim nations (Nash, 1989).

Sweatshops Return Sweatshop conditions, entailing long hours, no fringe benefits, harsh and unsafe working environments, and the use of child labor have reemerged in the textile and garment industry in North America and elsewhere in the world as a result of increased competitive pressures:

> Where ten years ago there were fewer than 200 garment factory sweatshops, there are now between 3,000 and 4,000 sweatshops in New York. . . . They employ between 50,000 and 70,000 persons and a large portion of their employees are illegal migrants from the Caribbean, Latin American, and Oriental countries. . . . This segment cannot be considered a mere aberration. . . . Sweatshop labour is a necessary condition of global competition. (Ross and Trachte, 1983:416)

Sweatshops have also sprung up on the West Coast, where they employ large numbers of Asian and Latin American immigrants. College students have recently become sensitized to these issues because of increased awareness of the use of sweatshop labor in the manufacture of popular clothing brands. Students have organized into a group called United Students Against Sweatshops (*www.usasnet.org*) to protest the conditions under which workers manufacture these garments.

Competition for Specialty Textiles Not all of the American market share in textiles has been lost to countries that offer cheap labor. A significant market share has also been lost to Germany and Italy, which pay wages comparable to, or higher than, those in the United States. These countries have been able to increase their market share by a steady stream of incremental innovations in manufacturing design. The innovations have allowed textile and apparel manufacturers in West Germany and Italy to customize their products, making significant inroads against mass-produced goods. Such innovations have also put these firms in a better position to take quick advantage of new market trends and developments. This technological progress has been encouraged by trade associations, cooperative banks, and apprenticeship programs under the joint sponsorship of unions and governments (Freeman, 1994). In the United States, with its tradition of more individualistic forms of competition, such practices would be viewed as collusive and as an unfair constraint of trade. In the European nations they are seen as cooperation for the sake of economic development.

GLOBAL COMPETITION
AND THE NEW WORLD ORDER

Many manufacturing industries in North America and in other industrially advanced nations have stagnated in recent decades. What are the reasons? One is increased international competition. Japan and Europe have been able to rebuild their industrial capacity, which was destroyed during World War II. The rebuilding occurred during the 1950s and 1960s; by the 1970s these

countries were providing substantial competition for goods made in the United States. In addition, many Third World nations have been pursuing the path of industrialization. Some nations have met with only limited success in this regard. Many others, however, such as Taiwan, South Korea, and Brazil, have secured sizeable market niches for themselves in the world economy.

The Wrong Policies at the Wrong Time

American firms have responded inadequately to this heightened competition. They have pursued three main strategies: (1) exporting jobs overseas in the search for cheaper labor, (2) attempting to drive down wages at home, and (3) manipulating balance sheets and profit margins through what has come to be known as **paper entrepreneurialism** (Reich, 1992). These strategies all rely on the increased worldwide mobility of capital allowed by the heightened availability of telecommunications and jet transport. Rapid transportation and virtually instantaneous communications allow the coordination of engineering and marketing among diverse facilities around the world. In addition, the control of capital has become more concentrated over time within larger and larger corporations. Large corporations have the financial power and political leverage to take advantage of these new worldwide networks of production and marketing. At the dawn of the twenty-first century, capital has simply become more mobile than labor. And this **capital mobility** has allowed American capital to increasingly leave the United States in the search for cheaper labor costs and less restrictive environmental regulations.

The strategy of driving down the wages of American workers has been carried out in part through demands for wage and benefit concessions from trade unions. Often threats of bankruptcy and plant closings have backed up these demands. As a result there has been strong downward pressure on the wages and benefits of American workers. This strategy has also been pursued through attacks on Social Security, education, and other social programs that were secured by middle- and working-class Americans during the post–World War II boom years (Levine, 1995).

Although the dual strategy of exporting jobs and reducing wages at home has secured profits for some firms, this strategy has been costly for the American economy as a whole. It has resulted in the decline of real income for average Americans as well as tremendous costs to local communities in the wake of large-scale layoffs and plant closings (Edwards, 1993); and the new jobs that have been created in the service sector are often much lower paying than the jobs that disappear in manufacturing. Clearly, a more rational policy, focusing on heightening productivity rather than exporting jobs and driving down wages, would be preferable. In this section we examine current policies and their consequences and look at some policy alternatives.

Financial Shell Games Some American manufacturers can be characterized as *hollow corporations.* Instead of investing in productivity-enhancing activities at home, such as technological innovation and job redesign, they subcontract production outside the United States. As we discussed in the section on automobiles, many of the goods we think of as being "American made" are now being made in foreign nations, with the American company providing little more than the packaging and the labeling. In this way some of the largest American corporations have attempted to maintain their profit rates by, as Norman Jonas puts it, "lowering their wage base, by outsourcing, by literally turning their companies inside out—in short by hollowing. But enduring prosperity will require investment in human and physical resources" (Jonas, 1986:59).

Plant Closings Many of the jobs lost in the United States were exported overseas by American corporations. For instance, "General Electric expanded its overseas payroll by 5,000 during the 1970s while it was diminishing its U.S. work force by 25,000. RCA increased its foreign work force by 5,000 while it cut domestic employment

by 14,000" (Lustig, 1985:123–124). These closings cost workers in terms of lost livelihoods, and they also cost communities and states in terms of lost tax revenues and increased social welfare expenses. Box 8.4 presents arguments for and against legislation regulating plant closings.

Externalization Plant closings are encouraged by the fact that large companies pay only part of the total costs entailed in moving from location to location. Other costs are **externalized** and are paid by taxpayers or by individuals. Such costs include tax write-offs negotiated by companies with the new localities in which they locate, accelerated depreciation allowances, and utility discounts. These costs reduce local revenues. Local services must then be curtailed or other taxes increased, such as personal property taxes or sales taxes. These adjustments create additional difficulties for communities that are trying to attract skilled labor and professional workers. Attracting and retaining skilled labor is made more difficult by the underfunding of social services, such as parks, police, and education.

From the company's standpoint this pitting of communities and regions against one another offers some attractive advantages. By complaining about pollution standards, taxes, and other regulations in one area, a company can induce a community into offering concessions on these issues to attract jobs. To entice a company to relocate, a city may offer tax breaks or offer to build an industrial park on city-owned property. The costs are absorbed by the city and by taxpayers and are *real costs* of relocation. Even moves that are economically irrational for the nation as a whole may thus be made because they are profitable to the individual company, which receives a subsidy from the new host community. Additionally, the costs of moving families, shutting down schools and rebuilding them in new locations, and building new streets and sewers are all real costs of relocation that are not paid by the company but by the taxpayers or by the individuals affected. In addition, federal tax codes allow much of the value of the physical plant and equipment of

---

BOX 8.4   Arguments for and against Plant Closing Legislation

Recent debates on plant closings suggest the following arguments for and against legislation to regulate such closings.

The principal arguments in *favor* of legislation:
1. Advance notice and income guarantees are necessary to mitigate the great economic and psychological burden of closure on employees.
2. Businesses get substantial economic incentives and rewards from government; workers deserve the same protection.
3. Communities also deserve advance notice, because they lose a major source of tax revenue when plants shut down and face the added burden of social welfare payments to displaced workers.
4. Plant closure has a multiplier effect within a community, causing a greater loss of jobs than those in the plant itself.
5. Because businesses usually plan shutdowns far in advance, prior notification requirements pose no great burden.

The principal arguments *against* legislation:
1. State plant closure laws would place unconstitutional restraints on interstate commerce.
2. State plant closure laws would create more, not less, local unemployment, because large firms would establish or increase operations in other states to avoid penalties.
3. Workers receive adequate economic protection through state unemployment benefits and job search services.
4. Businesses need to be free to close inefficient plants with obsolete equipment and replace them with new facilities.
5. Based on their educational qualifications and skills, manufacturing workers receive generous wages; as a quid pro quo they should assume the risk of closure.

SOURCE: Adapted from Paul D. Staudohar and Holly E. Brown, 1987, *Deindustrialization and Plant Closure.* Lexington, Mass.: D.C. Heath, pp. 275–276

---

closed factories to be counted as depreciation against income from new plants located elsewhere or even against earnings from newly purchased subsidiary companies. The loss of jobs and the loss of social safety nets because of plant closings have contributed to increased problems of poverty, homelessness, and marginality in the 2000s. These problems are discussed further in Chapter 14.

Downsizing and Flexibility   A key corporate strategy used in recent decades to adjust to increased competition is to **downsize.** Workforce downsizings have become increasingly common in the 2000s. By downsizing core employment and subcontracting many functions, corporations believe they can cut costs through increased **flexibility** in matching their resources to market needs. Some of the lost manufacturing jobs have been replaced by employment growth in smaller firms; however, many of these new jobs have a short life span. In addition, much of the growth in the small-firm sector has actually been produced by start-ups of subsidiaries of large corporations in the process of seeking out locations with cheaper labor costs (Vallas, 2001). The growth of the service sector has also provided new jobs, but these generally pay much less than jobs in manufacturing and contribute substantially less to the gross national product.

Displaced automobile workers report losing an average of 44% of their previous earnings during the first two years after being laid off. Workers in steel, meatpacking, and aerospace report similar losses. A laid-off shipyard worker reports: "In reality, I'm not just out of a job. I'm out of a city. My kids are out of school. My wife is out of her education, and I'm out of self-respect" (Lustig, 1985:123).

A Declining Middle Class Heightened pressures on jobs and wages have helped create a trend toward a **declining middle class** (Braun, 1997). Traditionally, in manufacturing and construction, wages clustered around the middle range of income. Employment in these sectors thus helped create a prosperous working class that was a bulwark of middle-class society. Today, real wages in these jobs are stagnant or declining. Average wages increased in the United States during the post–World War II period until the early 1970s when they peaked at about $12 per hour (in current dollars). Since then real wages have fallen by about 15% to near their level in the early 1960s (Census, 2000). Families have compensated by increasing their hours of work for both husbands and wives. In the most recent decade, these efforts netted a 1% increase in real family income for a 4% increase in hours worked (Mishel et al., 2001:18). Unmarried individuals have fared less well.

The economic strategies pursued by the largest American corporations have resulted in the loss of millions of jobs and have slowed growth in the gross domestic product. The reason is not that the heads of these corporations are ignorant or stupid. It is because what is good for large corporations in a global economy is often disastrous for national economies:

> These shutdown decisions are made for reasons that are internal to the needs of the companies, but which result in substantial social costs to the affected parties. Yet, workers, unions and communities have little say in the matter and virtually no influence over either the process or the outcome. Most of the closings had tactical significance relevant to the product market shares or labor market costs, or both. Without exception, the reasons are only vaguely communicated to local area residents and organizations. (Craypo, 1986:112)

Current federal policies do little to ameliorate these tendencies and nothing to prevent them. The Trade Adjustment Act provides some retraining benefits to workers who lose their jobs to international competition. But it has been interpreted very narrowly by federal administrative bodies and excludes workers displaced because of American corporations moving overseas. Similarly, Title 3 of the Job Training Partnership Act is earmarked for retraining redundant workers; however, only about 2% of the agency's budget has been allocated for this purpose (Hooks, 1987).

Unexplored Alternatives

A potentially important set of policy options that could address these issues center on establishing "fair trade" between nations instead of unregulated "free trade." The goal of **fair trade** policies is to establish common standards for the treatment of workers worldwide that will allow nations to compete with each other without doing so on the basis of lowering wages, working conditions, and environmental standards to the lowest common denominator.

Federal legislation could help address these issues. One possibility is to extend the Fair Trade and Tariff Act. This act provides duty-free privileges for certain imported goods from 114 developing countries. In the original act, countries became ineligible for these privileges if they expropriated American property without compensation, failed to prevent drug trafficking, or failed to meet several other conditions. In the 1984 renewal of this act, the AFL-CIO convinced Congress to include the requirement that these countries follow certain "internationally recognized worker rights." These rights, spelled out in the conventions of the International Labour Organization (*www.ilo.org*), include the right to organize and bargain collectively, a minimum age for the employment of children, and acceptable conditions of work with respect to wages, hours, and safety and health. This principle could be extended beyond the Fair Trade and Tariff Act (which covers fewer than 5% of imports) to the entire range of U.S. international trade (Gray, 1987). Such standards could help elevate the conditions of workers around the world through reducing competition between workers in different countries over who will accept the smallest benefits package and the most hazardous conditions in order to secure jobs. And they could have the long-term effect of heightening productivity and expanding markets for manufactured goods among workers worldwide.

Recent years have also witnessed increased public concern about the actions (and inactions) of the World Trade Organization (WTO, *www.wto.org*), World Bank (*www.worldbank.org*), and the International Monetary Fund (*www.imf.org*). These nonelected global institutions have acted to open markets and to promote the free flow of goods between nations. These actions have increased trade and development in some areas. These global institutions, however, have been much more reluctant to implement policies that provide protections for workers or for the environment. As a result, their actions in support of free trade have accelerated the competition of workers in different nations against each other without providing any framework for the establishment of minimum standards of child labor, safety, or workers' collective rights to form unions.

SUMMARY

Advanced industrial societies include a diversity of industries, which can be broadly classified into extractive, manufacturing, and service sectors. Productivity in extractive industries has increased to such an extent that only a relatively small part of the labor force is still employed in this sector. Similarly, a declining share of workers is employed in manufacturing. The decline in manufacturing jobs is based on rising productivity, renewed world competition, and the movement of American corporations overseas in the search for cheaper labor and more lenient safety, health, and environmental regulations.

Some jobs in manufacturing are highly skilled. Workers in these jobs experience a great deal of autonomy and pride in their work. Such jobs are similar to professional jobs in their training requirements and the autonomy they afford the worker. Other jobs in manufacturing are boring and repetitive.

Employment growth in advanced industrial economies is occurring mainly in the service sector. Some researchers argue that the growth of the service sector foretells an increasingly unequal society as middle-level positions in manufacturing decline in number and are replaced by poorer-paying service jobs.

As manufacturing productivity increases, a decline in its relative share of employment is inevitable. However, the sharp decline of employment in American manufacturing due to a loss of competitiveness is neither inevitable nor desirable. Investments in technology and in worker training can improve productivity and competitiveness and help keep manufacturing jobs in the United States.

KEY CONCEPTS

extractive industries

manufacturing industries

postindustrial society

advanced industrial society

aquaculture

occupational communities

skilled craft work

semiskilled work

unskilled work

working-class culture

mini-mills

paper entrepreneurialism

capital mobility

plant closings

externalization

downsizing

flexibility

declining middle class

fair trade

QUESTIONS FOR THOUGHT

1. Describe some of the problems facing farmers in North America today. What policies might help address these problems?

2. Why has employment in manufacturing in North America declined in recent years? Construct arguments for and against the following proposition: "The decline of manufacturing is an acceptable and desirable aspect of industrial growth and development in advanced economies."

3. Why do skilled craft workers generally resent being closely supervised? Is the autonomy of skilled workers a roadblock or a potential asset for economic growth in advanced economies?

4. If you were working as a machine operator in a factory, what would you like least about the job? What would you like most?

5. Identify three policies that you believe would both increase the competitiveness of North American manufactured goods in world markets and increase employment opportunities for North American workers. Rank order these three policies by their probable impact and defend your ranking.

MULTIMEDIA RESOURCES

Print

William Form. 1985. *Divided We Stand.* Urbana, Ill.: University of Illinois. An analysis of the diverse nature of the working-class experience and the political consequences of this fragmentation.

John Gaventa. 1980. *Power and Powerlessness: Quiescence and Rebellion in an Appalachian Valley.* Urbana, Ill.: University of Illinois. Moving insights into the lives, communities, and worldviews of Appalachian miners and their families.

David Halle. 1984. *America's Working Man.* Chicago: University of Chicago. Work life and community life among chemical plant workers in a New Jersey factory town.

Studs Terkel. 1974. *Working.* New York: Avon. Over a hundred short excerpts from the lives of working people in their own words. The selections are fascinating and moving.

Websites

Government publications. *infomine.ucr.edu/search/ govpubsearch.phtml* A comprehensive guide to online government information.

Upjohn Institute for Employment Research. *www.upjohninst.org* Innovative research on workplace issues.

Economic Policy Institute *www.epinet.org* Policy reports on contemporary economic and workplace issues.

Labor Posters from World War II. *americanhistory.si.edu/ victory/victory4.htm*

Exploring social inequality. *www.trinity.edu/%7emkearl/ strat.html* A comprehensive exploration of North American and global economic inequality using charts, figures, and statistics and including hundreds of useful links.

Thomas Register of American Manufacturers. *thomasregister.com* Online version of a core resource for finding business products, parts, and services.

RECOMMENDED FILM

The Perfect Storm (2000). Gripping true-life story of the trials of the crew of a fishing vessel caught in the storm of the century.

9

The High-Technology Workplace

With the availability of small computers that can be plugged into immense data banks, word-processing equipment, and other devices, more and more people are setting up what Alvin Toffler calls the "electronic cottage." . . .

Pat Lee, a consultant and special projects manager, likes the flexibility: "It means if you want to work on a rainy day for 12 hours, and spend 12 hours in the sun the next day, you can. If you feel like sitting in your chenille bathrobe at midnight writing an employee handbook, and you really get into it and want to stay up till 4:00 in the morning doing it, that's terrific." . . .

Author Jessica Lipnak, who also works at home with her husband, likes the opportunity for parenting: "We both really enjoy it in relation to our small children, because we've had a lot of access to the kids—they're seven months and three years. I'm nursing the baby, so it's made it possible for me to be one of those rare women who can work and nurse. That's a big issue for a lot of people with infants. So that's been wonderful. Being able to see them on and off all day and make choices about spending time with them has been great. You know, if something came up. . . . If I wanted to spend some time, or if Jeff wanted to take Miranda off on a bicycle or something, we can do all of those things, and that is terrific."

(APPLEGATH, 1982:46)

Maria, a twenty-six-year-old political refugee from Argentina, who chooses not to be known by her real name, found work in the Silicon Valley, but she did not strike gold. Maria quit her $4.10 an hour production job at

Memorex to have her first baby. For two years, she illegally stuffed and soldered thousands of printed circuit (PC) boards in her home. Her employer, a middle-aged woman she calls "Lady," subcontracted assembly work from big firms—so Maria was told—like Apple and Memorex.

Maria gladly accepted the low piece-rate work because child care would have eaten up most of her after-tax earnings at a full-time job. She quit, however, when Lady asked her to wash her assembled boards by dipping them into a panful of solvent, heated on her kitchen stove. Maria, unlike most Silicon Valley cottage workers, had studied chemistry before immigrating to the United States, and she knew that the hydrocarbon fumes could make her young son, crawling around on the kitchen floor, seriously ill.

(SIEGEL AND MARKOFF, 1985:138–139)

These workers' experiences illustrate the diverse consequences of high-technology production. The effects of high technology on the workplace are little understood despite the widespread attention that has been focused on its potential for spurring economic growth.

In this chapter we examine the consequences of the new technologies for workers and for organizations. We present competing views of the influence of high technology on work. We also examine the consequences for the skill content of jobs and the mix of job opportunities. In discussing the new high-technology job market we explore the hypothesis that the growth of high-technology work is increasing the gap between an upper tier of well-paid jobs and a bottom tier of poorly paid jobs. Finally, we look at how advanced technology affects the meaning of work and the dynamics of organizations. These complex and differentiated effects of high technology on work and workers highlight the importance of examining issues of power and the social organization of work simultaneously with the examination of technological change.

COMPETING VIEWS
OF HIGH TECHNOLOGY

What exactly is high-technology production? At the heart of high technology is the tiny silicon computer chip. With the widespread application of the microchip to a variety of work settings beginning in the 1980s, entirely new industries, such as robotics, have sprung up, and others, such as computer manufacturing, have expanded dramatically. Electronic technology has also altered the nature of production across a wide range of settings. These settings include both traditional blue-collar manufacturing and white-collar professional and clerical work. Our investigation of the high-technology workplace thus focuses both on high-technology industries and on more traditional jobs that have been transformed by microprocessor technology.

The U.S. Bureau of Labor Statistics defines **high-technology industries** as those employing one and a half times the average proportion of technology-oriented workers (engineers, life and physical scientists, mathematical specialists, engineering and scientific technicians, and computer specialists); having research and development expenditures twice the average for all industries; or being above average on both these criteria (Burgan, 1985:9). This definition identifies forty-eight high-technology industries, including electronics, machinery, ordnance, chemicals, instrumentation, pharmaceuticals, aerospace, genetic engineering, and communications equipment.

There are competing perspectives on the influence of advanced technologies on work. One perspective emphasizes the benefits resulting from technological advances. The work of American economist Faye Duchin (1998) is representative of this tradition. Duchin views technology as liberating individuals from the necessity of performing undesirable work. Gerald Piel (1994), publisher of the magazine *Scientific American,* considers technology to be the major vehicle bringing about higher living standards. A key characteristic of this optimistic view is its conception of technology as neutral and inevitable. As Mowshowitz (1976) suggests, "technological change is . . . an inevitable feature of human existence." Technologically based potentials for improved working conditions and skill upgrading also figure prominently in this view. Riche (1982:38) sums up this optimistic perspective when he writes, "There is general agreement that the benefits of new technology far outweigh the disadvantages and that innovation has led to economic progress, new job opportunities and a more prosperous society."

Other analysts, however, describe the role of technology in less glowing terms. These social scientists identify technological growth more as a cause of current problems, such as unemployment, than as a solution to these problems. A leading spokesman in this more pessimistic view of technology is sociologist David Noble (1997), who argues that there is nothing automatic about either the development or the consequences of technology. In his view, technology is neither neutral nor inevitable but is, instead, a tool to increase management's leverage in bargaining with workers. Noble argues that when options exist, employers systematically select the technologies that weaken workers' autonomy and solidarity. Systems that rely on expensive automated machine tools, for example, are selected over less-expensive machine tools that require more skill to operate. In this view the selection of new technologies reflects and reinforces the unequal distribution of economic and social power.

Another sociologist, Mike Cooley (1980), faults the new technologies for reducing workers'

skills and dissipating their motivation. Computers and other forms of automation in the workplace can have negative effects on the experience of work. Two negative changes may occur when control is removed from workers and placed in computers and other advanced forms of technology. The first is that the work may become less diverse and less rewarding. The second is that automation may actually decrease efficiency and quality because achieving high quality requires skilled human input, which is excluded in automated systems (Shaiken et al., 1997).

Although there are competing views about the consequences of new technologies for work and for workers, there is a reasonable degree of consensus on at least two aspects of high technology: (1) The major effect of new technologies is to transform existing jobs rather than to create new ones. (2) Technological innovations can increase productivity, product quality, and the ability to customize products thus improving the competitive position of those organizations that use them effectively (Ozaki, 1999).

In the following sections we incorporate a variety of sociological perspectives by looking at specific areas of agreement and disagreement about the consequences of high technology. Our discussion revolves around three core issues:

1. How are new technologies transforming the skill requirements of different occupations?

2. What new occupations are created by new technologies, and what jobs are eliminated?

3. How are working conditions being modified by new technologies, and how are new technologies affecting the meaning and nature of work?

MICROPROCESSOR TECHNOLOGIES AND SKILL REQUIREMENTS

As a first step in answering these questions, let us consider three divergent positions in the debate about the effect of advanced technologies on the central issue of skill. Recall that we first discussed skill upgrading and deskilling in Chapter 7.

The Skill-Upgrading Thesis

The most optimistic position argues that new technologies have increased skills. Among the oldest and most frequently cited empirical studies dealing with technology and skill upgrading is Blauner's (1964) study of continuous-process automation in the chemical industry. Continuous-process systems involve a continuous flow of the product (in this case chemicals) through the system. Human intervention is required only to monitor pressures, temperatures, flows, and so on. Blauner found that continuous-process automation requires a greater proportion of skilled maintenance workers than less automated manufacturing systems. In addition, machine operators in the chemical industry have greater responsibility for the care and proper functioning of expensive capital equipment than those in mass-production settings.

More recently, in a Communications Workers of America (CWA, *www.cwa-union.org*) membership poll, 78% of the respondents indicated that technological change had increased the skill requirements of their jobs. Automated systems often require workers to utilize both high levels of technical knowledge and skills acquired only through lengthy experience (Zuboff, 1988). It is easy to underestimate the depth of knowledge required by the technicians who press buttons on automated equipment or who click boxes on computer screens. Trained and experienced workers are necessary for the effective utilization of technologically advanced automated systems, even where managers devise systems that they believe are "foolproof."

Based on a case study of banks and bank tellers, Adler (1984) arrives at similar conclusions about skill upgrading. Focusing on worker responsibility, cognitive learning, and job interdependence (rather than mastery of a fixed set of tasks), Adler finds that the least skilled jobs are the ones most affected by automation within banks. Many of these jobs were eventually eliminated. Therefore, the overall effect of automation was to upgrade the average skill requirements of the remaining jobs. Similarly, in a study of six companies that made significant investments in microprocessor production techniques—computer-aided numeric control machine tools, computer-aided design, and management information systems—Francis and his colleagues (1981) find "no evidence of deskilling." Their study finds only one instance in which the number of skilled positions was reduced. They conclude that increased use of microelectronics produces no discernible reduction in skill requirements.

Several studies have emphasized the new skills that workers must acquire to operate technologically advanced production systems. Based on analysis of work in automated paper mills, Penn and Scattergood (1985) find that high-level maintenance skills requiring autonomous choices increase with advances in technology. In a review of technological change in three industries (printing, banking, and metalworking), Ozaki (1999) finds that the increasing use of high technology results in greater requirements for formal knowledge, precision, and autonomous decision making. Based on a study of thirty-six continuous process companies that are "technical leaders in their industries," Cross (1985) finds an expanding use of electronics in the control and monitoring of production. In interviews with more than a hundred workers in these firms, he discovered that the introduction of technology forced workers to learn important new skills, including the ability to use and maintain the new technology and also the ability to diagnose systems problems. He sees these increased skill requirements as arising from the use of more complex and expensive equipment, from greater integration of different production processes, and from greater demands for product quality. Box 9.1 describes some of the potentially empowering aspects of working with sophisticated new computer-assisted production systems.

BOX 9.1 Computer Technology and the Flow of Information

Based on her study of a variety of facilities undergoing computer-based automation, Shoshana Zuboff believes that the new technologies tend to disperse information to workers at the front lines of production.

Some believe that organizations in the future will achieve competitive advantage based on their ability to better understand their own businesses and apply imagination to newly available data in order to generate higher levels of innovation. A corporate vice president, reflecting on the emerging manufacturing environment, struggled to formulate such an alternative:

> "What has been managerial access to information is not as comfortable a notion as it may seem. There has been a fear of letting it out of our hands—that is why information is so carefully guarded. It could be misused or misinterpreted in a way that cannot be managed. Traditionally, we have thought that such data can only be managed by certain people with certain accountabilities and, I hesitate to say, endowed with certain skills or capabilities. But with the new technology it seems there is an almost inevitable kind of development if you have as a goal maximizing all business variables and maximizing the entire organization's ability to contribute to that effort. I don't think you can choose not to distribute information and authority in a new

way if you want to achieve that. If you do, you will give up an important component of being competitive."

Judgment means the capacity to ask questions, to say no when things are not right. It also creates the possibility of asking "why?" or "why not?" One of the managers most respected for his willingness to "pass the knowledge down," described what he called "the developmental learning process" that operators must go through if they are to become critically competent at the data interface.

> "At the first level, of course, people need to know how to keep the equipment running. But the next step is to ask, "Why am I doing what I am doing?" Only if people understand why, will they be able to make sense of the unknowns. The third step is process optimization and diagnostic problem solving. At that point, they can hone in on the real issues. . . ."

According to an operator at the factory:

> "We need to know the whys of this process. I can't just punch this button because I was told to. I have to do it because I know why and what happens. That's the only way I can run it better."

SOURCE: Excerpt from Shoshana Zuboff, 1988, *In the Age of the Smart Machine*. New York: Basic Books, pp. 288–290.

The Deskilling Thesis

Substantial evidence, however, also suggests that deskilling often accompanies new technology. Bright (1966), although noting that automation produces increasing levels of responsibility for some workers, concludes that, overall, it creates a tendency toward declining skill requirements. He argues that as mechanization progresses, initial changes that demand increased skills give way under automation to a progressive loss of skill resulting in an inverted U-shaped skill curve. According to this thesis, as automation

progresses, skills first rise and later decline. Moreover, Bright questions whether the growing demand for trained technicians to operate automated technology is not just a form of credential inflation rather than a true upgrading of skills. That is, degrees and other credentials may be required because so many workers have them, not because the skills they represent are actually required on the job. The printing industry provides an example of declining skills in a traditional craft occupation as a consequence of microprocessor-based automation. Increased automation in printing has radically reduced the need for skilled printers (Wallace and Kalleberg, 1982).

Two case studies describe in detail how new technologies can deskill workers (Boddy and Buchanan, 1981). The first study involves the transformation of copy typists into video display typists. Deskilling occurs because with automated equipment there is less need to type text correctly the first time. With the new equipment corrections are easier, and the printer positions the paper and takes over other functions formerly performed by copy typists. However, some new skills are required of video display typists. These include increased concentration and familiarity with codes for formatting and editing text.

The other case documented by Boddy and Buchanan involves the introduction of automated mixing equipment at a large cookie factory. The major consequence of technology was the transformation of the "doughman" into a mixer operator. Formerly, a master baker had held the position of doughman. But as the computer replaced the need for human intervention in the mixing process, the doughman suffered a loss of craft skills. The new automated equipment left the doughman with the residual responsibility of pressing a button to start the mixing cycle. Skilled maintenance positions at the factory also declined in skill as repair work became merely a matter of running a series of simple tests and replacing defective parts. No aspects of the automatic technology demanded that the mixer operators or maintenance workers acquire new skills or knowledge to perform effectively. Workers became bored, apathetic, and careless. They rejected responsibility for breakdowns of the new system. Further, they developed few new skills that would have made them promotable. The factory managers thus lost a valuable source of potential recruitment into supervisory positions.

Based on her analysis of several insurance companies in New York, California, and Pennsylvania, Appelbaum (1984) identifies deskilling as the main pattern associated with the introduction of new technology in insurance underwriting. Her work indicates that in 80% of the cases, computers make the underwriting decision on personal insurance lines. In short, as their work becomes increasingly standardized, insurance underwriters retain their skill in title only.

In a review of problems posed by the use of automated methods for industrial and office workers, Mowshowitz (1997) finds skill requirements for workers mainly declining after the introduction of new technologies. For office workers computerization makes work more standardized and formally defined (Frenkel et al., 1999). Workers also have less autonomy as more of the decision making involved in the production process is assumed by higher-level managers or built directly into automated information systems.

The Mixed-Effects Position

Finally, many analysts view the impact of new technologies on skill requirements as a dynamic process in which some skill requirements are increased and others are reduced. Milkman and Pullman (1991) argue that the introduction of robotics and other automated processes into manufacturing industries produces both deskilling and skill upgrading. They conclude that robotics affect workers' skills in a positive way because the jobs created in robotics and robotics maintenance require more technical background than did the manufacturing jobs they replace. A solid understanding of post–high school math and science is essential for robotics technicians. For the most part, this training can be obtained through technical programs offered at community colleges. However, they also warn that many of those displaced by robots may end up in low-skill, low-wage service jobs.

Automation also has certain inherent limits that necessitate the continuation of skilled positions. In the machine tool industry, for example, as lathes and other tools age and settle, tolerances spontaneously get out of whack. When machine tools are computerized or numerically controlled, problems arise in achieving precision cuts. Without a skilled craft worker in charge of the machine, such problems can be difficult to fix

Mark Belanger. 1983. *The Facts* 5, 7 (September): 36. Reprinted by permission of The Canadian Labour Congress.

Sociologist Kenneth Spenner (1990) and industrial relations expert Jeffrey Keefe (1999) argue that part of the problem in drawing firm conclusions about the effects of technology on skills is that skill is often a vague and undefined concept. What is meant by skill appears to change as technologies evolve. Thus, it may be difficult to determine whether skills have increased or declined as a result of technological change. According to Spenner (1985:126), "Skill changes have been uneven, offsetting, and are not attributable to any one single cause, including technology." The observed general stability in skill levels may thus be a result of offsetting changes. Upgrading of skills appears to be occurring through the creation of jobs that require more training. Simultaneously, deskilling appears to be occurring through the downgrading of job content over the lifetime of many jobs.

Variability in High-Technology Effects It is difficult to avoid the conclusion that the effects of high technology are highly diverse and vary from one setting to another. In some situations, automation demands that workers learn new skills to monitor sophisticated equipment. Automation may also increase the responsibility of workers for complicated and expensive integrated production systems. In other situations, manufacturing workers may be reduced from operating equipment that required advanced skills to simply loading and unloading an automated version of the same equipment. The latitude of clerical workers to make decisions about their work may be similarly restricted as they labor under increasingly routinized systems. There are also situations in which workers find that their old skills are obsolete but that they need new skills in mathematics, electronics, science, or programming to handle their new responsibilities. Rather than universally upgrading skills or deskilling jobs, advanced technologies appear to have mixed and contradictory effects. Some jobs are upgraded, some are deskilled, and some experience changing skill requirements (Liker et al., 1999).

(Shaiken et al., 1997). Many computer-driven systems may simply be so fragile and vulnerable to disruption that they inevitably require close maintenance by a skilled craft worker to be utilized effectively (Hirschhorn, 1997). To the extent that this is true, craft skills can never be eliminated from many manufacturing operations.

Related problems can occur in automated engineering and drafting systems in which the computer generates generic parts and fits them into a design. Although such systems facilitate the work of the drafting engineer, they also eliminate less-demanding parts of the task that provided essential "mulling" time for the engineer to survey and comprehend the broader goals of the project. Such limitations suggest that computers will never completely replace skilled workers.

In a study of computerization in the insurance industry, Baran and Teegarden (1984) find that the higher-level functions are most vulnerable to automation. Once automated, the work is then assigned to less-skilled and lower-paid workers. At the top, Baran and Teegarden find a shrinking group of professionals. For these workers the emphasis is on innovation and personalized sales relations. At the bottom, data-entry clerks perform high-volume standardized duties.

One important mechanism through which this transformation occurs is that older workers with more vintage skills are laid off and younger workers with newer training are hired. This works well for the companies involved because they avoid the costs of in-house training to update the skills of their existing employees. Plus, they may also be able to pay the new workers less than existing workers with seniority. This solution to upgrading the skills of a company's workforce, however, does not work well for middle-aged workers who are left without jobs and with few employment options. Many employers have also recruited technically trained immigrants from other nations with lower wage structures, such as Ireland and India. Some employers even subcontract work such as software production to offshore sites in the search for cheaper labor. These strategies further weaken the position of North American workers, especially older workers.

Skills appear to increase in settings where workers have the power to insist that new technology be introduced in a manner that upgrades their skills. Workers' power to demand skill upgrading may rest on their organization into a union or on their professional expertise. Skills also appear to be upgraded when managers perceive that deskilling the workforce is counterproductive because of the complexity of the production process. When worker power is absent and managers perceive no need to maintain or cultivate workers' skills, high technology has resulted in the deskilling of work.

The insights of researchers in the mixed-effects position allow us to see that the effects of high technology on skills are not determined solely by technological imperatives. They are often determined by the social context in which the technologies are introduced and by the relative power of the actors involved.

Training for Changing Skill Requirements

New technologies profoundly affect workers' needs for training. There is now "less demand for manual dexterity, physical strength, and for traditional craftsmanship. . . . Employers [are] stressing formal knowledge, precision, and perceptual aptitudes" (Riche, 1982:37). Some researchers maintain that existing workers can be retrained to fill the new jobs. In practice, however, this process can be very difficult. Who pays for a year's leave for a worker in mid-career to learn a new skill? Retraining the workers directly displaced by robots for new jobs in robotics may not always be realistic. For example, substantial training would be required to teach an assembly-line welder to repair and maintain the welding robot that will be doing the welder's job in the future (Bills, 1995). On the other hand, retraining skilled plant maintenance workers to maintain industrial robots is relatively simple.

Continuing Education Many researchers argue that the educational demands for new high-technology jobs will be substantial, but that they can best be met by continuing-education programs and on-the-job training rather than by additional training for young workers before they enter the labor market. They argue that the new technology changes so rapidly that it often requires highly specific training, which can most effectively be acquired on the job or through continuing education classes (Ozaki, 1999). As a result, training for high-technology jobs will become a lifelong endeavor, with new training being required as new technologies emerge. "Reduced reliance on a *single* vocational skill makes sense in as much as fewer workers follow lifelong occupations compared with the past" (Spenner, 1985:151).

Liberal Arts Education The increased focus on job-specific training also includes a continuing role for broader education in the liberal arts:

> De-emphasizing liberal arts may be short-sighted from the standpoint of future labor requirements in high-technology industries. That point is brought out in a study of computer services in California. The study points to the experience of one firm doing research on computer speech synthesis and

voice recognition—an area that is likely to grow with the rise of "user-friendly" computer systems. The company's employees are graduate students in linguistics and are "valued not only for their knowledge of the mechanics of speech, but for their general professional commitment to understanding communications, not just computing machinery." (Peltz and Weiss, 1984:279)

Much of the skill upgrading associated with new technologies involves cognitive and interactive skills, exactly the kind of skills taught in traditional liberal arts curriculums (Wolff, 2000). In addition, job seekers flooding highly focused technology training programs may create an overabundance of computer technicians and related specialists. Unfortunately, as we will discuss, the actual number of new jobs in these fields may be relatively small.

Training Options How is training for high-technology jobs best provided? One option is formal, degree-granting programs. For example, twenty-seven degree-granting programs in robotics maintenance exist in the United States. In addition, 343 robotics courses are taught at various other institutions in the United States (Lapan et al., 2000). The nondegree courses may be at least as important as the degree-granting programs, because they can be combined into flexible training programs to meet rapidly changing needs.

Another alternative is to institute such training programs as part of collective bargaining agreements. Starting in the 1980s, the CWA succeeded in bargaining for training programs to upgrade members' skills in working with sophisticated telecommunications equipment. The initial contract allocated $36 million for training programs for employees (Noble, 1986:10). Today the CWA, in cooperation with the regional telephone companies, runs a wide range of lengthy training programs and shorter courses targeted to meeting and anticipating specific skill needs in telecommunications. This commitment to training is an important part of the CWA's and the regional telephone compa-

nies' plans to remain competitive in the rapidly changing telecommunications industry.

The bulk of the evidence indicates that requirements for formal education in the labor force are increasing but that, simultaneously, this education is becoming less general and more specific. Bright (1966) resolves this seeming contradiction by arguing that increasing educational requirements simply represent credential inflation. There is another interpretation, however. Advanced education can be both formal and narrow at the same time. Indeed, this reality is at the heart of the current debate in colleges and universities between a traditional broad liberal arts education and more narrowly defined professional training. These developments should warn us that no unitary concept of skill is adequate for understanding the consequences of high technology for job requirements.

CHANGING JOB CONTENT

To better understand the consequences of high technology for the future of work, let us first look at how it has affected the distribution of occupations. Significant changes in the distribution of occupations occurred in the last two decades of the twentieth century and the beginning of the twenty-first century. Relative to the distribution in 1980, a significant increase occurred in professional jobs, from 15.6% of the labor force to 19.8%. Moderate increases also occurred for craft workers (from 13.3% to 15.0%), machine operatives (15.7% to 16.5%), and service workers (12.4% to 14.7%). Declines for managers (from 9.5% to 7.2%) and clerical workers (17.8% to 11.4%) offset these increases. Sales workers and laborers maintained their relative share of employment (Census, 2000).

These figures may come as a surprise to those who believe that in the future everyone will be a high-technology worker. This misconception arises from paying too much attention to which occupations will have the *fastest growth rates* and too little attention to which occupations will actually provide the *largest number of jobs*. The first fastest-growing occupations, all with growth rates

BOX 9.2   You Are Shooting a Lot—You Are Killing a Lot

I started making computer games when I was ten years old. My dad had a small PC business. . . . I was just lucky. I went to MIT, took a lot of science and liked it. . . . I first worked on *Mech Warrior 2* and then did an add-on to that—*Ghost Bear's Legacy*. . . . That add-on was probably the funnest thing I've ever done. It took two and a half months, and was very low budget, and we just had a great time. It was such a small team and all of us really dug the design we were working with. We put hundred-hour weeks into it, sleeping at the office a lot, and I was like a kid again, you know? It was just a joy to work on. . . . And it sold three hundred thousand units, which the company loved. We were all golden boys for a while. . . .

The company I work for now, it's a pretty small company. We just do development of games. We don't actually sell the game afterwards. My old company Activision does that for us. They're kinda like our client—they have to like the game we make, which means usually they pick the topic. . . . They give us the main idea, they distribute it, we do the fun stuff. . . .

This is a dream job for me. It's the best job in the world. It doesn't change the world for the better, but it's at least giving people some enjoyment for a couple of hours a day. And it's only going to get bigger in the future. Five years from now a lot more people are going to have games. You're going to have the game system just built into your television. The Playstation is pretty close to that now and Web TV is starting with that direction, too. So many more people are going to have games. Which means the audience is gonna get much broader. . . . When I started in this, we were all geeks who played games, so you could make games for geeks. Now you have to make games that have big action stars in them, or have cool music—like the Beastie Boys might be in one of the games. Something that really draws in a large crowd. . . .

There is definitely a violent aspect of it all. You are shooting a lot. You are killing a lot. But, that's kind of obviously the point, right? And it's more than that, too. . . . It's an experience. It's a release. . . . You're watching someone play and you see their eyes get wide. "Wow!" They're hurting their thumb playing the game. It's very good to see.

SOURCE: John Bowe, Marisa Bowe, and Sabin Streeter. 2000. *Gig: Americans Talk About Their Jobs*, pp. 305–309. New York: Crown Publishers. Reprinted with permission.

---

of 50% or more in the first decade of the 2000s, are expected to be database managers, systems analysts, and computer engineers. However, the *total number of jobs* created in these occupations will be fewer than the number of new positions created for cashiers, home care aides, truck drivers, or teachers (Census, 2000).

Large numbers of new jobs will *not* be created in the high-technology industries themselves. However, rapid technological change will dramatically transform many occupations in the next ten years. In this section we examine recent changes in the nature of work for some of the occupations most affected by advanced technology. Such occupations include engineering, assembly work, machine operation and maintenance, service work, clerical work, management, technical work, and work in the home.

Engineering

Engineering is the glamor occupation of advanced technology. From the back cover of Tracy Kidder's *The Soul of a New Machine* (1981) we learn: "Management thought it was impossible. But thirty of America's computer whizkids pushed themselves beyond the limits of ability and endurance to build a machine more complex than any one of them could understand, more powerfully advanced than anything like it. . . . And they did it in record shattering time!" Engineers frequently report that one of their most important motivators at work is the opportunity to work on intellectually challenging projects (Meiksins and Smith, 1996). It is highly unlikely that the creative aspects of engineering will ever be automated away. Box 9.2 describes the work of a young computer programmer involved in video game production.

Engineers make up 10% to 15% of the workforce in electronics, compared with about 3.5% in the U.S. economy as a whole. In robotics manufacturing, this figure runs as high as 23.7%, with an additional 15.7% of the workforce employed as engineering technicians. Well over 50% of the jobs in robotics require two or more years of college training, compared with fewer than 20% in the rest of manufacturing. Salaries for computer programmers averaged $36,000 in 2000 (Wright, 2000). These employment figures suggest a positive job outlook for many engineering specialties. Employment opportunities in these jobs, however, will tend to favor already privileged groups and college-educated immigrants, thus further intensifying existing class and racial divisions in the economy.

Assembly Jobs

At the other extreme are assembly jobs in high-technology industries. Frequently, these jobs are dull, repetitive, poorly paying, and hazardous. Many of them are exported overseas to countries with large pools of labor willing to work for what would be considered subminimum wages in North America and Europe:

> Roughly 200,000 Asians work in semiconductor plants along the eastern rim of Asia. More than 90% of these people are young women. And most do assembly—the bonding of chips to the caterpillar boards—and associated packaging.
>
> Most of them are young women because the employers, knowing that the women don't have families to support, can get away with paying poverty-level wages. Also, the employers don't have to worry about maternity benefits or the women's health because they are expected to drop out of the labour force to get married at 25 or so.
>
> Close monitoring and strict quotas keep the women at their microscopes for a full eight hours a day with just one 45-minute break. And this is while they are suffering from chronic conjunctivitis—eye inflammation—caused by toxic gases and dust in the factories. (Belanger, 1983:36)

Because partially assembled electronic components are easy to export by jet, North American and European workers employed in such jobs are placed in direct competition with lower-paid workers around the world. Working conditions in electronics assembly in the advanced nations are not as bad as those in the Third World, but the availability of cheap labor places downward pressure on the wages and conditions of electronics assembly workers in the industrially advanced nations.

Some aspects of high-technology assembly work are becoming increasingly automated, with workers being replaced by robots that will be untroubled by doing endlessly repetitive or dangerous tasks. The term *robot* is derived from the Czech word *robota*, meaning "servitude, drudgery, or work." There have been five generations of robots on a scale of increasing complexity: compressed air control (popular in Europe twenty-five years ago), drum can control, pin control, numeric control, and microprocessor control. The advent of the microcomputer-controlled robot makes robots flexible enough that their cost can now be justified across a potentially wide range of production operations. However, the introduction of such fully automated systems can create problems for workers remaining after their introduction. "If there is a malfunction of the robot, the system is designed so that the mechanical arm can be pulled off the line and a human worker 'inserted' in its place. The human would then be doing a job designed and paced for a robot while the robot itself was being repaired" (Shaiken, 1984:173). Obviously, such a system restricts freedom of movement and regiments the pace of work to a greater extent than even the most arduous of today's assembly lines.

Most assembly workers have somewhat ambivalent, though generally positive, attitudes toward robots. In a study made ten weeks *before* robots were introduced into a manufacturing plant

BOX 9.3   High-Technology Jargon

VDT—Video Display Terminal A video display terminal is a remote terminal attached to a computer. VDTs are the backbone of computerized word processing.

NC—Numeric Control Under numeric control, machine tools are directed by a set of instructions punched as numbers onto a computer card or tape. These instructions direct the positioning, depth, and speed with which drills and other cutting tools operate. These instructions replace control by a skilled machinist.

CNC—Computer-Aided Numeric Control Under computer-aided numeric control, the instructions are no longer punched on a card or tape but are programmed with software language into a microcomputer that directs the machine. This greatly increases the flexibility of the programming and eliminates the need to punch a new control card for every design adjustment.

MIS—Management Information System Management information systems are integrated systems of computer accounting. The information compiled may include daily reports by workers, sales personnel, or secretaries about production matters or about their work activities; information collected by computer monitors directly from the production line; information about workers' whereabouts and activities taken from pass cards that workers insert into control boxes when moving from one area to another; or any other type of information that can be collected and recorded on computers. This information is compiled by the computer, cross-checked between sources where possible, and tabulated into reports. MIS has the potential to displace many middle-level managers whose jobs have entailed tabulating this kind of information.

CAD—Computer-Assisted Design Computer-assisted design involves the use of computer-generated generic part designs to replace original designs developed by engineers. For example, a structural engineer might type in weight and size specifications for a bracket to hold part of an aircraft wing together. The computer would use existing descriptions of parts in its memory to generate a bracket to fit these specifications. The design could then be displayed on the engineer's terminal or printed as a blueprint. In more advanced systems the computer-generated design specifications can be directly transferred to a CNC machine tool and the bracket milled with little or no direct human participation at any stage of the process. CAD has greatly changed the jobs of some engineers. It also has the potential to virtually eliminate the occupation of engineering drafter.

employing 1,000 workers, 87% of the workers thought that the robots would make their company more competitive. However, 50% also worried that the robots would displace workers. Five months later (ten weeks *after* the introduction of robots) attitudes were decidedly less positive. Workers complained that robots had increased costs, increased accidents, and lowered the quality of the product. They were also no longer convinced that robots increased productivity (Argote, 1999).

Machine Work

The work of skilled craft workers is also being affected by computer technology. Skilled machinists provide one of the clearest examples.

These workers make the tools, patterns, molds, and machine parts that make modern industrial production possible. Today, these jobs are being transformed by the introduction of **numeric control (NC)** and **computer-aided numeric control (CNC).** Under NC and CNC systems, the metalworking lathes, drills, and cutting tools operate automatically. The machinist is left with the residual job of feeding and unloading the machine while the computer controls the cutting tools, their speeds, and their depths. Some of the most commonly used terms in high-technology production work are defined in Box 9.3.

In NC and CNC production systems the right to program the control device becomes a central issue. Is this work to be done by machinists after they receive training in programming, or is it to be done by computer programmers after they receive training in machine tooling? Shaiken et al. (1997) argue that in most cases the best programmers are machinists, because the programming software is "user-friendly" and is not hard to learn. The machinists' skills, however, are very complex, requiring a long training period and much experience to master, and this knowledge is essential for designing computer programs that will work.

Skilled Maintenance Work

High technology will also influence the work of those who maintain and repair machinery. Because of the rapid pace of technological change, maintenance workers will no longer be able to specialize in one narrow craft or type of application: "The single disciplined craftsman has no foreseeable future in most frontline maintenance situations" (Cross, 1985:203). Instead, the trend will be toward the multirole, multidisciplinary craft worker who is simultaneously machinist, electrician, and computer programmer (Mort, 2000).

Clerical Work

Some of the most dramatic effects of high technology are being felt in clerical work. Clerical jobs have been dramatically transformed because their basic task is handling information, and it is this task for which computers are most suited. For instance, the introduction of word processing is estimated to reduce the cost of producing a letter by a secretary from $7 to $2.

The development of computerized office automation can be divided into two stages. In the first stage individual workstations are automated through the introduction of personal computers. This stage dramatically increases individual productivity. In the second stage, as yet incompletely realized, individual workstations are eliminated, and the entire information accounting system becomes fully automated. Jobs in this stage may

be more highly skilled, but they will be largely programming and machine maintenance rather than clerical, and the reduction in clerical jobs is projected to be even more dramatic than in the first stage (Wright, 2000).

Capital outlay per worker will increase with the development of the automated office. Office work has traditionally been badly undercapitalized, with an average investment of only about $2,000 per worker. By contrast, factory work has an average investment of about $25,000 per worker. In automated office work, investment typically rises by fivefold or more to about $10,000 per worker (Wright, 2000).

Office automation will intensify the pace of clerical work and may also increase alienation among workers. One possible scenario is suggested by changes in the clerical support work for the New York Stock Exchange (Morgan, 1983). At Citibank, which transacts a large share of business on the exchange, check sorters operate machines that have twenty-eight slots. The clerks look at codes and then push buttons to determine the bank (slot) to which a check is to be routed. The machine measures the rate at which checks are routed, and the faster the operators work, the more they get paid. "A check sorter routing 900 checks an hour [one every 4 seconds] (the minimum) will get about $12,000 a year," says a Citibank vice president, Joseph Reddington. "One doing 2,000 an hour [one every 1.8 seconds] will get $23,000." The average is about 1,400 checks an hour [one every 2.6 seconds] for which the clerk earns $17,500.

Middle Management

Middle management is also being affected by computer technology. Traditionally, the job of middle managers has been to search out, compile, and digest production and marketing information and then pass this information on to top management. These are precisely the activities that are most easily automated with a computerized management information system, or MIS (see Box 9.3). Where computerized accounting systems are used, middle managers may face displacement and more pressured working conditions because they are in direct competition with the more cost-effective systems. Such systems also allow closer monitoring, not only of workers, but also of middle managers, thus eroding important aspects of their traditional power and autonomy. Researchers have observed increasing resistance to automated production systems among some middle managers, a group normally identified with willing compliance to organizational goals (Smith and Walter, 1997).

Technical Workers

In addition to affecting existing occupations, new technologies are creating entirely new occupational specialties and greatly expanding others. These occupations include computer programmers, health technologists, and engineering and science technicians. Many of these occupations require two-year college degrees and they are some of the fastest growing professional and semiprofessional occupations. Many of the jobs that are being created are in rapidly growing high-technology industries. For example, chemical and biological technicians work in laboratories evaluating medical and scientific specimens for the rapidly growing medical and biotechnical industries. Other technical workers provide technical support functions to more traditional professions that have experienced rapid technological transformations in the nature of their work (see Chapter 11). Nurse anesthetists, radiologists, dental hygienists, and biological and chemical laboratory technicians are examples of such rapidly growing technical support occupations (Rabinow, 1999).

Telecommuting

Because of the falling costs of personal computers, it has become economically feasible for workers to do increasingly more information processing work away from the office (Barker and Christensen, 1998). **Telecommuting** is doing

Mark Belanger. 1983. *The Facts* 5, 7 (September): 53. Reprinted by permission of The Canadian Labour Congress.

work that would normally be done in the office at home on a personal computer or remote terminal. It is estimated that by the year 2000, 10% of the labor force was telecommuting at least part time. Telecommuting can have both positive and negative consequences for workers.

Negative Consequences Negative consequences include reduced personal contact with co-workers and clients and exclusion from informal sources of feedback that may be essential for doing a job correctly and efficiently. Isolation and diminished visibility can also reduce an employee's opportunities for making contacts essential for promotion and upward mobility. Blue Cross–Blue Shield provides an example of the negative side of telecommuting. "The company pays clerical home workers piece rates, offers no paid vacation or benefits and charges $2,400 a year in equipment-rental charges. Their 'cottage keyers,' as the workers are called, process more than 200 medical claims a day and net only about $100 a week" (Moore and Marsis, 1984:13). Processing the average claim requires about two and one-half minutes, and the home workers net about ten cents for the task.

Telecommuting potentially generates a host of problems for clerical and data-entry workers.

Isolation is central among these, but also significant is the problem of working on a piece-rate basis. According to a developmental editor in the publishing industry:

> I find in my freelance work that most employers require an estimate, which allows little flexibility. Though they offer an hourly wage, they give me a ceiling for the entire job. Thus, I must either stay confined to the number of hours allotted to the job—even if the job needs more hours than estimated—or not get paid for the extra hours worked. And some employers work on a piece-rate basis alone. The rate offered is almost invariably based on fewer hours than the work requires.

Also, clerical workers who are forced to choose home-based work because of difficulty in securing affordable child care may find that their problems are compounded by trying to work at home and take care of children at the same time. These problems may be further intensified by difficulties in finding adequate work space in already cramped quarters. For clerical workers who telecommute, "home work could very easily become a new confining characteristic of 'women's work,' a new female job ghetto. . . . Women could find themselves losing much of the ground they've gained over the past fifty years" (Menzies, 1982:149–150).

Benefits of Telecommuting Positive consequences include increased flexibility of working hours and reduced commuting time. The greatest benefits of telecommuting are likely to occur for professional workers who use personal computers to do portions of their work at home during the evenings or on weekends. Telecommuting allows significantly increased flexibility in work schedules for these workers. Even for data-entry workers, however, positive as well as negative examples of telecommuting can be cited. In 1982, the American Express Company instituted a home work program for disabled workers on an experimental basis. These workers transcribe dic-

tation from a centralized system. The transcribed copy is then transmitted to a computer at the main office, where it is printed and distributed. Initially the workers were paid on a *subcontracting* (piece-rate) basis, but the program was so successful in terms of productivity and accuracy that the new workers were reclassified as regular, full-time, payrolled employees of American Express. Similarly, F International, which offers a comprehensive range of data-processing services and employs 850 people worldwide, has operated principally on the basis of telecommuting for twenty years. The average workweek for F International employees is twenty hours, and over 96% of its workers are women. Wages and benefits are good, and 34% of the workforce has been with the company for more than five years (National Research Council, 1985).

"Offshore" Telecommuting Perhaps the most ominous trend for the North American labor force is the growth of "offshore" telecommuting. Over forty companies in the United States, Australia, and Japan now use overseas offices for routine clerical work. The telecommuted work includes keypunching magazine and newspaper subscription lists, airline reservations, and survey questionnaires and preparing book manuscripts for electronic typesetting. One of the most widely noted examples of international telecommuting involves American Airlines:

> Every work morning an American Airlines jet arrives on the Caribbean island of Barbados and unloads a quarter-ton of used ticket coupons. The tickets are taken to the airline's Caribbean Data Services (CDS) subsidiary, where over two hundred women, each earning less than $3 an hour, key data from the coupons into a local computer system. Almost immediately, CDS transmits the information via satellite to the company's central data-processing operation in Tulsa, Oklahoma. (Siegel and Markoff, 1985:99)

Data can be electronically transmitted between countries even more readily than partially assembled electronic components, and the threat to the jobs and conditions of North American workers is at least as immediate.

Because of threats to the jobs and working conditions of American labor as a result of telecommuting, the AFL-CIO has made several recommendations about how such work should be organized. These recommendations include: (1) provision of pay and benefits for home workers equivalent to office workers doing comparable work, (2) automatic inclusion of telecommuters in existing unions at the workplace, and (3) restrictions on the import of clerical work and data-processing work done overseas for domestic employers (Chamot and Zalusky, 1983).

JOB DISPLACEMENT AND JOB CREATION

Some occupations are eliminated by new technologies while others are created or expanded. Is this process creating a two-tiered, or dual, occupational structure by eliminating middle-range jobs and expanding opportunities at the bottom and the top of the occupational distribution? Or does technological change lead to a more middle-class, egalitarian society? Again, there is a range of views on these issues. Many researchers focus on the creation of high-prestige jobs such as systems analyst and ignore the displacement of other workers. Others focus only on the potential displacement effects of new technologies and ignore the jobs that are being created. Many offsetting forces are involved, and the overall effect of high technology on employment remains an open question. The possibility of further significant unanticipated developments is also high. For example, the declining growth rate in clerical occupations because of word processing was not anticipated prior to the rapid expansion of microcomputers. Conversely, increased productivity due to new technology has lead to a growing economy and increased employment opportunities for many workers (Gore, 1993).

Job Displacement

Many researchers believe there has been a net displacement of jobs resulting from the introduction of new technologies, particularly microprocessor-based technologies and robotics. Such an elimination of jobs is called **technological displacement.** Clerical workers, especially typists, stenographers, and office-machine operators, are often identified as the group most affected (Hartmann, Kraut, and Tilly, 1986). In addition, robotics may make a wide variety of industrial labor redundant. Finally, there is likely to be some displacement of machinists, tool and die makers, and metalworkers resulting from the introduction of NC and CNC machine-tooling systems (Wright, 2000). Box 9.4 presents Kurt Vonnegut's futuristic vision of an automated society in which NC machine-tooling systems have made workers obsolete.

Is the Electronic Revolution Really Different from Prior Technological Advances? Two critical differences between the current wave of technological change and the wave of continuous-process automation that occurred in the 1950s may amplify current negative effects on employment: "The new 'automation' debate therefore differs from the 1950s debate in two crucial respects. First, the chip is a genuinely revolutionary device. Second, it has appeared at a time when economic growth in the industrial societies can no longer be taken for granted" (Forester, 1981:xvii). A key difference between electronic technology and the automation of the 1950s and 1960s is that the consequences of microcomputers are at least as significant for white-collar work as for blue-collar work. The potentially negative impact of computer technology on employment can be glimpsed by examining some of the industries in which its impact has been, or is being, felt.

The telephone industry employed 213,000 operators in 1960 but, in spite of tremendous expansion in the industry, it is projected to employ only 175,000 operators by 2005, due largely to the introduction of electronic switching. Railroads

BOX 9.4   Futuristic Automation?

In 1952 Kurt Vonnegut, Jr., wrote *Player Piano*. The novel portrays a future society in which automation has eliminated all jobs except those of a few select engineers. Unemployment is pervasive in spite of public works projects and a chronic state of manufactured war. Vonnegut's fictional scenario rests on the futuristic projection of data tape automation first used in the 1950s to automate machine milling work. This type of automation is almost identical to that used in a player piano.

In Vonnegut's story, engineers have captured the skills of master machinist, Rudy Hertz, by attaching recording instruments to his lathe. His movements are then reproduced, with variations, ad infinitum.

> Rudy hadn't understood quite what the recording instruments were all about, but what he had understood, he'd liked: that he, out of thousands of machinists, had been chosen to have his motions immortalized on tape.
>
> And here, now, this little loop in the box before Paul, here was Rudy as Rudy had been to his machine that afternoon—Rudy, the turner-on of power, the setter of speeds, the controller of the cutting tool. This was the essence of Rudy as far as his machine was concerned, as far as the economy was concerned, as far as the war effort had been concerned. The tape was the essence distilled from the small, polite man with the big hands and black fingernails. . . .
>
> Now, by switching in lathes on a master panel and feeding them signals from the tape, Paul could make the essence of Rudy Hertz produce one, ten, a hundred, or a thousand of the shafts.

SOURCE: Excerpt from Kurt Vonnegut, Jr., 1952, *Player Piano*. New York: Dell, pp. 9–10.

---

employed over 1 million workers in 1955, but only 232,000 in 2000 as a result of automated traffic switching, sophisticated machines for track maintenance, and mechanized warehouse loading operations (Census, 2000). Simultaneously, the tonnage of freight moved on railroads has actually increased.

The printing industry has also experienced advances in productivity coupled with a dramatic decline in employment:

> A magazine cover or a full-page ad in the period immediately after World War II would require about 40 hours of "dot etching" time. It also required a minimum of six hours of camera time. Today, that same job can be done in *seven minutes*, with electronic color scanners. (Chamot and Baggett, 1979:39)

Retail trade, hotel management, libraries, and many other white-collar, trade, and service industries are undergoing similar transformations as their operations and accounting systems are increasingly automated (Frenkel et al. 1999).

Employment Losses in White-Collar and Service Occupations The decline in employment growth in white-collar and service industries has occurred at a time when these industries are being counted on to absorb manufacturing labor displaced by advanced technology:

> Meanwhile, marketed services have stagnated, only partially compensated by growing demand for intermediate services in manufacturing and a growth in non-marketed (public) services. In short, innovations have opened up the prospect of "jobless growth" in the very sector expected to absorb labour displaced in manufacturing. (Standing, 1984:144)

Rough times ahead can also be projected for middle managers as more information processing is built into technological systems and centralized under the control of top management. Currently popular management theories pose further threats to middle managers. For instance, Tom Peters (1997), in his best-selling book, *The Circle of Innovation,* argues in favor of "lean" managerial structures with few levels and with few supervisory personnel. Such management theories both reflect and add to technologically based threats to employment opportunities for middle managers.

Robotics The employment impact of robotics is one of the most widely debated issues in the study of technological change. It is estimated that by 1990 robots had displaced 17% of welders and 32% of production painters in the automobile industry.

In 1980 there were about 30,000 robots in use worldwide (Schlesinger, 1984:27). This represented only about 1 robot for every 10,000 manufacturing workers. By 1990, the number of robots was estimated to have increased to about 100,000, but this still represents only about 1 robot for every 3,500 workers. Between 32,000 and 64,000 new jobs were created in the robotics industry during this ten-year period, but robots displaced between 100,000 and 200,000 production workers during this same period.

The cost of a programmable (servo) robot ranges from $30,000 to $120,000, and the purchase price does not include installation, site preparation, or maintenance, all of which can be very expensive (Chamot and Baggett, 1979:15). Bearing these costs in mind, the U.S. Congressional Research Service does not expect a wholesale displacement of manufacturing workers by robots. It would require the diversion of the entire gross private domestic investment in the United States to displace even a fourth of U.S. manufacturing workers (U.S. Congressional Research Service, 1985:28). However, it is also possible that the use of robotlike devices can be extended to nonmanufacturing settings. Mechanized warehousing and inventory control provide examples of these possibilities. Such computer-related mechanization in support services may cause further substantial employment losses.

Job Creation

Many researchers argue that relatively few employees are laid off as a direct result of technological change. They argue that such displacement is often moderated by techniques used to prevent layoffs, such as providing notice and retraining and reassigning displaced employees to new jobs. In addition, computer technology has introduced entirely new occupations such as systems analyst, programmer, data-entry clerk, and web master. Expanded job opportunities are also occurring for engineers and research-and-development scientists. Acknowledging that some workers are hurt by the introduction of new technologies, these researchers argue that technology is not the culprit. Instead, they see displacement problems as resulting from limits in adjustment, retraining, and relocation programs for workers.

Many possible effects of advanced technology may offset short-term job losses. New products and industries are being created; demand may be stimulated by cheaper, better products; and economic growth may occur through increased investment in new technologies. Increased demand for machine maintenance required by ever-greater investment in technology will also tend to increase employment. According to Blanchard (1984:275), "there is no reason at this point to conclude that the net long-term effect of new technologies will be heavily detrimental to the availability of productive jobs." In the 2000s, the potential for computer applications to expand employment has been demonstrated through economic growth brought about by the rapid expansion of the internet as a marketing tool. By 2000, the internet economy directly employed 2.5 million people, more than many older industries such as insurance, communications, or public utilities (*www.internetindicators.com*).

Long-Term and Indirect Effects Drawing final conclusions about job creation and displacement is difficult. The main problem in determining whether the dominant pattern is toward job creation or job displacement arises from the difficulty of calculating the **long-term and indirect effects** of new technologies. When the direct effects alone are examined, it is clear that job displacement outweighs job creation; but when indirect effects are considered, the analysis quickly becomes clouded. For instance, if the introduction of robotics leads to higher productivity and lower prices, it may stimulate demand and create a significant number of jobs throughout the economy. Technologically based productivity gains may even improve a country's competitive position in the world market, thus leading to even greater employment gains. These potentially important consequences, however, are difficult to verify. Conclusions about the effects of new technologies on job creation and job displacement must be made with great caution.

Increasing Segmentation?

As well as considering the question of whether job displacement or job creation is the prevailing consequence of new technology, sociologists are also concerned about the consequences of advanced technology for the relative prevalence of desirable and undesirable jobs. At first glance, job structures in many high-technology settings appear to suggest an increasingly middle-class occupational distribution. For example, professional and technical occupations make up 10% of jobs in manufacturing as a whole but 33% in the electronics industry (High Tech Research Group, 1984:14).

Examining the labor force as a whole, Levin (1995) concludes that positions requiring the highest level of skill and those requiring the lowest level have decreased, resulting in a broadening of the middle. Nevertheless, he is cautious in interpreting these results, because they rely solely on an analysis of general educational development while ignoring more specific vocational training, on-the-job training, and earnings. Levin also notes that the skill differences between certain occupations previously

regarded as "middle-range" may be increasing in importance. For example, assembly jobs may be fragmenting into more skilled and less skilled specialties.

Other researchers have found that new technologies sometimes create dual occupational structures in high-technology settings (Tolbert et al., 1998). The increasing use of educational credentials as a screening device can also heighten barriers to mobility between nonconnecting career lines (Burris, 1993). For example, secretarial and accounting career paths are typically segregated on the basis of a degree in accounting. Such limits on mobility can lessen motivation for all employees. Moreover, it is mostly women who are affected by these reduced opportunities for mobility (Glass, 2000). Minority women, concentrated in the lowest-level clerical jobs, face displacement. Better-educated white clerical workers confront deskilling and blocked mobility opportunities.

Even in high-technology industries, many of the newly created jobs will be in traditional *"low-tech" occupations* with less than average earnings. In 1980, only 25% of the jobs in high-technology industries could be classified as truly high-technology occupations. In Santa Clara County, California (popularly known as Silicon Valley), 28% of jobs in the semiconductor industry are in the top third of the national earnings distribution. The middle third is almost empty, however, with only 9% of the jobs, and 63% of the jobs are in the *bottom third* of the income distribution (Appelbaum, et al., 2000).

This earnings distribution is unlikely to improve with further advances in technology, because the profit imperative demands that highly paid labor be automated before poorly paid labor. For example, one of the most successful areas of automation in computer manufacturing is in computer-aided design, or CAD (see Box 9.3). This technological breakthrough has displaced large numbers of reasonably well-paid engineering drafters. Meanwhile, the development of automatic equipment for inserting microchips into printed circuit boards has advanced much more slowly, partly because the work is being done by poorly paid workers in the Third World.

Temporary Work   When examining the consequences of high technology for the occupational distribution, we need to look at the full range of jobs created by high-technology companies, including jobs that may be filled by workers not directly employed by the companies themselves. High-technology companies employ large numbers of temporary workers to help them adjust to the cyclical nature of production in these industries. Many workers are also employed on a subcontracting basis, and conditions for these workers are generally inferior to those of full-time employees of the parent company. It is also necessary to consider the plight of those *without* jobs when appraising the occupational distribution resulting from high-technology production systems. If technological changes increase unemployment, the occupational distribution resulting from new technological advances is more negatively skewed as a result. We discuss temporary work and other types of employment marginality in more detail in Chapter 14.

The International Division of Labor   A final group of high-technology workers to consider are those employed by North American companies operating abroad. The six major high-technology companies operating in Massachusetts, for example, employ 28% of their workforce overseas (High Tech Research Group, 1984:46). Hourly wages for these workers range from an average of $6.56 in Singapore, to $1.26 in Mexico, to as low as $0.21 in India (United Nations, 2000). If these workers are brought into the equation, the resulting distribution of jobs is even more negatively skewed.

The available evidence suggests a strong possibility that advanced technology may contribute to an increasingly unequal occupational distribution made up of a few highly paid jobs and an increasing share of poorly paid, relatively alienating jobs. We discuss this possibility, and the opportunities for avoiding such a future, in Chapter 17.

Public Policy and Employment

Because of concerns about technological unemployment, various types of full-employment programs have gained increasing public attention. The programs include such diverse proposals as advance notice of plant closings, more extensive retraining, reductions in the workweek, early retirement, and expansion of public services, especially for young workers. If technological unemployment in the face of advancing productivity is indeed to be the fate of the industrially advanced nations, then systematic planning and adjustment needs to take place.

Local Impact of High-Technology Development   One important aspect of technology-related employment changes is their impact on local communities. Many community planners believe that development policies that stress the recruitment of high-technology industries are ill-conceived. For example, even after an aggressive (and expensive) state program to attract high-technology firms, fewer than 7% of all jobs in North Carolina are in high-technology businesses (Luger, 1996). Programs narrowly targeted to attract high-technology businesses may actually be detrimental to states' long-run economic interests. Many high-technology industries have limited employment-creating potential, and a substantial outlay of public funds is often required to entice them to move to a specific location (Colclough and Tolbert, 1992).

It is likely that development funds could more effectively be targeted to help *existing* businesses become more technologically advanced. Such development plans would be comprehensive rather than targeted toward high-technology businesses per se, and they would stress integrated economic development. "State initiatives for technological and industrial innovation are said to be the makings of a national policy for industrial competitiveness; but without greater consideration of their human and spatial implications, they could very well turn into a national policy of regional and social abandonment" (Peltz and

Weiss, 1984:278). An important component of more integrated development programs could be the requirement that any company receiving publicly subsidized financing give substantial advance notice of plant closures or major layoffs and agree to refund public investments made to attract the company.

WORKING WITH HIGH TECHNOLOGY

Technological changes are capable of increasing both productivity and job satisfaction. To what extent are these possibilities being realized? In this section we explore the consequences of high technology for the meaning of work and for organizational dynamics. We also examine union responses to the challenges of technological change.

Computer Technology and the Meaning of Work

Considerable evidence suggests that new technologies often increase job satisfaction (Lincoln and Kalleberg, 1996). Reviewing the literature on job satisfaction and computer technology, Danziger et al. (1993) argues that workers tend to experience technological change as mildly benign. Japanese workers appear to be even more favorably disposed toward technological change. The majority welcome such change, including robotics, and only small numbers have negative attitudes.

In contrast, other observers see advanced technology as having less than ideal consequences for job satisfaction. Examining the work of machinists, Shaiken et al. (1997) find that NC systems create heightened alienation and stress. Specifically, they see increased alienation stemming from workers' loss of control over the production process. This heightened stress is partly due to growing isolation at work, as proportionately more machines and fewer

workers are involved in production. Shaiken et al. are not completely negative in their evaluation of NC, however, noting that it can also mean less noise, more accuracy, and more cleanliness. In particular, they note that NC and CNC in the machine tool industry have created a situation in which workers are less frequently confronted with chronically "cranky" machines that are difficult or impossible to operate within prescribed tolerances.

Computer technologies can also have negative effects for white-collar workers. These include lessened variety, reduced physical mobility, computer-paced pressure to produce, reduced interaction with other workers, and "feelings of being zombie-like" because of prolonged, intense interaction with a computer while being physically and psychologically isolated from other workers (Buchanan, 1997). For example, one of the most significant negative effects faced by copy typists when their work is automated is the loss of the personal, informal working relationship between authors and typists.

Some researchers have even gone so far as to suggest that diminished mental capacity is needed to operate some of the new automated equipment. Based on his own work experiences, Cooley (1986) asserts that the ideal worker for NC equipment would have a mental age of twelve. More characteristic of the literature as a whole is Dyer-Witheford (1999), who observes that isolation and constant monitoring can create stress but that, simultaneously, new technologies are eliminating much tedious and dangerous work.

Computer Technology and Organizational Dynamics

How are organizations being influenced by computers and what will high-technology organizations be like in the future? Many researchers argue that the introduction of computer technologies is fundamentally altering the nature of organizations. Some argue that the new technologies are increasing the availability of information and thus dispersing power throughout the organization. Researchers find that alienation is lessened under these circumstances. Others argue that computers are leading to a centralization of control.

Dispersion of Information A key component of the thesis that technological change leads to improved working conditions is the argument that advanced technology tends to **disperse information** and authority more broadly throughout the organization. This thesis has gained popular recognition in Naisbitt's (1999) slogan that "computers destroy hierarchy" and in Cleveland and Anderson's (1999) thesis concerning the "twilight of hierarchy." This thesis, ultimately based on technological determinism, is given further support by current management schemes that attempt to promote productivity through the involvement of all employees (Peters, 1997).

Many researchers also argue that technological advances have resulted in greater interdependence among tasks and have made it easier for groups to determine the nature and pace of their work. Because high-technology industries make a large capital investment per worker, they depend heavily on workers' understanding of the job and on their good will, motivation, and commitment. Technological advances may thus empower workers and improve the experience of work for many. In addition, quality-control functions are sometimes reintegrated into production rather than being allocated to separate divisions. This reintegration results partly from management decisions about how best to pursue quality control. But it also stems from the increasing ability of the workers who use advanced technology to monitor their own work and produce consistently high-quality products (Hirschhorn, 1997).

Centralization of Control In direct contrast to those who believe that advanced technology lessens organizational inequalities, other researchers observe a connection between technology and the centralization of authority. These researchers argue that the natural tendency of automation is to concentrate the functions of control and decision making in the upper levels of management (Daday and Burris, 2001).

In the early 1980s, the Bell telephone system installed a computerized Line Maintenance Operating System (LMOS), which is basically a computerized filing system into which operators record customers' complaints. The system then dispatches a repair crew to the problem site and sets a "commitment time" when it is supposed to be there. Michelle Brooking, a shop steward for repair attendants in Washington, D.C., describes how the system actually works:

> "LMOS may sound efficient, but it isn't. It's easier in getting information, but as far as getting your phone fixed, it takes a lot longer now than it ever did before." Before LMOS, repair service attendants worked in local repair bureaus linked to the central offices. They were in frequent telephone contact with repair and installation crews. Often, they would relay messages between the customers and the crews—changes in commitment time, last minute repair jobs, etc. "You knew how many men you had on the street," says Brooking, "how many troubles you could handle." With LMOS, all the attendants are in centralized bureaus. Their only contact with the repair crews is one way, via the computer. They do not know if the crew actually makes it to the customer's house at the appointed time (though, often, they have to listen to the customer's complaints if it does not). "Now, you're doing it blindly." (Howard, 1981:13)

Organizational analysts have long noted a connection between advances in technology and organizational centralization (Pfeffer, 1998). It is important to be aware that the flexibility of computer-aided production systems is primarily flexibility in *information retrieval* and *product design*. This flexibility does not necessarily translate into greater organizational, task, or interpersonal flexibility.

Electronic Surveillance Advances in technology also create the specter of **electronic surveillance** at work. In Austin, Texas, Productivity Innovations, Inc., markets an electronic time card that employees pick up every morning and use whenever they change activities or locations. A centralized computer keeps detailed records of their activities throughout the day. The computer creates a report on each worker's activities for the day, week, and month (Sullivan, 1985). In the trucking industry, electronic monitors have been installed to track drivers' activities. These devices record the duration of stops for fuel, rest, pick-ups, and drop-offs and even control maximum and minimum driving speeds. "Drivers, predictably, are unhappy with the new controls" (Salpukas, 1984).

Electronic devices are also used to closely monitor the work of many clerical workers, especially in large companies. This monitoring may include regular printouts for managers on keystroke rates, error rates, and break times. Even the work of grocery checkout clerks has become closely monitored with the advent of electronic scanning of canned and packaged grocery items (Marx, 1999). In combination with chemical surveillance through urine tests, electronic surveillance significantly increases the ability of corporations to intrude into the lives of their employees both on and off the job.

In Canada, electronic monitoring of work is considered an invasion of privacy and a violation of the Canadian Charter of Rights:

> The Labour Canada Task Force regards close monitoring of work as an employment practice based on mistrust and lack of respect for human dignity. It is an infringement on the rights of the individual, and an undesirable precedent that might be extended to other environments unless restrictions are put in place now. (Long, 1984:288)

Even more stringent restrictions on the electronic monitoring of individual workers have been implemented in many European countries (Ozaki, 1999). Box 9.5 describes a recent effort to implement such safeguards in the United States.

Management from the Rear A variety of other organizational factors besides increasing hierarchy and the possibility of electronic surveillance also suggest at least the possibility of more stressful working conditions under high-technology systems. One commonly cited problem concerns the stresses caused by working on accelerated project schedules (Rabinow, 1999). Perhaps more importantly, many high-technology companies have incompetent managers. This problem arises for two reasons. First, even if managers are promoted from within the production staff, rapid product changes may quickly make their knowledge obsolete. Without continuing hands-on involvement in design and production, even engineering managers rapidly begin to lose touch with the new technology and with its problems, possibilities, and limitations. Second, managerial orientations toward short-run profit may come into direct conflict with the efficient operation of new production systems. Managers may rely excessively on cost-cutting techniques or may make poorly considered demands for getting the product out too fast or on too tight a schedule. Whatever its cause, working under chronically incompetent managers can devastate worker morale and long-run productivity (Hodson, 2001).

The High-Technology Life Cycle A rapid cycle of corporate birth and death is characteristic of high-technology industries. High-technology companies frequently produce a wide variety of spin-off companies, often started by engineers leaving the parent company. Spin-off companies generally specialize in products either competing directly with or complementing the products of the parent company. For example, Massachusetts Computer is an offshoot of Digital, Stratus is an offshoot of Data General, Automatix is an offshoot of Computervision, and Precision Robotics is an offshoot of Teledyne.

BOX 9.5 Electronic Surveillance

Plugged in, booted up and logged on, more and more workers in the Information Age are finding out that Big Brother is no distant fiction. According to a recent issue of *Macworld*, employers "may view employees on closed-circuit TV; tape their phones, e-mail, and network communications; and rummage through their computer files with or without employee consent—24 hours a day." Some 20 million American workers, including mail sorters, word processors and data-entry clerks, may be subject to electronic eavesdropping through their terminals.

Though employers use electronic monitoring in the name of productivity, studies show that it actually contributes to employee tension, anxiety, depression and other stress-related illnesses. . . . "The technology keeps advancing and it keeps getting less expensive," said American Civil Liberties Union (ACLU) spokesperson Jonathan Anderson. "Without some kind of protective legislation, the problem will get worse."

Because of a loophole in the 1986 Electronic Communications Privacy Act, employers have virtually unlimited rights to monitor workers. In May, Senator Paul Simon introduced "The Privacy for Consumers and Workers Act," which would require employers to alert workers about possible monitoring and how the collected information might be used. It would also prohibit covert, periodic or random monitoring, including video surveillance in bathrooms and locker rooms.

Simon introduced the bill in Congress by recounting a number of recent cases, including a lawsuit filed this year against the Sheraton Boston Hotel for secretly videotaping the male employees' changing room. When Sheraton employee Franklin Etienne, who emigrated to the U.S. from Haiti, saw a tape of himself changing into his uniform, he said, "Things like this used to happen in my country. My dream was to come to this country and be free to express myself. This is not the America I was thinking of." Other cases cited by Simon involved video cameras in women's locker rooms, e-mail interceptions and on-the-job electronic monitoring that records an employee's key strokes and the length of bathroom breaks and phone calls.

SOURCE: Aushra Abouzeid, 1993, "Spies On-Line: Computers Keep Tabs on American Workers," *In These Times* (June 28): 8–9. Reprinted with permission.

Just as high-technology companies come into existence quickly, so do they quickly pass out of existence, often through corporate mergers. These mergers may occur when larger corporations seek to buy out smaller firms with compatible product lines. Acquisitions may also occur when highly diversified conglomerates move into high-technology fields:

> Conglomerate buy-outs [in the high-technology field] should be viewed suspiciously because of the tendency to redeploy the assets of acquired firms in an attempt to increase short-term profitability. In the Massachusetts machine tool industry, for example, numerous conglomerate acquisitions resulted in "milking" the cash flow (profits plus depreciation) from the Massachusetts operation for reinvestment elsewhere. (High Tech Research Group, 1984:41)

The nature and consequences of the accelerating pace of mergers are discussed further in Chapter 15.

A final characteristic of high-technology industries is a quickened cycle of boom and bust. Tens of thousands of workers in the electronics industry were laid off in the early 1980s, an event unheard of in the 1970s. However, after layoffs in the 1980s, high-technology firms began hiring again in the 1990s and 2000s. Such booms and busts occur because of the rapid movement of new firms into high-technology fields, which produces periodic gluts in the market. These problems are increased by rapid changes in product lines as new products quickly eclipse older ones.

Union Responses

Unions are concerned about a variety of issues associated with the introduction of new technologies (Ozaki, 1999). These concerns include the following issues: (1) Advance notice of technological changes is necessary to give unions time to study their impacts and develop reasonable strategies for accommodating to these changes. (2) Unions need to be included on a consultation basis from the very earliest planning stages so that they can have a role in determining which new technologies will be selected. (3) It is important to initiate technological changes on a trial basis so that their unintended effects can be examined. (4) Workers need to be protected from reclassification to lower grades or pay scales. (5) Training programs for teaching existing workers the new skills required are strongly preferred over hiring new workers to displace existing ones. (6) Job security against technological layoffs is a particularly important issue and may be based on reduction of the workweek, voluntary early retirements, or redeployment to other facilities. (7) Workers need to be protected from potential health hazards associated with new technologies and from electronic surveillance at work.

The above list of contractual provisions are what unions would like, but what have they actually been able to get? Fewer than 5% of workers in the electronics industry in the United States are unionized, compared with 25% in manufacturing as a whole (Mort, 2000). The American Electronics Association reports only ninety collective bargaining contracts among its more than 1,900 member companies. Even when a high-technology company has a union, contractual provisions dealing explicitly with technological change are not common. For example, the AFL-CIO reports that only about 14% of agreements among its member unions contain provisions for worker retraining after the introduction of new technology.

New Technologies and Union Power One reason why unions have not been more successful in negotiations over technological change is that union power is often reduced in the situations where technological change is occurring. Automation of production makes it increasingly possible for managers to run production operations without workers, at least until repair and maintenance problems mount. As a result, the effectiveness of labor's major bargaining weapon, the strike, is reduced. In addition, there are well-grounded fears that too many demands from North American workers may encourage companies to move their production operations overseas.

In spite of these adverse conditions, union organizing and bargaining activity does occur in high-technology industries. In Massachusetts the CWA, the UAW, the Machinists, and the Electrical Workers have joined in a coalition with several other unions and are seeking to organize workers at the scores of plants and industrial laboratories along Boston's Route 128. Probably the greatest union successes have been in the area of demanding training for existing workers so that they can successfully utilize the new technologies (Ferman et al., 1990).

Union organizing activity also occurs in the high-technology plants exported to less developed countries:

> In April, 1985, workers at Silicon Technology, a subcontractor in the Philippines, struck against layoffs of nearly a quarter of its work force. The union won the reinstatement of some of the workers, but one strike leader, Jemoschick Paul, was reportedly shot dead by a company security officer. . . .
>
> In Bombay, India, this spring, 900 women employed by Tandon Magnetics—a subsidiary of U.S.-based Tandon, which is run by an Indian emigre—struck for better pay and conditions. The plant, located in the Santa Cruz Electronics Export Processing Zone, is not subject to Indian minimum wage laws, and the workers complained that

they were receiving pay even lower than other Tandon affiliates in the Zone. Working mothers also objected to rotating shifts, and some employees accused male supervisors of sexual harassment. ("Workers Active," 1985:1)

Because of the ability of high-technology companies to move production jobs around the world, improved conditions for electronics assembly workers may increasingly depend on the ability of workers in different countries to coordinate their demands for better conditions.

International Agreements In Canada unionized workers have been able to negotiate stronger agreements than those in the United States. Advance notice and consultation are required in 30% of contracts. Provisions for retraining are specified in 19% of contracts. Even in Canada, however, 60% of unionized workers and 85% of the labor force as a whole have no contractual rights concerning technological change (Belanger, 1983:79).

European unions have been actively engaged in negotiating technological change longer than North American unions and generally bargain on at least two additional issues. First, they frequently bargain for advance training for workers' representatives so that selected shop stewards can be fully trained in all aspects of the new technology before its introduction. Second, they generally specify that electronic devices cannot be used to collect personal or production data on individual employees.

The greatest strides in agreements in Great Britain have been made in white-collar industries, particularly banking, where strong job security provisions are the norm. These agreements typically contain provisions concerning safety and health, as well as job security. Britain has also been a leader in worker retraining with its state-financed Training Opportunities Programs (Cressey and Williams, 1990).

The most comprehensive technological agreements have been reached in the Scandinavian counties. Sweden's largest and most dynamic sector is advanced machinery and machinery parts, an industry requiring the very latest technology to maintain its competitive international position. Sweden has the largest number of industrial robots per capita in the world. Many of the Swedish provisions concerning technological change are legislative rather than contractual, including a complete prohibition against the electronic monitoring of individual workers' output. Additional provisions are included in collective bargaining agreements. In Norway the primary innovation is the creation of **data stewards.** Data stewards keep abreast of the latest technology being considered by the company, consult and negotiate with the company concerning its deployment, and protect workers' legislative and contractual rights.

SUMMARY

New technologies transform skill requirements, destroy entire occupations, and create new ones. They also produce changes in the occupational distribution, working conditions, and organizational dynamics. There are few consistent, undisputed findings concerning the consequences of technological advances.

Nowhere are these contradictory outcomes more apparent than in the area of skill requirements. In some cases new technologies destroy requirements for traditional craft skills; in others they create new skill requirements. Often the prevalence of deskilling or skill upgrading has little to do with the nature of the new technology itself. Instead, it may depend on the power of workers to demand the preservation and extension of their skills and on managers' perception of the importance of preserving a skilled workforce.

The majority of the studies favor the notion that a dual occupational structure emerges after the introduction of new technologies. A dual occupational structure appears to be a prominent feature in the electronics industry. In fields such as the manufacturing of microprocessors, computers, and scientific instrumentation, large numbers of lower-level production workers are required along with smaller numbers of highly trained engineers and technicians.

There is substantial agreement that advanced technologies are producing new sources of alienation and stress for workers. Increasingly, control and discretion are removed from workers and placed directly in automated equipment or in computer-assisted management information systems.

Many analysts cite the role of new technologies in eliminating dirty, tedious, and dangerous jobs as an important element in improved working conditions. But this point tends to ignore the fact that displaced workers may not have any job after the new technologies are introduced. Because new technologies create few jobs in contrast to the number they eliminate, acquiring new skills may be of little benefit to these workers. Long-term and indirect employment generating consequences of high technology may help offset these employment losses, but these effects are hard to verify.

Union responses to technological change focus primarily on the issue of job security for workers threatened by technological displacement and on the availability of new jobs for displaced workers. Unions also encourage the development of programs that provide training for workers in new skills and that protect workers from electronic monitoring.

KEY CONCEPTS

high-technology industry

numeric control (NC)

computer-aided numeric control (CNC)

video display terminal (VDT)

management information system (MIS)

computer-assisted design (CAD)

telecommuting

technological displacement

long-term and indirect effects

dispersion of information

electronic surveillance

data stewards

QUESTIONS FOR THOUGHT

1. Do you think the jobs in the year 2025 will be more skilled or less skilled as a result of widespread use of computer technology?

2. What aspects of the current computer-based technological revolution generate the greatest potential for job displacement? Do you think other factors will be able to offset this job displacement and, if so, what will these factors be?

3. Describe the potential benefits and problems of telecommuting. Do you think you will be telecommuting in your future career? What about this experience do you expect will be positive? What will be negative?

4. Computers can allow both greater access to information and greater centralization of control. Which of these do you think will be their dominant effect? Will some people benefit from greater access to information while others suffer from additional control and surveillance?

5. It is often said that the younger generation, which has grown up with computers, is more comfortable with their use than older workers. It is also the case that in developing nations few people at all have access to computers or computer skills. Will these realities lead to a world divided between young computer literate workers in developed nations versus everyone else?

MULTIMEDIA RESOURCES

Print

Heidi I. Hartmann, Robert E. Kraut, and Louise A. Tilly. 1986. *Computer Chips and Paper Clips: Technology and Women's Employment*. Washington, D.C.: National Academy Press. An in-depth analysis of the impact of electronics on the work of women, with a special focus on clerical work.

David Noble. 1997. *The Religion of Technology: The Divinity of Man and the Spirit of Invention*. New York: Knopf. An analysis of the causes and consequences of technological change and its role in human development.

Shoshana Zuboff. 1988. *In the Age of the Smart Machine*. New York: Basic. Presents the argument that the essential character of computer technologies is their ability not only to automate work but also to provide information about the production process, which then serves as a source of potential power for workers.

Websites

U.S. Department of Education. *www.ed.gov* A valuable site for information on high-technology training opportunities.

National Center for Education Statistics. *nces.ed.gov* Comprehensive information on educational programs and educational attainment of the labor force.

Journal of Industrial Teacher Education. *scholar.lib.vt.edu/ejournals/JITE* A core site for information on vocational-technical training.

Public Broadcast System (PBS) Online. *www.pbs.org* High-quality information (and entertainment) on a wide range of social issues.

Internet Resources for Sociologists. *www.umsl.edu/~sociolog/resource.htm* A huge site with links to information and research on a wide range of topics of sociological interest.

Internet Indicators. *www.internetindicators.com* Key indicators on the range and growth of the internet economy.

RECOMMENDED FILM

Star Trek television series and all its progeny. A thought provoking window on the future of technology.

10

Services

A post-industrial society is based on services. Hence, it is a game between persons. What counts is not raw muscle power, or energy, but information.

(BELL, 1976:127)

. . . [T]he reasons for the rapid growth of service occupations in both the corporate and government sectors of the economy [are]: the completion by capital of the conquest of the goods-producing activities; the displacement of labor from those industries, corresponding to the accumulation of capital in them, and the juncture of these reserves of labor and capital on the ground of new industries; and the inexorable growth of service needs as the new shape of society destroys the older forms of social, community, and family cooperation and self-aid.

(BRAVERMAN, 1974:359)

The great majority of the North American labor force already works in service jobs, and most new jobs are being created in the service sector. Students reading this book are likely to be employed in service occupations or industries for most of their working lives. As the two quotations above indicate, however, observers disagree on the implications of the transformation of the economy from goods-producing to service-producing. In this chapter, we examine the nature of service work, the problems that face service workers, and the challenges that lie ahead for individuals and firms that produce services.

WHAT ARE SERVICES?

Members of the labor force provide either goods (tangible items) or services to be sold. In Chapters 7, 8, and 9 we discussed the production of such goods as agricultural products and manufactured products. *Services* are acts provided in return for payment. The distinction between goods and services may be based on how close in time or space the work is to consumption by the user. Pressers in a clothing factory are counted as manufacturing workers, but pressers doing exactly the same job in a dry-cleaning establishment are considered service workers (Braverman, 1974:360–361). A worker in a frozen-food factory who prepares and freezes a casserole for later sale is performing a manufacturing task. The restaurant cook who microwaves the casserole for a seated customer would be considered a service worker. So would the supermarket cashier who rings up the frozen casserole for purchase by a shopper. But the consumer who microwaves the casserole at home performs that activity outside the market economy, and so this food preparation would be considered home production. The difference is that the frozen-food worker is preparing a product for sale. The cook and the salesperson are compensated for behaviors (cooking, selling) that make the product accessible to the consumer. The unpaid actions undertaken by the final consumer are not considered work.

Service workers may be self-employed, employees in private enterprises, or government employees. Their customers may be individuals, organizations, or governments. The workers may interact directly with their customers, they may only hear their voices or read their correspondence, or they may never interact directly with a customer. Some service workers receive direct payment on the basis of their services—for example, automobile salespeople typically receive a commission for each car they sell. Some service workers receive payment for the services they are *ready* to provide during a certain period of time, even if there are no customers. For example, a cook in a restaurant would be paid even if no customers ordered food on a particular evening. We pay for some services that we hope never to need, such as emergency medical services or national military defense. The insurance industry is a service of this sort: the policyholders pay for the protection and peace of mind, and hope that others than themselves will need the benefits (Jureidini and White, 2000).

Characteristics of Services

Time-Boundedness The many types of services share some common characteristics. Most services are **time-bound,** in the sense that there are culturally defined times for providing the services. Some services, such as electrical utilities, hospitals, or police protection, are offered twenty-four hours a day. Other services are offered at the times most convenient to customers or providers. Restaurants do most of their business, and must have most of their workers present, at the traditional mealtimes. Health clubs are open evenings and weekends when their members have leisure time. Some services must be offered according to regularly occurring deadlines. Accounting firms, for example, require longer hours of their workers as the April 15 tax deadline nears. Youth camps employ counselors only during the summer.

Services are also time-bound in the sense that their production and consumption occur nearly simultaneously. Services cannot usually be stockpiled, as goods can be produced and stockpiled in a warehouse. An airlines reservations clerk may have no calls during one hour of her shift, and then dozens of calls during a brief period of time. The earlier idle time, however, cannot be put to use later when she is busier. Fast-food restaurants may try "banking" of hamburgers, which means cooking additional hamburgers in anticipation of a rush at lunchtime, but this tactic is limited by the need to forecast accurately the customers' orders and by the need to provide fresh, hot food. The limited ability to stockpile services reinforces the need to match workers' schedules closely to the customers' convenience.

Low Productivity Services have traditionally been characterized as having *low productivity,* although there is much variation among the different types of services (Fuchs, 1968). There is also concern that the measurement of productivity appropriate for goods production is less meaningful for services. Suppose that a child-care worker who has been caring for four infants must now care for five. From one perspective the worker's productivity had increased. Others, however, might conclude instead that the quality of the service provided to the babies will decline (Malambre and Clark, 1992).

For many service jobs, technology has enhanced productivity only to a limited extent. In part low productivity derives from the time-boundedness of services, and in part it results from the requirement to work with clients or customers individually. The most efficient airlines reservations clerk can handle only one call at a time, and will not be productive at times when there are no calls. Telephone and computer technology may aid the clerk's efficiency—for example, call-waiting and automated messages may pacify the customers waiting for service—but the productivity gains are limited by the nature of the work. In a hospital, electronic monitors may substitute for a nurse's or an aide's direct observation of a patient, but other tasks require direct interaction with one patient at a time—giving baths, administering medication, or assisting with physical therapy.

Restaurants provide a second example of the difficulties in increasing productivity. Technology may reduce the time required to prepare food in a restaurant, but a waitperson is still needed to bring the dishes to the diners' tables. In the fast-food industry, innovations in organization have increased productivity and kept prices low. Lower prices have other causes as well. Salad bars, buffets, and cafeterias substitute a customer's labor for the waitperson's, thus reducing costs.

Important consequences flow from the lower relative productivity of service work. As agriculture and manufacturing become more productive, they require fewer workers to maintain equivalent levels of output. The displaced workers seek employment in the service industries. Managers in the service industries seek higher productivity as a way to increase their profits; to the extent that they are successful in raising productivity, there may be a further movement of workers within the service industries from more productive to less productive jobs. Economists usually identify low productivity as one reason why compensation in service jobs is lower than in manufacturing jobs. Other factors that contribute to lower compensation for service workers are their lower levels of unionization, the greater likelihood that they work in small firms, and the greater probability their work is part time or seasonal.

Insatiable Demand Finally, there may be an *insatiable demand* for services, in the sense that people can always consume more services, or more expensive versions of the services they now consume. The demand for agricultural and manufactured products is more limited. There are only so many bowls of cereal or television sets that a consumer will buy. For services, by contrast, the consumer with more disposable income will find increasingly more attractive services to purchase. As long as consumers can afford to pay for services, service employment will grow. Figure 10.1 shows the increase in service employment during the 1990s and the corresponding unemployment rate.

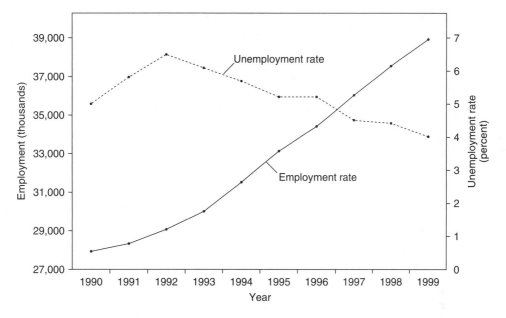

FIGURE 10.1 Service Employment versus Unemployment Rate in 1990s

SOURCE: U.S. Department of Labor, Bureau of Labor Statistics, *http://stats.bls.gov/iag/iag.service.htm.*

Sources of the Demand for Services

The demand for services can grow in several ways. First, service industries can arise from *new manufactured products.* The widespread use of computers generated many new service jobs. Wholesalers and retailers sell computers, peripheral equipment, and software. Consultant firms advise companies about purchasing computers; other firms maintain and repair the equipment; and employee training firms develop programs to help employees use new hardware and software. In a similar fashion, other new agricultural and manufacturing products generate new service jobs.

Second, service industries grow with the shift from unpaid production to *paid production of services.* Sick people were once tended at home. Today workers from many health occupations provide medical services in hospitals, clinics, outpatient centers, and medical offices. Families once provided most food preparation, child care, gardening, cleaning, and repair services needed

by their households. Today many families purchase these services, either regularly or occasionally. Because these services can still be produced in the home, during times of hardship some families once again produce these services for themselves, and the sellers of such services may be especially vulnerable to the business cycle. For example, household budgets for many families are balanced by reducing the number of meals eaten away from home.

Third, many service industries arise in response to *higher disposable income* of consumers. For most households, leisure and entertainment services, such as tourism, professional sports, and the theater, are discretionary purchases as budgets permit. Retailers rely for much of their profits upon consumers' discretionary holiday spending for clothing, gifts, and special foods. Because these expenditures may also be curtailed when consumers are worried about the economy, retailers of these items are especially vulnerable to recessions.

Table 10.1 The Sectoral Shift of the U.S. Labor Force,[a] Women Workers, and Black Workers, 1900–2000

	1900	1910	1930[b]	1950	1970	1980	1986	2000[c]
Percent of Labor Force in:								
Agriculture[d]	41.0	33.2	23.4	12.2	4.4	3.4	2.9	2.5
Manufacturing[e]	28.2	30.5	31.1	35.9	33.1	29.3	26.6	22.2
Services	30.8	36.3	45.5	52.0	62.5	67.3	70.5	75.3
Percent of Female Labor Force in:								
Agriculture[d]	18.4	22.4	10.4	4.1	2.2	1.6	1.3	1.4
Manufacturing[e]	24.7	22.6	21.5	24.2	22.3	17.9	15.4	11.8
Services	56.9	55.0	68.1	71.7	75.5	80.5	83.3	86.8
Percent of Black[f] Labor Force in:								
Agriculture[d]	65.4	55.5	37.0	19.8	4.7	2.3	1.4	21.0
Manufacturing[e]	5.1	13.9	20.1	25.1	30.5	27.7	24.4	18.0
Services	29.5	30.6	42.9	55.1	64.8	70.0	74.2	81.0

[a]Employed civilians only.

[b]The data for women and blacks in 1930 omit clerical workers, who accounted for 18.5% of the female workforce and 0.7% of the black workforce.

[c]Preliminary.

[d]Includes farming, forestry and fishing.

[e]Includes manufacturing, mining and construction.

[f]The term "Negro" was used until the 1970 census; the 1970 figures refer to non-whites. Data for blacks in 1900 exclude occupations that accounted for 18.3% of the black workforce.

SOURCE: Teresa A. Sullivan, 1989, "Women and Minority Workers in the New Economy." *Work and Occupations* 16, 4 (November) and calculated from BLS, 2001a, Employment and Earnings (January 2001).

Historically, the emergence of a service society occurred over a long period of time and was influenced by many developmental factors that you have already studied in earlier chapters. In the next section, we turn to the issue of how so many workers came to be employed in services.

THE RISE OF THE SERVICE SOCIETY

Table 10.1 shows the transformation of the American labor force in terms of industrial sector. (Industrial sectors were discussed in Chapter 8.) In earlier centuries, many service workers were household servants. The stigma of servitude, and the expectation of subservience from the workers, still affects much service work (Paules, 1991). As Table 10.1 indicates, however, there has been a secular trend in employment from agricultural work to industrial work and finally to services. Services are projected to grow both in terms of the numbers of workers and in terms of the proportion of the labor force that delivers services. Shortly after World War II, half of the U.S. labor force was employed in services. Today, three-fourths of the labor force works in services (BLS, 2001a).

Sectoral Transformation

The shift from agricultural through manufacturing to service work is called the **sectoral transformation of the labor force.** The shift proceeds in two ways. First, positions in agriculture, mining, and manufacturing have not been filled when workers retire and other jobs have been eliminated through reorganization and plant closings. Second, new jobs have disproportionately been service jobs. Teenagers and young adults are especially likely to enter the service sector. Many new service jobs are knowledge based and found in fields such as telecommunications or transportation, but others are found in personal services such as restaurants.

Tertiarization

In the developing countries the process of gradual sectoral transformation is often short-circuited by what is called **tertiarization,** or the development of a service economy without a manufacturing base. The term comes from the tripartite division of the economy into primary (agricultural and extractive), secondary (manufacturing), and tertiary (service) sectors. In the United States and many other advanced countries, services did not come to dominate the labor force until after the rise, maturation, and decline of manufacturing. In the developing countries, by contrast, the labor force has frequently shifted from primarily agricultural employment to an urban, service-oriented economy without the intermediate step of manufacturing. In such countries, manufacturing may be limited to the first-step processing of raw materials (for example, rice mills or sugar processing plants), with most manufactured products imported from abroad.

Developing countries face a difficult task if they seek to increase their labor force employed in manufacturing. To duplicate the heavy industry of industrially advanced countries—such as steel mills and automobile plants—requires huge capital investments and successful entry into the heavily competitive world market for those products. Protecting their new plants from foreign competition by imposing restrictive tariffs may backfire: Their country may lose favorable trading terms for its raw products. Another way to industrialize is to engage in joint ventures with large multinational corporations. This strategy may bring only limited investment, often for intermediate-stage assembly plants that export both their products and their profits to the international corporation's headquarters. Often, these are "run-away plants," so that their opening in another country follows a plant closing in the United States or elsewhere. (The global economy is discussed in greater detail in Chapter 16.)

Even strenuous manufacturing efforts may be insufficient to absorb much of the labor force. Tertiarization—the rapid growth of the third, or tertiary (service) sector—is the consequence. In the Bahamas, for example, about 76% of the labor force works in some service area, compared with only 4% in manufacturing (calculated from ILO data available at *http://www.ilocarib.org.tt/digest/countries.html*). Tertiarization rarely results from a deliberate employment policy. Rather, employment in a low-paid service job is often the last resort of the worker. In the cities of developing countries, many workers labor in the most marginal of service enterprises. Urban service workers are often recent arrivals from the countryside who migrate because of the lack of agricultural land, the push of growing population, or the hope for urban opportunities. Some workers find employment as household servants for better-off urban dwellers. Box 10.1 describes Peruvian women who work in households in Lima, Peru.

Other people employ themselves as petty entrepreneurs. Many large cities in developing countries have cadres of self-employed workers who swarm over cars stopped at red lights, offering services such as wiping the windshield, selling flowers or newspapers, or selling single cigarettes from a package. Street vendors serve as indigenous fast-food outlets, offering cooked foods that are not generally available in the home. Tertiarization in the developing world does not mean the growth of the knowledge-based and business services; instead, the term implies the expansion of poorly paid service work so tenuous that it is sometimes referred to as casual or informal labor. This type of service employment is also found in U.S. cities, where homeless people

BOX 10.1   Domestic Service in Lima, Peru

As in many other developing countries, women migrants to Lima are disproportionately recruited to work as servants in private households. This work status is often the first full-time job of their adult lives, but it is also often a transitional stage in their lives. Street vendors are another important occupation for women in Lima, but whereas 90% of the servants are single, nearly half of the street vendors are married. Anthropologist Margo Smith was curious to know whether former domestic servants later became street vendors. In 1982 she reinterviewed women whom she had first interviewed twelve to fifteen years earlier. Although she did not find them to be street vendors, she recorded the changes she learned about in the reinterviews.

Señora F. still works as a servant, a trusted *ama de llaves* (housekeeper) for an affluent family. She has been employed by three households in the same family for more than thirty-five years. Starting as the teenaged *ama* caring for the two young daughters of her employers, she has improved her position to the point of being completely responsible for managing the household on a daily basis, and she supervises the work of two other servants. Approximately thirty years ago, right after her marriage to a cabinetmaker, she worked as a market vendor for about a year but returned to domestic service in response to her former employers' (and marriage godparents') appeal. She and her husband have been separated for many years, and she has been entirely self-supporting. For several years she worked half-days in the knitting factory owned by her brother-in-law, in addition to her employment as a servant, in order to earn more money, but she no longer does so. In the mid-1960s she purchased a small apartment for herself in a modern low-income building in Lince, a modest, middle-income neighborhood of Metropolitan Lima; she is now purchasing a

second unit there for one of her nieces, a single mother. Her living conditions in the home of her employer are much more comfortable than they were in 1970; as her employer prospered economically, her situation also improved. She has no children but has contributed to the support of her six nieces and nephews. . . .

Señora M. M. is a housewife. She had a five- or six-year career as a servant prior to her marriage and quitting her job. One of her employers had promised to set her up in a small shop selling meat, but never did so. When her first child died as an infant, she was so grief-stricken that she took a job as a servant for a European family and lived with them for a year in Europe before returning to her husband in Lima. Since then she has had two children and is determined not to have any more because they are so expensive. Both children are in elementary school. Her husband had studied accounting at a Lima university, but dropped out before graduating. He works at two low-level white-collar jobs. They are building their own brick home in a new lower-income development in Metropolitan Lima. Her home is more completely furnished than those of her mother or sisters; it has indoor plumbing, a refrigerator, a gas stove, and television. She proudly displays an album full of the photographic *recuerdos* of a few vacations she and her husband and children have taken around Peru. When the back bedrooms of the house are completed, one of the front rooms will be converted into a small store for her to operate, the same goal she had spoken of in 1969. She complains about the plight of the poor in Lima, a group with which she identifies.

SOURCE: Margo L. Smith, 1989, "Where Is Maria Now? Former Domestic Workers in Peru." In *Muchachas No More: Household Workers in Latin America and the Caribbean,* edited by Elsa M. Chaney and Mary Garcia Castro. Philadelphia: Temple University Press, pp. 137–138.

may scavenge castaway articles or recycle aluminum cans to eke out a living (Snow and Anderson, 1993). (In Chapter 14, we discuss peripheral employment in greater detail.)

TYPES OF SERVICE INDUSTRIES

Sociologists identify six types of services classified by the type of service produced and by the customer to whom it is delivered (Browning and Singelmann, 1975).

Professional Services

Professional services provide knowledge-based services that enhance the well-being of individuals or firms who are clients. Examples of professional services are medical services, legal services, and counseling services. Not all workers in these industries are themselves professionals. Nonprofessional workers in legal services and architectural services, for example, are employed as aides, helpers, and assistants to the professional staff. In 2000, about 17.1 million workers were employed in professional services (BLS, 2001a).

Professional workers are often able to provide services to only one client at a time. Because it is difficult to gauge in advance how much time each client will need, both professionals and their clients often express frustration with the quality of their interactions. One physician reported:

> You can spend an entire career trying to help the health and mental health of one patient, but the realities of practice are that you have hundreds of patients and I would make an appointment with someone for twenty minutes for a problem that should have taken five minutes and an hour later I would be sitting there listening to her complaints. (Ebaugh, 1988:55)

Clients, of course, experience the reverse frustration of lengthy waits and truncated visits. We discuss professionals in more detail in Chapter 11.

Business Services

Business services assist individuals, firms, and organizations in carrying out their economic functions. Examples of business services are temporary worker firms, advertising, financial services, accounting, real estate, and insurance. Nearly 8 million people are employed in the finance, real estate, and insurance industries. Almost 10 million people work in other business services, such as computer and data-processing services (2 million) and temporary-help agencies (3.4 million) (BLS, 2001a).

Automation is causing rapid change in many business services. The insurance industry, which handles massive amounts of data, is a good example of an industry in which automation has changed jobs.

> Premium billing and collection, for example, is often performed through the "turnaround" billing system. After a machine-readable notice and payment are returned by the policyholder, the computer stores the payment data for the accounting department, calculates the agent's commission on each premium, and credits the policyholder's account. These computerized billing systems handle about 5000 remittances an hour and the capacity of new equipment is over 40,000 remittances an hour. . . . In general, clerical occupations which involve a variety of tasks, individual decision-making, and face-to-face communication have been least affected by office automation. Those which require routine and repetitious work have been proven most vulnerable. (Cornfield et al., 1987:118–119)

Producer Services

Producer services help other industries create their products or services. Utilities are a good example of producer services. Providing electricity, gas, water, sanitation, and communication are important services needed by every business and household. Some producer services, such as municipal garbage collection or water treatment, are provided by the public sector and paid for with user fees and taxes. Communications employ 1.6 million workers, and utilities and sanitary services employ another 900,000 workers (BLS, 2001a).

Utilities jobs are often well compensated and may require considerable skill with new technology and equipment. The services are so necessary that even brief disruptions in service will inconvenience thousands of people. The development of complex utilities networks can increase pressures for workers because even small errors on the job can have major consequences.

> Even changing light bulbs has its dangers in these highly engineered, complex systems [nuclear power plants]. In 1978 a worker changing a light bulb in a control panel at the Rancho Seco I reactor in Clay Station, California, dropped the bulb. It created a short circuit in some sensors and controls. . . . The loss of some sensors meant the operators could not determine the condition of the plant, and there was a rapid cooling of the core. (Perrow, 1984:44)

Distributive Services

Distributive services bring about the geographic dispersion of goods. Goods produced in New Jersey reach a consumer in Oregon because of transportation, warehousing, and retail marketing, each of which is an example of a distributive service. Sales workers are discussed in greater detail in Chapter 13. The distributive-service industries include 4.6 million workers in transportation, 7.1 million workers in wholesale trade, and 23.2 million workers in retail trade (BLS, 2001a). Among the transportation workers, about 1.8 million workers are in trucking or warehousing.

> You sit in a truck, your only companionship is your own thoughts. Your truck radio, if you can play it loud enough to hear—you've got the roar of the engine, you've got a

transmission with sixteen gears, you're very much occupied. You're fighting to maintain your speed every moment you're in the truck. . . .
>
> You have to get all psyched up and keep your alertness all the time. There's a lot of stomach trouble in this business, tension. Fellas that can't eat anything. Alka-Seltzer and everything. There's a lot of hemorrhoid problems. And there's a lot of left shoulder bursitis, because of the window being open. And there's a loss of hearing because of the roar of the engine. . . . There has been different people I've worked with that I've seen come apart, couldn't handle it any more. (Terkel, 1974:283–284)

Besides truck drivers, transportation services include airlines, passenger and freight railways, and taxi driving. Box 10.2 describes the work of an urban bus operator.

Social Services

Social services are often financed by the government and benefit not only individuals, but the society as a whole. An important subset of these services is protective services: the military, police, prison workers, and other law enforcement officers, and also the firefighters, rescue squads, and emergency response services. Education and social welfare are other examples of social services.

Government employees in general are service workers in the sense that they produce services for the entire society. There are 20.5 million government workers, most of whom are engaged in providing social services. Others provide producer services such as utilities and sanitation. There are 4.8 million state workers, 13.1 million local government workers, and 2.6 million federal employees (BLS, 2001a). Important government services include national security and international affairs, justice, public order, postal service, and human resource programs. In addition, nearly 1.8 million people are on active duty as members of the armed forces.

BOX 10.2 Bus Driver in Los Angeles: Lupita Pérez

I'm a "bus operator." They don't like calling us bus drivers, they like to call us "bus operators." I have no idea what the difference is. "You're not bus drivers, you're bus operators." Okay, no problem.

I'm thirty-eight years old and I've been doing this for three years. I used to work at an elementary school as a teacher's assistant. And I liked that okay, but the money wasn't there, you know? And I was looking for a part-time job for the summer and a neighbor of mine told me, "Hey, the Transit Authority is hiring for part-time. What have you got to lose?" So that's how it all started. And it turned out to be a career! [Laughs]

It's not a bad job at all. It really isn't. I mean like anything, you have your good days, you have your bad days. But it's a good opportunity for me and my family. The benefits. The insurance. I'm a single mom so I have to think of this stuff, you know? This is just a good opportunity. I'm going to stick it out for the long haul, and take advantage of this good retirement. When I hit sixty, it's gonna be beautiful. . . .

The route I prefer—everybody thinks I'm nuts—is the 81. The 81 is the Figueroa and it goes from Eagle Rock Plaza all the way to 117th and Imperial—the middle of Watts. You go through some interesting parts on the 81. It's one of the busiest lines in the country. I like the ones that are very heavy and busy. They're always on the go.

Always. And before you know it, your day is done. I hate the routes that drag—that you hardly pick up anybody and you got to go really slow—and it's so boring! I don't like those. Not at all.

I really like the people on the 81, too. Because let's face it, the 81, basically we pick up the poor people. The 81 is basically what they consider the low-class passengers. But for being low class, they pay. And they don't give me any hassles. And they always say, "Hi, how are you?" Or "Good afternoon," or "Good morning.". . . .

Of course, the biggest thing is learning about how to deal with so many different types of people. You deal with different nationalities. Different kinds of everything. You gotta know the rules. In this part of town, they treat me good. Maybe because they see me as the same nationality. But, like, when I go to South Central, umm, it's okay for them to treat me like garbage, but I can't go around and treat them like garbage. That's the way it is. And then like when I go into San Marino or Beverly Hills, I get treated different and I have to treat them different, because hey, people act different. And every day is something new. Fortunately, for me, I enjoy that aspect. Most of the time, I love these people. They make the job for me.

SOURCE: John Bowe, Marisa Bowe, and Sabin Streeter, eds. *Gig: Americans Talk about Their Jobs.* New York: Crown Publishers, 2000, pp. 151–153. Reprinted by permission of Random House, Inc.

Besides government-provided social services, about 3 million private-sector workers are employed in social services, including child day-care services and residential care. A private nursing home attendant reports her round of daily work activities:

My baby here has cerebral thrombosis. She is ninety-three years old.

I get in this morning about eight-thirty. I shake her, make sure that she was okay. I took her tray, wipe her face, and give her cereal and a cup of orange juice and an egg. She's unable to chew hard foods. You have to give her liquids through a syringe. . . .

The first thing in the morning, after breakfast, I sponge her and I give her a back rub. And I keep her clean. She's supposed to be turned every two hours. If we don't turn her every two hours, she will have sores. Even though she's asleep, she's got to be turned.

I give her lunch. The trays come up at twelve-thirty. I feed her just the same as what I feed her in the morning. In the evening I go to the kitchen and pick up her tray at four o'clock and I do the same thing again. About five-thirty I leave here and go home. (Terkel, 1974:651)

Police work is a major form of public employment in the social services, and private security personnel are a fast-growing area in the private sector. The police receive substantial training in their work. This training has increased in recent decades, as have the efforts of police departments to provide continuing education for their employees. Policemen and policewomen have substantial autonomy and discretion in deciding how to do their work in the field. Their autonomy results both from their work being spatially dispersed, and therefore difficult to supervise, and also from the fact that they have the training necessary to make quick judgments in a crisis. Private security personnel vary widely in terms of their training and duties. Both police and private security forces encounter danger, many rules prescribing appropriate behavior and proscribing inappropriate behavior, and often harsh discipline for infractions of rules. Because these characteristics are at odds with traditional roles for women, women officers and guards may encounter problems (Appier, 1998, Britton, 1997). A Toronto home alarm seller reported:

> I've had weird cases, where I will go to the home and the male, he'll say, "Oh, I can't believe we've got a female doing this job," which really bothers me. . . . They don't think females understand the technical side of things. I do . . . (Erickson, Albanese, and Drakulic, 2000:315).

As a consequence of the mixed professional and nonprofessional nature of police work, employees have sought to increase their power and prestige through both professional associations and collective bargaining. They have succeeded increasingly in both endeavors, and the status and rewards of police work have risen accordingly (Lundman, 2002).

Personal Services

Personal services are produced principally for a family or for individuals, and include restaurants, hotels and motels, entertainment, tourism, and repair services. A subset of personal services are the fewer than 1 million private household workers, employees of individual households who clean, cook, garden, and care for dependent family members. Over 3.7 million workers are employed in personal service, entertainment, and recreational industries. Over 7 million workers are employed in eating and drinking places, but are counted with retail trade services in government statistics. About 1.9 million work in hotels, and another 1 million are employed in industries such as laundries, beauty shops, and barber shops (BLS, 2001a).

The defining characteristic of personal service work is that the worker delivers a service directly to the final consumer. This service may be operating a movie projector, carrying luggage to a hotel room, or sewing alterations for a suit. Service workers are often expected to address their customers as "Sir" or "Madam" or by their title and last name, but customers usually address the service workers in more familiar terms or by their first names.

> A barber, he has to talk about everything, baseball, football, basketball, anything that comes along. Religion and politics most barbers stay away from. . . .
>
> Usually I do not disagree with a customer. If there is something that he wants me to agree with him, I just avoid the question. (Laughs.) This is about a candidate, and the man he's speaking for is the man you're not for and he asks you, "What do you think?" I usually have a catch on that. I don't let him know what I am, what party I'm with. The way he talks, I can figure out what party he's from, so I kind of stay neutral. That's the best way, stay neutral. Don't let him know what party you're from cause you might mention the party that he's against. And that's gonna hurt business. . . .
>
> When I leave the shop, I consider myself not a barber any more. I never think about it. When a man asks me what I do for a living, I usually try to avoid that question. I figure that it's none of his business. There

are people who think a barber is just a barber, a nobody. If I had a son, I'd want him to be more than just a barber. (Terkel, 1974:315–317)

Some service jobs carry a stigma (Saunders, 1981). Like the barber, workers in these occupations find little to brag about concerning their jobs. When asked what they do for a living, their response may be guarded.

Janitorial work is an example of an occupation that is stereotyped as low-status, dirty work. Janitors, however, often play important roles, protecting their buildings from illegal entry, fires, and other hazards. In addition, they are free from direct supervision for most or all of the day. These job characteristics give many janitors a basis for taking pride in their work:

> I make a pretty good buck. I figure if I do my work and do it honestly I should be entitled to whatever I make. For high-rise buildings, head man makes a thousand dollars a month and his apartment. You never heard of that stuff before. I've turned down high rises by the dozens. I can make more money on the side on walkup buildings. . . .
>
> Today I can walk in the boiler room with clean trousers and go home with clean trousers. You check the glass, you're all set. That's the first thing you do. I check my fires and bring my garbage down right away. I take one of those big barrels on my back and I bring it up the flight of stairs and back down. I do this on three buildings and two have chutes. . . .
>
> I enjoy my work. You meet people, you're out with the public. I have no boss standing over me. People call me Mr. Hoellen. Very respectable. If I'm a good friend, they say Eric. I'm proud of my job. I've made it what it is today. Up in the morning, get the work done, back home. Open the fires and close 'em. (Terkel, 1974:119–125)

Such positive aspects of janitorial work are counterbalanced by less desirable duties, such as handling garbage and cleaning rest rooms. Some janitors are also expected to be on call for emergencies seven days a week. These aspects of the occupation are the basis for the negative stereotype.

COMPENSATION IN SERVICES

The conditions of workers who provide services vary a great deal. Figure 10.2 shows the trend in average hourly earnings of workers in service industries through the 1990s. These averages, however, conceal important differences in competition. The *high-paying service industries* include transportation, communication, public utilities, wholesale trade, finance, insurance, real estate, professional and related services, and public administration. The *low-paying service industries* include retail trade, repair services, personal services, and entertainment and recreation services (BLS, 2000a).

Table 10.2 shows the wages and earnings by broad industry group for goods-producing and for the two tiers of service-producing industries. Not every job in the high-paying service-producing industries will pay a high salary, just as not every job in the low-paying service-producing industries will necessarily pay low wages, but the average wages in the table show a clear pattern of differences. Table 10.2 indicates goods-producing jobs paid more than services, and the low-paying service sector paid even less.

Various circumstances explain the differences in compensation for different types of service workers. As we have already mentioned, *productivity* tends to be related to higher pay. Telecommunications is an example of an industry in which electronic technology has made workers more productive, and this is a higher-paying service industry. The nature of the *client* is also an issue in how much payment the service provider can demand. In general, professionals such as lawyers accord higher prestige to clients who are corporations rather than individuals (Heinz and Laumann, 1982). Corporations are also better

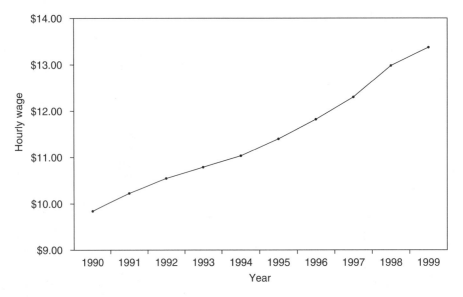

FIGURE 10.2 Average Hourly Earnings of Nonsupervisory Workers in Service Industry, 1990s
SOURCE: U.S. Department of Labor, Bureau of Labor Statistics, *http://stats.bls.gov/iag/iag.service.htm.*

Table 10.2   Average Hourly Earnings by Broad Industry Group, 2000

Total, private employment	$13.83
Goods-producing industries	$15.46
Service-producing industries	$13.33
High-paying services	
Transportation and public utilities	$16.30
Wholesale trade	$15.32
Finance, insurance, real estate	$15.19
Low-paying services	
Retail trade	$9.54
Services (other)	$13.97

SOURCE: U.S. Department of Labor, Bureau of Labor Statistics, 2000a *http://www.bls.gov*, data for September, 2000.

able than individuals to pay high fees and to keep lawyers on an indefinite financial retainer. In these fields the worker is often described as working with a *client;* in the lower-paying service fields, the worker is more often described as working with a *customer* (Rothman, 1987). The *nature of employment* is a third issue. A waitperson

cannot be rude to a customer without jeopardizing the tip, which represents a large proportion of total earnings. A salaried service worker may have more leeway to be rude to customers.

SERVICE INTERACTION

Standards

The essence of service work lies in the interaction between the service provider and the service recipient, who may be called the client or the customer. A successful interaction requires that the employer, the customer, and the worker all meet standards for a successful interaction.

Employer Standards   There are several criteria for judging the success of this interaction from the point of view of the *employer.* First, how many customers were served? Serving a number of customers is an indicator of higher productivity and efficiency. Second, how well were they served? Did the customers feel satisfied with this interaction? Did they spend as much money as the employer hoped? Are they likely to be repeat customers?

Customer Standards There are also criteria by which the *customer* judges the success of the interaction. For the most part, customers want to be treated quickly, courteously, and efficiently, but they also want to be treated as individuals. These may be contradictory criteria. The service provider may be able to interact more quickly with more people by being relatively impersonal. Speedy interactions meet the customer's desire for quick service, but they do not allow for individualized treatment. Handling special requests, explaining the product, or in other ways personalizing the encounter may lengthen the interaction, thus reducing the worker's efficiency. In some service industries (for example, among information operators), the length of interactions is monitored, and an operator who talks too long with customers may be admonished. Operators learn to keep most interactions as brief as possible, so that they can spend more time on the occasional complicated request without raising their average time-per-call to an unacceptable level.

Some customers also employ other criteria for judging the success of the interaction. Because service workers do serve the public, some people wish to treat them as servants, demeaning them if possible and expecting subservience in return. Service workers are often prohibited from responding in an angry or reproachful tone to the customer, and so they must react meekly and according to company procedures, even if the customer is rude or belligerent.

Worker Standards *Service workers* also have criteria for the success of an interaction, but these criteria vary depending on the nature of the interaction. Most interactions involve customers who are voluntarily present as shoppers, callers, or diners. Service workers in these situations expect to be treated with at least minimal respect, and they expect the customer to play the accustomed "customer role." Thus, while a certain amount of joking or teasing might be tolerated from a diner, a waitperson would not expect insulting language or physical assault from a customer. Part of the expected customer role in the restaurant also involves paying a tip.

In other situations, the "customer" is present involuntarily, and then the service workers may excuse the behavior of the customer. An ambulance attendant or an emergency room trauma team might make allowances for unusual behavior among patients who are in a great deal of pain. A police officer would not expect a suspect to be pleased about being arrested, but the officer does expect respectful treatment in ordinary interactions with citizens.

Because each of these three parties has different criteria for the success of an interaction, the result may be a struggle for the control of the interaction. As the opening quotation from Daniel Bell suggests, there is a "game among people" that occurs in service work. In the following sections, we examine how employers and workers seek to control the interaction.

The Role of Employers

Most service workers are employees. Management tries to control service interactions in three ways: (1) establishing standards for the conduct of its employees; (2) training employees so that they understand how, from management's perspective, the ideal interaction proceeds, and what to do when the interaction deviates from this norm; and (3) establishing controls to ensure conformity to the standards and practices of the company. Even service workers who are self-employed will often adhere to similar standards because customers come to expect a comparable level of service.

Setting Standards Setting standards may be formal or informal. Many service occupations require workers to wear a uniform that readily identifies their occupation. Often such uniforms reflect the subordinate status of the worker. For instance, waitresses and barmaids may be required to wear costumes reminiscent of the clothing of maids or clothing that is sexually revealing. Companies deploy slogans such as "the smile comes before the sale" to remind employees that friendliness is part of the ideal interaction. To reinforce this friendliness, management may mandate that employees wear name tags bearing only their first names, thus forcing customers to address the worker familiarly, a practice associated with friendliness.

Companies may also require certain rituals, such as introducing oneself to the customer by name, greeting customers as they enter the store, or asking certain questions in a required manner. One objective of these rituals may be fairness, or treating all customers equally. Service workers may be required to ask all customers the same questions or present them the same information, even if customers are uninterested. This could be a matter of fulfilling legal formalities. For example, the admitting personnel in a hospital may be required to ask questions to determine the patient's financial eligibility. Pharmacists must counsel customers concerning potential side effects of drugs.

Alternatively, the service workers' required "script" may be merely a means to increase the amount of money the customer spends. Telemarketers may receive actual written scripts to repeat over the telephone. Waitpersons may be required to recite the evening's specials even to a diner who has already announced an order. An insurance sales agent may make "cold calls" to prospects who have not expressed interest in life insurance. In each of these cases, one of the employer's standards for a successful interaction is following the script, even if both the worker and the customer would prefer to omit it.

Training Employers' expectations for performance are set forth in more or less explicit training sessions. Box 10.3 presents an account of the training sessions in two such settings: McDonald's Hamburger University and the training program of Combined Insurance (Leidner, 1993). Aside

from the explicit details that apply to the particular firm, such training usually seeks to achieve two goals. The first goal is the **routinization** of the interaction. In fast-food restaurants, the exact procedures for preparing each type of food are minutely detailed (Reiter, 1991). Procedures are also spelled out for the ways in which the service worker is to interact with the customer. According to Hamburger University instructors, McDonald's franchisees should buy ten to twelve videotapes a year for training their employees (Leidner, 1993:65). The standardization in training helps ensure not only that the customer tastes the same kind of sandwich, but also that the customer receives consistent service.

Routinization may ensure observance of health or safety rules. Airline mechanics and pilots have checklists to complete before every flight to verify that the airplane is in full working order. The pilot and copilot must complete a separate checklist before each takeoff. Hospitals have for years prescribed certain routines of care. Nurses in a surgical ward must follow a prescribed routine before sending a patient to surgery; this routine is designed to make sure that the necessary preliminary tests have been done and that the correct surgery is performed on the patient. The work of telephone operators was routinized soon after telephones became widespread (Norwood, 1990), partly to guarantee comparable standards of service.

The more recent wave of routinization in sales and personal service, however, is rarely to achieve goals of safety or performance, but instead to rationalize service delivery so that it is as efficient as possible. Rationalization reduces the interaction to its minimum elements and speeds up each transaction, but a routinized interaction may be unsatisfying to the customer because of its impersonality. So a second important goal of the training is to teach service workers to program and manage their emotional responses to the customer. **"Face-work"** refers to the management of the facial expression and, by extension, of the worker's emotional reactions (Goffman, 1955). The process of managing one's

BOX 10.3   Fast Food and Insurance: Training for Insurance Work

Sociologist Robin Leidner did fieldwork in two large service firms, McDonald's and Combined Insurance. She recounts some details of the training for both groups of workers. In both cases, the worker is instructed to routinize the interaction to achieve the company's goals.

"Window class" . . . lasted a whole morning. My training group included two other newly hired window workers and a grill worker who was being cross-trained to work window. The training began, as usual, with a videotape. It emphasized the importance of the window crew's work, telling us that to guests (McDonald's word for customers), "You ARE McDonald's." Interactive work is only part of the window crew's job, though. In addition to learning about dealing with people, we had many details to learn about dealing with things. The videotape provided instructions on what the various-sized cups and bags were used for, how to stock the counter area, how to work the soda and shake machines, and how to load a bag and set up a tray properly.

Interactions with customers, we were taught, are governed by the Six Steps of Window Service: (1) greet the customer, (2) take the order, (3) assemble the order, (4) present the order, (5) receive payment, and (6) thank the customer and ask for repeat business. . . . The videotape provided sample sentences for greeting the customers and asking for repeat business, but it encouraged the window crew to vary these phrases. According to a trainer at Hamburger University, management permits this discretion not to make the window crew's work less constraining but to minimize the customers' sense of depersonalization:

"We don't want to create the atmosphere of an assembly line," Jack says. They want the crew people to provide a varied, personable greeting—"the thing that's standard is the smile." They prefer the greetings to be varied so that, for instance, the third person in line won't get the exact same greeting that he's just heard the two people in front of him receive.

Training for door-to-door insurance salespeople was at least as standardized.

In addition to scripting the life insurance agent's words, Combined Insurance tried to standardize how the agents held themselves, how they delivered their lines, how they gestured, and how they used the physical setting. Consider, for example, my field notes from a class lecture on getting through the door:

After ringing the doorbell and opening the screen door, wait for them to answer with your side to the front of the door. Do a half-turn when they open it. It's almost as though they catch you by surprise—nonconfrontational. Be casual.
Lean back a little when they open the door. Give them space.
Attitude: "Be as loose as a goose," Mark says. Be able to respond to what they say.
To get inside, use The Combined Shuffle. It has three steps:
1. Say, "Hi, I'm John Doe with Combined Insurance Company. May I come in?" Handshake is optional. Break eye contact when you say, "May I come in?"
2. Wipe your feet. This makes you seem considerate and also gives the impression that you don't doubt that you'll be coming in.
3. START WALKING. Don't wait for them to say yes. Walk right in. BUT—be very sensitive to someone who doesn't seem to want you to come in. In general, act like a friend; assume you'll come in. Mark demonstrates one effective technique: he keeps shaking hands while he's walking forward, which makes it hard to stop him.

SOURCE: Robin Leidner, 1993, *Fast Food, Fast Talk: Service Work and the Routinization of Everyday Life.* Berkeley: University of California Press, pp. 68, 111.

BOX 10.4   Emotional Work: Flight Attendants

Flight attendants typically work in teams of two and must work on fairly intimate terms with all others on the crew. In fact, workers commonly say the work simply cannot be done well unless they work well together. The reason for this is that the job is partly an "emotional tone" road show, and the proper tone is kept up in large part by friendly conversation, banter, and joking, as ice cubes, trays, and plastic cups are passed from aisle to aisle to the galley, down to the kitchen, and up again. Indeed, starting with the bus ride to the plane, by bantering back and forth the flight attendant does important relational work: she checks on people's moods, relaxes tension, and warms up ties so that each pair of individuals becomes a team. She also banters to keep herself in the right frame of mind. As one worker put it, "Oh, we banter a lot. It keeps you going. You last longer."

It is not that collective talk determines the mood of the workers. Rather, the reverse is true: the needed mood determines the nature of the workers' talk. To keep the collective mood stripped of any painful feelings, serious talk of death, divorce, politics, and religion is usually avoided. On the other hand, when there is time for it, mutual morale raising is common. As one said: "When one flight attendant is depressed, thinking, 'I'm ugly, what am I doing as a flight attendant?' other flight

attendants, even without quite knowing what they are doing, try to cheer her up. They straighten her collar for her, to get her up and smiling again. I've done it too, and needed it done."

Once established, team solidarity can have two effects. It can improve morale and thus improve service. But it can also become the basis for sharing grudges against the passengers or the company. Perhaps it is the second possibility that trainers meant to avoid when in Recurrent Training they offered examples of "bad" social emotion management. One teacher cautioned her students: "When you're angry with a passenger, don't head for the galley to blow off steam with another flight attendant." In the galley, the second flight attendant, instead of calming the angry worker down, may further rile her up; she may become an accomplice to the aggrieved worker. Then, as the instructor put it, "There'll be *two* of you hot to trot."

The message was, when you're angry, go to a teammate who will calm you down. Support for anger or a sense of grievance—regardless of what inspires it—is bad for service and bad for the company.

SOURCE: Arlie Russell Hochschild, 1983, *The Managed Heart: Commercialization of Human Feeling.* Berkeley: University of California Press, pp. 115–116.

---

emotional responses in interactions with customers is called **"emotional work."**

Emotional work plays an important role in the customer–worker interaction, but whether it has positive or negative effects on the workers depends in part on how much autonomy the worker has in the situation (Wharton, 1993). A flight attendant's primary duty is to ensure the safety of passengers. As Box 10.4 indicates, however, an important part of the job is the face-work required to keep passengers feeling satisfied with the airline and its meal and beverage service. Bill collectors, on the other hand, are encouraged to dominate the interaction, even at the risk of appearing nasty (Hochschild, 1983). The bill collector can freely express negative feelings to the customer, because

the bill collector's objective is to induce enough fear, shame, or guilt that the customer will pay an overdue bill. Many companies hire debt collection agencies to collect overdue debts so that their own workers do not have to spoil their customer relations with unpleasant interactions.

Because feminine gender roles have often been associated with nurturing and soothing others, emotional work and service work more generally have sometimes been labeled "women's work" (Fishman, 1978). Some customers may hassle service workers and seek to redefine the service role as a sexualized encounter (Williams, Giuffre, and Dellinger, 1999). These situations require the service worker to negotiate the boundaries of appropriate interaction with the customer, a situation that intensifies the difficulties of emotional work. (Sexual harassment was discussed in Chapter 5.)

Social Control Managers need to assure themselves that their service workers are upholding the company's standards for interaction. Managers may monitor employee behavior. For example, the store manager may observe the greeter in a discount store welcoming customers to a store, or the front-end manager of a grocery store may check to see that cashiers greet every customer and inquire whether the customers found everything that they needed. Managers may electronically eavesdrop on telephone operators' conversations. Airline reservations managers may check electronic logs of the length of calls received by their reservation clerks.

Managers also check directly with the customers to see if service has been adequate. Airlines, restaurants, and hotels distribute surveys to their customers, inquiring about the adequacy of service or particular areas that require further attention. A customer who places a telephone order for merchandise, or who answers a telephone consumer survey, may be contacted later by a supervisor checking on the courtesy and accuracy with which the clerk completed the interaction.

Managers even set up systems whereby customers may report, favorably or unfavorably, on the behavior of specific employees. In effect, customers are recruited to take on part of the monitoring task normally fulfilled by managers. Because customers also have a desired outcome for the interaction, they may welcome the opportunity to become involved in evaluating the service worker. This evaluation may take the relatively benign form of inviting customers to nominate a particularly helpful worker for "employee of the month." Sometimes there is an incentive for the customer to report employee infractions; for example, a grocery store customer may receive a free quart of milk if the cashier fails to provide a receipt. The general public is sometimes invited to send in anonymous reports on employee behavior. Trucks, buses, delivery vans, and taxis often bear a message such as "Am I driving safely? Call 1-800-RATONME." By this tactic, managers encourage even passers-by to help the company enforce its driving standards.

From management's perspective, who should control the interaction between worker and customer? The answer to this question is equivocal. On the one hand, the customer may not always be right, but the customer should always feel well cared for. On the other hand, the service worker needs to control the interaction sufficiently to achieve smooth and efficient functioning of the enterprise. Thus, a prevalent management strategy is to provide the worker with the training to control the interactions, and to give the customer adequate opportunities to complain.

The Worker's Perspective

Workers respond in a variety of ways to the pressures of interacting with the public day after day. A worker's response may even vary from day to day, depending upon the characteristics of the customers, the pace of the workday, and how well the rest of the worker's life is going. But several responses are sufficiently commonplace to deserve special commentary.

Manipulating the Interaction One response of workers is to seek the control of the interaction as completely as possible, with the objective of keeping to a work schedule, making the sale, or perhaps increasing the size of a bill and therefore the size of a tip. Keeping to a schedule is likely to be important when there are many people to be served, and there is no financial advantage to lingering longer with some customers than with others. Workers who are unsuccessful at manipulating the interaction may find themselves caught between company policy and the demands of clients (Troyer, Mueller, and Osinsky, 2000).

The ability to exercise some control over the timing of interactions is often crucial for members of the helping professions. Many patients would like to spend longer with their physician, asking questions and telling the physician more about their health than the physician feels is necessary. Workers learn to manipulate the interaction so that they maintain greater control over the interaction. A doctor may arrange to be paged after spending a few minutes with a patient or may arrange the timing of a laboratory test so that it will effectively end the interview. Doctors, counselors, and others who often hear the confidences of customers also face the "therapist's dilemma": after hearing the problems of others all day, they need to express their own emotional reactions. Because the implicit terms of the interaction do not permit them to share their emotional state with clients—this is one aspect of "face-work"—they must make other arrangements in the round of their workday to regroup themselves before helping another client. Social workers, police officers, and health professionals may find that the time necessary for paperwork, documenting the details of each interaction, provides a respite from the emotional work that is otherwise expected of them.

Manipulating the interaction may also help to increase the size of a sale. Automobile salespeople are taught to channel the conversation with potential car buyers to increase the chance of making a sale. Entertainers may sacrifice their artistic preferences to play "square" music requested by the audience, and so increase the likelihood of receiving a tip or a longer engagement. Tip-increasing tactics used by waitpersons include suggesting a round of drinks before the dinner is ordered, inviting the diners to order appetizers, salads, or a la carte items, or suggesting dessert (Butler and Snizek, 1976).

Manipulation makes the interactions less personally satisfying for both the workers and the customers. Many service workers prize their regular customers because they have the opportunity to build a personal relationship with the customer. In this case, a genuine relationship may replace the struggle to control the relationship.

"Losing It"   Occasionally a customer will be so offensive that the worker will "lose it," or fail to maintain the face-work expected in the interac-

tion. Managers may see this as a serious lapse in behavior, but workers often justify it by pointing out that the customer's behavior has violated the role expectation for a customer. Describing a customer who had tried to embarrass her, one waitress reported:

> I stood up to her regardless of whether they were customers, regardless of whether I lost my job. Nobody's going to degrade me like that because I'm a waitress. And then she started getting loud. And boisterous . . . I said, "Look," and then I put my book down. I slammed my book down. I put my pen on the table. . . . I said, "Look. If you can do a better job than me, you write your damn order down yourself and I'll bring it back to the kitchen." (Paules, 1991:153)

Workers "lose it" because they perceive the customer as indifferent or hostile. The depersonalizing of the interaction, which may also affect the customer's satisfaction, is interpreted by the worker as a specific, personal affront from the customer. For example, calling a waitperson "Girl!" or "Boy!" is seen as depersonalizing or demeaning. The terms "boy" and "girl," which imply that the worker is not yet an adult, are especially demeaning. These terms have also been used by some white people to emphasize the lower status of African-Americans or other minorities in service positions. Calling the worker by the job title ("Waitress!" "Cabbie!") may also be interpreted as depersonalizing. One waitress reported the rejoinder "Don't call me Waitress. I don't call you Customer or Eater" (Howe, 1977:104). "We are people too, and we have feelings" is a common response among service workers.

Burnout   *Burnout* refers to chronic stress in a service job, often as a result of interactions that are too frequent, too repetitive, or too upsetting. Service workers of all types can burn out, with the result that many of them leave their jobs and seek a different type of work. Burnout seems most likely to occur as a response to an inability to control the number or content of interactions.

Burnout may be a temporary response to the daily difficulties of service work, or may reflect a long-term reaction to service work. Workers who are headed for burnout may seek to reduce their commitment to the job, perhaps by working fewer hours, seeing fewer customers, or refusing extra duties (Ebaugh, 1988). Many workers will seek refuge in another part of the company, in management positions, or in a related field that they perceive as less stressful. For example, some school teachers respond to the stress of the classroom by seeking positions as administrators in the school system. Administrators still have to deal with the public, but they may have more control over their time and their schedule of interactions.

Many service jobs report high levels of turnover, partly because the work conditions are stressful to workers. The personal costs of turnover are always high for the worker. For jobs that require relatively low levels of training, turnover may be a manageable cost of doing business. Turnover among highly trained and experienced workers, however, entails substantial organization costs, especially in agencies whose purpose is to offer social or personal services. Finding ways to avoid burnout is an especially important objective in organizations that need to retain highly trained service workers.

THE FUTURE OF SERVICE WORK

Although service provision may change in some respects—for example, there may be greater appli-cation of electronic technology in some service—services will continue to provide employment to the great majority of workers in the advanced economies, and probably in the developing economies as well.

Moreover, most of us encounter service workers regularly in our everyday lives. Especially in large, urban areas, most people will have many transitory encounters with service workers providing many different types of service. Whether these encounters are satisfying or unpleasant will constitute an important part of everyday existence. The concern about the productivity of the service sector has led to greater interest in routinizing and controlling interactions, leading in some large enterprises to "people processing," an application of assembly-line techniques to *customers* instead of to products. These encounters may leave the consumers feeling that they have never received any human attention. These encounters may also prevent service workers from experiencing the satisfactions of assisting others.

In the next four chapters, we discuss specific nonmanufacturing occupations in which people work. Chapter 11 discusses the professions and related occupations. Most of these workers are employed in the high-paying service industries, and their jobs rely on the application of knowledge to serve clients. Chapter 12 discusses managers and administrators. Chapter 13 discusses clerical and sales workers. In Chapter 14 we consider marginal employment, including private household workers and personal service workers.

SUMMARY

American workers are increasingly employed in the service industries; many of them are employed in service occupations. The service industries are heterogeneous, serving both individuals and organizations with a variety of professional, business, producer, distributive, social, and personal services. The conditions of work among these industries vary widely between high-paying services and low-paying services. The process of tertiarization in some developing societies represents a growth of the low-paying service industries without a manufacturing base.

Service work differs from goods production in that the service production is time-bound. Services cannot be stockpiled, and the production of the service often occurs nearly simultaneously with the consumption of the service. Service work is often characterized by low productivity, and one response to the low productivity has been routinizing the interactions between the customer and the service worker. Although this routinization has advantages, it also makes the encounters more impersonal and may lead to dissatisfaction by the customer and to burnout among the workers.

KEY CONCEPTS

time-boundedness

sectoral transformation of the
labor force

tertiarization

professional services

business services

producer services

distributive services

social services

personal services

routinization

face-work

emotional work

burnout

QUESTIONS FOR THOUGHT

1. Think of a business or store that you patronize and that you would characterize as having good service. What are the characteristics that lead you to describe the service as "good"? Are these characteristics widespread in the service field?

2. Think of an interaction that you have had recently with a service worker or with a customer (if you work in services). Who controlled the interaction, and how? Did you play your role as the other person expected? Was emotional work required?

3. The founders of Federal Express offered a new service to the marketplace. Imagine that you are an entrepreneur. What new service could you provide to the economy? How would you classify this service—as a professional, business, producer, distributive, social, or personal service?

4. Suppose that you manage a service industry. What steps could you take to ensure that customers receive satisfactory service? What steps could you take to prevent burnout among the employees?

5. To what extent is productivity the critical variable in the sectoral transformation of the labor force? Is your answer true only for advanced economies, or does it apply to developing countries as well?

6. What factors bring about the different rates of pay in the high-paying service industries and the low-paying service industries?

MULTIMEDIA RESOURCES

Print

Elsa M. Chaney and Mary Garcia Castro, eds. 1989. *Muchachas No More: Household Workers in Latin America and the Caribbean.* Philadelphia: Temple University Press. Household service has been a common form of work for women migrants in developing countries. This series of chapters explores the conditions of household service for workers in Latin America and the Caribbean, and it charts the development of household workers' unions and other forms of organization.

Arlie Russell Hochschild. 1983. *The Managed Heart: Commercialization of Human Feeling.* Berkeley: University of California Press. This study of airline attendants and bill collectors examines the role of "emotional work" among service workers. She argues that an occupational hazard for such workers is becoming estranged from their own emotions.

Robin Leidner. 1993. *Fast Food, Fast Talk: Service Work and the Routinization of Everyday Life.* Berkeley: University of California Press. Based on fieldwork at McDonald's and Combined Insurance, this book documents how training of service workers seeks to regulate the workers' words, looks, attitudes, and demeanor.

Greta Foff Paules. 1991. *Dishing It Out: Power and Resistance among Waitresses in a New Jersey Restaurant.* Philadelphia: Temple University Press. The author was a participant observer in a New Jersey restaurant, and she documents the strategies waitresses use to negate the stereotype of servitude and to successfully maximize tips while preserving their dignity.

Websites

Newspaper Guild. *http://www.newsguild.org/* A resource for media workers affiliated with the Communication Workers of America.

"Justice for Janitors." *http://www.seiu.org/j4j/j4j2000.cfm* This website has information about efforts to improve janitors' pay and working conditions.

Flight attendants. *http://www.flightattendant-afa.org/* This website has access to statements and news releases about air rage, cabin safety, and other issues of concern to flight attendants.

RECOMMENDED FILM

Cable Guy. A dark comedy directed by Ben Stiller and starring Jim Carrey as the cable installer who eventually stalks a client. Rated PG-13.

❖❖

PART IV

Occupations and Professions

Workers who perform the same tasks belong to the same occupation. As you learned earlier, occupation refers to what a worker does. In Part IV we examine occupational groups. The chapters in this part emphasize how occupational groups vary in the characteristics of their incumbents, their tasks, and their collective power.

In examining the characteristics of the workers, we describe how many workers the occupation has, the education or skills they possess, and their ascriptive characteristics (for example, gender or ethnicity). Job characteristics include how much judgment and autonomy are associated with each job, how frequently the job is a rung on a long-term career ladder, how secure the job is, and how well it is compensated. *Collective power* refers to the control that members of an occupation exercise over the selection and training of new members, the definition of high-quality work, and compensation. Occupations vary greatly in how much control their members exercise over employers, clients, their working conditions, members of other occupations, and one another.

Just as the industrial composition of the developed countries has shifted from agriculture to manufacturing and then to services, so the occupational composition of the labor force has changed. Table A shows the occupational transformation of the United States from 1900 to the changes projected to occur

Table A   Occupational Distribution of the United States Labor Force, 1900–2008

Occupational Group[a]	1900	1930	1960	1970	1980	1998	2008
White-collar workers	**17.7%**	**29.4%**	**43.4%**	**48.3%**	**52.2%**	**57%**	**59%**
Professional, technical	4.3	6.8	11.4	14.2	16.1	18	20
Managers, administrators	5.9	7.4	10.7	10.5	11.2	11	11
Clerical workers	3.0	8.9	14.8	17.4	18.6	17	17
Sales workers	4.5	6.3	6.4	6.2	6.3	11	11
Blue-collar workers	**35.9**	**39.6**	**36.6**	**35.3**	**31.7**	**24**	**24**
Craft and kindred	10.6	12.8	13.0	12.9	12.9	11	11
Operatives	12.8	15.8	18.2	17.7	14.2	13	13
Nonfarm laborers	12.5	11.0	5.4	4.7	4.6		
Service workers	**9.0**	**9.8**	**12.2**	**12.4**	**13.3**	**16**	**16**
Farmers and farm workers[b]	**37.6**	**21.2**	**7.9**	**4.0**	**2.8**	**<3**	**<3**

[a]Data for 1993 and projected data for 2005 use new occupational codes that do not correspond exactly to the codes for previous years. Detail may not add to 100% because of rounding.

[b]Includes forestry and fishing workers after 1980.

SOURCES: 1900 and 1930 data from Philip M. Hauser, 1964, "Labor Force." In *Handbook of Modern Sociology*, edited by Robert E. L. Faris, Chicago: Rand McNally, p. 183, 1960, 1970, and 1980 data from U.S. Department of Commerce, Bureau of the Census, 1981, *Statistical Abstract of the United States*. Washington, D.C.: U.S. Government Printing Office, p. 401, 1998 data and projected data for 2008 from U.S. Bureau of Labor Statistics, *Futurework, http://www.dol.gov/dol/asp/public/futurework.*

by 2005. Farm workers are projected to decrease dramatically from more than 37% of all workers to less than 3%. The manual (blue-collar) occupations that are often associated with construction and manufacturing first increased and then decreased. The proportion of professional/technical, clerical, and service workers increased substantially, and that is expected to continue. In the following chapters, we will examine the dynamics that accompanied these shifts.

Some occupations are widely dispersed among industries, and in the following chapters we will discuss such occupations. Chapter 11 examines professionals, a group of occupations that are distinctive in worker characteristics, job characteristics, and collective power.

Professionals are well educated, tend to work in jobs that require judgment and autonomy, and exercise control over members of their occupational group and other occupational groups.

Chapter 12 examines executives, managers, and administrators. This category includes a large number of occupational specialties that are found in every industrial setting, because every industry needs at least some specialized workers who direct and oversee the production of goods and services. The restructuring of many corporations has focused attention on how many such workers are needed.

Chapter 13 discusses clerical and sales workers, who are required in many industries to manage information and to help market the product or service. Chapter 14 examines workers in marginal jobs. Some of these workers are located in declining occupational specialties such as farm work and household work. Others are located in low-paid jobs, often in small firms, or they are freelancing as temporary workers. We will discuss ways in which workers may find themselves in peripheral parts of the occupational structure.

These four chapters do not include occupations that are closely tied to specific industries, such as the manufacturing, high technology, and service industries. Such occupational groups were already considered in Chapters 8 through 10. The chapters in Part IV do discuss many occupational specialties within the broad occupational categories, however. In Appendix Table 1 you will find a listing of these specialties along with the proportion of the workers who are women or members of minority groups.

11

Professions and Professionals

[The professional complex] has already become the most important single
component in the structure of modern societies. It has displaced first the
"state," . . . and, more recently, the "capitalistic" organization of the
economy. The massive emergence of the professional complex . . . is
the crucial structural development in twentieth-century society.

(PARSONS, 1968:545)

I propose that we name the mid-twentieth century The Age of Disabling
Professions, an age when people had "problems," experts had "solutions"
and scientists measured imponderables such as "abilities" and
"needs." . . . It will be remembered as the age of schooling, when
people for one-third of their lives had their learning needs prescribed and
were trained how to accumulate further needs, and for the other two-
thirds became clients of prestigious pushers who managed their habits.

(ILLICH, 1977:11–13)

As these quotations indicate, social scientists disagree on a single view of
the professions, but they agree that the professions are important. In this
chapter we will explore the characteristics of professions and how professions
are changing. Almost everyone recognizes a few occupations as being professions. Other occupations have some characteristics of a profession, and still

others seek collectively to acquire more of the characteristics of a profession. Occupations can be ordered along a continuum from more professional to less professional, and in this chapter we will identify some of the social forces that push occupations toward one end of the continuum. In particular, we will examine the semiprofessions and the paraprofessions as groups that aspire to professional status.

HOW SOCIOLOGISTS RECOGNIZE PROFESSIONS

People often use the word *profession* when sociologists would use the terms *occupation* or *career*. (Recall that in Chapters 2 and 3 we discussed the sociological use of these terms.) Likewise, the term *professional* is used in many different ways. For example, someone may say that a job "was really professional" to indicate that it was well done or refer to "professional athletes" to distinguish them from amateurs. Sociologists, however, assign a specific meaning to *profession*. A **profession** is a high-status, knowledge-based occupation that is characterized by (1) abstract, specialized knowledge, (2) autonomy, (3) authority over clients and subordinate occupational groups, and (4) a certain degree of altruism. We will refer to these four characteristics as the hallmarks of a profession. A professional is a person who is qualified and legally entitled to pursue a specific profession (Hughes, 1965:2).

Thus, we can see how the popular usage parallels the sociological definition. A profession is a certain type of occupation. A professional is an expert, and we usually associate expertise with a job well done. Finally, a professional is somewhat like the professional athlete, who is distinguished from the amateur both by higher status and by higher pay. To understand this chapter, it is important to remember that we will always use the terms *profession* and *professional* in the sociological sense of these characteristics. University professors, for example, are professionals who devote their work careers to becoming experts in a particular area, such as sociology. Although others may read sociology books in their spare time,

the sociology professor is paid to study society and to teach students about it.

The characteristics we have just presented do not perfectly describe many occupational specialties. Most experts agree that law, medicine, and the ministry possess the characteristics of a profession. Other occupations, including military officer, scientist (Ben-David, 1984), and university professor (Freidson, 1986:15), are widely accepted as fitting the definition of a profession. The members of other occupational specialties may self-consciously seek to elevate the status of their occupation by adopting the characteristics of professionals. **Professionalization** is the process by which an occupational specialty seeks to emulate a profession by demonstrating the four hallmarks of a profession. For now, we will discuss the four hallmarks. We discuss the trend toward professionalization later in the chapter.

Identifying professions by a set of hallmarks is also called the structural-functional approach, the traits approach, or the characteristics approach. In this chapter we will refer to it as the hallmarks approach. Not all sociologists agree that this approach is the best way to understand the professions. Instead, they contend, the professions are merely the *powerful* occupations that are currently winning in the constant struggle among occupations to control preferred types of work. Therefore, the problem is not *which* occupations are recognized as professions but, rather, the *process* by which they gained their recognition. Over time, people may come to change their perceptions about which occupations are powerful (Haber, 1991). In a preindustrial society, for example, the most powerful occupations might have been hunter and shaman, whose members controlled, or were believed to control, the food supply and the spirit world. Hallmark theorists would reply that hunter and shaman were the occupations with the greatest claim to the four hallmarks. Here we elaborate the four hallmarks, and then in a later section evaluate them using different approaches.

Abstract, Specialized Knowledge

Our definition states that professions are knowledge-based occupations. But every occupation has its lore, a body of knowledge that its members master. What distinguishes the professions is the *type* of knowledge they master. All societies have both common knowledge (generally known by many) and esoteric knowledge (known by only a few). The traffic laws are common knowledge in most contemporary societies, whereas specific treatments for cancer are more esoteric knowledge, usually known principally by physicians. Not all esoteric knowledge is equally valuable, and people and societies vary in the value accorded different knowledge. For example, astrology is still considered an important body of knowledge in some countries, but in the advanced industrial countries astrology is commonly viewed as merely entertaining. In general, the esoteric knowledge commanded by professionals is considered important, even a matter of life or death, for the well-being of individuals or groups.

In preindustrial societies with a meager accumulation of knowledge, the "professionals" of a tribe or village were sages or shamans, who knew the important lore concerning health practices, weather, the spirit world, genealogies, and the history of the group. It was possible for a single person in every village to know all of these things, and that person was consulted by the entire community. Although others might have aspired to such knowledge, they were often denied it. This specialization met a simple survival need because most members of the tribe were needed to provide food. But the sage also gained an advantage in keeping the information mysterious and inaccessible. Mystification added much to the prestige of the village sage. In addition, the sage had greater discretion to choose his or her successor if everyone believed that the knowledge was important, hard to learn, and entrusted to only a few worthy followers. The selection of apprentices and the teaching of esoteric knowledge were accompanied by rituals, which underscored the importance of the sage and of the knowledge. Later on, monasteries and other formal religious institutions preserved knowledge, maintaining the link between knowledge and religion.

In an industrial society the number and variety of such experts, or professionals, has greatly expanded, and their knowledge has been secularized. As you saw in Chapters 7 and 8, increased productivity in agriculture and manufacturing has shifted the opportunity structure so that more workers can enter the knowledge-based occupations. In addition, the base of human knowledge has exploded and continues to grow exponentially. Not all of this knowledge will be considered central or important, but most of it can be considered esoteric in the sense that only a few specialists will be aware of it. The very volume of knowledge means that those who transmit knowledge must become increasingly specialized and that advanced industrial societies will require more numerous and more varied knowledge-based workers. Knowledge-based fields have fueled much of the growth of the service sector, especially that part called *professional services* (for example, educational, legal, and medical services). Some authors define the postindustrial society as one that is dominated by professional experts (Bell, 1976).

Knowledge Base The knowledge base of a profession consists of three parts. The first part is *theoretical knowledge,* which may seem rather far removed from the day-to-day activities of the professional. This knowledge is often acquired in college. For example, physicians must have a base of theoretical knowledge in biochemistry, mathematics, and physics, which they learn before medical school. Although such bodies of knowledge continue to grow, many physicians keep up with them only in limited ways. Physical and biological scientists, on the other hand, actively maintain and extend this abstract knowledge.

The second part of the knowledge base is *detailed, practical information* that can be applied in serving a client. Physicians, for example, know specific information about diseases and treatments. This part of the knowledge base also expands quickly, but the professional must stay abreast of these developments to provide the best service to clients. One advantage of specialization—for example, in cardiology or pediatric cardiology—is that the professional must keep pace with new information only in a relatively specialized area. Bar associations, medical associations, and other professional organizations frequently require their members to update their practical information through annual continuing education.

The third part of the knowledge base, *technique,* is the application of the knowledge base (Ellul, 1964). Knowing that something must be done is not enough; one must also know how to do it. A physician may know that a certain condition will respond to intravenous medication, but he or she must also be able to inject the medication into the vein. Techniques are learned in an applied or clinical portion of a professional training program. The later years of medical school include clinical rounds during which students learn techniques and skills in many fields of medicine. Technique may also be learned or perfected during an apprenticeship to a more experienced professional. Internships and residencies are opportunities for new physicians to learn techniques. The internship or residency after basic medical training is devoted to learning techniques specific to an area of specialization—for example, orthopedic surgical procedures.

Both the professional associations and the professional schools seek to expand and refine the profession's knowledge base (Hoffman, 1989). The associations may lobby for public funding of research to be conducted by the faculty of the professional schools. The results of the research—new knowledge and techniques—reach the members through professional journals. Professional associations or professional schools publish these journals, available to members through subscriptions and specialized libraries. Formal continuing education is provided through conferences, video and audio cassettes, compact disks, web-based systems, and electronic mail.

Professionals regard some specialties as "hot" or "exciting" if many discoveries are being made in them. Their intellectual excitement may draw disproportionate numbers of the new professionals to these fields. The very proliferation of knowledge encourages further division of professional knowledge into subspecialties to enable workers to keep up with the volume of knowledge. Specialization encourages a further division of labor among professionals just as it does in other occupational groups.

Professional Culture   Every profession has its characteristic jargon, behaviors, and lifestyles—all elements of the **professional culture.** (More technically, it should be thought of as a subculture within the national culture.) Professional schools convey not only knowledge but also the norms, values, and lifestyles of the profession. Older, established professionals become role models who demonstrate how to dress and how to interact with clients and peers. Some of this preparation is quite explicit, such as the requirement that students complete courses in professional ethics.

Other information is conveyed informally. Students learn from the faculty in professional schools to accord prestige to researchers in the profession, and they learn which specialties have high prestige and which do not. Many law students are initially surprised to learn that criminal law has relatively low status compared with corporate law (Heinz and Laumann, 1982), and some students change their goals accordingly. Learning the professional culture helps neophytes blend in with more experienced professionals.

Professional associations play an important role in the professional culture (Halliday, 1987;

Powell, 1989). In addition to representing the interests of the profession to outsiders, they also shape consensus within the profession about norms for practice and the social organization of the work (Halliday, Powell, and Granfors, 1993).

Autonomy

Autonomy means that professionals can rely on their own judgment in selecting the relevant knowledge or the appropriate technique for dealing with the problem at hand. Professionals justify their autonomy by their mastery of the knowledge base. Lay people often accept this autonomy because they assume that professional training is necessary to make decisions. Professional standards limit autonomy to some degree. For example, a physician can use experimental drugs or treatments only within certain well-defined limits but is free to choose among accepted therapies (Freidson, 1970).

Autonomy in decision making generates issues of accountability. The client is sometimes unable to judge whether the professional is providing the correct remedy, and the client must trust either the standards of the profession itself or make judgments based on the personal bearing of the professional—for example, the bedside manner of the physician. When knowledge is important, highly specialized, and inaccessible, its misuse is an important concern to professional and layperson alike. If a physician or a lawyer makes a mistake, the results for the client may be death or severe punishment. Yet clients cannot fully protect themselves against such mistakes, because they do not have all the requisite knowledge to judge the medical or legal services they have received. Even a client who seeks the opinion of a second professional may have little basis for judging the value of the second opinion.

The professions' response to the accountability issue is to claim that they police their own membership in the interests of protecting the public. We discuss this issue further in the section on altruism. For professionals who practice in an organizational setting, such as doctors in a hospital, bureaucratic rules attempt to establish accountability to the organization and, presumably, to the clients. These bureaucratic rules may also infringe on the autonomy of the professional, a topic we will discuss later in this chapter.

Authority

Authority, the third hallmark of a profession, means that a professional can expect compliance with his or her orders from clients and subordinate occupational groups.

Authority and Clients Authority is usually understood to be part of the relationship between a professional and a client. In the doctor–patient relationship, for example, the doctor is usually assumed to have the authority to advise the patient on the proper treatment of illnesses and to prescribe medication. The professional relationship also justifies behaviors that would not otherwise be allowed. For example, the doctor is permitted to view and manipulate the unclothed body. The patient is expected not only to comply but also to trust the professional, even revealing personal secrets (Hughes, 1965:3). A patient with a sexually transmitted disease, for example, will be asked questions about intimate relationships. The expectation that patients will reveal secrets reaches its height in the specialties of psychiatry and psychology.

As the body of professional knowledge expands, professionals seek to expand the scope of their authority over clients. By redefining many aspects of life as health related or by redefining problems (such as alcoholism) as "medical" problems, physicians claim greater authority over patients and over some other occupational groups. As physicians have become more concerned with maintaining health and preventing disease, they have increasingly offered their patients advice on diet, exercise, smoking, child rearing, family relationships, and many other topics. The doctor–patient relationship implies the doctor's authority to expect compliance with the advice.

Lawyers have defined many transactions of everyday life as containing legal pitfalls or the potential for lawsuits, and they encourage legal audits (analogous to the routine physical health examination) to prevent lawsuits or liability. Other professions seek to increase their authority by claiming dominion over additional areas of life, asserting that people need additional expertise. Members of the clergy propound moral principles that affect many aspects of life, offer counseling to members of their congregations, and explicitly advise exhibiting or avoiding some behaviors. Various professions now offer expert advice upon such issues as child rearing, care of elderly family members, and investing money.

Just because professionals assert authority does not mean that their clients will accept their authority. Clients may ignore advice on issues that they consider to be beyond the professional's domain. Patients may take a medication that has been prescribed, but refuse to stop smoking. Professionals consider "compliance," or getting clients to follow professional advice, to be a major problem.

Authority and Other Occupations Professional associations play an important role in maintaining the authority of the profession. They seek laws establishing licensing and preventing the practice of the profession without a license. Legal sanctions sometimes reinforce the profession's claims to authority and prevent other occupations from competing in their area of expertise.

Most professions interact with a variety of occupational groups, some of which are subordinate to them. In some cases the subordinate worker is an employee of the professional, such as a dental hygienist employed by a dentist. In other cases, however, the subordinate nature of some occupations is established by a larger organization. In a hospital, for example, physicians have some authority over registered nurses, therapists, dietitians, pharmacists, and other health occupations. The physician leaves orders or prescriptions that these workers handle. Registered nurses, in turn, have autonomy in certain nursing duties.

Frequently, for instance, only nurses are permitted to give particular medications. They also have some authority over other occupational groups, including licensed practical nurses, vocational nurses, nurse's aides, and orderlies.

The dominant profession often delegates what is considered its dirty work to subordinate occupations. Physicians, for example, delegate routine patient care, such as feeding, bathing, and bedpan handling, to the nursing staff. Attorneys delegate routine legal work, such as the completion of standard forms or letters, to paralegals or to legal secretaries. The rationale for this delegation is that it frees the professionals for the more highly skilled work for which they alone were trained. The delegation does entail certain risks for the professional. The professional directs the work, at least in theory, and takes responsibility for it (Cassell, 1991). If a paralegal makes a mistake, it is ultimately the lawyer who is responsible to the client.

Where these relationships of authority and delegation are accepted, there may be relatively little conflict among the groups. One study that compared two hospitals found less friction among the various occupations than might have been expected, although there was friction *within* some groups (such as nurses) based on their differing conceptions of what the nurse's relationship to the doctor should be (Guy, 1985). Nurses were more likely to disagree among themselves over the proper way to respond to a doctor whose orders were vague than they were to disagree openly with the physicians.

When two or more professions are expanding their spheres of authority into new areas, there are often struggles for domination. If there is no clearly dominant professional group, a struggle may ensue for the authority to claim expertise. In the United States there was a historic battle in which medical doctors sought to establish their profession and prevent the recognition of other health practitioners. The medical doctors fought in state legislatures to prevent the licensure and acceptance of osteopaths, chiropractors, herbalists, and lay midwives (Gevitz, 1982). Lawyers have struggled to maintain their professional authority against what they see as infringements of credit counselors (in bankruptcy proceedings), real estate agents (in property transactions), and nonprofit counseling services (in uncontested divorces).

When professionals declare that more areas of human life need their expertise, new "turf" is created that is not squarely within the province of the existing professions. Nutritionists, dietitians, physicians, and university scientists all offer advice on nutrition. Although charges of "charlatan" and "quack" are sometimes heard among the disputants, there is as yet no clear public acceptance of whose authority should prevail. For many clients the advice given by a nutritionist does not yet carry the weight of advice given by a physician. Other clients, however, may doubt the physician's competence in nutrition. These are areas in which the struggle among occupations for authority can be observed (Zhou, 1993).

Altruism

Altruism means concern for others. No one doubts that professionals seek an income from their practice, but the hallmark of altruism implies that they officially see themselves as having additional objectives. Most professions have codes of ethics that express the ideal relationship among the professional, the client, and the community. Altruism implies that the professional will incur some self-sacrifice to help the client. Altruism also involves the profession's duty to use its knowledge for the public good. On the one hand, because the knowledge is important and is monopolized, the profession has the duty and responsibility to preserve, enhance, and transmit it and to use it in the public interest. On the other hand, the confidential knowledge gained about an individual client must *not* become public; to violate a client's confidentiality is a breach of ethics. The courts uphold some professional–client confidences as privileged; that is, they do not have to be revealed in court.

The first aspect of the profession's altruism is its self-policing. The expression "only professionals can judge professionals" refers to the inability of those without specialized knowledge to evaluate professional competence. The self-regulation may begin before admission to professional school, when there may be efforts to determine whether the prospective student is "of good character." Further character investigations may take place before formal admission to the profession. Hospital review boards, professional grievance procedures, peer review panels, ethics panels, and disciplinary committees of professional associations police the working professionals. One aim of these procedures is to reassure members of the public that the profession is truly acting in their interest. Many people, however, remain cynical about the effectiveness of such procedures, believing that professionals would rather cover up the incompetence of a fellow worker than admit that one of their number had made a serious error.

The second aspect of the profession's altruism is its advocacy of community service, called *pro bono publico* work among lawyers. This is professional work that is volunteered or performed for a lower fee than is usually charged. Local professional associations often arrange for volunteer services to be provided by their members. A local legal society may provide a legal aid clinic, or a local medical society may offer free health screenings or vaccinations to low-income people. Professional associations may also sponsor hot lines, tape-recorded educational messages, and informational materials with professional advice. The association encourages individual professionals to donate their services to the poor or to nonprofit organizations.

Professionals engage in public service in other ways as well. The professional culture inculcates a sense of *noblesse oblige,* an expectation that professional elites will be visible contributors to community life. Part of the professional lifestyle includes providing leadership to civic and voluntary endeavors. Thus, physicians often spearhead drives to raise money for new facilities at the local hospital, and attorneys often promote local good-government movements. Clergy members provide significant leadership in public affairs. An example is the leadership of the Reverend Martin Luther King, Jr., in the civil rights movement. Although such activities do not require professional training, the prestige of the professional is believed to add to the success of the efforts.

Such community activities may not be merely altruistic, for they also develop important contacts and referrals. Many professionals become important members of the local business community or candidates for elective office. The prestige they receive from their community service is also a symbolic reward in addition to the monetary rewards of professional work (Barber, 1965:19). Such considerations lead some observers to identify these activities as more self-serving than altruistic.

EVALUATING THE FOUR HALLMARKS

The four hallmarks of a profession are a useful starting point for understanding the professions, but they are incomplete. Taken together, they compose an idealized model that imperfectly describes reality. The type of analysis on which they are based, often called structural-functional analysis, has been criticized by sociologists for ignoring much of the reality of professionals' lives. Many critics use the conflict approach to provide a different perspective on the profession. In this section we will review some of their criticisms.

From the perspective of the conflict approach, the four hallmarks are incomplete explanations of how the professions retain a position of power. The knowledge base, which is monopolized by the professional, is one key to maintaining professional power. Their associations, their lobbying efforts, and their use of state laws to enforce their privileges serve to maintain their power while denying others a share in it. Maintaining professional power also implies maintaining autonomy and authority. From the power perspective, the

notion of altruism is mostly a smoke screen to prevent the public from investigating the internal workings of the profession.

How Powerful Are the Professions?

The professions are powerful occupational groups. One criticism of the structural-functional approach has been its neglect of professional power (Geison, 1983:46). Professions are attractive occupations because their members, unlike production workers, clerical workers, and many workers in other occupational specialties, wield real power over their own work lives. They decide what they will do and how, direct other workers in their tasks, and act to prevent other occupations from doing the same sort of work (Halliday and Karpik, 1997).

Monopolizing Knowledge

The professional knowledge base and the esoteric lore of the village shaman both represent a monopoly over knowledge (Larson, 1977). Just as the village sage monopolized the possession and transmission of esoteric knowledge, only the professionals command the knowledge of their field. They earn their living by applying their professional knowledge in service to paying clients. Contemporary professionals monopolize knowledge by erecting three basic barriers to laypeople: professional schools, licensure, and mystification.

Professional Schools Admission to the professional schools is intensely competitive. Just as the village sage had to find and train a successor, so the profession must select and train its new members. In advanced industrial societies this training is nearly always accomplished in formally organized professional schools, such as medical schools, law schools, and divinity schools, which are often part of larger universities. There are usually many more applicants than available spaces. Leaders of professional schools reason that by limiting the number of students, the schools will produce more competently trained graduates. Aside from maintaining quality controls over the students, the professions also fear producing a glut of graduates who will compete against one another for clients. The restriction of access helps maintain a higher price for professional services and higher incomes for the professionals.

The professional association is intensely interested in the organization, curriculum, and graduates of the professional schools. The American Bar Association, for example, accredits law schools. In some states, law graduates are eligible to become lawyers only if they received a law degree from an ABA-accredited school. By accrediting, the professional association seeks to ensure that the school teaches a core body of knowledge and skills and that only qualified faculty conduct the training. Professional associations are often reluctant to accredit new schools or to permit schools to enlarge their student bodies, usually citing concerns over maintaining the quality of graduates. The professional association may be compared to a gatekeeper, deciding whom to admit. Other occupations also have gatekeepers; for example, the unions that conduct the apprenticeship programs in the skilled trades have power over their occupational group because they control access.

Licensing Even if laypeople were to acquire professional knowledge, they would not be able to put it at the service of clients. Nearly all professions require licensing by the state, and an applicant can qualify only with educational credentials and a passing grade on a licensing examination. There may be additional requirements for periodically updating the license or for completing further training. Most professions are protected by laws that make it illegal to practice without a state license (Zhou, 1993). Other professionals such as scientists, although unlicensed, are essentially unemployable without formal schooling credentials, because employers will not hire them. Either through licensing or through convincing employers to hire only trained graduates, the profession can act collectively to restrict access to its knowledge base.

To be certified in a professional specialty within the profession requires further examinations and licensing restrictions. Specialty certification is most widespread in medicine, though some states certify legal specialties. Specialists maintain their own professional organizations (such as the American College of Obstetricians and Gynecologists) as well as the general professional organization (such as the American Medical Association).

Mystification The mystification of professional knowledge comes about because the learning is so specialized and often couched in such difficult jargon that even after being exposed to the information, a layperson might not understand it. Despite their efforts to communicate, the professionals may be so absorbed in their distinctive lingo that their advice is incomprehensible. Laypeople then perceive the knowledge as esoteric. Just as the shaman enjoyed increased prestige from possessing esoteric knowledge, a client today may leave a professional's office thinking, "That person must be smart, because I didn't understand a word of it."

Mystification reinforces the profession's claim to knowledge and discourages efforts by laypeople to take upon themselves the tasks that professionals wish to monopolize. In theory these factors that restrict the supply of new professionals should protect the jobs of the existing professionals and help them extract higher fees for their services. Even in the professions, however, employment is not guaranteed. As a rule, the unemployment rate for professional workers has been low, under 2% in 2000 (see the most current data at *http://www.bls.gov*). Despite the restrictions on supply, however, in some fields there has been an oversupply of professionals.

Power within the Professions

The structural-functional view of the professions is principally concerned with distinguishing the professions from other occupations. It neglects inequality within the professions, except in its observation that some specialties enjoy greater prestige than others. The conflict approach pays more attention to the differential distribution of power within the professions. The most powerful professionals are likely to be those with the greatest control over the sources of knowledge (such as the deans of the professional schools), those who possess the most important and specialized knowledge (Abbott, 1988; Halpern, 1992), those with the most prestigious clients (Heinz and Laumann, 1982), those who have the most challenging and complex intellectual tasks (Cullen, 1985), and those with the greatest organizational or personal resources to maximize their personal autonomy and authority.

Changes in the Professions

The structural-functional approach best describes the so-called free professions, those in which individuals open their own practice and charge their clients fees for services rendered. This view of professional practice has been criticized because it leads to a romanticized vision of the professional–client relationship and ignores the fact that today there are relatively few independent professionals. Many professionals today are also employees. The recent movement of physicians into health maintenance organizations points to the new prominence of professional firms and group practices, even within the traditional professions.

An important business advantage of group rather than independent practice is that overhead expenses such as rent, furniture, equipment, support staff, insurance, and accounting can be pooled. Fees for services are also pooled, and the professionals are usually compensated with salaries, although partners may also receive a share of profits. Only licensed professionals can be shareholders (Freidson, 1986:126–129). Some advocates argue that group situations offer better service because the client may be served by whomever of the professionals is available ("on call"). Not all clients prefer such arrangements, for many still favor the personal attention of one professional who is familiar with their case. Moreover, at least some professionals have ethical questions about the advisability of rotating clients among professionals (Starr, 1982:198–232).

The group practice retains control by the profession, in the sense that someone with the requisite professional knowledge makes decisions. The members are not necessarily equal, however. There may be a managing partner or director, and some professionals will be full members or partners with financial equity in the firm, whereas others will be provisional, salaried employees who may not ever achieve partnership. Other arrangements, however, including the so-called "McDoctor Clinics," represent the reorganization of medical practice as franchises or conglomerate organizations. "McDoctor Clinics" are small clinics that offer rapid medical attention to minor medical situations, just as fast-food outlets offer quick meals to hungry customers (see Box 11.1).

A related change has been the increased employment of professionals in other organizations, such as universities, research organizations, corporations, and government agencies. Most big corporations, for example, employ house counsel, lawyers who work as staff members, even though outside law firms may also be engaged for specific tasks. This arrangement subjects the professional to bureaucratic or organizational authority that may supersede their professional authority. Authority is eroded further when the only "client" is also the employer. The house lawyers must provide the best legal advice possible at the risk of losing their employment, but then the top managers of a corporation are free to ignore their advice. A law firm might suggest to a client who consistently refuses its advice that it is time to seek new lawyers; the house counsel cannot easily make such a suggestion to the employing corporation.

An additional major change has been the entry of third parties, especially the government, into the professional–client relationship. This role is most evident when a third party pays for the professional services, as when medical care is paid for by an insurance company or by the government through Medicaid or Medicare. Although the professional is still engaged by the client, the economic relationship has changed because the link between receiving and paying for services has been severed. The professional and the client may both be affected by the third party's judgment about whether the

BOX 11.1   The Rise of Corporate Medicine

One important influence on the professions is the organizational climate in which they work. In this selection, sociologists George Ritzer and David Walczak describe the ways in which bureaucratic initiatives are changing the practice of medicine.

[A] significant development is the rapid growth of Health Maintenance Organizations (HMOs). While these organizations take a variety of forms, all of them are business organizations in which the objective is to work within budgetary constraints, or in some cases to turn a profit. As business organizations, HMOs are under tremendous pressure to rationalize operations in order to minimize costs and maximize income. Physicians, and their norms of autonomy, authority and altruism, are affected by these pressures whether they control these HMOs or are employed in them.

Also significant is the "conglomeration" of medicine. Whether it be through expanding product lines or product-line diversification, these large medi-giants aim to integrate healthcare facilities, medical services, and insurance functions on a regional or national basis. Large corporations are not only coming to own for-profit hospitals, McDoctors, and HMOs, but they are also controlling health insurance firms, buying out hospital supply companies, and generally moving in the direction of controlling as many components of the "medical-industrial complex" as possible (e.g., nursing homes, laboratories, emergency-room services, home care, etc.). To compete, large, non-profit organizations like Blue Cross–Blue Shield are expanding in a similar manner. In either case, ever larger conglomerates are increasingly impelled to rationalize their operations. This means that they are going to seek ways of becoming more efficient, predictable, and calculable, of replacing more and more workers with nonhuman technologies, and of exerting increasing control over employees (including physicians).

Those physicians who resist moving into various large-scale capitalistic medical organizations are going to have to be even more entrepreneurial than they have been in the past. Physicians have, of course, always combined professionalization and entrepreneurialism, but in order to compete in today's more market-oriented medical system, physicians are going to be forced to emphasize business practices more with the result that professionalization may suffer. Further, physicians in private practice are coming to emphasize entrepreneurialism more and professionalism less by opening an array of free standing or noninstitutional health care centers including pathology labs, health promotion centers, obesity and substance abuse clinics, chains of local emergency rooms, proprietary hospitals, and nursing homes.

While good empirical data do not yet exist on the impact of entrepreneurialism on patient care decisions, there is suggestive evidence that "is adequate to confirm the common sense conclusion that investments and economic arrangements that reward physicians financially for making certain patient care decisions (e.g., ordering lab tests) will bias physicians in favor of making such decisions." In any case, it is clearly harder for primarily entrepreneurial physicians to continue to claim to adhere to substantive rationality, especially the value of altruism. Furthermore, the demands of the capitalistic marketplace will exert greater control over an entrepreneurial physician.

One result of this control by the market is the expansion of advertising by the various agencies involved in the medical business, including even physicians in private practice. While there are still strong negative sentiments among some physicians toward advertising, there is evidence to suggest that advertising is gaining acceptance in the medical profession. Here is one of the clearest examples of the erosion of the distinction between professional and capitalist, between substantive and formal rationality. Again, with physicians advertising their services on television and in the newspapers, it will be harder for them to continue to put forth an altruistic image. Patients are more likely to question the authority of physicians selling their services side-by-side with used car salespeople. [Notes and references omitted.]

SOURCE: Excerpt from George A. Ritzer and David Walczak, 1988, "Rationalization and the Deprofessionalization of Physicians." *Social Forces* 67,1 (September):10–11. Copyright © The University of North Carolina Press. Used with permission of the publisher and the authors.

services are needed or appropriate and whether the professional is charging too much for them.

The reorganization of professional practice creates new hierarchies within the professions, with some professionals exercising managerial control and commanding higher salaries than other professionals who are salaried employees (Bok, 1993). These newer forms of professional practice may, however, improve the availability and affordability of professional services.

Are the Professions Meritocracies?

The hallmark view implies that the professions are *meritocracies*—that is, groups in which rewards are based on achievement. In the profession the achievement is presumably mastering the core knowledge, contributing to new knowledge or techniques, or practicing the profession well. Because these tasks are difficult, high monetary and symbolic rewards are used to attract the most capable individuals into the profession. The achievements are presumably independent of such ascribed characteristics as race, gender, and social class, and the selection criteria of the professional schools presumably identify actual abilities and potential accomplishment.

In fact most professions have been dominated by white men, often with origins in the upper middle class. As Figure 11.1 indicates, in elite professional specialties the proportion of women and minority group members remains small. This lack of balance has raised concerns that the professions are not really recruiting the "best and brightest" but instead relying on occupational inheritance. One study of lawyers in the United States showed that the major determinant of whether one became a lawyer was whether one's father had been a lawyer (Carlin, 1966).

Since the 1960s and the changes generated by the civil rights movement, minority group members and women have sought better representation in the professions. The exclusion of women from the professions was not only customary but also had legal sanction. At one time the courts upheld the rights of the state legislatures and of the bar associations to exclude women from spe-

cific professions. In 2000, women accounted for nearly 30% of all lawyers and judges and 28% of physicians, though the median weekly earnings of women lawyers were 73% those of male lawyers and the median weekly earnings of women physicians were 58% those of male physicians (based on full-time workers only; data from *http://www.bls.gov*). Women are now better represented in the professional schools; nearly half of medical, law, and dental students are women.

The number of minority professionals has also increased in recent years; in 2000 there were 1.8 million African-American professionals, and over 900,000 Hispanic professionals. But the *proportion* of professionals who are minorities has changed little in recent years. One reason for the underrepresentation of racial and ethnic minorities in the professions is the cost of professional school. African-Americans, Hispanics, and other minorities are often reluctant or unable to incur heavy debts to finance professional schooling.

Once in the professions, moreover, women and minorities tend to have lower incomes and to occupy the less prestigious specialties. Many female and minority-group professionals do not specialize but maintain general practices. One reason for this may be a preference to serve in a minority community whose members have fewer resources for professional services (Weinfeld, 1999).

Women who specialize within medicine are often steered to such "appropriate" specialties as pediatrics, gynecology, family practice, and psychiatry (Lorber, 1984). Specialties such as dermatology or radiology, which rarely involve emergency or weekend work, are also thought to be more compatible with marriage and a family. Within law, female and minority professionals are overrepresented in the specialties of family law and real estate law (Epstein, 1970:160–161). They are also more likely to work with individual clients, although work for corporations is more lucrative and prestigious. Women and minorities entering professions may find themselves isolated and even harassed by colleagues (Booze, 2000; Davey and Davidson, 2000).

A 1986 U.S. Supreme Court decision made it illegal for the partners in a law firm to discriminate

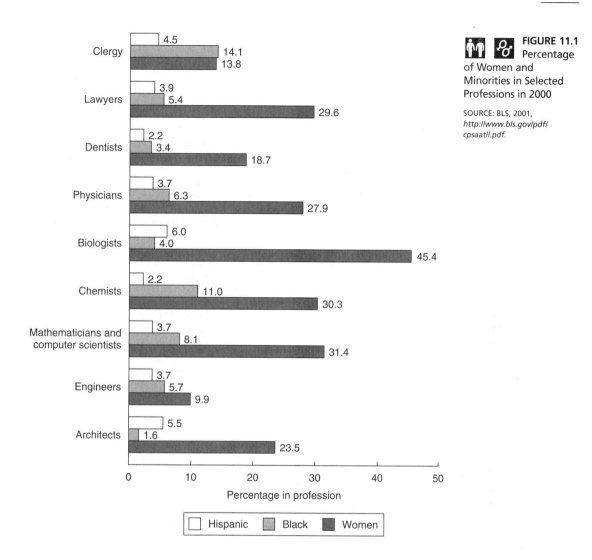

FIGURE 11.1 Percentage of Women and Minorities in Selected Professions in 2000

SOURCE: BLS, 2001, *http://www.bls.gov/pdf/cpsaatll.pdf.*

against a qualified woman lawyer by preventing her from becoming a partner in the firm. (A new lawyer is usually an associate of a firm for six or more years before being considered for a partnership.) Even so, many women and minority professionals experience discrimination. Sometimes this takes the form of direct verbal aggression, as when women physicians are referred to as "hen doctors" or when male law school professors use hypothetical examples that depict women as stupid (Benokraitis and Feagin, 1986). Women and minorities are somewhat more likely to have their authority challenged, both by

clients and by co-workers from subordinate occupations (MacCorquodale and Jensen, 1993). Within the profession they may be less likely to receive referrals of potential clients from other established professionals (Epstein, 1982:65). Perhaps for these reasons many female and minority professionals do not maintain traditional solo practices nor even group practices, but seek employment in government agencies and other large, bureaucratic workplaces. Box 11.2 discusses some issues that arise for professionals in a multicultural information technology workplace in Great Britain.

BOX 11.2   How Racism Persists in Professional Workplaces: A Case of Information Technology in Britain

Journalist Roisin Woolnough interviewed nonwhite workers in Britain's information technology (IT) industry, and found that they experienced racial prejudice from clients who expected to work with white consultants. In addition, minority workers were sometimes excluded from important business-related networking.

The IT industry is perceived as a particularly open industry where race is not an issue. Look around most IT departments and they are invariably made up of a diverse workforce. But take a closer look, and it often becomes apparent that white IT professionals are the ones who are making it to executive positions of authority.

Rene Carayol is a man who's made it. Formerly the managing director of IPC Electric, the 42-year-old is now chief executive of an e-business consultancy. Yet he says he is still shocked by the amount of racial prejudice that he comes across and how it affects his career. "When I go to new offices with my team, receptionists instantly turn to a member of my team to see who is the boss," he says.

"At IPC Electric, many of the suppliers didn't expect to be working for a black IT director and they had to get used to working with me. It took some time."

Carayol thinks it took him longer to get where he has in the IT industry than it should have done, because of the colour of his skin. "I found that I stayed a junior for what appeared to be a little too long. It is a persistent complaint that I hear from ethnic minority people. It's taking minorities longer to reach top positions than it should."

Sarabjit Ubhey, head of operational control at Bupa, thinks institutional racism has to be at the root of it. "People expect racism not to be an issue in IT because it's a new industry," she says. "But the so-called glass ceiling seems to be there as there are so few senior non-white IT people. The numbers don't match up and that shows there is an issue.

"The fundamental problem is awareness. For example, my team has a very diverse make-up. There are a couple of Muslims and people from other parts of the world. As a manager, if I am organising an event I have to make sure I don't arrange things that will exclude some of them. For example, if I invite people to the pub I'm sending out the wrong message because I could be excluding the two who don't drink. So, I make sure we go out to eat instead."

A lot of networking goes on at work social events, and if someone is not party to that interaction, it could mean they are missing out on opportunities.

Paul Riddell, the Commission for Racial Equality's (CRE) head of legal strategy and private sector in London and the South, says one of the most common complaints from people who approach the CRE is that they feel excluded from these networks and that this has hindered their career progression.

Riddell agrees it is the responsibility of managers to make sure people are not being excluded. "They shouldn't use the informal stuff as a forum for career advancement," he says. "Instead, they need to do performance assessments. The ideal is career appraisals every six months with a yearly in-depth one."

Some companies have support groups for ethnic minority groups. While these can be a good forum, they can also have the negative effect of compounding a "them and us" situation.

Some companies have also appointed diversity managers. But Carayol thinks companies should not need to employ a diversity manager: "Diversity should be on every manager's list."

The more role models that non-white people have, Carayol says, the easier it becomes for them to progress up the career ladder. The good news is that he thinks it has already begun.

"Get a couple of people through and it opens the floodgates. The race issue is in the same place as women were 10 years ago. It is taking longer than I expected and it is more painful, but it has started to change."

SOURCE: Roisin Woolnough, November 4, 2000, "Racism Reinforces the Glass Ceiling," *The Guardian* (London), p. 301. Used by permission.

CHANGING DEGREES OF PROFESSIONALIZATION

Regardless of the sociological approach taken, virtually everyone agrees that professionals play a key role in society. Because of their monopoly of knowledge, the desirability of their knowledge, and the barriers to entry into their professions, many professionals have been able to earn high incomes and to retain substantial control over their daily work lives. They also are able to delegate some unpleasant parts of their jobs to others (Stevens et al., 2000). Many workers would like to have these generally pleasant work conditions, and many occupations have sought to develop similar work conditions by emulating the professions. As noted earlier, this process is called professionalization.

Professionalization

Professionalization can be understood as the effort by an occupational group to raise its collective standing by taking on the characteristics of a profession. Professionalization needs to be distinguished from **professionalism,** which is the competence and effectiveness of workers in their job performance. Professionalization often occurs in occupations with considerable heterogeneity in the training and background of their members (Barber, 1965:23). It often takes the form of seeking to adopt the four hallmarks of the professions to a greater or lesser extent, and it is usually initiated and maintained by the leaders of the occupation.

Steps in Professionalizing The first step in professionalizing is often forming an organization or strengthening an existing one. The professional association tries to convince the general public, the state legislature, and perhaps other professions that its claims to professional status are legitimate. Some occupational groups are divided over the best way to achieve greater status, and they may have several competing organizations,

not all of which adopt professionalization as a strategy. In the efforts to improve the status of elementary and secondary teachers, for example, some organizations are oriented toward a union model, and others explicitly follow a professionalization strategy.

A second, very important step is standardizing the body of knowledge that the members of the occupation should have. This step requires developing courses of training, and most professionalizing groups seek to locate the training in universities. One step toward the professionalization of nursing was to develop university schools of nursing to supplement the traditional training provided in hospital nursing schools. Although the university-trained nurses would also undergo clinical training in hospitals and take the examination to become registered nurses, they would have a bachelor's degree in nursing. A professionalizing occupation usually develops its own research program, journals, and continuing education. Registered nursing is an occupation that has done all of these things.

Developing the body of knowledge is not sufficient (Torstendahl and Burrage, 1990). The professionalizing group must convince the public that the knowledge is important and that only those graduates who complete the certification process should be assumed to possess the knowledge. The association must develop a certification process or specify a credential required to be in good standing. This credential might be a degree from an accredited school, passage of an examination, or a state-issued license (Freidson, 1986:63–88). Ideally, the professionalizing association would like to convince a legislative body to mandate that workers in the occupation be licensed or certified.

A field that is currently seeking some type of licensure is marriage counseling; in many states marriage counselors need not have any training nor any certification. Certification laws would protect those already in the field from further competition. In addition, certification would protect members of the public from unqualified persons whose marriage counseling might be of poor quality or even be emotionally damaging. In this particular example certification would probably encourage heightened competition for "turf" among social workers, clinical psychologists, clinical sociologists, clergy members, and others who offer marriage counseling.

Most occupational groups must negotiate their status with the public and with legal authorities for many years. Even more formidable opposition, however, is likely to come from the established professions, especially if the "newcomer" occupation potentially threatens the territory of a well-established profession. Established professional associations often oppose efforts by new professions to achieve a measure of autonomy or authority (Smith, 1993). For example, dentists have challenged the right of dental hygienists to open their own offices. Medical societies have prevented the granting of hospital privileges to chiropractors and have sometimes argued that the work of nurse-midwives constitutes practicing medicine without a license.

Some new professional specialties have been allowed a toehold of recognition only because they have agreed from the beginning that their members will be supervised by a licensed member of an existing profession. For example, nurse practitioners usually work under the direct supervision of a physician, although they have advanced training and are able to take on some tasks once reserved for physicians. We will discuss such workers further in the section on paraprofessions.

Professionalizing occupations often adopt codes of ethics and internal disciplinary procedures, but the codes may be vague, general, and hard to apply in concrete cases, and the enforcement mechanisms usually work poorly. These codes and procedures often seem to be part of the apparatus of convincing others that misuse of the knowledge base could endanger the public. Professionalizing associations may encourage their members to be publicly visible and to donate their time and talent, both because that is something that many professionals do and because it again brings their occupational group before the public eye.

Most professionalizing groups achieve autonomy and authority slowly, if at all (Forsyth and Danisiewicz, 1985). Sometimes they succeed because of a lack of personnel in the established professions. For example, osteopathic doctors (also called osteopaths) have achieved standing in many small, rural communities because they provide medical services where medical doctors are unavailable (Gevitz, 1982). In rural communities pharmacists may have higher standing than do their counterparts in large cities, because people must rely on them for advice and not just for filling the prescriptions of a medical doctor.

Obstacles to Professionalizing Perhaps the most significant obstacle to professionalization is the opposition of existing professions. The established professions not only monopolize knowledge but are also gatekeepers for other functions. Physicians, for example, control access to prescription drugs, hospitals, and key government health agencies. These gatekeeper functions may frustrate the efforts of other would-be health professions to professionalize.

A second obstacle is the fact that most would-be professionals do not practice independently but work for large organizations. Teachers, for example, are employed by school boards. The claims of such workers to professional authority clash with the bureaucratic authority exercised by administrators. The employing organization is especially likely to resist claims of professional expertise that disrupt its operating procedures. For example, although teachers contend that their expertise should influence curriculum revision and textbook adoption, these functions are often performed by state and local educational bodies that give the teachers' opinions little weight.

A third obstacle to professionalization is public skepticism. At a time when many occupational groups argue that the public needs more experts to provide advice on more subjects, some consumers have come to resent the professions. Claims to expert knowledge, autonomy, and authority are likely to be taken with a grain of salt by both the public and elected officials. This skepticism is one element of a countermovement called "deprofessionalization."

Deprofessionalization

Deprofessionalization, sometimes called post-professionalism, is the process of weakening or eliminating the professional characteristics of an occupational group (Krtizer, 1999). It is rare for an occupational group to deprofessionalize itself, although current efforts to unionize such professionals as veterinarians or dentists are sometimes interpreted in that light. Rather, deprofessionalization usually originates with the general public, the government, or organizational employers (Ritzer and Walczak, 1988).

Demystifying and Empowering One form of deprofessionalization is demystifying the esoteric knowledge of the profession so that it is more widely shared among the general public. This process could also be described as *empowering the consumer,* because consumer knowledge redresses the imbalance in the professional–client relationship. Publicized cases of malpractice may induce consumers to learn more about professional knowledge so that they will make better choices in seeking professional advice. Sometimes professionalizing occupations that are competing for the same clients will seek to inform the public about their area of expertise, bolstering their own public-service claims.

Noncompeting occupations may also be responsible for demystifying knowledge. For example, journalists and broadcasters now disseminate medical and health information through specialized websites, health newsletters, general magazines, newspapers, and radio and television programs. Even without consulting physicians or other health professionals, consumers can become more informed about improving their health through lifestyle changes and preventive medicine. Marketing developments and changes in government regulation may also help demystify knowledge. Home medical screening kits, the reclassification of some prescription drugs for over-the-counter sales, and new courses in home health care and disease prevention have also empowered clients to take more responsibility for their own health. Do-it-yourself kits for wills, divorces, and bankruptcies may serve the same function in deprofessionalizing some aspects of the law.

The response of the professional associations to disseminating information has been ambivalent. Although most professional associations encourage clients to learn more, they also warn against the possibility of misusing information or being misled by the do-it-yourself approach. Physicians and lawyers alike warn about the dangers of self-diagnosis. Doctors worry that home births, unauthorized cancer or AIDS treatments, and similar experiments outside the sphere of professional medicine may increase the risks to patients. Attorneys worry that uninformed clients actually worsen their legal situations because they do not understand the potential legal ramifications of such actions as writing their own wills or representing themselves in court.

One response by the professional associations has been to undertake their own programs to demystify their knowledge and to empower the client. Medical associations often sponsor hotlines with information and referrals, and many hospitals offer courses on medical self-help. Physicians have begun some of the health-related websites such as *http://www.drkoop.com* and *http://www.healthcentral.com.* Legal associations offer "ask-a-lawyer" services that feature brief, often free, consultations, and they provide publications that describe the circumstances under which a consumer needs to consult a lawyer. But the professional associations remain vigilant to see that other occupational groups are not taking advantage of consumer empowerment in ways that might constitute practicing law or medicine without a license.

Regulation A second form of deprofessionalization is regulation, which attacks both the autonomy and the authority of a professional. Regulation does not refer here to the self-regulation of the professional association but rather to rules imposed by government agencies, insurance companies, or other third parties.

Government agencies indirectly regulate physicians through such measures as approving drugs, regulating hospitals, and placing some restrictions on medical schools that receive tax dollars. Government regulations also mandate additional self-regulation by physicians on such issues as the length and appropriateness of hospital stays. Medicare and HMOs reduce fees by refusing to pay charges that they consider excessive. Many HMOs further restrict their members to physicians who have agreed in advance to accept lower fees. The courts, too, have been active in changing the professions. The U.S. Supreme Court ruled in 1978 against professional associations' bans on advertising and price competition among their members.

Private insurance companies may require second opinions before surgery or prior approval of hospital stays. Many insurance carriers will not reimburse patients for treatments that are experimental or considered inappropriate. Many proposals for health-care reform advocate more extensive use of such regulation.

Managerial Control A third form of deprofessionalization comes from managerial control, which constitutes a potential threat to the autonomy and the authority of the professional. Managerial control may compromise the right of the professional to be judged by others in the professional community. In bureaucratic organizations the professional employee has a supervisor who is likely not to be a professional and not to have access to the same body of knowledge. In making decisions, the supervisor is likely to consider many organizational factors and not just the recommendations based on professional knowledge. Far from prescribing what a client should do, the professional in the bureaucratic organization is reduced to making suggestions or offering recommendations.

In addition, the employing organization will have its own set of rules and procedures that may occasionally conflict with the ethics or sound practice of the profession. In Chapter 3 you learned about the problems of role conflict between work and home. Many professionals may find themselves in a conflict between two work roles if they are expected to behave in one way as a professional and in a different way as an employee. A potential case of role conflict faces a company physician. The company wishes to minimize time lost to sickness, to avoid worker compensation claims, and certainly to avoid liability for permanent damage to employees' health. The physician may be pressured to determine that a worker's symptoms might be caused by nonjob factors, such as lifestyle. A physician may feel pressure, for example, to find that lung cancer was caused by smoking, not by the worker's involvement in the production of a carcinogenic chemical.

Box 11.3 recounts the story of two European scientists who deceived their superiors about their research for fear that the superiors would order them to do different work. Scientists usually expect to be able to set their own agendas for research. A common source of role conflict for scientists employed in industry is that their professional judgments of what issues are important to investigate may not be the issues the company wishes to have studied. In this case the company, which was IBM, ended up quite pleased with their employees' work, for they won the Nobel Prize!

To resolve role conflict, many large organizations assign some managerial functions to experienced professional scientists (Shenhav, 1988). For the individual such assignments may constitute another aspect of deprofessionalization, because they gradually draw him or her away from the professional community, with its norms and values, and into a different occupational group with different norms and values. Professionals in large organizations may find their loyalties shift from their profession to the company and, perhaps, to trade associations with which their company is affiliated. In Chapter 12 we will discuss why the concerns of managers are likely to be different

BOX 11.3   IBM Scientists: A Prize for Insubordination

In 1987 the Nobel Prize for physics was awarded to two IBM physicists for their research into superconductors. Corporate laboratories have been the scene of many important scientific discoveries, and so it is not surprising that J. Georg Bednorz and K. Alex Mueller were working for a large corporation. What is surprising is that they conducted the work "in spare moments" on the job after "telling a supervisor a half-truth."

Rueschlikon, Switzerland—Four years ago, two scientists at International Business Machines Corp. began work on an idea so big that they were loath to share the details even with IBM.

The researchers, at IBM's Swiss laboratory here, wanted to solve one of the toughest problems in physics: finding a cheap, simple substance to conduct electricity without resistance. But they were searching where experts least expected this new "superconductor" to be found—in a ceramic that normally conducts electricity poorly.

Their idea made as little sense as making a picture window out of solid steel. "We were sure anybody would say, `These guys are crazy!' " recalls J. Georg Bednorz, one of the scientists. Indeed, when his partner, K. Alex Mueller, first mentioned the idea to an IBM official, he got a skeptical reaction. So the two scientists hunted quietly, telling a supervisor a half-truth and steering a curious visitor off the track.

Thus, it was quite a surprise when the scientists came up with one of the most significant electronics discoveries since the invention of the transistor in 1947. After 2-1/2 years of lonely lab work conducted in spare moments, Messrs. Bednorz and Mueller discovered a new class of ceramic, superconducting materials. The announcement of the finding triggered the current worldwide frenzy about superconductors. In time, the materials may affect industry as profoundly as the transistor did, making computers smaller, electric motors more efficient and electricity far cheaper.

The scientists' unlikely breakthrough was part painstaking work, part crazy thinking and part luck. Prof. Mueller, 60 years old, who also teaches physics at the University of Zurich, and Mr. Bednorz, 37, made a good team: a brainy, irascible Swiss physicist and his German protege, a whiz at the lab bench. Their idea, based on a former colleague's theory, was "a far-out notion," says Brown University Prof. Leon Cooper, a 1972 Nobel laureate in superconductivity. "It would never have occurred to me to look" where they did. . . .

A veil of secrecy dropped over the project. . . . The issue is a sore point now for IBM brass. The company cultivates a public image of a tightly run ship. Clandestine freelancing in its lab clashes with that image.

"You shouldn't stylize it (in print) that management doesn't know what's going on," says Martin Reiser, the Zurich lab director. "The general thrust of their work was known."

But exactly what was known is still unclear. In a June interview, Prof. Mueller at first volunteered that "we did [superconductivity research into metal] oxides without the knowledge of management." . . .

As for Mr. Bednorz, he says that he kept his involvement in the project from his supervisor. . . .

So the two men labored alone, squeezing the project in between regular duties. . . . Meanwhile, though, Mr. Bednorz was becoming increasingly uncomfortable about the secrecy. At first, his work didn't attract attention because his main job involved metal-oxide experiments for other projects. But the new IBM employee was spending 30% of his time on something his boss didn't even know about. "It could give the impression you're sleeping" on the job, Mr. Bednorz says.

In a compromise, Prof. Mueller finally told Mr. Bednorz's boss, Belgian physicist Eric Courtens, that they were doing experiments based on Prof. Thomas's theory. But he omitted the crucial detail: that they were using the theory to invent a new superconductor.

SOURCE: Excerpt from Richard L. Hodson, 1987, "How Two IBM Physicists Triggered the Frenzy over Superconductors," *Wall Street Journal* (August 19):1, 9. © Dow Jones & Company, Inc. 1987. All rights reserved. Used with permission of Dow Jones & Company, Inc

from those of the staff professionals. Deprofessionalization is an important counterprocess to professionalization. Whether the professions of the future will have more or less prestige and influence than they have today will depend in large part on the relative progress of the two movements.

THE SEMIPROFESSIONS AND THE PARAPROFESSIONS

The Semiprofessions

The term **semiprofession** refers to an occupational group that has achieved some of the characteristics of a profession or possesses the hallmarks of a profession to an attenuated degree. Many people accord workers in these occupations professional status, and studies conducted by government statistical agencies classify them as professional specialty workers. Sociologists, however, do not always classify them as true professions. Most of these occupations are professionalizing and may ultimately succeed in gaining professional status.

Among the major semiprofessions are elementary and secondary schoolteachers, librarians, social workers, and registered nurses. These occupations are sometimes called the female semiprofessions because they are dominated by women. Figure 11.2 shows both the sex and minority composition of these semiprofessions. The few men in these occupations may earn higher salaries and receive rapid promotions, unlike the women who dominate the occupation (Williams, 1992). Because women generally exercise less social power than men, the female semiprofessions face obstacles in gaining collective power.

Members of minority groups have the greatest relative representation among social workers, dietitians, and nursery school and kindergarten teachers. Their levels of representation are lower among librarians and pharmacists. In 2000, for example, only 3.3% of pharmacists were African-American and only 3.8% were Hispanic.

Other occupations that could qualify as semiprofessions are opticians, human resources officers, and systems analysts. Some sociologists would add other occupations to the list of semiprofessions. Rather than comparing various lists, however, it is probably more useful to talk about distinctions between semiprofessions and other occupational types.

The semiprofessions have a body of knowledge, but they usually do not monopolize it, nor do they erect effective barriers to entry. Elementary teachers have a college degree and usually have completed specialized work in education; in addition, they have a body of knowledge on child development, teaching methods, and many other fields. But the basic information they teach in class is known to most adults. Most adults can count, add, subtract, and read. Although some adults accord authority to the elementary teacher who has mastered the specialized knowledge of teaching, many others do not, presuming that "anyone could do that." Most parents would probably give greater deference to the advice of pediatricians than to the advice of elementary teachers concerning a child's performance. Teachers do have to be licensed by the state, however, and so there are barriers to entry.

Librarians pose a somewhat different case (Winter, 1988). Although librarians with a degree in library science are trained in a variety of techniques that others will not know, many libraries (especially smaller ones) will hire people without librarian credentials. In this case the credential and the specialized knowledge it represents are not considered necessary. There are few barriers to entry for many librarian jobs (Abbott, 1988).

A still different problem occurs when the semiprofession's body of knowledge closely overlaps that of an established profession. For example, both registered nurses and pharmacists share many aspects of professional knowledge with physicians. But because physicians are an established, powerful profession, they can limit nurses'

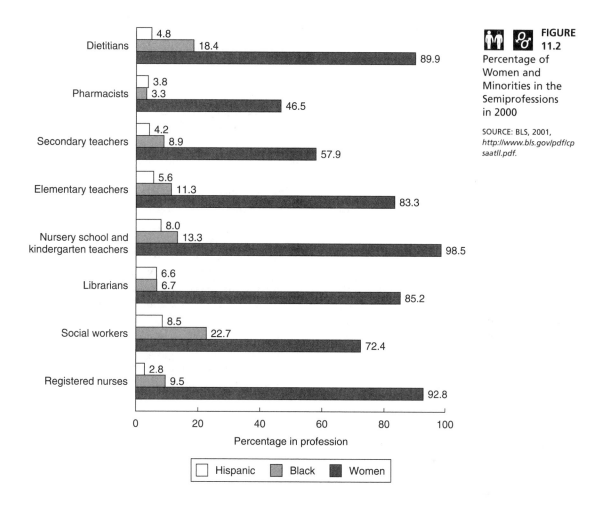

FIGURE 11.2

Percentage of Women and Minorities in the Semiprofessions in 2000

SOURCE: BLS, 2001, *http://www.bls.gov/pdf/cp saatll.pdf.*

and pharmacists' abilities to use the information. Both nurses and pharmacists may know what medication would normally be effective for a specific condition, but only the physician can diagnose the condition and prescribe the drug.

The semiprofessions enjoy only partial autonomy because most of their members work in bureaucratic settings. For example, a teacher has guidelines about what must be taught, what books to use, and how students may be disciplined. A social worker has agency policies and rules to follow concerning how outreach is to be done, how new cases are written up, and what reports must be filed. A registered nurse has hospital policies and physician orders to follow. Even

so, within these constraints the semiprofessional has discretion in dealing with individual clients. Teachers diagnose the learning problems of specific students and work to overcome them. Physical and occupational therapists typically work under a physician's direction, but with autonomy to apply their specialized knowledge of rehabilitation techniques to a patient's needs.

Some semiprofessions have adopted the trappings of the professions, including those associated with altruism. Many semiprofessions have codes of ethics and disciplinary rules that are enforced by the members of the professional associations. Unlike the established professions, however, their associations are less likely to be politically powerful. The semiprofessions' associations are often cautious for fear of offending major employers or the more powerful professions. They must also continue to educate the public and negotiate with state agencies to improve their position regarding licensure and recognized authority. To make the process even more difficult, in some semiprofessions several associations compete for the allegiance of the members.

In summary, semiprofessions have eagerly sought professionalization. Many semiprofessions are accorded status and legitimacy. Some have considerable authority over clients, and most have some form of licensing or credential. But they still lack the autonomy and organized power of the professions. From the power perspective, the lack of collective power is the single most important difference between the semiprofessions and the professions.

The Paraprofessions

A different group of near-professions are the **paraprofessions,** which might be considered occupational agents of the major professions. Members work closely with professionals, usually doing tasks delegated and supervised by them. The dominant profession dictates the conditions of training and certification, if any, and provides the guidelines mandating the point at which a member of the profession must be consulted.

Paramedics and paralegals are examples of paraprofessionals. Paramedics provide emergency medical care until a patient can be transported to a hospital or other site for treatment by a physician. In many large American cities the paramedic is linked by radio to physicians who can give precise instructions for emergency care. Paralegals work under the direct supervision of an attorney. They may do some legal research, prepare and file forms, interview clients on routine matters, and perform other necessary tasks, but always under the direction of the supervising attorney. Paralegals are con-

sidered to be cost-effective because they can perform routine duties and free the attorneys for more specialized tasks.

The paraprofessions, unlike the semiprofessions, do not have their own body of knowledge, although they share some of the knowledge of the dominant profession. Paramedics, for example, know how to treat accident victims for shock, and they are able to use a number of medical techniques on an emergency basis. Paraprofessionals have substantial responsibility but no real autonomy because of the close supervisory relationship with a professional. They may have some authority over the client, but it is sharply circumscribed and is derived from the authority of the supervising professional. Finally, they have limited ethical codes and lobbying efforts and are not generally self-disciplining, because the supervising professional is presumably responsible for the quality of their work.

Workers in many occupations share some characteristics of paraprofessionals. These workers include physician's assistants, teaching assistants and teacher's aides, library aides, caseworkers without formal social work credentials, and various technical assistants who work with engineers, architects, or accountants. Many occupational coding systems do not list them as professional workers; instead, they may be listed as technical workers, clerical workers, or service workers. What these occupations have in common is their close association with one of the professions or semiprofessions (Creighton and Hodson, 1995).

As with the semiprofessions, many of the paraprofessions are predominantly female. In the United States, over 98% of dental hygienists and nearly 85% of legal assistants, for example, are women. The representation of minority group members in these fields is still relatively small, however. Among dental hygienists, fewer than 3% are black and 1% are Hispanic, and among legal assistants, 8% are black and nearly 10% Hispanic. In Canada, dental hygienists are also predominantly female and seeking professionalization (Brownstone, 2000).

THE FUTURE
OF THE PROFESSIONS

The two passages that open this chapter summarize two views of the future. The quotation from Talcott Parsons suggests that the future will be the age of professions, with professionals exercising great social control over work and knowledge. The quotation from Ivan Illich suggests that deprofessionalization will accelerate so that we will come to wonder why we accorded so much authority to the professions. There is support in this chapter for both of these views.

The professions constitute the backbone of the **"new middle class"** that is based on knowledge rather than on the possession of capital (Mills, 1951; Gouldner, 1979; Brint, 1984). Among sociologists there is considerable disagreement about the importance of this new class (Bruce-Briggs, 1979; Kellner and Heuberger, 1992), but some aspects of the importance of professionals are well established. Professionals who are in private practice constitute an important segment of small-business owners, and they tend to be respected members of local business communities. Many professionals are also in demand to play important roles in large organizations either because their expertise is central (for example, a physician who is a hospital administrator) or because they perform important functions on the staff (for example, a lawyer who is house counsel to a corporation). Both individually and collectively, professionals have gathered considerable political and economic power (Polsky, 1991). Through their professional associations they have become adept at maintaining their power through promoting favorable legislation and educating the public about their importance. Their self-organization and substantial control over their work environment are often envied by other workers with little such control.

But the professions are changing. The relentless creation of new knowledge has many consequences. One consequence is *intellectual obsolescence* among professionals who cannot keep up with the new knowledge (Leventman, 1981). A second consequence is the *routinization of professional judgment*. New computer applications in the professions provide a good example. Interactive computer terminals can be used to take medical histories, and computers can search databases for illnesses consistent with a cluster of symptoms. Databases of legal opinions can identify, within seconds, opinions that once would have required a lawyer many hours of library research. Although these developments may make more information accessible to professionals, they may also deskill the individual professional by putting more control in the hands of managers, technicians, and bureaucrats and leading to a lower value being placed on professional knowledge and judgment (Derber, 1982:196–198; Ritzer and Walczak, 1988).

A third consequence of the new knowledge is *renewed pressure for specialization*. Specialization, in turn, has created the need for integration. The knowledge needed to solve a particular problem may require that the efforts of several specialists be integrated by a generalist. This integration has often taken place in large professional firms, where the professional values may remain paramount but the individual professional has much less influence. Another alternative is to train more professionals who are generalists rather than specialists (Winslow, 1993). A key unanswered question for the future is who will exercise the social control over esoteric knowledge (Rueschemeyer, 1983).

Many professionals no longer work in independent practices or even in large-scale professional practices; instead, they are employees of a large organization or government agency. They may often find that much of the self-regulating nature of their profession is lost, because for them the bureaucratic rules of the employing organization often take precedence (Leventman, 1981). Physicians who manage HMOs or hospitals may find themselves in conflict with physicians who remain in practice; professions are not so cohesive under such conditions as they once were. Globalization, information technology, and other developments challenge the traditional practices of the professions (Greenwood and Lachman, 1996).

Other pressures may eventually diminish the authority and autonomy of professions. New bodies of knowledge may challenge the old claims of the professions, as "alternative medicine" has done for physicians. Acupuncture, herbs, and various forms of therapy now compete with traditional Western medicine and its practitioners. Semiprofessions and other occupations seeking professionalization add to the pressures against the established professions. Deprofessionalization affects both the established professions and the occupations seeking to professionalize. There is a renewed demand to demystify professional knowledge so that clients can be more intelligent consumers. Court decisions and third-party regulation have affected the autonomy of the professional to set fees. Professions with adjacent bodies of knowledge often do battle with one another to claim the unique right to advise clients in their areas of expertise.

Nevertheless, the professions retain considerable strength. The fact that their knowledge bases continue to grow leads to more specialized services. Most have been successful in convincing consumers that their professional services are needed and worth the price. Legislatures still defer to professional authority. Individual professionals are accorded high prestige within the community, and most continue to command incomes well above the median income for all workers. Despite their problems, professionals remain among the most powerful of workers (Halliday and Karpil, 1997).

SUMMARY

Professions are high-status, knowledge-based occupations. According to the hallmark or structuralist-functionalist approach, they are characterized by a monopoly over advanced, specialized knowledge; by autonomy; by authority over clients and often over subordinate occupations; and by a degree of altruism. The professions attempt to be self-regulating by using associations to maintain their legal perquisites, erecting barriers to entry, and policing the behavior of their members. The associations monitor professional schools, promulgate codes of ethics, and educate the public about the importance of the profession. The professional association also seeks to maintain the authority of its members against the competing claims of other occupations whose members may have mastered similar bodies of knowledge.

The conflict approach emphasizes that the four hallmarks of the professions are more the result of professional power than its cause. In contrast to the structural-functional approach, the conflict approach stresses the struggle for power among members of a profession, among professions, and between the professions and other occupational groups.

Professionalization is a process by which occupational groups seek to improve their collective status by more closely resembling a profession. Semiprofessions are lower-status, knowledge-based occupations that are seeking to emulate the professional model. Most semiprofessionals are employed by organizations, and both bureaucratization and unionization may further weaken their drive for professional status. Some of the semiprofessions may also be considered of lower status because they are primarily female in composition and have traditionally been rather low paid. Paraprofessionals are more closely tied to a particular profession, which oversees their training, day-to-day performance, and evaluation. Although the paraprofessions may have a professional organization and credentials, they are really agents in the service of the dominant profession.

Deprofessionalization poses an important challenge to the existing professions and to the professionalizing occupations. Individual professionals may lose autonomy when they are employed in large, professional firms. Autonomy is also lost when they are employed in bureaucratic organizations or in government agencies. Professions as a group face challenges from the general public, many of whom are no longer content to be passive clients. Professions are also challenged by new government regulations, the demands of third-party payers, and the unionization of their own members.

Professionals are significant models for other workers because they have achieved so much autonomy and authority. Being a member of a profession confers considerable prestige and usually higher income. Many of these individuals are considered members of the so-called new middle class, which achieves its position based on knowledge and achievement and not on the acquisition or inheritance of capital. The future of the professions may take either of two paths. The process of deprofessionalization challenges the position of the professions and of the other occupations that seek to emulate them. As knowledge expands and the economy is increasingly transformed by the need for greater knowledge and more services, the collective prestige of the professions is likely to grow.

KEY CONCEPTS

profession	professionalism	paraprofession
professionalization	deprofessionalization	new middle class
professional culture	semiprofession	

QUESTIONS FOR THOUGHT

1. According to the definitions used in this chapter, is a professional football player a professional worker? Why or why not? Would your answer to this question differ if you adopt the conflict approach instead of the hallmark (structural-functional) approach?

2. The professions are based on knowledge. How does the spread of information technology affect the professions? What is the effect of the internet on the monopoly of professional knowledge?

3. Choose an occupation you have considered for your own career. Describe how through this occupation you might try to develop abstract knowledge, authority, autonomy, and altruism in an effort to professionalize or to maintain a professional status.

4. Suppose an occupation did not require a state-issued license for its practice. Could it still be considered a profession? Can you think of any examples? What are the alternatives to professionalization for an occupation that wishes to elevate its status?

5. Many semiprofessions are predominantly female. How might the gender of the workers affect the ability of an occupation to professionalize?

MULTIMEDIA RESOURCES

Print

Andrew Abbott. 1988. *The System of Professions: An Essay on the Division of Expert Labor.* Chicago: University of Chicago Press. A systematic discussion of the role of professionals and their work.

Gale Miller. 1997. *Becoming Miracle Workers: Language and Meaning in Brief Therapy.* Hawthorne, NY: Aldine de Gruyter. An analysis of how psychotherapists approach short-term therapeutic contacts, and how their use of language shapes the interactions.

Paul Starr. 1982. *The Social Transformation of American Medicine.* New York: Basic Books. A Pulitzer Prize–winning account of the struggles of the medical profession to gain recognition in the United States.

Websites

American Bar Association. *http://www.abanet.org/* Because the ABA is one of the most powerful professional associations, many parts of this site are password protected for ABA members only. Visitors to the site can read articles in the association's *Journal* and find information about legal practice and related links.

American Medical Association. *http://www.americanmedicalassociation.org/* This site provides information on finding a doctor, recent medical news, and other information for both members and nonmembers.

Aboriginal Nurses Association of Canada. *http://www.anac.on.ca/* This professional association seeks to encourage nursing careers among aboriginal peoples and is also concerned with improving health for aboriginal peoples.

RECOMMENDED FILM

Patch Adams. 1998. Based on a real-life story, Robin Williams plays a medical student who confounds professors at his medical school with his belief that laughter is the best medicine. He eventually founds the Gesundheit Institute to further his cause. Rated PG-13.

12

Executives, Managers, and Administrators

First-line managers and supervisors—those who supervise direct production and service workers—have always had a difficult job: exhorting workers to live up to standards and demands thrust upon them by higher levels. The new corporate era, which has brought new work systems, only increases the pressures on them.

(KANTER, 1983:56)

Workers have willingly offered themselves complete with their personal and family lives to the organization. Like children, they have allowed the organization to determine what is best for them and how they should best be used. They expected no power and that is precisely what they got. Such childlike passivity is fading fast. The divine right of kings or managers no longer exists.

(KELLEY, 1985:93)

Ask most workers who has power in their workplace, and they will answer "the boss" or "management." Executives, managers, and administrators form the occupational group that organizes and coordinates work and makes decisions about production, finances, and the hiring, firing, and deployment of other workers. These responsibilities do not imply, however, that managers can act unilaterally. As the quotations above emphasize, even the boss faces limitations. In this chapter we examine the types of jobs managers hold, the changes they are experiencing in their roles, and the ways in which managerial performance is evaluated.

TYPES OF MANAGEMENT ROLES

Many kinds of workers can be classified as executives, managers, or administrators, from the cabinet-level head of a government agency to the self-employed proprietor of the corner candy store. In this chapter, we will use the generic term **manager** to refer to all three groups. Managers may have responsibility for millions of dollars in a budget, or they may manage the evening shift in a round-the-clock restaurant. In labor relations, management is distinguished from labor by its responsibility for the maintenance and growth of the employing organization. Managers are defined principally in relation to their employing organization and not in relation to their occupational group. Unlike a professional, for example, who shares common training and credentials with other members of the profession, managers do not necessarily share a common base of knowledge, nor is entry into management restricted by licensure or certification.

What is common to management jobs is the need to provide leadership and coordination, one of the oldest specializations in work roles. (See Chapter 1 on the emergence of leader roles in preindustrial societies.) Leadership roles have especially been associated with political and government work, and the terms **executive** and **administrator** are likely to be used for such workers. In the private sector, and especially in corporations, the terms *executive* and *manager* are more common. All such work involves integrating, synthesizing, and coordinating information, but different terms are used depending on the type of organization and on how highly placed the worker is within the bureaucracy.

Executives

The term *executive* refers to a person who is at the very top of a workplace bureaucracy. The president of the United States is called the chief executive of the United States. Within the government, the executive branch refers to all the agencies that carry out government programs. Even very small tribes typically have a chief who serves the executive function. Agrarian societies developed hereditary monarchies to provide leadership. Later, executive roles in the modern political states were filled by government officials. The top officers of most private corporations and public agencies are also called executives.

Managers

Although management as a field of study for specific occupational preparation is relatively new, for centuries managers ran farms and factories in the absence of owners. Even a large household might have had a steward as manager. These prototype managers had foremen, supervisors, or servants reporting to them. The functions of these managers included purchasing and accounting; directing which products were produced and how; hiring, monitoring, and disciplining workers; and providing for the maintenance, repair, and replacement of equipment. These functions are similar to those of the managers of today's bureaucratic enterprises.

But today's management is different for two basic reasons. First, today's enterprises are far larger and more geographically dispersed than the early farms and factories were. Second, today the ownership is also more dispersed (Hansmann, 1996). The old-style managers reported directly to the owners, but the modern corporation is owned by many shareholders who elect a board of directors. The board in turn appoints corporate executives. The rest of the management team, who are appointed by the executives and their subordinates, ultimately receive their authority from the board of directors.

Administrators

Government administrators assisted rulers in developing and enforcing policies, and today they serve the same functions in much more elaborate and complex government agencies. In Europe, the medieval church developed separate offices to perform specific functions, and each office had a cadre of administrators. The local bishop served as the executive, in the sense of having final authority, and other members of the clergy administered parishes, charitable institutions, or church offices. The top managers of other nonprofit firms are often called administrators as well. For example, at your college or university there are probably specialized officers who are recognized as administrators. Hospitals, nursing homes, and similar institutions are often headed by an administrator.

Staff and Line Managers

Executives, managers, and administrators can also be distinguished by whether they have line or staff responsibility. A line manager supervises workers directly involved in the production process. **Line managers** typically have substantial numbers of "direct reports"—lower-level supervisors and workers who report directly to them. A **staff manager,** by contrast, provides or supervises a service that is necessary to the organization but is not the organization's major product or service. Examples of staff specializations that may have managers include accounting, personnel, purchasing, maintenance, and legal services. In restructuring, many companies have sought to outsource some of these staff functions. For example, instead of maintaining an account-

ing staff, the company may contract with an accounting firm to provide this service.

EXECUTIVES, MANAGERS, AND ADMINISTRATORS AT WORK

Management attracts workers because its duties are varied and carry some measure of autonomy and authority. Managers also tend to be well compensated, although compensation depends upon experience and the employing organization. In 2000, the median weekly earnings of all full-time workers were $576. For men employed full time in executive, administrative, or managerial specialties, the earnings were $1,014 and for women they were $686 (BLS, 2001a). Top corporate executives are among the best compensated workers in the United States, with salaries, fringe benefits, and perquisites that easily make them multimillionaires.

Demand for Managers

Within a firm, the demand for managers is a function of technology and organization. Technology may alter a manager's **span of control,** the number of workers a manager supervises, and this in turn will affect the number of managers required. For example, a manager who supervises workers through telecommunications links to computer workstations can supervise a greater number of workers, and workers dispersed over a geographically larger area, than can the manager who must supervise in person. This is a case in which technology might increase the span of control.

The bureaucratic complexity of the work organization also affects the number of managers who are needed. The **steepness of hierarchy** refers to how many levels of supervision are required. A firm that has many establishments needs a set of managers at each location. If the divisions or product lines of a company are autonomous, the company needs to hire a set of managers for each one.

Changes in organizations also affect the number and type of managers needed. Reorganizations, especially acquisitions and mergers, are often followed by the layoff of some managers in the acquired firms. In a merger, where both firms already have a full set of managers, there is usually lengthy negotiation over how many managers will be retained. Executives, or top management, may negotiate in their hiring contracts for compensation if a merger, acquisition, or reorganization leads to the loss of their jobs. These severance pay provisions are sometimes so lavish that they are called "golden parachutes." Lower-level managers and ordinary production workers receive much smaller severance benefits, if they receive severance pay at all.

Within the labor force as a whole, the demand for managers is a function of the number of firms as well as of technology and organization. Even very small firms must have managers. The term *manager* appears in the job title of many workers—office managers, sales managers, apartment complex managers, restaurant managers, and so on—even though these jobs may seem to have relatively little in common with managing a large corporation.

Managers' unemployment rates have traditionally been very low. In 2000, when overall unemployment was 4%, the rate for managerial and professional specialties was 1.7% (BLS, 2001a). As an occupational category, managers are also less likely to suffer from involuntary part-time employment or very low pay (Clogg and Sullivan, 1983). Managers, who are responsible for personnel decisions, tended to lay off themselves and their fellow managers last.

More recently, however, *organizational changes* have affected the long-term job security of managers. Mergers and acquisitions often result in laying off "redundant" managers (Newman, 1993). Foreign competition, new technology, deregulation, and other financial issues have led some organizations to reduce their managerial staff. During the 1990s, "downsizing" was the term often used to justify flattening the company hierarchy by laying off some middle managers. These layoffs followed decisions by many executives that their companies were too fat and needed to become "lean and mean."

Interestingly, the willingness to lay off more workers has come to be identified as a sign of managerial talent. In the first two months of 1994, 144,000 jobs were lost in U.S. staff cutbacks, a process that is now termed *job shedding*. One observer comments that "job shedding has become fashionable—the mark of a good manager" (Uchitelle, 1994). Middle managers continue to be targets for job shedding, but upper-level executives have been fairly safe.

The Self-Employed Worker

The self-employed proprietor is often classified as a manager, even if there are no employees and the proprietor is managing accounts, inventory, and the daily business rather than people. A little less than 8% of the U.S. labor force is self-employed, but the proportion of self-employed is much higher among certain groups, such as immigrants.

The self-employed tend to be concentrated in a few industries. About 15% of the self-employed are in professional services, the industry that provides legal, educational, and medical services, and other similar services to consumers. In this industry, the self-employed person may be both a professional, in the sense we used the term in Chapter 11, and the manager of a self-owned accounting service, law firm, or clinic. The self-employed are also commonly found in retail trade, in which the manager-proprietor operates a small retail store. The most common retail stores run by the self-employed are grocery stores and small clothing stores. A third common industry for the self-employed is personal services, in which the proprietors may own bars or restaurants, dry-cleaning establishments, or small motels.

Some ethnic communities have developed specializations as small-business owners (Butler, 1991; Woodard, 1997). Examples include Koreans who specialize as greengrocers (vegetable sellers) in New York City, or Indians who specialize as small motel owners (Kim, Hurh, and Fernandez, 1989; Yoon, 1991; Dong, 1992). The ethnic community often loans the capital needed by these small-business owners (Light, 1972). The term *ethnic enclave* is used to describe a cluster of ethnic business in which the owners provide substantial employment to other members of their ethnic community. Cuban Americans in Miami represent an example of an ethnic enclave (Portes and Jensen, 1989). Geographic clusters of small businesses often provide tacit knowledge and social ties that allow new entrepreneurs to enter an industry (Sorenson and Audia, 2000).

The failure rate of small businesses is very high. In 1995, the small-business "birth rate" was 10.8 new businesses per 1,000 existing businesses, and the "death rate" was 9.1 business failures for every 1,000 existing businesses (calculated from Census, 1999:557). Although many small businesses fail, some entrepreneurs become extremely successful and begin large companies.

One way that small-business owners can share some of the risks of ownership is by becoming the *franchisees* of establishments affiliated with large chains of restaurants, convenience stores, motels, and other businesses. Franchises are also available in personal services, such as weight-loss groups or house-cleaning services. Although some franchise establishments are owned by the company, in over 80% of the cases the franchisee buys the business or a share in the business by investing a large sum of money. The franchisee agrees to run the business according to the conditions set down by the national organization. The franchisee receives the know-how and advertising of the national organization. This is a great advantage in building a clientele because customers will recognize the nationally advertised product. The national organization may sell its training, require the franchisee to buy its products, and receive a portion of the profits as well, all without risking much capital. The franchisee provides the capital; hires and supervises the workers; arranges purchasing, accounting, and

other services; and oversees the daily operations of the business. Although many franchisees hire additional workers, they also commonly work long hours alongside other members of their families.

Supply of Managers

Management was historically the preserve of white Protestant men. One reason given for this exclusivity is that managers must work in a climate of uncertainty, and homogeneity among themselves is believed to reduce one source of miscommunication and uncertainty (Kanter, 1977). This view cannot explain, however, why the preferred group is white, Protestant, and male. Even white Protestant men sometimes miscommunicate and may hold different values, however, so the workplace culture may be more important than recruitment in maintaining homogeneity. The fundamental reason is simply that a group with power tends to maintain its power. Figure 12.1 shows the percentage of women and minorities in selected managerial specialties.

Partly because of legislative pressure and partly because of popular sentiment, women in recent years have made notable advancement into management positions (Blair-Loy, 1999). About 45% of all executive, managerial, and administrative workers are women. Firms that use formal recruitment procedures for managers, as opposed to using more informal networks, tend to have higher proportions of female managers (Reskin and McBrier, 2000). Some firms have a two-track system of promotion to management, with the "mommy track" (or *ippanshoku* in Japan) that has less complicated jobs, lower mobility requirements, and a ceiling on promotions (Strober and Chan, 1999).

Management remains a field with relatively little minority representation. In 2000, only 7.6% of all executive, managerial, and administrative workers were black, and only 5.4% were Hispanic. Black managers had their greatest representation among inspectors and compliance officers (13.9%)

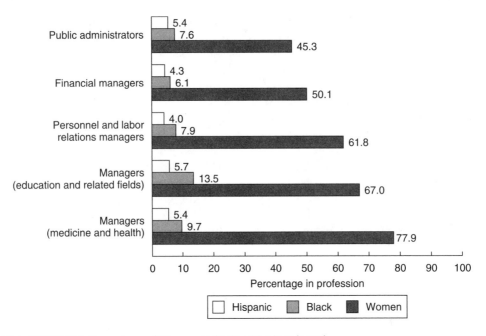

FIGURE 12.1 Percentage of Women and Minorities in Selected Managerial Specialties, 2000.

SOURCE: BLS, 2001a, 178.

and education administrators (13.5%). By contrast, only 4.2% of the nation's 755,000 marketing, advertising, and public relations managers were black. Hispanics had their greatest representation (9.2%) among managers of food serving and lodging establishments (BLS, 2001a). Box 12.1 reports the experience of some black executives during a period of corporate downsizing.

Social class plays a role in recruitment to management. Virtually all executives have a college education; controlling for education, people with upper-class origins are also more likely to be within the top ranks of corporate management. Top managers are more likely to have a bachelor's degree from a top-ranked college and a master's degree in business administration (MBA) or a law degree from a leading institution. Holding a law degree and having an upper-class background are useful for joining formal and informal networks inside and outside the corporation; these networks are useful for career moves within and between corporations (Useem and Karabel, 1986).

Middle managers are more likely to have come from middle-class or even working-class origins and to have graduated from state colleges or denominational colleges. There are also industries, such as auto repair services and retail sales, in which the majority of manages do not have college degrees (Verma and Boyer, 2000). Their position within management may represent social mobility from their parents' positions. Both they and their spouses often feel proud of their success, and this keeps them committed to the corporation even though the conditions of work may be grueling (Newman, 1988). Middle managers often experience a "ceiling effect," reaching a level within the firm above which they cannot pass.

Business proprietors and other self-employed workers often have backgrounds similar to those of the middle managers, although they are not so likely to have college degrees. Many of the self-employed do have specialized training and skills, however; examples include independent craft workers and professionals.

BOX 12.1 Downsizing, Flattening, and Affirmative Action

Sociologist Sharon Collins studied the effects of corporate response to affirmative action. She argued that black managers were disproportionately affected by the efforts to downsize firms and flatten hierarchies, and that slackening enforcement efforts made the situation worse.

> African Americans attribute their career stagnation to processes they associate with corporate downsizing and restructuring, such as the trimming of functions and excess personnel. Some black executive careers are casualties of corporate attempts to meet new competition from domestic and international businesses. For instance, a company once considered the leader in its industry went through a period of flat sales brought on by stiffer competition, followed by several phases of staff reductions, departmental reorganizations, and new standards of profit accountability from middle management. In this process the size of a black executive's staff had been reduced from three people reporting to him in 1986 to one in 1992. Other individuals stagnated because of a relatively new business trend in which hierarchical, vertically integrated corporations are "flattening." In Chicago, for example, one nationally known professional service firm implemented a team-leader strategy, dismantling the hierarchical framework that more typically defines promotions in corporate structures. Such organizational changes make it harder to move up in a company and easier to fail. One executive who had worked for the company for twelve years said, "I don't think I'll make twenty." Nineteen of the fifty nonracialized executives in this study (38 percent) left companies altogether between my first and second interviews and went out on their own.
>
> African American sentiment mirrors that of nonblack managers. The displacement of executives, administrators, and managerial professionals due to cutbacks or plant closings jumped by 50 percent between 1987 and 1992 (U.S. Bureau of Labor Statistics, unpublished tables). In the theoretical framework linking blacks' attainments to political pressure, however, one may plausibly identify the brunt of such changes as inequitably borne by African American executives.

SOURCE: Sharon M. Collins, 1997, *Black Corporate Executives: The Making and Breaking of a Black Middle Class.* Philadelphia: Temple University Press, pp. 144–145.

The Managerial Career

In large firms the managerial career is marked by a number of transitions. In her interviews with managers at a firm she calls "Global Products, Inc.," Diane Margolis (1979) identified critical stages in the managerial career. During the *initiation,* the period immediately after hiring, the manager comes to adopt a self-identity as a manager and, equally important, an identification with the firm. The initiation is intended to be a training period, and the firm uses every expedient to train the new manager as quickly as possible. The new manager works long hours and often long weeks; to refuse these assignments, without exceptionally good cause, makes superiors wonder aloud if the manager is really committed and "has what it takes." (This period of overwork also characterizes some professional workers, including medical interns, assistant professors, and young associates in law firms.)

Frequent transfer is a technique firms use to train managers and to initiate them into the company. Ostensibly, the transfer provides adequate staffing for the firm and teaches the manager new skills. As one executive put it:

> We have somebody in the Philippines we're going to replace. This situation has triggered five moves. The man in the Philippines is being let go because he's not doing the job. A man in Hong Kong will be moved to the

Philippines. A man in Singapore will go to Hong Kong. One in Thailand will be transferred to Singapore and a man somewhere else in Asia, we haven't decided who yet, will replace the man in Thailand. (Margolis, 1979:48–49)

But the frequent moves also serve other functions. They make it difficult or impossible for the manager's family to become involved in the community or even to have many friends outside the company. When moves come as often as every six months, the family relies on itself or on other corporate families for companionship (Newman, 1988). Limiting ties with other groups increases the manager's reliance on the corporation as a source of income, identity, and prestige.

Most successful managers have **mentors** during their early years in the company (Kanter, 1977). The typical mentor is two levels higher than the manager and serves as a role model, protector, adviser, and cheerleader. Many management jobs are ambiguous, and companies often act in a climate of uncertainty. In such a climate, the mentor can provide a road map for understanding. Mentors may warn their young protégés of interdepartmental "dirty tricks" and teach them how to protect themselves while giving every appearance of being good team players. Eventually, as a part of developing the protégé's separate identity, the protégé and mentor part ways, sometimes with hard feelings on both sides. But as the protégés becomes better established, a rapprochement and lasting friendship sometimes develop between the former protégé and the former mentor (Harriman, 1985).

On the other hand, management careers also encounter barriers. Table 12.1 indicates the percentage of male and female managers who reported experiencing various career barriers. Typically, managers endure a period of *pruning* during which some managers are promoted and others are encouraged to leave (Margolis, 1979). Many firms do not actually fire managers, but they manage to convey their sense that the manager "doesn't fit in" or "would be happier elsewhere." Many large corporations now have their own outplacement offices for helping to relocate

Table 12.1   Percentage of Respondents Mentioning Principal Barriers in Their Corporate Career, by Sex

	Men	Women
Inflexible working patterns	6	14
Family commitments[a]	44	43
Lack of adequate childcare	3	8
Lack of career guidance	45	50
Lack of training provision	9	5
Prejudice of colleagues	11	23
Lack of personal motivation/confidence	19	21
Senior management seen as a "club"	32	54
Social pressures	1	10
Sexual discrimination/harassment	1	17
Insufficient education	11	13

[a] Here the data refer only to respondents with children.

SOURCE: Judy Wajcman, 1998, *Managing Like a Man: Women and Men in Corporate Management.* Cambridge, England: Polity Press, p. 88. Used by permission.

displaced managers. Some corporations also have a group of smaller client firms to whom they refer surplus managers. These smaller firms, often unable to afford the training that large firms provide, are eager to cooperate because they can receive managers who will be better trained than if they had been promoted from within the smaller company. Short of being pruned, managers may also find themselves *rechanneled* into the by-waters of the corporations, often into staff positions that are not rungs on a career ladder.

The manager who survives and is promoted walks a fine line between too obviously advancing his or her own career and being perceived only as a team player. The manager who is too interested is suspected of being insufficiently loyal; the one who is too little interested in promotion is insufficiently ambitious. Promotion to higher levels within the firm is often acknowledged to be "political," in the sense that it is based on something other than technical qualifications. The corporate

BOX 12.2   Whose Idea Was It?

Sociologist Calvin Morrill studied conflicts and their resolution among executives. Successful subordinates learn not to outshine their superiors, and yet they still need to achieve recognition to advance. In this excerpt, a younger executive explains how he led his boss to a good decision.

Asymmetries between superiors and subordinates also occur in written communication. First, more memos (almost ten times as many) travel upward than downward. Second, the qualitative content of executive communication differs among superiors and subordinates. Upward communication tends to be informative *and* covertly rhetorical. Conversely, superiors are overtly rhetorical in their downward communication and less informational. An example of the differences in the contents of upward and downward communication was related by this senior vice president of branch management in dealing with his superior, an executive vice president division manager:

We had a branch that I wanted to close. The place was in an older part of [a city] and wasn't doing very well. We had a newer branch nearby that could serve the needs of the older branch's customers. It would be too expensive to make the older branch

state-of-the-art. I could have written a memo that argued for closing the old branch, but I didn't. I had my staff make up a report using several sets of data—cost accounting, depreciation of the building, customer profile—you name it, it was in there; lots of pretty graphics from our new graphics software, nicely bound cover, the works. I sent that up to the EVP. It said what I wanted it to say—Close the old branch down—without saying it in so many words. [Smiling as he tells this part of the story.] Led the EVP right to the decision to close the sucker. It's a different story when he wants to get something done. One time he wanted to close a branch that had been a flagship at one time for a small bank we acquired a few years ago. He writes a one-page memo telling me to start disentangling ourselves from the branch after the new year. No figures, no nothing, just do it. Obviously, if I know what's good for me, I don't write a memo like that to him. It's simple here, really, at Old Financial; different rank, different way to get things done.

SOURCE: Calvin Morrill. 1995. *The Executive Way: Conflict Management in Organizations.* Chicago: University of Chicago Press, pp. 102–103.

---

ideology, however, is that the firm is a meritocracy and that those who hold the highest positions have achieved them by their accomplishments.

This means that for the many managers who never make it to the top, the ideology encourages the belief that they themselves are to blame for their failure (Newman, 1988).

Box 12.2 presents a situation in which a subordinate had to convince his superior that an appropriate decision was the superior's idea.

Many questions have been raised in recent years about the ethics of managers in American corporations. Revelations of managers' bribery of overseas officials, their tolerance for unsafe prod-

ucts and production facilities, their efforts to evade regulation and statutes, and their conspiring to fix prices have led many to ask whether the lengthy process of initiation into management does not destroy the moral sensibility of managers (Jackall, 1988).

CONTINUITIES AND DISCONTINUITIES IN MANAGEMENT ROLES

The jobs done by village chiefs, medieval bishops, plantation overseers, and ships' captains are still recognizable as management tasks. There are also, however, important differences between the types of work historically done by managers and administrators and the types of work that they do today. These differences stem from changes in *scale, environment, specialization,* and *technology.*

Changes in Scale

Compared with the governments and companies of the past, today's agencies and corporations are larger and more bureaucratized. They are likely to have several layers of managers, each of which coordinates and monitors the layers below. This leads to the distinctions among **top management** (the executives), **middle management,** and **first-line supervisors** (foremen and the like). The first-line supervisors are not considered part of management by some organizations. The larger the organization, the more likely it is to develop a more complex bureaucratic structure. In large, bureaucratized workplaces, the average manager must think about both the **horizontal** and **vertical dimensions of bureaucracy.**

The horizontal dimension refers to the development of functional units that perform discrete tasks. For example, a factory may have separate departments for purchasing, personnel, production, maintenance, and shipping. It is the manager's job to ensure not only that his or her own unit is doing its job but also that its job is coordinated with those of other units. If purchasing has failed to buy needed raw materials or if completed products cannot be shipped, the production manager may still receive part of the blame. In many corporations, considerable friction exists between functional units over budgets, resources, and perceived importance within the company.

In a conglomerate (a company made up of companies), the functional units may be separate corporations. If these units are autonomous or nearly so, coordination among them may be a key function of top management. One accusation leveled at American corporations is that their quality control suffers because of uneven attention given to the functional units. As more and more chief executive officers are recruited from

among lawyers and financial experts, their attention shifts to those areas and away from the basic products of the company (Hayes and Abernathy, 1980).

The vertical dimension refers to the bureaucratic hierarchy of the workplace. Before workplaces became large and functionally specialized, different functions were often performed by small, separate firms and the coordination was done by impersonal market forces. As workplaces have become larger, the coordination has been done by managers and has been accomplished by adding layers to the hierarchy (Chandler, 1980). The typical manager reports to a superior who in turn reports to other superiors. The number of layers in a hierarchy is referred to as the *height* of an organization.

Changes in Environment

Managers, and especially executives, must be concerned not only with what is happening within their organization but also with the relationships between their organization and other organizations. The organization will not survive unless it is successful in its exchanges with others in its environment. These exchanges can be cooperative, as are those between suppliers and purchasers. The exchanges can also be competitive, particularly between organizations producing the same goods or services. Relationships with regulatory agencies, consumer organizations, lenders, and possible merger partners may range from quite friendly to fiercely adversarial. Top levels of management must pay serious attention to each of these relationships. Managers also try to monitor what changes are being made within other organizations, especially if similar innovations might benefit their own company.

Monitoring and responding to changes in the environment are among the most important tasks of contemporary managers. These tasks are especially pressing in industries in which the environment is changing rapidly because of technology, organizational realignments, expansion of the global market, or other developments (Madu,

BOX 12.3 Staying Competitive

By 1966, as a result of a string of technical innovations, Corning [Glass Works] held the dominant position in the U.S. market for television picture tubes, and the television business provided nearly three-fourths of Corning's earnings. Yet by 1975, in less than a decade, the television business was almost all gone. Corning no longer made black-and-white tubes, and its color tube business was losing money. What happened? The Japanese. In the late 1960s they had started entering the U.S. market for televisions, first for black-and-white and then for color sets. Gradually they began running American manufacturers like Corning's customers out of the market. By the mid-1970s, imported televisions accounted for over 60 percent of the market for black-and-white sets and over 40 percent of the market for color sets. Corning's base of customers for its tubes had shrunk from twenty-eight manufacturers to only five.

With the decline of its television business and setbacks in other areas, Corning set out in the late 1960s and early 1970s to do what it had always done: it launched a wide-ranging search for new products, hoping to accomplish the next big

breakthrough. During this period, Corning pursued development of such diverse products as safety windshields, glass-based lasers, ceramic roofing shingles, ceramic heat exchangers for turbine engines, optical fibers, medical instruments, optical waveguides for telecommunication, glass razor blades, integrated circuits, and even computer terminals. Many of these development efforts, such as the safety windshields, ceramic components for glass turbines, and glass lasers, built upon Corning's historical core competency in glass-ceramics technology. Others, such as efforts to establish a position in medical equipment, integrated circuits, and biotechnology, did not.

There was a marked difference in Corning's success with these different types of ventures. When Corning pursued ventures that were consistent with its core competencies, it was usually successful. When it ranged far afield and abandoned its expertise, it wasn't.

SOURCE: Joseph H. Boyett with Himmie T. Boyett, 1995, *Beyond Workplace 2000: Essential Strategies for the New American Corporation*. New York: Dutton, pp. 44–45.

1996). Box 12.3 discusses how Corning Glass Works successfully changed in response to international competition for one of their products.

Changes in Specialization

In small firms, a manager may handle all decision making and day-to-day administration. Many management tasks require controlling and organizing the flow of information about accounts, products, timetables, productivity, and so on. These tasks may be barely distinguishable from clerical work. Management also requires a great deal of work with other people. In the small firm, the manager handles all personnel decisions in addition to interacting with customers, suppliers, the owners, and others. Managers in small firms know the entire business, and their job is varied and requires a range of skills.

In large firms, however, no single manager can be expected to be the jack-of-all-trades. As you saw in Chapter 11, the increasing base of knowledge forces professionals to specialize. Increased size and complexity of the organization cause managerial functions to become more specialized. Whether in the professions or in management, specialization always implies acquiring a particular type of knowledge. The two major ways to acquire this knowledge are through training and experience. Some managers are now *trained* to assume one particular management function. For example, personnel training prepares a person to recruit, test, train, assign, and evaluate workers. The public relations manager needs knowledge of press relations, advertising, community service activities, and so on. The training to perform these functions goes on in specialized programs, often within university business schools. The graduates could perform these functions, with minor modifications, in other firms. The phrase *professional manager* refers to someone who is occupationally prepared to take on such management functions within any workplace.

Alternatively, managerial specialization comes about through *experience* in one or more specialized departments within the organization. Experience implies specific, detailed knowledge of how things are done and how people work within a particular organization. Managers' experience depends on the functional departments they have managed during their careers. For example, the circulation manager of a large metropolitan newspaper develops expertise in setting up distribution networks, but not in gathering news. An account manager in an advertising firm becomes very familiar with the products and plans of a specific clientele, but might know very little about other clients. An administrator in a state highway department performs some of the same functions as an administrator in the state human services agency, but the specific decisions made in the two agencies are quite different. Managers' importance to their firms depends partly on their experience within the specific organization.

These two types of specialized knowledge, training and experience, presuppose very different types of preparation for management. The former approach assumes that the professional manager is one who has been trained, perhaps in an MBA program, to perform tasks needed in every organization. Such a manager should be able to move laterally from one organization to another, applying the expertise to the new situation. The latter approach assumes that the most important managerial knowledge comes from experience in a specific firm: knowledge of its production practices, its clientele, and its workers.

These two types of knowledge are not incompatible; someone trained as an MBA with specialized knowledge of one management function may also gain valuable experience in a particular organization and have both types of knowledge. But more typically, the professional manager is seen as one who moves laterally from one firm to another, joining the company's labor force through distinct entry ports. The manager with local experience is more often seen as one who stays with a single company for long periods, often being promoted within the company's internal labor market. In recruiting new managers, the existing management must choose between the options of lateral recruitment and internal promotion.

These two types of knowledge are related to the traditional distinction between staff and line managers. The line managers exercise bureaucratic authority over production of goods or services. The staff managers provide ancillary or auxiliary services, often of a skilled or technical nature, to the line managers. So, for example, line managers are knowledgeable about the technology and equipment used in production. Staff managers may be responsible for accounting and other financial services, legal services, personnel management, safety, marketing, and public relations. Staff managers may be blocked in seeking promotions because they lack knowledge about the central production processes. For this reason, they may seek advancement by moving laterally to similar (but perhaps better-paid) positions in other firms. Line managers are more likely to be promoted from within; moreover, their specialized knowledge might not be transferable to other firms.

If line managers can be promoted, a first-line supervisor could theoretically rise to chief executive officer because the first-line supervisor is most likely to have actual experience both with the production process and with managing workers. At one time, first-line supervisors were promoted within the ranks, at least in large manufacturing firms. Today, a sharp distinction is frequently made between first-line supervisors, who are often already at the top of their job ladder, and middle managers. Middle managers are recruited directly from the outside, often because they have credentials, such as college degrees, that foremen do not have. The middle management positions, because they are filled from the outside, are entry ports and the first rung on a separate career ladder reaching higher within management.

Changes in Technology

Managers manage people, production, and information. Technology affects how managers perform all three functions. It typically does so by increasing the amount and type of information available to managers. For years, a manager's *knowledge of workers* was a function of direct observation and perhaps rudimentary productivity records. New technology makes it possible for the manager to electronically eavesdrop on the speed and accuracy of clerical workers, technicians, and operatives using machinery of any sort. Managers potentially possess far more information about workers now than they ever did (Zuboff, 1988). On the one hand, the new technology may also expand the span of control of the managers, which means that fewer managers may be needed. On the other hand, the increased volume of information implies the need for hiring more managers. One reason for the succeeding rounds of employment and layoffs in management suites is resulting indecision about the appropriate staffing ratios of managers to other workers.

Technology affects a manager's *knowledge of production* in many ways. New machines, procedures, and methods of organizing work require the manager to learn new techniques. Technology may lead to changes in the size and characteristics of the company's workforce. What form that transition takes will depend on managerial decisions to purchase and deploy new equipment. Managers sometimes decide that a different number of workers, with a different skill mix, will be needed (McElroy and Hazzard, 1994). For example, the adoption of electronic data-processing equipment may lead to fewer clerical workers and more equipment operators. Jobs may be deskilled (as discussed in Chapter 8). Alternatively, the jobs may be upgraded to require new or different skills.

The staffing ratio (the ratio of workers to capital) is a management decision, and it has a major effect on the number of managers. Upgrading may involve giving new supervisory skills to some workers so that the overall number of managers can be reduced. Managers may also decide that each production process can now use fewer workers. In the short run, this will result in a smaller workforce, although in the long run new jobs may be created if the product market expands or the company becomes more competitive (Osterman, 1986). Alternatively, managers may decide to maintain the same staffing ratio but to reorganize the work into two shifts or in other ways redeploy the existing workforce.

Finally, technology affects a manager's *information base,* usually by dramatically expanding the available information. Managers may receive so much information that they simply cannot use it and suffer from "information overload." If managers receive conflicting types of information, they must decide which data are reliable. For example, a manager may demote or discipline all slow or inaccurate workers to boost average speed and accuracy. Another manager who relies on "eyes and ears" may judge a worker to be valuable to a company, even though electronic monitoring may reveal the worker to be only average or even below average in speed or accuracy.

Many service firms now use customer responses as a source of information for managers.

> A customer relations manager in a hotel chain reported that when top management received the results of customer surveys they wrote to middle managers, informing them of problems and complaints; middle managers had to respond within 30 days, notifying upper-level management of the ways in which they had rectified any faulty conditions. If no response was made by 30 days corporate management instigated formal proceedings against the manager . . . (Fuller and Smith, 1996:83).

How information will be used within various organizations is at least partly a function of the

---

BOX 12.4   A Checklist for Managers

1. Planning-coordinating
   a. setting goals and objectives
   b. defining tasks needed to accomplish goals
   c. scheduling employees, timetables
   d. assigning tasks and providing routine instructions
   e. coordinating activities of different substitutes to keep work running smoothly
   f. organizing the work
2. Staffing
   a. developing job descriptions for position openings
   b. reviewing applications
   c. interviewing applicants
   d. hiring
   e. contacting applicants to inform them whether they have been hired
   f. "filling in" when needed
3. Training-developing
   a. orienting employees, arranging for training seminars, and the like
   b. clarifying roles, duties, job descriptions
   c. coaching, acting as a mentor, "walk- ing" subordinates through tasks
   d. helping subordinates with personal development plans
4. Decision making–problem solving
   a. defining problems
   b. choosing between two or more alternatives or strategies
   c. handling day-to-day operational crises as they arise
   d. weighing trade offs, making cost–benefit analyses
   e. deciding what to do
   f. developing new procedures to increase efficiency
5. Processing paperwork
   a. processing mail
   b. reading reports, emptying the "in box"
   c. writing reports, memos, letters
   d. routine financial reporting and bookkeeping
   e. general desk work
6. Exchanging routine information
   a. answering routine procedural questions
   b. receiving and disseminating requested information
   c. conveying the results of meetings
   d. giving or receiving routine information over the phone
   e. attending staff meetings of an informational nature

---

prevalent models of management. In the next section we discuss different distinctions among management models.

TRACKING MANAGEMENT PERFORMANCE

Because managers are found in every type of establishment and industry, the actual content of their jobs ranges widely. As we have just discussed, four factors that affect management jobs are scale, environment, specialization, and technology. Each of these factors, however, is little more than an external constraint on how man-

agers play their roles. In practice, even a manager with a very specialized job will find that the job contains many different tasks. The manager must simultaneously seek good relationships with subordinates, peers, and superiors, without neglecting the specific tasks of directing, coordinating, and planning. The manager must pay attention to day-to-day operations without forgetting about the company's long-range goals and objectives. High-quality job performance depends on juggling many tasks. Depending on the company, the season, and the business climate, the aspects of the job that must be emphasized will change daily.

Sociologists have developed several ways to investigate the job performance of managers, and

7. Monitoring-controlling performance
   a. inspecting work
   b. walking around and checking things out, touring
   c. monitoring performance data (computer printouts, production, financial reports)
   d. preventive maintenance
8. Motivating-reinforcing
   a. allocating formal organizational rewards
   b. asking for input, participation
   c. conveying appreciation, compliments
   d. giving credit when due
   e. listening to suggestions
   f. giving feedback on positive performance
   g. increasing job challenges
   h. delegating responsibility and authority
   i. letting subordinates determine how to do their own work
   j. sticking up for the group to superiors and others, backing a subordinate
9. Disciplining-punishing
   a. enforcing rules and policies
   b. nonverbal glaring
   c. demotion, firing, layoff
   d. any formal organizational reprimand or notice
   e. criticizing a subordinate for negative performance

10. Interacting with others
    a. public relations
    b. contacting customers
    c. contact with suppliers, vendors
    d. external meetings
    e. community service activities
11. Managing conflict
    a. managing interpersonal conflicts between subordinates or others
    b. appealing to higher authority to resolve a dispute
    c. appealing to third-party negotiators
    d. trying to get cooperation or consensus between conflicting parties
    e. attempting to resolve conflicts between a subordinate and oneself
12. Socializing-politicking
    a. chitchat about family or personal matters
    b. informal joking around
    c. discussing rumors, hearsay, grapevine
    d. complaining, griping, putting others down
    e. politicking, gamesmanship

SOURCE: Fred Luthans and D. L. Lockwood, 1984, "Toward an Observation System for Measuring Leadership in Natural Settings." In *Leaders and Managers: International Perspectives on Managerial Behavior and Leadership,* edited by James G. Hunt, D. Hosking, C. Schriesheim, and R. Stewart. New York: Pergamon, p. 122. Reprinted by permission of the publisher.

we will describe two of them. One is *behavioral,* and it is based on the observation of individual managers; thus, it is a micro-level approach. The second emphasizes *organizational cultures,* and it generalizes from studies of large numbers of managers to develop more abstract models of management styles. It is a macro-level approach that is concerned more with the social structures that make managers behave similarly than with variations among individuals.

The Behavioral Approach

The behavioral approach to managerial performance emphasizes how managers choose to perform their roles from among the many activities available to them. Mintzberg (1973) observed five chief executives over five-day periods. As a result of this work, he reports that managers play ten major roles: figurehead, leader, liaison, monitor, disseminator, spokesperson, entrepreneur, disturbance handler, resource allocator, and negotiator. Each of these roles requires specific activities. Box 12.4 provides a checklist for specific activities that managers perform. So many activities are part of the managerial role that there can be substantial variation in how two managers perform the same job. The checklist in Box 12.4 can be used by trained observers to record the actual behavior of managers at time intervals throughout the day. Another way to gather such data would be to ask managers to recall their activities or to log them during the day. Using trained observers is probably better, however, because it avoids the problems of recall or of subtle distortion of one's performance.

Several studies have observed the actual time managers allocate to these tasks. Mintzberg (1973) found that top-level managers spend relatively little time on the activities traditionally thought to be central to the managerial role, such as long-range planning. Instead, their work consists mostly of superficial and reactive encounters. Later studies have confirmed that top managers spend much of their time interacting with others through short conversations and jokes that seem unrelated to work, but actually help to build workplace communication (Kotter, 1982).

One study of successful and unsuccessful managers used the checklist in Box 12.4 to identify what successful managers really do on the job. In this study, the successful managers were identified through their relatively rapid promotion records. They were more likely than others to engage in "networking" and other forms of informal contact with their subordinates and superiors. Although this networking often did not involve exchanging any substantive information, it did serve to maintain good horizontal and vertical relationships. Then when there was important information to be exchanged, the successful manager was in a good position to receive or transmit it. Successful managers were also more oriented toward conflict management and spent more time on decision making, planning, and coordinating (Luthans, Rosenkrantz, and Hennessey, 1985). Studies of successful managers also emphasize that beyond the horizontal and vertical relationships within the firm, managers must also maintain good relationships with outsiders who affect the environment of the firm.

The *simulation* is a type of study in which a group of managers is asked to solve a hypothetical problem. The managers are observed during this exercise to better understand how they reach

a decision, which factors they consider most important, and how they interact with one another. This technique can also be used to encourage team development, which has become more important with the advent of such practices as matrix supervision (see Chapter 7). It can also be used to diagnose problems. For example, one simulation revealed that the managers were too deskbound, wrote too many memos, and did not communicate well along the horizontal dimension because they preferred vertical communications (Kaplan, Lombardo, and Mazique, 1985).

American managers, when compared with their counterparts in other industrialized countries, have been accused of failing to make good decisions or failing to emphasize the most important roles in their job. Box 12.5 presents one criticism of the behavior of American corporate managers.

The Organizational Culture Approach

The organizational culture approach to managerial behavior identifies unique features of organizations that affect the behavior of managers as a group. The organization, like a society, is said to have a distinctive culture (Trice and Beyer, 1992). The major components of the organizational culture that have been identified are climate, supervisory leadership, peer leadership, and group process. These components affect production costs and financial outcomes, and they also affect individual workers and their behaviors, including absenteeism and job satisfaction (Hodson, 1999; Steele and Hubbard, 1985). These analyses of management behavior are closely tied to the sociological study of organizations. We will review three classifications of organizational culture: scientific management, human relations, and the so-called new school. These ideas were presented in Chapter 7 as Theories X, Y, and Z. Here we examine the effect of these theories on managers.

Scientific management (Theory X) has already been discussed as one stage in the development of labor control. It also has important implications for how managers behave. The scientific management school views workers as motivated principally by economic rewards, and it views management's task as distributing rewards and punishments. This version of management is paternalistic, in that managers make virtually every decision about workers' performance and the views of workers are rarely sought. This approach is also very hierarchical, and lower-level managers often find themselves treated paternalistically by those above them in the hierarchy. The successful manager pays close attention to vertical relationships, watches productivity carefully, and emphasizes rewards and punishments. The successful manager also looks for ways to simplify tasks and to make the workers as interchangeable as possible—processes that minimize the costs of turnover. High turnover is not by itself a sign of poor management but is attributed to worker characteristics.

The *human relations approach* (Theory Y) views workers as motivated both by extrinsic rewards, such as money, and by intrinsic rewards, such as job satisfaction. In this approach, management's task is to make work more pleasant while still maintaining productivity. This might mean redesigning the jobs to make them more interesting, but it is more likely to mean that the company provides amenities such as an attractive lunchroom, lighted parking areas, and so on. Managers are expected to pay close attention to their relationships with their subordinates. High turnover is considered a warning signal.

The *new approach* (Theory Z) emphasizes a much closer, almost fraternal relationship between managers and workers (Kelley, 1985; Ouchi, 1981). This school views workers as highly trained and skilled partners in production; it is often recommended for managers who supervise knowledge workers. Some examples of appropriate work sites include research and development departments, high-technology firms, organizations that hire many professionals (hospitals or universities), or organizations composed chiefly of professionals (law firms or engineering firms). This organizational culture recognizes that managers have responsibilities to allocate resources and otherwise maintain the organization but that efficient production requires teamwork by both managers and workers. Because the workers have specialized knowledge, their cooperation and performance are essential to the firm's success. In this model, high turnover is a sign of management failure.

The argument has been made that this model is appropriate for all work sites, even those in which most workers are not highly trained, because all workers have developed some expertise and may have insights into improving the production process. Because this theory emphasizes worker participation in some types of decision making, it is associated with such organizational practices as climate surveys and quality circles. Climate surveys assess workers' perceptions of the workplace. Quality circles are informal discussion groups composed of workers from various levels of the organization who meet periodically to discuss improving the work process and product.

Studies of organizational climate are very important for understanding the managerial occupations. A good climate may boost productivity and morale, just as a poor climate may lead to deterioration in the quality of products and in the relationships among the workers. No single manager, however, can change the organizational climate to any significant degree. Instead, new managers learn to work within an organizational climate, and their success usually requires being able to reproduce that climate. Studying organizational climate illuminates the roles managers play in maintaining the climate, and the ways in which they organize to influence change in the climate.

BOX 12.5   How American Managers Fail

Many reasons have been offered for the decline in competitiveness of American industries. In this selection, two professors of business administration at the Harvard Business School argue that American managers should claim some responsibility for the decline.

In the past, American managers earned worldwide respect for their carefully planned yet highly aggressive action across three different time frames:

   Short term—using existing assets as efficiently as possible.

   Medium term—replacing labor and other scarce resources with capital equipment.

   Long term—developing new products and processes that open new markets or restructure old ones.

   The first of these time frames demanded toughness, determination, and close attention to detail; the second, capital and the willingness to take sizable financial risks; the third, imagination and a certain amount of technological daring.

   Our managers still earn generally high marks for their skill in improving short-term efficiency, but their counterparts in Europe and Japan have started to question America's entrepreneurial imagination and willingness to make risky long-term competitive investments. As one such observer remarked to us: "The U.S. companies in my industry act like banks. All they are interested in is return on investment and getting their money back. Some times they act as though they are more interested in buying other companies than they are in selling products to customers."

   We refuse to believe that this managerial failure is the result of a sudden psychological shift among American managers toward a "super-safe, no risk" mind set. No profound sea change in the character of thousands of individuals could have occurred in so organized a fashion or have produced so consistent a pattern of behavior. Instead we believe that during the past two decades American managers have increasingly relied on principles which prize analytical detachment and methodological elegance over insight, based on experience, into the subtleties and complexities of strategic decisions. As a result maximum short-term financial returns have become the overriding criteria for many companies.

   For purposes of discussion, we may divide this new management orthodoxy into three general categories: financial control, corporate portfolio management, and market-driven behavior.

Financial Control

As more companies decentralize their organizational structures, they tend to fix on profit centers as the primary unit of managerial responsibility. This development necessitates, in turn, greater dependence on short-term financial measurements like return on investment (ROI) for evaluating the performance of individual managers and management groups. Increasing the structural distance between those entrusted with exploiting actual competitive opportunities and those who must judge the quality of their work virtually guarantees reliance on objectively quantifiable short-term criteria.

   Although innovation, the lifeblood of any vital enterprise, is best encouraged by an environment that does not unduly penalize failure, the predictable result of relying too heavily on short-term financial measures—a sort of managerial remote control—is an environment in which no one feels he or she can afford a failure or even a momentary dip in the bottom line.

THE FUTURE OF EXECUTIVES, MANAGERS, AND ADMINISTRATORS

The Bureau of Labor Statistics predicts continued growth in the number of executives, managers, and administrators from 8.4 million in 1992 to 10.4 million in 2005 (Silvestri, 1993). Every organization will continue to need workers who provide leadership and coordination. But as organizations change, the number of managers will change and the tasks they perform will also change. As this chapter indicates, alterations in the characteristic forms of technology will also

Corporate Portfolio Management

This preoccupation with control draws support from modern theories of financial portfolio management. Originally developed to help balance the overall risk and return of stock and bond portfolios, these principles have been applied increasingly to the creation and management of corporate portfolios—that is, a cluster of companies and product lines assembled through various modes of diversification under a single corporate umbrella. When applied by a remote group of dispassionate experts primarily concerned with finance and control and lacking hands-on experience, the analytic formulas of portfolio theory push managers even further toward an extreme of caution in allocating resources. "Especially in large organizations," reports one manager, "we are observing an increase in management behavior which I would regard as excessively cautious, even passive; certainly over-analytical; and, in general, characterized by a studied unwillingness to assume responsibility and even reasonable risk."

Market-Driven Behavior

In the past 20 years, American companies have perhaps learned too well a lesson they had long been inclined to ignore: businesses should be customer oriented rather than product oriented. Henry Ford's famous dictum that the public could have any color automobile it wished as long as the color was black has since given way to its philosophical opposite: "We have got to stop marketing makeable products and learn to make marketable products."

At last, however, the dangers of too much reliance on this philosophy are becoming apparent. As two Canadian researchers have put it: "Inventors, scientists, engineers, and academics, in the normal pursuit of scientific knowledge, gave the world in recent times the laser, xerography, instant photography, and the transistor. In contrast, worshippers of the marketing concept have bestowed upon mankind such products as new-fangled potato chips, feminine hygiene deodorant, and the pet rock. . . ."

The argument that no new product ought to be introduced without managers undertaking a market analysis is common sense. But the argument that consumer analyses and formal market surveys should dominate other considerations when allocating resources to product development is untenable. It may be useful to remember that the initial market estimate for computers in 1945 projected total worldwide sales of only ten units. Similarly, even the most carefully researched analysis of consumer preferences for gas-guzzling cars in an era of gasoline abundance offers little useful guidance to today's automobile manufacturers in making wise product investment decisions. Customers may know what their needs are, but they often define those needs in terms of existing products, processes, markets, and prices.

Deferring to a market-driven strategy without paying attention to its limitations is, quite possibly, opting for customer satisfaction and lower risk in the short run at the expense of superior products in the future. Satisfied customers are critically important, of course, but not if the strategy for creating them is responsible as well for unnecessary product proliferation, inflated costs, unfocused diversification, and a lagging commitment to new technology and new capital equipment.

change the day-to-day functioning of the manager. Many of the information tasks that were formerly done by managers are today delegated to clerical workers, who also serve an important function in coordinating information. We will consider clerical workers in greater detail in Chapter 13.

SUMMARY

Managerial occupations vary a great deal among firms and even within firms. Self-employed proprietors are considered to be managers, but so are the top executives of multinational corporations. What the jobs have in common is the necessity to make decisions, control and allocate resources, and provide for the maintenance and growth of organizations. Concern about management behavior and organizational climates is directed toward understanding how managers can improve on juggling the many roles that are part of their job. Larger issues are being raised about the ill effects on society of having managers' identities closely tied to those of their organizations. References to "the organization man" or "the man in the gray flannel suit" underline the sense in which the manager is socially, intellectually, and morally dominated by the firm in which he or she is employed. More recently, women and members of minority groups are beginning to join white Protestant men in the management suite.

KEY CONCEPTS

manager	span of control	first-line supervisor
executive	steepness of hierarchy	horizontal dimension of bureaucracy
administrator	mentor	
line manager	top management	vertical dimension of bureaucracy
staff manager	middle management	

QUESTIONS FOR THOUGHT

1. Layoffs of managerial personnel have tended to target middle managers and staff managers. Why are these positions in particular jeopardy?

2. Is there such a thing as a "dead-end job" in management? What kind of job might it be?

3. A researcher discovers that minority and women managers are more likely to be in staff positions than in line positions. How would you explain this finding?

4. How do managers' responsibilities differ when considering the vertical dimension of bureaucracies as opposed to the horizontal dimension of bureaucracies?

5. How is technology likely to change the span of control and the steepness of hierarchies?

MULTIMEDIA RESOURCES

Print

David M. Gordon. 1996. *Fat and Mean: The Corporate Squeeze of Working Americans and the Myth of Managerial "Downsizing."* New York: Free Press. An economist's analysis of the role that bloated corporate management plays in squeezing the wages of ordinary workers.

Sherryl Kleinman. 1996. *Opposing Ambitions: Gender and Identity in an Alternative Organization.* Chicago: University of Chicago Press. A case study of corporate culture in a medical center and its consequences for the men and women who work there.

Anne E. Kingsolver, ed. 1998. *More Than Class: Studying Power in U.S. Workplaces.* Albany: State University of New York Press. Each chapter of this edited volume offers an ethnographic glimpse of a workplace and how power is distributed within it.

Ann M. Morrison. 1996. *The New Leaders: Leadership Diversity in America.* San Francisco: Jossey-Bass. An examination of women and minority executives and the factors that lead to their success.

Rosabeth Moss Kanter. 1983. *The Change Masters.* New York: Simon and Schuster. An analysis by a leading sociologist of management's role in bringing about change in the economy and society.

Websites

Academy of Management. *http://www.aom.pace.edu/* This website belongs to the professional group that studies management and publishes some of the leading journals reporting research on business.

Forbes. *http://www.forbes.com/tool/toolbox/forbes500s/* The website for *Forbes Magazine's* annual listing of the 500 largest corporations and their characteristics.

http://www.lycos.com/business A website for small-business owners, with information on a wide variety of topics.

RECOMMENDED FILM

Breakfast of Champions. 1999. In this adaptation of the Kurt Vonnegut novel, Bruce Willis plays Dwayne Hoover, the most important man in Midland City, the owner-manager of a large automobile dealership, but someone who is experiencing internal turmoil. He encounters an impoverished writer named Kilgore Trout (Albert Finney) and their two worlds collide. Rated R.

13

Clerical and Sales Workers

Secretaries, clerks and bureaucrats were once grateful for having been
spared the dehumanization of the factory. . . . They had higher status
than blue-collar [workers]. . . . But today . . . such positions offer
little in the way of prestige . . . imparting to the clerical worker the
same impersonality that blue-collar workers experience in the factory.

(U.S. DEPT. OF H.E.W., *WORK IN AMERICA,* 1973:38)

Selling is when it matters how you and your customers get to know each
other, how you help her find what's best, how you understand what she
needs. Here you no sell. Lady comes in, asks for something, you go find it,
she buys or no buys, that's it, you never see her no more. That's not
selling. That's showing. But most of the time, you don't do that even.
You're putting labels on corsets, or emptying boxes.

—MARLENE, A DEPARTMENT STORE SALES WORKER (HOWE, 1977:87)

Clerical and sales occupations are sometimes labeled the lower-level white-
collar jobs. Figure 13.1 shows the growth of the clerical and sales workforce. The clerical occupations, categorized in U.S. government publications
as "administrative support, including clerical," have grown rapidly in recent
years. By 2001, the 14.4 million full-time clerical workers accounted for over

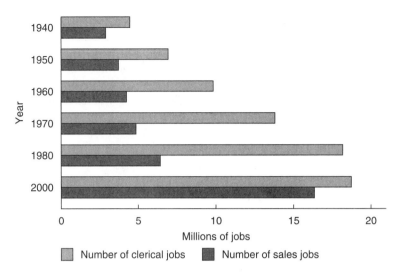

FIGURE 13.1 Number of Clerical and Sales Jobs, 1940–2000 (millions of jobs)

SOURCES: Census, 1952, *Statistical Abstract of the United States, 1952.* 73d edition. Washington, D.C.: U.S. Government Printing Office, table 221, p. 186. Also, Census, 1981, *Statistical Abstract of the United States, 1981.* 102d edition. Washington, D.C.: U.S. Government Printing Office, table 673, p. 401; and BLS, 2001, *http://www.bls.gov/pdf/cpsaatll.pdf.*

14% of full-time workers, and the 10.1 million full-time sales workers accounted for about 10% of the full-time workers (BLS, 2001).

Size alone makes clerical and sales work important to study. In addition, the two groups have in common a rapid transformation driven by the use of technology in offices and stores. The outcome of this transformation is by no means certain. Demand for workers might increase, or new machinery might displace them. Newly created jobs might be more interesting and have more complex tasks, with each worker assuming greater autonomy. Or the transformation might result in even more specialized, boring, and repetitive jobs that reproduce factorylike conditions in offices and stores. The outcomes depend on the choices made by different firms. Furthermore, the new technology might blur the distinction between sales and clerical jobs. In this chapter, we will examine the origins of clerical and sales work, the demand for and the supply of clerical and sales workers, and the ways that technology is changing these jobs.

HISTORY OF CLERICAL WORK

Clerks have been needed since written records were developed. When most Europeans were illiterate peasants, work that required literacy was performed by clerics—that is, those in priestly orders. The terms *clerk* and *clerical* originated at that time. Bishops, abbots, and other high church officials required secretaries and other clerks for such tasks as reading and writing correspondence and keeping business accounts and inventories. Thus began a close association between clerical work and the professions; in this case the profession was the clergy. We preserve this older usage of the term *clerk* today when we speak of the law clerk, who is usually a law student or a new attorney performing tasks under the close supervision of a judge or an experienced attorney. Unlike many other clerical workers who work with professionals, most law clerks will eventually assume full-fledged professional positions. In general, however, clerical workers cannot expect to become professionals without leaving their jobs and acquiring further training. Remember that we discussed the training of professionals and entry into the professions in Chapter 11. The work of clerks is subordinated to that of professionals, although most professionals require a support staff of clerical workers in addition to any semiprofessionals or paraprofessionals whom they employ.

Clerks were also needed in secular pursuits. Monarchs and their representatives needed clerks to transcribe rulings and laws, send messages, and maintain records. Feudal lords required at least rudimentary records on land use, agricultural products, taxes, and levies. With the rise of mercantile capitalism, merchants and shippers needed clerks who could keep accounts, maintain correspondence, and generally assist the owner. This was the origin of the close association between clerical work and managers that is retained today in job titles such as confidential secretary.

Clerks replaced neither professionals nor managers in making decisions. Nevertheless, when reading and writing were skills in relatively short supply, the clerical worker was well regarded and had relatively high prestige. A good clerk might expect raises and perhaps promotion, because the clerkship served as a training ground and apprenticeship for management. The clerk knew intimately how the business was run and what its financial situation was, so that the clerk often became the confidant of the business owner and might eventually become a partner in the business.

Work that required reading and writing was generally perceived to be clean; hence, the clerical worker could wear a white collar and dress in a style that approached that of the gentleman.

Before the U.S. Civil War, the clerical occupations were almost entirely male. Among other reasons, men were more likely to have received the education needed to perform clerical tasks, and they were viewed as more appropriate apprentices. Although some women completed high school or even college at this time, they were much less likely to be working outside the home. As we shall see, the male domination of the clerical occupations was to be completely reversed, and that reversal was accompanied by a number of other changes.

Demand for Clerical Workers

The changes began when the number of clerical jobs increased. Between 1870 and 1930, the clerical workforce grew from 76,000 to 3.8 million (Census, 1943). The last half of the nineteenth century saw the growth of large, bureaucratic organizations, the forerunners of today's giant corporations. Government agencies, too, proliferated. As we saw in Chapter 7, the growth of these organizations resulted in a heightened need for more efficient organization. Just as there was a division of labor among workers, there came to be a division of labor within companies. Departments and other organizational groups specialized to perform certain tasks (Chandler, 1980). For example, a typical company developed divisions for sales, orders, production, billing, and shipping. Both the larger scale and the new organizational structures increased the requirements for coordination and record keeping (Mills, 1951). There were also changes in the kind of records kept. After 1910, the widespread use of cost accounting led to a greatly increased volume of financial records (Strom, 1987). These changes required additional workers to prepare, send, receive, and file the records.

With the greater numbers of clerical workers came increased specialization. During the mid-nineteenth century there were four basic clerical jobs (Davies, 1982). Before telephones the messenger (or office boy) was needed to run errands; he also did housekeeping chores around the office and was often studying bookkeeping or some other office skill for possible promotion within the firm. Before the invention of carbon paper or photocopying, copyists were required to write and rewrite documents. Some copyists could eventually aspire to become clerks, but many never advanced. Most businesses kept only rudimentary accounts, but later there was a need in larger firms for bookkeepers to keep track of sales, debts, and credit extended. Some bookkeepers might eventually be promoted, but as bookkeeping became more complex, their mobility was often limited to the accounts department. Their training was so specialized that some business owners preferred not to lose bookkeepers to supervisory or management positions. The clerk, the fourth specialist, often had minor management authority and had probably acquired all or most of the other skills. In small firms one person did the tasks of all four specialties.

Greater occupational specialization arose from two sources. One source was the *reorganization of work*. Some occupations, such as personnel worker, file clerk, or record clerk, arose from the consolidation of similar tasks previously performed in each division of the organization. A second source of specialization was the development of *office equipment*. New occupations were developed and named according to the technology that was used. Typists operated typewriters, and there were billing-machine operators and switchboard operators; today there are word processors. Appendix Table 1 indicates the variety of clerical occupations.

In smaller firms the clerical worker (often called a secretary) was not so specialized and often performed all sorts of tasks, including transcribing dictation, keeping records, filing papers, and running personal errands for the boss. A small office had neither enough workers nor enough office machines to justify a higher degree of specialization. For the all-purpose secretary, clerical work involved variety and often a good deal of autonomy. Specialized clerical workers in large firms, however, were more likely to find their work monotonous and closely supervised.

The demand for clerical workers remains strong. One indication of strong demand is their relatively low unemployment rate, which in January 2001 was 3.4%, compared with the overall average of 4.2% (BLS, 2001b).

Supply of Clerical Workers

The increased demand for clerical workers during the latter half of the nineteenth century led employers to look for new sources of workers. During the Civil War, the U.S. Treasury Office became one of the first employers to experiment with female clerks. The young males who were the usual clerical workers were in uniform, and women were sought as their replacements. Box 13.1 discusses this experiment, which the government evaluated and generally considered to be successful. Other employers repeated the experiment and a major change was under way— the change in clerical occupations from mostly male to mostly female.

The Feminization of Clerical Work Today's clerical labor force has been largely *feminized*. In 2000, 98% of all secretaries were female, and 79% of all clerical workers were female. Over 25% of all women workers were in the clerical fields (BLS, 2001a). How did this transformation occur? During the last quarter of the nineteenth century, the demand for clerical workers increased sharply. Employers were concerned about finding a good supply of skilled workers while minimizing their wage costs. Native-born white women became the clerical labor supply for four basic reasons: (1) they often had the necessary training; (2) they accepted lower pay than men; (3) clerical work was attractive to them because of its relatively high status; and (4) competing sources of labor lacked the necessary language skills and education.

First, because of educational reforms, women had the *necessary training* to perform clerical tasks. Universal public education meant that girls as well as boys had the necessary literacy skills for clerical work. By 1890, the number of women

BOX 13.1   The Civil War Experiment with Female Clerks

Today most clerical workers are female, but once they were mostly male. This selection discusses an experiment made by the U.S. government in hiring women for clerical work. This passage hints at some of the reasons for the "feminization" of clerical work: there was a willing supply of well-trained women workers available, and they were willing to work for lower pay than the male clerks received.

During the Civil War the U.S. Treasurer General, Francis Elias Spinner, confronted a severe labor shortage caused by the large numbers of men in Union uniforms. He decided over considerable opposition to hire some female clerks, who worked at relatively mechanical tasks such as sorting and packaging bonds and currency. This "experiment" was continued after the war and in 1869 Spinner declared "upon his word" that it had been a complete success: "Some of the females doing more and better work for $900 per annum than many male clerks who were paid double that amount." Such wage figures indicate one of the reasons Spinner thought so

highly of his experiment: female labor was cheaper than male. A contemporary claimed that most of these early female clerks got their positions through political patronage, with the result that some of them were not well trained for their jobs and had to take writing lessons after they were employed. But a study of federal government clerks from 1862 to 1890 has found that, by and large, the women did have sufficient education for clerical work, most of them having remained in school at least until the age of sixteen. They came overwhelmingly from white, native-born, middle-class families and were the daughters of men with jobs that ranged from clerks to judges; almost none of them were the daughters of craftsmen, much less unskilled laborers. Whether they were single, widowed, or, less frequently, married, these women sought clerical jobs out of economic necessity.

SOURCE: Margery W. Davies, 1982, *Woman's Place Is at the Typewriter: Office Work and Office Workers, 1870–1930*. Philadelphia: Temple University Press, p. 51. © 1982 by Temple University Press. Reprinted with permission of Temple University Press.

---

receiving high school diplomas had outstripped the number of men receiving diplomas (Davies, 1982:57). When clerical work required skills beyond literacy, "commercial courses" such as shorthand and bookkeeping were added to the high school curriculum. Many proprietary trade schools offered secretarial training to high school graduates, and by about 1895 a third of their students were women (Davies, 1982:73). By 1924, women composed two-thirds of the enrollment in high school business courses (Strom, 1987:85).

The provision of high school and post–high school business education for young women meant that employers provided minimal training for new workers. This result was important because they expected a female labor force to have high turnover, which increases training costs because workers must be trained even if they stay on the job for only a few months. Before being hired, a new secretary was expected to have

already mastered the basic office skills, and that mastery could be verified with tests of typing, spelling, and dictation.

Second, women workers traditionally received *lower pay* than men. Employers expected women to accept less pay because they expected the women to work only a short time before getting married. Employers also justified lower pay because typewriters were thought to make the job easier by reducing its physical demands. When bookkeeping machines were introduced, they were usually operated by women, who were paid less than the more skilled male bookkeepers who knew how to do all of the bookkeeping tasks by hand (Strom, 1987). Thus, even though the numbers of clerical workers grew, substituting women for men helped contain total labor costs.

The rate of substitution varied, however, depending on how intensively the firm used clerical workers. In industries with large numbers of clerical workers, large savings resulted from hiring women. In industries with only a few clerical workers, men tended to keep the clerical jobs because the differential in the wage bill was relatively small (Bridges, 1982; Cohn, 1985a, 1985b; Lowe, 1987). Some clerical specialties continue to be predominantly male; these are often in workplaces in which the clerical specialty is closely related to production and in which production workers may be promoted into clerical jobs. Examples of such jobs are meter readers and clerks who do traffic routing, shipping, and receiving.

Third, office work had relatively *high status,* which attracted women from both the middle class and the working class. Although men of all classes and working-class women had other job opportunities, few appropriate jobs were open to middle-class women who wanted to work. Offices were seen as "clean," suitable workplaces for women workers. Middle-class women were not interested in the many jobs available as factory hands and servants, but they perceived office work as consistent with their position in the community. For working-class women, office work provided a channel of upward mobility from the farm or the factory. In addition, although women were paid less than men, a woman could still earn more in clerical work than she could in almost any other line of work. In 1883, when the average weekly wage for women in manufacturing was about $5, copyists earned $6.78, and cashiers $7.43 (Davies, 1982:64). The difference in hourly wages was even more pronounced because women in offices often worked shorter hours than the women in manufacturing. Ironically, however, as women came to dominate the clerical fields, these occupations often lost status, at least in part because they became identified as "women's work."

Fourth, *language skills and education requirements* generally disqualified the competing sources of cheap labor. At the time that the clerical labor force was feminizing, immigrants from Europe were the major source of new workers. The immigrants, however, were less likely than the native-born to be literate and much less likely to have an excellent command of English. Good written and spoken English remained an important qualification for virtually every clerical job. But prejudice against the immigrants also was strong. Even immigrants who spoke English, such as the Irish in Boston, were far less likely than native-born women to be hired for clerical work (Handlin, 1973).

Another possible pool of competing labor, African-Americans migrating from the South, were generally not considered by employers for clerical work. One rationalization for this discrimination was the belief that the schooling of African-American women had been so inferior that they could not perform clerical tasks adequately. It is equally likely that both the employers and their white employees resented the idea of sharing the relatively high status and clean white-collar office work with African-Americans. After World War II, however, minority women became an important source of clerical labor.

Minority Women as Clerical Workers
African-American women were first hired as clerical workers in businesses owned by other African-Americans. When these women were hired in other firms, they tended to be in "back-office" jobs that required few face-to-face interactions with the public. As an example, one relatively early employer of African-American women was the Montgomery Ward mail-order house (Strom, 1987:85). These workers filled mail orders and never interacted with the customers. The Depression in the 1930s reinforced the disadvantaged position of African-American women who wanted clerical jobs, because competition for the jobs with white women was intense.

The continued growth of clerical employment after World War II drew minority women into the field, but they entered the least skilled and most poorly paid jobs. White women are more likely to be found among secretaries and receptionists, and minority women are more likely to be found among typists, file clerks, and mail clerks. Nevertheless, the number of minority women has been rising rapidly, even in the predominantly white specialties: Between 1970 and 1980, the number of black female secretaries increased by 148% and the number of Hispanic female secretaries by 131% (Murphree, 1986:100). By 2000, nearly 14% of all clerical workers were black and nearly 10% were Hispanic.

Opposing Trends in Clerical Work Continued occupational specialization within the clerical fields tended to follow one of two trends, each of which affected the supply of workers. One trend was *professionalization,* which we explored in Chapter 11. One objective of professionalization is to "reserve" jobs for members of a profession and to protect them from competition. Members generally seek to upgrade entry requirements and to define an area of knowledge that only specialists master. By restricting the supply of workers only to those with the appropriate skills, professionalizing fields are able to command higher wages. Among clerical workers, professionalization predominated in areas dominated by men. Such areas as accounting, personnel administration, and actuarial science became men's fields. After women started entering bookkeeping in large numbers, male accountants professionalized the work of bookkeepers, which had the effect of protecting accountants' positions and limiting the access of women.

Professionalization also had its limiations, however. The newly professionalized specialists were often passed over for promotion to management jobs because their knowledge was narrow. Furthermore, the specialized training required for professionalization was provided by colleges and universities that did not necessarily restrict the entry of women into the field. Women have now gained access to these fields through university training. By 2000, nearly 57% of accountants and auditors were women, and women accounted for over 66% of personnel,

training, and labor relations specialists and 72% of underwriters (BLS, 2001a).

The second trend was the development of *clerical operatives,* who specialized in specific pieces of office equipment. Their jobs were defined by the machinery they operated rather than by their knowledge. This knowledge was available in vocational education courses or in commercial business schools. Many office workers learned to use only one or two pieces of equipment. New technology could render them and their machinery obsolete. Because they were often perceived as relatively short-term workers, however, their narrow specialization was a distinct advantage from management's perspective. If they left for marriage or family reasons, their replacements could be quickly trained. Figure 13.2 shows the continued rise and fall of various clerical occupations between 1950 and 2000. Typists originally rose, declined, and then rose again a bit as word processing became more prevalent. Data entry, which rather clearly involves specialization in one type of equipment, has always been done mainly by women in positions such as keypunchers and word processors.

The changing gender composition of the occupation of computer operator represents an interesting mix of both trends. Computer work was originally seen after World War II as a rather limited clerical specialty, and the first computer operators were women. Within a few years, however, men were being hired in these jobs. In 1960, the occupation was 65% female, but by 1970 it was only 29% female. The expansion of computer science, an engineering-based and hence stereotypically male area of endeavor, may have stimulated the influx of men. The trend again reversed, however; by 2000, 49% of computer operators were women.

What is common to both the professionalization and specialization trends is that the costs of training are usually borne by the workers themselves as part of their job preparation. The employer provides relatively little training except when new technology is being introduced.

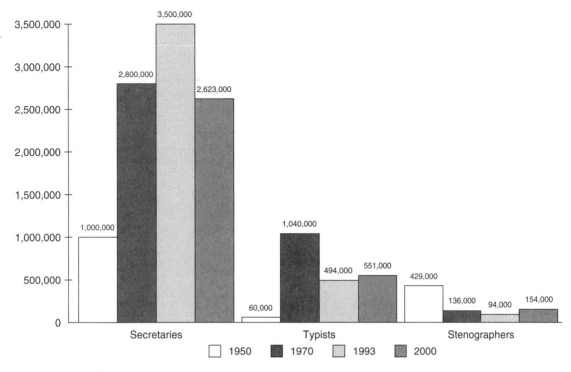

FIGURE 13.2 Rise and Fall of Clerical Occupations, 1950–2000

SOURCES: Allan H. Hunt and Timothy L. Hunt, 1987, "Recent Trends in Clerical Employment: The Impact of Technological Change." In *Computer Chips and Paper Clips,* edited by Heidi I. Hartmann. Washington, D.C.: National Academy Press, pp. 228–229. Also, BLS, 1994a. *Employment and Earnings* 41.1 (January): 206–207, table 22; BLS, 2001, *http://www.bls.gov/pdf/cpsaatll.pdf.*

TRANSFORMING THE CLERICAL OCCUPATIONS

The office is sometimes portrayed as playing the same role for an organization that the brain plays in a living organism. The office receives information flowing in from all parts of the organization (or organism) and from the external environment, processes that information and sends back responses, instructions, and commands through an extended nervous system—established channels of communication. In every enterprise, there is a need for an office, or a central location for information management. Two developments, new technology and work reorganization, are responsible for dramatic recent changes in office work.

Office Technology

For the last 125 years, technology has had an important impact on the nature of clerical work. Figure 13.3 indicates the relative timing of the introduction of new office technology. The typewriter and the telephone, both introduced between 1870 and 1880, revolutionized even the small office, where the all-purpose secretary had to master them. In larger offices the new machines became the basis for displacing some workers, changing the job content for others, and developing new specialties. The telephone made the office messenger nearly obsolete, changed the job of receptionist, and created jobs for switchboard operators. The typewriter made the copyist obsolete, changed the job of secretaries, and

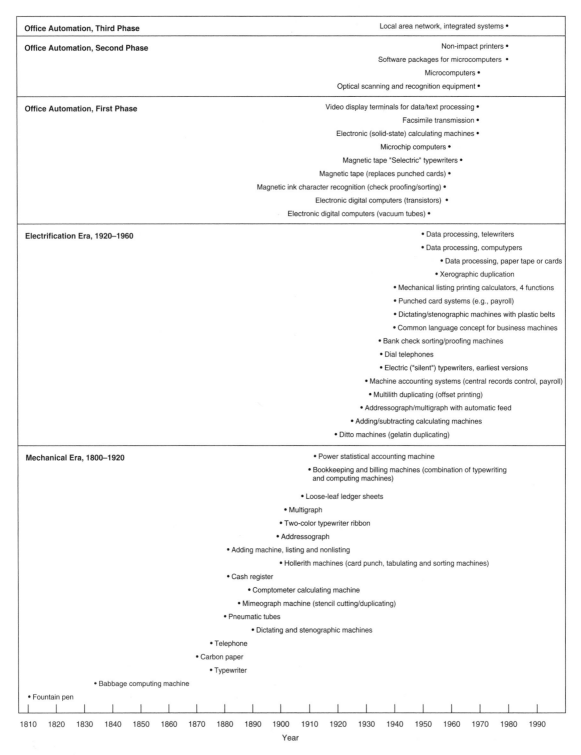

Office Automation, Third Phase
Local area network, integrated systems •

Office Automation, Second Phase
Non-impact printers •
Software packages for microcomputers •
Microcomputers •
Optical scanning and recognition equipment •

Office Automation, First Phase
Video display terminals for data/text processing •
Facsimile transmission •
Electronic (solid-state) calculating machines •
Microchip computers •
Magnetic tape "Selectric" typewriters •
Magnetic tape (replaces punched cards) •
Magnetic ink character recognition (check proofing/sorting) •
Electronic digital computers (transistors) •
Electronic digital computers (vacuum tubes) •

Electrification Era, 1920–1960
• Data processing, telewriters
• Data processing, computypers
• Data processing, paper tape or cards
• Xerographic duplication
• Mechanical listing printing calculators, 4 functions
• Punched card systems (e.g., payroll)
• Dictating/stenographic machines with plastic belts
• Common language concept for business machines
• Bank check sorting/proofing machines
• Dial telephones
• Electric ("silent") typewriters, earliest versions
• Machine accounting systems (central records control, payroll)
• Multilith duplicating (offset printing)
• Addressograph/multigraph with automatic feed
• Adding/subtracting calculating machines
• Ditto machines (gelatin duplicating)

Mechanical Era, 1800–1920
• Power statistical accounting machine
• Bookkeeping and billing machines (combination of typewriting and computing machines)
• Loose-leaf ledger sheets
• Multigraph
• Two-color typewriter ribbon
• Addressograph
• Adding machine, listing and nonlisting
• Hollerith machines (card punch, tabulating and sorting machines)
• Cash register
• Comptometer calculating machine
• Mimeograph machine (stencil cutting/duplicating)
• Pneumatic tubes
• Dictating and stenographic machines
• Telephone
• Carbon paper
• Typewriter
• Babbage computing machine
• Fountain pen

1810 1820 1830 1840 1850 1860 1870 1880 1890 1900 1910 1920 1930 1940 1950 1960 1970 1980 1990
Year

FIGURE 13.3 History of Technology Used in the Office

SOURCE: U.S. Congress, Office of Technology Assessment, 1985, *Automation of America's Offices,*
1985–2000. OTA-CIT-287, December, Washington, D.C.: U.S. Government Printing Office, p. 9.

336

created a whole new occupational specialty, the typist. As you learned in Chapter 7, these three outcomes—displacement, transformation, and the emergence of new occupational specialties—are typical responses to new technology.

The typewriter initially sparked controversy, and its adoption was relatively slow (Davies, 1982). Managers were skeptical that typing would ever be faster than handwriting. There were also concerns that typed correspondence, instead of the handwritten letter, was inappropriate business etiquette. Today, some managers express parallel concerns about the propriety of electronic correspondence, but a typed letter is likely to be seen as more businesslike than a handwritten one. Eventually, the typewriter was widely adopted and typing courses were instituted in many schools.

Typist training programs, such as those offered by YWCAs beginning in 1881, were aimed at young women and specifically oriented toward teaching them a useful job skill. Offering such courses was controversial, for several reasons. Giving women job skills was seen as encouraging them not to marry and have families; it was also seen as a way to deprive men of jobs needed to support their families. Related concerns were expressed that working at a typewriter might be dangerous to the health and constitution of young women and, by implication, to their childbearing. An equally ideological counterresponse was soon forthcoming: Women had more manual dexterity and endurance for repetitive tasks, and so they were better "suited" to be typists. Typing, it was argued, reduced the stamina required for clerical work and made clerical jobs more suitable for women.

Much like the copyist, who was unlikely to experience upward mobility in the office, the typist was expected to do only one thing and might not learn any other office skills. Many offices sought to reduce the number of skilled office workers, such as stenographers, while increasing the number of typists. Typists, in turn, were vulnerable to further technological changes. Electric typewriters were faster and made better carbon copies than manual machines. Duplicating machines, and then later photocopying equipment, eliminated the retyping of multiple copies. The development of typewriters that were self-correcting or programmable reduced the need for typing accuracy and spelling skills. Audio recording equipment made shorthand obsolete. Figure 13.3 shows the years these new technologies were first introduced.

Recent Innovations Word processors, microcomputers, and electronic workstations are more recent office technologies. Although most of these innovations do not require the clerical worker to be "computer literate" in any sophisticated sense, they do require skills that many clerical workers, especially older ones, have not had the opportunity to master. Some older workers are reluctant to learn the new machines, thus endangering their jobs. Others complain that the skills they worked for years to master have been rendered obsolete, putting them on a par with the newest young workers hired. Some studies indicate that women experience more negative effects of technology introduction than men, perhaps because they are less often invited to participate in the decision-making process (Zaucher et al., 2000).

Although the demand for workers with these new skills is rising, the increased demand may not translate into a better competitive position for the existing workers. Because different firms use different processing systems and software programs, the horizontal mobility of clerical workers may be impaired. For example, a word processor who is proficient in one system might not be hired in a firm that uses a different system.

Another possibility is that the skill levels in some jobs may be reduced. For example, bank clerks once received up to six months' training in back-office functions before becoming tellers. The back-office functions included processing canceled checks, checking records, and performing other functions that do not require personal interaction with customers. With the computerization of many of these functions, the needed training can now be provided in four months (Rajan, 1985). Automated teller machines (ATMs) have further reduced the need for tellers, although tellers now perform more nonroutine activities.

More innovations in office technology lie ahead. One trend is the linking of office machines into a continuous-process system. (Recall that this term was introduced in Chapter 9.) The "capture" of data at the point of origin eliminates the need for repeated data entry. For example, an order can be electronically tracked from receipt to delivery, producing automatically along the way the mailing labels, invoices, and inventory adjustments required by the transaction. Electronic files and records have largely replaced paper records in many offices. Prototype voice recognition systems can produce correspondence that is virtually untouched by human hands, while electronic mail systems have already made some correspondence obsolete. It is difficult to completely forecast the effect of these innovations on clerical workers because of the key role played by work reorganization in the adoption and implementation of new technology.

Work Reorganization

In a majority of institutions, clerical jobs were extremely routine in the precomputer days. Far from deskilling clerical workers, new technology can be viewed as their liberator. However, this is not to imply that technology's impact is predetermined—far from it. Much depends upon specific decisions on the part of management (Rajan, 1985:413).

Many studies have shown that clerical workers feel powerless and experience considerable stress (Reeves and Darville, 1986). The adoption of new technology and accompanying restructuring are often stressful for both clerical workers and management (Fox and Sugiman, 1999) Whether technology can change this situation is a function not only of what devices are developed but also of how they are used. Technology can degrade and deskill clerical workers when it

forces narrow specialization with accompanying loss of variety, autonomy, and job mobility. By one estimate, 90% of clerical workers using a computer have little autonomy or discretion in their work (Crompton and Jones, 1984). **Work reorganization** affects three major dimensions of the job: job content, supervision, and relations among workers.

Job Content For a quarter of a century following World War II, secretarial work was commonly organized by assigning one secretary, who was usually a woman, to a particular "boss" or perhaps a group of "bosses," who were nearly always men. Secretaries had many jobs to perform in a day, but they could schedule them as they wished, subject to the boss's priorities. A trusted secretary made decisions about screening calls and mail and bringing urgent matters to the attention of the boss. Thus, the job had both variety and some degree of autonomy. Upward mobility was possible because successful managers usually took their secretary with them to a more responsible position (Kanter, 1977). Secretaries gained prestige from the position of their boss, and the nature of the supervisory relationship was a very personal one, with the boss and the secretary often protecting each other from the demands of superiors or office rivalries among competing departments (Glenn and Feldberg, 1977).

The secretary remained dependent on the goodwill of the boss. Many bosses viewed the secretary role as an extension of the wifely role and expected personal services (Pringle, 1988). Although secretaries made many decisions during the course of a day, they worked in a gray area that made them liable to accusations both of having exceeded the limits of their authority and of not having taken sufficient initiative. The secretary's role under such conditions required considerable skill.

When clerical work was **rationalized,** the constituent parts of secretarial work were identified as discrete tasks, and workers specialized in those tasks. Thus, a receptionist greeted all visitors.

Switchboard operators answered all phone calls and took messages. Typing was centralized in a typing pool, where all work to be typed was delivered. Any typist might be assigned the work of any manager. This system resembled an assembly line. Today in many offices a word-processing pool has replaced the typing pool.

The purpose of rationalization is to make the office more efficient, but that is not always the result. In theory, specialization makes clerical work more productive; in practice, this does not always happen. Receptionists and switchboard operators often lack sufficient knowledge to screen out nuisance calls or alert managers to urgent ones. Because of their lack of full information, they sometimes annoy or frustrate potentially important customers. Typists, because they work for many managers, never learn to read all their different handwriting idiosyncrasies, and they never gain the overall sense of projects or priorities that the individual secretary had.

Supervision The rationalization of office work also changed the supervision of clerical workers. Some clerical workers in banking and retail trade, because they deal with money, have always been closely supervised, and the new technology has merely changed the means of control. Electronic monitoring of workstations has replaced record checking by a supervisor. But in other cases the nature of supervision has changed more dramatically. Whereas many clerical workers once experienced one-to-one supervision by the boss, today clerical workers are more likely to be supervised as a group. First-line supervision is often provided by another clerical worker who has been promoted to supervisor. The supervisor is often forbidden to socialize with the former co-workers but is also unlikely to be promoted further (Glenn and Feldberg, 1977).

The old-style manager knew whether the work was being satisfactorily done, and whether secretaries made good use of their time. In the rationalized office, supervisors rely on more abstract, measurable criteria of productivity, many of which are generated by the new technology.

For example, typists can be rated in terms of keystrokes per minute. Telephone company supervisors can "listen in" or have machinery automatically measure how long operators spend in responding to each call. Timing devices attached to a workstation can monitor how long a clerical worker is absent to collect supplies, change positions, or take a break. Some workplaces block certain websites during business hours and forbid games on company computers.

Relations among Workers When office technology reproduces factorylike conditions, the relationships among clerical workers also change. Under the old system, secretaries had sufficient autonomy to schedule breaks and lunches together and to cover for one another when the office was particularly hectic or when one of them needed to run an errand (Strom, 1992). It was also true that the secretaries of higher-ranking managers often did not associate with the secretaries of lower-ranking managers and that the rivalries among managers at the same level might be carried over to their secretaries. But for many clerical workers, one attraction of the job was being part of a work group.

By contrast, supervision using the new office technology often makes clerical work lonelier by preventing employees from socializing. Some offices stagger breaks and lunch periods, so that workers cannot eat together or even drink coffee together. Some firms also stagger shifts, so that there can be more or less continuous typing, photocopying, and word processing available, with the result that clerical workers have fewer chances to meet and know other workers. Workers cannot add variety to their jobs by covering for a busy co-worker or trading jobs with another worker.

In some firms, the clerical workers are forbidden even to talk to one another during working hours. Some managers justify these practices by claiming that socializing on the job (or even during breaks) reduces productivity. But such practices also reduce the flexibility of the clerical workers to respond to busy periods, and they eliminate the opportunity for workers to learn from one another. Further, the depersonalization of their jobs may contribute to lower work commitment and less cooperation with management (Costello, 1984:119).

Alternatives to Rationalization When efficiency becomes the overriding issue in adopting new technology, many of the more pleasant aspects of clerical work are eliminated, and the work becomes more monitored, monotonous, and meaningless. How technology is adopted is a management *decision,* however, not an outcome predetermined by the equipment. When clerical work has been rationalized, the clerical worker spends the time saved by doing still more of the same specialized tasks. Nevertheless, it is possible to adapt the technology so that the quality of work life is improved for clerical workers and their work has more variety and autonomy.

Many technological changes have the potential to make clerical jobs more interesting and less tedious. For example, the development of photocopying equipment freed typists from retyping or handling carbon paper. Using liquid correction fluid and later self-correcting typewriters freed typists from tedious erasing and retyping. Because of word processing, typists no longer type the same form letters over and over, or retype all of the drafts of a report or manuscript. Many secretaries report that these innovations make it possible for them to use the time saved to perform tasks that need to be done but for which there was never time. Telecommuting clerical jobs may allow workers more flexibility in combining work with other responsibilities (Steward, 2000).

In general, less routinized work is associated with greater job control by workers (Simpson, 1985:421). As alternatives to routinization, **job enlargement** and **job enrichment** can change the content of clerical work when technology is introduced. These two strategies take advantage of the time saved by the technology to make the job content more interesting and challenging. Job enlargement makes a job "bigger" by adding on previously separate tasks. In banking, a teller may

now handle many different types of transactions that were once handled by separate departments or specialized tellers. A routinized solution, by contrast, would route all transactions of one type to just one teller, who would repeat the tasks over and over. With job enlargement, the bank teller learns to perform a wider variety of transactions.

Job enrichment, a strategy that has been used in the insurance industry, is similar to job enlargement except that it gives the worker a fuller cycle of sequential tasks. For example, a single worker in an insurance firm may handle a customer's account from initial inquiry through underwriting and claims processing (Baran, 1987). In the routinized solution, all inquiries would be referred to one worker, who would channel them to different workers who would write and process policies, adjust claims, and so on. Job enrichment, like job enlargement, requires workers to acquire a wider range of skills and knowledge to perform their jobs. Job enlargement and job enrichment are alternatives to routinization that have applications to many occupations and not just to clerical work.

The Persistence of the Secretary The preceding discussion may seem to imply that the all-purpose secretary is extinct, but such is not the case. Smaller firms employ clerical workers who do all clerical tasks around the office, from greeting visitors to typing (Simpson, 1985). In these settings new technologies have been most welcome, because these workers have many uses for the time they have saved. For example, legal secretaries are highly skilled clerical workers who prepare legal documents for an attorney's signature. With technological innovation, the documents have become permanently available in the word processor, and the blanks can be quickly filled in. In many law firms, this technology has freed the legal secretary to do more skilled tasks, some of which may be considered paraprofessional tasks of the sort we discussed in Chapter 11.

In other law firms, however, the secretary's job description now requires relatively fewer skills. The specialized knowledge of the legal secretaries is no longer needed, but their jobs persist because having a personal secretary is a sign of prestige among the top lawyers in a firm. The personal secretary is still found working for the highest-ranking lawyers, but the secretary's job description has often changed dramatically The secretaries' functions have come increasingly to revolve around welcoming clients, serving them coffee, and running errands for the attorney (Murphree, 1981).

THE FUTURE OF CLERICAL WORKERS

Technology can change not only the *content* of jobs, but also the *number* of clerical jobs. One issue that is inevitably raised is whether new technologies will displace workers or will lead to the creation of jobs. This issue is of particular interest because, as mentioned earlier, clerical occupations are a large segment of the labor force.

The pessimistic view is that the new office technology will increasingly displace workers, so that the job of record keeping and coordination currently done by many clerical workers will be done mostly by machines. In general, the occupations that work with obsolete equipment, such as typists and switchboard operators, are more likely to decline than the occupations that require interpersonal interactions, such as hotel clerks (Silvestri, 1993).

A more optimistic view is that the new office technology will eventually lead to the creation of clerical jobs, although these jobs may differ in content from the older jobs. Better coordination, resulting from both human and machine contributions, may improve efficiency, leading to lower prices and increased demand for the goods or services produced. In this case, the increased demand may require the hiring of new workers (Osterman, 1999).

There are not enough studies so far to let us choose between the pessimistic and optimistic views. One study of U.S. data between 1972 and 1978 suggests that both displacement and new jobs will result from new technologies. In the short run, some clerical workers are displaced when computers are deployed. Shortly thereafter, employment expands because coordination needs increase after the technology is adopted. Jobs are created to meet these needs (Osterman, 1999). Automation will affect clerical specialties differently. According to the Bureau of Labor Statistics, two of the fastest growing occupations between 1998 and 2008 will be office clerks and computer support specialists (*http://www.bls.gov/emplr99.htm*). Whether the future contains more elements of the optimistic scenario or of the pessimistic scenario, changes in clerical work will continue to be guided by the three factors that have so far transformed it: the technology, the work organization, and the size of the firm. These three factors, in turn, affect the degree of specialization, variety, and autonomy on the job.

HISTORY OF SALES WORK

Goods and services were originally sold by their producers, who were often farmers or craftworkers. Learning to price, display, and sell the wares was part of the training for most occupations. Even merchants were not principally sales workers, and their apprentices often functioned mainly as clerical workers, although attending customers was one of their duties. Just as sales work was not necessarily distinguished from management, farming, and craftwork in the past, so sales work today is not always clearly distinguishable from clerical work. This is especially true when the sales worker has responsibility for shipping, maintaining inventory, and keeping track of orders. Cashiers are an example of workers whose jobs have aspects of both clerical work and sales work.

Sales work became a separate occupation when some workers began to sell what others had produced and traded. The earliest sales workers were probably itinerant peddlers. In rural areas the peddler was a source of entertainment and news as well as goods. The peddler might carry products that were imported or otherwise hard to find, such as spices or salt. Later, peddlers carried small manufactured goods. Specialized sales jobs accompanied the expansion of mercantile capitalism and manufacturing, extensive export–import trade, and large-scale distribution systems. Large-scale trade created tiers of merchandising, with both wholesale and retail levels. The traveling salesman solicited orders for wholesalers or manufacturers, much as the peddler retailed goods to individual consumers. The new forms of merchandising, in turn, created specialized jobs for brokers, agents, and salespeople. Mail-order houses and telemarketing have created new sales jobs.

DEMAND FOR SALES WORKERS

Sales work grew because of the proliferation of trade and the elaboration of distribution systems. Worldwide improvements in transportation, communication, and credit arrangements widened markets. Specialized sales workers became more important in reaching and developing new market areas.

Sales work is a heterogeneous occupational category. Table 13.1 shows some of the sales occupations, which the U.S. Department of Labor classifies by the product sold. In addition to the variety introduced by different products, sales work is organized in many different ways. Sales workers may work for either wholesale or retail concerns; they may be paid on commission, on salary, by the hour, or in some combination of ways; they may work either as regular employees or independent contractors. Many factors contribute to this diversity, including the marketing history of the product, the type of firm, and the knowledge expected of the sales worker. These factors affect earnings and benefits, the full-time or part-time nature of the work, job security, and mobility.

Product Marketing

Several innovations in marketing have affected the organization of sales work. The development of the department store was accompanied by the specialization of sales workers into *departments*. Department stores first appeared in the 1860s and 1870s. Some began as single-line retail outlets, while others were part of wholesale operations. By 1900, the largest department stores employed as many workers as the largest manufacturing establishments (Carter and Carter, 1985:587). Employment growth was rapid and specialized (Benson, 1986). By the 1980s, department stores accounted for 80% of the general merchandise market (Hartmann, Kraut, and Tilly, 1986:48).

Many store departments adopted the sales practices of the specialty stores they partially replaced. For example, furniture stores usually paid their sales workers on commission, while clothing and fabric stores paid their workers an hourly wage. In the department store, departments that sold "big-ticket items," such as appliances and furniture, followed the custom of paying the sales worker a commission on the products sold. In the clothing departments, it was customary to pay an hourly wage. This practice was justified by the argument that customers required more individual attention when they were buying expensive items, and so the sales worker had to work harder for fewer sales.

These differential payment practices came to be reflected in gender segregation within sales work. In department stores men worked in such departments as appliances, furniture, and automotive sales, where they were paid on commission. Women worked in women's and children's clothing, housewares, cosmetics, stationery, and similar departments, and they were usually paid hourly wages. Men's clothing represented an intermediate case. For the customers' comfort and privacy, men worked in the men's clothing department, and selling clothing was usually compensated on an hourly basis. But selling the more expensive menswear, such as suits, sometimes brought the

Table 13.1 Total Sales Workers, 1900–2000, with Proportions in Specific Sales Occupations (numbers in thousands)

Date	1900	1910	1920	1930	1940	1950	1960	1970	1983	1993	2000
Total employed in sales	1,307	1,755	2,058	3,059	3,450	4,025	4,801	5,625	11,818	14,245	16,340
Insurance sales	6.0%	5.0%	5.8%	8.4%	7.3%	6.9%	7.7%	8.2%	4.7%	4.1%	3.5%
Real estate sales	2.6	4.4	4.3	4.9	3.4	3.6	4.1	4.7	4.8	5.0	4.8
Securities and financial services sales	0.3	0.3	0.5	0.7	0.5	0.3	0.6	1.8	1.8	2.5	3.7
Advertising and related sales	0.9	0.6	1.2	1.3	1.2	0.8	0.7	1.2	1.0	1.1	1.0
Street and door-to-door sales workers	6.0	4.6	2.4	1.9	1.6	0.6	1.2	2.2	NA	NA	1.9
News vendors	0.5	1.7	1.4	1.3	1.7	2.5	4.1	1.2	NA	NA	0.7
Retail sales workers[a]	83.2	82.8	83.8	81.1	83.9	63.0	56.7	50.6	46.7	44.1	41.5
All other sales	0.5	0.6	0.6	0.4	0.4	22.3	24.9	28.3	41.0	43.2	42.9
Total	100%	100%	100%	100%	100%	100%	100%	100%	100%	100%	100.0%

NA = not available.

[a]Until 1940, includes sales workers and sales clerks in manufacturing, wholesale trade, and other industries as well as retail trade.

SOURCES: Adapted by the authors from the Census, 1975, *Historical Statistics of the United States, Colonial Times to 1970.* Bicentennial Edition, Pt. 1, Washington, D.C.: U.S. Government Printing Office, p. D 358–441; U.S. Census, 1994, *Statistical Abstract of the United States,* 1994, p. 408, Table 637; U.S. Bureau of Labor Statistics, 2001, *http://www.bls.gov/pdf/cpsaatll.pdf.*

sales worker a commission. In 2000, male sales workers earned $684 weekly, compared with $407 for women (BLS, 2001a).

A second innovation was the development of the *self-service store*. The modern supermarket is an achievement of this innovation; customers roam the aisles looking for the merchandise they want, and they rarely request sales personnel to help them. In fact, even if the customer needs help in locating an item, no sales personnel may be available. The self-service concept has been adopted widely throughout the retail trade industry. The old-style "five and dime store," which had a salesperson at every counter, has been replaced by discount department stores that feature self-service and check-out counters. New self-service stores have also proven a major economic challenge to the older, established department stores, in part because fewer sales workers are needed and the workers are often paid lower wages than the department stores pay.

Self-service systems work best if the consumers receive sufficient information to select products for themselves. The more information provided directly to the customer, the less important it is for the customer to find a sales worker. Fixed prices, price tags, location maps and signs, and alphabetical arrangements of goods or departments are ways to make the customer self-sufficient inside the store. Depending on the goods they sell, different stores will seek new ways to inform their customers. Food stores often post information on the price per unit (such as the price per ounce) and on nutritional composition. Ready-made clothing carries tags with information on fabric care and cleaning. Many retail stores use advertising both to lure the customer to the store and to extol the virtues of various products. Consumer-oriented magazines and newspaper articles also help take the place of the knowledgeable salesperson.

A third innovation with an impact on retail sales workers is the creation of large *chains* of stores, located in separate cities or parts of the city but all carrying the same products and following similar personnel policies. Chains offer considerable advantages that locally owned stores have trouble matching. Because of their size the chains can buy in large quantities at discounted prices and advertise widely. Customers are attracted by national brand names such as K-Mart or Wal-Mart. Chain and discount stores account for 89% of the department store trade.

A fourth change is the *longer operating hours* kept by many retail establishments. Routine retail shopping has usually been a task for women, but working women often find it impossible to shop during traditional business hours. In response, retail stores have extended their closing times and added weekend and holiday hours. Many convenience stores and some grocery stores are open twenty-four hours a day. In shopping centers and malls, tenant stores may be required to stay open during all the hours the mall is open, perhaps as many as twelve hours a day. These expanded hours require retail establishments to hire more workers and arrange work shifts. One solution is hiring part-time workers, especially for peak days or hours. Another solution is to hire temporary workers for especially busy seasons such as Christmas. More than a third of department store employees work only three months of the year, usually around Christmas (Hartmann, Kraut, and Tilly, 1986). The temporary and part-time workers are usually poorly paid, which is sometimes a source of friction between full-time and part-time workers.

A fifth change is the resurgence of *long-distance merchandising*. Retailers such as Sears, Roebuck Company have for years sold merchandise through their catalogs. Catalogs were a convenience for people living in rural areas or small towns. As more Americans settled in large metropolitan areas, many observers assumed that the mail-order business would become obsolete. Instead, consumers today have many options for shopping over the internet or by phone and mail order. These options attract busy workers who have little time to shop. Catalog companies are easy to access because of toll-free telephone numbers and fax numbers, with payments made by credit card. The Home Shopping Network and similar sales networks on cable television, and the eventual development of interactive video, promise additional types of retailing and new sales jobs.

Type of Firm

Sales work is organized differently in different firms. In general, the larger the employing organization, the more secure the sales job. Large corporate employers usually have a sales department or division; if there are several product lines, there may be a sales division for each product. Sales workers in these companies have relatively secure full-time positions with salaries and fringe benefits, and they may earn commissions in addition. For example, sales of securities usually carry a commission. In many companies, upward mobility is possible from sales work to sales management and even into general management.

At the other extreme are the relatively insecure commission jobs with locally owned firms. In sales positions with real estate agencies and car dealerships, for example, earnings depend on one's social networks (to find customers) and skill. The significance of networks is emphasized by this insurance salesperson:

> This may sound silly, but my manager considered it very important for me to join community organizations and boards: Little League, PTA, school boards, hospital boards. You get your face in the local paper. You know: "Ralph Williams and what he's doing now." It tells people you're with Pilgrim Mutual. You're well-established, you're a family man, you're part of middle class America . . . you can be trusted. This is the guy you go to for insurance. So networking yourself into all these small

community organizations is very important. (Oakes, 1990:22–23)

In effect, these workers are independent contractors who are offered a desk in the firm but who have few of the other characteristics of steady employment. These jobs often cannot be categorized as "full time" or "part time," because some workers will work up to sixty hours a week, while others work fewer hours.

There are other independent contractors in sales work, too, many of whom think of themselves as self-employed workers. In many cases their sales work is a second job and is called *direct sales,* meaning that they have no store or other permanent place of business. Direct sales jobs include distributing household products and cosmetics, often through a merchandising technique called a *house party* held in customers' homes. Door-to-door sales workers still exist, although with the decline in full-time homemakers and the fear of crime in urban neighborhoods, this technique is becoming less common. *Telemarketing,* or the use of the telephone to make sales calls, is becoming the technique of choice for many workers in direct sales. On the one hand, direct sales work is financially risky if the worker is expected to buy the inventory in advance. The work carries uncertain earnings and no benefits; the pay for telemarketing, for example, is often based entirely on commission. On the other hand, direct sales companies argue that their workers have the advantage of selling known products that are nationally advertised and that they provide sales training and incentives. Box 13.2 describes one direct sales firm, Mary Kay Cosmetics.

Intermediate in working conditions are the many local retail stores. Most of them hire many part-time workers who have few benefits, little job security, and no prospects for promotion. Except for the smallest firms, however, most retail stores also have full-time employees who run the store when the proprietor is absent. These full-time jobs may carry fringe benefits;

BOX 13.2   Direct Sales and Mary Kay Cosmetics

Entrepreneurs provide a new product or service or innovative marketing for old products, and, in the process, they provide jobs for themselves and others. Some entrepreneurs become executives of their own companies, and a few become extremely wealthy. Mary Kay Ash, the founder of Mary Kay Cosmetics, is a successful entrepreneur who was recently listed among the 100 wealthiest Texans. In 1985, Mary Kay Cosmetics, Inc., posted revenues of $250 million (Anderson, 1989). Despite her wealth, Ash encourages people to use her first name. "I'm called Mary Kay. If people call me Mrs. Ash, I think they either don't know me or they're unhappy with me" (Ash, 1981:168).

Mary Kay Cosmetics manufactures cosmetics and sells them through direct sales. Mary Kay products are not sold in store, but through "beauty shows" to which hostesses invite five to six friends. Mary Kay's 185,000 sales workers, called beauty consultants, earn commissions on their sales of cosmetics. Their sales directors also earn commissions from the sales of other beauty consultants whom they have recruited. In effect, the organization is made up of many entrepreneurs, and the success of each consultant's own business adds to the success of the company. Such arrangements are sometimes called *pyramids.*

Some sales directors have earned over $1 million in commissions, and many are enthusiastic and loyal members of the organization. Most consultants, however, do not earn very much money; for many of them, Mary Kay Cosmetics is a second job or a part-time job (Coughlin, 1989). The organization provides consultants with training, products, and sales incentives, such as diamond bumblebee pins, furs, and pink Cadillacs. Especially successful sales workers are termed "royalty" at annual seminars, which have been described as "a combination of the Academy Awards, the Miss America Pageant, and a Broadway opening" (Ash, 1981:191).

Mary Kay Ash's own career included working in direct sales for Stanley Home Products and World Gifts. As a divorced mother with three children, she had to hold three Stanley house parties a day to make ends meet. Mary Kay's mother had worked long hours as a restaurant manager to support Mary Kay's invalid father and the children, and Mary Kay credits her mother's example with giving her the encouragement she needed. But

she found it irritating that she would be asked to train men who would then become her superiors, and she wrote of this time, "It seemed to me that a woman's brains were worth only fifty cents on the dollar in a male-run corporation" (Ash, 1981:24).

Mary Kay began her cosmetics business in 1963 after buying a formula for skin care products from the daughter of a hostess for a Stanley Home Products party. By the second year, her company's sales had reached $800,000. During her work in direct sales for other organizations, she developed an upbeat, enthusiastic philosophy about direct sales work and its benefits for women workers. Stating her belief that a woman's priorities should be "God, family, and work," in that order, she notes that direct sales work is flexible in scheduling. Mothers can be home when their children arrive from school or are sick. Husbands are also considered a part of the Mary Kay team; Mary Kay has been quoted as saying "A woman who has her husband with her is a woman and a half. A woman who doesn't have her husband with her is half a woman" (Ash, 1981:70).

Mary Kay Cosmetics also tries to maintain a personal relationship with consultants. They receive birthday greetings, newsletters, and frequent praise from their sales directors and from the national organization (Ash, 1984). This philosophy differs sharply from that of many corporate employers, but it is very attractive to many potential workers. The charismatic personality of entrepreneurs like Mary Kay has been a key to the success of direct sales organizations (Biggart, 1989).

Not every direct sales worker is happy. Turnover is high, and sales workers become discouraged. Beauty consultants are not employees in the usual sense of the word. They receive no compensation other than their commissions, and they receive no fringe benefits. Consultants must pay cash to keep an inventory on hand. Because cosmetics are available in every drugstore and supermarket, Mary Kay's beauty consultants must convince their customers that Mary Kay's products are worth the extra expense and trouble of direct sales. Despite the drawbacks, however, entrepreneurship offers them the opportunity to be their own boss and to succeed through hard work.

they do provide a measure of autonomy and variety, and they may lead to supervisory positions or to positions such as store manager, shift manager, or department manager. They also provide useful experience for the worker who wishes eventually to open a small business. Although these jobs may carry the title of manager, sales remains an important part of their job content.

Knowledge Base

Except in large companies, most sales workers experience little upward occupational mobility. The pay in many of the jobs is minimum wage, and many jobs are part time. Turnover is high, especially in retail trade, so employers invest little in training. With high turnover, hiring is nearly continuous, and in recent years the sales force has become younger and often better educated. In general, those sales workers with greater knowledge of the business and the product are the ones most likely to find upward mobility with the employer.

Sales jobs that require detailed or specific knowledge are more likely to be secure and well compensated, often with a combination of salary and commission. Some sales workers are highly trained and are expected to know and understand the products they sell. One sales director says, ". . . give the sales force two or three key points about each product to differentiate the product. . . . If your company is more focused, like a manufacturer, then provide your sales team with more in-depth information. Give them the manufacturing process, the benefits and the features of their product, and compare it to competitiors' products" (Selling Power, 2000: 166). In areas that are rapidly advancing, the sales worker may provide management with important information concerning the inroads that competitors are making in the market and the new features that the customers want. Because of their knowledge of production and of the market, these workers may be promoted, either within the sales department or to management positions.

Most sales jobs, however, require little or no training, and the sales worker may not be expected to know much about the products. Indeed, many managers believe that good sales skills are not learned and therefore cannot be taught; thus, there is little value even in having experienced workers train new workers. Sales skills are viewed as deriving from one's personality; workers either have it or they don't. Sales workers are expected to have generic knowledge about maintaining displays and inventories, keeping the store tidy, and opening and closing cash registers. Because sales work is a common part-time job for students, many young workers have already learned these generic skills and are employable in a variety of sales jobs. Because these skills are considered to be widespread, they do not command high wages or job security.

One competitive technique used in some lines of retail trade is hiring well-trained and knowledgeable sales personnel to assist the consumer (Carter and Carter, 1985). For example, more expensive clothing stores often feature salespersons who are well versed in fashion, alterations, and other relevant areas. These shops may even provide such special services as wardrobe consultation, closet coordination, and home visits. *Personal shopping* is another service that can be offered to the busy consumer. The personal shopper has a list of the customer's needs and knows the customer's preferred styles, colors, and price range. The personal shopper provides the customer with a selection of possible purchases, which are often brought to the customer's home for approval. These additional services may lead to higher wages for the sales worker, even in competitive product markets with slim profit margins. Box 13.3 discusses the development of customized services in the case of supermarkets.

BOX 13.3 Grocery Sales

Grocery stores employ 133,000 workers, most of whom do retail sales work for at least part of their jobs. Grocery store workers are expected to do many other things, however, including stocking shelves, carrying groceries to the customers' cars, and cleaning the stores. These jobs are often strenuous, and may help explain why grocery stores have relatively high levels of occupationally related injuries (Company and Personick, 1992).

Grocery stores are extremely competitive and have small profit margins. One innovation in grocery sales has been the expansion of the stores into full-service stores that offer lines of sundries, toys, video rentals, flowers, and specialty foods. Many larger stores feature prepared foods as a direct response to fast-food measurements. One interesting feature of the prepared foods is the new autonomy it has allowed some of the workers.

Q: Have you got a grill there?
DELI CLERK: Yes, we've got everything, a grill, stove, oven, deep fryer. We make doughnuts, fried chicken. We have hot lunches. We can make whatever we want.
Q: Do you have recipes?
DELI CLERK: No, whoever works the morning, just makes what she likes, what will sell. You try to make something simple, brats and sauerkraut, fried chicken, meat loaf. . . . You get two vegetables and a potato, or roll. It's about $2.00. We sell out every day. When you're cooking, you decide what you want, and then take it off the shelves. You can make anything, because we have all the equipment and we have a whole grocery store. . . . We also make all our own salads. We make pasta salad, antipasto, potato salad, ambrosia, stuffed shells.

There is a great deal of worker autonomy at the store level. While the division office issues a marketing plan, the store employees have a lot of flexibility within that plan. . . . Given a bundle-of-tasks perspective, in-store production seems to make more sense for many types of foods. Because service work is intermittent, adding production tasks to the bundle allows workers to spend their down time on production. Not only is their labor used more effectively, but the complexity of the jobs also increases, and therefore, presumably, job satisfaction increases. Similarly, the move into catering allows the firm to diversify in a way that takes advantage of both the production skills of employees (who must be able to cook for in-store production) and the down time in their schedule (because the meals can be made between waiting on customers during the morning and still be ready for an even that afternoon or evening). Employees of specialty shops can also use this time to prepare goods for self-serve, such as pre-made sandwiches or fresh pizzas. In-store production not only allows better time management by employees, it also allows a better fit to the variation in local demand, since employees know what sells in their market. . . . Finally, in-store preparation allows firms to take advantage of mistakes in other parts of the store. For example, the deli can run a special on meat loaf if the meat department grinds too much meat or make a Polynesian stir-fry if the produce department overorders pineapples. Thus, while interdependencies in the store are limited (the different departments are not forced to interact very often), there are synergies among the departments that can increase productivity or buffer the store from uncertainties.

SOURCE: John P. Walsh, 1993, *Supermarkets Transformed: Understanding Organizational and Technological Innovations*, pp. 119–120. Copyright © 1993 John P. Walsh. Reprinted by permission of Rutgers University Press.

SUPPLY OF SALES WORKERS

The low-paid sales jobs that require little knowledge and skill are disproportionately filled by women, minority workers, and young people. By 1900, women were already one-fourth of the retail sales force (Carter and Carter, 1985). Today, about half of all sales workers are women. Women are more likely than men to be counter clerks and to work in apparel sales. Women make up 76% of all cashiers. Men still dominate jobs selling higher-priced items, such as motor vehicles, furniture, appliances, and parts (BLS, 2001a). Most insurance and securities sales workers are men.

The trend to part-time jobs has certainly affected the labor supply for sales work. About 28% of all sales workers are part time, but 38% of women sales workers are part time. The development of telecommuting for mail-order and internet sales opens up more flexibility in hours and the possibility of working from home for some sales workers (Webster, 2000).

Nearly 9% of sales workers are black, and 8.5% are Hispanic. Both groups are overrepresented among cashiers, of whom 16.5% are black and nearly 13.5% are Hispanic. In other sales occupations, however, minority group members are underrepresented. This situation is long-standing, and some employers justify it by arguing that sales workers from a minority group might alienate white customers. Because white customers have accepted minority group members in clerical and technical specialties, however, it seems likely that they would also be accepted as sales workers. Fears about "acceptability," because some employers assume that racism is a constant, help perpetuate racist stereotypes in the workplace.

Because of the high turnover in sales jobs, it is relatively easy for youths to find sales work, but they are unlikely to seek a sales career. As more part-time sales jobs are created, there are more opportunities in sales for students in school, but part-time work is less attractive to many other workers. Hourly wages are often only marginally above the minimum wage. It seems unlikely that these jobs will become more attractive as careers because of the changes being wrought by new technologies.

THE FUTURE OF SALES WORKERS

As with clerical workers, two important issues affecting sales workers are how technology is changing the content of their work and how it will affect the number of available jobs. New technology is simplifying sales work. Many retail establishments already use "intelligent" cash registers that simultaneously check stocks, control inventory, check customers' credit records, and record the transactions. Sales assistants do not even need to know the price of items if they use scanners that read product codes or icons that picture the product (Sullivan, 1996). In stores that already employed low-skilled workers, even the routine sales skills will become less important. Sales workers may become a *service* that the owners of the new high-technology stores may not wish to offer.

Technology has changed the traditional work of the traveling sales worker. Telecommunications, computer networks, and modern transportation have diminished the amount of time that many sales workers must spend on the road while also providing better means for keeping track of sales contacts, orders, and so on. One effect of these changes may be to reduce the size of the sales force, even when a national or international market is being served.

As technology transforms sales work, sales jobs come to resemble clerical jobs because the sales worker also performs many of the tasks of information management and processing. The job of cashier, one of the fastest growing, is a good example of this trend. The cashier's job is to complete a sales transaction, but today the cash register may also perform a number of other information-processing tasks at the same time. The cashier may not even be aware of the many uses of the data being recorded by the register.

Despite these changes, however, the U.S. Bureau of Labor Statistics predicts that sales jobs will be one of the fastest growing occupational categories. Retail trade is projected to add 563,000 jobs by 2009, and there will be 556,000 new cashier jobs (*http://www.bls.gov/emplr99.htm*). The most rapid growth rate is expected among travel agents, real estate brokers and appraisers, and securities and financial services sales workers.

SUMMARY

Clerical and sales jobs have grown rapidly in the recent past and can be expected to grow in the near future, but the content of these jobs is likely to be changed by developments in work organization and technology. Among clerical workers, job content has been changed by the imitation of assembly-line production techniques. The introduction of new technology may worsen this trend, or it may reverse it through the processes of job enlargement and job enrichment. Among sales workers, work reorganization has led to a proliferation of part-time jobs, and the new technology has made possible greater use of relatively unskilled and inexperienced workers. On the one hand, the business of "selling" the consumer has often been taken over by advertising, self-service retailing, and other innovations that have reduced the need for skilled sales workers. On the other hand, where products require sales workers with specialized or technical knowledge, the workers are likely to retain greater control and influence.

KEY CONCEPTS

feminization of clerical work	rationalization	job enrichment
work reorganization	job enlargement	

QUESTIONS FOR THOUGHT

1. A number of analysts have tried to explain the feminization of clerical jobs. Can you think of other jobs that might be "feminized" or "masculinized" in the future?

2. Some experts think the number of clerical jobs will increase, while others expect a decrease. With which position do you agree, and why?

3. Under what conditions does technology improve clerical and sales jobs?

4. Think of a recent technological innovation. How could this innovation improve or worsen the work life of a clerical or sales worker?

5. Under what conditions is a clerical or sales job a dead-end job? Under what conditions is it part of a job ladder?

MULTIMEDIA RESOURCES

Print

Margery W. Davies. 1982. *Woman's Place Is at the Typewriter: Office Work and Office Workers, 1870–1930*. Philadelphia: Temple University Press. An excellent history of how clerical work came to be dominated by women.

Evelyn Nakano Glenn and Roslyn L. Feldberg. 1977. "Degraded and Deskilled: The Proletarianization of Clerical Work." *Social Problems* 25,1. (October):52–64. A classic description by two sociologists of how the factory model of clerical work comes to affect the workers.

Heidi I. Hartmann, Robert E. Kraut, and Louise A. Tilly, eds. 1986. *Computer Chips and Paper Clips,* Vol. 1. Washington, DC: National Academy Press. A readable and interesting report on how modern technology may change women's occupations, especially clerical occupations.

C. Wright Mills. 1951. *White Collar.* New York: Oxford University Press. A classic work in sociology by a leading American analyst. Examines the transformation of the occupational structure and the nature of white-collar work.

Websites

Office Workers Career Centre. *http://www.clericalworkerscentre.org/* This nonprofit agency in Toronto serves office workers with career planning, workshops, and other services.

Monster.com. *http://www.monster.com/* A large website with information for job seekers. Visitors may look for jobs by occupational title.

RECOMMENDED FILM

The Big Kahuna. 2000. Stars Kevin Spacey and Danny DeVito as sales reps going after the biggest contract their company could ever have—the Big Kahuna. Rated R for language.

14

Marginal Jobs

Many employers carefully select a core group of employees, invest in them, and take elaborate measures to reduce their turnover and maintain their attachment to the firm. Many of these same employers, however, also maintain a peripheral group of employees from whom they would prefer to remain relatively detached, even at the cost of high turnover, and to whom they make few commitments. The strength of these employers' desire to avoid attachment to this peripheral group is such that a broad spectrum of institutional arrangements has emerged to facilitate this relationship.

(MANGUM, MAYALL, AND NELSON, 1985:599)

Now the delegation of dirty work to someone is common among humans. Many cleanliness taboos, and perhaps, even many moral scruples, depend for their practice upon success in delegating the tabooed activity to someone else.

(HUGHES, 1958:2)

Most of us carry a mental image of a bad job. In this chapter, we use the term **marginal jobs** to refer to jobs that are undesirable because they are boring, low paid, intermittent, lacking autonomy, and dead end. The term *marginal* describes only the job; in many circumstances, the workers in such jobs are hardworking and capable, but unable to find better jobs.

As the quotations above suggest, marginal jobs originate in different ways. In this chapter, we will discuss types of marginality, marginal occupational positions, the marginal jobs that arise from employer decisions, and some characteristics of workers found in marginal jobs. We will also discuss trends in the numbers and types of marginal jobs.

WHAT IS A MARGINAL JOB?

A marginal job departs significantly from the employment norms of the society in which it is located. Because jobs are governed by many norms, there are also many ways in which jobs can depart from these norms. Four important norms commonly apply to jobs: (1) the job content should be legal; (2) the job should be institutionally regular; (3) the job should be relatively stable; and (4) the job should provide adequate pay with sufficient hours of work every week to make a living. We will discuss marginal jobs in terms of their departure from one or more of these norms. It is important to note the subjective evaluation involved in identifying marginal work; for example, two people might disagree whether a wage level is sufficient to provide an adequate level of living.

Illegal or Morally Suspect Occupations

> The longer you strip, the harder it is to retain a positive view of men. Long-term dancers—at least the ones I met—were all bitter. . . . Be a waitress."—former stripper. (Bowe, Bowe, and Streeter, 2000:368–369)

Some jobs are marginal because the content of the work is socially defined as improper, immoral, or illegal. Drug pushers, pimps, and prostitutes can be considered members of **deviant occupations** (Miller, 1978). Deviant work includes criminal behavior, such as the production and distribution of illegal goods or services. Criminologists sometimes speak of criminal "occupations" and criminal "careers."

Which occupations are considered deviant varies, depending on which groups have the power to define deviance, and the labels may change with time and place. During Prohibition, legislators defined as deviant the people who sold alcoholic beverages, because the law prohibited the production and sale of alcohol. Today, although regulated by federal and state law, producing and selling whiskey, wine, and beer are legal. Liquor stores alone employ 116,400 workers (BLS, 2000a). Nevertheless, some people might view these jobs as deviant if they regard alcohol consumption as a moral evil.

There are other occupations such as stripping that, although legal, violate some people's norms of modesty. Some occupations, such as prostitution and playing professional poker, are legal in some geographic areas but not in others (Hayano, 1982). Box 14.1 discusses a bordello in Guatemala (where prostitution is legal) that leaves its owners with little profit.

Workers may be recruited into deviant occupations because other, more legitimate economic opportunities are blocked. Some observers argue that minority youths are susceptible to recruitment into deviant occupations because other jobs have migrated to the suburbs, far from the ghetto homes of many minority families (Wilson, 1987). Deviant jobs may also be more lucrative than regular jobs. Even if they are not officially in the labor force, those in deviant occupations can be considered workers.

Unregulated Work

> They pay us once a week by check. You can't open a bank account without a Social Security number so it's a little difficult.

BOX 14.1 Prostitution and the Informal Economy in Guatemala

This selection from an anthropological study of Guatemala highlights one of the many strategies that the residents of San Pedro use to make a little extra money. San Pedro's economy is characterized by many small enterprises and an informal economy. Prostitution is legal in Guatemala, but the business failed anyway, and the anthropologists try to explain this failure.

The local house of prostitution, known as *La casa de las mujeres* or *la casa,* is a business owned and operated by a woman and her three daughters. None of these women are prostitutes. They employ four women from out of town for that work . . . the women have no control over the prostitutes, who only work when they want to. If the employees leave for the coast for a few days, the owners cannot open and thus lose money. They complain quietly, but have neither the leverage nor the powers of persuasion normally attributed to a male owner, madam, or pimp. This lack of authority is not simply a matter of physical strength, nor of sexual exploitation, but is as well a result of the absence of female experience as employers. Women have traditionally worked with relatives alone. As minimally capitalized corporate enterprises made up of the labor of family members, by their very nature these businesses make employment of outsiders difficult. Where *muchachas* or apprentices are hired, authority may be expressed almost as an option of status or class. With a group of independent, strong-willed prostitutes, however, this is an improbable response for *Sampedranas* [women residents of San Pedro] who do not have the experience of men as

employers. Rather than being bosses and strong authority figures, *la casa* owners are as kind and considerate to the prostitutes as they would be to any houseguests. They are not coercive. To the contrary, Doña Tila tries to take good care of them by feeding them lunch and dinner, with sizeable portions of meat as often as she can. She pities the prostitutes for their way of life and in small ways attempts to ease their situation by doing things like caring for the son of one of them while his mother works. In another sense, their being women is fiscally limiting as they have little access to capital to invest in the business. They have only debts. . . .

La casa is both like other female family businesses and different from them. It does contain many of the common elements of female business systems such as minimal capitalization, donated labor of educated and uneducated women, reciprocal responsibilities. On the other hand, it was begun a year ago as a desperate investment by a group of inexperienced women. It has no history in their family nor do any of the members express loyalty or dedication to it. It has been a difficult and unsatisfying business experience which has barely sustained them financially. Interestingly, the pending disappearance of the eldest daughter from *la casa* is typical of the trend away from female family businesses and toward modern employment among *Sampedranas,* as is the closing of the business that will doubtless accompany her resignation.

SOURCE: Tracy Bachrach Ehlers, 2000, *Silent Looms: Women and Production in a Guatemalan Town.* Austin: University of Texas Press, pp. 91–93. Used by permission.

You can cash the check in the company bank. . . . There are also a lot of thefts—people break into our houses and steal our money, because they know we can't keep our money in banks.—an undocumented immigrant working in a poultry processing plant (Bowe, Bowe, and Streeter, 2000:190)

A second norm, that the job be covered by government work regulations, recognizes that the government is interested both in protecting workers and in collecting taxes. Especially in industrialized countries, state regulation is supposed to ensure that "regular jobs" meet minimum standards for pay, working conditions, and safety. The employer is also expected to withhold employees' income and Social Security taxes and to pay taxes on its own payrolls (such as unemployment compensation insurance). Thus, a second dimension of marginality characterizes jobs that lie outside this system of regulation.

In the **shadow economy,** jobs are not necessarily illegal, but they are institutionally irregular because the income is not reported, there is no official monitoring of health, safety, or working conditions, and other institutional regulations of work are missing. In developing countries, terms such as the *informal* or *submerged economy* are more commonly used (Portes and Sassen-Koob, 1987). What is common to all the definitions is that the shadow economy is neither accountable to state authorities nor subject to economic measurement. Examples include maids or gardeners who work for cash or whose employers refuse to withhold taxes. The shadow economy is illegal in the sense that it evades normal regulation; the goods and services produced may not be illegal. In the case of the undocumented worker quoted above, working in the chicken processing factory is certainly legal, but the worker has no Social Security number and so there is an irregularity involving the payroll tax, if not other things as well.

The shadow economy may or may not use money as a medium of exchange. Workers may receive their wages in cash or in goods, services, discounts, or even barter. However, even if the workers are paid in money, one goal of the shadow economy is to evade the monetary costs of regulation. The head of one French firm remarked: "If I were to declare the work of a translator it would cost me 5,000 francs and she would get only 3,500 francs net. I'd rather give nothing to the taxman and pay her about 4,200 francs" (DeGrazia, 1980:550). No one can be certain how large the shadow economy is, but estimates run into the billions of dollars.

Contingent Work

The worst part about temping is probably when people ask you what you do. It's, like, "Oh, you're doing temp work still, Lillian? Can't you get a job?" That's a big one. My parents are, like, "Lillian, let's get a job." (Henson, 1996:145)

The third norm, stability, implies that a job should represent a relatively long-term relationship between the employer and the employee, assuming that both of them perform their contractual responsibilities. A corresponding dimension of marginality is **instability** or **contingency,** terms that imply that the job inherently carries no job security.

Unstable jobs are not the same as probationary jobs. Many employers initially hire workers for a probationary period that may vary in length from a few weeks to several years. Typically, the probationary period is relatively short for unskilled or semiskilled jobs, but often quite long for professional workers. Among university professors, for example, the usual probationary period is six or seven years. During the probationary period, the professor receives no job security. But a professor who is given a tenured position has a secure job unless the school faces severe financial problems, eliminates a program, or finds the professor guilty of serious misconduct. By contrast, an unstable job never offers job security.

Employers have many devices for getting work done without making a long-term commitment to the worker, including the use of consultants, short-term contracts, and temporary workers. In a business that is about to fail, all of the jobs in the firm may become unstable jobs. A job may be unstable because the work is seasonal or because intense foreign competition makes it difficult for the employer to guarantee continued work. Even in successful businesses, an employer may for strategic reasons refuse to guarantee continued work, reasoning that workers who are fearful for their jobs will be more docile and compliant. Some employers are willing to provide stable employment for those workers whom they consider central to the workplace, but they are unwilling to provide similar stability to other workers.

Berry's World

In 1999, there were 5.6 million workers who held contingent jobs. A majority of them (53%) report that they would have preferred a permanent job, and only 39% say that they prefer their current arrangement. Even full-time contingent workers earned only 77% of what full-time workers earned in permanent positions (BLS, 1999).

Underemployment

[I had] a definite decrease in weekly hours. Used to work a lot of hours, 50 to 75 hours per week. Regular work week is 40 hours. For the past year my hours were cut dramatically. Work never picked up again.— Julie Keys, filing for bankruptcy (Sullivan, Warren, and Westbrook, 2000: 95)

A final norm is that work be adequate in terms of the hours of work available and the compensation offered. **Underemployment,** or inadequate work, describes jobs that may be permanent but that provide fewer hours of work than the worker would like (a condition called *involuntary part-time*

employment) or that provide very low wages. Involuntary part-time work may occur in several ways. Workers may have their hours of work reduced, as Julie Keys did. A company may reduce every worker from five work shifts per week to only four, with pay reduced proportionately. Such partial layoffs may also be imposed because of bad weather, shortages of needed materials, and other conditions beyond the worker's control. When the weather turns rainy, construction workers may be sent home until conditions are better for working. Involuntary part-time work also occurs if a worker who wants a full-time job can find only a part-time one. Very-low-wage workers are called the working poor. We will use the term *inadequate work* in this chapter to refer to these types of underemployment. The word also describes the situation of a worker who has greater qualifications or skills than the job requires (Sullivan, 1978; Clogg, 1979).

Many jobs are characterized by more than one dimension of marginality. In the following sections, we examine entire occupational groups in which the jobs are marginal. Later we discuss the expected growth of marginal employment and the possible division of employment into a stable sector and a growing marginalized sector.

HOW DO JOBS BECOME MARGINAL?

Some jobs are marginal because of the content of the work. Undesirable or "dirty" work is likely to be delegated by more powerful occupational groups to less powerful groups, and the resulting jobs are also likely to be unstable, poorly paid, and in other ways marginal. Other jobs become marginal because new technology changes the job content, making the workers' skills obsolete. While it is not possible to predict with certainty just which skills will become obsolete, we will examine some of the major effects of technology on marginality. Finally, some employers create marginal jobs, regardless of the job content. Thus, given two identical jobs, the job at one firm might be secure and stable while the job at the other firm might be marginalized.

Marginal Occupational Groups

> In every advanced capitalist society, somebody has to haul sheetrock, clean toilets, can fruits and vegetables, harvest crops, cut up and package animals and fish to be trucked to supermarkets. Turks sweep Switzerland's streets. Jamaicans pour concrete in London. In central Texas, illegal immigrant Mexicans carry turkey semen from barn to barn to inseminate female birds. Puerto Rican women working in the tuna canning plants of Mayaguez smell so bad that the local taxi drivers refuse to carry them in their cars, and North Carolina women working in crab houses throughout the eastern part of the state acquire a rash they simply call "crab rash" for lack of a medical term. (Griffith, 1993:4)

In terms of the dimensions of marginality we have discussed, some farm work, private household work, and some manufacturing and service jobs can be defined as marginal. These major occupational groups share three related characteristics: (1) Their jobs are generally considered unattractive, (2) they work in relatively small, unbureaucratized workplaces, and (3) they have little occupational power.

Unattractive Work Nearly every job contains at least some tasks that are dull or unpleasant. People in more powerful occupations, however, are often able to delegate unpleasant tasks to other occupations. Physicians delegate personal care tasks to registered nurses; nurses in turn may delegate the tasks of making beds, giving patients baths, and emptying bedpans to nursing assistants or orderlies. A similar process occurs in many industries, with some jobs consisting mostly of dirty, repetitive, or boring work. Because others often perceive these jobs as unskilled and poorly paid, these occupations have relatively low occupational prestige. It is important to note, however, that these generalizations are being made about the occupation overall. It is not necessarily the case that every individual within the group will have few skills and receive low pay.

Small, Unbureaucratized Workplaces Members of these occupational groups are among the least likely to work for large bureaucratized firms. Their employers are likely to be farm owners, small-business owners, or homeowners—people who may not even think of themselves as employers because they employ only one or a few workers. As a result, there is little commitment to provide permanent jobs or full-time work. To make ends meet, each worker may have to work part time or part of the year for several employers, ensuring that every employer has even less incentive to improve job conditions. Even if a large bureaucracy is the employer, such as agribusiness corporations that employ farm workers, marginal work may be treated differently from other work. For example, many farm workers are excluded from state unemployment coverage and minimum wage laws.

Low Occupational Power Marginal occupations tend to have little occupational power. The workers are rarely in a good position to bargain for better wages. Workers who have small, unbureaucratized employers find collective organization difficult (see Chapter 6). The workers' knowledge is not sufficiently specialized to claim the sort of monopoly that professionals or craftworkers have. Nor are their skills believed to be in great demand. Further, the workers rarely own the capital necessary for doing these jobs, such as the farm land and equipment, the restaurant equipment, or the cleaning equipment and appliances. As a result, the workers are in a poor position to improve their jobs.

Because of poor working conditions, employers often claim an inability to find enough people to fill marginal jobs. Large growers petition Congress to admit immigrant farm workers. Owners of restaurants and hotels complain of high turnover among their employees. Homeowners complain that "there's no good help available." Their comments suggest that there is actually high demand for marginal jobs, but only if wages are kept low and job security provisions are avoided.

Farm Laborers   Farm laborers are employees, permanent or casual, of farm owners. Their principal duties include the planting and harvesting of crops and the other tasks needed to bring the crop to harvest, such as weeding, fertilizing, irrigating, and crop dusting. Farm laborers also care for livestock and help prepare animal products (such as eggs or milk) for market. As you saw in Chapter 8, agriculture is declining in terms of the numbers of workers employed. In 2000, there were 768,000 farm workers, and their numbers are projected to decline further (BLS, 2001a). This decline is attributed to the increased use of machinery for planting, fertilizing, and harvesting. But farm workers are still needed to harvest the soft fruits and vegetables that would be damaged by mechanical pickers, and they are needed to run farm equipment.

Farm work is considered low-skilled work, but it would be more accurate to say that farm work does not require formal schooling credentials but instead is learned on the job (Smith, 1987). The work is often dangerous because of exposure to pesticides, the use of heavy equipment, and other potential hazards. Farm work involves heavy manual labor and exposure to all kinds of harsh weather.

Farm workers can be defined as marginal in terms of job regularity, stability, and adequacy of employment. The organization of agriculture has affected the *regularity* of farm work. There are several types of farm laborers, but each of them is irregular in some aspect of the job. The **migrant farm worker** follows the crops as they mature and works for a succession of employers. Different migration patterns follow the seasonal harvests through continental climatic zones, up the East Coast, the West Coast, or the Midwest. The **sharecropper,** a decreasingly common type of farm laborer, lives on the land owned by another and is paid partly with a portion of the harvest. Between harvests, sharecroppers often borrow money, seed, and fertilizer from their landlords, and pay them back with interest after the harvest (Jaynes, 1986). Because of high interest rates and occasional crop failures, many sharecroppers were once so far in debt to their landlords that they were virtually captive labor. The most "regular" farm labor jobs are those of the **hired hands,** including cowboys. These are often relatively permanent positions. Hired hands receive wages, but part of their pay is sometimes room and board.

More recently, farm workers have been employed by large-scale agribusiness firms. Three types of corporations now dominate agricultural production. The first is the family corporation engaged in farming. This is the "family farm," but usually on a very large scale, often with thousands of acres under cultivation. The second type is the publicly owned corporation that is primarily engaged in food processing, such as Del Monte and the Ralston-Purina Company. The third is the publicly owned conglomerate whose major operations are in nonfood industries. Despite having large, corporate employers, however, farm laborers still do not have completely "regularized" jobs, because they are explicitly excluded from many types of labor legislation. Most laws regulating pay and working conditions exempt farm workers. Thus, a farm worker employed by a subsidiary of Chevron might not receive the fringe benefits available to other Chevron workers.

Farm jobs are often *unstable*. Some instability arises because farm work is seasonal; in addition, if the crop fails or is poor, there will be little or no work at all. Some instability results from the economics of agriculture, which encourage each farm owner to hire workers only for the peak periods. To make ends meet, each worker may have to work part time for several employers, and the worker is often legally defined as an independent contractor rather than an employee. Independent contractors are ineligible for fringe benefits. Migrant workers often make a customary annual circuit of employers. The employment relationship, however, is asymmetrical. The workers may be bound to the employer through contracts and sometimes through loans made to provide for the workers during the off-season. But the employer has no reciprocal requirement to rehire the workers. Most farm jobs offer no opportunities for career advancement, except perhaps for promotion to crew leader.

Many farm jobs would also be considered *inadequate employment*. Full-time farm workers in 2000 earned median weekly wages of only $309, compared with $446 for operators, fabricators, and laborers and $469 for administrative support and clerical workers (BLS, 2000a: Table 39). The compensation is often based on piecework rates, with a sum of money paid for each bushel harvested or each acre planted. In practice, farm laborers may literally work from sunrise to sunset without earning enough money to support themselves. The farm worker may lose wages if a piece of equipment breaks down or if a thunderstorm begins. If the farm worker is paid an hourly wage, an additional problem is finding and keeping full-time work. Many farm workers are not even covered by Social Security, let alone pension plans, so that when they become too old to work, they must rely on public assistance. Migrant workers may not even qualify for public assistance because of ambiguities about their legal residence (Nordheimer, 1988).

Given the prevailing work conditions, recruiting workers to farm laborer jobs is difficult. Many farm workers are members of minority groups; 36% of them are Hispanic and 7% are black. Teenagers are recruited for work during school vacations. Large growers have persistently sought to recruit unskilled immigrant workers as temporary or permanent farm laborers (Majka and Majka, 1982). The late civil rights leader, Cesar Chavez, helped organize the United Farm Workers, and the union has been able to negoti-

ate improved wages, working conditions, and some pension coverage for its members (Nordheimer, 1988). Only 5.5% of all farm, foresty, and fishing workers are represented by a union (BLS, 2001c).

Marginal Service Workers As noted in Chapter 10, the service industries and the service occupations are growing rapidly. Many service jobs are not marginal. Professional services, including medical, legal, social, and educational services, are rarely marginal (see Chapter 11). Similarly, business services, such as finance, insurance, and real estate, are rarely marginal, nor are such protective services as the police and fire departments.

Three groups of service workers more likely to be marginal are personal service workers, food service workers, and cleaning service workers. Personal services are provided directly to individuals, and include porters, dry cleaners, shoeshiners, beauticians and barbers, and repairers. Food service workers include cooks, dishwashers, waiters and waitresses, and counterpersons. Cleaning services, another major category, need to be distinguished from private household workers who clean homes. The principal difference is that cleaning service workers are paid by cleaning companies and have no direct employment relationship with the people whose dirt and mess they clean (Hood, 1988).

Service work may not be *regular* in terms of its legal protections. In 2000, there were 1.7 million workers who were paid hourly rates below the federal minimum wage, most of them in services such as restaurants, personal services, and other service industries (BLS, 2001a: Table 45) In other ways, however, greater regulation is beginning to characterize these jobs. For example, cleaning companies that hire maids and send the maids to work in private households provide more regular employment than do private household employers, because the wages and working conditions are established by the cleaning company and not through individual negotiation with the homeowner. Illegal aliens have often been employed in positions such as hotel porters, dishwashers, and janitors. With the passage of the Immigration Reform and Control Act of 1986, it became illegal for employers to hire aliens without work authorization.

Service jobs are often *unstable,* in the sense that the work contracts are rather informal and employers easily breach them. Many service jobs are seasonal, especially in the resort industry and in some restaurants. Other service jobs are intermittent, with the employer calling on workers when they are needed. When business is slack, the workers are furloughed.

Finally, service jobs are often *inadequate* in wages and hours. In 2000, the median weekly wage of full-time service workers was $355, compared with $576 for the entire labor force (BLS, 2001a: Table 39). Both male and female service workers earned only about 64% of what their counterparts earned in other fields. Moreover, about 61% of female service workers and 52% of male service workers worked part time or part year (Census, 1999: Table 703).

Service work differs from the other marginal occupational groups in that the number of jobs is expected to grow. Demand for services such as food preparation increases as home meal preparation diminishes. Precisely because these jobs were once in home production, however, the knowledge and skills needed are often widespread. Workers who are technologically displaced from manufacturing or clerical work are available to take the service jobs. Thus, even though demand will remain strong, wages and conditions of employment in many of the services may not improve.

Some service workers are highly skilled. Some cooks may have learned their trade on the job, but others are graduates of culinary arts institutes or chefs in four-star restaurants. Barbers and beauticians are skilled and must hold a state-issued license. Unlike professionals, however, service workers have difficulty limiting access to their occupations. The higher wages that skilled service workers might otherwise command are often undermined if the service can be mass produced.

For example, a fine restaurant can seat relatively few diners in an evening, compared with a fast-food outlet. A barber can cut hair for relatively few customers in a day, but a chain-operated discount haircut store can employ more barbers and serve dozens of customers every hour. Because of such open competition, even skilled service workers may not be highly compensated.

There is a growing unionization movement among some service workers, including hotel workers, food preparation workers, and cleaning service workers. As mentioned in Chapter 6, organized labor has targeted the service industries as a major site for future organizing.

Private Household Workers Private household workers are distinguished from other service workers because they are never employed by corporate employers. There are about 820,000 private household workers in the United States. Private household workers include launderers, cooks, maids and housekeepers, gardeners, babysitters and nannies, chauffeurs, butlers, and other household servants. The occupational group is dominated by women; only 3% of private household workers are men.

Private household workers are marginal in terms of job regularity, stability, and adequacy. Their jobs are *irregular* because, like farm workers, they are excluded from many forms of labor legislation. The 1938 Fair Labor Standards Act extended minimum-wage coverage to nearly 87% of all nonsupervisory employees as of 1985, but private household workers were exempt (Levitan and Shapiro, 1987). When many household workers were covered by minimum-wage legislation for the first time in 1974, employers often responded by reducing their hours of work (Rollins, 1985:68). Live-in maids may be expected to work for ten or twelve hours a day and to accept part of their pay in room and board. Another form of **payment in kind** that flourishes is the giving of gifts, such as old clothes, cast-off furniture, and leftover food. Private household jobs rarely have job descriptions. Workers complain that their employers increase the amount of work expected from week to week. Even lunch hours and coffee breaks may be frowned on (Rollins, 1985).

Even if some benefits are mandated, employers often fail to comply. For example, employers of private household workers are supposed to pay Social Security taxes, but many do not. One private employer said, "No, I've never taken out for Social Security or made arrangements about sick pay or vacation. In this kind of job, there is no formalized arrangement. I can't imagine what the [vacation arrangement] would be, but certainly time without pay" (Rollins, 1985:78). The workers themselves may resist deductions from their already meager wages.

Private household work is also *unstable*. There is little or no job security. If the employer is out of town or on vacation, the worker is effectively laid off. Workers who want full-time work must often arrange to work in several different households. There are no career opportunities. Employers can discipline or fire their household workers at will, and the workers have no grievance procedures or other protection beyond the goodwill of the employer (and perhaps the difficulty of "finding good help"). The work may be heavy, and it is frequently lonely and isolated.

Private household work is often *inadequate* in terms of the hours worked and the pay received. Most of the workers are part time, although about 12% of them would prefer full-time work (Sullivan, 1978). Only 23% of private household workers had full-time jobs (Census, 1999). Median weekly earnings in 2000 were $264 (BLS, 2001a).

Private household employment is similar to unpaid housework in that it is generally devalued. The work is described as being unskilled, and it has typically been relegated to the least powerful members of the society: women in general and women of lower-class or minority origins in particular. Private household workers were traditionally recruited from among immigrant groups or from racial minorities (Romero, 1992). In 2000, 15.7% of private household workers were black, and another 30% were Hispanic (BLS,

2001a: Table 18). Box 14.2 discusses the association of domestic work with gender, race, and class inequalities.

There is a small unionization movement among private household workers. Its objectives are to provide better and more standardized working conditions and to improve wages and benefits (Chen, 1993). A major factor limiting unionization has been the isolation of domestic workers, which makes it easier for individual employers to fire one worker and hire another.

EMPLOYERS WHO MARGINALIZE JOBS

> INTERVIEWER: Do you feel like you get a message from higher up in terms of what you should be doing with part-time and full-time? STORE MANAGER: Oh, without a doubt. They want to decrease the full-time ratio and increase the part-time ratio. . . . We would want at least a three-to-one ratio here, [three] part-time to one full-time. (Tilly, 1996:110)

Many workers in the three marginal occupational groups just discussed will have a marginal job. In other occupational groups, *some* workers will have marginal jobs. For example, many factory workers are unionized and paid well. They have monetary and other benefits and at least some promotion opportunities. But others work in sweatshops with unstable, poorly paid jobs and without grievance procedures or other protections.

We have already seen that the members of the same occupation may be employed in many different industries and in many different firms. Because industries and firms may have independent effects on which jobs are marginal and which are not, there are occupations in which a large proportion of workers hold very good jobs, but in which another large proportion could be classified as holding marginal jobs. In terms of job

BOX 14.2 Race, Class, and Gender among Household Workers

The [employer's] use of domestics' first names, calling domestics of all ages "girls," the encouragement of performances of subservience, demands of spatial deference, perceiving domestics as childlike, giving domestics used household articles—all of these conventions of domestic servitude have in common the quality of affirming the employee's inferiority. . . .

The low regard for this sphere of labor—whether paid or unpaid—has been well documented. The female employer, regardless of the degree to which she may have chosen to buy her way out of it, knows that she is seen as responsible for all household maintenance and that this is devalued work. She perceives the person she hires to do such work as doing *her* work in a way the male employer does not. The domestic is something more than an employee; she is an extension of, a surrogate for, the woman of the house. And she operates in what is increasingly the least prestigious realm of women's activities. This view of the domestic on the part of the employer—as an extension of the more menial part of herself rather than as an autonomous employee—may help to explain why the women tend to see domestic service as a more informal arrangement than other occupations. . . .

The employer's low regard for this "women's work" can combine with her own sexism, racism, and class prejudice to further degrade the work and the groups already subordinate in the "three structures of power" in the United States (women,

people of color, and the lower classes). For some employers, like Alberta Putnam, it is incongruous to hire a man to do such work: "I would feel uncomfortable with a man in that position. I wouldn't feel right giving him orders like that. I even feel funny asking my husband to clean the dishes." For some, like Holly Woodward's husband, it is incongruous to hire a middle-class person: "Then there was Patricia, a fascinating British girl. Her father was an actor and she wasn't sure what she wanted to do. My husband was against hiring her. He told me, 'You don't want help like that around.' " And it may be assumed that for some employers—particularly in the South, Southwest, and Far West, where the servant population has been almost exclusively black, Mexican-American, Native American, and Asian-American—it is incongruous to hire a white. One can begin to see why the lower-class woman of color, just *because* of this society's sexism, racism, and class prejudice, might be psychologically the most desirable "type" for a position of servitude and why being associated with this archetypal "women's work" further degrades her—even, or perhaps especially, in the eyes of her female employer. The employer benefits from the degradation because it underscores the power and advantage (easily interpreted as the rightness) of being white and middle class.

SOURCE: Judith Rollins, 1985, *Between Women: Domestics and Their Employers.* Philadelphia: Temple University, pp. 183–184, 194. Copyright © 1985 by Temple University Press. Reprinted with permission of Temple University Press.

stability and permanence, probably no occupations are "safe" from marginality. Rather, some form of economic vulnerability is possible in every occupation as a function of the industry, firm, or employment contract.

Employers create marginality by dichotomizing workers, jobs, firms, and even industries into categories that are "more important" or "less important." A single employer can divide jobs into **core and peripheral jobs,** carefully rewarding the core job holders to minimize their

turnover. The peripheral workers, by contrast, may be laid off, intermittently laid off, or even temporarily hired from an outside agency. Firms can be classified along similar lines. The core firm employs a permanent workforce, but may also subcontract to peripheral **client firms** when there is too much work or if it wants to hire temporary workers to expand production. Even industries can be divided into those that are most central to the economy and those that are more peripheral.

This argument suggests a symmetry such that the preferred workers are able to secure core jobs within core firms and within core industries. The preferred workers are usually able-bodied, well-educated white men between the ages of twenty-five and sixty-four. The remaining workers are more marginal—not because they are necessarily poor workers, but because they have the highest probability of filling marginal jobs. The symmetry between core and peripheral economic structure and the division of labor force into "preferred" and "unpreferred" workers is sometimes called the **dual labor market.**

In this section, we consider three sources of marginality: industry, firm, and temporary employment contracts. In later sections, we will consider in more detail the workers who become marginalized, and we will also return to the dual labor market hypothesis.

By Industry

Some industries are associated with high proportions of marginal jobs, and the common characteristics of these industries can help us understand why the jobs are often marginal. In general, modern industrialized economies can be thought of as having certain *core,* or essential, *industries* (Averitt, 1968). These industries are responsible for a disproportionate share of the wealth produced, and, given the international division of labor, they also tend to be the industries in which the home country is most specialized. *Peripheral industries,* by contrast, are considered less important. They might have unstable product markets, perhaps because they produce a seasonal product or because the market is subject to sudden, intense price fluctuations. Three additional characteristics of peripheral industries seem to be associated with marginal jobs: competitiveness, low productivity, and a management orientation toward low-wage, low-skill production practices.

Marginal jobs tend to be more common in industries that are highly competitive. Within a single country, very competitive industries often have many small firms in economic competition

with one another. In industries that are dominated by only a few firms—industries called **oligopolies**—individual firms often face much less competition for domestic markets. Even oligopolies, however, may now face stiff foreign competition, so that the number of firms within an industry in any given country is no longer a sufficient index of competitive pressures. Competitive conditions can lead to more marginal jobs in two ways.

First, a firm may respond to competition by creating low-wage, unstable jobs and laying off its workers as a result of even minor economic setbacks. This strategy gives management some additional flexibility in meeting competitive conditions; it is easier to move the company in a new direction, modify the product lines, or simply conserve capital. But it is costly in terms of employee morale and loyalty. When managers follow this strategy, competition results in increased numbers of marginal jobs.

In the second case, the firm may decide simply to export its jobs to another region or even another country, often a developing country, where workers will accept lower wages. In this case, the effect of competition is to decrease total employment in the first location by eliminating some jobs, although a number of lower-paid jobs will be created elsewhere. It is difficult to predict which strategy will be used more frequently, and so it is difficult to assess whether the effect of competition is more marginal jobs or a decreased total number of jobs. A combination of both effects may occur.

Some foreign competition is real. Multinational corporations may, however, use the *fear* of foreign competition to justify lower wages and poorer working conditions within the industrialized countries. The multinational corporation can produce its products, and create jobs, in a number of countries. Its profit margin may not depend on the locale of production. Its managers may threaten workers with the possibility of closing its North American plants and moving them to Taiwan, the Dominican Republic, or Malaysia (Adler, 2000). In such cases, workers or their union representatives may agree to lower wages and poorer working conditions to keep their jobs. By participating in this process, workers may seem to acquiesce to their own marginality, but they may see a marginal job as preferable to unemployment.

The *less productive* industries are also more likely to create marginal jobs. Lower productivity may result because the work itself is labor-intensive and makes little use of technology. Many of the service industries, for example, are less productive than manufacturing industries because no technological process is available to improve the delivery of the service. In other cases, productivity-enhancing technology exists, but firms have lacked the ability or willingness to invest in it.

There are also industries that have a tradition of creating low-income, unstable jobs, usually as a result of *management practices.* The textile industry, which was once thought to be a low-wage industry only because of intense competition, actually employs quite productive workers who are, however, paid less than they would be paid in other industries (Galle et al., 1985). They often cannot seek higher wages because the textile mill is the only major employer in the town. It appears that the practice of paying textile workers low wages, and of seeking lower-wage labor markets for textile mills, is, in part, a result of management tradition. Textile mills have traditionally sought to hire low-wage workers, first native-born women, then immigrant women, then minority women, and now workers from developing countries. The practice of always seeking out lower-wage workers keeps prices (and wages) low.

By Firm

Firms were once ready to keep extra, trained workers on their payrolls so that the workers would be available to progress through the internal labor market. This practice ensured a ready supply of skilled labor when there was a lot of work. More recently firms appear oriented to shrinking rather than to maintaining employment levels (Osterman, 1999). When markets are slack, employers cut their wage bills rather than support a large number of employees. When there is an abundance of work, employers want to get the work done without permanently expanding the size of their workforce. This means that a large firm may identify some of its workers as being more important than others. They are thought of as the core workers in the firm, and they often have job security, good pay and benefits, and a career line. Less important workers, perhaps those with lower skill levels, may have unstable jobs, be subject to layoff and recall, or in other ways be deemed expendable in times of slack demand.

Firms sometimes seek institutional mechanisms, such as subcontracting the work to client firms (Averitt, 1968), for dealing with extra work. In 1992, for example, Aetna Life & Casualty Company decided to "reengineer" by cutting 2,600 jobs and then contracting out the work (Zachary and Ortega, 1993). As mentioned, client firms are smaller, more specialized firms that contract to perform work as needed when demand is high at the larger, patron firms. For example, a local mailing service may be called in to address and mail an unusually large number of notices for a bank, although the bank would normally use its own clerical staff to prepare mailings. Client firms tend to have weak market positions because they are dependent both on the level of demand and on larger firms. As a result, the client firm often creates marginal jobs and resorts to even more stringent layoff policies than those that characterize the patron firms.

Small firms may be more likely to rely on part-time or seasonal workers and to offer lower wages and fewer benefits. Client firms, and small firms more generally, are often legally exempted from providing some fringe benefits. Small firms are also less likely to be bureaucratized, to have established personnel policies, to have well-trained supervisors, and to have grievance procedures. This can have the advantage of a friendly, less bureaucratized work environment. Alternatively, it can mean that a worker is relatively powerless and must continually appease a capricious employer or boss. For all these reasons, small size is one important characteristic of firms that is associated with marginal jobs.

By Employment Contract

Another way the large firm may deal with peak periods of work is to contract with temporary workers. Temporary workers may also be hired to fill in for permanent core workers who are sick or vacationing. In one survey, the leading reason that employers gave for hiring temporary workers was "to alleviate an overload of work" (70% of employers); the second most frequent response was "to cover for workers on leave" (52% of employers) and "to cover for vacationing workers" (51% of employers) (McCarthy, 1988). The percentages overlap because employers could give more than one reason in this survey.

Temporary workers move from one employer to another. Temporary workers who perform manual labor are often called *day laborers,* and most cities have an area where homeless people congregate to be hired for manual day labor. This hiring is often called casual hiring because the job will last for only one day. Many companies maintain a roster of workers who can work as substitutes. Most school systems, for example, have a group of substitute teachers who can be called on as needed, usually to fill in for a full-time teacher who is ill. Unions and some professional groups also maintain rosters of members who are available for temporary work.

In a more bureaucratized form, temporary hiring agencies serve as intermediaries between the workers and the employers. Once popular chiefly in clerical work, now there are temporary workers in many types of jobs, including nursing and other health professions, computer programming, retail sales, and manual labor (Henson, 1996). Box 14.3 recounts the experience of Robert Parker, now a sociology profes-

sor at the University of Nevada at Las Vegas, whose doctoral dissertation research consisted of participant observation as a temporary manual worker.

"Temps" differ from others in the same occupation not primarily in their skills or knowledge but in their job contract. Their contract is with the temporary agency rather than with a specific employer. The pay is not high, and a portion of the pay goes to the agency. Although some temporary agencies provide training, most provide few benefits beyond Social Security coverage, some worker's compensation coverage, and, after completion of a minimum number of work hours, unemployment compensation (Henson, 1996).

Temporary workers may work a full workweek, but they may also work only one or two days a week, depending on demand. In 2000, there were 862,000 workers employed by "personnel supply services," of whom 52% were women, 16% were black, and 27% were Hispanic (BLS, 2001a: Table 18). Many "temps" are available only for part of the year. School teachers and students, for example, may hold temporary jobs during school vacations. Many retail stores offer temporary employment during the holiday rush season.

Temporary workers can be in a variety of occupations and may work in a variety of industries. We can consider them marginal to the extent that their work is involuntarily intermittent and their pay and hours of work are substantially below what they might obtain in a more conventional, permanent job. Some "temps," however, choose this type of work because it is compatible with their other home or school responsibilities and because it provides useful job experience and exposure to a variety of workplaces and potential employers. In one survey, the most important reason given for working for a temporary help service was "freedom to schedule my work in a flexible manner." Fewer than 10% reported that temporary work was "a stopgap measure until I can obtain a permanent job" (Gannon, 1984).

BOX 14.3 Working as a Temporary

On the morning of the assignment, I phoned the temporary company's office at approximately 7:20 A.M. to check on work availability. The firm's job dispatcher said there would be no assignments available for me that day. About twenty minutes later, however, she called back with an offer for a light industrial assignment. This quick turnaround of circumstances illustrates both the uncertainty surrounding the availability of work and the short notice commonly given to and expected of temporary employees. Taking note of the employer's address and other details, I accepted the temporary firm's offer and headed for the warehouse.

On this assignment I had been hired by a major U.S. retailer that needed so-called special-project workers. The retail company's regional management was attempting to reorganize a local warehouse more efficiently. In addition to myself, more than a dozen other temporary workers were assigned for this special project. I later learned that I had been called in as a last-minute replacement for a temporary employee who, according to another worker's account, decided that yet another mid-August day in a warehouse was intolerable. His reluctance to return was easy enough to comprehend; by the time I arrived at 8:30 A.M., the temperature had surpassed ninety degrees.

The job site on this occasion was easy to find, but the pervasive uncertainty that characterizes temporary employment was evident in my inability to locate the appropriate personnel upon arriving. Furthermore, the temporary company that had arranged the assignment did not know to whom I should report. For roughly ten minutes I toured

the inside of the warehouse before finding the supervisor. After being introduced to the first of several managers from whom I would take instructions that day, I began my day-long task— rotating appliances from one part of the warehouse to another. At first I moved water softeners, then tool kits, and later drill presses, but it was all the same kind of manual work.

Despite the mundane activity, this assignment was particularly valuable to me in terms of understanding light industrial temporary employment from the workers' perspective. It allowed me to interact informally with as many as twelve other temporary workers at one time. In addition, it eventually yielded several in-depth interviews. One of the interviews that stemmed from this assignment was with Ron, a recently divorced, black Vietnam veteran. Compared with most temporary employees I met, Ron displayed a significant level of assertiveness at the workplace. For example, he remarked to one of our supervisors that "we don't need any of your help" (in organizing the warehouse). Also on this assignment, I worked alongside, and interviewed, Tim, a white thirty-two-year-old who had recently quit working as a drummer for a local band. Another light industrial temp in my immediate work group was Tom, a white forty-year-old who had recently been operating a small engraving shop in the central city. The four of us had lunch at my house that day and became better acquainted.

Back on the job that afternoon, rotating stock grew increasingly disagreeable, difficult, and dangerous as the temperature eventually climbed to 103 degrees. Much of the work we had earlier

WHY ARE SOME WORKERS CONSIDERED MARGINAL?

Certain segments of the labor force are more likely than others to fill marginal jobs. It is safe to say that these are the workers who have fewer options than other workers, but why their options are limited may vary. Workers in marginal jobs are more likely than other workers to have character-

istics of geographic isolation, low education, disability, and recent job displacement, and to be distinctive in terms of demographic characteristics such as age, race and ethnicity, and gender.

Geographic Isolation

Some workers may find that because of their place of residence the only available job opportunities are marginal jobs. For example, workers in accomplished solely with our backs was increasingly relegated to forklifts and dollies. In midafternoon, as a collective expression of our discomfort, we took a half-hour break, a period three times longer than officially allowed. When we returned to work, one manager began to convey his disapproval about the time our idleness had consumed. Before he could finish, however, he was cut off by Ron, who relayed his own displeasure about this particular supervisor's "rigid" managerial approach. Given the heat and arduous labor, none of the temps expressed feeling particularly guilty about slowing the work pace or taking extra time off. Even the head supervisor was sympathetic. And despite our transgression, the retailer's management encouraged all of us to return the following day.

After the late afternoon break, we were instructed to perform tasks typically forbidden by temporary help firms. Specifically, the work involved disassembling two sections of scaffolding that rose more than thirty feet. All the storage components were made of long, awkward, heavy pieces of metal and required at least two workers to handle them safely. Accomplishing this task genuinely aroused concerns within me about safety, especially because of the lateness of the day and the fact that the work required all the stamina and alertness that could be mustered. Nonetheless, we successfully completed the last-minute chore without incident, signed out, and left for home by 5:00 P.M.

Two general points are noteworthy about this temporary assignment. First, the work itself was dead-end, offering no opportunities to learn new skills or hone existing ones. With the exception of operating a forklift, the work required no formal skills and no ongoing supervisory instruction or on-the-job training. Several co-workers that day expressed the belief that temporary employees are often assigned dead-end work because permanent employees cannot handle its stifling character.

Second, there were several modes of worker resistance that temporary employees exercised to protest the numbing assignments. As intimated earlier, temporary workers have several methods at their disposal to resist the exploitative working conditions they encounter. The most fundamental form is to simply resign from temporary work arrangements or routinely decline assignments when temporary firms make an offer. This is not a practical strategy, however, since most temporary employees are working out of economic necessity. Another form of worker resistance I witnessed was the deliberate restriction of output. At the warehouse assignment workers cajoled one another not to give more than $4.00 an hour worth of effort (the wage they were being paid that day). Taking frequent breaks is yet another method. On the warehouse assignment described here, workers left early for lunch, extended allotted breaks beyond specified time limits, and took numerous other informal breaks. When resistance surfaced, temporary workers often discussed the issue and justified their behavior in terms of the lack of employer commitment and the generally dubious working conditions they confronted.

SOURCE: Robert E. Parker, *Flesh Peddlers and Warm Bodies*, pp. 116–119. Copyright © 1994 Robert E. Parker. Reprinted by permission of Rutgers University Press and author.

rural areas or in inner-city ghettos may find that only unstable, low-paid jobs are available within their immediate vicinity. Better jobs may be available in other parts of the same metropolitan area, but workers who do not already have good jobs may not be able to afford the transportation costs. Even in large cities with good mass transit systems, ghetto residents often find that the good jobs are being created in suburban areas and that regular outbound transportation to the suburbs is available only at the end of the day when suburban commuters return to their homes.

Geographic isolation also occurs in small towns, where the only available jobs may be unstable or poorly paid. Because employers do not have to compete to attract laborers, there may be little improvement in the jobs, even over a long period. In areas that are economically depressed, high levels of unemployment may be chronic. In such circumstances, workers or their organizations may turn to the political process to lobby for economic development funds or for other efforts to bring employment to the community.

People with limited geographic mobility may be in a difficult employment situation even if they are not actually isolated. As you learned in Chapter 3, two-career couples are a notable group with limited mobility. Married women, in particular, are likely to find their job prospects limited by their husbands' jobs. Better jobs might be available to these women if they were free to move, but there may be a poor match between their skills and the jobs available in their local labor market.

It is difficult to estimate how many workers have marginal jobs because of their isolation. Many workers do not know that their wages would be higher in another location, and, indeed, they cannot be certain that they would be hired for more attractive jobs even if they moved.

Educational Level

Workers' education and skills are highly correlated with their earnings. In 2000, 90% of all adult workers had at least a high school diploma. Those who had not completed high school earned on average only about 62% as much as the average for all workers. Earnings increase with level of schooling. In 1997, a full-time, full-year male worker with at least one college degree earned an average of $66,393, compared with $32,611 for high school graduates (Census, 1999: Table 758). Workers with some college, but no degree, were in an intermediate position with annual earnings of $39,367.

Years of schooling is not a sufficient indicator of education, however, because as many as 10% of high school graduates are functionally illiterate, and as many as 44% of the foreign-born in the United States may be functionally illiterate in English. Productive workers need to be literate even for marginal jobs; farm workers, service

workers, and household workers are exposed to toxic pesticides and chemicals daily, and at a minimum, they need to be able to read the labels of the products they use. In one reported instance, a herd of cattle had to be destroyed because an illiterate worker unknowingly added a poisonous compound to their feed (Levitan and Shapiro, 1987:70). National surveys of new workers, young adults aged twenty-one to twenty-five, show a decline in literacy over time ("Lowdown on Literacy," 1994).

Disabling Conditions

Although many people suffer some work disability, about 12 million American adults report themselves unable to perform an instrumental activity of daily living, such as using the telephone or doing light housework (Census, 1999: Table 235). These workers tend to be overrepresented among the underemployed (Sullivan, 1978; Vickers, 2000). Only 23% of adults with a severe disability work at all (Census, 1994a); of those who do work, many are part-time workers. The disabled may not be able to compete well for the most desirable jobs. On the one hand, they may be disqualified from some jobs because of their specific disability; for example, a paralytic might not be able to move heavy objects. On the other hand, there are many jobs that the disabled could handle but that are not offered because the employer has many able-bodied applicants. Some disabled people are forced to accept marginal work because no better job is available.

Job Displacement

Displaced workers are defined as workers who are at least 20 years old who lost their job because their company closed or moved away, because there was insufficient work for them to do, or because their position or shift was abolished. In February 2000, the Bureau of Labor Statistics conducted a survey of the period 1997 through 1999, and found that 7.6 million workers had experienced displacement during this time period. Of those workers, 3.3 million had worked for their employer for three or more years. A similar survey covering the years 1995 through 1997 showed that a similar number of workers had been displaced during those three years (BLS, 2000).

Figure 14.1 shows the work situation of the displaced workers in 2000. About 75% of the displaced were reemployed, 10% were unemployed, and the remainder were no longer in the labor force. The most likely workers to be reemployed were those younger than 54. Older workers had much lower reemployment rates. Men were more likely to be reemployed than women, and white workers had somewhat higher reemployment rates than blacks and Hispanics. Only four in ten of the displaced workers received advance written notice that their jobs would end; workers who received this notice were more likely to be reemployed.

Technicians were the most likely to be reemployed (86%) and machine operators, assemblers, and inspectors were the least likely to be reemployed (66%). Manufacturing accounts for a disproportionate number of the displaced workers, although service industries also displace workers. As Figure 14.1 shows, a substantial proportion of the displaced workers are earning less than they made on their last job, although another substantial proportion are earning as much or more than they had on their lost job.

Age

Age tends to be correlated with skills and experience, and for this reason young and inexperienced workers are most likely to be recruited for marginal jobs. Some observers believe that it is an advantage for teenagers that many farm labor and service jobs require little training, because these jobs are then available to provide job experience to teenagers. Some legislators have proposed the **subminimum wage,** a wage lower than the current minimum wage, as a way to increase the number of jobs available to young people. To be sure, these jobs would be poorly paid. But because the jobs would provide work experience

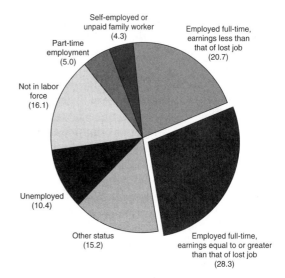

FIGURE 14.1 Percent Distribution of Displaced Workers Who Lost Wage and Salary Jobs between January 1997 and December 1997, by Labor Force Status in February 2000

SOURCE: Bureau of Labor Statistics, 2000, calculated from Tables 1 and 7.

and some on-the-job training, proponents of the subminimum wage hope that young workers would be able to move into more responsible and better-paid jobs.

Child labor laws have limited the types of work and the hours of work for teenagers. In general, any worker aged 18 or older may do any job for unlimited hours, but 16 and 17 year olds may not do jobs declared too hazardous by the U.S. Department of Labor. Such jobs include logging, slaughtering livestock, work with explosives, and mining. There are also state labor laws that affect what type of work children may do, but the level of protection varies. In Oregon, children as young as 9 may pick berries and beans (Mendoza, 1997).

Elderly workers whose experience might be considered obsolete may also be hired for marginal jobs, although older workers may also respond to job loss by retirement. As we mentioned in Chapter 3, the decline in births during the 1970s meant fewer entry-level workers in the 1990s, and so the firms and industries that traditionally hired teenagers began to view the rapidly growing elderly population as a potential source of workers (Pereira, 2000). Advertisements for several fast-food chains depict older employees. These ads are part of a campaign to convince elderly and retired people to consider part-time work. Although elderly workers might be good for the firm, it is not clear that the jobs will be attractive to them. Low-wage jobs are less likely to have job-related medical insurance and pension coverage, benefits of particular interest to elderly workers.

Race and Ethnicity

The overrepresentation of minority groups in marginal jobs stems directly and indirectly from discrimination (DeAnda, 1996). Discrimination directly affects minority workers through a cultural division of labor that relegates them to the less desirable jobs. Members of minority groups are sometimes denied access to better jobs and to further education and training on the grounds that they prefer or are naturally attracted to the marginal jobs they hold. These marginal jobs may even become a sort of occupational ghetto, with employers refusing to hire members of the majority group. This dynamic was described in the case of household service in Box 14.2, with some employers unwilling to hire majority white women as maids.

Discrimination may also indirectly influence the job opportunities available to members of minority groups. Inferior schools may keep many minorities channeled into less attractive jobs. Residential segregation, and especially the concentration of minority workers in inner-city neighborhoods, isolates many minority workers from attractive jobs in the suburban ring. In addition, urban transportation systems are better developed for bringing suburban dwellers downtown than they are for conveying inner-city residents to jobs in the suburbs (Wilson, 1997).

Immigrants are a special case of the general marginality of minority groups (Lamphere,

1992). Generations of immigrants have taken relatively menial and marginal jobs because they were the only jobs available. This is still true of immigrants who have relatively low levels of education or job skills. (Immigrants with professional skills, on the other hand, often earn on a par with native-born workers within a few years.) Immigrants continue to be recruited as temporary farm workers and household workers. Discrimination, poor ability in the language of the host country, and unfamiliarity with host country regulations may all be reasons why immigrant workers tend to be channeled toward marginal jobs. Employers often assume that the native-born will not take a marginal job. For example, some U.S. agricultural growers recruit foreign workers because they assume that no American workers would take the jobs. Such employers may prefer to recruit aliens instead of making the job more attractive to the native-born. More attractive jobs may not be open to immigrants because of legal barriers. For example, jobs with the government or within certain core industries may be limited to citizens, or the licenses required for some jobs may specify that the credential be obtained from training schools within the host country.

Gender

Women have traditionally been more likely to be involuntary part-time workers and to be among the working poor (Sullivan and Mutchler, 1985). Constraints unrelated to the job market, such as family responsibilities, may influence women workers to take temporary or part-time jobs that are marginal. Women are disproportionately found, for example, in household service, other personal service jobs, and temporary work. This circumstance ironically reinforces the stereotype that women work only for "pin money," that their work roles are secondary to their family roles, or that they are unreliable. Such stereotypes reinforce gender discrimination if female workers are relegated to lower-paying, marginal jobs within a firm.

Interacting Characteristics

Some of these characteristics interact with each other. For example, young African-Americans are more likely to be geographically isolated and somewhat more likely to be school dropouts than are other workers. Having several of these characteristics increases the probability that a worker will be employed in marginal jobs. As we have discussed in Chapter 5, some employers use a type of statistical discrimination in hiring for core jobs. To minimize the time spent in interviewing and hiring candidates, these employers simply hire according to their stereotype of the appropriate worker for the job. Because the gender, race, education, and other easily verifiable characteristics of productive workers are already known, employers prefer to hire more workers who have these same characteristics. When they are hiring for their more peripheral jobs (such as cleaning staff), women, minority workers, disabled people, immigrants, and others may be considered appropriate.

Such stereotypes may become self-fulfilling prophecies. Some observers fear that marginal jobs "train" people to be bad workers who cannot take initiative, move aimlessly from job to job, and fail to develop good work habits. For example, if the employer shows little interest in the jobholder and the job offers no chance for advancement, the employee may likewise see little reason to be punctual or to work very hard. Employers who expect little from their workers often find their expectations fulfilled. Such workers may quit frequently to seek better jobs (Brown, 1982). Unfortunately, the record of frequent resignations and job changes may be interpreted as indicating that the worker is unstable and unwilling to commit to a job. This may mean that the worker who begins to work in a series of marginal jobs may have difficulty finding a more permanent, core job. A vicious cycle develops in which good habits are not rewarded and bad habits are reinforced. As bad as this situation is for the individual worker, it also perpetuates stereotypical thinking about whole groups of workers. This stereotypical thinking eventually leads to the occupational steering of some people toward marginal jobs.

MARGINAL WORKERS AND SOCIAL CLASS

> There were times when people who worked for GM, in the factory, might have been good [role] models. But those factory jobs are gone. So what you have to impart to a young guy is that they have to have diverse skills and flexibility. My own experience is that I had to confront a very painful need to figure out how to exist economically without having to go and apply for what is considered a "job."—Hakim, a street vendor of books (Duneier, 1999:40–41)

Jobs and occupations are important indicators of social class. In Chapter 3, we discussed this important influence on workers. Many social scientists are concerned about what the persistence or perhaps increase of marginal jobs will mean for the class structure of advanced industrialized societies. They express this concern in two basic ideas: the development of an **underclass,** and the **shrinking middle class.**

The underclass is a social group that cannot achieve social mobility because of lack of personal resources and opportunities (Wilson, 1987; Auletta, 1982). To be a true social class there must be a mechanism to transmit the job status of parents to their children. Children of marginal workers are less likely to finish school than are children of nonmarginal workers. Teachers and employers may believe that these children cannot achieve much in the workplace and may steer them away from college preparatory curricula and other opportunities that would prepare them for better jobs. In addition, the low income levels of the families may mean that the teenage children must take available jobs—often marginal ones—and then never advance into good jobs. In addition, the children are less likely to have a network of contacts who might help them get a better job

(Granovetter, 1995). The occupational inheritance of marginal jobs from generation to generation is thought to be part of the formation of an underclass.

Welfare reform has required a number of former welfare recipients to seek employment. The employment most often available to them consists of relatively low-paid jobs, often in the service sector (Edin and Lein, 2000; Newman, 2000). The former welfare recipients who go to work often face much higher costs for transportation, child care, and other expenses, and the jobs available to them are not always sufficient to cover the additional expenses. This group of the "working poor" competes with teenagers, recent immigrants, and inner-city residents for the available low-level jobs.

The hypothesis of a shrinking middle class suggests that the traditional middle-class jobs are declining in relative numbers or, at least, are becoming more unstable (Levy, 1987). Some jobs are being reduced to marginal jobs, perhaps because they become part time or temporary. Some jobs are being deskilled because of technological advances. Other jobs are being lost altogether to overseas competition, perhaps because they are being exported to other countries. Whereas the underclass argument suggests that class position is inherited, the shrinking middle-class argument suggests that middle-class children may *not* be able to inherit their parents' class position and may slip below their parents' class position. At the same time, mobility into the middle class will become even more difficult (Rosenthal, 1985; McMahon and Tschetter, 1986).

There is as yet little convincing evidence to support or refute these two hypotheses, although some social scientists have argued that public policy tends to emphasize the problems of the poor and the issues of the rich, to the exclusion of the middle class (Skocpol, 2000). As you saw in Chapters 12 and 13, technological advances can lead, in some situations, to improved jobs. The creation of a marginal job is not inevitable. Moreover, at least some marginal jobs may still provide an opportunity for young people and immigrants to gain experience before moving into more permanent employment. Social scientists still disagree whether workers in marginal jobs will become a permanent underclass or whether they can "cross over" into better employment. Plainly, the class arguments hinge on the future number and characteristics of marginal jobs.

THE FUTURE OF MARGINAL JOBS

Every society has some work that is defined as "dirty work," and dirty jobs will probably always have low status and may also have low pay. But as we have seen, many marginal occupational categories are declining in numbers. Thus, the important issue in assessing the future of marginal jobs would seem to be marginality that results from the organization of employment. Two types of theories are useful in understanding this type of structural marginality: *dual labor market* theory and *internal labor market theory.*

Dual Labor Markets

Sociologists have hypothesized that labor markets are increasingly demarcated between good jobs and bad jobs (Hodson, 1983). In developing countries, the distinction is sometimes made between formal and informal sectors or between modern and traditional sectors; in industrialized countries, the distinction may be expressed as one between primary and secondary economic sectors or between core and peripheral industries. Productivity and profits are expected to grow in the primary sectors, permitting the substitution of more capital for labor. Thus, the number of good jobs may not grow rapidly, although the existing good jobs might become even better in terms of pay and fringe benefits. Meanwhile, there is concern that the more labor-intensive secondary sectors, which are poorly paid and less secure, will absorb disproportionate numbers of workers. Such theories attribute job creation to the overall trajectories of economic growth.

Internal Labor Markets

According to **internal labor market** theories, the decisions of employers are crucial in determining the relative proportion of good versus bad jobs. Even an employer in a profitable, core industry may have two kinds of jobs to offer. Employers may provide good jobs for the workers with many skills or with rare skills and marginal jobs for unskilled workers. In times of economic recession, the workers holding the more peripheral jobs are laid off. Even if the workers in the marginal jobs have high turnover, the production process will be little affected. According to internal labor market theories, even substantial economic growth might not be translated into better jobs. The critical issue will be the decisions of employers or, possibly, the countervailing power of workers.

Additional Factors Two more factors affect the future number of marginal jobs. International competition and technology may lead either to the creation of more marginal jobs or to the creation of better jobs. It is not yet certain which of these is the more likely outcome. It may happen that the adoption of technology and the impact of competition will lead to better jobs in the core industries and firms and to more marginal jobs in the peripheral ones. If this is the case, competition and technology would tend to aggravate the duality of labor markets. It might also happen that competitive pressures will be met, and technology will be deployed, to reinforce existing divisions between internal labor markets and secondary labor markets. If this happens, competition and technology will aggravate duality, even within workplaces.

SUMMARY

A marginal job is not defined by job content alone; it can also be defined by its organizational and economic context. The job content, job regularity, job stability, and adequacy of employment in some occupational groups mark them as marginal jobs. These jobs include farm labor, service work, and household work. Other occupational groups may be employed either in marginal jobs or in stable, "good" jobs. For these occupational groups, the industry, the firm, and the nature of the labor contract determine whether their jobs are marginal. For example, temporary clerical workers may do the same job as others in the same clerical occupation, but their jobs are intermittent because of the nature of their employment contract.

The workers most often hired for marginal jobs tend to be distinctive in terms of their race,

gender, nationality, schooling, disability status, previous work history, and geographic location. Employers may use statistical discrimination as a means of channeling workers into the good jobs and the marginal jobs.

Social scientists do not yet know whether the relative number of marginal jobs is increasing. Dual labor market theory suggests that economic growth and concentration encourage the creation of marginal jobs. Internal labor market theory emphasizes the labeling of jobs within the firm. International competition and technology will surely affect the eventual outcomes. If the relative number of marginal jobs increases, the middle class is likely to shrink, and the underclass may grow.

KEY CONCEPTS

marginal jobs

deviant occupations

shadow economy

job instability

contingent employment

underemployment

migrant farm worker

sharecropper

hired hand

payment in kind

core and peripheral jobs and
   industries

client firm

dual labor market

oligopolies

subminimum wage

underclass

shrinking middle class

internal labor market

QUESTIONS FOR THOUGHT

1. If my salary is not as high as the salary I want, is my job a marginal job? Why or why not?

2. Why are female and minority workers more likely to find employment in marginal jobs?

3. What are the differences between a secondary sector and an internal labor market? Why are these concepts important for a study of marginal jobs?

4. Suppose that a temporary worker receives employment on every day that he or she wishes to work. Could this person's job still be called a marginal job? Why or why not?

5. How is it that an industry could be peripheral in one country and core in another country? Are there marginal jobs even in core industries?

MULTIMEDIA RESOURCES

Print

David Card and Rebecca M. Blank, eds. 2000. *Finding Jobs: Work and Welfare Reform.* New York: Russell Sage Foundation. An analysis of how former welfare recipients have fared in the employment market.

Mitchell Duneier. 1999. *Sidewalk.* New York: Farrar, Straus and Giroux. An ethnographic account of New York street vendors and the network of relationships within their urban neighborhood. It provides good examples of intersecting formal and informal labor markets.

Katherin Newman. 2000. *No Shame in My Game.* New York: W.W. Norton. An analysis of the working poor in a fast-food outlet. Although they earn little money, they find dignity in their work.

William Julius Wilson. 1997. *When Work Disappears: The World of the New Urban Poor.* Reprint ed. New York: Vintage Books. An analysis of how the loss of the industrial base in American cities has contributed to an urban underclass without access to good jobs.

Websites

http://www.altavista.com/cgi-bin/query?rnsp+20& AltaVista.com is a search engine that provides this site for small-business owners who may need temporary workers.

http://www1.umn.edu/ihrc/ The Immigration History Research Center site, maintained by the University of Minnesota, with information about various immigrant groups that originated in Europe and the eastern Mediterranean.

RECOMMENDED FILM

The Grapes of Wrath. 1940. Starring Henry Fonda, Jane Darwell, John Carradine, and Doris Bowden, this classic movie is based on the John Steinbeck novel. It depicts hard times among poor farmers who left the Dust Bowl to seek a better life in California. Not rated.

PART V

❖❖

Work in the
Twenty-First Century

The three chapters in Part 5 discuss the forces that are shaping the future of work. Chapter 15 examines large corporations and their movement toward increased complexity and diversification. Chapter 16 shows how events in any one part of the global economy have consequences for all the other parts. Chapter 17 projects current developments into the future in an attempt to understand the face of work for future generations.

Chapter 15 focuses on the emergence of the large diversified corporation as the dominant organizational form in the modern economy. We discuss both positive and negative consequences of large corporations for workers and for society. Workers are frequently paid better in large firms than in smaller firms and have better benefit and retirement packages. Their work, however, may be more alienating and repetitive. The buying and selling of subsidiary companies by large corporations also creates new vulnerabilities for workers as jobs appear and disappear overnight for reasons that may be quite remote from how productive the workers are. Mergers, acquisitions, and divestitures also absorb financial and human resources that might be better invested in innovation and in activities that directly increase productivity. What is good for any given conglomerate may be bad for the health of the economy as a whole. In the concluding section of Chapter 15 we examine the continuing viability of the small-firm sector and its importance for the future of the economy.

Chapter 16 explores the origins of a world economy in the imperial system developed by Britain and other European states. We examine the replacement of political imperialism after World War II by a system of economic neo-imperialism with the United States at its head. Political independence and economic development in the Third World have greatly increased the complexity of the world economy. In conjunction with the rebuilding of Europe's industrial base after World War II, these economic and political developments have created a world economy with greater diversity than ever before. In the concluding section of Chapter 16 we examine some of these diverse systems of industrial organization, including those typical of Japan, Sweden, Germany, and the former Soviet Union.

Chapter 17 discusses possible futures for advanced industrial societies. These futures include both the possibility of a large innovative sector with rapid growth and good jobs and the possibility of a large marginal sector with low pay and unstable jobs. We argue that the choice between these futures depends on the concerted actions of all of us as producers and citizens.

15

The World of the Large Corporation

"It's mine. I built it. You bump it down—I'll be in the window with a rifle. You even come too close and I'll pot you like a rabbit."

"It's not me. There's nothing I can do. I'll lose my job if I don't do it. And look—suppose you kill me? They'll just hang you, but long before you're hung there'll be another guy on the tractor, and he'll bump the house down. You are not killing the right guy."

"That's so," the tenant said. "Who gave you the orders? I'll go after him. He's the one to kill."

"You're wrong. He got his orders from the bank. The bank told him, 'Clear those people out or it's your job.' "

"Well, there's a president of the bank. There's a board of directors. I'll fill up the magazine of the rifle and go into the bank."

The driver said, "Fellow was telling me the bank gets orders from the East. The orders were, 'Make the land show a profit or we'll close you up.' "

"But where does it stop? Who can we shoot? I don't aim to starve to death before I kill the man that's starving me."

"I don't know. Maybe there's nobody to shoot. Maybe the thing isn't men at all. Maybe, like you said, the property's doing it. Anyway I told you my orders."

(STEINBECK, 1939:51)

In the world of work you may find that your life is strongly influenced by economic forces that are outside your control. You may not be able to see who or what these forces are or to personalize them so that you can understand them. As an individual, you may be nearly powerless in the face of these forces. At some point in your career you may well work for a large, seemingly impersonal corporation. Such corporations exercise immense power because of their size. They exercise this power in relation to their employees, their customers, and smaller firms.

In this chapter we discuss the increasing size of corporations and its implications, focusing on the issue of power. We also examine the ways in which large corporations have expanded their influence over time. These ways include mergers, diversification across product lines, overlapping boards of directors, and the use of subcontracting. We conclude the chapter with a consideration of the continuing importance of small firms and their role in the economy of the future. In conjunction, the giant corporations and the constantly reemerging small-firm sector set the organizational stage for the economy of the twenty-first century.

THE POWER OF THE LARGE CORPORATION

Some workers are self-employed. Some work for small employers. Others work for huge corporations with holdings around the world. As one moves along this continuum, the worker and the owner of the enterprise are increasingly distant. When the worker labors alongside the owner in a small shop, there is often a personal link between them (though it may be a paternalistic link and entail very unequal power). In the large corporation workers may not even know who the owner or owners are. They may know who their immediate boss is or the plant manager's name, but they typically have little knowledge of who actually owns and controls the corporation.

Public Concerns about Corporate Power

Large corporations have immense power in society. Concerns about their role in society were enunciated during the earliest stages of industrial capitalism by Adam Smith (1776) who believed that monopolies were detrimental to the free development of productive forces. These concerns have led the governments of the United States and other industrialized nations to enact legislation to prevent or regulate monopolies. Most businesses would like to be monopolies if they could and it has become one of the roles of the government to prevent this. In spite of these concerns and regulatory actions, corporations have continued to grow in size and their large size often gives them many of the powers of monopolies (Carruthers and Babb, 2000).

The power of large corporations creates the possibility of corporate actions that can have grave consequences. Public concerns about the abuse of corporate power involve six major issues: (1) the exercise of concentrated economic power, (2) the exercise of concentrated political power, (3) the creation of highly bureaucratized organizations that are inflexible and resistant to change, (4) the dehumanization of work, (5) the exploitation of consumers, and (6) the degradation of the environment (Jacoby, 1997). In a later section we also discuss some of the positive aspects of work in large corporations.

Public concern about the **economic power** of large corporations arises from their powerful influence on smaller businesses and on aggregate employment levels. Large firms have the economic power to prevent the entry of smaller firms into their industries and to undermine the viability of small firms already operating there. This power rests on economies of scale and on control of technologies and markets. Such exclusionary power reduces price competition, innovation, and economic growth. It also displaces economic resources from other segments of society to large corporations (Marchak, 1991).

The *downsizing* of large companies through cutting back employment has also been a source of public concern in the 2000s. Such downsizing can entail plant closings (discussed in Chapter 8), technological displacement (discussed in Chapter 9), the movement to "lean" managerial structures (discussed in Chapter 12), and the intentional increase of marginal employment positions (discussed in Chapter 14). On a worldwide scale, domestic downsizing can also be achieved through moving production to lower-wage locations outside North America (see Chapter 16). In aggregate, these tendencies have increased chronically high unemployment rates and eroded the quality of available jobs. In the 2000s, the largest U.S. firms have achieved increasing profits but are employing fewer workers.

Large corporations exercise **political power** first and foremost through lobbying to promote favorable legislation and through campaign contribution to influence elections. Exercise of this power results in subsidies for favored industries and indirectly leads to the control of regulatory bodies by the corporations they were intended to regulate (Davis and Mizruchi, 1999). There is a general public awareness that large corporations are often able to shape national economic policy in their own interests. For instance, the news media, legislators, and citizens' groups frequently accuse military contractors of promoting expanded military budgets in order to make greater profits rather than to meet national security needs. Institutions of higher learning may also come under corporate influence because of gifts, grants, and contracts. Finally, many are concerned that large corporations damage the American image abroad by their exploitation of workers in weaker nations. (We discuss the realities behind these concerns about the world economy in Chapter 16.)

Fears about the **inflexibility** of large corporations focus on the possibility that they may become inefficient, bloated with too many highly paid managers, and overly bureaucratic. Such organizations are unable to change rapidly, adapt to new situations, or innovate. In some ways large organizations may be less efficient than smaller ones.

Many people also fear the power of large corporations to *exploit workers* and to reduce the quality of working life. Large corporations may undermine the sense of community among workers. As you saw in Chapter 4, workers dislike large corporations because of alienating and rigid procedures and reduced opportunities to use their abilities.

Finally, many people are concerned that large corporations *exploit customers* through high prices and the production of shoddy goods and that they *damage the environment*. These outcomes may occur through a disregard for the long-run consequences and side effects of economic activity. For example, a corporation may lower its costs of production by avoiding environmental regulations and allowing the public to absorb these costs in the form of increased pollution, noise, and ugliness. Many people believe that large corporations stress profit to the point of neglecting moral, social, and aesthetic values (Ritzer, 2000). The twenty largest U.S. corporations, ranked by sales, are listed in Table 15.1 along with the dollar values of their profits, assets, and employment. Note that the largest employer among U.S. corporations is Wal-Mart.

Table 15.1 The 20 Largest U.S. Corporations (ranked by sales) $

2000 Rank	Company	Sales	Profits	Assets	Employment
1	General Motors	$189,058,000,000	2,956,000,000	257,389,000,000	594,000
2	Wal-Mart Stores	166,809,000,000	4,430,000,000	49,996,000,000	910,000
3	Exxon Mobil	163,881,000,000	6,370,000,000	92,630,000,000	790,000
4	Ford Motor	162,558,000,000	22,071,000,000	237,545,000,000	345,000
5	General Electric	111,630,000,000	9,296,000,000	355,935,000,000	293,000
6	IBM	87,548,000,000	6,328,000,000	86,100,000,000	291,067
7	Citigroup	82,005,000,000	5,807,000,000	668,641,000,000	173,700
8	AT&T	62,391,000,000	6,398,000,000	59,550,000,000	107,800
9	Philip Morris	61,751,000,000	5,372,000,000	59,920,000,000	144,000
10	Boeing	57,993,000,000	1,120,000,000	36,762,000,000	231,000
11	Bank of America	51,392,000,000	7,882,000,000	632,574,000,000	8,500
12	Southwest Bell	49,489,000,000	8,189,000,000	83,215,000,000	204,500
13	Hewlett-Packard	48,253,000,000	2,945,000,000	33,673,000,000	124,600
14	Kroger	45,351,600,000	411,000,000	6,700,000,000	213,000
15	State Farm Insurance	44,637,200,000	2,449,000,000	69,442,000,000	22,200
16	Sears Roebuck	41,071,000,000	1,048,000,000	37,675,000,000	324,000
17	American International	40,656,100,000	5,055,000,000	268,238,000,000	55,000
18	Enron Oil & Gas	40,112,000,000	56,000,000	3,018,000,000	1,190
19	TIAA-CREF	39,410,200,000	840,000,000	102,216,000,000	6,500
20	Compaq Computer	38,525,000,000	2,743,000,000	23,052,000,000	90,000

SOURCE: *www.fortune.com* and *www.primark.com*. Also, *Moody's Industrial Manual*, 2000, New York: Mergent.

Types of Corporate Market Power

The power of corporations is ultimately based on their economic power. Large firms are able to exercise their economic power in a variety of ways. Social scientists conceptualize economic power in terms of large firms having either a **monopoly** or **oligopoly** position in relation to their environments. In addition, large size, in and of itself, confers many elements of power.

Monopoly Power A monopoly firm holds a dominant position in the market for the goods it produces. When a few firms dominate an industry, these firms can also be said to hold monopoly power. For instance, if the top four firms in an industry control more than 70% of industry sales, we would consider these firms to have monopoly power (Scherer, 1999). Technically, such a situation is called an oligopoly (dominance

by a few firms), but the economic power these few firms exercise is virtually identical to that of a true monopoly. Steel manufacturing in the United States was organized in this way before the introduction of foreign competition in the 1970s and 1980s. On the basis of market domination, monopoly firms are able to restrict competition and charge higher prices for their goods. As a result "since less output is produced and is sold at higher prices, real incomes fall for the consumers who purchase the monopolized goods" (Thurow, 1975:217). In a competitive economy such industries would be flooded with potential producers. What keeps this from happening in monopoly industries?

Barriers to Entry in Monopolies A variety of barriers keep competing firms out, thus protecting the monopoly firm's profits. One important barrier is provided by *economies of scale,* which make it difficult for small firms to compete with large firms. Large batches are generally less expensive to produce per unit than small batches. Monopoly firms may also own important sources of supplies and may make it difficult for competing firms to secure access to these supplies; this is called a **backward linkage.** They may also control wholesale or retail outlets for goods, a **forward linkage.** Companies with forward and backward linkages are said to be **vertically integrated.** Monopoly firms may also have privileged access to credit. A firm can also achieve monopoly power through advertising. If the firm is able to differentiate its products from competitors' products through advertising, customers may come to prefer these products and consider them superior to otherwise equivalent goods. Monopoly firms may even engage in **predatory pricing,** in which they intentionally underbid competitors until they drive them out of the market. At that point, no longer facing any competition, the monopoly firm recoups its losses by charging higher prices.

Industries with monopolistic structures typically have several distinctive characteristics (Averitt, 1968). These industries operate independently of suppliers, wholesalers, and distributors because they own their suppliers and control their own sales networks. They engage in a larger-than-average share of research and development. They set wage standards against which other firms have to compete. And they have a strong influence on the overall health of the economy, because so many subordinate and supporting firms depend on their economic vitality.

Monopoly control is unlikely in some industries. For example, residential construction precludes monopoly structure because the work is relatively unstandardized, production sites are geographically dispersed, and entry is relatively easy. Monopolies are also difficult to secure and maintain in personal services and in other services that cannot be standardized, such as repair. In the United States, antitrust laws have retarded the growth of monopolies in other industries such as interstate trucking.

Perhaps surprisingly, concentration within industries has not increased greatly over time. In 1935, the average four-firm **concentration ratio** in manufacturing industries was 37%. In other words, on average, the four largest firms in a typical manufacturing industry controlled 37% of industry sales. By the 1970s this figure had risen only to 39% and has remained relatively stable since then (*Census Quarterly Financial Report for Manufacturing,* 2000). One reason for this relative stability is that antitrust laws have limited the growth of concentration in key industries. As a result, large companies have grown by moving into related (and sometimes unrelated) industries rather than by further increasing their share of their primary market. In this way antitrust laws may have actually promoted the growth of large **conglomerate** corporations sprawling across several industries. Such corporations are the focus in the section on mergers.

Oligopoly Power An industry is considered dominated by *oligopolistic* corporations if its output is controlled by few enough firms that they can engage in collaborative price setting. Such price fixing would be illegal if done openly; however, collaborative pricing more typically involves informal agreements to hold prices at a specified level and is rarely caught or prosecuted. The maximum number of companies that can effectively coordinate such collaborative practices varies from industry to industry but may be as high as a dozen or more in some industries. This practice creates market conditions similar to those in monopoly industries. Cigarettes and processed foods are familiar examples of industries with oligopolistic structures.

Two types of activity are particularly distinctive of oligopoly corporations. First, oligopolies tend to divide the available market into spheres of influence. These spheres may be defined by geographic regions or by different product lines. Second, oligopolies take advantage of reciprocal buying agreements in which suppliers or purchasers are pressured into allowing the oligopoly company privileged access to their markets

(Scherer, 1999). For example, a wholesaler may be pressured to sign an agreement to buy high-priced microwave ovens from an oligopoly manufacturer as a prerequisite for obtaining their popular washing machines at the volume and price desired. The oligopoly company is thus able to use its power base in one industry to increase its profits in other industries.

Size as Power Large firms, regardless of their market share, have a great deal of power. This power is similar in many respects to monopoly or oligopoly power. For example, before Japanese firms began competing in the automobile industry, a third of the labor force in the United States was engaged in manufacture, sale, or distribution of automobiles or automobile parts. This gave the industry tremendous leverage in Washington:

> [If a large firm were to fail,] the effects on workers, shareholders, creditors, suppliers, and distributors of and tangent to the stricken organization would be intolerable. . . .
>
> Moreover, the dominant position of the giant firms demands that public policy-makers avoid at all costs decisions which could conceivably jeopardize the financial integrity of the mammoth. As a practical matter this means that most doubts in divestment proceedings, in antitrust prosecutions, or levies on unreasonable accumulations of surplus, on the vigor with which certain regulatory statutes will be applied . . . will be resolved in favor of the corporation in question. In a real sense the giant enterprises have achieved many of the privileges accorded the publicly regulated utilities without at the same time being burdened even by formal restraints imposed on the latter. (Baratz, 1971:151–153)

Thus size, in and of itself, confers tremendous power on large corporations, both in relation to the government and in relation to their competitors, customers, and employees.

In the 1980s, the Chrysler Corporation, after decades of bad management and bad decision making, used its influence to secure low-interest loans from the government and wage concessions from its employees in order to restructure its activities. Less powerful firms in less important industries would simply have been allowed to fail. In the 1990s, subsequent to its publicly financed revival, Chrysler was acquired at great profit to its shareholders by the Daimler-Benz Corporation of Germany. Although microeconomic theory focuses on a large market share as the sole basis of monopoly power, sociologists and institutional economists point out that many aspects of economic power result directly from large size alone.

The Legal Status of Corporations

Business corporations in the United States, including the very largest ones, have had legal protection as "persons" for over a century (*Santa Clara County v. Southern Pacific Railroad Co.,* 1886). This legal status serves at least two important functions. First, the owners of the corporation (stockholders) are protected from bearing full legal or fiscal responsibility for the actions of the corporation. Because the corporation is recognized as a person, debts, bankruptcy, and lawsuits are restricted to the corporate entity itself and do not carry over to its owners. This concept is known as **limited liability.**

The Cloak of Privacy A second consequence of their legal status as persons is that corporations receive certain rights granted to people, including the right to privacy. This right is particularly important for the large modern corporation:

> The nature of oligopolistic competition generates pressure on management to maintain secrecy. In the oligopoly, product price is a collective decision reached overtly or covertly through "oligopolistic pricing mechanisms." . . . Competition in such industries occurs through non-price mechanisms (e.g., advertising) and rivalry

over product quality. . . . Strategies for non-price competition (e.g., choice of product lines or marketing policies) must be secret until they are implemented. The need for secrecy in oligopolistic firms is heightened by their clear visibility among each other and before the public. (Cornfield and Sullivan, 1983:258–259)

Such secrecy, however, is not granted to *public* bodies; indeed, it would be considered a breach of democratic principles.

Antitrust Regulations In order to prevent single companies or small groups of companies from using their monopoly power to eliminate competitors and exploit consumers, the government attempts to regulate the existence and behavior of monopolies through the activities of the Federal Trade Commission (FTC, *www.ftc.gov*). This agency performs important functions in reducing unfair practices of large companies against both their competitors and against consumers. It is a bulwark in maintaining the effectiveness of a competitive market system through preempting its tendency to self-destruct through the excessive concentration of economic power. For instance, in 2000, the FTC found Microsoft guilty of anti-competitive practices against its competitors by putting "technological roadblocks" in its Windows operating system that prevented competing programs from operating effectively.

Owners versus Managers The owners of a large corporation (the stockholders) do not, in general, directly control the daily activities of the corporation. Rather, hired managers control daily activities. Management control has been alternatively interpreted as encouraging lesser and greater corporate responsibility. The economists Adolf Berle and Gardiner Means coined the term **management-controlled firms** in 1932 to describe corporations in which the stockholder with the largest holding owns *less than 20%* of the corporation's stock. Using data from 1929, they found that 44% of the 200 largest U.S. nonfinan-

cial corporations were management controlled. They were concerned that this management control might make corporations less responsible (in relation to their stockholders). Berle and Means (1932:9) argued that management control "destroys the basis for the old assumption that the quest for profits will spur the owner of industrial property to effective use."

These findings have been intensely debated since Berle and Means put them forward over half a century ago. According to a reanalysis by the sociologist Maurice Zeitlin (1974), only 22% (rather than 44%) of the corporations studied by Berle and Means were actually management controlled under their own definition. Further, researchers have argued that management-controlled corporations may, in fact, be just as oriented toward maximizing owners' profits as owners ever were (James and Soref, 1981). If profit rates are not as high as possible in a management-controlled corporation, owners may withdraw their holdings from the company and invest in other companies. This will cause a decline in the price of the corporation's stock and jeopardize the interests of management. Under this thesis the profit motive is built into the system of ownership and investment. It does not reside in individual owners' desires for accumulation.

Other researchers argue that managers are more interested in maximizing revenues than profits because maximizing revenues enhances their own power, prestige, and earnings. This thesis may help explain the drive toward increased size. Corporate executives are paid salaries unthinkable to the average wage or salary worker. Table 15.2 reports the salaries of the twenty highest paid chief executives in the United States. The highest paid of these chief executives makes over 25,000 times as much as the average full-time employee. Further, the salaries of chief executives increased at about four times the rate of inflation in 2000 while inflation-adjusted salaries for workers barely increased at all (*Forbes*, 2000). We delve more deeply into the nature and the consequences of the drive toward ever larger corporate size in the following section.

Table 15.2 The 20 Highest Paid Executives in U.S. Publicly Held Corporations $

Company	Industry	Chief Executive	Compensation
Computer Associates	Computer Software	Charles B. Wang	$650,048,000
Foundry Networks	Telecommunications	Bobby R. Johnson, Jr.	230,544,000
CBS	Entertainment & Information	Mel Karmazin	201,939,000
Gap	Retailing	Millard Drexler	172,816,000
Cisco Systems	Computers & Electronics	John T. Chambers	121,700,000
America Online	Internet	Stephen M. Case	117,090,000
IBM	Computers & Electronics	Louis V. Gerstner, Jr.	107,216,000
General Electric	Manufacturing	John F. Welsh, Jr.	106,855,000
Colgate-Palmolive	Household Products	Reuben Mark	97,150,000
Compuserve	Computer Software	Peter Karmanos, Jr.	87,521,000
Citigroup	Insurance	Sanford Wells	85,876,000
Conseco	Insurance	Stephen C. Hilbert	75,100,000
Qwest Communications	Telecommunications	Joseph P. Nacchio	69,317,000
Charles Schwab	Financial Services	Charles R. Schwab	69,111,000
Comcast	Entertainment & Information	Ralph J. Roberts	54,347,000
Walt Disney	Entertainment & Information	Michael D. Eisner	50,660,000
Enron	Energy Extraction	Kenneth L. Lay	49,812,000
El Paso Energy	Energy Extraction	William A. Wise	46,016,000
American Express	Financial Services	Harvey Golub	45,099,000
Bristol-Meyers Squibb	Health Care	Charles A. Heimbold, Jr.	44,037,000
Median earnings of employees in the United States (2000):			$25,540

SOURCE: *Forbes* (May 15, 2000):225. Also Bureau of the Census, 2000, *Statistical Abstract of the United States.* Washington, D.C.: U.S. Government Printing Office.

MERGER MANIA

Corporations have increased in size and concentration in a halting and irregular fashion patterned by periods of rapid activity and relative quiet. Corporate growth occurs not only through expansion of a firm's operations but also through the acquisition of other firms, and these **mergers** tend to occur in distinct cycles. The history of these cycles reveals not only dramatically changing merger rates, but also the emergence of different kinds of corporations during six distinct periods.

The First Five Merger Waves

The *first* wave of mergers occurred in the early 1890s in metallurgy, chemicals, and electrical machinery. Technological breakthroughs in these industries made large-scale mass production more efficient than smaller-scale production. Large firms also benefited from their domination of emerging national distribution and marketing systems based on transcontinental railway and telegraph lines. With these advantages large firms succeeded in forcing smaller ones out of business by using practices that would clearly be illegal under today's antitrust legislation. Such practices included buying up railroads and charging competitors higher railroad freight rates in order to put their products at a price disadvantage. Large firms also increased profits by using new mass-production technologies to replace skilled workers with less well paid semiskilled workers. This merger wave resulted in a dramatic increase in the size of firms in the affected industries (Zunz, 1998). Westinghouse, International Harvester, and Standard Oil emerged as industrial giants at this time.

The *second* merger wave occurred between 1897 and 1905 and was again based on economies resulting from large-scale production, marketing, and transportation systems. This wave had a wider industrial base than the first one and created monopolies in industries such as tobacco and food processing that are still in existence today. Companies emerging as giants at this time include American Can, United Fruit, and American Tobacco. The first billion-dollar merger occurred in 1901 when the Carnegie Steel Corporation combined with its leading rivals to form United States Steel (USX, *www.usx.com*) (Davidson, 1985:xiii).

Vertical Integration During the first two merger waves, large companies acquired smaller companies engaged in similar lines of production. The *third* merger wave occurred in the 1920s and introduced a new corporate form: the *vertically integrated* company. Vertically integrated corporations come into being with the acquisition of firms supplying raw materials or component parts (backward linkages) and firms engaged in further processing or in selling the manufacturer's products (forward linkages). Steel companies were central actors in this wave of vertical integration. They acquired ownership of mines, ore transportation networks, finishing mills, metal fabrication plants (which turn raw steel into pipe, construction girders, and other products), and even transportation and distribution systems for finished steel products. Bethlehem Steel and Republic Steel emerged at this time.

Diversification The *fourth* merger wave stretched across two decades, starting in the Great Depression and extending into the early 1950s (Chandler et al., 1997). **Diversification** was the keynote in this merger wave, and the first multi-industry companies emerged at this time. During the Depression, large companies were scrambling for opportunities to remain profitable. Many smaller companies were facing bankruptcy and

could be purchased for a song. Mergers tended to follow lines of technologically similar products or production processes and resulted in the expansion of companies into related product lines. The movement continued through the boom period of World War II when the larger and more diversified companies were able to take advantage of the heightened demand for new military weapons and supplies more effectively than were smaller companies. This merger wave created E.I. du Pont and many other corporate giants of today's economy.

Industrial Conglomerates The *fifth* merger wave, which occurred in the late 1960s and early 1970s, created a new type of company, the *industrial conglomerate*. These mergers occurred between firms engaged in completely distinct product lines; for example, in 1973 Mobil Oil purchased the Montgomery Ward retail chain, which it subsequently sold as unprofitable. Later Montgomery Ward was acquired by General Electric, which dismantled the company and liquidated its assets through bankruptcy proceedings in 2001. These mergers were motivated by the desire of large firms to move into profitable areas regardless of the field. Vastly different levels of economic power between the acquiring companies and their acquisitions allowed many "hostile takeovers" to take place. In such takeovers, the acquiring company subsequently exploits the cash reserves and market position of the newly acquired company for its own profit without making even minimal investments to maintain the acquired company's productive capacity.

The history of corporate mergers has witnessed a movement from growth through expansion within product lines, to forward and backward linkages, to expansion into related product lines (diversification), to expansion into totally different product lines (conglomeration). Large companies today operate across a variety of related and sometimes unrelated product lines. Diversification and conglomeration are essential concepts for understanding the modern corporation and the modern economy.

The Current Megamerger Frenzy

The fifth merger wave ended with the recession of 1974 and 1975. However, after a very brief respite, a *sixth* wave took off in the early 1980s and quickly accelerated to even higher levels of merger activity. The sixth wave is distinguished by what have come to be known as megamergers between some of the largest companies in the economy, such as Conoco Oil and E.I. du Pont. By the early 2000s, a high level of merger activity had become an established fact of economic life in the United States and Canada.

The contemporary period is one of intense merger activity. By the early 2000s, top managers and rank-and-file employees alike also had become increasingly nervous about the future of their companies as corporations came to be defined as salable bundles of assets. Takeover bids assumed an increasingly hostile character. In this section we explore the nature, causes, and consequences of this megamerger frenzy.

Why has a wave of megamergers occurred at this time? One reason is that it had become cheaper to buy a company in financial trouble than to build new plants and facilities. Many such troubled companies exist in North America because of increased international competition. The value of the acquired company on the stock market may greatly underestimate the actual value of its assets. Tax laws also encouraged conglomerate mergers. Part of what may make a company an attractive acquisition is its package of tax losses, write-offs, and depreciations. These are also acquired, along with the plant and equipment, and can be used to offset the taxes of the acquiring company. In some cases it has even been possible to "buy a company for less than the value of its tax losses and then deduct its losses from one's income" (Davidson, 1985:206).

Alternatively, a successful company with a secure market niche, but a low growth potential (and therefore a low price), may provide a desirable target for a corporation interested in growth through conglomerate expansion. Such a company can serve as a **cash cow.** The acquiring cor-

poration can use the regular profits and depreciation allowances of the acquired company to fund expansion in other areas rather than setting them aside to maintain or upgrade the productive capacity of the acquired company (Bluestone and Harrison, 2000).

A Permissive Legal Environment   Changes in the interpretation and enforcement of antitrust laws have also encouraged the megamerger frenzy. In the 1980s, the Federal Trade Commission stopped targeting specific industries for antitrust investigation and instead only went after companies "known to have engaged in illegal activities" (Blau, 1993). Antitrust cases were left to the Justice Department, where presidential appointees gave low priority to enforcement of this legislation:

> There is no doubt that the administration of President Ronald Reagan brought with it a different attitude toward antitrust legislation. . . . [The] case against IBM, the world's largest computer company, was dropped. A case against the eight largest oil companies in the United States was abandoned just before the trial was set to begin. And an investigation of the major American automobile companies was closed. In their place, the FTC sued a small group of attorneys who represent indigent criminal defendants in Washington, D.C. and started investigations of state boards that license taxicabs and optometrists and of labor unions that represent actors. (Davidson, 1985:125)

The fact that mergers between large corporations were no longer legally contested greatly accelerated their occurrence.

Merger Targets   How are targets identified for corporate mergers? One criterion is technological compatibility between two companies. Other criteria include shifting consumer preferences or a desire to control resources critical to the acquiring company's survival. In conglomerate mergers between two companies in unrelated industries, the target company is selected by financial advisers who appraise it as a "good buy" (Davis and Mizruchi, 1999). Conglomerate firms interested in acquiring subsidiaries will often pay a premium for their purchase, typically buying large blocks of the company's stock at between 40% and 50% more than their current market price. This provides a lucrative, almost irresistible situation for shareholders in the company being acquired who are interested in quick profits. Even if the firm wishes to avoid such a takeover bid, its only option may be to search out a "white knight," a more congenial firm with which to merge. A highly colorful language has emerged to describe the takeover process: " 'White knights' rush to rescue 'sleeping beauties' from 'black knights'—or end up 'paying ransom' as a last defense" (Hirsch, 1986:814). (Some contemporary merger jargon is decoded in Box 15.1.)

At the beginning of the megamerger frenzy, many of the corporations acquiring other companies were in the oil industry. These corporations had amassed huge stockpiles of profits as a result of rapidly rising energy prices. American companies received windfall profits from the sale of cheaply produced American oil at prices leveraged upward by the Middle Eastern oil cartel. With this money the oil companies expanded into nonenergy areas. In 1981, Mobil, Seagram, and du Pont entered into a bidding war for the purchase of Conoco, the fourteenth largest industrial corporation in the United States. By mid-year du Pont had won the battle but had to borrow the $3.9 billion it needed to buy out Conoco shareholders, thus increasing its long-term debt nearly fivefold. The total deal cost $7.2 billion and was the largest corporate acquisition to that point in history. The 1993 acquisition of McCaw Cellular Communications by American Telephone and Telegraph set a new record at $12.6 billion (Sikora, 1993:6).

In the early 1980s, the airlines also began to move toward a diversified structure as they bought hotel chains, restaurant chains, and car rental companies. By the mid-1980s, the merger

frenzy seemed to be consuming the industry. In 1985, United Airlines agreed to buy the Pacific routes of Pan American Airways for $750 million. In 1986, Southwest Airlines bought Muse Air. People's Express bought Frontier. Texas Air bought People's Express. Piedmont acquired Empire Airlines. Northwest agreed to buy Republic. By the summer of 1986, the sizes of the acquired companies had gotten even larger. Texas Air agreed to buy Eastern for $800 million, and this deal was quickly followed by TWA's offer to purchase Ozark Airlines. By the 1990s, virtually all of these recombined airlines had entered bankruptcy proceedings at least once.

The impact of these mergers extends beyond the immediate companies and workers involved. The fifty to one hundred hostile takeover attempts of large firms each year indirectly affect virtually *all* companies, because companies must spend a great deal of time, money, and energy to make themselves unlikely or difficult takeover targets. This pressure moves their investment decisions toward more conservative options in the companies that are fearful of being acquired (Taggart and McDermott, 1993). It also drives up the price of credit, making productivity enhancing investments more difficult.

In 1998, two of the world's largest automakers combined when Chrysler and Daimler-Benz merged. In 2000, the largest merger ever occurred when the two telecommunication giants, Time-Warner and America Online combined their resources. The combined company was valued at $350 billion, nearly ten times the value of the Chrysler-Daimler conglomerate.

In this context of heightened megamergers, workers, shareholders, and communities have come to fear that top managers are more motivated by greed and ambition than they are with the security and prosperity of the organizations they head. Top executives are in control of important information that they use to plan their own career strategies but that they may choose to share only selectively with stockholders or employees. They may even engage in whitewashing the facts and in stonewalling (withholding

---

BOX 15.1 Merger Jargon $

The following terms describing key merger events were identified by the sociologist Paul Hirsch. His research attempts to unravel the symbolic social meanings through which those involved in mergers make sense of the merger events and their aftermath.

afterglow postmerger euphoria of acquirer and/or acquiree, usually soon lost.

ambush swift and premeditated takeover attempt.

bear hug hostile tender offer, usually with considerable muscle behind it.

big-game hunting plotting and executing takeovers of large companies.

black book a point-by-point antitakeover plan that every potential target is presumed to have ready at all times.

black knights unfriendly acquirers drawn to a target by news that the company is already being propositioned by others.

bring to the altar consummate a merger, usually friendly.

chain letter effect apparent growth, through acquisition of other companies rather than high internal performance and productivity.

Chinese wall internal procedures to prevent communication of privileged information between bank lending officers and trust department staff. Gaps in the Chinese wall are objects of fear and resentment by takeover targets.

confetti stock traded by acquirer for that of acquiree, particularly if thought of as having little value (see also "Russian rubles").

courtship merger discussions, relatively friendly, between top executives of two firms.

cyanide pill antitakeover finance strategy in which the potential target arranges for long-term debt to fall due immediately and in full if it is acquired (see also "scorched earth").

double Pac-Man strategy target firm makes tender offer for the stock of its would-be acquirer.

dowry outstanding assets that the target may carry into a merger (e.g., low-interest loans, long-term contracts).

friendly offer merger proposal cleared in advance with the target company's board and top management; usually leads to the firm's recommending it favorably to shareholders for approval.

golden parachutes provision in the employment contracts of top executives that assures them a lucrative financial landing if the firm is acquired in a takeover.

greenmail a firm's purchase of its own stock, at a premium, from an investor who it fears will otherwise seek to acquire it or else initiate a proxy fight to oust its present management.

hired guns merger and acquisition specialists, other investment bankers, and lawyers employed by either side in any takeover.

hot pursuit warfare image for strenuous campaign by aggressive, would-be hostile acquirer to obtain shares of the target firm.

junk bonds high-risk, high-yield debt certificates traded publicly, so called because they are rated below investment grade, either

---

information or comment) while working furiously to feather their own nests.

Increased Diversification

Many large firms have increasingly diversified across a variety of industries. This tendency intensified dramatically during the megamerger wave of the 1990s and 2000s. Corporations view diversification as providing protection from vulnerabilities in any one industry and as leading to greater stability for their assets and profits. Business schools have also promoted this view and trained two decades of managers in its tenets. This distinctly American view of corporate strategies has encouraged a shift of attention away from production and toward the financial manipulation of assets.

by Moody's or by Standard & Poor's; junk bonds are often used to help finance hostile takeovers.

mushroom treatment postmerger problems from an acquired executive's standpoint: "first they bury us in manure, then they leave us in the dark awhile, then they let us stew, and finally they can us."

on the rocks failed, incompatible merger, often leading to divestment.

pigeons highly vulnerable targets.

pirates hostile acquirers.

poison pill see "scorched earth."

raiders hostile acquirers.

rape forcible, surprise hostile takeover, sometimes accompanied by looting of acquiree's profitability.

rescue party a company that provides any substantial aid to the target firm in a takeover battle (see also "white knight").

safe harbor antitakeover advantage: company owns a subsidiary in a heavily regulated industry (such as broadcasting or interstate trucking). The long time required for government approval to transfer its ownership delays the acquisition process and decreases the chances for successful takeover.

Saturday night special a fast and predatory merger.

scorched earth policy whereby the target company would rather self-destruct than be acquired (e.g., all personnel threaten to quit).

seed partner big investor committed to the firm as is, whose large ownership protects a potential target company from acquisition.

sex without marriage extended negotiations for a friendly merger that never takes place.

shark repellent protective strategies for preventing or combating a hostile tender offer.

sharks takeover artists.

shoot-out climax of a takeover battle, usually conducted by "hired guns."

sleeping beauties vulnerable targets (see also "pigeons").

studs aggressive suitors.

summer soldier executive of target company offering only token resistance against a takeover.

takeover the purchase of majority ownership in a corporation; usually resisted by the target company but accomplished nonetheless by paying a premium above the current market price for the firm's shares.

tender offer proposal to purchase a firm's stock from its shareholders for an amount higher than its current market price.

white knight acceptable acquirer sought by a potential acquiree to forestall an unfriendly takeover; the preferred suitor.

wounded list executives of an acquired firm who develop health or career problems from the deal.

SOURCE: Adapted from Paul M. Hirsch, 1986, "From Ambushes to Golden Parachutes: Corporate Takeovers as an Instance of Cultural Framing and Institutional Integration." *American Journal of Sociology* 91,4 (January):830–835. Used with permission of The University of Chicago Press and the author.

Growth through increasing conglomerate diversification is largely responsible for the increasing domination of the economy by a few large firms. Industrial concentration has two distinct meanings. In many industries a few key firms hold a dominant market share but operate solely or primarily within that industry. This type of concentration has been relatively stable since World War II. The second type of concentration results from a few huge firms, which operate across a variety of industries, controlling a large share of the economy. This type of concentration has increased dramatically. Economic theory has largely ignored the growing importance of the concentration of economic power in diversified conglomerates (Scherer, 1999).

Growth within an industry (and growth through vertical integration) occurs either through *expansion* of the core firm or through the acquisition of enterprises doing related work that then become **divisions** of the larger corporation. Growth through conglomerate mergers occurs through the incorporation of acquired companies as semiautonomous *subsidiaries.* This autonomy may only mean having a separate corporate name, or it may entail substantial latitude in decision making for the subsidiary company.

The Effects of Increasing Size and Concentration

What effects have the increasing size and concentration of enterprises had on workers and on organizations? Overall, the effects have been enormous. One need only reflect for a moment on the difference between working in a small corner grocery and a new multiservice shopping center that is part of a national retail chain. These changes can have both positive and negative aspects.

Effects on Organizations Organizational size has a huge impact on the workplace for two major reasons. First, specialization (division of labor) and complexity increase as more activities have to be coordinated (Carruthers and Babb, 2000). Second, bureaucratic rules and procedures replace direct control by owners, as you saw in Chapter 7. Increasing specialization, formalization of rules, and the prevalence of impersonal authority may result in reduced employee identification with the enterprise. Reduced identification makes employees reluctant to take on the values and goals of their increasingly impersonal employer. In addition, workers in large corporations often have fewer opportunities to develop their abilities across a wide range of activities than do workers in smaller firms (Hodson and Sullivan, 1985).

On the positive side, large firms typically pay their employees more than smaller firms. This difference allows large firms to compete successfully for the best workers and also helps motivate these workers. Large firms also typically have fairer and more equitable compensation schemes than smaller firms, resulting partially from the prevalence of bureaucratic rules and procedures that help limit favoritism. The unionization of large firms and the role of unions in negotiating compensation contribute to pay equity. Large firms also typically have better affirmative action records. Again, this results from their formalized procedures, which limit blatant forms of discrimination, and also from their greater public visibility. Finally, workers in large firms are typically more highly rewarded for their formal education than are workers in small firms. This difference results from the premium placed on educational credentials in bureaucratic settings.

Effects on Society Mergers between economic organizations can have both positive and negative consequences for society. The buying and selling of companies may help move capital from stagnant industries into growing ones. Conglomerate mergers can even intensify competition in an industry that has been dominated by a few giant firms. Large, multi-industry conglomerates may be able to compete on the home turf of monopoly companies in a way not possible for smaller companies.

Profit-making organizations, however, do not undertake mergers with the public interest in mind. Mergers are rarely peaceful, and they may have many negative consequences for society. Conglomerate mergers can hurt the economy because they divert resources from productive investment to the buying and selling of corporate properties. Managers who are focused on these transactions may overlook opportunities to increase productivity more directly. This preoccupation retards innovation, and it results in skyrocketing prices for companies that look like good merger targets, thus adding to inflationary pressures. Most researchers believe that excessive mergers damage the health of the economy (Stearns and Allan, 1996).

Conglomerate mergers can also damage the social fabric of local communities. If a conglomerate buys a local plant and then shuts it down after a few years, the loss of jobs and revenue can devastate a community. (Recall our discussions of layoffs and plant closings in Chapters 5 and 8.) Outside ownership of local enterprises also undermines the participation of local business leaders in civic affairs. The welfare of the large, outside-owned corporation and its managers does not depend on the cultivation of local contacts. Local civic participation declines accordingly. Outside-owned firms are also less likely to look to local sources for supplies, labor, and expertise than are locally owned firms.

An additional long-term consequence of mergers is that they damage public confidence in the economy. The movement toward mergers is based on the pursuit of profits through financial manipulation rather than through producing products or services of economic value. In addition, a great many mergers are obvious failures. Employment is often reduced, and stock prices of the acquiring company sometimes plummet. For the company being acquired, there is little indication that negative outcomes are reversed at any point in the future (Marchak, 1991).

Effects on Employees Top managers often fare very well in merger situations: "Mergers provide power and legitimacy to the acquiring management team, while at the same time effortlessly creating the conditions from which they can justify higher compensation. The larger unit you command, the more stars on your epaulets" (Thackray, 1982:86). Managers in the acquired firms are increasingly protected by **golden parachutes**

(see Box 15.1) that guarantee compensation to top corporate executives if the firm is taken over: "The provisions are golden because they are generous. A $4 million parachute protected William Agee, chief executive officer of Bendix. Ralph Bailey, head of Conoco, had a slightly softer landing on $4.1 million" (Davidson, 1985:63).

The potential effects on lower-level employees, however, can be devastating. Both material and psychological losses are likely. Employees may be laid off or transferred against their wishes, often abruptly. Seniority acquired over years or even decades may mean nothing in such situations. Benefit programs are also likely to be disrupted. New work rules and disciplinary procedures are likely to be implemented. Pensions, sick leaves, vacations, holidays, and employee stock purchase plans may all be subject to change. Pension funds have been a frequent target in mergers. Reorganization of these funds often makes large amounts of money available to the new owning firm at the expense of workers who had counted on these funds for their retirement.

A merger can also result in reduced trust among all levels of employees (Lorenz, 1992). Employees are likely to doubt that the company has their best interests at heart and may adjust their own commitment downward in response. Ambiguity in role expectations is also likely to increase after a merger, and the company may seem relatively aimless during the transition period. Morale and motivation typically decline

DILBERT reprinted by permission of United Feature Syndicate, Inc.

(Taggart and McDermott, 1993). Mergers threaten employees' relations to their jobs and their economic security. Fear spreads.

Even contractual labor relations may be disrupted by mergers. The National Labor Relations Board recognizes only the obligation of a new parent company to bargain with an existing union, not an obligation to recognize existing labor contracts. The "obligation to bargain" gives legal recognition to the union as the workers' legitimate representative. It does nothing, however, to help secure a new contract. The terms of any new contract are completely open for renegotiation, and the increased power of the enterprise as part of a larger organization may substantially alter the terms of the agreement.

A Slowdown of Mergers?

By the 2000s, forty states had enacted some form of antitakeover legislation (Stearns and Allan, 1996). These laws provide safeguards against the most financially irresponsible and predatory takeovers and have slowed the growth of mergers. In addition, unions and consumer and environmental groups have increasingly publicized illegal and morally questionable practices of large corporations, ranging from the employment of child labor in foreign nations to the capture of dolphins in tuna nets. Inside large corporations, whistleblowing by employees has also increased (Rothschild and Miethe, 1999). Increased scrutiny and demands for accountability by the government, by employees, and by the public have forced companies to become increasingly interested in corporate ethics. As a result, corporations have had to reconsider many of their activities. Many large corporations have hired experts in ethics to help internally monitor their own behavior; books and business school courses on ethics have flourished. The lasting impact of these developments is yet to be evaluated. Past history suggests that the driving force of corporate activity—profits—is strong but that it can be channeled and regulated if there is sufficient public will toward this goal. Box 15.2 reports on a partially successful lawsuit establishing the personal liability of corporate directors for financially irresponsible mergers.

INTERCORPORATE LINKAGES

We have observed that large corporations have tremendous power in relation to customers, competitors, and workers. If a corporation is spread across several industries, its employees may even have difficulty understanding for whom they actually work. Interlocking directorates, bank ownership of majority shares, and subcontracting provide other sorts of linkages between firms that may be even more difficult for employees and communities to perceive and comprehend (Useem, 1996). These links, however, may be critical factors in determining investment decisions and, thus, in determining the fates of employees and communities (Keister, 2000).

Interlocking Directorates

Many executives sit on the boards of directors of several different corporations, creating **interlocking directorates.** For example, many members of the Rockefeller family, and other wealthy families, are incumbents of the boards of directors of a variety of corporations. Such interlocking directorates can coordinate the policies of large corporations. Thus, they are important in facilitating various oligopolistic practices. The networks defined by such interlocking directorates are often quite intricate. Banks or enterprises that compete in one country may be parts of international networks that collaborate with one another in markets elsewhere in the world. Employees may find it extremely difficult to understand and support the goals of their company when those goals are in part determined by its role in an international network of corporations. The companies originating from Standard Oil of Ohio originally owned by John D. Rockefeller form one such interconnected group.

BOX 15.2 Sanctions Levied against Executives Involved in Financially Irresponsible Mergers

Directors of large corporations have begun to be held accountable for some of the financial consequences of poorly considered mergers. The following discussion is from a book intended as a guide for managers to help them avoid such accountability. The book also reports, somewhat ironically, on the careers of the corporate directors subsequent to the sanctions against them.

> For the Directors of the corporations that dominate or influence the lives of most U.S. citizens, January 19, 1985, was a black day. It was the day the Supreme Court of Delaware found several of Chicago's best and brightest business leaders guilty of breaching their duty to the company's shareholders. Nine directors of the billion-dollar blue-chip Trans Union Corporation were held liable for agreeing to sell the company without careful review of its value and ordered personally to pay the difference between the per share selling price and the "real" market value of the company's shares.
>
> A motion by the directors for a rehearing summed up the crisis: The court's decision "has shocked the corporate world in its unprecedented holding that knowledgeable

directors of a Delaware corporation, performing their statutory managerial function, may be exposed to catastrophic liability . . . where there were no charges or proof of fraud, bad faith or self-dealing." . . .

What became of Trans Union? Less than a year after the merger, Bruce Chelberg, Trans Union's president and chief operating officer, became senior vice president, international, for IC Industries, Inc. A few months later, Jack Kruizenga, once head of Trans Union's tank car division, was recruited to run Pullman Standard Inc., a railcar industry giant that Kruizenga proceeded to overhaul. On the recommendation of Van Gorkom's good friend George Shultz, President Ronald Reagan appointed Van Gorkom Under Secretary for Management of the State Department. The *New York Times* reported that Shultz wanted a "trusted confidant" in this State Department senior management position.

SOURCE: Arthur Fleischer, Jr., Geoffrey C. Hazard, Jr., and Miriam Z. Klipper, 1988, *Board Games: The Changing Shape of Corporate Power.* Boston, Mass.: Little Brown and Company, pp. 15, 37.

These companies and some of their interconnections are listed in Box 15.3.

The Role of Banks

Bank control of large corporations removes ownership and investment decisions yet another level from the actual site of production (Fligstein, 1996). The Chase Manhattan Bank alone has predominant control in more than 10% of the 200 largest industrial corporations in the United States. According to economists Edwards, Reich, and Weisskopf (1978:156), "Leading bankers have the power to determine or influence the allocation of capital over a significant portion of the economy, and to influence many other aspects of corporate behavior as well." The banking industry is concentrated much as industrial enterprises are. As of 2000, the 400

largest commercial banks in the United States held approximately 83% of all the deposits in the 8,774 commercial banks in the country (Census, 2000). The following list identifies three of the largest interconnected groups of banks and corporations in the United States (Mizruchi, 1982:156–157):

1. J. P. Morgan & Co. (New York), Aetna Life (Hartford), Exxon (New York), Coca-Cola Co. (Atlanta), Western Electric (New York), Procter & Gamble (Cincinnati), United Aircraft (Hartford), Southern Railway (Richmond), Prudential Insurance (Boston), TRW (Cleveland), Travelers Insurance (Hartford), Eastman Kodak (Rochester), Delta Airlines (Atlanta).

2. Celanese (New York), Bankers Trust (New York), Anaconda Copper (New York),

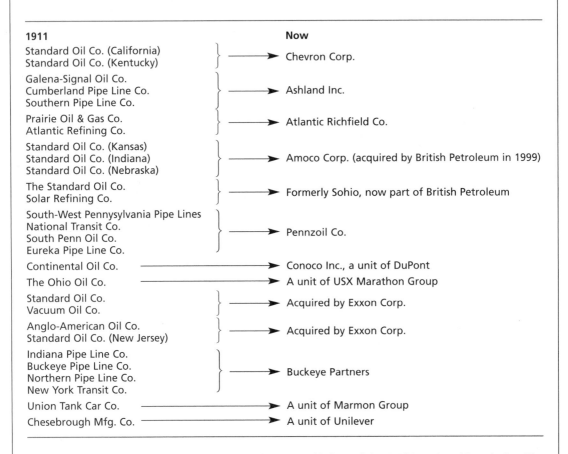

BOX 15.3   The Companies of Standard Oil

John D. Rockefeller formed Standard Oil of Ohio on January 10, 1870. By the early 1900s, he controlled up to 85% of the U.S. oil industry, inspiring the first antitrust laws. In 1911, the Supreme Court ordered the Standard Oil Co. to dissolve into thirty-four companies.

1911	Now
Standard Oil Co. (California) Standard Oil Co. (Kentucky)	Chevron Corp.
Galena-Signal Oil Co. Cumberland Pipe Line Co. Southern Pipe Line Co.	Ashland Inc.
Prairie Oil & Gas Co. Atlantic Refining Co.	Atlantic Richfield Co.
Standard Oil Co. (Kansas) Standard Oil Co. (Indiana) Standard Oil Co. (Nebraska)	Amoco Corp. (acquired by British Petroleum in 1999)
The Standard Oil Co. Solar Refining Co.	Formerly Sohio, now part of British Petroleum
South-West Pennysylvania Pipe Lines National Transit Co. South Penn Oil Co. Eureka Pipe Line Co.	Pennzoil Co.
Continental Oil Co.	Conoco Inc., a unit of DuPont
The Ohio Oil Co.	A unit of USX Marathon Group
Standard Oil Co. Vacuum Oil Co.	Acquired by Exxon Corp.
Anglo-American Oil Co. Standard Oil Co. (New Jersey)	Acquired by Exxon Corp.
Indiana Pipe Line Co. Buckeye Pipe Line Co. Northern Pipe Line Co. New York Transit Co.	Buckeye Partners
Union Tank Car Co.	A unit of Marmon Group
Chesebrough Mfg. Co.	A unit of Unilever

SOURCE: Adapted from Bernard Wysocki, Jr., 1998. "Oil Megadeal Signals a World of Low Inflation." *Wall Street Journal*, December 2, p. 112, with permission.

Metropolitan Life (New York), Mobil Oil (New York), Union Pacific (New York), R.J. Reynolds (Winston–Salem), Crocker National Bank (San Francisco), American Can (New York), Continental Airlines (Los Angeles).

3. Manufacturers Hanover Trust (New York), Kraftco (New York), B.F. Goodrich (New York), Goldman, Sachs & Co. (New York), Chrysler (Detroit), LTV (Dallas).

Subcontracting

A final type of corporate linkage is based on **subcontracting** (also called **outsourcing**).

Subcontracting arrangements exist in two forms. In *industrial subcontracting* the subcontractor manufactures parts that will be incorporated into a final product made by the principal company. This type of subcontracting is common in the manufacturing of equipment and machinery, including automobiles. In *commercial subcontracting* the subcontractor manufactures an entire finished product that the principal company markets, typically under its own brand name. Commercial subcontracting occurs in a wider range of industries, including clothing, footwear, toys, plastic articles, and many other consumer goods. Many large firms also subcontract out peripheral support functions, including food service, security, and janitorial services.

Lower Wages and Reduced Benefits The motivation for subcontracting is often to avoid the higher wage and benefit packages secured by unionized workers in the principal firm. Subcontracting sets up relationships of subordination between companies. The job security of workers in the subcontracting firm may be highly vulnerable to decisions made by the principal firm. Workers in subcontracting firms face problems resulting from their employer's weak position in relation to the principal firm. Subcontracting firms often achieve a competitive price for their product by cutting back on wages and by failing to pay Social Security, medical insurance, pensions, unemployment insurance, and other worker benefits. The small companies involved in subcontracting arrangements are frequently exempted from legislation protecting workers and working conditions. Subcontracting may also be a tool for undermining a union in the principal company. The union may find that its membership is being eaten away by employment losses as work is subcontracted outside the company. Unions generally bargain for some say in the subcontracting arrangements entered into by their employing company. Unless the union is very strong, however, it rarely succeeds in securing a contract that excludes subcontracting.

The growth of subcontracting erodes the prevalence of the sort of permanent employment relations that once typified most jobs in the American economy. Insecure employment contracts have contributed to the decline of real income for American workers. Many of those who have lost jobs they previously considered permanent have only been able to secure employment at much reduced wage rates (Newman, 1999). This tendency has contributed to the growth of contingent and peripheral employment discussed in Chapter 14.

The organizational innovation of "just-in-time" delivery systems, first developed in the Japan automobile industry, has also increased the impetus toward subcontracting. In just-in-time delivery systems, suppliers are required to deliver specified lots of parts to the principal manufacturer on a very exact but rapidly changing time schedule. This allows the manufacturer to drastically reduce inventory and related costs. The subcontracting supplier absorbs these uncertainties by paying lower wages and utilizing flexible employment levels.

Even when the subcontracting arrangement is relatively favorable, the subcontracting firm still exists in a very precarious relationship to the market for its product. In the face of an economic downturn, orders from subcontractors can be cut rapidly to preserve available work for the principal firm. Subcontracted production is relatively easy to enter into because of the specialized nature of the production process involved and because of the guaranteed market. But, for these same reasons, it is also extremely vulnerable to market downturns. In the early 2000s, many developing countries have encouraged the rapid development of semiconductor component subcontracting as a route to industrial development. However, cyclical downturns in the semiconductor industry have caused widespread cutbacks in this industry, and many nations, most notably South Korea, have been left with large, expensive plants standing idle.

Subcontracting relationships are not necessarily exploitative. In the international context such arrangements can have positive consequences for developing nations when they encourage raw material production and component production in the developing nation (Berthomieu and Hanaut, 1980). Such arrangements may provide an opportunity for the developing nation to manufacture items it was previously importing. Because relationships between international corporations and domestic subcontractors in developing nations generally involve a strong imbalance of power, Berthomieu and Hanaut argue that the governments of developing nations should closely regulate such agreements in order to ensure that they are economically beneficial.

THE SMALL-FIRM SECTOR

Our discussion in this chapter has so far focused on the increasing importance of corporate entities that may swallow up individuals in large, monolithic organizations. This focus should not lead us to assume, however, that the small-firm sector is on the verge of extinction. Small companies come into existence every day, and though many of them also quickly pass out of existence, their aggregate effect on employment is quite significant. This is especially true in the service sector, which employs an increasing proportion of the labor force. The subcontracting relationships that we have just discussed are also important in the ongoing reproduction of a small-firm sector. One factor that heightens the significance of small firms is that employment is relatively stagnant in large firms, in spite of the increasing proportion of production that is taking place there. Large firms typically use capital investment to increase productivity, and the employment-generating consequences of this strategy can be marginal or even negative (O'Connor, 1998).

Small firms are generally run by a single individual or family. Profits and retained earnings are lower than those in larger firms. Long-term borrowing is difficult, and the firms are heavily affected by local economic conditions. Produc-

tion and marketing strategies are frequently outdated, and the firm generally maintains its competitive position by cost cutting rather than by brand-name recognition or by the other forms of competition available to large firms. Such small firms, however, employ a significant share of the labor force.

Satellites, Loyal Opposition, and Free Agents

Three types of small firms exist in the economy: satellites, the "loyal opposition," and free agents (Averitt, 1968). **Satellite firms** engage in subcontracting relationships with larger firms, supplying either components or distribution and marketing services. Satellite firms may either be attached to a particular principal company or may "float," being dependent on sales to a certain industry but not to any particular firm in that industry. Companies supplying automobile glass, upholstery, and electrical components are good examples of satellite firms. **Loyal opposition firms** provide competition to larger firms in their own industries. Such companies rely heavily on local sources of supply, have technically inferior equipment, have limited access to foreign markets for their goods, surrender the power to set prices to the larger firms, and as a group have lower profits than larger firms. American Motors occupied such a position relative to Ford, Chrysler, and General Motors for decades. In the early 1980s, the remnants of American Motors were divided between the French automobile manufacturer Renault and Chrysler. **Free agent firms** are a diverse group that spring up in the nooks and crannies between large firms. In manufacturing, free agents often specialize in small-batch production. In retail trade and services they serve markets in which it would be difficult for larger firms to achieve economies of scale. The economic niches occupied by free agents typically have too low a profit rate to attract larger firms. Specialty tool shops, repair services, and residential construction firms are all examples of free agents.

How Important Is the Small-Firm Sector?
Simply counting the number of firms in the
economy would indicate that small firms are of
overwhelming importance: "Eighty-five percent
of all U.S. businesses have fewer than 20 employees. . . . [S]mall firms, particularly the young
ones, harbor a great reservoir of jobs and innovation. For distressed communities, research shows
that these enterprises may be the best hope of
revival" (Peirce and Steinbach, 1981:4). However, recall that in earlier sections on the increasing concentration of corporate assets we arrived
at different conclusions. Large firms control the
greatest share of productive assets in industrially
advanced societies and are responsible for most
economic production.

How are we to resolve this seeming anomaly?
What are the economies of industrially advanced
nations actually like? Are they dominated by a
few large corporations or by many small firms?
By looking at the distribution of employment,
we will be able to arrive at a clearer answer to this
important question.

The proportion of manufacturing employees
in establishments smaller than 100 employees
declined from 39.2% in 1904 to 27.4% in 2000.
An establishment is a plant or other place of
employment. But as you have learned, employment growth has increasingly shifted toward services and wholesale and retail trade over this
period. In these sectors small firms are more typical, and the decline in their role has been slower.
In services, for example, the proportion of
employees in establishments with fewer than 100
employees decreased from 83.5% in 1948 to
62.4% in 2000. In the economy as a whole
approximately 50% of the labor force is employed
in establishments with fewer than 100 employees,
and 50% is employed in establishments with more
than 100 employees.

Different Places, Same Company? These
figures are for employment in different **establishments** and do not take into account the
spread of **multiplant companies.** For example,
individual Wal-Mart discount stores may have
only a few dozen employees at any one establishment, but Wal-Mart has tens of thousands of
employees nationwide. Nor do these figures take
into account the type of concentration in which
one company purchases the production facilities
of other companies. Thus, in manufacturing only
16.2% of employees actually work in *firms* with
fewer than 100 employees. In the economy as a
whole, while approximately 50% of the labor
force is employed in *establishments* with fewer
than 100 employees, only 40% work in firms
with fewer than 100 employees (Granovetter and
Swedberg, 1992).

How many workers are employed in the very
largest industrial corporations? The list of the
largest U.S. corporations published in *Fortune*
magazine (*www.fortune.com/fortune/fortune500*)
provides a common frame of reference. If we
include the top 1,500 *Fortune* companies—made
up of the top 1,000 industrial companies; the top
100 firms in diversified services, commercial
banks, and diversified financial companies; and
the top 50 firms in life insurance, retail, transportation, and public utilities—approximately a
third of the labor force is employed in the 1,500
largest industrial, financial, and service firms
(Hodson, 1983).

What do these figures tell us about the organizational locus of work in modern society?
There are a great many small firms in industrially advanced societies. Assets and sales, however, are concentrated in the very largest corporations. Approximately a third of the labor force
is employed in small firms (those with fewer
than 100 employees). Another third is employed
in intermediate-size firms (those with between
100 and 5,000 employees). The final third is
employed in large firms (such as those identified
in the *Fortune* list). Thus, employment positions
are spread relatively evenly across the three
types of organizations. Many workers are
employed by one of the huge corporate giants.
Many others work in one of the many small
firms that fill the niches between these giants.
And many others are employed in intermediate-
size firms.

The Birth of New Jobs

In recent years the news media have given a great deal of attention to the importance of small firms in creating jobs. Much of this discussion was sparked by research by economist David Birch (1981). His most widely cited conclusion was that firms with fewer than twenty employees create two-thirds of the new jobs in the U.S. economy. We must interpret this conclusion with a great deal of caution, however. Many of these jobs disappear just as rapidly as they are created. Accordingly, some of these jobs represent a circulation of workers among small firms as new firms are born, struggle, and die rather than a lasting contribution to the supply of available jobs (Bednarzik, 2000).

Small firms do create many jobs. Some of these jobs, however, are of short duration; many others are relatively marginal and offer only low wages and meager or nonexistent benefits. Seasonal hiring in small retail stores of part-time or temporary workers provides a good example of this latter type of job.

Economic Revitalization

As researchers and policymakers have increased their awareness of the continuing importance of small firms in creating and maintaining employment, many have expressed hopes that this sector will provide a solution to the lingering economic doldrums of the early twenty-first century. Charles Sabel and Jonathan Zeitlin (1997), for instance, argue that production strategies pursued in the past by large firms in the industrially advanced nations are no longer viable. The cornerstone of past strategies was the production of large numbers of standardized products through the use of highly product-specific equipment. This strategy is sometimes called **Fordism** because of its successful use by Henry Ford to revolutionize the automobile industry. Over time, however, this strategy tends to deskill workers, rendering industries based on Fordism attractive to low-wage competitors such as South

Korea, Taiwan, Brazil, and Mexico. International competition over markets for such mass-produced goods has eroded the once dominant position of the North American and West European nations in many industries.

The solution proposed by Sabel and Zeitlin and others is that the industrially advanced nations begin to concentrate on specialized products that imitators find impossible or unprofitable to copy. Sabel and Zeitlin offer an example of an existing system of this type in central and northeastern Italy. Starting in the 1980s, many small factories employing from five to fifty people sprung up in this region. These shops specialized in the production of textiles, automatic machines, machine tools, automobiles, buses, and agricultural equipment. They have been able to escape the role of being subcontractors dependent on larger firms through diversifying their marketing operations to include other small producers in the local area and in the broader European market. As a result they have altered their status to become independent innovators, developing new products for a wider market. In the past, the subcontractor's customers arrived with a blueprint to execute, now they arrive with a problem for the subcontractor to solve.

Size and Innovation Part of the optimism concerning small firms results from their demonstrated ability to innovate more quickly and less expensively than large bureaucratic organizations. The National Science Foundation finds that small firms are about six times more effective in creating technological innovations than are large firms (U.S. Senate, 1986). A frequently cited example of this phenomenon comes from the steel industry:

> The oxygen converter is one of the most important advances ever developed in steelmaking. It was invented and used in Europe prior to the Second World War. It was introduced in the United States in the 1950s—not by big firms, but by McLough Steel, a small independent firm based in

Table 15.3 Some Important Innovations by Independent Inventors and Small Organizations in the Twentieth Century

Invention	Inventor	Invention	Inventor
Xerography	Chester Carlson	Continuous hot strip rolling of steel	John B. Tytus
DDT	J. R. Geigy and Co.	Helicopter	Juan De La Cierva, Heinrich Focke, and Igor Sikorsky
Insulin	Frederick Banting		
Vacuum tube	Lee De Forest	Mercury dry cell	Samuel Ruben
Streptomycin	Selman Waksman	Power steering	Francis Davis
Penicillin	Alexander Fleming	Kodachrome	L. Mannes and L. Godowsky, Jr.
Cyclotron	Ernest O. Lawrence		
Cotton picker	John and Mack Rust	Air conditioning	Willis Carrier
Shrink-proof knitted wear	Richard Walton	Polaroid camera	Edwin Land
		Ball-point pen	Ladislao and Georg Biro
Dacron polyester	J. R. Whinfield	Cellophane	Jacques Brandenberger
Catalytic cracking of petroleum	Eugene Houdry	Tungsten carbide	Karl Schroeter
Automatic transmission	H. F. Hobbs	Oxygen steel-making process	C. V. Schwarz, J. Miles, and R. Durrer
Jet engine	Frank Whittle and Hans von Ohain	Video games	Noland Bushnel
		Artificial heart	Robert Jarvik
Frequency modulation radio	Edwin Armstrong	Solar-powered car	Hans Tholstrup and Larry Perkins
Self-winding wristwatch	John Harwood		

SOURCES: U.S. Department of Commerce, 1967, *Technological Innovations: Its Environment and Management.* Washington, D.C.: U.S. Government Printing Office. Also, Kevin Desmond, 1986, *Inventions Innovations Discoveries,* London: Constable.

Detroit. A comparable case is continuous casting, a revolutionary process in steel production, which was introduced by ninth-ranked Allegheny Ludlum. (Reid, 1976:50)

Some other important innovations made by individuals and small firms are listed in Table 15.3. Public policy agendas supporting small firms as a source of innovation have gained increasing acceptance in recent years. Great hopes are being placed on this sector and on the belief that "decentralization is the great facilitator of social change" (Naisbitt, 1999). A contemporary example is provided by steel mini-mills that convert scrap metal into specialty steel products. These are being proclaimed as at least a partial answer to the decline of the U.S. steel industry. (See Chapter 8 for a discussion of the role of these mills in the U.S. steel industry.)

SUMMARY

Large corporations exert great power in relation to workers, their host communities, and smaller companies. In recent years they have further increased their power through mergers, expansion into diverse product lines, interlocking boards of directors, and subcontracting.

The prevalence of large, diversified corporations and the rapid pace of merger activity have many potential negative consequences for workers and communities. Workers may lose jobs without warning and on the basis of decisions in which they had no role and about which they have little or no knowledge. Similarly, the use of subcontracting may lessen the power of workers and communities and increase their vulnerability to changes in corporate strategies.

Large-scale production in huge corporations will continue to be the dominant way in which most of the manufactured goods used around the world are produced. However, it also seems likely that the small-firm sector will continue to be reproduced in the niches between these corporate giants. Despite its limited productivity, this sector employs a significant number of workers.

Whether it can produce technological and organizational innovations that will increase its relative share of production seems less certain. Its significance, however, should not be underestimated; it will continue to play an important role in providing jobs, goods, and services. The fact that a third of the labor force works for firms with fewer than a hundred employees should caution us not to assume that the experiences of workers in the largest corporations are the only or most significant type of work experience in industrially advanced nations.

KEY CONCEPTS

corporate economic power	concentration ratio	interlocking directorates
corporate political power	conglomerate	subcontracting (outsourcing)
corporate inflexibility	limited liability	satellite firms
monopoly	management-controlled firms	loyal opposition firms
oligopoly	corporate mergers	free agent firms
backward linkage	diversification	establishments
forward linkage	cash cow	multiplant companies
vertical integration	corporate divisions	Fordism
predatory pricing	golden parachutes	

QUESTIONS FOR THOUGHT

1. Identify a large economic enterprise in your community. Research this enterprise and find out as much as you can about who owns it and about the ownership and control linkages between this enterprise and other enterprises. Is this enterprise a subcontractor to some other enterprise? Does it subcontract out any of its own work or support services?

2. What factors encourage the increased size of organizations? What factors, if any, limit organizational size?

3. In what ways do large corporations exercise power over their employees? Over competitors? Over the government? Over communities?

4. Draw a time line depicting the history of corporate mergers. Note the different types of mergers occurring during the different periods.

5. What purposes are served by limited liability for the owners of corporations? Write an essay either defending or criticizing limited liability. How do the sanctions levied against the directors of Trans Union relate to limited liability?

6. How important do you think the small-firm sector will be in the future of advanced industrial society? Why do you think it will be more or less important? Would you prefer to work in a small firm, a medium-size firm, or a large firm?

MULTIMEDIA RESOURCES

Print

Barry Bluestone and Bennett Harrison. 2000. *Growing Prosperity: The Battle for Growth with Equity in the Twenty-First Century.* Boston: Houghton Mifflin. A critique of the economic inefficiencies caused by large corporations. Includes a variety of proposals for revitalizing the economy based on distributing control to workers and communities.

Eric Mann. 1987. *Taking on General Motors.* Los Angeles: Center for Labor Research and Education, Institute of Industrial Relations, University of California. A firsthand case study of the union and community effort to keep the GM Van Nuys plant open.

Jacoby, Sanford M. 1997. *Modern Manors: Welfare Capitalism Since the New Deal.* Princeton, NJ: Princeton University. A revealing analysis of the rise of the modern economy dominated by large corporations.

Michael Useem. 1996. *Investor Capitalism: How Money Managers Are Changing the Face of Corporate America.* New York: Basic Books. A penetrating analysis of the influence of money managers on the economy and on modern society.

Websites

Fortune Magazine List of 500 Largest Corporations. *www.fortune.com/fortune/fortune500*

Business History. *www.batnet.com/prologue* Provides links to fascinating histories of some of the largest North American companies.

Small Business Administration. *www.sbaonline.sba.gov* The core federal agency providing support for small businesses.

Financial Scandals around the World. *www.ex.ac.uk/~RDavies/arian/scandals* Reports both business and government fraud and financial scandals.

Power Structure Research. *darkwing.uoregon.edu/~vburris/whorules* An internet guide to contemporary social science research on the concentration of economic power and its consequences.

RECOMMENDED FILM

Coca-Cola Kid. (1995). Coca-Cola sends its best operative to eliminate competition from one of the last independent soft-drink bottlers in a remote part of Australia with unexpected results.

16

Work in a Global Economy

Three times a day scores of small buses make the six-mile trip north from
Saltillo's working-class neighborhoods to Ramos Arizpe where both
General Motors and Chrysler operate huge assembly plants. At Chrysler,
three shifts per day assemble motors for shipment to the U.S. and other
Chrysler plants in Mexico. At General Motors, less than a mile away, shift
workers turn out complete cars, also for shipment north. Autoworkers
here buy reduced priced meals in the plants, ride free company buses,
labor their way through the seven-step plant hierarchies, and enjoy up
to two weeks of vacation per year. They study family planning,
collaborative working, and a variety of technical mini-courses offered
within the plants. It's almost like working in Detroit. But here the air is
clear and the wage scale about three dollars for a ten-hour day.

(DUBOSE, 1987)

Today we live in a vast, integrated world economy. Workers in one part of
the world make products consumed by workers in others parts of the
world; and because production facilities are increasingly mobile, workers also
compete for the same jobs. This interdependence has been growing. Before
the eighteenth century most goods were produced and consumed locally. By
the nineteenth century, increasing numbers of people were producing goods
destined for national, or even international, markets and were consuming

goods that had been produced in other parts of the world. In the twenty-first century these trends have become an overwhelming reality.

This chapter provides a brief history of the world economy. It also highlights the emergence of world economic dominance by the United States following World War II and the rebirth of international competition in the latter decades of the twentieth century. In the last section of the chapter we examine international differences in how countries organize their economies to confront the rapidly changing world economy. The themes we have developed in previous chapters concerning the division of labor, inequality, technology, trade unions, gender, race, and marginality provide important building blocks for understanding the modern world economy.

HOW HAS THE GLOBAL ECONOMY DEVELOPED?

Three major theories have been advanced to explain industrialization—the process through which societies change from a primary reliance on agricultural production to include a greater role for manufacturing and services. In this section we briefly present these theories and evaluate their contributions. We also review the historical process leading to today's world economy.

Theories of Industrial Development

The process of industrialization has fascinated social scientists since the beginnings of the Industrial Revolution. Many scholars trace the birth of sociology to a concern with the social disturbances produced by the Industrial Revolution. Sociologists have developed three major theories to explain the process of industrial development: modernization theory, dependency theory, and world systems theory.

Modernization The earliest theory of development is called **modernization theory.** This theory was formalized following World War II and argues that societies advance from agricultural production to industrial production, based largely on *internal dynamics* (Rostow, 1998). These internal dynamics include growing consumption needs and technological advances. These dynamics work to transform feudal economies grounded on agricultural production for local consumption into industrial economies grounded on urban manufacturing for regional or world markets. In this view, countries that specialize in a certain commodity do so because they have some comparative advantage in that commodity, such as naturally occurring resources or a conducive climate or location.

This theory of development points out the important role of domestic needs and technological advances. Its greatest relevance, however, is for the earliest nations to industrialize—the Netherlands and England. And even for these nations, the theory overlooks the crucial role of international markets in the expansion stage of industrialization. All later developing societies have had to pursue the process of industrialization in competition with existing industrialized societies. Modernization theory tends to overlook the early emergence of a world economy and the incorporation of virtually all nations into that economy, either as manufacturing nations or as suppliers of raw materials.

Table 16.1 lists some of the largest nations of the world ranked by their gross national product (GNP) per capita. A nation's ranking on this list provides at least a partial indication of its level of industrial development. Note the extreme range of the inequality between nations and the important role of the timing of industrialization in determining a nation's wealth. Note also that the United States is now ranked sixth among nations in GNP per capita. You might also want to compare this table with Table 15.2, which reports the salaries of the highest paid U.S. executives.

Table 16.1   Per Capita GNP around the World

Low-Income Economies	U.S. Dollars 2000	Upper Middle-Income Economies	U.S. Dollars 2000
Ethiopia	110	Turkey	3,130
Tanzania	210	South Africa	3,210
Nigeria	280	Poland	3,590
Sudan	290	Mexico	3,700
Vietnam	310	Venezuela	3,840
Uganda	330	Croatia	4,060
Kenya	340	Czech Republic	4,060
Bangladesh	360	Hungary	4,510
India	370	Malaysia	4,530
Ghana	390	Brazil	4,790
Pakistan	500	Chile	4,820
Honduras	740	Argentina	8,950
Sri Lanka	800	**Richest Economies**	
China	860	Portugal	11,010
Burma	970	Greece	11,640
Lower Middle-Income Economies		Spain	14,490
Indonesia	1,110	Israel	16,180
Syria	1,120	Ireland	17,790
Bulgaria	1,170	Canada	19,640
Egypt	1,200	Italy	20,170
Philippines	1,200	Australia	20,650
Morocco	1,260	United Kingdom	20,870
Romania	1,410	Hong Kong	25,200
Algeria	1,500	Netherlands	25,830
Jamaica	1,550	Sweden	26,210
Iran	1,780	France	26,300
Belarus	2,150	Belgium	26,730
Colombia	2,180	Austria	27,920
Peru	2,610	Germany	28,280
Russia	2,680	United States	29,080
Costa Rica	2,680	Singapore	32,810
Thailand	2,740	Denmark	34,890
		Norway	36,100
		Japan	38,160
		Switzerland	43,060

SOURCE: U.S. Department of Commerce, Bureau of the Census, 2000, *Statistical Abstract of the U.S.,* p. 841

Dependency The unequal relationships between industrialized and less industrialized countries are the central focus in a theory of economic development called **dependency theory.** This theory notes that many less industrially developed nations have long been incorporated in the world economy but in a dependent role. Thus, Cuba was incorporated in the world economy for nearly 300 years as a sugar-growing colony and has only recently begun to develop a more diversified industrial structure. Similarly, Honduras has long participated in the world economy, but in the limited role of a banana-producing economy.

According to dependency theory, less industrially developed nations do not have the opportunity to advance from agricultural societies to industrial societies as dictated by their internal dynamics. Rather, their role as less developed, **peripheral nations** is maintained by their relationship to the industrially advanced, **core nations,** which buy raw materials from them and sell them manufactured goods in return (Frank and Gills, 1993). This dependency operates in a variety of ways. Along the west coast of Africa, for example, railroad lines were built under the direction of European colonial rulers. These rail lines, however, all run inland rather than along the coast. This network facilitates the exploitation of the interior of Africa for export trade but limits the development of regional trade ties along the coast. The industrial infrastructure of railroads, highways, and urban centers in West Africa thus supports the dependent relationship of this area to the world economy while contributing only minimally to local development.

Additional evidence for dependency theory comes from analysis of data on world capital investments, exports, and economic growth. Export-oriented development has a positive short-term effect on economic growth rates in developing nations. However, on average, this effect is reversed when the country has large amounts of foreign investment, when it is engaged in exporting a raw material, or when the prices of its exports fluctuate drastically (Jaffee, 1998). These negative effects occur because foreign investment distorts economic infrastructure, as in the example of railroads in West Africa. In addition, education and professional services often remain underdeveloped. Thus, being involved in the world export trade may help countries develop, but this one-sided "development" maintains their dependent status as suppliers of raw material or as subcontractors in larger manufacturing processes.

World Systems The most recent theoretical advances in the study of economic development have come from **world systems theory.** This theory builds on the notion of dependency between core and peripheral nations by adding the insight that the world economy is and has long been an *integrated economic system.* Thus, this theory focuses on the stability of the world system and its continuing relations of inequality. Proponents of this theory reject any notion of a necessary development of nations from agriculture to manufacturing that is either inevitable (as proposed by modernization theory) or actively blocked by already industrialized countries (as proposed by dependency theory). Instead, they focus on how the roles of nations in the world system are constantly reproduced by existing trading relations (Wallerstein, 1999). World systems theory also highlights the emergence of a "semiperiphery" of developing nations involved in the manufacture of relatively simple mass-production goods and components to be assembled into final products in more developed nations. (In a subsequent section we refer to the countries in the semiperiphery as "developing nations.") World systems theory also identifies low levels of capital investment as central to understanding continuing underdevelopment, especially in Africa and South Asia.

The final major insight of world system theory is that the world economy involves not only the movement of capital but also the large-scale migration of people. The European migrations to the East Coast of the Americas in the 1880s and 1920s and the Chinese migration to the West Coast of the Americas in the 1870s are well-known historic examples of such economically inspired population movements. Two major flows of population have been occurring in the world system in more recent decades. One of these flows is between the periphery and the semiperiphery and involves the migration of workers to countries with large oil exports and to cities with rapidly growing industrial bases producing for the world market, such as Singapore and Djakarta, Indonesia. The second major migration is from

Asia, South America, and the Caribbean Basin directly to the Western industrialized nations, particularly to Great Britain and the United States (Portes et al., 1997).

Emergence of the Contemporary World Economy

The first elements of an integrated world economy emerged in the 1600s with the development of the Dutch trading empire. The principal markets for Dutch goods were other European countries and the civilizations of North Africa. The Netherlands became a manufacturing center at this time, with over 50% of its population living in urban places (Wallerstein, 1999). The Dutch not only engaged in trading but also, along with Spain, England, and the other imperial nations, seized immense amounts of wealth from their colonial empire in Africa, the Americas, and the Pacific in the form of precious metals, trade goods, and human slaves. This wealth and these new resources helped underwrite the expansion of European manufacturing. Marx (1967 [1887]) called this process **primitive accumulation** and considered it a necessary step in the development of capitalism.

The British Empire The British colonial empire held the center stage in the world economy of the 1800s. Britain's colonies were incorporated into this economy as producers of raw products (such as cotton, sugar, rum, tobacco, and rubber) for the homeland and as consumers of industrial products from British factories (textiles and, later, steel and other manufactured products). The subordinate role of new territories in the British empire frequently sparked resentment among colonists and indigenous populations because of the unfavorable terms of trade imposed. These terms included high taxes, inflated import prices or high import duties, and low prices paid for commodities exported from the colonies. The Boston Tea Party and related episodes leading to the American Revolution provide early examples of angry resistance to unfavorable terms of trade.

Not only did Britain and the other imperialist nations impose unfavorable terms of trade on their colonies, but they also dismantled indigenous industries to open up markets for their own manufactured goods. For instance, loom operators in India who had supplied textiles to Asian and European markets in the 1600s and 1700s were put out of business when the British Parliament in the early 1800s restricted Indian producers to growing raw cotton and stipulated that textiles could be manufactured only in Britain. Later, when this restriction was removed, Indian textiles were subject to high tariffs when exported to Britain, whereas goods made in Britain were admitted into India with little or no import duty (Hobsbawm, 1996).

Specialization among Less Developed Nations In their roles as suppliers of raw materials, the colonies of the European nations often became highly specialized. Subsistence agriculture for local consumption was displaced by production of a more restricted range of agricultural products for export. Sometimes whole regions specialized in providing just one product. The American South became primarily a cotton-producing area, and Jamaica and the West Indies primarily produced sugar. As railroads advanced into the interiors of these countries, agricultural production for local consumption was displaced by agricultural production for the export market. Thus, from the very earliest stages of the Industrial Revolution these nonindustrialized regions played a major role in the world economy, but it was a subordinate role based on the greater economic and military might of the imperial powers.

The transformation from subsistence agriculture to the production to agricultural commodities for the imperial powers was accompanied by a removal of small producers from the land and the concentration of land ownership in large plantations. In Central and South America these plantations were known as *latifundia*. Peasants were transformed by this process from independent producers of agricultural products for personal or local consumption to dependent hirelings (or slaves) of landowners with large holdings. Just as the enclosure movement in England, which we described in Chapter 1, was a long and violent process, so too was the process of centralizing land ownership in the colonies. The labor force for the production of export crops was secured primarily through subjugating native people to various forms of peonage, such as tenant farming, sharecropping, and wage labor, and through importing slave labor. Colonists migrating from the imperial homeland in the search for new land also played a major role in the production of export commodities.

A Captive Market The Industrial Revolution resulted in a dramatic increase in manufacturing productivity in England. The British colonial empire played an essential role in this industrial revolution. Without world markets for the goods mass-produced in England, there would have been much less demand spurring on the development of modern industrial techniques and machines (Hobsbawm, 1969). British industrial productivity rose most dramatically around the mid-1800s. British manufactured goods were the essential medium through which the world economy came to penetrate all the continents of the globe. At the height of Britain's colonial power in the second half of the nineteenth century, its armies and navies controlled the world's seas and a sizable proportion of its inhabited territories. These military forces ensured the ability of Britain to sell its manufactured goods around the world and to purchase raw materials at favorable prices.

Dominance by Britain and the other European nations over their Asian and African colonies continued until the end of World War II. The military destruction of the industrial base of Japan and the European nations during the war resulted in a temporary decline in the economic and military power of these nations. No longer able to police worldwide empires, the industrialized nations began to lose control of their previously dependent colonies. Liberation movements demanding national autonomy sprang up in Africa, Asia, the Caribbean, and the Pacific Islands. Almost immediately after World War II, France, attempting to hold on to its colonies, became embroiled in disastrous wars in Vietnam and Algeria. Similarly, Britain became involved in conflicts in its former colonies of Malaysia and India and Belgium became embroiled in conflict in Zaire (formerly the Belgium Congo). Other European nations, realizing their inability to police colonial empires, allowed the transition to political independence to take place more peacefully. The next two decades saw the sometimes peaceful, but often violent, transformation of former colonies into politically independent nations.

The Rise of the United States and Neo-Imperialism In terms of military and civilian casualties and economic devastation, the United States fared better in World War II than any other major combatant. At the end of the war it had the only intact industrial apparatus in the world and possessed the largest standing army and navy. The United States was also a creditor to most of the industrial nations, who were heavily in debt for the purchase of American military goods, foodstuffs, and manufactured goods during the war.

The United States was thus in a unique position in the world economy in the 1950s. In some ways its position was similar to that of England at the beginning of the Industrial Revolution. However, the global U.S. empire was based more on economic dominance than on direct political or military control over dependent areas, though these more direct means were used as well. Historians and sociologists give the name **neo-imperialism** to this new version of the world economy in which less industrialized nations possess at least nominal political independence (Emmanuel, 1984).

The United States also worked selectively to encourage reindustrialization following World War II in nations it identified as potential allies against the spread of Communism. Economic development policies, such as the Marshall Plan, facilitated the reindustrialization of Germany and

Japan, as well as many other nations, with the indirect consequence of sharply limiting the period of uncontested American economic dominance in the world economy.

Neo-imperialism, with the United States at its head, remained unchallenged for only about two decades. During these decades the economies of Europe and Japan were rebuilding, and they began to compete with the United States. Also, some less developed countries had at least limited success in undergoing their own processes of industrialization and provided additional competition in the world market for manufactured goods. Many less developed countries also began to supply some of their own local needs for manufactured goods, thus reducing their dependence on the world market for these goods. Other regions organized into cartels to reduce competition among themselves over the pricing of their exports. A **cartel** is a combination of political or economic organizations that act to limit price competition among themselves. The Organization of Petroleum Exporting Countries (OPEC) provides the most successful and widely known example of a cartel pricing arrangement. Similar cartels emerged for regulating the price of such commodities as coffee and cacao. Thus, the world economy at the beginning of the twenty-first century is more complex and diverse than during any prior period in history. It includes large numbers of *industrialized nations, industrializing nations,* and an even larger number of *less industrialized nations,* many of which are either actively trying to industrialize or are at least trying to protect the prices of their raw export products.

THE WORLD ECONOMY TODAY

In this section we discuss the nature of the contemporary world economy and highlight the role of multinational corporations in the global economy. We also discuss the causes of slowed rates of economic growth in the industrialized nations and current problems of development in the less developed nations.

The Role of Multinational Corporations

So far we have discussed the world economy as if international trade were merely a matter of relations between nations. In fact, the world economy is to a large extent orchestrated by the heads of **multinational corporations** (MNCs) rather than by the heads of governments. A MNC is a firm that operates in many countries. MNCs typically have their headquarters in one of the industrially developed countries but many of their plants are located in less developed nations. For example, IBM's popular personal computer, retailing at $860, contains $625 worth of components made outside the United States, including $360 in components made by overseas plants of U.S.-based MNCs. These plants are located principally in East Asia. Nevertheless, because the computer is made by an American company and undergoes final assembly and testing in the United States, it carries a prominently displayed sticker reading "Made in the U.S.A." (Judis, 1987:7). Similarly, the Ford Escort is assembled primarily from parts made outside the United States, including parts made in Austria, Belgium, Britain, Canada, Denmark, France, West Germany, Italy, North Ireland, Japan, Netherlands, Norway, Spain, Sweden, and Switzerland (Bluestone and Harrison, 2000). By the early 2000s, the sales of U.S.-owned firms abroad were three times the value of U.S. exports. That is, three-fourths of production by U.S. firms for the world market takes place in branch plants located outside the United States.

The United States is not the only home base for large multinational corporations eager to expand their operations. Japanese automobile makers have also dispersed their production and assembly operations around the world. The United States and Canada have been an important destination for Japanese automobile assembly and parts plants, many of which were started as joint ventures with North American automakers

(Perrucci, 1994). The chief location for the development of new Japanese automobile plants, however, has been Southeast Asia.

Power without Boundaries Many social scientists are concerned about the intense concentration of economic power in the hands of the leaders of a few global corporations. By 1985, the 200 largest global corporations controlled over 80% of the assets of the Western world. "The men who run the global corporations are the first in history with the organization, technology, money, and ideology to make a credible try at managing the world as an integrated unit" (Barnet and Cavanagh, 1994:13). The centers from which the world economy is managed are the headquarters of the huge multinational corporations located in such cities as New York, Tokyo, London, Munich, and Paris.

Their large size gives the multinational firms the ability to influence the actions of governments in ways unavailable to other parties. In the 1970s, for example, International Telephone and Telegraph (ITT) was accused of having offered $300,000 to $400,000 to U.S. Attorney General John Mitchell for his influence in getting the Justice Department to arrive at an antitrust decision that was agreeable to ITT (Sampson, 1995).

According to many sociologists, the danger that these unimaginably large firms present is based not only on their size but also on their ability to make decisions independently of the nations in which they operate. "The capacity of any government to command a particular firm to undertake a specified task in support of a public policy, such as settling in a backward region or holding down a key price, has been reduced; large firms now have a capacity that they never had before for choice between competing nations" (Vernon, 1998:136–137). Table 16.2 lists the twenty-five largest MNCs. Note that the very largest companies are American but that Japan has more companies listed in this elite group of corporate giants than the United States. Germany ranks third on this list of largest corpo-

rations with the United Kingdom, France, and the Netherlands also having at least some companies of this magnitude.

Multinational corporations sometimes benefit less industrialized nations by transferring technology and expertise to them. The technology transferred, however, is often outdated. In addition, managers and professional workers are more likely to be transferred from the home country than to be recruited locally. Technology transfers thus often fail to significantly improve the employment prospects or the rate of economic growth in the host country:

> By means of administered prices for intermediate goods, license fees for patents, technology and management services, all of which are transferred to headquarters, the MNCs' customers in peripheral countries pay for the organizational superstructure of the MNC without receiving any of the employment benefits. . . . Such flows of resources contribute to the further development of innovation and expertise, and therefore to the expansion of clean and well-paid jobs in the expanding expert and command classes in the core, whereas routinized, dirty, and simple tasks with low average incomes are increasingly transferred to the periphery. (Bornschier, 1983:256)

The potential advantages of the technology and skill transfers that do occur are often overshadowed by the loss of economic autonomy suffered by the less industrialized nation, by increased dependence on foreign technology, and by lost opportunities to cultivate the development of local infrastructure, technology, and skills (Dicken, 1998).

In the industrially advanced nations many researchers fear that the actions of multinational corporations create a more segmented job structure. A small increase in managerial and professional jobs in an industrialized nation may occur as a result of the expansion of MNCs headquartered in that nation. As these companies ship jobs overseas in the search for cheaper labor, however,

Table 16.2   The World's 25 Largest Multinational Corporations

2000 Rank	Corporation	Country of Origin	Sales
1	General Motors	United States	$189,058,000,000
2	Wal-Mart	United States	166,809,000,000
3	Exxon Mobil	United States	163,881,000,000
4	Ford Motor	United States	162,558,000,000
5	General Electric	United States	111,630,000,000
6	DaimlerChrysler	Germany	154,615,000,000
7	Mitsui	Japan	109,373,000,000
8	Itochu	Japan	108,749,000,000
9	Mitsubishi	Japan	107,184,000,000
10	Toyota Motor	Japan	99,740,000,000
11	Royal Dutch-Shell	Netherlands	93,692,000,000
12	Marubeni	Japan	93,569,000,000
13	Sumitomo	Japan	89,021,000,000
14	IBM	United States	87,548,000,000
15	Citigroup	United States	82,005,000,000
16	AXA Insurance	France	78,729,000,000
17	Volkswagon	Germany	76,307,000,000
18	Nippon Telephone	Japan	76,119,000,000
19	British Petroleum-Amoco	United Kingdom	68,304,000,000
20	Nissho Iwai	Japan	67,742,000,000
21	Nippon Life Insurance	Japan	66,300,000,000
22	Siemens	Germany	66,038,000,000
23	Allianz	Germany	64,875,000,000
24	Hitachi	Japan	62,410,000,000
25	Matsushita Electric	Japan	59,771,000,000

SOURCE: *www.fortune.com/fortune/global500,* May 2000.

manufacturing workers lose employment and suffer deteriorating conditions (Milkman, 1997). Even in the industrially advanced nations, manual workers may come to work under conditions not unlike those of workers in less industrially developed countries. In Geismar, Louisiana, for example, chemical workers at a plant owned by BASF, a huge multinational company headquartered in Germany, complain of union-busting activities, substitution of lower-priced labor, unsafe working conditions, and policies that encourage high turnover (Streeck, 1996). These conditions are particularly worrisome because the plant makes pesticides similar to those that killed over 3,000 residents of Bhopal, India, in 1984 after a leakage

at a Union Carbide plant (see Chapter 5). Even workers in industrially advanced nations can be peripheralized by powerful MNCs.

Slowed Growth in the Industrialized Nations

The emergence of a more tightly integrated world economy at the beginning of the twenty-first century has brought with it a number of problems. In the industrialized world, growth rates have slowed appreciably. The annual growth rate in the United States fell from 3.8% in the 1960s to 2.8% in the 1970s but rose again slightly in the 1980s to 3.1% and in the 1990s returned to 3.8%. Industrial growth in Japan was once more than double that of the United States, but it, too, has dropped: from 10.5% in the 1960s to 4.6% in the 1970s to 4.1% in the 1980s to 1.8% in the 1990s. Growth rates in Germany and France have also steadily declined. Growth rates in the United Kingdom have been the lowest of any of the great powers in the post–World War II period but have rebounded to 4.0%—a level more comparable with the average for other industrial powers during this period.

These irregular and declining growth rates result from increased international competition and from the ability of MNCs to move around the world in the search for cheaper labor and more lenient environmental regulations. Figure 16.1 illustrates the rates of economic growth among some of the leading industrialized nations.

The End of U.S. Economic Dominance

The United States is still the largest producer of industrial goods in the world. However, its position has slipped dramatically since the 1960s. It is no longer the leader in automobiles, steel, railroads, or electronic goods. Between 1963 and 1995, the U.S. share of world manufacturing output dropped from 40.3% to 23.0%. Meanwhile, Japan's share increased from 5.5% to 20.4% (Dicken, 1998).

Why did the United States decline as the leading industrial nation in the world, especially given all its advantages of market power and technological leadership? One important reason is simply that the United States no longer dominates world markets as a sole supplier. The European nations and Japan have successfully recovered from the devastation of World War II and have become major competitors with the United States.

A second reason for the relative decline of the United States as an industrial leader is that its government spends a huge share of its research and development money on military research. Although military research has spin-offs that may increase productivity in some industries, they are

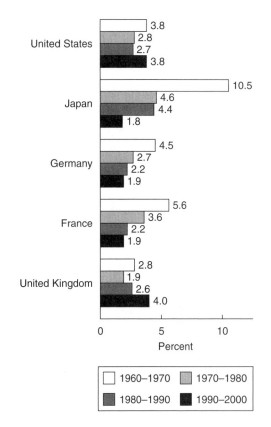

FIGURE 16.1 Declining World Economic Growth Rates (Average annual rates of change for gross domestic products)

SOURCE: U.S. Department of Commerce, Bureau of the Census, 2000, *Statistical Abstract of the U.S.*, Table 1362, p. 841.

likely to be far less significant than the economic benefit of using R&D monies directly for industrial research. In 2000, the United States spent 3.2% of its GNP on the military. West Germany spent 2.6%; Sweden, 2.4%; Switzerland, 2.1%; and Japan, 1.0% (Census, 2000). All of these countries have achieved growth rates in excess of that of the United States. Conversely, the former Soviet Union spent 11.7%—a burden that contributed to the Soviet Union's economic and political problems and to its eventual collapse.

A third cause for the declining international dominance of U.S.-made products is that American labor has been undercut by lower-priced labor in less industrialized countries. Even if workers are less productive in these countries, the difference between the wages of workers in less industrially developed counties and those of U.S. workers can make products manufactured in these countries competitive on international markets. In many less developed nations wages are as little as a tenth or a twentieth of what they are in the United States and other industrialized nations. One reason wages remain low in these nations is that workers there lack trade unions and have limited ability to strike for higher wages. Unions are typically outlawed or closely controlled by the government. MNCs also typically employ large numbers of female workers, who are paid even less than their male counterparts. Such extreme wage differentials pose a serious threat to the competitiveness of U.S.-made products and to the jobs of workers in the United States.

Even professional workers in the United States and other industrialized nations run the risk of being undercut by lower-priced labor located elsewhere (Marshall and Tucker, 1992). For example, in the early 1990s U.S.-based Intel Corporation built a new microchip manufacturing plant in Ireland and hired exclusively Irish engineers at salaries only a fraction of those it pays its engineers in the United States (Zachary, 1993).

Protectionism, Free Trade, and Fair Trade

Protectionism—protecting domestic industry through high tariffs on foreign imports—is often proposed as a solution to the challenges of international competition. High tariffs decrease or eliminate the competitive advantage of products made elsewhere in the world. Such a policy would seem to have great potential for protecting troubled industries in the industrialized nations. Japan, China, and what have sometimes been called the "four tigers"—Hong Kong, Taiwan, South Korea, and Singapore—used tariff barriers to shelter their domestic industries from the world economy during crucial periods of development. High tariffs on imported manufactured goods were an important component in these

countries' successful efforts to promote sustained economic development.

Many people fear, however, that increasing tariffs on foreign goods could provoke other nations to establish similar tariff barriers. High tariffs can thus protect some industries from international competition, but they can also negatively affect the health of other industries that are dependent on foreign markets for the sale of their goods. Indeed, many historians believe that tariff wars in the 1930s increased the intensity and duration of the Great Depression. Starting in 1947, the General Agreement on Tariffs and Trade (GATT), involving most of the industrialized nations of the world, has slowly reduced tariffs on key goods traded around the world. The seventh round of the GATT negotiations, called the Uruguay Round, concluded in 1994 and resulted in average reductions of 33% on manufactured goods and 36% on agricultural products ("World's Biggest Trade Deal," 1993).

In 1995, the World Trade Organization (WTO, *www.wto.org*) was established by representatives of the governments of the leading industrial nations as a permanent body to replace the periodic Gatt meetings. Its mandate is to reduce tariffs worldwide, thus increasing trade between nations as a mechanism to promote growth and development. The goal of the WTO has been to encourage **free trade**—trade unrestricted by tariffs and by other governmental restrictions. This goal has been broadly successful and has resulted in increased global trade and in increased development and employment worldwide. Unfortunately, this goal has sometimes meant the elimination of restrictions against goods manufactured with child or prison labor or under unsafe conditions. The actions of the WTO have in general undermined the ability of nations to use their purchasing power to pressure other nations to adopt safer and more humane labor conditions and better environmental practices (Tonelson, 2000). As a result of this mixed record of achievements the WTO has often been the object of criticism. Opposition to the actions of the WTO center on the need for fair trade as well as more open markets. **Fair trade** is typically defined as bringing workers' rights, workplace safety, and environment regulations up to the standards enjoyed in industrialized nations rather then reducing them to the lowest common denominator (see also Chapter 8 on the global restructuring of manufacturing).

In addition to the reduction of global barriers to trade, many nations have also entered into regional trading partnerships as a mechanism for promoting trade and development. The emergence of North American and European integrated trading blocks as a strategy for reducing trade barriers is discussed in the next section. These regional agreements provide additional opportunities to encourage trade and development and also to establish uniform and fair conditions for trade. As we will see, however, increased trade is often easier to negotiate than are protections for workers and the environment.

Trading Blocks: Regional Solutions to Lagging Growth

The governments of Europe and North America have sought to create regional trading blocks to lower tariffs and promote the exchange of goods. The European Economic Community (EEC) and the North American Free Trade Agreement (NAFTA) are the outcome of these efforts. Responses to these efforts at integration have been mixed on both continents and have included both great optimism and great trepidation. Proponents of these measures argue that the industrially dominant nations will find expanded markets for their goods with the lowering of trade barriers and that less industrialized nations will likewise find expanded markets for their products, leading to growth for all nations in the block. Opponents argue that, with the lowering of trade barriers, corporations will quickly move to the less industrialized nations in the block to lower their wage bills and escape environmental regulations at home. This will cost jobs in the core nations but not necessarily raise the standard of living in the peripheral nations.

The primary motivation for the development of the EEC and NAFTA has been the desire to stimulate economic growth, but they have also been motivated by a desire to reduce the flow of immigration from less industrialized nations to more industrialized nations within trading blocks. Proponents of the trading blocks argue that by stimulating growth in the peripheral nations, immigration to the core nations will be reduced. The EEC, operating under this theory, has developed explicit policies to promote economic convergence among EEC members (Metcalfe, 2000). In NAFTA such outcomes are left up to the invisible hand of the market, in spite of a much wider divergence in GNP per capita between the United States and Mexico than between, for example, Germany and Portugal. Because NAFTA addresses only trade issues and fails to include safeguards for working conditions or the environment, unions in both the United States and Mexico have generally opposed it (Dreiling, 2000). The economic, political, and environmental consequences of the EEC and NAFTA will take decades to be realized and positive outcomes may depend on safeguards still waiting to be introduced.

Combined and Uneven Development in Less Developed Nations

With increased involvement in the world economy, many less industrialized nations have experienced **combined and uneven development,** in which the export sector is highly developed but the domestic economy is badly underdeveloped. One important cause of combined and uneven development is the tendency of multinational corporations to borrow most of their investment capital locally. It has been estimated that in Latin America, U.S.-based global corporations secure 87% of their financing locally (Barnet and Cavanagh, 1994). Borrowing by multinational corporations absorbs resources that could otherwise be used for economic development of the domestic economy. Thus, borrowing by MNCs contributes directly to the underdevelopment of the domestic economy. This process creates a few jobs in the capital-intensive export sector but at the expense of many more jobs that might have been created in the labor-intensive domestic consumption sector. For example, a few job openings may occur in automobile parts manufacturing but at the expense of opportunities lost in bicycle manufacturing, food processing, and residential construction.

Combined and uneven development can thus contribute to inequalities in the economies of the nations involved (Boswell and Chase-Dunn, 2000). A small elite class consumes imported luxury goods in the face of mass poverty and the underdevelopment of production for domestic consumption. Partly as a result of such patterns of combined and uneven development, world unemployment levels are at an all-time high, exceeding 30% in many developing nations (Bradshaw and Wallace, 1996).

The **commodification of agriculture—**that is, the change from agricultural production for local consumption to production for international commodity markets—also contributes to combined and uneven development. This transformation creates scarcities in local markets and drives up the price of the foodstuffs that are available. Commodification of agricultural production has also brought with it increased mechanization and more centralized ownership, which have resulted in the displacement of large numbers of peasants. This has fueled the swelling ranks of homeless people living on the outskirts of major cities in less industrialized nations, such as Mexico City and Bombay, India.

A final cause of combined and uneven development is the use of **transfer prices** that do not accurately reflect the value of the goods produced in less developed nations. Transfer prices are the prices that large corporations charge when a product is transferred from one division to another. For example, a U.S. automobile manufacturer may build its engine blocks in Mexico and then export them to Detroit for final assembly. The transfer price is the amount that the

Ford Motor Company pays its Mexican plant for the engine blocks shipped to Detroit. These prices may bear little or no relationship to the actual value of the component. Instead, transfer prices may be an accounting mechanism for shifting profits wherever desired in order to avoid taxes, leverage nations or communities into greater economic concessions, or simply conceal the export of profits to be reinvested elsewhere. Before the creation of OPEC, the giant oil companies successfully transferred profits to their domestic operations in this way by undervaluing crude oil exports while simultaneously overvaluing refined oil products, such as gasoline and heating oil (Claes, 2000). Unrealistic transfer prices can seriously deplete the resources of less developed nations desperately in need of capital to finance economic development.

Developing nations involved in global markets thus often fill very specific and vulnerable niches. Often their role involves extractive activities based on mining or agriculture. Other involvements include the manufacture of subcomponents that are later assembled into finished products elsewhere in the world. In this way, developing nations become involved in **commodity chains** that may stretch around the world (Gereffi, 1999). The accounting of where and how much value is added at each stage of the commodity chain is often open to administrative manipulation because the different components are often manufactured by subsidiaries of the same company rather than by separate companies competing in an open market. The administrative decisions that set these transfer prices thus often include nonmarket considerations that may result in the developing nation's role being economically undervalued (Kessler, 1999).

Border Plants The plants on Mexico's northern border, known as *maquiladoras,* provide a good example of the contradictory effects of international investment on the economies of developing nations. More than 1,150 factories employing 500,000 workers have opened along the border since 1990 (Lowe and Kenny, 1999). Overwhelmingly, these plants are subsidiaries of U.S.-based MNCs, manufacturing everything from television sets to automobile parts to Levi pants to brassieres. These plants generate about $7 billion a year in exports for Mexico, second only to oil. Over 90% of the products are exported directly back to the United States. Wages average about $1.75 per hour, and over 80% of the labor force is female. The plants have brought large numbers of Mexican workers into the international economy, and the *maquiladora* sector is responsible for an increasing share of Mexico's export trade.

There is also evidence, however, that the *maquiladoras* have many negative consequences for the Mexican economy. For example, unemployment has increased in the host cities, partly because more workers are drawn to these areas in the search for jobs than actually find employment. Few manufacturing or service facilities have been spun off, because most of the components and services are imported directly from the United States. Even more damaging is the vulnerability of the plants to international fluctuations in the demand for their products and in the cost of manufacturing them elsewhere. Border plants have a notoriously short life expectancy. Plant closings leave the Mexican economy with unused industrial capacity, unemployed workers, and a history of distorted investment decisions (Bacon, 2000).

The Debt Crisis Many less developed nations have borrowed heavily to fuel industrialization and domestic consumption. Because of the dynamics of combined and uneven development, however, these debts have been very difficult to repay. These developments have produced a **debt crisis** in the world economy. The less developed nations owe the industrially developed nations over $1 trillion. As the costs of manufactured goods have risen in recent years and the costs of raw commodities have fallen, many of these nations have fallen into seemingly unrepayable debt (Edgar, 2000).

HOW DO WORK PRACTICES DIFFER AROUND THE GLOBE?

The modern world economy includes a great variety of work practices and working conditions. To adequately understand these differences, we need to take a closer look at variations in how work is organized in different countries. These differences are the focus of this section.

The level of industrialization in a country is one of the most important determinants of working conditions. Accordingly, we discuss working conditions first in the **least developed nations** and then in somewhat more advanced **developing nations.** Industrial relations in developed economies also vary widely. We discuss examples from state-regulated capitalism (France, Canada, and South Africa), codetermination (Germany), autonomous work groups (Sweden and Norway), macroplanning (Japan), and post-communist societies (China, Eastern Europe, and the former Soviet Union). For each type of economy we begin with the question "Who makes the key economic decisions?" and examine the consequences of the answer for economic growth and working conditions.

Least Developed Nations

Pakistan, Mozambique, Uganda, Kenya, Honduras, and El Salvador are among the least industrially developed nations in the world. In such countries economic power is held to a large degree by agricultural interests—often by owners of *latifundia* and other large plantations producing for the international market. The patron of such an estate holds enormous power over the peasants who till the land. These relations are sometimes considered feudal in nature, because they may entail obligatory labor on the patron's land in exchange for the peasant's right to till a small plot for private consumption (Scott, 1990). Besides such export-oriented agricultural enterprises, the economies of less developed nations also include substantial amounts of subsistence agriculture.

They are also frequently typified by large migrations of workers from rural to urban areas as subsistence agriculture is displaced by export-oriented agriculture.

In such societies many goods consumed locally are still produced in the **informal sector** by local artisans and sold by small shopkeepers and street vendors (de Soto, 2000). One striking feature of less industrialized nations is the stark contrast between the poverty of the rural poor and the consumption of Western consumer goods by urban elites. Western clothes, food, and soft drinks are promoted on billboards, radio, and television. The transformation of local tastes to include these expensive commodities provides a market for Western consumer goods but drains significant amounts of money from the market for local goods.

Developing Nations

Venezuela, Algeria, Turkey, Brazil, and Mexico have made greater advances along the road to industrialization. Such countries are often referred to as developing nations. In developing nations the power shifts from landowners who produce raw materials for export to wealthy industrialists who have ties to multinational capital (Toledo, 2002). The Philippines provides a good example of this type of dependent industrial system. It was a U.S. colony between 1898 and 1946. During that time the economy slowly moved from agricultural exports of sugar and rubber to a manufacturing base. In the 1970s and 1980s, the Philippines became a major exporter of textiles, apparel, and electronic components, mainly to the United States (Shalom, 1993). This pattern of **dependent development** has had many negative consequences for the economy and the people of the Philippines. The Philippines has one of the lowest average wage rates in the Pacific Basin. Under the late President Ferdinand Marcos it also gained a reputation as one of the most graft-ridden economies in the world and has a history of political repression and unrest. Partly as a consequence of these factors, the Philippines has also been one

of the most important sources of immigration from the Pacific Basin to the United States in the post–World War II period.

Working conditions in many factories in developing nations are reminiscent of conditions in the now industrialized nations at the beginning of the twentieth century. The labor of children is commonly used in industrial pursuits and unsafe working conditions are commonplace (Moberg, 1998). For example, on May 11, 1993, a fire in a toy factory 15 miles west of Bangkok, Thailand, resulted in the deaths of 200 workers. The factory had only two narrow fire escapes for *over 4,000 employees.* The fire was uncannily reminiscent of the infamous Triangle Waistskirt fire in the United States 82 years earlier in 1911, which we discussed in Chapter 6. Table 16.3 reports average hourly wages for manufacturing workers in a variety of least industrialized, industrializing, and industrialized nations.

Unions often play an important role in the process of development. In nations first undergoing industrialization and in developing nations trapped in dependent development, labor movements are important organizers of resistance against low wages and oppressive working conditions. Once independence is achieved, however, national leaders often limit the power of unions as new objectives emerge that may not include sharing power with trade unions. During this period trade unions are frequently repressed by the very governments and political parties they helped bring into power. If the society succeeds in sustaining industrial development, there is a good possibility that unions will again gain an increased role as important actors in a more open and pluralistic society (Frenkel, 1993).

Consumers in the industrially advanced countries can also be a source of pressure toward improving working conditions in developing nations. Consumer boycotts of goods manufactured with child labor or under harsh and exploitative conditions bring attention to these issues and can be an important force for change. This is especially the case when highly visible retail chains in the industrially advanced nations wish to avoid the

Table 16.3 Hourly Wages for Manufacturing Workers in Various Countries

Country	U.S. Dollars 1999
India	.21
Sri Lanka	.30
Pakistan	.32
China	.34
Mexico	1.26
Portugal	3.23
Poland	3.43
South Korea	4.18
Taiwan	4.38
Brazil	4.69
Hong Kong	4.80
Greece	5.60
Singapore	6.56
France	8.94
Israel	10.07
Ireland	10.41
Australia	11.56
Canada	11.82
Germany (unified)	12.03
Belgium	12.62
Italy	12.71
United Kingdom	12.79
United States	12.80
Japan	15.59
Sweden	17.16
Norway	17.67
Denmark	17.88

SOURCE: United Nations, *Statistical Yearbook,* 44th edition, 2000. New York: United Nations.

negative publicity associated with having goods produced under exploitative conditions.

State-Regulated Capitalism

In industrial capitalism the owners and managers of large corporations hold the economic power. The state, however, regulates the activities of corporations so that they do not undermine the interests of the nation as a whole. Government regulations cover such issues as minimum wages, fair hours, child labor, how retirement accounts can be invested, monopolistic practices, and environmental protection. Although these societies frequently proclaim an ideology of laissez-faire (unregulated) capitalism, all of the industrially advanced capitalist nations can be more correctly regarded as examples of **state-regulated capitalism.**

Canada Labor relations in Canada are broadly similar to those in the United States. Both countries rely on collective bargaining between unions and companies in preference to direct government regulation of working conditions. However, the Canadian economy holds a different place in the world economic system than the United States. Canada, a sparsely populated nation, has never been a major world economic power and has long struggled against economic domination by the United States. It has pursued what some economists call an **import substitution** development strategy, in which a nation attempts to develop industries that produce goods it has previously imported. This policy has made Canada a self-sufficient producer of many of the manufactured goods consumed in the country (Krahn and Lowe, 1998). More recently, Canada has also been successful in becoming a successful member of the world *export* economy. Prior to the 1980s, its exports were primarily those of field and forest. However, in the 2000s it has also become a major exporter of automobiles, aircraft, and advanced telecommunications equipment (Giles and Belanger, 2002).

France The French economy is dominated, on the one hand, by a few industrial giants and, on the other, by a great many small producers. This curious organization can be traced both to the early bourgeois revolutions in France, which created political structures designed to support the bourgeoisie (small producers), and to later state-sponsored efforts to industrialize through creating giant modern corporations. Thus, the French economy is simultaneously both more centralized and more decentralized than the U.S. economy.

The overriding importance of a few large industrial giants in France has led to a cartel relationship among these enterprises as they have organized to divide available markets among themselves. Large French firms include Renault, Peugeot-Citroen, and Thompson Electronics (which now owns the U.S. firm RCA). The Economic Ministry institutes regular five-year plans. Although these plans are not blueprints for production, they do outline the general directions in which the government will encourage the economy to develop. The government exerts its control of the economy through state ownership of the banks, with resulting control of investment decisions. Many wage rates also fall under legislative control.

Collective bargaining in France differs from that in the United States and Canada. Multiple unions exist within every major company, and a comprehensive written agreement is not the final product of collective bargaining. Instead, a process somewhere between consultation and collective bargaining takes place. If the outcome is not agreeable to both the unions and the company, there will be a strike or lockout as they attempt to pressure each other toward their respective goals. This bargaining system is responsible for a paradox: France has one of the numerically weakest labor movements in the industrially advanced societies and yet it has one of the most volatile and strike-prone systems of industrial relations (Desmarez, 2002). Current economic policy in France includes an attempt to increase the role of collective bargaining in hopes of making labor relations more consistent across enterprises and industrial sectors.

South Africa   In South Africa a strict system of **apartheid** (separation) historically defined relations between *indigenous blacks, white colonists* (originally from Holland and England), and immigrants from India (designated under apartheid as "*coloreds*"). This system provided for rigid racial segregation in the workplace and in other social settings such as restaurants, beaches, and public facilities. The system of apartheid entailed sharply lower wages for blacks than whites and intermediate wages for coloreds.

Strikes by blacks in the South African railroad and mining industries in 1987 were the largest ever in the country and were a turning point in apartheid. These workers failed to win their immediate demands because of violent repression and because high unemployment among blacks provided pools of ready replacements for striking workers. These strikes, nevertheless, were a sign of growing unrest and their failure only further intensified the underlying conditions that gave rise to resistance and conflict.

Black resistance to apartheid succeeded in eliminating the system through parliamentary elections in the mid-1990s. Changes in the economic system, however, have been slow to follow. The legacies of apartheid include an impoverished black working class and grave inequalities in areas of literacy, education, training, home ownership, and personal wealth. Improvements in the employment situation of black South Africans are occurring but have been slower than many initially hoped and provide a continuing source of tension between the races (Webster, 2002).

German Codetermination

In unified Germany the principal holders of economic power are the owners and managers of large corporations; however, they share power with the government, unions, and **works councils.** Works councils are comprised of worker representatives (typically selected through the union) who sit on management boards and represent the workers in the day-to-day affairs of the enterprise. Working together, owners and workers collectively **codetermine** production practices and employment policies. German workers have a much greater say in the management of their enterprises than do North American workers. In large enterprises workers have 50% representation on the management board. The works councils, which date back to the violent social revolutions of 1848, are a unique form of worker participation and have attracted a great deal of interest outside of Germany (Muller-Jentsch, 2002).

Works councils have a strong influence on personnel matters such as scheduling, vacations, and pensions. They also influence, though they do not have the final say on recruitment, training, job transfers, and discharges. Works councils have the least say in economic matters, such as manufacturing techniques, the introduction of new technology, plant layout, and plant closings and openings. In the area of economic policy, their role is bolstered by union-operated banks, which influence investment decisions and, therefore, directions for economic growth. Unions also exert influence on the German economy through training programs collectively organized by unions, management, and the state. These programs include rigorous apprenticeship systems, as well as ten-month paid sabbaticals for workers selected for retraining in needed areas of technological growth (Rogers and Streeck, 1995).

Based on its strong apprenticeship program and highly skilled and committed workforce, Germany has been able to control world markets for high-quality goods and to produce these goods in volume. The archetype for this achievement is the Mercedes-Benz line of cars and trucks, which are recognized for quality throughout the world. Quality standards are ensured by the utilization of highly trained and skilled workers. This strategy also allows German manufactures to introduce new design features quickly and effectively—little new training is required because the workers are already highly skilled and are actively involved in ongoing training (Streeck, 1996).

The system of codetermination is responsible for a low strike rate in Germany compared with that in the United States. This low strike rate is based on German workers having additional effective channels to voice their concerns. The most important such concern over the last decade has been the demand for a shorter workweek with limited reduction in pay. Significant movement in this direction has been made especially in the important metal trades where it has been defended as a strategy for preserving employment in the face of technological and market changes (Seifert, 1991).

In the economic slowdown experienced by the advanced industrial nations in recent decades, the German industrial system has fared relatively well. The existence of a stable, mutually agreed-on system in which workers have a regular voice in the operations of enterprises is an important ingredient in this success. Early retirements have also been used to maintain employment levels for younger workers—a strategy made possible by historically high wages and a well-funded and secure retirement system. These institutional structures have also helped ease the disruptions associated with the reunification of East and West Germany (Jurgens, Klinzing, and Turner, 1993).

Scandinavian Autonomous Work Groups

The most distinctive aspect of labor relations in Sweden and Norway is the large degree of power held at the shop-floor level by **autonomous work groups.** Because these countries have capitalist economies, the owners and managers of large corporations hold a large share of economic decision-making power. As in Germany, however, they do so in combination with the government, trade unions, and workers. In the Scandinavian countries workers in many companies are organized in small shop-floor groups rather than in councils operating at the level of the enterprise as in Germany. Legislation in the 1970s substantially strengthened workers' roles in economic decisions in Sweden. The 1976 Codetermination Law was particularly important in this regard. This law extended the areas that could be legally covered by collective bargaining to "any decision that affects employees" (Haas, 1983). Partly as a result of this and related laws, Sweden has been a leader in innovative work practices that have employee well-being and job security at their center (Berggren, 1992).

The Scandinavian system is in some ways similar to the German system but it includes greater power for autonomous work groups. These work groups focus on their own well-being as well as on product quality. The German and Scandinavian systems are similar in that both are backed by strong, independent worker organizations operating at the enterprise and national levels (Clayton and Pontusson, 1998). A classic example of the Swedish system is provided by the Volvo automotive plant at Kalmar, where groups

---

BOX 16.1   Work Groups at the Kalmar Volvo Plant

Throughout the company, we have a hierarchy of works councils, with representatives from both management and the employees. Some of these councils—one per plant—have been required by collective agreement since 1946. Others, like the Corporate Works Council, have been created on a voluntary basis to meet our own needs for consultation. . . .

Job rotation . . . started in the upholstery department, for very practical reasons. Employees complained of sore muscles from doing the same operation over and over. They discovered that if they paired off and traded jobs every day or so, they were able to use different sets of muscles. From that ergonomic, down-to-earth beginning have grown most of the other changes that focus on the quality of work life more generally. Soon the upholsterers (mainly women) were trading among three or four, and the relief from wrist, arm, shoulder and back pains was far greater than anything they had been able to get from the small army of supervisors, doctors, safety specialists, and industrial engineers who had developed methods or mechanical aids to relieve their complaints. . . .

Eventually the workers in the upholstery department were given responsibility for planning all their own work, a fairly complex task because

the blue seat with the special fabric has to leave the upholstery group in time to reach the assembly line at the right moment to meet the blue car with the special headliner. The planning had been a white-collar job in the past, but the group members, who alternated the paperwork task, had no trouble absorbing the planning. They get information on the assembly requirements from a teleprinter, and have to translate this into individual assignments. They have also taken over checking the quality of incoming upholstery materials, a job that used to require four inspectors. All the operators attended a short training program to learn how to find and identify all the kinds of defects and what to do about them.

As these changes took place, rather gradually, there was a concomitant increase in team spirit and individual commitment. Employee turnover and absenteeism changed dramatically, and the upholstery quality improved because team members understood the entire process and felt much more responsible for the product of their work.

SOURCE: Pehr G. Gyllenhammar, 1977, *People at Work*. Reading, MA: Addison-Wesley Publishing Company, pp. 81–82, 85. Reprinted by permission of the author.

---

of fifteen to twenty workers build entire cars (or major subsystems) rather than each worker being assigned a highly specific task on a rapidly moving assembly line. Some of the roles of work groups in the Kalmar Volvo assembly plant are described in Box 16.1.

The Scandinavian countries were late to industrialize but have been a major success story in the world economy. Only a century ago these countries were primarily agricultural and the majority of their industrial development has occurred since World War II. The economies of Sweden and Norway depend heavily on export trade; over 30% of the gross national product of these countries comes from exports (Berggren, 1992). Most of these exports are high-quality machine parts, machine tools, lumber, and finished wood

products. Scandinavian exports have a reputation for quality and workmanship.

In Sweden 91% of the eligible workers, including most white-collar employees, are members of trade unions. This high level of participation has brought labor stability and peace to the Scandinavian countries. Workers have successfully gained access to economic power through the political process, trade unions, and autonomous work groups. In this context most workplace issues become matters of information, negotiation, and cooperation rather than issues to be resolved by dispute and confrontation. The proportion of hours of work lost because of strikes in Sweden is only about 1.4% of that lost in the United States.

The Social Democratic Party in Sweden has been a powerful representative of workers' interests at the national level and has contributed several unique features to Swedish industrial relations (von Otter, 2002). These include national labor courts that review workers' grievances. In the United States only unionized workers, government workers, and some professional workers, such as academics, have access to this sort of judicial review of workplace complaints. The Swedish government has also developed the most comprehensive full-employment, job-retraining, and relief-work programs of any nation in the world. There seems little doubt that the labor peace and job security provided by these programs have been important contributors to the success of Swedish industry in producing high-quality products for the world export market. In Sweden, the average hours worked per week has dropped in the twentieth century from 64 to 36, and wages have risen almost fourfold in this same period.

Macroplanning in Japan

In Japan economic power is held by large corporations, especially those involved in the export sector, and by government planning ministries. In 1980, Japan overtook the United States as the largest producer of cars in the world (Branscomb et al., 1999). This and other successes are attributable to the unique economic and industrial system that Japan has developed since World War II. The Japanese system of **macroplanning** includes economic planning at the economy level and job security and teamwork at the enterprise level.

Long-Term Planning The Ministry of International Trade and Industry organizes the Japanese economy to compete effectively in the international market for a variety of key goods, including steel, automobiles, and electronics. The ministry has identified each of these areas as a target for development. It coordinates the efforts of different industries, as well as the educational system and the financial system, to facilitate these success stories. To protect selected domestic sectors, Japan has also vigorously upheld restrictive import practices. These trade restrictions have been an important component in the policy of sheltering domestic

industry to a point at which it can effectively compete in the international market (Kono, 1992).

Work Practice Innovations Japan is also an innovative leader in workplace practices. Central to these innovations is a strong reliance on work groups. Work groups, organized into **Quality Control Circles,** take initiative for developing incremental changes in production techniques that help increase productivity through **continuous improvement** in production techniques (Deming, 1982). Such systems of teamwork depend on first establishing trust between the worker and the enterprise. This is achieved through the assurance of lifetime employment for many workers, something virtually unheard of in the United States. Once loyalty and trust have been established, workers are more willing to cooperate with their teammates and supervisors to promote productivity. The system is in some ways similar to the German and Scandinavian systems of worker attachment and commitment to the firm and has many of the same positive consequences for productivity and labor relations. The Japanese system differs from the German and Scandinavian systems, however, in that workers have less of an *autonomous* voice in enterprise affairs. Box 16.2 presents a corporate song and recitation that highlight differences between Japanese and American work orientations.

The teamwork system in large corporations has had many positive consequences both for productivity and for the quality of work life in Japan. Large enterprises are more concerned with employee welfare than is typical in most Western state-regulated capitalist societies. Japanese firms spend about 8.5% of total labor costs on housing, medical facilities, food service, transportation services, sports, social activities, and special welfare grants. In contrast, British firms, for example, spend only about 2.5% of their payrolls on these employee services (Dore and Sako, 1998).

The Japanese system of employment security and team organization has made it possible for enterprises to integrate quality control operations into production processes. Because workers are committed to the success of the firm, they are

---

BOX 16.2   Japanese Corporate Spirit

The following song and recitations start every working day at the large Uedagin Bank in Japan:

> A falcon pierces the clouds
> A bright dawn is now breaking.
> The precious flower of our unity blossoms here.
> Uedagin Uedagin
> Our pride in her name ever grows.

> Smiling in our hearts with glory,
> For we carry the responsibility for tomorrow's Independent Japan.
> Our towns and villages prosper
> Under our banner of idealism raised on high.
> Uedagin Uedagin
> Our hopes inspired by her name.

> Marching forward to the new day
> With strength unbounded,
> We continue forward step by step.
> Oh, the happiness of productive people.
> Uedagin Uedagin
> Brilliantly radiates her name.

They remain standing to recite in unison the "Uedagin Principles" (*Uedagin koryo*):

> Constantly abiding by the ideas of cooperative banking, we will, together with the general populace, advance in our mission to serve as an instrument of small and medium business enterprises.

> Intent in the spirit of service, we will contribute to public welfare and social prosperity. Emphasizing trust and possessing an enterprising spirit we will advance scientific administration.

> With mutual respect and affection, we will work with diligence, employed in maintaining systematic order.

Possessing a spirit of love for Uedagin, we pledge to plan for the prosperity of the bank and for the public welfare and to make the bank the greatest in Japan.

And finally, before sitting down, a second catechism is recited, this one known as the "President's Teachings" (*shachokun*):

> Harmony (*wa*). The bank is our lifelong place of work, let us make it a pleasant place, starting with our greetings to each other each morning.

> Sincerity (*seijitsu*). Sincerity is the foundation of trust, let us deal with our customers with a serious and earnest attitude.

> Kindness (*shinsetsu*). Have a warm heart. Be scrupulously kind.

> Spirit (*tamashii*). Putting our heart and soul into it, let us work with all our strength.

> Unity (*danketsu*). Strong unity is the source of energy for our business.

> Responsibility (*sekinin*). Responsibility makes rights possible; first let us develop responsibility.

> Originality (*soi*). In addition, let us think creatively and advance making each day a new day.

> Purity (*seiketsu*). Have a noble character and proper behavior.

> Health (*kenko*). With ever growing pride let us fulfill the Uedagin dream.

SOURCE: Thomas P. Rohlen, 1974, *For Harmony and Strength*, p. 36. Copyright 1974, The Regents of the University of California. Used with permission of The University of California Press.

---

more eager to improve quality than workers in other advanced capitalist economics, who are rewarded only for the work they directly perform. American and British firms do most of their quality control checks after production is complete. They then have to go through the costly process of fixing mistakes that the Japanese system encourages workers to avoid making in the first place. In addition, guarantees of employment security mean that workers no longer have to fear

technological innovation and can instead work to implement innovations as effectively as possible.

The Underside of the Japanese Miracle   The much-acclaimed system of lifetime employment security in Japan applies only to part of the Japanese industrial system. Workers in such enterprises make up only about a third of the Japanese labor force. The other two-thirds is made up of workers in peripheral industries, which are often

BOX 16.3   Japanese Workers Protest Workload

The Plight of Labor in Japan:

WHO SAID,
"YOU'VE COME A LONG WAY, BABY!"
?

Deceptive new bill now before Diet

A bill to renew the "Labor Standards Law" is now being proposed by the Japanese Government for legislation during the present Diet session. It calls for a 40-hour work week, instead of the 48-hour work week under the current law. This sounds like an advancement, but in fact it is deceptive.

To start with, no timetable is given for the enforcement of the 40-hour work week. On top of that, an unbelievably generous exemption is being granted to all businesses with less than 300 employees, which account for 85% of the total labor force in Japan, for an unspecified period of time.

This means that the new bill does not guarantee equality under the law, one of the principal rights guaranteed by our Constitution.

We haven't really.
And someone is making our way ahead longer and tougher.

How many of you would look at the short vacations taken in Japan and say, "A reasonable way to reduce labor costs. Let their employers enjoy further immunity so that they can strengthen their competitiveness." ?

A minimum of 10 days of paid annual leave sounds like an improvement from the 6 days under the current law. Deceptive again, because the paid annual leave that an employee can take at his or her own exclusive discretion without losing pay will be reduced to only 5 days. The remaining "leave" may be allocated at the convenience of the employer.

Such a system may simply eliminate the need for replacement personnel, eliminate additional employment even though the nominal "vacation" may increase, all at the sacrifice of the rights of the employees.

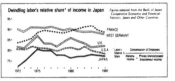

How many of you would look at the long working hours in Japan and say, "A great way to win in international competition. Let them have another decade of this." ?

We believe that these long hours, which have been a feature of the Japanese working environment for decades, constitute one of the major causes behind the current trade imbalance between Japan and the U.S., and we fear that this new legislation may well worsen the situation. Management will be entitled to override the principle of up to 8 hours work per day, **without overtime pay.**

Legalizing longer normal working hours (even 10 to 12 hours a day) at peak periods is sure to make it difficult to override the principle of up to 8 hours work per day, particularly career women, to retain their jobs.

With all these privileges, Japanese business and industry are likely to be in an even more competitive position with respect to international trade.

After all, can you stand yet another deceptive move by the Japanese Government that will affect you through industrial competition ?

We are by no means born-to-be workaholics. We are totally dissatisfied with the fact that we are deprived of pay and time for ourselves and our families which we believe we deserve. The compensation we receive as a percentage of corporate income, for example, is the smallest among the five major advanced countries.

Voice Your Concern, Now

Not against the people of Japan, but in support of the people of Japan, and to protect your own established rights for fair competition.

Express your view on this issue in your own way to the Japanese Embassy and through your own channels to the government and the Diet (congress) of Japan.

We call for international solidarity to rectify the labor policy of the Japanese Government so that we will have

NO MORE UNFAIR COMPETITION

and a solution to the Japan-U.S. trade imbalance.

CONCERNED LABOR OF JAPAN
c/o Japan Federation of Newspaper Workers' Unions
5-6, Misakicho 3-chome, Chiyoda-ku, Tokyo, 101 Japan

This message is brought to you by the committee formed and supported by individuals and organizations covering a broad spectrum of professions and industries, including the print media, broadcasting, publications, advertising, printing, medical and clinical services, transportation, distribution, education, national and municipal public services, agricultural cooperatives, legal services and jurisprudence, together with contributions from private citizens.

SOURCE: Used with permission of Concerned Labor of Japan.

subcontractors to the core firms, and women, who work as secondary workers in all sectors of the economy (Lo, 1990). These workers do not have lifetime employment or any of the other benefits of workers in the core manufacturing sector. The exploitation of these workers has been an essential, though hidden, aspect of the Japanese "economic miracle" (Brinton, 1993). In many ways the Japanese economy is a contradiction, containing elements of both highly advanced industrial economies and developing economies. High levels of efficiency, the latest innovations in computer technology, and lifetime employment security exist alongside a large sector of underpaid and highly vulnerable workers.

In addition, Japanese workers, as a whole, work longer hours and have fewer holidays than workers in other industrially advanced nations. Pressures for unrecorded and unpaid overtime are also widespread (Roberts, 1994). Death by overwork (*karoshi*) has been recognized as a national health problem in Japan. An organization of Japanese workers airs some of these complaints in a newspaper advertisement presented in Box 16.3.

Japanese unions are generally **company unions.** Membership rose steadily after World War II, reaching a peak of 35% of the labor force in 1970 (Shigeyoshi and Bergmann, 1987). It has since declined to about 24%. These unions are typically dominated by management and do not provide an autonomous voice for workers. On the positive side, enterprises are often willing to share planning information with their unions because of the cooperative spirit that pervades Japanese labor relations. Because many workers have a guarantee against being laid off, the union can work effectively with management to devise the best plan for adjusting the company's labor force to meet changing economic circumstances. Strikes in Japan are infrequent. They generally address highly specific issues and are of short duration. The Japanese system of industrial relations, somewhat ironically, demonstrates that worker participation can encourage productivity without necessarily entailing increased worker power, autonomy, or satisfaction (Marsh, 1992).

China

China is the largest country in the world with a population of over 1.2 billion. In recent decades it has experienced an annual economic growth rate in excess of 10% in contrast with the 2% to 4% growth rates typical of Western industrialized nations. This rapid growth rate results from internal changes in the Chinese political and economic system as well as from external forces.

Internally, China has moved toward privatization of parts of its economy, although this movement has been slow and only partially realized in many sectors. In China many hybrid enterprises exist that are partly public and partly private. Market relations have emerged between these enterprises but even these relations often operate within networks that are defined by past administrative linkages (Keister, 2000). In addition, many small private firms have emerged to fill the niches between these larger enterprises.

External forces adding to these changes include heavy direct foreign investment in China by multinational corporations. This investment has been motivated by a desire to take advantage of China's cheap labor and, even more importantly, by a desire to secure access to China's immense potential as a market for goods. These markets are both for imports and for goods made in China by foreign firms producing directly for the Chinese market (Lee, 1998).

With the advance of privatization, however, China's commitment to equality appears to be eroding. Free college education for the most qualified students, regardless of ability to pay, is now a thing of the past. Rapidly rising tuitions have increasingly made college an option only for the children of the newly emerging entrepreneurial class. Similar trends are taking place in housing and health care. Such trends may result in many Chinese rethinking the particular mix of planned and private enterprise that will eventually be the China of the twenty-first century. Currently, however, these trends toward increasing inequality are occurring without significant gains in democratization that would be needed in order for a serious public debate about China's future to take place.

The Four Tigers

The Pacific Rim is also home to what are sometimes called the **four tigers** of Asia: Hong Kong, Singapore, South Korea, and Taiwan. These nations have all experienced dramatic economic growth in recent decades. The rapidly rising productivity of these nations, coupled with that of Japan and China, is the basis for projections that the twenty-first century will be economically defined by events in Asia rather than in Europe or North America.

All four nations have succeeded in growing in spite of limited natural resource bases. Instead of relying on natural resources, these nations have invested heavily in their educational systems and succeeded through cultivating the development of their people's human capital. Given their relatively small size and limited domestic markets, they have all had to succeed through the manufacture of export goods for the world market (Krahn and Lowe, 1998).

Development strategies among the four tigers have been diverse, suggesting that there is more than one path to economic success in the 2000s. Hong Kong has relied on family enterprises, inherited from its Chinese cultural tradition, as the basis for economic growth. Similarly, Taiwan has successfully developed on the basis of small and medium-size manufacturing establishments combined in networks through extensive subcontracting systems (Hsiung, 1996). South Korea, in sharp contrast, has succeeded in its development project by investing in a few giant corporations that compete on the world market in basic commodities such as steel and automobiles. Singapore, in contrast, has relied heavily on direct investment by foreign multinational corporations and serves as a major subassembly center for world manufacturing. Singapore is also characterized by a strong role of the central state in directing development. Although all four of these nations are nominally democratic, none enjoy the full range of freedoms taken for granted in the industrialized West.

Eastern Europe and Russia

In the past, the power to make economic decisions in the Soviet Union (as well as Communist China) was held jointly by the Communist Party and by government administrative bodies. The government developed annual plans and five-year plans for the economy as a whole. At the national level these plans specified only broad goals for each sector of the economy. It was left to local and regional government agencies and enterprises to develop detailed blueprints for achieving these goals. The Soviet model of industrial organization emphasized clear lines of top-down authority and incorporated no formal role for workers' councils at the enterprise level or for autonomous work groups at the shop-floor level. Industrial relations at the

plant level were thus similar to those in many state-regulated capitalist economies; however, **central planning** rather than market forces drove the decisions about what to produce (Lengyel and Neumann, 2002).

Trade unions existed in every Soviet industry and were an important mechanism through which workers' voices were heard on enterprise management boards and in government ministries. Trade unions were important in the areas of labor recruitment, training, safety and health, and cultural activities. They were also actively involved in welfare functions for workers such as managing health care and retirement systems. Although trade unions in the Soviet Union represented workers' interests to the government and to enterprise managers, they were not autonomous in the same sense as unions in the state-regulated capitalist economies. The leaders of Soviet unions were members of the Communist party (as were the majority of people in leadership positions in all aspects of Soviet society). In this way the party exerted a strong influence over union activities and union demands rarely fell outside party agendas. In spite of this strong influence, strikes were not unknown in the Soviet Union, including walkouts by thousands of workers (Silverman and Yanowitch, 2000). Strikes, however, tended to address highly specific local issues and to be of short duration for fear of government reprisal.

Under the leadership of President Mikhail Gorbachev, the Soviet Union instituted reforms calling for *glasnost* (openness) and *perestroika* (restructuring) in the late 1980s. This increased openness set the stage for a political devolution of the Soviet Union into independent states. Today many republics of the former Soviet Union are fully independent and others have significant autonomy. The remaining core of the old Soviet Union, composed of Russia, Siberia, and many contiguous republics, is today organized politically as the Russian Federation.

Economic reforms have created a climate for change and have set the groundwork for many subsequent changes in Russian life and in the Russian economic system. Privatization of the economy has become a central focus of these reforms. Little knowledge existed in the Soviet Union, however, about how to proceed with privatization. As a result, many early reforms were exploited as opportunities for illicit gain. The result might more aptly be described as *pirate capitalism* than as industrial capitalism. Unemployment, poverty, and shortages have increased dramatically.

At present, the economies of the states of the former Soviet Union and those of its former satellites in Eastern Europe combine capitalism entrepreneurialism (especially in the small firm and service sectors), limited elements of state ownership retained from the communist period, and some worker collective ownership of factories sold off by the state (Standing, 1996). Backlashes have occurred in response to the excesses and failures of the transition period and the future economic structure and health of the economies of the states of the former Soviet Union, including the Russian Federation, remain uncertain (Stark and Bruszt, 1997). In many regions, despair has become widespread as the promises of privatization have been replaced by the realities of rising unemployment and falling incomes. Alcoholism has increased and suicides have risen to record levels.

In regions where peaceful political transitions can be achieved without civil war between rival nationalities, there appear to be reasonable opportunities for the ferment of competing strategies of economic development to yield improvements in the standard of living. Chronically high unemployment rates and increased poverty for many, however, also appear likely for the foreseeable future.

Competing Organizational Forms

Japanese automobile plants are being transplanted to the United States and Canada. German automobile plants are being transplanted to the United States and vice versa. British petroleum corporations are opening retail gasoline stations around the world. French pharmaceutical companies are building factories worldwide with a special focus on former French colonies. As large economic organizations become increasingly global in their operations, they take their organizational cultures with them to their new host nations. Thus, in the world of the twenty-first century, there is a regular commerce in organizational forms as well as in the products produced by these organizations (Perrucci, 1994; Rinehart et al., 1997).

Do global corporations take on the organizational features, human relations policies, and other characteristics of their new hosts, or do they impose their own organizational structures and human relations policies on their host nations and on their new labor forces? These questions are crucial for the future of work in many nations.

Probably the most widespread organizational transplants in the world today are those of Japanese automobile and electronics companies, which have opened operations in South Asia, North America, and Europe. These plants generally seek to reestablish the Quality Control Circles and other innovations that are core components of the Japanese system of production based on continuous improvement. However, these plants also have to make adjustments to local conditions and cultures (Besser, 1996). Resistance to Japanese corporate cultures has been significant in North America and Europe where workers expect to maintain greater independence and autonomy from their employers than do Japanese workers (Delbridge, 1998).

The organizational forms that produce the highest-quality products most cheaply and that are most flexible in adjusting to rapid market changes will win in the world of heightened global competition. Organizations that respect workers' rights and motivate them to give their best efforts to the success of the enterprise will have a significant advantage in the increasingly competitive environment of the twenty-first century (Pfeffer, 1998).

International Labor Solidarity

Workers have many resources of power, even in the modern interconnected world economic system. In the face of increasing economic complexity and interconnections, workers have attempted to develop new strategies for securing a share of the benefits of economic development. Labor unions in the United States have successfully supported federal legislation requiring the Overseas Private Investment Corporation to refuse insurance to U.S. corporations operating in countries that fail to protect "internationally recognized workers' rights." These rights include "the right of association; the right to organize and bargain collectively; a prohibition on any form of forced or compulsory labor; a minimum age for the employment of children; and acceptable conditions of work with respect to minimum wages, hours of work, and occupational safety and health" (International Labor Organization, *www.ilo.org*). Such legislation is intended not only to promote more humane working conditions in less developed countries but also to keep workers in the United States from having to compete against workers who labor under exploitive conditions. Such legislative efforts are an important way in which workers in the industrially advanced nations can find common ground with workers in developing nations. Box 16.4 presents an appeal by the United Automobile Workers to support additional legislation of this sort.

The International Labour Organisation (ILO) The United Nations conducts research and develops policy guidelines on labor issues through the International Labour Organisation. The ILO does not involve itself in contract negotiations or other direct interventions in labor relations. Instead, it serves as a research institute on international labor issues and as a clearinghouse

for information of value to international unions and other workers' groups.

World Labor Confederations Three major **international labor organizations** are directly involved in supporting and coordinating union activities around the world. They are the World Federation of Trade Unions (WFTU), the International Confederation of Free Trade Unions (ICFTU), and the World Confederation of Labour (WCL). The WFTU (*www.wftu.cz*), the largest of these groups, has a membership of over 200 million. About two-thirds of these members are in Eastern Europe and the former Soviet Union. The ICFTU (*geneva.ch/icftu.htm*) claims an affiliated membership of about 105 million, mostly in Western Europe and in former European colonies. Both of these organizations have grown in recent decades and have memberships that dwarf that of the American labor movement. The WCL (*www.cmt-wcl.org*) was founded in 1920 as a Catholic labor organization. It aggressively opposes the current role of MNCs in the world economy. Although the WCL has lost membership in Western Europe, it has grown in many less industrialized nations and has maintained a large membership in Canada. Its largest membership today is in Latin America where it enrolls about 19 million workers—similar to the number of North American workers enrolled by the AFL-CIO.

These international labor confederations have developed a number of strategies for improving the position of labor in the increasingly complex world economy. They have promoted United Nations guidelines for labor standards. They have promoted codetermination arrangements in which union representatives have voting rights on the boards of directors of enterprises. International unions have also encouraged unions' efforts to use their financial assets, held largely in pension funds, to promote better conditions for workers by investing in companies with favorable labor policies, disinvesting in companies with unfavorable labor policies, and encouraging worker ownership of

BOX 16.4   UAW Support for Trade Bill

Many "Cheap" Foreign Imports Come At Too High A Price.

It's a simple fact that many Americans don't want to face. Workers' rights are being trampled upon abroad to keep labor and other production costs down. The price tags we see on goods from those countries may be lower, but that's because someone else has already paid dearly.

Jail terms, beatings, kidnappings, torture are all-too-common features of life in nations like Chile, Paraguay, South Korea, and Taiwan. Workers are paying for their efforts to form unions, bargain collectively, or seek some small measure of safety and health protection.

These violations of basic human rights should not enable foreign nations or multi-national corporations to prosper. That's why we need strong worker rights provisions in our trade laws.

The Pease Worker Rights provision in the House Trade Bill—H.R. 3—would define such practices as unfair under U.S. law. It would allow action to stop employers and governments from profiteering at the expense of basic worker rights. Contact your U.S. Representative and urge support for the Worker Rights provision in the House Trade Bill.

Don't let them trade on repression.
Support the Worker Rights Provision in the House Trade Bill—HR 3.

International Union, UAW

Owen Bieber, President

enterprises wherever possible. Finally, these organizations have attempted to achieve their aims by encouraging more active involvement by workers in national politics (Bergsten, 1996).

The Challenges of Global Coordination for Unions Attempts to coordinate workers' demands on a world scale face significant problems that have limited the effectiveness of coordinated international efforts by workers to improve their conditions. In brief, workers are more immediately concerned with their own wages than with those of workers in other nations. Problems also arise from trying to coordinate the activities of workers in societies with different levels of unionization. Where only a small proportion of the labor force is unionized, union recognition and survival are central concerns. Where a majority of the labor force is unionized, concerns for job security, safety and health, and worker training take precedence. Workers in different countries may also face different legal frameworks, some of which are very repressive.

An example helps illustrate the potential successes and limitations of internationally coordinated trade union activities. In 1978 Akzo, the Belgian-based chemical company, fourteenth largest in the world, announced the layoff of between 5,000 and 6,000 workers in plants located mainly in Western Europe. Outraged workers in Belgium and Germany staged sit-down strikes in the plants to be closed. In addition, the International Federation of Chemical Workers worked directly with Akzo to develop alternative plans to maintain employment. The Dutch government was heavily pressured to sanction the company in some way. Although the government did not act, observers believe that the possibility of such action was important in forcing the company to withdraw its plans for layoffs in all but a Swiss plant (Turner, 1998).

This example suggests that in spite of the difficulties involved, workers in different nations can support one another and coordinate their efforts to improve their working conditions. Direct involvement by international unions can have a significant positive influence on the outcome of local or regional labor negotiations. Workers can also exert influence through pressuring their governments to support improved working conditions in other countries and through mobilizing public opinion in support of fair working conditions.

SUMMARY

The world has become increasingly integrated in an interconnected economy. Theories of modernization, dependency, and world systems each contribute to our understanding of the global economy. This economy is typified by a dominant role for multinational corporations, reduced growth rates in the industrialized nations, increasing privatization in former Communist nations, uneven development in less developed countries, and a high degree of diversity among nations in the organization of work.

Britain established a colonial empire stretching around the world. The colonies supplied raw materials for British factories and markets for British manufactured goods. As the Industrial Revolution spread to other European nations, they, too, established colonial empires to support their industrialization. This system lasted until World War II, after which it was replaced by a new system of neo-imperialism with the United States as its head. This system was based on economic control of the underdeveloped areas rather than on direct political control. The same patterns of dependent development continued under this new system of imperialism, with nonindustrialized nations supplying raw materials to the industrialized nations and providing a market for manufactured goods.

Multinational corporations occupy a central role in the world economy at the beginning of the twenty-first century. MNCs are the central institutional arrangement through which economic transactions occur in the world economy. They have encouraged the development of export-oriented economies in less industrialized nations. This type of industrialization encourages underdevelopment in the domestic manufacturing sector by absorbing resources that could have been used for the production of goods to be consumed locally. Export-oriented industrialization has created a development process called combined and uneven development. Such development strategies have as yet been unable to lift the majority of people in these developing societies out of poverty. Some nations, such as China, Japan, and South Korea, have achieved greater successes. They have encouraged the development of industries producing for domestic consumption by protecting domestic markets from foreign goods. Simultaneously, they have developed their own export sectors. Meanwhile, the economies of the advanced industrial nations have experienced reduced rates of growth because of increased international competition.

The world economy at the beginning of the twenty-first century is more diverse than at any previous time. Not only have less industrialized nations begun to take a variety of paths to development, but great diversity in industrial organization also exists within the industrially developed nations. Individuals, governments, or workers may have ownership over economic enterprises. Economic planning decisions may be made by the managers of large enterprises, by the state, by a planning body of enterprises and the state (with or without union participation), or by some form of codetermination in which workers and owners are both represented. One particularly bright spot in the world economy is the economic successes associated with increased levels of worker participation. These successes suggest that the most effective systems of economic production in advanced industrial societies may also be those that are the most democratic and humane. Such settings allow human beings to apply the full range of their creative talents.

The diversity of successful forms of economic organization suggests that many ways of organizing industrial society are possible. The continuation of such diversity is probably good for the future of industrial society. No single way of organizing is best for every situation; each has its own problems and potentials. Only through accepting and encouraging such experimentation and innovation can we hope to improve our systems of industrial production and the quality of work life.

KEY CONCEPTS

modernization theory

dependency theory

peripheral nation

core nation

world systems theory

primitive accumulation

neo-imperialism

cartel

multinational corporation

protectionism

free trade

fair trade

combined and uneven development

commodification of agriculture

transfer prices

commodity chains

maquiladoras

debt crisis

least developed nations

developing nations

informal sector

dependent development

state-regulated capitalism

import substitution

apartheid

works councils

codetermination

autonomous work groups

macroplanning

Quality Control Circles

continuous improvement

company unions

four tigers

central planning

international labor organizations

QUESTIONS FOR THOUGHT

1. Consider your daily round of activities. Make a list of where the products that you use are made. Be aware that the components of a product assembled in one country may be made in other countries.

2. Describe the manner in which some less industrialized nations are dependent on the export of one or more raw agricultural or extractive products. How might a nation move out of this role? What difficulties would the nation be likely to encounter in trying to move out of this role?

3. What factors have contributed to a reduced rate of economic growth in North America and Europe in recent decades? What growth rate do you think we can expect in coming decades? What might be done to increase this growth rate? Is an increased growth rate even desirable?

4. Describe some of the positive and negative roles played by MNCs in the world economy. What policies might be effective in limiting the negative behaviors of MNCs while retaining their positive effects?

5. Compare and contrast Scandinavian autonomous work groups, German codetermination, American collective bargaining, and Japanese Quality Control Circles as systems for incorporating workers into workplace decisions. Which systems do you think will be most likely to emerge as dominant in the next twenty-five years?

6. What is the role of international labor organizations in the modern world economy? Do you think that such organizations will have a significant effect on working conditions in the twenty-first century? Why or why not?

MULTIMEDIA RESOURCES

Print

Daniel B. Cornfield and Randy Hodson (editors). 2002. *Worlds of Work*. New York: Plenum. Overviews of the nature of work and its study in the major industrialized and industrializing nations by leading experts in each country.

Terry Boswell and Christopher Chase-Dunn. 2000. *The Spiral of Capitalism and Socialism: Toward Global Democracy*. Boulder, Colo.: Lynne Rienner. An examination of the tension between market dynamics and social justice. Advances the thesis that this tension is useful for the pursuit of democracy and desirable social outcomes.

William M. Adler. 2000. *Mollie's Job: A Story of Life and Work on the Global Assembly Line*. New York:

Scribner. Mollie runs a metal stamping machine in a factory in Patterson, New Jersey, making ballasts for fluorescent lights. This book chronicles the movement of her job from New Jersey, to rural Mississippi, to Matamoros, Mexico, and tells the stories of those who hold it.

Ronald P. Dore and Mari Sako. 1998. *How the Japanese Learn to Work*. London: Routledge. The best available examination of Japanese training strategies in the workplace and their role in producing quality products.

Kathryn Ward (editor). 1990. *Women Workers and Global Restructuring*. Ithaca, New York: Cornell University School of Industrial and Labor Relations. An overview of women's economic roles around the globe.

Websites

United Nations Human Development Report. *www.undp.org/hdro* Key data and insights on the economic condition and quality of life of people around the world.

Multinational Monitor. *www.essential.org/monitor* Private watchdog group reporting on violations of human, civil, and workers' rights by multinational corporations.

Earthrights International. *www.earthrights.org* Dedicated to defending human rights and the environment. Includes current news items from around the world.

U.S. Department of Labor, Bureau of International Labor Affairs. *www.dol.gov/dol/ilab* Comprehensive international statistics on employment, benefits, working hours, compensation, injuries, and related topics.

Institute of International Studies, University of California at Berkeley. *globetrotter.berkeley.edu* A core site for investigating global social issues of all kinds.

Southeast Asia web links. *www.gunung.com/seasiaweb* A great site for learning more about some of the Pacific Rim nations.

RECOMMENDED FILM

Mi Familia (1995), starring Edward James Olmos. A multigenerational epic of the Sanchez family starting in the 1920s with grandfather Sanchez's long walk from Mexico to Los Angeles.

17

The Future of Work

The media provide mounting evidence of "time poverty," overwork, and a squeeze on time. Nationwide, people report their leisure time has declined by as much as one third since the early 1970s. Predictably, they are spending less time on the basics, like sleeping and eating. Parents are devoting less attention to their children. Stress is on the rise, partly owing to the "balancing act" of reconciling the demands of work and family life.

The experts were unable to predict or even see these trends. I suspect they were blinded by the power of technology—seduced by futuristic visions of automated factories effortlessly churning out products. . . . To understand why forty years of increasing productivity have failed to liberate us from work, [it is necessary] to abandon a naive faith in technological potential and analyze the social, economic, and political context in which technology is put to use.

(SCHOR, 1992:5–6)

Changes of great significance are occurring in the nature and organization of work. For the first time in history we can meaningfully speak of a global workplace. Telecommunication networks link distant work sites. Night shifts in North America communicate instantaneously with morning shifts in Europe and with afternoon shifts in Asia. Multinational corporations link production sites around the world with little regard for geographic or political boundaries.

It is tempting to view these transformations in terms of technological change alone. But each aspect of change—new products, new technology, and new work organizations—has its own unique dynamics. Innovations in the organization of work, such as the assembly line, have historically contributed at least as much to productivity as have mechanical inventions.

It is also tempting to view these changes as beyond the control of the average worker. Even chief executives may feel helpless in the face of "world competition" or "the system." The welfare of workers and of all of us depends on who controls the nature and direction of change. Workers around the world are deeply involved in this process and are far from powerless. **Workers' power** is based on their knowledge of the technology and the product and on their collective organizations, such as unions and professional associations.

The difficulty, for workers and analysts alike, is to anticipate the consequences of these changes in the nature of work. All changes have multiple effects. Some are likely to be benign, others not so benign. In this chapter we examine three master trends that will be of crucial importance in determining the nature of work in the twenty-first century. We project, based on these trends, the possibility of a future society in which there is an increasing divide between a highly innovative and productive sector and a marginal sector. The emergence of such a divided society is a possibility but not a certainty. Therefore, we also discuss mechanisms for increasing innovation and reducing marginality.

PIVOTAL WORK TRENDS

The beginning decades of the twenty-first century are witnessing many important changes in the nature of work. Three trends are particularly important: (1) the spread of electronic technology throughout the workplace, (2) increased competition in the world economy, and (3) the increased movement of women into paid labor. In this section we review each of these trends and discuss how they are shaping the nature of work in the twenty-first century.

Computer Technology

During the twentieth century advances in technology and organization greatly increased productivity in manufacturing. Initially, these advances involved a finer division of labor into more and more minute tasks and the organization of jobs along moving assembly lines. New manufacturing technologies in the twenty-first century involve the use of automated machines and computer-controlled robotics. These electronic technologies are having an even greater impact on a broader range of industries and occupations than did the mass-production technologies of the twentieth century.

On the positive side, computer technologies have increased productivity and created skilled jobs. They have also shortened the time needed to develop new products. Thus, the pace of change has increased, giving rise to potentially faster economic growth. Microprocessor technologies have also encouraged greater worker participation, which is required to ensure the successful use of sophisticated technologies. On the negative side, new technologies have created new stresses because of the deskilling of some workers, the displacement of others, and the electronic monitoring of many others.

In sum, an acceleration in the pace of technological change associated with the extensive use of microprocessors has created a period of rapid change in the workplace and in workplace relations. These changes have destabilized older patterns of work and relationships among workers and between workers and managers. Technological change has created a period of tremendous flux in the workplace. Technology by itself, however, does not determine the direction of these changes. Other factors are equally important for determining changes in the nature of work.

An Integrated World Economy

Starting in the 1950s, an integrated world economy emerged following the devastation of World War II (see Chapter 16). Today, this integrated world economy includes tremendous diversity among nations. Countries vary in their level of industrialization and in how they are integrated into the global economy. Some are dependent suppliers of raw materials or partially finished components to more industrialized nations; others are autonomous producers of finished goods and services.

Recent developments in the world economy have greatly increased competition. After World War II, Japan and the European nations rebuilt with newer, more modern factories and reentered the world market. Many developing nations have also industrialized and have become important producers of manufactured goods. For example, South Korea now sells cars in the American market, something undreamed of twenty years ago. In addition, much of the work in low-wage, labor-intensive industries is now performed in less-developed nations, further intensifying the competitive pressure on workers in the industrialized countries.

Thus, changes in the world economy have also contributed to a period of change and flux in the workplace. Increased international competition has heightened threats to the jobs of workers in the industrialized nations. Greater competitive pressures require organizational change and innovation for economic survival.

Similar technological, organizational, and market forces have caused a convergence between capitalist and socialist nations. Formerly socialist nations today incorporate market forces in significant parts of their economies. And the most successful capitalist nations have moved toward greater economic planning. Such planning includes targeting specific industries for expansion, identifying the training needs of the labor force for these industries, and developing programs to meet these needs. The next stage of industrial society will transcend the distinction between capitalism and socialism. Other distinctions, such as those between more innovative and less innovative industrial systems and those between developed and less developed nations, however, are becoming more salient. Both capitalist states and the former communist states of China, Eastern Europe, and the Soviet Union are likely to face problems of high unemployment and may need to encourage greater worker participation to facilitate technological change.

Female and Minority Workers

Women have always worked; however, the transformation of their work into paid labor outside the home occurred somewhat later for women than it did for men, and it is still continuing today. The decline of high fertility and the reduction of child-rearing duties have been crucial in this process. In addition, employment opportunities have expanded in areas that have traditionally employed female workers, especially in clerical and service occupations. In combination, these factors have resulted in a rapid increase in the proportion of female workers in the labor force. This process has had tremendous consequences for the family. It can no longer be assumed that women will remain at home for child-care and home-tending duties. It would be incorrect, however, to assume that women's and men's paid jobs are similar. Women are still segregated by occupation and industry, and they are much more likely than men to be part-time and part-year workers.

Since the passage of the Civil Rights Act of 1964 and related legislation, female and minority workers in the United States have made important strides in the workplace, as well as in other spheres of society. Opportunities for female and minority workers have increased significantly. Full equality, however, is still far away. For minority workers, centuries of oppression have become embedded in class inequalities that make it difficult for many to take advantage of increased opportunities. Similarly, continuing assumptions on the part of both men and women that home and child-care duties should fall more on women's shoulders make it difficult for many women to take advantage of increased opportunities. In addition, for both female and minority workers, blatant and subtle forms of prejudice and discrimination continue to create barriers. Thus, in spite of improved conditions, women continue to suffer segregation into lower-status and lower-paying jobs, and blacks and other minorities still differentially occupy the lowest positions in society.

THE FACE OF WORK
IN THE TWENTY-FIRST CENTURY

Technological change, increased world market competition, and the increased proportion of women and minority workers in the labor force are major changes that will determine the nature of work in the future. What will work be like in the twenty-first century? It appears likely that the economy will be typified by two very different employment sectors. In one, which we call the **innovative sector,** the response to heightened international competition and technological change will be the development of technological and organizational innovations leading to

increased productivity. In the innovative sector, technological innovation will be continuous, jobs will be reasonably secure, pay will be adequate, job conditions will be more or less pleasant, and, perhaps most importantly, workers will have an increased say in determining the conditions of their work.

In the other sector, which we call the **marginal sector,** employers will respond to heightened international competition and technological change by reducing labor costs through lowered wages and benefits. In the marginal sector, innovation will be slow, jobs will be insecure, pay will be low, and conditions will be unpleasant and even hazardous. The organizing principle of the marginal sector will be to achieve economic viability by driving down wages rather than by promoting technical and organizational innovation. Workers will have little say in determining the conditions of their employment or the policies of their organizations. It is likely that female and minority workers will be disproportionately employed in the marginal sector. Thus, we do not believe as some have proposed that work will disappear in the high-technology future (Rifkin, 1995). Rather, work is here to stay, but the nature of work appears to be diverging between two increasingly distinct sectors of employment.

In this section we discuss the reasons for a divergence between these two distinct ways of organizing work and the characteristics of work in each sector. Box 17.1 depicts the three trends in work that we have described and possible future scenarios to which these trends may lead.

The Innovative Sector

What factors encourage the growth of an innovative sector? On what basis do we project its continued and increasing importance? The growth of innovation results from the development of microprocessor technologies that facilitate such innovation and from increased international competition and worker pressure that demand it in order to protect profits and jobs (Bluestone and Harrison, 2000). Many recent workplace innovations can be seen as attempts to use new technologies to adapt to increased competition in the world economy while maintaining high-wage employment. Another impetus toward workplace innovations has been the desire to reduce the inefficiencies of bureaucratic and hierarchical arrangements of work (Vallas, 2001). In this sense, workplace relations in the innovative sector can be better described as **postbureaucratic** than as postindustrial. Because human beings are central to the process of production, innovations in the organization of work will continue to be at least as important for the success of this sector as technological innovations.

The defining characteristics of work in the innovative sector are increased **worker education** and **participation.** Increased worker education and participation create the conditions for **continuous learning** and continual **job redesign** (Appelbaum et al., 2000). Continual job redesign will be necessary because of the pace of technological change and the highly competitive and rapidly changing global economy. The specific ways in which jobs will be redesigned are impossible to predict in great detail because they will be unique to each setting and each technology. However, some general principles are relatively clear (Szell, 1992).

The Centrality of Participation Workers hold the power to make organizational and technological advances succeed or fail. Some observations by industrial sociologist Robert Guest provide a good example. Guest was a visitor at a steel mill in the process of making a major technological change. A new steel process was delayed for six months because of differences between the company and the union on the incentive plan that would distribute part of the benefits of the new technology to the workers. After the incentive question was finally settled, Guest got a call from a worker and was told that he would see something interesting if he came down to the mill at the start of the midnight shift. Guest reports the following events:

> At precisely midnight a loud klaxon sounded. The lead man raised his arm and in a loud voice called out, "Let 'er roll!" The red hot billets spit out of the helical rolls at a speed I have never seen before. There were no delays or breakdowns on the shift and within a month capacity had gone up over twenty percent. (Guest, 1987:5)

The moral of the story is that workers hold the key to the success of programs of technological and organizational redesign. To engage workers' fullest abilities requires that they have a piece of the action—not just a share of profits but a share in decision making. The sectors of the economy that succeed in introducing competitive innovations will be the sectors that include a leading role for worker participation at every stage of innovation.

Work Groups One important form of worker participation occurs through small work groups of eight to twelve workers who are given collective responsibility for a task. Work groups offer an important role for worker participation, though they often limit the topic of discussion to product quality or to minor aspects of the work environment. Many college students work in settings that involve some aspects of team organization, or at least claim team organization by calling employees "associates" or "partners." Work groups can be an important source of innovation (Pfeffer, 1998). They can also be important for improving the **quality of work life.** However, as we will see, they can also be used to intensify work and heighten pressures on the job.

Team systems of production based on significant degrees of self-management by work groups have become increasingly important in contemporary organizations. Team-based production systems, however, actually have a long history in the workplace. Miners, seafarers, and other skilled trades have long relied on teams to coordinate work in situations involving complex and difficult tasks.

The increased importance of teams in the modern workplace reflects many forces, including increased skill demands associated with sophisticated technologies, new management theories about how best to organize production,

BOX 17.1   A Crystal Cube of the Workplace

Social scientists do not have crystal balls and cannot predict the future. They can, however, examine trends and describe what might happen if certain trends continue. Throughout this book we have tried to identify trends that may affect the workplace of the future. Three master trends that we have discussed are the role of technology and organization in the workplace, the competitive climate, and the changing composition of the labor force. These three trends may combine in many ways to produce many possible outcomes.

The cube in Figure A shows how the three master trends might interact. Imagine North American jobs right now as centered somewhere in the middle. A choice to use organization and technology in a way that increases the utilization of workers' skills would represent a shift to the left; a choice to use organization and technology to simplify or eliminate jobs would represent a shift to the right. Similarly, if increased productivity alleviates the competitive threat to North American industry, jobs would shift downward in the cube. As we have noted, women and members of minority groups will come to be a larger proportion of the labor force, but whether their labor power is adequately utilized depends on the types of jobs available to them. A "nearer" point in the three-dimensional space represents greater equality for female and minority workers. This would imply more access for female and minority workers to full-time, year-round jobs in occupations and industries comparable with those of white men. It is possible that women and minorities could be less adequately utilized than they are today; this is the "farther" end of the third dimension.

These options could, of course, be combined in many, many ways, but there are two points in our hypothetical space that we wish to discuss further. One is Point I on the figure; at this point technology and organization are used to complement workers' skills, competitive threats are lessened by productivity increases, and women and minorities move closer to equality. Even if most jobs in North America fail to move toward this point, some jobs will probably approach Point I. In this chapter, we refer to such jobs as comprising the innovative sector.

On the other hand, organization and technology may be used to deskill or to eliminate jobs, especially in response to competitive pressure. Under such conditions, pressures will persist to use women and members of minority groups as reserve workers, calling them up for part-time or seasonal work as needed or paying them low wages to undercut the wage demands of higher-paid workers. These conditions are represented schematically as Point II in the diagram. Even if most jobs do not move toward Point II, some jobs will. In this chapter we refer to such jobs as representing the marginal sector.

It is difficult to predict with any accuracy the relative size of the marginal sector and the innovative sector. Most jobs will probably still lie somewhere between these extremes. But there are reasons to think that these two sectors will represent significant numbers of jobs in the twenty-first century. It is useful to discuss these possibilities as "best-case" and "worst-case" scenarios representing endpoints in the master trends we have identified.

and worker demands for increased voice at the workplace (Ortiz, 1998).

Japanese companies and their affiliates around the world have lead the way toward increased utilization of team-based production systems. Under Japanese team production systems, employees organized in *Quality Control Circles* are expected to be ever vigilant for opportunities to work more effectively by identifying and eliminating underutilization of time and resources.

An important underpinning of Japanese Quality Control Circles and other initiatives to improve productivity and increase quality has been the tying of the worker to the company through lifetime employment and through finely graded systems of seniority-based pay (Dore and Sako, 1998). The

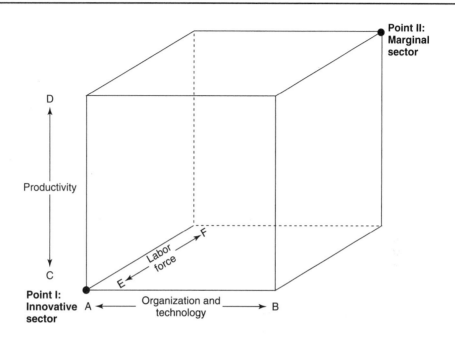

Organization and technology:

    A. They are used to increase productivity by complementing the abilities of skilled workers.

    B. They are used to replace workers and deskill jobs.

Productivity:

    C. Innovation and productivity increase.

    D. Productivity stagnates.

Labor force:

    E. Women and minority groups approach equality with majority men.

    F. Women and minority groups disproportionately occupy low-wage and part-time jobs.

FIGURE A Three Master Trends in the Workplace

tying of the employee and the firm together in a lifelong partnership encourages workers to use their skills to improve productivity and thus ensure the firm's future (Delbridge, 1998).

Other researchers note, however, that Japanese workers are not necessarily enthusiastic about involvement in team-based production (Lillrank and Kano, 1989). Rather, they see participation in Quality Control Circles and related team activities as a requirement for the economic success of their enterprise. Japanese workers thus participate in problem-solving activities with honesty and candor, but not generally with great enthusiasm or a sense of personal gratification.

In many workplaces, Japanese-style teams have been associated with work intensification (Endo,

1994), increased pressures for production (Rinehart et al., 1997:27,78), employee monitoring of peers (Roberson, 1998:78), and antiunion campaigns (Grenier, 1988:47,132). It is also a mistake to assume that managers and supervisors disappear in team production settings. Under Japanese team-based systems, frontline supervisors continue to play an active role in controlling and evaluating workers. In many ways, workers are more tightly controlled in team settings than in traditional supervisory settings. The power of the supervisor is not removed; rather, it is extended through allocating additional supervisory functions to the team as a whole (Rinehart et al., 1997:86).

In summary, team organizations of work are an important potential source of innovation, creativity, and heightened productivity (Smith, 1996). However, when used as the sole form of participation, they can force employees to ever greater efforts with no compensating gains for employees.

Codetermination and Joint Union–Management Programs Worker participation can also occur through formal consultation with workers at every level of the organization, from the shop floor to the boardroom. In Western Europe, various forms of worker participation are widespread. These forms include the workers' councils in Germany, which act as an autonomous board to review management policy, and technology stewards in Norway, who review and advise on technological change. Additional forms of participation occur through joint union–management initiated programs.

Joint union–management programs are based on explicit collectively negotiated agreements between unions and management to jointly sponsor programs that include employee involvement. In the United Kingdom, such programs are relatively commonplace across a wide range of industries (Marks et al., 1998). In North America such programs are concentrated in the automobile and telecommunications industries (Cooke, 1990). Other well-established programs exist in steel, construction, and the public sector. The key focus of many of these programs is on improved worker training to meet the challenges of automation and global competition (Milkman, 1997:160).

In joint union–management programs, the issues to be discussed are not necessarily restricted to management defined agendas. Workers in the automobile industry have successfully bargained for various forms of accelerated training under joint union–management programs and voice a great deal of satisfaction with these programs (Ferman et al., 1990). In these programs, workers receive additional training as part of an exchange for their greater involvement in the workplace and their increased contributions to productivity. The programs often involve supplemental training both on and off company time.

Increased communication and direct consultation with workers are also hallmarks of joint union–management programs. A joint program at an American car manufacturer includes the following principles:

- establish effective lines of communication among all employees,

- encourage participation of all employees who desire to become involved,

- strive for expeditious resolution of mutual problems,

- treat all employees with dignity and respect, and

- recognize the contributions of each individual (Milkman, 1997:161).

Note that these principles include a focus on employees and their rights and contributions rather than focusing solely on production-related issues as is typical of Japanese Quality Control Circles.

Workers in joint union–management programs are also increasingly allowed to go on purchasing and sales trips previously reserved for management and sales personnel. Workers provide valuable hands-on information in negotiations to secure the best components and new technologies. They also work directly with customers to learn how to improve quality and meet customer needs. The new knowledge and flexibility that such programs generate provide workers with opportunities to develop better relationships with their co-workers and with workers up and down the production chain. The opportunities provided by joint programs thus encourage employees to construct their organizational roles more actively. This active orientation generates new roles and new ideas that are often missing when work roles are unilaterally prescribed by management. In general, workers have been very enthusiastic about joint union–management programs and about participating in decision-making processes historically reserved for management (Milkman, 1997; Pfeffer, 1998).

The bilateral nature of joint initiatives provides a legitimacy to these programs that is sometimes missing when programs are initiated unilaterally by management. This legitimacy has been identified as a significant foundation for the success of joint union–management programs in stimulating productivity and improving working conditions. The initiatives emerging from joint union–management programs are also often more complementary with the public purpose than unilateral management initiatives because they include a focus on the preservation of employment and on the quality of employment as well as on increased productivity (Ferman et al., 1990:187).

Joint union–management programs are widely acknowledged to have played a leading role in stabilizing employment in the U.S. automobile industry and improving its competitive position in world markets. The resurrection of the Gary Works steel plant as a result of a worker participation program is described in Box 17.2.

Worker Ownership An additional type of worker participation is based on **worker ownership.** Worker ownership can be either total or partial through **employee stock ownership plans (ESOPs).** In 2000, 13 million U.S. workers, or about 10% of the labor force, participated in ESOPs (Census, 2000).

Employee ownership generally results in improved productivity and improved employee satisfaction (Pendleton et al., 1998). A core underlying reason for these improvements is that worker-owned enterprises are simply more concerned with the well-being of their employees than organizations owned by outside shareholders (Tucker, 1999). They are able to solicit high levels of worker involvement and participation because of the genuine overlap between the goals of the enterprise and those of the employees (Bradley, Estrin, and Taylor, 1990). Improved communication, teamwork, and participation are important underpinnings for the relative success of worker-owned enterprises.

The greatest participation occurs in worker-owned cooperatives in which workers not only own the firm but also actively manage its day-to-day affairs (Logue and Yates, 1999). Worker buyouts have provided an important counterbalance to the tendency of conglomerate companies to shut down or reduce their labor forces in the search for profitability. For instance, in 1994, the 78,500 unionized employees of United Airlines purchased the company, making it the largest employee-owned company in the United States (Moberg, 1994).

Worker buyouts of existing companies have often been initiated in an effort to preserve jobs. Worker-owner enterprises thus often face precarious circumstances because of external factors. Worker ownership often results from an employee buyout of a plant in a last-ditch effort to save the plant and the jobs it represents. In such situations, market forces may already be working against the enterprise. The market niche it serves may be shrinking or its technology and equipment may be outdated. Employee buyouts thus often face a precarious future because of the circumstances of their birth (Keef, 1998).

Worker ownership offers no necessary panacea to troubled firms, but even tested in this harsh environment, it has had a good record of success. In the Boston Harbor area, for example, about ninety boats make a living off lobsters. However, these boats were being priced off the docks by condominiums, office complexes, and

---

BOX 17.2   Competitiveness through Participation

U.S. Steel's Gary Works was all but banished from General Motor's supplier rolls. Ford was threatening the same. "Find a new way of doing your business," was the blunt mandate. . . .

Not only was the steel bad, it arrived late. "And we were arrogant," recalls [manager] Robert Pheanis. Gary Works—the biggest mill in the steel company—was alienating customers. It was also losing more than $100 million a year. The heart of Big Steel was about to stop beating.

Revival came from a small team of gritty union hardhats in the dreary 6-mile-long steel mill. . . . First one, than another—eventually five steelworkers were freed from mill jobs to visit automotive customers' plants and see problems for themselves. . . . The union crew was generally free to change how steel was made, stored and shipped so that the 50-ton rolls and quarter-ton sheets arrived at customers' plants in better shape.

They demanded rubber pads on flatbed trucks to cushion the steel. They created plastic rings to protect the rolls from crane damage. They persuaded workers who package and load each roll or stack to take responsibility for its condition by signing a tag attached to the shipment.

The result: automotive customers, who buy almost half of Gary Works' steel, now reject just 0.6%, down from an industry worst 2.6%. . . .

Says [consultant] Thomas Johnson, "The average guy working in most American companies today is a hell of a lot smarter than top management is willing to give him credit for." . . .

Gary Works' quality-improvement program differs from many because it was created from, and continues to be built on, the experience and intuition of hourly workers battling to save their plant, their jobs, their way of life. . . . But neither is it a by-guess and by-gosh effort. In the best tradition of statistical process control, computers measure every variation and tolerance at the plant so workers know when something is too hot, too thick, or too slow.

Ford [which nearly canceled its contract the previous year with Gary Works] gave it the Q1 award as a high-quality supplier. "We'd had instances of breakage where we'd get the managers, the engineers, the metallurgists working on it. Then the hourly guys come in and get their heads together, and the problem would go away," says a manager at Ford's nearby Chicago Heights stamping plant. . . .

[In one instance] hourly worker Bill Barath was pulled from his galvanizing job applying anti-corrosion coating at the end of the Gary steel line and was sent to Ford's Chicago Heights plant to eyeball the steel.

He found a galvanizer's nightmare: flaking zinc. Steel he so carefully coated at Gary was shedding its anti-corrosion skin like a snake when Ford formed it into fenders and doors.

Barath knew instantly: too much zinc buildup on the edges of the steel. The rods that trimmed it off at the mill were out of whack. Barath took that intelligence back to the mill, and an amazing thing happened. The problem got fixed. Right now. No tangled bureaucracy, no scapegoating.

Word spread, and other auto plants demanded Gary Works' liaisons. Even GM came around and has increased Gary Works purchases fivefold.

"These are line workers in the steel industry, supplying the auto industry. I can't think of two more battered industries. They've really pushed this idea of empowerment down to the workers where it belongs," says [plant manager] Goodwin. "That's the spirit of American industry that's coming back, and people don't know it. To have hourly people with that kind of power is not typical in this industry or this country, but it has to be."

SOURCE: James R. Healey, 1992, "U.S. Steel Learns from Experience," *U.S.A. Today* (April 10):B1–B2. Copyright 1992. *USA Today.* Reprinted with permission.

---

yacht marinas. With the help of the Boston archdiocese, these fishing families leased their own dock and established a cooperative enterprise that is flourishing today (McManus, 1987).

Job Security An essential foundation for all forms of heightened worker participation is **job security.** Without job guarantees, both on paper and in a history of commitment, workers are reluctant to give their best efforts to increasing productivity. This reluctance is especially strong in areas of active technological change, where the possibility of displacement for large numbers of workers is very real. Only when there is a strong commitment by the organization to maintain employment levels will workers give their full support to overcoming the inevitable problems associated with technological and organizational innovations. Loyalty by the company to its workers is thus an essential precondition for the realization of the full benefits of worker participation.

Training Training programs for employees are essential if they are to have the knowledge necessary to take a leading role in a more innovative workplace. Such programs have grown dramatically in the 2000s in community colleges and training institutions around the country. Community colleges provide flexible course sequences tailored to the needs of local industry. In addition, there has been increased interest in expanding traditional apprenticeship programs for the skilled trades. This interest has been sparked by unfavorable comparisons between the U.S. system of apprenticeship and the more developed German system, which many observers credit with making German products world renowned for their quality (Streeck, 1996). In-house training programs have also been expanding as employers and unions seek to expand the skills of workers as a means of increasing productivity, saving jobs, and increasing profits.

Distributing Profits A final key to successful job redesign is the development of mechanisms for distributing some of the profits of innovation back to workers (McHugh, Cutcher-Gershenfeld, and Polzin, 1999). The redistribution of profits is important for maintaining worker enthusiasm and commitment. The most innovative organizations are aware of this and have developed a variety of means to redistribute increased earnings to workers. These include higher pay, production bonuses, and profit sharing. The Ford-UAW bonus plan discussed in Chapter 6 provides a good example of one such plan.

Many innovative organizations also seek to tailor their compensation schemes to the needs of workers. One strategy is called the **cafeteria approach.** In this approach, workers are offered a variety of benefits and are free to choose the ones best suited to their own needs. Such packages may include supplemental retirement savings, supplemental life insurance, child-care vouchers, supplemental medical or dental coverage, or other benefits that appeal to specific workers.

The spread of such cafeteria-style benefit packages has also gained momentum from the increased presence of women in the labor force. Female workers bring different needs to the workplace than male workers. Because many men and women expect women to take greater responsibility for children, female workers often take a greater interest in workplace provisions for child care.

The increased attention to workers' needs in innovative organizations may result in an improvement in women's relative position in this sector. Because of reduced discrimination and less traditional career choices on the part of women, female workers may move increasingly into traditionally male-typed jobs, which are more likely to be in the innovative sector. The greater flexibility of women in their career choices as a result of lessened child-rearing duties may further facilitate these trends (England and Folbre, 1999).

Barriers to Innovative Job Redesign Job redesign also faces certain barriers that may limit its effectiveness. Chief among these is the problem of limited commitment by large corporations to their workers. Worker participation programs in the United States have often been superficial and have been accused of being more window dressing than substance (Parker and Slaughter, 1994). U.S. corporations have often acted as if *proclaiming* their allegiance to worker participation somehow constitutes an adequate solution to lagging productivity and to lack of management loyalty and commitment to their workers. Developing programs that actually incorporate workers in active roles at all levels of decision making and becoming committed to the long-term interests of employees requires more than just a public relations announcement of commitment to such goals (Graham, 1995). In addition, large organizations may experiment with job redesign in one plant with great success only to cancel the experiment because of changes in organizational strategy initiated from the top (Rothschild, 2000).

A second potential point of conflict can arise because increased worker participation often reduces the need for managers by incorporating managerial and supervisory activities within shop-floor groups. Employees may even be asked to evaluate their manager's performance. These changes may threaten the jobs of supervisors and middle-level managers and may stimulate resistance on their part (Smith, 1990). Middle-level managers are in a strong position either to facilitate job redesign programs or to sabotage them through subtle noncompliance and other tactics that workers themselves occasionally use with great effectiveness. (Recall our discussion in Chapter 4 on the tactics used by alienated workers.) When combined with agendas of corporate restructuring and theories of "lean production" that stress downsizing, job redesign programs put middle managers at significant risk of being laid off.

The ways in which worker participation can be incorporated into job redesign and technological innovation are extremely diverse. In this chapter we have described some of the forms of worker participation. The major forms of worker participation are enumerated in Figure 17.1, with the corresponding issues that are open to negotiation at the various levels of worker participation. At one extreme, workers are involved only in decisions about how to improve product quality and efficiency. At the other extreme they are involved in investment decisions about when and where to build new factories and what new lines of endeavor to pursue. Workers have shown themselves able to participate effectively in deci-

sions about their own working conditions, in decisions about the production process, and in decisions about investment (Gittleman, Horrigan, and Joyce, 1998). All of these forms of worker participation are important, all have been proven to be effective in some circumstances, and all have problems. No one form or level of participation is right for all circumstances. In industries with rapidly changing technologies, worker participation in job design may be most important. In industries with rapidly changing market situations, worker participation in investment decisions may be essential for continued economic viability.

The persistence of enthusiasm for worker participation and job redesign demonstrates that these programs are here to stay. Greater worker participation can make important contributions to productivity and competitiveness. Such programs are precursors to new systems of industrial relations that will increasingly characterize a significant share of employment positions in the twenty-first century.

The Marginal Sector

In the economy of the twenty-first century, it is also likely that a large sector of marginal employment will exist. The existence of such a sector alongside a highly innovative sector with increased worker participation suggests a more divergent economic structure in the next century. Why might a marginal sector grow in the future? Driving down wages is one possible response of organizations to competition. By cutting wages, enterprises can remain competitive, at least for a time. In sectors where technological change is slow, such as services, such a strategy may appear more attractive than strategies for increasing productivity.

Low Pay and Few Benefits   An increase in the marginal sector would cause an intensification of current social problems. Workers in this sector would have reduced buying power, thus limiting the demand for the goods produced by other workers and slowing the growth of the economy as a whole. Without a national health

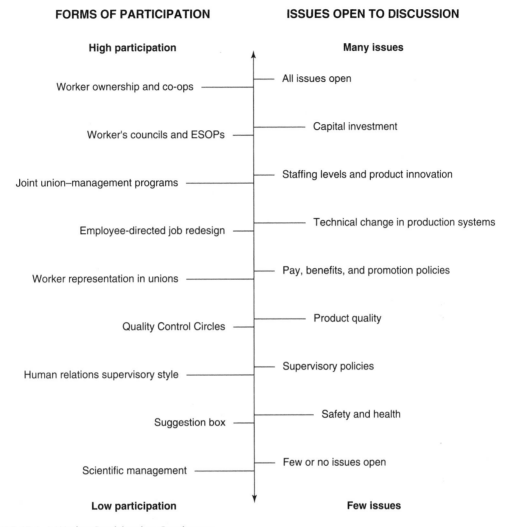

FIGURE 17.1 A Worker Participation Continuum

care plan, workers and families would lose health insurance and other benefits, thus putting additional pressure on already pinched social services and welfare programs. The growth of marginalized jobs is deeply implicated in increasing inequality in the United States over recent decades. It has been estimated that the highest earning 1% of U.S. families (those with incomes above $500,000) received 60% of the growth in after-tax income during the 1990s (Braun, 1997). Meanwhile, joblessness and poverty remained at high levels. Continued

expansion of the marginal sector would further increase poverty and homelessness.

What will work be like in the marginal sector of the future? Based on the characteristics of the current marginal sector, as explored in Chapter 14, we can anticipate that many aspects of long-term, stable employment will be missing. Pay will be low, and part-time and temporary work will be common. This sector will be typified by extensive subcontracting and frequent use of temporary workers (Tilly, 1996). Employers will have little interest in or commitment to employees. Probationary periods of employment with reduced rights will also be common. Two-tier pay scales with lower pay for newer employees might be used to bring in lower wage levels, with senior employees on the higher scale being rapidly phased out. Workers in the informal economy of unreported (and untaxed) work will also be part of this expanded marginal sector. While some informal work can be quite lucrative, basic benefits and employment security are largely missing from this employment sector.

Modern Sweatshops Some aspects of work in the marginal sector of the twenty-first century will replicate those of preindustrial work and the factories of the early nineteenth-century Industrial Revolution described in Chapter 1—but with some new twists. There will be fewer safeguards against hazardous working conditions than in the innovative sector. Enthusiastic commitment to work will be rare. Discontent, subtle forms of noncompliance, and even sabotage will be relatively common. Many workers in this sector will be underemployed, and many others will experience periodic unemployment as temporary jobs come and go. Young people will have a particularly hard time locating permanent jobs. Similarly, disregard of minimum wage restrictions and fair time and hours standards will be all too common. Mandatory drug testing, electronic surveillance, and other forms of monitoring may be widespread. Workers in the marginal sector will represent a transient **underclass.** It is an unfortunate but realistic projection that such an underclass may be a sizable component of the North American workforce in the twenty-first century (Wilson, 1997).

An Absence of Employer Commitment In what parts of the economy will marginal employment grow? According to sociologist Dan Cornfield, marginal jobs with unilateral managerial control and without formal worker participation will be most likely to occur in situations:

> with little unionization and/or with favorable macroeconomic conditions. In

these industries, management encounters little organized worker resistance to technological innovation, either because, in the absence of unionization, workers have low attachment to their employers and lack sufficient bargaining strength to demand participation in managerial decision making; or because favorable macroeconomic conditions lessen the threat of technological displacement for workers. (Cornfield, 1987:332)

The marginal sector is thus likely to grow in the absence of organized worker power and of competition based on technological change or foreign producers. Service industries, such as restaurants, and retail trade provide examples of such situations. It is also possible that the marginal sector will grow where advanced technology is used to reduce the need for skilled workers.

Without technologically or market-based competition there is little reason to initiate job redesign and worker participation. Without organized worker power there is often inadequate reason for managers to respond to heightened competition with workplace innovation rather than with lower wages and reduced **employer commitment** to workers. In situations where either condition is absent, it is likely that marginal jobs will increase. Such situations may represent a sizable or even majority share of new employment positions in the economy of the twenty-first century. Indeed, if the trend continues for new areas of job growth to be outside traditional union strongholds, elements of marginality may typify many areas of new employment. In areas of growing employment, such as services and microprocessor manufacturing, marginality can be expected to increase. Box 17.3 describes a hypothetical marginal job of the future.

Continuing Disadvantages for Female and Minority Workers How will female and minority workers fare in the economy of the twenty-first century? This question is difficult to answer. The recent past has seen improvements in

BOX 17.3   A Marginal-Employment Scenario

It had snowed all night, but the snow-jets had not yet cleared the lot. No matter, thought Debra as she braked her bike. Like most other parking lots built in the 1990s, this one was too big. Way too big, she thought, for my bike. If today was typical, hers would be the only vehicle parked there through the ten-hour shift. Marge Henry's little van was there now, but it was nearly shift-change time. Debra eased her right leg over the bike and into the ankle-deep snow. This was no time to slip and break something: the company had already cut back the health benefits to the government-required minimum, and Debra had nothing extra for the 50% coinsurance.

The snow muffled all sounds; it was almost as quiet here in the parking lot as it would be inside. The winter sun was nearly up over the horizon, and Debra looked over her shoulder at the lightening gray clouds in the east. In the winter she saw the sun only on Sunday. Her suntan, like her marriage, was a casualty of the company's shift policy. Ten hours a day, six days a week, take it or leave it. Many, many would take it if she left it. Certainly poor Jim. It was hard having an unemployed husband. It *had been* hard, she corrected herself. She pulled off a mitten and pushed her thumb against the glass for print recognition.

As the glass doors rolled back, she stepped across the laser beam on the threshold. It was easier than punching a clock but unrelenting in its accounting of her time. All personal areas of the shop had been fenced in with laser beams, so any "nonproductive" time was deducted from your paycheck. No tolerance, she thought wryly, if you had morning sickness, a touch of the "bug," or just too much spicy food last night. That was one reason for showing up a little early—it gave you a cushion of time if you needed an extra stop in the personal room. She crossed the corridor to the control room and pressed her thumb against another print-reader. Marge, she knew, would be locked inside. The company kept tight security, mainly for the sake of the robots (which were expensive).

"Any problems, Marge?" Debra asked.

"Number thirty-two is acting up again. Quiet night otherwise."

Marge crossed the laser beam to the dressing room and sat down to pull on her snow boots. She resented Debra. Both women worked ten hours a day, but Marge worked the night shift, and so she received only minimum wage. Debra made ten cents an hour more. Education had nothing to do with it; both jobs required a degree in robotics. The difference, the company claimed, was in the volume of work. This warehouse filled orders for toys—or rather, its forty-five robots did. The orders came in on the electronic LINCOM, and the robots located the proper storage bin, selected the proper number of boxes, issued address tags, and loaded the boxes into huge shipping containers. Twice a day, trucks came to pick up these containers. During the day shift, orders came in from across the United States and Canada, a few from Mexico. During the night shift, most of the orders came from Europe, Japan, and Africa. Although the toy market was growing there (in inverse proportion, it seemed, to declining birthrates), the volume of night business was much lower. Ample reason, according to the company, for Marge's paycheck to be a dollar less every day than Debra's even though the same forty-five robots needed routine and preventive maintenance and there were still occasional foul-ups.

Then there was the matter of the swing shift. Between 6 P.M. and 10 P.M. a young college graduate came in to work the short evening shift. It was so hard for new graduates to find jobs these days that some companies viewed these little part-time jobs as favors they did the worker, not as jobs that required a paycheck. This company paid an honorarium—half of minimum wage—for the decided advantage of gaining work experience. The evening shift wasn't all that easy, either, what with the boom in the Pacific Basin.

Marge was ready to leave. Debra saw her out, then checked the security system. The LINCOM was already humming, and she saw two more robots move off to fetch toys. Thirty-two was idle and would bear watching. As she punched up the maintenance record on thirty-two, she saw the reflection of a blue light blinking overhead. The blue light was the "spy in the sky," the electronic monitoring device that ensured that employees did not alleviate their boredom with transistor videos, personal message units, or other distractions. She was glad that someone, somewhere, knew she was here, even if only to check up on her. The workday had begun, but except for the humming of the LINCOM and the soft whir of the robots' wheels, you might never have known.

SOURCE: Adapted from Teresa A. Sullivan, 1989, "Women and Minority Workers in the New Economy: Optimistic, Pessimistic, and Mixed Scenarios." *Work and Occupations* 16,4 (November):393–395. Used by permission of Sage Publications, Inc.

employment conditions for female workers but the situation for minority workers has been more mixed. If women's earnings continue to advance against men's earnings, due to reduced discrimination in hiring and promotion and less traditional occupational choices by women, the prognosis for the future of women's employment is relatively good (England, 1992). If minority workers continue to make occupational advances into fields previously dominated by whites, the prognosis for minority employment is reasonably good. However, any remaining prejudices and discrimination against women and minority workers will result in their disproportionate representation in the marginal work sector of the twenty-first century. If the burden for child care continues to fall disproportionately on women and if the divorce rate continues at a high level or even increases, then poverty associated with family dissolution will continue to plague many women and will seriously constrain their employment options. Similarly, the reemergence and growth of a black underclass reminds us that, even if some minority workers have made progress in recent decades, many others have been left behind (Wilson, 1997).

We have predicted the emergence of a divided economy with large innovative and marginal sectors. Social and political forces, some of which are outside the scope of this book, however, will influence the relative size of these two employment sectors. Similarly, the relative representation of minority groups and women in each sector will be contingent on the ability of these groups to further reduce discrimination through mass movements and legislative agendas. Nevertheless, there are many relevant factors that will influence the relative development of these two sectors that we can consider.

ACHIEVING A BRIGHTER FUTURE

The emergence of an economy divided between innovative and marginal sectors is not inevitable. The problem before society is how to increase the size of the innovative sector and how to decrease the size and diminish the negative consequences of the marginal sector. How might these goals be achieved?

Increasing Innovation

What conditions foster innovative job redesign and worker participation? Workers in industrially advanced nations today are more highly educated than ever before in history. They are also more interested in safety and health issues and in improving the conditions of their employment. They have come to have high expectations about the satisfactions and rewards that work can provide. These expectations are perhaps the most important precondition for the growth of a highly innovative sector based on heightened participation and continuous learning.

Organizational and Market Imperatives Not all industries, however, are equally likely to pursue a strategy of workplace innovation. Innovation has tended to emerge when "both labor and management perceive a necessity for technological innovation; . . . and workers have the bargaining strength to demand participation in managerial decision making" (Cornfield, 1987:333). When these conditions are missing, innovation is unlikely to occur, and work practices will more closely resemble those of the mass-production industries of the past or those of the marginal sector.

Intensified competition also provides an impetus to workplace innovation. Labor and management do not typically cooperate unless competitive pressures force them to do so. Without these pressures they will be more likely to pursue their separate goals by engaging in adversarial bargaining over their respective shares of the economic pie.

Increased Worker Power An important precondition for innovation is that workers have sufficient power to demand participation and innovation. When workers are unable to demand participation, managers may favor a strategy of reducing wages, benefits, and job security in order to remain competitive rather than following the more uncertain, complex, and demanding route of workplace innovation. In the long run, however, the option of remaining competitive by reducing wages may be unrealistic in many industries (even ignoring the ethical hazards involved in such a strategy). As a result there is reason to hope that a substantial portion of organizations in industrially advanced societies will opt for workplace innovation.

The Role of Unions Worker participation is most likely to take place in unionized firms. The concepts of direct worker participation and worker cooperation with management, however, are to some extent antithetical to traditional union approaches to industrial relations. On topics ranging from work rules to grievance procedures, unions have increased workers' rights by negotiating explicit rules and ensuring their enforcement. This approach has often translated into a formalistic and legalistic approach to workplace issues, exactly the sort of approach that can be stifling to innovation (Freeman and Rogers, 1999).

Unions have also been cautious in advocating innovative workplace changes because they fear that altering the status quo may undermine their ability to deliver the package of benefits and protections that have been their traditional offering to workers. To survive in innovative settings, however, unions will have to stop *reacting* to changing circumstances and begin to take an *active role* in developing proposals for increasing productivity. In the past, union leaders have often dismissed worker participation as "a ruse to increase worker productivity" (Parker and Slaughter, 1994). Continuing to dismiss worker participation in this manner will undermine the role of unions in the workplace of the future. The emergence of joint labor–management groups at various levels of the organization is necessary to increase productivity in rapidly changing and highly competitive industries. If

the innovative sector is to grow and prosper, it will have to include the unionized industries, but unions will have to change to accommodate and promote innovation. These changes are threatening to established union practices in many ways. Unions in the industrially advanced European nations, however, have taken a leading role in promoting increased worker participation. Unions and worker participation are not inherently antithetical; indeed, they may be dependent on each other.

Education and Training Educating the workforce for innovation will require an increased commitment not only to college education but also to **vocational education** and to ongoing **retraining** for workers already in the labor force. Many European nations have instituted programs for mid-career retraining. Observers credit a substantial part of these countries' economic vitality to such programs. North Americans must move beyond retraining programs designed to remedy the plight of the currently unemployable. Such programs include the Comprehensive Employment Training Act and Job Training Partnership Act. Although these programs are an important safeguard for workers otherwise condemned to work in the marginal sector, they are largely inadequate for training the highly skilled labor force necessary for the innovative sector (Slessarev, 1997). Besides providing college for the middle classes and job training for marginal workers, North Americans must provide skilled workers with the education and continuing training necessary to compete effectively in the global economy.

Workplace Experimentation Increased **workplace experimentation** is also needed to stimulate the growth of an innovative sector. Successful job redesign comes only from experimentation. It cannot be fully specified in abstract formulas—it requires a constant process of review and change. Unfortunately, sustained programs of experimentation in job redesign and worker participation are not widespread in the United States and Canada.

To increase the innovative sector, the preconditions for its existence must be broadened. These preconditions include workplace experimentation, increased worker power, and technological innovation. Greater worker power can come about either on an individual basis, through increased education, training, and professional development, or on a collective basis, through increased representation in unions and professional associations. Encouraging education and unions fosters the growth of an innovative sector and decreases the prevalence of marginality. The impetus for these changes must come from workers, unions, and professional associations. Reforms of this sort will not occur unless workers and their organizations demand them (Freeman and Rogers, 1999).

Reducing Marginal Employment

Besides underwriting workers' abilities to demand technological innovation and competitiveness and providing a labor force educated to meet the challenge of innovation, society can also set up roadblocks against encroaching marginality. These roadblocks could include an increased minimum wage, restrictions on mergers and plant closings, and standardization of tax, labor, and environmental laws across the country so that states are not forced to compete with one another to offer companies the most minimal employment standards (Phelps, 1997). In many cases higher minimum wages have been shown to have significant positive effects on worker stability and the acquisition of additional training without substantial negative effects on new hiring (Card and Krueger, 1995). Roadblocks to capital flight would keep capital at home and in place where it can be used to foster innovation and increased productivity as preferred responses to competitive pressures. Programs encouraging worker buyouts, ESOPs, and cooperatives might also reduce marginality. Banks could be encouraged to lessen their resistance to extending needed credit to worker-owned enterprises.

Such forms of collective ownership, though they do not necessarily eliminate marginality, at least reduce some of its worst consequences (Cornforth, 1992).

Expanded worker training programs are important, not only for facilitating the spread of innovation, but also for reducing marginal employment. The need for such programs is increased by the heightened pace of technological change. Fewer and fewer workers can expect to work at just one job throughout their careers. With more rapid job changes, improved job placement and matching services also become increasingly important if jobs are to be filled by qualified applicants, if workers' abilities are to be used effectively, and if workers are to avoid periods of marginality. Employment of the labor force at nearer its full potential would increase productivity and reduce welfare expenses immensely.

Reducing the marginal aspects of available jobs would help lessen underemployment, unstable employment, and blocked mobility. In conjunction with education and training programs and the expansion of the innovative sector, a reduction of marginal employment positions would facilitate the incorporation of all Americans in the mainstream of society. The success of such agendas will also depend on the vigorous enforcement of laws protecting the rights of workers.

Expanding Leisure

Expanded leisure hours could also make an important contribution to improving the quality of work life and distributing available employment. The average hours worked per week dropped from near seventy in 1850 to about forty by the 1930s. Starting in the 1970s, however, this trend began to be reversed. Fully employed U.S. workers found that their workweek increased by about 5% during the 1970 to 2000 period. Simultaneously paid holidays and vacation days fell by 15% (Schor, 1992). This change occurred simultaneously with rising unemployment. In the 2000s, full-time work has become harder to get, and it is often highly demanding when it is finally secured. Reducing the hours of work would be an effective way to distribute available employment in a period of rapidly increasing productivity and technological change. Reduced hours of work would also help prevent polarization of society between those with too much work and those with inadequate work or no work at all (Hunnicutt, 1996).

A related possibility is that people will combine work, leisure, and education throughout their lives rather than completing education before starting work and saving their leisure years for retirement. As we discussed in Chapter 3, such a pattern would deviate from the so-called normative career pattern. A career pattern with greater integration of work, education, and leisure would include periodic breaks for retraining and renewal, breaks that would be extremely important for sustaining the innovative spirit in workers. Lifestyles that allow for a greater integration of work, education, and leisure across the life cycle would also help resolve the dilemma of fewer workers being needed as technologically based advances in productivity continue to accumulate.

Expanding Public Goods

Over the history of industrial society the provision of "public goods" has increased. **Public goods** are products or services to which the citizens or residents of a society are entitled without direct payment or at a nominal fee. Such public goods include education, clean air, public parks, retirement income, and, increasingly, health care. If the provision of public goods expands, workers who remain in the marginal sector of employment will at least be spared some of the most debilitating consequences of poverty and marginality. The provision of public goods can thus be an important preventive against the reproduction of a marginal sector. The provision of public goods also creates jobs, providing additional escape routes from marginality. Alternatively, if such services are provided only privately as parts of benefit packages restricted to the innovative sector, many in society will have to do without them. The provision of public goods and the training of workers for employment in the innovative sector are among the most important strategies for avoiding a deeply class-divided society in the twenty-first century. Whether we pursue such agendas as a society depends on the importance that we attach to equality and the creation of a just society (Kuttner, 1996).

SUMMARY

The rapidly changing economic, technological, and organizational realities of today's global economy are setting the stage for the world of work in the twenty-first century. Alternatives are available, but these depend on the political and social actions we take today. The dilemmas described in this book can be resolved in a positive manner. Workers and other members of society can pressure organizations to respond to competition by increasing innovation rather than by cutting wages and fostering marginality. They can also pressure labor unions and other workers' associations to adapt to the new conditions of a competitive environment and to provide leadership in the areas of technology, organizational innovation, and worker training. Finally, they can pressure the government to enact laws and programs that will foster innovation, job creation, worker involvement, expanded leisure, and a better quality of life for all.

Industrial societies have the capacity to constantly increase productivity through technological and organizational innovation. This capacity has never been greater than it is today. However, this very capacity creates the dilemma of how to distribute available work when more and more goods and services can be produced by fewer and fewer workers. These, then, are the central challenges of contemporary industrial society: How do we increase productivity? How do we distribute available work? And how do we distribute the goods and services produced? The manner in which we resolve these dilemmas will determine the political, social, and economic landscape of the twenty-first century.

KEY CONCEPTS

worker power	job redesign	underclass
innovative sector	quality of work life	employer commitment
marginal sector	joint union–management programs	vocational education
postbureaucratic	worker ownership	worker retraining
worker education	ESOPs	workplace experimentation
worker participation	job security	expanded leisure
continuous learning	cafeteria-style benefit packages	public goods

QUESTIONS FOR THOUGHT

1. Consider your chosen (or likely) field of work. Identify one way in which you think each of the factors of technological change, increased world competition, and increased worker participation will influence the nature of your work and career.

2. .What developments, other than those discussed in this chapter, do you think will significantly influence the nature of work and society in the future? (HINT: Environment? Population? Immigration? Something else?)

3. What factors might continue to channel female and minority workers into the marginal sector in the twenty-first century? Do you think women will gain full equality in the workplace and in society? Will minorities gain equality? Why or why not?

4. To what extent do you thing the United States will be successful in expanding the innovative sector? Which industries are likely to play a leading role in this process?

5. How optimistic are you about the possibilities for reducing marginality in the twenty-first century? What might be some of the consequences of failing to address the issue of marginality?

MULTIMEDIA RESOURCES

Print

James R. Barker. 1999. *The Discipline of Teamwork: Participation and Concertive Control*. Thousand Oaks, CA: Sage. An ethnographic account of the use of a team-based approach to increase productivity. Reveals both the benefits and the pressures of team-based systems.

Eileen Appelbaum, Thomas Bailey, Peter Berg, and Arne L. Kalleberg. 2000. *Manufacturing Advantage:*

Why High-Performance Work Systems Pay Off. Ithaca, NY: Industrial and Labor Relations Press. A systematic analysis of work systems involving greater employee involvement and their benefits for productivity and enhanced worker well-being.

Gyorgy Szell (editor). 1992. *Concise Encyclopedia of Participation and Co-Management*. A comprehensive international tour of worker participation programs and their successes and limitations.

Websites

Self-directed work teams. *users.ids.net/~brim/sdwth.html* A great resource for those involved in developing, managing, or working in self-directed teams. Many examples are provided, including Saturn and Federal Express.

Employee Involvement Association. *www.eia.com* Dedicated to promoting workplace change through employee involvement.

Canadian Human Resources Department. *labour-travail.hrdc-/drhc.gc.ca/wip* A workplace innovations support site with great links.

Canadian Labour–Management Partnership Program. *labour-travail.hrdc-drhc.gc.ca/doc/fmcs-sfmc/eng/lmpp.cfm* Support for both labor and management as they seek new ways of working together.

RECOMMENDED FILM

Office Space. (1999). A comedy about computer workers who cannot take their jobs anymore and seek redress through a computer accounting scam.

Appendix Table 1 Employed Civilians by Detailed Occupation, Sex, Race, and Hispanic Origin, 2000

Occupation	Total employed (thousands)	Percent of total		
		Women	Black	Hispanic origin
Total, 16 years and over..	135,208	46.5	11.3	10.7
Managerial and professional specialty	40,887	49.8	8.2	5.0
Executive, administrative, and managerial	19,774	45.3	7.6	5.4
Officials and administrators, public administration	651	52.7	13.1	7.0
Administrators, protective services	62	27.8	16.1	3.4
Financial managers	784	50.1	6.1	4.3
Personnel and labor relations managers	226	61.8	7.9	4.0
Purchasing managers	123	41.3	7.0	3.2
Managers, marketing, advertising, and public relations	755	37.6	4.2	4.2
Administrators, education and related fields	848	67.0	13.5	5.7
Managers, medicine and health	752	77.9	9.7	5.4
Postmasters and mail superintendents	55	58.5	5.4	3.8
Managers, food serving and lodging establishments	1,446	46.8	9.9	9.2
Managers, properties and real estate	552	50.9	8.2	7.2
Funeral directors	58	18.3	10.9	7.2
Management-related occupations	4,932	56.5	9.5	5.4
Accountants and auditors	1,592	56.7	8.9	5.1
Underwriters	104	71.9	7.7	3.8
Other financial officers	837	51.3	9.4	4.9
Management analysts	426	38.5	7.2	2.9
Personnel, training, and labor relations specialists	628	66.6	11.6	6.3
Buyers, wholesale and retail trade, except farm products	224	51.3	5.9	6.5
Construction inspectors	72	9.5	4.7	4.3
Inspectors and compliance officers, except construction	255	40.1	13.9	6.8
Professional specialty	21,113	53.9	8.7	4.6
Engineers, architects, and surveyors	2,326	11.1	5.3	3.9
Architects	215	23.5	1.6	5.5
Engineers	2,093	9.9	5.7	3.7
Aerospace engineers	78	9.7	5.4	3.6
Chemical engineers	85	10.4	5.1	1.0
Civil engineers	288	9.7	6.1	2.7
Electrical and electronic engineers	725	9.8	6.3	3.6
Industrial engineers	244	15.3	6.4	4.0
Mechanical engineers	342	6.3	4.7	3.7
Mathematical and computer scientists	2,074	31.4	8.1	3.7
Computer systems analysts and scientists	1,797	29.2	8.0	3.6
Operations and systems researchers and analysts	227	45.5	10.9	4.4
Natural scientists	566	33.5	5.4	3.2
Chemists, except biochemists	153	30.3	11.0	2.2
Agricultural and food scientists	53	28.2	6.1	3.9
Biological and life scientists	114	45.4	4.0	6.0
Medical scientists	84	49.5	4.6	4.6
Health diagnosing occupations	1,038	27.1	5.2	3.4
Physicians	719	27.9	6.3	3.7
Dentists	168	18.7	3.4	2.2
Veterinarians	55	30.6	3.4	1.5
Health assessment and treating occupations	2,966	85.7	9.0	3.4
Registered nurses	2,111	92.8	9.5	2.8
Pharmacists	208	46.5	3.3	3.8
Dietitians	97	89.9	18.4	4.8
Therapists	478	74.7	8.1	5.0
Respiratory therapists	78	62.4	10.8	5.3
Occupational therapists	55	91.4	3.5	5.8
Physical therapists	144	61.1	6.5	6.8
Speech therapists	102	93.5	4.5	2.0
Physicians' assistants	72	57.6	5.6	7.8
Teachers, except college and university	5,353	75.4	10.4	5.2
Prekindergarten and kindergarten	626	98.5	13.3	8.0
Elementary school	2,177	83.3	11.3	5.6
Secondary school	1,319	57.9	8.9	4.2
Special education	362	82.6	9.2	3.2
Counselors, educational and vocational	258	70.2	17.1	5.3
Librarians, archivists, and curators	263	84.4	6.0	5.8
Librarians	232	85.2	6.7	6.6
Social scientists and urban planners	450	58.9	7.8	4.1
Economists	139	53.3	6.3	4.4
Psychologists	265	64.6	8.1	4.0

Occupation	Total employed (thousands)	Percent of total		
		Women	Black	Hispanic origin
Social, recreation, and religious workers	1,492	56.4	17.4	6.4
Social workers	828	72.4	22.7	8.5
Recreation workers	126	71.0	9.5	4.9
Clergy	369	13.8	14.1	4.5
Lawyers and judges	926	29.7	5.7	4.1
Lawyers	881	29.6	5.4	3.9
Writers, artists, entertainers, and athletes	2,439	50.0	6.9	5.6
Authors	138	54.1	7.7	2.2
Technical writers	70	64.2	2.1	1.7
Designers	738	57.2	4.0	6.3
Musicians and composers	161	34.1	13.5	6.0
Actors and directors	139	41.5	12.8	6.1
Painters, sculptors, craft artists, and artist printmakers	238	46.5	6.8	4.2
Photographers	148	32.6	5.7	5.9
Editors and reporters	288	55.8	5.0	3.0
Public relations specialists	205	61.1	10.8	5.5
Announcers	54	10.7	10.5	6.0
Athletes	90	19.8	10.9	5.5
Technical, sales, and administrative support	39,442	63.8	11.4	8.9
Technicians and related support	4,385	51.7	11.2	6.9
Health technologists and technicians	1,724	80.5	15.0	8.2
Clinical laboratory technologists and technicians	342	75.0	18.0	7.5
Dental hygienists	112	98.5	2.4	1.7
Radiologic technicians	161	69.2	10.8	7.7
Licensed practical nurses	374	93.6	20.0	5.0
Engineering and related technologists and technicians	1,002	20.4	10.0	6.1
Electrical and electronic technicians	468	16.9	11.0	7.1
Drafting occupations	219	23.4	6.2	4.7
Surveying and mapping technicians	79	7.4	6.5	7.8
Science technicians	270	41.4	8.7	8.4
Biological technicians	108	59.5	7.1	8.2
Chemical technicians	71	21.2	7.2	7.6
Technicians, except health, engineering, and science	1,389	40.5	7.9	5.7
Airplane pilots and navigators	129	3.7	1.9	4.3
Computer programmers	699	26.5	8.1	3.5
Legal assistants	387	84.4	8.4	9.8
Sales occupations	16,340	49.6	8.8	8.5
Supervisors and proprietors	4,937	40.3	6.6	7.3
Sales representatives, finance and business services	2,934	44.5	7.6	4.9
Insurance sales	577	42.5	6.5	4.4
Real estate sales	787	54.3	5.3	5.0
Securities and financial services sales	600	31.3	8.2	3.4
Advertising and related sales	165	61.9	9.2	5.7
Sales occupations, other business services	805	42.7	9.9	6.2
Sales representatives, commodities, except retail	1,581	27.5	2.8	6.4
Sales representatives, mining, manufacturing, and wholesale	1,549	27.9	2.8	6.4
Sales workers, retail and personal services	6,782	63.5	12.3	11.4
Sales workers, motor vehicles and boats	329	11.0	9.1	8.4
Sales workers, apparel	411	77.1	14.2	13.8
Sales workers, shoes	114	55.4	16.1	14.6
Sales workers, furniture and home furnishings	185	50.7	6.6	7.7
Sales workers, radio, television, hi-fi, and appliances	258	27.0	7.6	6.2
Sales workers, hardware and building supplies	328	22.2	4.0	7.4
Sales workers, parts	186	8.9	5.8	15.2
Sales workers, other commodities	1,428	66.4	9.6	9.0
Sales counter clerks	185	68.0	9.2	8.7
Cashiers	2,939	77.5	16.5	13.5
Street and door-to-door sales workers	311	76.0	7.7	13.5
News vendors	110	44.7	6.8	7.5
Sales-related occupations	107	69.1	9.3	2.2
Demonstrators, promoters, and models	71	73.6	8.3	1.8
Administrative support occupations, including clerical	18,717	79.0	13.7	9.7
Supervisors, administrative support	710	60.3	17.0	9.4
Supervisors, general office	404	71.2	17.4	7.0
Supervisors, financial records processing	73	80.9	8.7	6.2
Supervisors, distribution, scheduling, and adjusting clerks	217	34.2	17.4	15.2
Computer equipment operators	323	48.6	16.6	7.4

Occupation	Total employed (thousands)	Percent of total		
		Women	Black	Hispanic origin
Computer operators	321	48.7	16.6	7.4
Secretaries, stenographers, and typists	3,328	98.0	9.9	8.6
Secretaries	2,623	98.9	8.5	8.7
Stenographers	154	94.7	4.7	4.4
Typists	551	94.6	17.8	9.3
Information clerks	2,071	88.0	11.3	10.4
Interviewers	212	78.9	16.1	9.2
Hotel clerks	130	76.3	15.9	8.2
Transportation ticket and reservation agents	287	71.9	11.7	9.4
Receptionists	1,017	96.7	9.7	11.6
Records processing, except financial	1,119	81.5	16.9	10.6
Order clerks	305	77.1	24.4	12.4
Personnel clerks, except payroll and timekeeping	84	82.5	18.7	4.8
Library clerks	152	87.0	10.8	6.5
File clerks	338	80.2	15.3	12.0
Records clerks	227	85.9	13.4	10.7
Financial records processing	2,269	91.8	9.2	7.3
Bookkeepers, accounting, and auditing clerks	1,719	92.2	7.8	6.1
Payroll and timekeeping clerks	174	91.3	8.7	8.4
Billing clerks	198	92.2	16.3	12.4
Billing, posting, and calculating machine operators	134	91.7	16.8	12.3
Duplicating, mail and other office machine operators	55	54.2	16.8	9.6
Communications equipment operators	167	84.3	21.8	10.7
Telephone operators	156	83.9	22.9	10.4
Mail and message distributing	978	41.2	21.9	7.7
Postal clerks, except mail carriers	304	54.4	32.4	6.2
Mail carriers, postal service	340	30.6	14.7	5.9
Mail clerks, except postal service	178	54.0	22.6	11.2
Messengers	157	23.9	16.7	10.4
Material recording, scheduling, and distributing clerks	2,052	46.7	15.3	12.8
Dispatchers	269	51.7	15.1	9.0
Production coordinators	227	58.5	12.0	6.6
Traffic, shipping, and receiving clerks	661	33.8	16.1	17.5
Stock and inventory clerks	460	44.9	15.1	13.1
Weighers, measurers, and checkers and samplers	64	51.9	19.3	16.7
Expediters	310	66.5	13.4	10.5
Adjusters and investigators	1,818	75.5	17.5	10.1
Insurance adjusters, examiners, and investigators	451	73.9	14.6	7.0
Investigators and adjusters, except insurance	1,097	76.0	17.0	11.5
Eligibility clerks, social welfare	94	89.2	16.1	9.5
Bill and account collectors	176	69.4	28.2	9.5
Miscellaneous administrative support	3,826	83.9	14.3	10.4
General office clerks	864	83.6	12.9	10.5
Bank tellers	431	90.0	13.7	8.2
Data-entry keyers	749	83.5	18.8	11.2
Statistical clerks	104	88.5	15.8	8.4
Teachersí aides	710	91.0	12.8	14.4
Service occupations	18,278	60.4	18.1	15.7
Private household	792	95.5	14.9	31.7
Child care workers	275	97.5	11.6	19.9
Cleaners and servants	500	94.8	16.9	37.7
Protective service	2,399	19.0	19.6	8.7
Supervisors	201	15.1	13.9	7.8
Police and detectives	116	14.3	10.5	3.0
Guards	53	23.5	28.4	18.5
Firefighting and fire prevention	248	3.8	8.7	5.4
Firefighting	233	3.0	9.0	5.0
Police and detectives	1,060	16.5	18.3	8.4
Police and detectives, public service	560	12.1	13.0	10.1
Sheriffs, bailiffs, and other law enforcement officers	156	19.2	20.2	5.8
Correctional institution officers	344	22.5	25.9	6.9
Guards	889	27.0	25.7	10.0
Guards and police, except public services	745	20.1	27.5	10.6
Service occupations, except private household and protective service	15,087	65.1	18.0	16.0
Food preparation and service occupations	6,327	57.7	11.9	17.2
Supervisors, food preparation and service	434	68.6	11.8	12.0
Bartenders	365	51.8	2.0	13.2
Waiters and waitresses	1,440	76.7	4.4	11.0

Occupation	Total employed (thousands)	Percent of total		
		Women	Black	Hispanic origin
Cooks	2,076	43.3	17.6	21.6
Food counter, fountain and related occupations	357	67.9	12.6	11.8
Kitchen workers, food preparation	317	71.1	13.0	12.3
Waiters' and waitresses' assistants	670	51.4	10.5	18.8
Miscellaneous food preparation	668	52.1	16.9	26.0
Health service occupations	2,557	89.5	31.4	10.1
Dental assistants	218	96.4	5.1	10.6
Health aides, except nursing	356	82.6	26.4	8.7
Nursing aides, orderlies, and attendants	1,983	89.9	35.2	10.4
Cleaning and building service occupations	3,127	45.0	22.2	23.4
Supervisors	166	38.0	21.9	20.2
Maids and housemen	650	81.3	27.7	28.3
Janitors and cleaners	2,233	36.3	20.9	22.5
Pest control occupations	71	5.0	15.0	11.7
Personal service occupations	3,077	80.5	14.8	10.8
Supervisors	119	63.2	17.8	8.8
Barbers	108	25.3	27.8	12.7
Hairdressers and cosmetologists	820	91.2	10.9	10.7
Attendants, amusement and recreation facilities	246	39.4	9.9	6.0
Public transportation attendants	127	80.9	12.3	7.9
Welfare service aides	99	87.2	30.3	12.7
Family child care providers	457	97.7	14.8	13.3
Early childhood teachersí assistants	480	95.2	17.4	10.6
Precision production, craft, and repair	14,882	9.1	8.0	13.9
Mechanics and repairers	4,875	5.1	8.2	10.7
Supervisors	223	8.9	5.4	8.4
Mechanics and repairers, except supervisors	4,652	5.0	8.3	10.8
Vehicle and mobile equipment mechanics and repairers	1,787	1.6	7.1	13.1
Automobile mechanics	860	1.2	7.3	15.6
Bus, truck, and stationary engine mechanics	345	.9	6.6	9.1
Aircraft engine mechanics	126	6.1	8.3	9.0
Small engine repairers	60	1.7	8.8	5.1
Automobile body and related repairers	186	1.3	8.7	16.1
Heavy equipment mechanics	162	.8	3.8	10.0
Industrial machinery repairers	524	4.2	9.3	8.0
Electrical and electronic equipment repairers	999	11.5	10.7	7.8
Electronic repairers, communications and industrial equipment	192	9.3	13.8	9.8
Data processing equipment repairers	342	15.4	9.8	4.8
Telephone line installers and repairers	53	3.1	3.8	9.9
Telephone installers and repairers	295	13.1	11.6	9.5
Heating, air conditioning, and refrigeration mechanics	371	1.2	6.3	11.9
Miscellaneous mechanics and repairers	949	6.3	8.2	10.7
Office machine repairers	53	9.8	6.2	5.8
Millwrights	78	2.4	9.0	3.9
Construction trades	6,120	2.6	7.0	16.4
Supervisors	967	2.2	6.3	7.5
Construction trades, except supervisors	5,153	2.7	7.2	18.1
Brickmasons and stonemasons	242	.7	13.0	18.6
Tile setters, hard and soft	94	1.3	2.0	34.4
Carpet installers	125	1.3	7.4	18.9
Carpenters	1,467	1.7	6.0	16.3
Drywall installers	206	5.6	6.1	39.2
Electricians	860	2.7	7.7	9.1
Electrical power installers and repairers	132	1.4	8.0	9.9
Painters, construction and maintenance	624	5.8	7.4	24.4
Plumbers, pipefitters, and steamfitters	540	1.3	6.1	13.5
Concrete and terrazzo finishers	99	1.1	11.0	36.1
Insulation workers	58	6.1	12.3	21.5
Roofers	215	1.6	7.3	30.1
Structural metalworkers	79	2.3	11.8	8.9
Extractive occupations	128	1.9	3.6	7.8
Precision production occupations	3,759	25.0	9.5	14.4
Supervisors	1,129	19.7	9.0	10.6
Precision metalworking	865	7.5	6.3	8.9
Tool and die makers	121	2.7	3.8	3.9
Machinists	488	6.3	7.3	8.8
Sheet-metal workers	121	4.2	5.6	10.0
Precision woodworking occupations	127	11.0	8.3	10.0
Cabinet makers and bench carpenters	89	.8	6.6	9.6

Occupation	Total employed (thousands)	Percent of total		
		Women	Black	Hispanic origin
Precision textile, apparel, and furnishings machine workers	192	54.2	7.8	25.2
Dressmakers	77	92.7	8.6	20.8
Upholsterers	64	20.8	6.9	25.1
Precision workers, assorted materials	554	52.6	12.2	14.8
Optical goods workers	77	58.5	7.7	9.3
Electrical and electronic equipment assemblers	336	62.4	15.2	15.7
Precision food production occupations	481	39.9	13.1	32.9
Butchers and meat cutters	265	27.3	14.5	38.3
Bakers	154	45.6	12.6	26.1
Food batchmakers	62	79.5	8.1	26.5
Precision inspectors, testers, and related workers	148	24.4	11.3	8.2
Inspectors, testers, and graders	136	24.5	11.6	7.7
Plant and system operators	264	5.4	10.2	11.8
Water and sewage treatment plant operators	69	5.5	7.1	7.6
Stationary engineers	118	6.7	12.0	11.5
Operators, fabricators, and laborers	18,319	23.6	15.4	17.5
Machine operators, assemblers, and inspectors	7,319	36.9	14.7	19.3
Machine operators and tenders, except precision	4,546	36.6	14.9	20.7
Metalworking and plastic working machine operators	349	18.4	10.7	10.8
Punching and stamping press machine operators	94	25.7	11.8	10.4
Grinding, abrading, buffing, and polishing machine operators	98	15.2	10.2	17.2
Metal and plastic processing machine operators	150	23.5	15.1	17.4
Molding and casting machine operators	84	34.0	10.6	14.7
Woodworking machine operators	137	14.8	8.9	11.1
Sawing machine operators	78	12.6	9.0	11.8
Printing machine operators	369	23.7	7.4	12.1
Printing press operators	292	17.6	7.9	13.4
Textile, apparel, and furnishings machine operators	854	69.2	18.3	33.0
Textile sewing machine operators	425	78.4	16.3	40.6
Pressing machine operators	81	66.6	13.9	49.9
Laundering and dry cleaning machine operators	214	56.9	18.0	25.6
Machine operators, assorted materials	2,665	32.2	15.8	19.9
Packaging and filling machine operators	345	60.7	17.3	34.0
Mixing and blending machine operators	112	8.1	17.9	11.7
Separating, filtering, and clarifying machine operators	62	3.7	10.1	15.8
Painting and paint spraying machine operators	187	12.4	9.0	20.3
Furnace, kiln, and oven operators, except food	53	5.8	8.9	4.2
Slicing and cutting machine operators	149	29.1	14.6	25.9
Photographic process machine operators	103	62.6	9.8	11.4
Fabricators, assemblers, and hand working occupations	2,070	33.5	13.9	17.1
Welders and cutters	594	4.9	10.2	15.3
Assemblers	1,299	44.6	16.4	18.2
Production inspectors, testers, samplers, and weighers	703	48.5	16.3	17.3
Production inspectors, checkers, and examiners	497	48.0	14.9	13.2
Production testers	64	29.6	9.8	10.7
Graders and sorters, except agricultural	134	58.9	25.3	36.1
Transportation and material moving occupations	5,557	10.0	16.5	11.9
Motor vehicle operators	4,222	11.5	16.7	11.8
Supervisors	77	18.6	15.0	10.8
Truck drivers	3,088	4.7	14.4	12.5
Drivers-sales workers	167	10.5	11.2	6.8
Bus drivers	539	49.6	26.1	8.0
Taxicab drivers and chauffeurs	280	10.8	26.0	14.0
Parking lot attendants	60	16.2	27.8	16.6
Transportation occupations, except motor vehicles	183	3.5	13.7	3.6
Rail transportation	127	4.2	15.8	5.1
Locomotive operating occupations	63	1.7	15.6	6.9
Water transportation	56	1.8	9.1	.2
Material moving equipment operators	1,152	5.4	16.0	13.7
Operating engineers	253	1.7	12.8	10.4
Crane and tower operators	70	.9	7.6	4.6
Excavating and loading machine operators	98	2.9	4.6	9.7
Grader, dozer, and scraper operators	52	5.5	4.8	4.5
Industrial truck and tractor equipment operators	569	7.0	22.1	18.4
Handlers, equipment cleaners, helpers, and laborers	5,443	19.8	15.3	20.7
Helpers, construction and extractive occupations	120	5.9	7.8	29.6
Helpers, construction trades	111	4.2	8.4	30.4

Occupation	Total employed (thousands)	Percent of total		
		Women	Black	Hispanic origin
Construction laborers	1,015	3.8	11.2	27.7
Production helpers	75	23.8	18.5	20.7
Freight, stock, and material handlers	2,015	22.4	17.7	14.6
Garbage collectors	54	3.8	44.0	11.6
Stock handlers and baggers	1,125	30.0	12.8	14.1
Machine feeders and offbearers	82	30.3	15.4	11.6
Freight, stock, and material handlers, n.e.c.	739	11.6	23.3	15.9
Garage and service station related occupations	184	7.7	8.0	16.1
Vehicle washers and equipment cleaners	313	13.8	17.6	28.2
Hand packers and packagers	366	63.2	15.9	34.7
Laborers, except construction	1,307	20.8	15.5	18.4
Farming, forestry, and fishing	3,399	20.6	4.9	23.7
Farm operators and managers	1,125	25.4	.9	3.0
Farmers, except horticultural	879	27.3	.6	1.7
Horticultural specialty farmers	69	12.8	4.5	11.9
Managers, farms, except horticultural	149	22.3	.7	5.6
Other agricultural and related occupations	2,115	18.9	7.1	36.1
Farm occupations, except managerial	847	19.5	4.7	47.6
Farm workers	768	18.7	4.7	47.4
Related agricultural occupations	1,268	18.5	8.7	28.4
Supervisors	174	9.3	2.4	15.9
Groundskeepers and gardeners, except farm	870	7.4	10.9	30.9
Animal caretakers, except farm	148	70.7	3.7	9.0
Graders and sorters, agricultural products	68	68.3	8.6	70.9
Forestry and logging occupations	109	8.4	4.4	7.8
Timber cutting and logging occupations	66	4.0	6.3	4.8
Fishers, hunters, and trappers	51	11.9	3.6	2.0

NOTE: Generally, data for occupations with fewer than 50,000 employed are not published separately but are included in the totals for the appropriate categories shown. Beginning in January 2000, data reflect revised population controls used in the household survey.

SOURCE: U.S. Department of Labor, Bureau of Labor Statistics, 2001, *http://www.bls.bov/pdf/cpsantll.pdf.*

Glossary

achieved characteristics   characteristics of workers that are developed through a worker's choices and experiences, such as level of education

advanced industrial society   a society with highly productive extractive and manufacturing sectors and a large labor-intensive service sector; also called "postindustrial society"

affirmative action   government-mandated hiring procedures that attempt to compensate for past discrimination through hiring goals, preferential consideration among otherwise equal candidates, or active recruitment of women or minority workers

alienating work   work that provides inadequate sustenance, meaning, and identity

American Federation of Labor (AFL)   oldest existing union organization in North America; serves as an umbrella organization enrolling member unions and coordinating their activities

apartheid   a strict South African system of racial segregation in the workplace and in other settings such as restaurants, beaches, and public facilities; ended in 1994

apprenticeship   a combination of classroom instruction and supervised on-the-job training that typically lasts from two to six years

aquaculture   the raising of fish in ponds or in large floating cages in the sea

artisan   a specialist in a skilled trade; used especially to refer to skilled workers in medieval cities

ascribed characteristics   characteristics with which workers were born and therefore cannot easily alter, such as gender or race

assembly line   a power-driven system that moves objects past workers as each does a small repetitive task toward the assembly of a finished product

autonomous work groups   Swedish industrial system in which significant power over procedures is held by groups of 12 to 15 employees in each area of the factory

backward linkages   control by a firm of important sources of supplies

bureaucracy   a hierarchical power arrangement based on fixed areas of expertise and written rules

CAD   "computer-assisted design" involves the use of computer-generated part designs to replace original designs by engineers

cafeteria-style benefit packages   a range of benefits from which workers can select a package best suited to their needs

capital mobility   the easy movement of money to fund investments around nations and around the globe

career   the sequence of events within a person's work history

carpal tunnel syndrome   a chronic inflammation of the nerves in the wrist due to repeated overuse

cartel   a group of political or economic organizations that act to limit price competition among themselves

case study   a study of a workplace that uses multiple sources of information and seeks to understand the perspectives of multiple actors in the workplace

cash cow   a profitable company acquired as a subsidiary to generate cash for expansion in other areas

centralization of control   concentration of power by the top executives in an organization and by the largest companies in the economy

central planning   detailed planning for economic output goals by the central government

chemical sensitization   heightened allergic reactions due to extended exposure to toxic chemicals at work

chronic stress injuries   injuries of the bones and soft tissues due to using improper equipment, assuming unnatural postures, or performing repetitious tasks; common varieties include back injuries and carpal tunnel syndrome

client firm   a firm that is dependent for most of its business on another, usually larger firm

codetermination   the sharing of economic power in Germany by large corporations, the government, unions, and works councils of employees in each factory

collective bargaining   bargaining between an employer and workers over wages, benefits, and conditions of employment through use of a union as a representative of the workers

combined and uneven development a highly developed export sector in combination with underdevelopment of the domestic consumption sector

commitment an attachment between workers and their employer that develops when workers perceive that the goals and values of their employer are compatible with their own goals and values

commodification of agriculture the movement from agricultural production for local consumption to crops sold on commodity markets

commodity chains the geographic, financial, and social linkages that trace raw products and subcomponents as they are combined into finished goods

company unions unions that are dominated by management and do not provide an autonomous voice for workers

comparable worth discrimination occupational differences in pay that include unequal rewards for equal qualifications and job demands between jobs that have typically been filled by men and those that have typically been filled by women

comparison group a specific group used as a standard of comparison; important for evaluating one's own level of job satisfaction

complexity the level of manual, interpersonal, or mental skills required by a job

concentration ratio the percentage of sales in an industry by the largest firms

conglomerate a firm operating in several nonoverlapping industries

Congress of Industrial Organizations (CIO) splintered from the AFL in 1935 to organize semiskilled factory workers in the newly emerging mass-production industries; reintegrated in 1955 to form the AFL-CIO

contingency theory argues that organizational structure results from the environment that the organization faces

contingent employment nonpermanent employment whose continuity depends upon the availability of work

continuous improvement the key goal of Japanese Quality Control Circles

continuous learning ongoing education and training throughout people's work lives

continuous-process technology a production process involving a continuous flow of product through the manufacturing stages, such as in oil refineries or chemical plants

core and peripheral jobs and industries in dual labor market theory, the principal division that separates stable employment (core) from unstable employment (peripheral)

core nation industrialized nations that sell manufactured goods and buy raw products from peripheral nations

corporate accountability the effort to hold corporations accountable for illegal activities such as price-fixing, environmental pollution, false advertising, bribery, tax evasion, political payoffs, and the sale of unsafe products

corporate campaign a union pressure tactic that targets financial backers of the company, consumers, and the public through informational campaigns about the corporation's undesirable behaviors

corporate divisions semi-independent product lines within a large corporation

corporate economic power the power of large companies to exclude smaller competitors from the market and influence overall rates of economic growth

corporate inflexibility an inability to change rapidly, adapt to new situations, or innovate resulting from excessive management hierarchy and bureaucracy

corporate mergers corporate growth through the acquisition of other firms

corporate political power corporate power exercised through campaign contributions to promote legislation favorable to the industry; also frequently exercised through industry control and staffing of the governmental bodies intended to regulate the industry

craft technology skilled handicraft production based on the manufacture and use of specialized tools for each trade

craft union association of skilled workers in a given craft or trade

cultural division of labor historical and cultural patterns that channel minorities and majorities and men and women toward different occupational roles

cyclical unemployment the gap between available work and people needing work during a temporary economic downturn

data stewards union show stewards in Sweden whose job is to keep abreast of the latest technology, consult with the company concerning its deployment, and protect workers' legislative and contractual rights

dead-end jobs entry-level jobs that require relatively little skill, often have a high turnover, and rarely lead to promotions

debt crisis the heavy burden of debt acquired by less developed nations in frequently unsuccessful attempts at industrialization; high interest payments further undercut development

declining middle class a reduction in the proportion of middle-income earners due to decreased job security and stagnant or falling wages

demographic characteristics the characteristics of workers, such as age, gender, race, or level of education

dependency theory argues that many less industrially developed nations are thoroughly incorporated in the world economy but in a dependent role

dependent development a pattern of industrialization in which a nation's manufacturing activities fill niches as determined by the needs of stronger nations and multinational companies

deprofessionalization the process of weakening or eliminating the professional characteristics of an occupational group

developing nations nations with some industrialization; often serve as subcomponent manufacturers for large multinational corporations

deviant occupations occupations whose job content is defined as illegal or immoral

deskilling a reduction in needed job skills due to automation and job simplification

direct personal control the company owner shows the workers how to do the work, dictates the appropriate pace, and evaluates the workers' performances

discouraged workers people who give up actively seeking employment because of discouragement and a perceived absence of opportunities

dispersion of information the thesis that computer technology tends to make information more widely available

diversification movement of a company into new but technologically similar product lines

division of labor the division of work into its component tasks and the specialization of workers in these different tasks

downsizing reduction in a company's workforce; sometimes achieved through increased outsourcing or the movement of work to cheaper locations

dual labor markets job markets that operate to preserve jobs with better pay and greater security for dominant groups

economies of scale cost savings for large producers resulting from lower per-unit costs for large batches

electronic surveillance the monitoring of the activities of individual workers through their interface with a computer

emotional work the process of managing one's emotional responses in interactions with customers or supervisors

employer commitment a sincere employer pledge to respect the rights of workers, their job security, and their benefits

enclosure movement the fencing of grazing land previously held in common by peasants and landlords for the exclusive use of the landlord in raising sheep for wool or other cash crops

entry ports entry-level jobs that offer the worker the possibility for training and promotion

environmental degradation erosion of the quality of the air, water, or landscape by dangerous industrial practices and insufficient attention to reducing or containing waste products

ESOPs employee stock ownership plans which entail partial worker ownership, sometimes with only limited rights of participation and control

establishment the physical building and location of an organization; one company may have many different establishments

establishment survey a survey in which the resonding units are workplaces

ethnography (as used in this text) a careful analysis of a work situation by a knowledgeable observer who may be a participant or nonparticipant in the workplace

expanded leisure reducing the hours of work as a way of improving the quality of work life and distributing available employment

experimental bias a change in the results of an experiment or an observation that results from workers changing their behaviors in response to the presence of the experimenter or observer

externalization of costs the ability of companies to lower their expenses by unlawfully dumping dangerous chemical wastes, continuing dangerous work practices, producing unsafe products, and bargaining with local governments for tax abatements

extractive industries industries such as agriculture, forestry, fishing, and mining in which a raw product is taken from the environment

extrinsic rewards rewards that are realized off the job, including such things as pay, fringe benefits, and job security

fair trade a doctrine that argues that "free trade" must also include common standards for the rights of workers and for environmental protection

family life cycle the stages of formation, growth, and dissolution of the nuclear family

feminization of clerical work process by which clerical work changed from predominantly male to predominantly female workers

feudal society agricultural societies in which peasants were obligated to stay on the land and to give a share of their crops to the wealthy land owners

firm an employing organization, generally incorporated as a legal entity

flexibility corporate strategy based on downsizing to facilitate more rapid matching of resources to market opportunities

Fordism the production of large numbers of standardized products using power-driven equipment and assembly lines; named after Henry Ford, a pioneer in this production strategy

foreman's control a hired foreman performs the duties of recruiting, training, and supervising workers

forward linkages control by a firm of the next manufacturing stage for its product, potentially including wholesale or retail marketing of its products or services

four tigers the nations of Hong Kong, Taiwan, South Korea, and Singapore in recognition of their rapid industrial development

free agent firms small firms that occupy less profitable market niches unexploited by larger corporations

free trade international trade unrestricted by tariffs or quotas

frictional unemployment unemployment that results from unavoidable delays between jobs, such as when a person moves to a new location

fringe benefits nonwage compensation and perquisites, such as pensions, health and other forms of insurance, sick leave, and paid vacation

gender-typing the categorization of certain occupations as appropriate for women (or men) because women (or men) have historically been responsible for these types of activities

general strike coordinated simultaneous strikes involving several trades

general union a union that enrolls all workers, regardless of occupation or industry

golden parachutes employment contracts that guarantee generous severance packages to top executives if their firm is acquired

grievance procedure a set of procedures for handling workers' complaints about violations of their rights under an employment contract

guilds medieval associations of skilled workers for regulating the hours and quality standards of an occupation; precursors to modern trade unions and professional associations

Hawthorne effect the observation that job redesign tends to increase short-term productivity, regardless of its content, because it entails added social attention; see *experimental bias*

high-technology industry industries with high proportions of engineers and scientists, such as electronics, ordnance, chemicals, pharmaceuticals, and genetic engineering

human capital the valuable productive skills possessed by workers

human relations school a tradition that argues that a supportive social context maximizes workers' productivity

ideology a system of ideas that specifies a given economic and political arrangement as appropriate and desirable

import substitution a development strategy in which a nation attempts to develop industries that produce goods it has previously imported

income squeeze the phenomenon that families often have greater income needs at the career stage where the worker's salary increases tend to flatten out

indentured labor workers under contract to work for a certain amount of time—typically eight to ten years—for a set price or as part of their penalty for being found guilty for a crime such as petty theft or vagabondage

indoor air pollution dust, chemicals, gases, molds, and bacteria trapped inside airtight or poorly ventilated buildings

Industrial Revolution the often violent transition between subsistence agriculture and the widespread production of manufactured goods for sale

industrial union a union that organizes all the various trades in a given industry

Industrial Workers of the World a radical general union active in North America in the early 1900s; supported replacement of capitalists by committees of workers

industry a branch of economic activity that is devoted to the production of a particular good or service

informal sector small independent artisans, shopkeepers, and street vendors who produce and sell goods for local markets

informal work culture social relationships and expectations that emerge among the people who work together

injunction a court order to cease and desist a specific action, such as a strike; frequently used against unions well into the twentieth century and even occasionally today

innovative sector industries in which heightened international competition, technological change, and increased worker participation lead to organizational innovations and increased productivity

inside game union pressure tactic that operates within the establishment during normal operations to pressure management; includes rigidly following rules and conspicuous displays of symbols of worker solidarity

instrumental orientation a view that the primary purpose of work is to provide for family and leisure needs

intergenerational mobility a child's attainment of an occupation of greater or lesser status than that of his or her parents

interlocking directorates a set of companies with overlapping boards of directors; facilitates oligopolistic practices

internal democracy a competitive party system for determining union officers; relatively uncommon in unions

internal labor markets job ladders within organizations or occupations

international labor organizations umbrella organizations involved in supporting and coordinating union activities around the world

intragenerational mobility a change in occupations or jobs within the career of a single worker

intrinsic rewards　rewards that are realized while at work, including such things as the freedom to plan one's own work and the chance to use one's abilities

inverted U-curve of technological development and alienation　the observation that alienation increases following the displacement of craft organizations of work by machine pacing and assembly-line technologies, but that it decreases again under more advanced continuous-process technologies

job autonomy　extent to which workers control their own job tasks and organization of work

job diversity　the number of different tasks and responsibilities required by a job

job enlargement　process of varying job content through additional tasks, often requiring additional skills

job enrichment　process of improving jobs through developing higher-level skills in workers

job instability　fluctuating wages or hours and periods of unemployment associated with specific jobs

job redesign　change in job content and organization; key current components include worker participation and team production systems

job satisfaction　the summary evaluation that people make of their jobs

job security　a reasonable expectation that workers can keep their jobs over time

job stress　chronic tension resulting from job conditions such as noise, time pressure, abusive or incompetent bosses, excessive responsibilities, and long hours

job–worker fit　the degree of fit between a worker's values and needs and the content of his or her job

joint union–management initiatives　jointly sponsored union–management programs that include employee involvement, such as new training programs

journeyman　a skilled craftworker who works as an employee rather than opening his or her own business

Knights of Labor　a progressive general union in North America active in the decades before 1900; supported immigrants and women

labor force　all of the employed and unemployed workers in a geographic area at a specific time

labor force participation rate　the percentage of people eligible to be counted as part of the labor force who are either employed or unemployed

layoffs　large-scale temporary or permanent employment terminations because of falling sales, downsizings, or plant closings

least developed nations　nations characterized by subsistence agriculture and raw material production for export

life cycle　the ordering of roles in a human life beginning with infancy and ending with death

lifetime employment　guaranteed job security for life for workers as a strategy for securing commitment and extra effort; popular in Japan

limited liability　corporate liabilities such as debts, bankruptcy, and lawsuits are limited to the corporation itself, rather than extending to its owners

long-term and indirect effects　indirect consequences of technology that may generate job growth through increased productivity and economic expansion

loyal opposition firms　smaller firms that provide competition to larger corporations in their own industries

macroplanning　coordination between corporations, banks, governments, and educational institutions to target certain industries for development

major occupational group　broad categories of occupations

making out　creative efforts by workers to organize their tasks to get the work done without exhausting themselves in the process

management-controlled firms　corporations without large controlling stockholders; differences in firm behavior between management-controlled firms and owner-controlled firms are an object of debate

manufacturing industries　industries that process raw products into finished goods

maquiladoras　U.S. factories on Mexico's northern border that manufacture goods for export to the United States

marginal jobs　jobs generally viewed as inadequate because they do meet the community norms for wages, hours, or use of worker skills; see *underemployment*

marginal sector　industries in which a lack of worker power allows employers to respond to heightened international competition and technological change by reducing labor costs

mass-production technology　production of large batches of products through the integration of specialized tools and equipment in a mechanized production operation

mass strike　a coordinated general strike of workers with support from the general population through demonstrations, marches, and insurrections

matrix organization　a bureaucratic system in which each worker reports to a division supervisor (e.g., electrical engineering) and to a project supervisor (e.g., a certain aircraft project)

merchant capitalism　the earliest form of capitalism based on buying and selling finished goods and agricultural products in intercity market networks

meritocracy　a system in which rewards are based on achievement

microchip technology　computer technology based on circuits chemically etched on miniaturized silicon chips

middle managers　layers of management above first-line and below executives

migrant farm workers farm workers who move with the agricultural cycle and work for many employers in the course of the season

mini-mills small steel mills that use electric furnaces to turn scrap into basic steel products

MIS a computerized "management information system" that records production information and potentially monitors individual output and workers' activities

modernization theory argues that societies advance from agricultural production to industrial production based largely on internal dynamics

monopoly exclusive or dominant control of a market

multinational corporation a firm that operates in many countries

multiplant companies companies with establishments in many different locations

National Labor Relations Act (1935) lays groundwork for union–management relations in the United States; provides for secret ballot vote by workers for or against union representation

NC and CNC systems that operate metalworking lathes, drills, and cutting tools automatically from numeric codes (NC) or computer-aided numeric control (CNC)

neo-imperialism the use of economic dominance, rather than direct military subjugation, to control and exploit poorer nations

NILF abbreviation for "not in the labor force"

nonparticipant observation a type of ethnography in which a trained researcher observes but does not actually become a part of the work group

norm a rule that a group develops for thinking, feeling, or behaving

occupation a cluster of job-related activities constituting a single economic role that is usually directed toward making a living

occupational communities distinct occupational subcultures that provide their members with a shared identity

occupational disease a disease resulting from improper working conditions or exposure to toxins at work

occupational segregation the channelling of minority groups or women into restricted sets of occupations

oligopoly a market controlled by few enough firms that they can engage in collaborative price-fixing

on-the-job training formal and informal training that occurs at the workplace

opportunity structure the occupations and jobs that are available within a society at a particular point in time

organizational structure the architecture or pattern of formal and informal relations in an organization

paper entrepreneurialism buying and selling companies and manipulating balance sheets to realize profits from the sale of productive assets rather than from the sale of products

paraprofession an occupation that assists an established profession

parent company a firm that owns and controls another firm

payment in kind payment in goods instead of money

peripheral jobs and industries see *core and peripheral jobs and industries*

peripheral nation nations that sell raw materials and buy manufactured goods from industrialized nations

plant closings plant shutdowns due to lost competitiveness or to the movement to cheaper locations

population the group to which a researcher wishes to generalize the results of a randomly selected sample

postbureaucratic possible organizational structure of the future in which hierarchy and bureaucracy are replaced by worker participation and innovation

postindustrial society a society in which high manufacturing productivity allows more and more workers to be engaged in service provision

predatory pricing intentional short-term underbidding of competitors by large corporations to drive them out of the market

primitive accumulation the seizure of wealth from the European colonial empires in Africa, the Americas, and the Pacific in the form of precious metals, trade goods, and slaves; important for financing European industrialization

profession a high-status occupation marked by abstract knowledge, authority, autonomy, and altruism

professional culture the characteristic jargon, behaviors, and lifestyles associated with the practitioners of a profession

professionalization a process by which an occupation seeks to more closely resemble a profession

protectionism the protection of domestic industry through high tariffs or restrictive limits on foreign imports

Protestant work ethic a religious view that identifies success in the pursuit of one's occupational calling with spiritual grace

public goods products or services to which the citizens of a society are entitled without direct payment or at a nominal fee, such as public schools and parks

putting-out system a system of production in which a merchant capitalist "puts out" raw materials to be worked up by people in their homes and later collects the finished products to be sold

Quality Control Circles Japanese system in which work groups identify production techniques that can be changed to improve quality and productivity

quality of work life a set of factors that make for satisfying and fulfilling work, including adequate pay, safety, and autonomy

rationalization   the process by which ends are more closely fitted to means

reliability   a characteristic of research methods that produce the same results repeatedly or that produce the same results when used by a different researcher

response error   errors in sample surveys that result from vague or erroneous answers

right-to-know legislation   laws allowing workers to be informed about the chemical hazards in their workplaces

role   a set of behaviors or expectations associated with a particular position in society

sabotage   intentional acts of disruption and destruction against employers for vengeance or leverage

sample   a mathematically selected subgroup that is representative of a population

sample survey   a research technique that uses questionnaires to seek information from a representative group of people

satellite firms   dependent firms engaged in subcontracting relationships with larger firms

scientific management   the idea that there is one best way to do every task and that this way can be discovered through scientific techniques of observation and experimentation

secondary boycott   unions supporting each other by refusing to handle struck goods

sectoral transformation of the labor force   the shift over time in the principal industry of employment from agricultural through manufacturing to services

self-actualization   the realization of one's broadest human needs through work

semiprofession   an occupation that has achieved some of the characteristics of a profession

semiskilled work   jobs requiring less than two weeks of formal training

seniority   allocating protection from layoffs, first choice of shifts, access to training programs, and so forth, on the basis of years of service

service industries   industries that provide useful activities instead of goods, including personal service (waitressing, hairdressing, entertainment), educational services (teaching), medical services (nurses, doctors), and business services (accounting)

services   acts provided in return for payment

severance pay   a lump-sum payment to workers at the time of a permanent layoff; paid only if mandated by an union agreement or other employment contract

sexual harassment   repeated, unwelcome behavior with a sexual content when submission to such behavior is explicitly or implicitly a condition for the person's hiring or for other employment decisions, or when such behavior creates a hostile, intimidating, or offensive working environment

shadow economy   production of goods and services that is not in cash, is not taxed, or is unregulated by the government

sharecropper   an agricultural worker whose pay consists of a portion of the harvest

shop steward   a worker who is allowed under a collective bargaining agreement limited time away from work to inform fellow workers of their contractual rights and to handle their grievances about violations of these rights

shrinking middle class   hypothesis that the class structure tends to increase the wealthy and the poor at the expense of the middle of the class structure

sit-down strike   a strike in which workers stay in their places but stop working; prevents the use of replacement workers

skilled craftwork   trades requiring extensive training and experience such as machinists, mechanics, and electricians

social classes   groups in society defined by the possession of different amounts of wealth and power

socialization   the process of learning norms, roles, and skills through either formal or informal instruction

social relations of production   the rights and obligations of people toward each other at every level of production

social stratification   the unequal distribution of rewards and power in society

social structure   the architecture of social relations in formal and informal groups and in the larger society

soldiering   intentionally working well below one's capacity

solidarity   mutual defense and support in times of crisis or challenge

span of control   the number of persons that one manager supervises

staff and line positions   bureaucratic positions entailing ancillary expertise (staff positions) versus chain of command (line positions)

state-regulated capitalism   an economic system in which large corporations hold economic power but the state regulates their activities to protect the interests of the nation as a whole; includes the United States, Canada, and most industrialized nations

statistical discrimination   discrimination that arises when an employer bases hiring or promotion decisions on the *average* qualifications of a group, rather than on an individual's qualifications

steepness of hierarchy   the number of levels of supervision

structural unemployment   the chronic gap between the number of jobs the economy provides and the number of people seeking work

subcontracting (outsourcing) purchasing goods and services that could be made by the company from outside sources

subminimum wage a wage below the minimum wage that may be legally paid to workers in specified categories (such as teenagers)

subsidiary a firm that is owned by another firm

Taft-Hartley Amendments (1947) amendments to the National Labor Relations Act that outlaw secondary boycotts, allow states to outlaw union shops, and allow the president to declare a state of emergency and force workers and employers to resume production

technical control control of work tasks by machine pacing and job design as on an assembly line

technological determinism a theory that organizational structure and the nature of work are rigidly determined by technology

technological displacement elimination of jobs resulting from the introduction of new technologies; a current concern because of the widespread application of microprocessor technologies

technology knowledge, skills, and tools used for the achievement of practical purposes

telecommuting doing information processing work at home on a personal computer or remote terminal

tertiarization the development of a service economy without a manufacturing base, usually in a developing country

Theory X argues that employees work hard only if coerced (by threat of firing) or bribed (by promises of pay raises) into doing so

Theory Y argues that workers will be more productive if they receive more humane consideration and attention

Theory Z views productivity as embedded in workers, in their skills, and in their cooperative attitudes

tokenism the hiring of a few minorities or women in an occupation as token adherence to diversity goals

trade association an organization of firms within the same industry

transfer prices the prices that large corporations charge when a product is transferred from one division to another

underclass an economic class with access only to marginal employment at best

underemployment employment that is inadequate for the worker in terms of income, hours or work, or use of the worker's skills

unemployed persons labor force participants who are not at work for even one hour during the reference week, who are actively seeking work or awaiting results of a recent search, and who are available to work

unemployment compensation limited-term payments (typically twenty-six weeks) to workers by states following layoffs

unemployment rate the proportion of the labor force who are unemployed at a specific time

unskilled work jobs requiring little or no formal training

validity a characteristic of research methods that produce information that fits closely with reality

VDT a video display terminal for a personal computer or operated as a remote station from a mainframe computer

vertical integration control by a firm of both its supplies (backward linkages) and marketing (forward linkages)

vocational education education specifically targeted toward usable workplace skills; typically one or two years in duration or as part of a longer apprenticeship program

worker education ongoing adult and on-the-job education necessary for the innovative workplace of the future

worker ownership majority ownership of a company by its employees

worker participation the inclusion of workers or their representatives in all levels of organizational decision making

worker power worker participation in economic decisions within the firm; essential for generating change and innovation

worker resistance intentional withdrawal of cooperation by workers ranging from soldiering to sabotage

worker retraining additional education for workers already in the labor force to train them in new areas or occupations

working-class culture key elements include pride in doing quality work, fears of economic insecurity, and solidarity with co-workers

working-class ideology a worldview common in working-class occupations that stresses solidarity with other workers against the more powerful capitalist and managerial classes

workplace experimentation a sustained process of developing and evaluating new forms of work organization

works councils worker representatives in Germany who sit on management boards and represent workers in the day-to-day affairs of the enterprise

world systems theory argues that the world is an integrated economic system with difficult-to-change roles for peripheral nations

References

Abbott, Andrew. 1988. *The System of Professions: An Essay on the Division of Expert Labor.* Chicago: University of Chicago Press.

Abbott, Andrew. 1989. "The new occupational structure: What are the questions?" *Work and Occupations* 16 (Fall): 273–291.

Abouzeid, Aushra. 1993. "Spies on-line: Computers keep tabs on American workers." *In These Times* (June 28): 8–9.

Adler, Paul A. 1984. "Tools for resistance: Workers can make automation their ally." *Dollars and Sense* 100 (October): 7–8.

Adler, Paul S., and Bryan Borys. 1996. "Two Types of Bureaucracy: Enabling and Coercive." *Administrative Science Quarterly* 41,1 (March): 61–89.

Adler, William M. 2000. *Mollie's Job: A Story of Life and Work on the Global Assembly Line.* New York: Scribner.

AFL-CIO. 1985. *The Changing Situation of Workers and Their Unions.* Washington, D.C.: AFL-CIO.

Aminzade, Ronald. 1993. *Ballots and Barricades.* Princeton, N.J.: Princeton University.

Anderson, John. 1989. "The Texas 100." *Texas Monthly* 17,8 (August): 111–171.

Apostle, Richard, and Gene Barrett. 1992. *Emptying Their Nets.* Toronto: University of Toronto Press.

Appelbaum, Eileen R. 1984. "The impact of technology on skill requirements and occupational structure in the insurance industry, 1960–90. Cited in Philip Kraft, "A review of empirical studies of the consequences of technological change on work and workers in the U.S." Washington, D.C.: National Research Council.

Appelbaum, Eileen. 1987. "Restructuring work: Temporary, part-time, and at-home employment." Pp. 268–310 in Heidi I. Hartmann (editor), *Computer Chips and Paper Clips: Technology and Women's Employment.* Vol. 1. Washington, D.C.: National Academy Press.

Appelbaum, Eileen, Thomas Bailey, Peter Berg, and Arne L. Kalleberg. 2000. *Manufacturing Advantage: Why High-Performance Work Systems Pay Off.* Ithaca, N.Y.: Industrial and Labor Relations Press.

Appelbaum, Eileen R., and Rosemary Batt. 1994. *The New American Workplace.* Ithaca, N.Y.: Industrial and Labor Relations Press.

Appier, Janis. 1998. *Policing Women: The Sexual Politics of Law Enforcement and the LAPD.* Philadelphia: Temple University Press.

Applebaum, Herbert. 1981. *Royal Blue: The Culture of Construction Workers.* New York: Holt, Rinehart and Winston.

Applebaum, Herbert. 1998. *The American Work Ethic and the Changing Work Force.* Westport, Conn.: Greenwood.

Applegath, John. 1982. "What's good about the home office?" *The Futurist* 16,3 (June): 46.

Argote, Linda. 1999. *Organizational Learning.* Boston: Kluwer.

Argyris, Chris. 1973. "Personality and organization theory revisited." *Administrative Science Quarterly* 18: 141–167.

Aronowitz, Stanley, and William DeFazio. 1994. *The Jobless Future.* Minneapolis, Minn.: University of Minnesota Press.

Ash, Mary Kay. 1981. *Mary Kay.* New York: Harper and Row.

Ash, Mary Kay. 1984. *Mary Kay on People Management.* New York: Warner.

Attewell, Paul. 1990. "What is skill?" *Work and Occupations* 17,4 (November): 422–448.

Auletta, Ken. 1982. *The Underclass.* New York: McGraw-Hill.

Averitt, Robert T. 1968. *The Dual Economy: The Dynamics of American Industrial Structure.* New York: Norton.

Bacon, David. 2000. "Tijuana troubles: NAFTA is failing Mexican workers." *In These Times,* August 21: 23.

Baker, Ted, and Howard E. Aldrich. 1996. "Prometheus stretches: Building identity and cumulative knowledge in multiemployer careers." Pp. 132–149 in Michael B. Arthur and Denise M. Rousseau (editors), *The Boundaryless Career.* New York: Oxford University Press.

Baldi, Stephanie, and Debra Branch McBrier. 1997. "Do the determinants of promotion differ for blacks and whites?" *Work and Occupations* 24: 478–497.

Baratz, Morton S. 1971. "Corporate giants and the power structure." Pp. 146–155 in Richard Gillam (editor), *Power in Post-War America.* Boston: Little Brown.

Barbash, Jack. 1975. "Unionizing low-paid workers." *Challenge* 18,3 (July–August): 39–43.

Barber, Bernard. 1965. "Some problems in the sociology of the professions." Pp. 15–34 in Kenneth S. Lynn and the editors of Daedalus (editors), *The Professions in America.* Boston: Houghton Mifflin.

Barker, James R. 1999. *The Discipline of Teamwork: Participation and Concertive Control.* Thousand Oaks, Calif.: Sage.

Barker, Kathleen, and Kathleen Christensen. 1998. *Contingent Work.* Ithaca, N.Y.: Industrial Relations Press.

Barnet, Richard J., and John Cavanagh. 1994. *Global Dreams: Imperial Corporations and the New World Order.* New York: Simon and Schuster.

Barnett, William, James N. Baron, and Toby E. Stuart. 2000. "Avenues of attainment." *American Journal of Sociology* 106,1 (July): 88–144.

Baron, James N., Brian S. Mittman, and Andrew E. Newman. 1991. "Targets of opportunity." *American Journal of Sociology* 96,6 (May): 1362–1401.

Baugher, John E., and J. Timmons Roberts. 1999. "Perceptions and worry about hazards at work: The U.S. petrochemical industry." *Industrial Relations* 38,4 (October): 522–541.

Baxter, Vern K. 1994. *Labor and Politics in the U.S. Postal Service.* New York: Plenum.

Beck, E. M., Patrick M. Horan, and Charles M. Tolbert, II. 1978. "Stratification in a dual economy: A sectoral model of earnings determination." *American Sociological Review* 43,5 (October): 704–720.

Becker, Howard S., Blanche Geer, Everett C. Hughes, and Anselm L. Strauss. 1961. *Boys in White: Student Culture in Medical School.* Chicago: University of Chicago Press.

Bednarzik, Robert W. 2000. "The role of entrepreneurship in U.S. and Canadian job growth." *Monthly Labor Review* 123,3 (July): 3–17.

Belanger, Jacques, Paul K. Edwards, and Larry Haiven. 1994. *Workplace Industrial Relations and the Global Challenge.* Ithaca, N.Y.: Industrial and Labor Relations Press.

Belanger, Marc. 1983. "The new electronic cottage." The Facts 5,7 (September). Toronto: Canadian Union of Public Employees.

Bell, Daniel. 1976. *The Coming of Post-Industrial Society.* New York: Basic Books.

Ben-David, Joseph. 1984. *The Scientist's Role in Society: A Comparative Study.* Chicago: University of Chicago Press.

Benokraitis, Nijole V., and Joe R. Feagin. 1986. *Modern Sexism: Blatant, Subtle and Covert Discrimination.* Englewood Cliffs, N.J.: Prentice Hall.

Benson, Susan Porter. 1986. *Counter Cultures: Saleswomen, Managers, and Customers in American Department Stores, 1890–1940.* Urbana, Ill.: University of Illinois Press.

Berggren, Christian. 1992. *Alternatives to Lean Production: Work Organization in the Swedish Auto Industry.* Ithaca, N.Y.: Industrial and Labor Relations Press.

Bergmann, Joachim, and Tokunaga Shigeyoshi. 1987. *Economic and Social Aspects of Industrial Relations.* Frankfurt: Campus Verlag.

Bergsten, C. Fred. 1996. *Dilemmas of the Dollar: The Economic and Politics of U.S. International Monetary Policy.* Armonk, N.Y.: M.E. Sharpe.

Berle, Adolf A., and Gardiner C. Means. 1932. *The Modern Corporation and Private Property.* New York: Harcourt Brace Jovanovich.

Berman, Daniel M. 1986. "The official body count." Pp. 124–135 in Lauri Perman (editor), *Work in Modern Society.* Dubuque, Iowa: Kendall-Hunt.

Bernstein, Dennis, and Connie Blitt. 1986. "Lethal dose." *The Progressive* 50,32 (March): 22–25.

Bernstein, Paul. 1997. *American Work Values: Their Origin and Development.* Albany: State University of New York Press.

Bersoff, David, and Faye Crosby. 1984. "Job satisfaction and family status." *Personality and Social Psychology Bulletin* 10,1 (March): 79–83.

Besser, Terry L. 1996. *Team Toyota: Transplanting the Toyota Culture to the Camry Plant in Kentucky.* Albany, N.Y.: State University of New York Press.

Best, Fred, and Barry Stern. 1977. "Education, work, and leisure: Must they come in that order?" *Monthly Labor Review* 100,7 (July): 3–10.

Bianchi, Suzanne M. 1995. "The Changing Demographic and Socioeconomic Characteristics of Single-Parent Families." *Marriage and Family Review* 20, 1–2, 71–97.

Biggart, Nicole Woolsey. 1989. *Charismatic Capitalism: Direct Selling Organizations in America.* Chicago: University of Chicago Press.

Bijker, Wiebe E., Thomas P. Hughes, and Trevor Pinch (editors). 1987. *The Social Construction of Technological Systems.* Cambridge, Mass.: MIT Press.

Bills, David B. (editor). 1995. *The New Modern Times: Factors Reshaping the World of Work.* Albany, N.Y.: State University of New York Press.

Birch, David. 1981. "Generating new jobs: Are government incentives effective?" Pp. 10–16 in Robert Friedman and William Schweke (editors), *Expanding the Opportunity to Produce: Revitalizing the American Economy through New Enterprise Development.* Washington, D.C.: Corporation for Enterprise Development.

Bittman, Michael, and Judy Wajcman. 2000. "The rush hour: The character of leisure." *Social Forces* 79,1 (September): 165–189.

Blair-Loy, M. 1999. "Career patterns of executive women in finance." *American Journal of Sociology* 104,5: 1346–1397.

Blake, Robert R., and Jane Srygley Mouton. 1982. "Theory and research for developing a science of leadership." *Journal of Applied Behavioral Science* 18,3: 275–291.

Blanchard, Francis. 1984. "Technology, work and society: Some pointers from ILO research." *International Labour Review* 123,3 (May–June): 267–276.

Blau, Francine D., Marianne A. Ferber, and Anne E. Winkler. 1998. *The Economics of Women, Men, and Work.* Upper Saddle River, N.J.: Prentice Hall.

Blau, Judith R. 1993. *Social Contracts and Economic Markets.* New York: Plenum.

Blau, Peter M., and Otis Dudley Duncan. 1967. *The American Occupational Structure.* New York: Wiley.

Blau, Peter M., and Marshall W. Meyer. 1971. *Bureaucracy in Modern Society,* 2nd ed. New York: Random House.

Blauner, Robert. 1964. *Alienation and Freedom.* Chicago: University of Chicago Press.

Bloom, David E., and Adi Brender. 1993. "Labor and the emerging world economy." *Population Bulletin* 48,2 (October).

Bluestone, Barry, and Irving Bluestone. 1992. *Negotiating the Future.* New York: Basic Books.

Bluestone, Barry, and Bennett Harrison. 2000. *Growing Prosperity: The Battle for Growth with Equity in the Twenty-First Century.* Boston: Houghton Mifflin.

Blumberg, Paul. 1976. *Industrial Democracy.* New York: Schocken Press.

Bohen, Halcyone H., and Anamaria Viveros-Long. 1981. *Balancing Jobs and Family Life: Do Flexible Work Schedules Help?* Philadelphia: Temple University Press.

Bok, Derek C. 1993. *The Cost of Talent: How Executives and Professionals are Paid and How It Affects America.* New York: Free Press.

Booze, Regena Michelle. 2000. "Protecting our essence: Coping strategies of African American female faculty at predominantly white colleges and universities." *Dissertation Abstracts International, The Humanities and Social Sciences* 60,12 (June): 4618A–4619A.

Bornschier, Volker. 1983. "The division of labor, structural mobility, and class formation." Pp. 249–268 in Donald J. Treiman and Robert V. Robinson (editors), *Research in Social Stratification and Mobility.* Greenwich, Conn.: JAI Press.

Bornschier, Volker, and Christopher Chase-Dunn. 1985. *Transnational Corporations and Underdevelopment.* New York: Praeger.

Boswell, Terry, and Christopher Chase-Dunn. 2000. *The Spiral of Capitalism and Socialism: Toward Global Democracy.* Boulder, Colo.: Lynne Rienner.

Bowe, John, Marisa Bowe, and Sabin Streeter. 2000. *Gig: Americans Talk about Their Jobs.* New York: Crown Publishers.

Bowes, J. M., and J. J. Goodnow. 1996. "Work for home, school, or labor force." *Psychological Bulletin* 119,2 (March): 300–321.

Bowie, Chester E., Lawrence S. Cahoon, and Elizabeth A. Martin. 1993. "Evaluating changes in the estimates." *Monthly Labor Review* 116,9 (September): 29–33.

Bowles, Samuel, and Herbert Gintis. 1976. *Schooling in Capitalist America: Educational Reform and the Contradictions of Economic Life.* New York: Basic Books.

Bowles, Samuel, David M. Gordon, and Thomas E. Weisskopf. 1990. *After the Waste Land: A Democratic Economics for the Year 2000.* Armonk, N.Y.: M.E. Sharpe.

Boyer, Richard O., and Herbert M. Morais. 1955. *Labor's Untold Story.* New York: United Electrical, Radio and Machine Workers of America.

Bradley, Keith, Saul Estrin, and Simon Taylor. 1990. "Employee Ownership and Company Performance." *Industrial Relations* 29,3 (Fall): 385–402.

Bradshaw, York W., and Michael Wallace. 1996. *Global Inequalities.* Thousand Oaks, Calif.: Pine Forge.

Bradwell v. Illinois. 1872. 83 U.S. 130.

Branscomb, Lewis M., Fumio Kodama, and Richard Florida (editors). 1999. *Industrializing Knowledge: University-Industry Linkages in Japan and the United States.* Cambridge, Mass.: MIT Press.

Braun, Dennis D. 1997. *The Rich Get Richer,* 2nd ed. Chicago: Nelson-Hall.

Braverman, Harry. 1974. *Labor and Monopoly Capital.* New York: Monthly Review Press.

Brecher, Jeremy. 1972. *Strike!* Boston: South End Press.

Bregger, John E., and Cathryn S. Dippo. 1993. "Why is it necessary to change?" *Monthly Labor Review* 116,9 (September): 3–9.

Bridges, William P. 1982. "The sexual segregation of occupations: Theories of labor stratification in industry." *American Journal of Sociology* 88,2 (September): 270–295.

Bright, James R. 1966. "The relationship of increasing automation and skill requirements." Pp. 203–221 in National Commission on Technology, Automation and Economic Progress, *The Employment Impact of Technological Change.* Vol. 2, *Technology and the American Economy.* Washington, D.C.: U.S. Government Printing Office.

Brint, Stephen G. 1984. "New class and cumulative trend explanations of the liberal attitudes of professionals." *American Journal of Sociology* 90,1 (July): 30–71.

Brinton, Mary C. 1993. *Women and the Economic Miracle: Gender and Work in Postwar Japan.* Berkeley: University of California Press.

Britton, Dana. 1997. "Gendered organizational logic: Policy and practice in men's and women's prisons." *Gender and Society* 11: 796–818.

Brodeur, Paul. 1985. *Outrageous Misconduct: The Asbestos Industry on Trial.* New York: Pantheon.

Brody, David. 1980. *Workers in Industrial America.* New York: Oxford University Press.

Bronfenbrenner, Kate. 1992. "Successful union strategies for winning certification elections and first contracts." New Kensington, Penn.: Department of Labor Studies and Industrial Relations, Pennsylvania State University.

Bronfenbrenner, Kate (editor). 1998. *Organizing to Win.* Ithaca, N.Y.: Industrial Relations Press.

Brown, A. Radcliffe. 1922. *The Andaman Islanders.* Cambridge, England: Cambridge University Press.

Brown, Charles. 1982. "Dead-end jobs and youth unemployment." Pp. 427–444 in Richard B. Freeman and David A. Wise (editors), *The Youth Labor Market Problem.* Chicago: University of Chicago Press.

Browne, Irene (editor). 1999. *Latinas and African American Women at Work.* New York: Russell Sage Foundation.

Browning, Harley, and Joachim Singelmann. 1975. *The Emergence of a Service Society.* Report to the U.S. Department of Labor, Manpower Administration.

Brownstone, Ellen Gail. 2000. "A qualitative study of the occupational status and culture of dental hygiene in Canada." *Dissertation Abstracts International, The Humanities and Social Sciences* 60, 12 (June): 4627-A.

Bruce-Briggs, B. (editor). 1979. *The New Class?* New Brunswick, N.Y.: Transaction Books.

Bryant, Clifton D., and Kenneth B. Perkins. 1986. "Containing work disaffection: The poultry processing worker." Pp. 155–165 in Lauri Perman (editor), *Work in Modern Society.* Dubuque, Iowa: Kendall-Hunt.

Buchanan, David A. 1997. "The limitations and opportunities of business processing re-engineering in a politicized organizational climate." *Human Relations* 50,1 (January): 51–72.

Budry, Grace. 1997. *When Doctors Join Unions.* Ithaca, N.Y.: Cornell University Press.

Buhle, Paul (editor). 1985. *Labor's Joke Book.* St. Louis: W.D. Press.

Burawoy, Michael. 1985. *The Politics of Production.* London: New Left Books.

Burawoy, Michael (editor). 1991. *Ethnography Unbound.* Berkeley: University of California Press.

Burgan, John U. 1985. "Cyclical behavior of high tech industries." *Monthly Labor Review* 108,5 (May): 915.

Burris, Beverly H. 1983a. "The human effects of underemployment." *Social Problems* 31,1 (October): 96–110.

Burris, Beverly H. 1983b. *No Room at the Top.* New York: Praeger.

Burris, Beverly H. 1993. *Technocracy at Work.* Albany, N.Y.: State University of New York Press.

Burris, Beverly H. 1998. "Computerization of the workplace." *Annual Review of Sociology* 24: 141–157. Palo Alto, Calif.: Annual Reviews.

Butler, John S. 1991. *Entrepreneurship and Self-Help among Black Americans: A Reconsideration of Race and Economics.* Albany: State University of New York Press.

Butler, Suellen, and William Snizek. 1976. "The waitress–diner relationship: A multimethod approach to the study of subordinate influence." *Sociology of Work and Occupations* 3,2: 209–222.

Cain, Pamela S., and Donald J. Treiman. 1981. "The Dictionary of Occupational Titles as a source of occupational data." *American Sociological Review* 46,3 (June): 253–278.

Card, David, and Alan B. Krueger. 1995. "Time-series minimum-wage studies." *American Economic Review* 85,2 (May): 238–243.

Carey, Max L., and James C. Franklin. 1991. "Industry output and job growth continues slow into next century." *Monthly Labor Review* 114,11 (November): 45–63.

Carlin, Jerome. 1966. *Lawyers in the Making.* Chicago: National Opinion Research Center.

Carruthers, Bruce C., and Sarah L. Babb. 2000. *Economy/Society: Markets, Meanings and Social Structure.* Thousand Oaks, Calif.: Pine Forge.

Carter, Michael J., and Susan B. Carter. 1985. "Internal labor markets in retailing: The early years." *Industrial and Labor Relations Review* 38,4 (July): 586–598.

Casey, C. 1996. "Corporate transformations: Designer culture, designer employees and post-occupational solidarity." *Organization* 3,3 (August): 317–339.

Cassell, Joan. 1991. *Expected Miracles: Surgeons at Work.* Philadelphia: Temple University Press.

Cattan, Peter. 1991. "Child-care problems: An obstacle to work." *Monthly Labor Review* 114,10 (October): 3–9.

Cavendish, Ruth. 1982. *Women on the Line.* London: Routledge and Kegan Paul.

Chamot, Dennis, and Joan M. Baggett. 1979. *Silicon, Satellites and Robots: The Impacts of Technological Change on the Workplace.* Washington, D.C.: Department for Professional Employees, AFL-CIO.

Chamot, Dennis, and Michael D. Dymmel. 1981. *Coopération or Conflict: European Experiences with Technological Change at the Workplace.* Washington, D.C.: Department for Professional Employees, AFL-CIO.

Chamot, Dennis, and John L. Zalusky. 1983. "The electronic sweatshop: The use and misuse of work stations in the home. " Paper presented at the meeting of the National Academy of Science, November 9–10.

Chandler, Alfred D., Jr., Franco Amatori, and Takashi Kikino. 1997. *Big Business and the Wealth of Nations.* New York: Cambridge University Press.

Chaney, Elsa M., and Mary Garcia Castro (editors). 1989. *Muchachas No More: Household Workers in Latin America and the Caribbean.* Philadelphia: Temple University Press.

Chang, Clara, and Constance Sorrentino. 1991. "Union membership statistics in 12 countries." *Monthly Labor Review* 114,12 (December): 46–53.

Chase-Dunn, Christopher, Yukio Kawano, and Benjamin D. Brewer. 2000. "Trade globalization since 1795: Waves of integration in the world-system." *American Sociological Review* 65,1 (February): 77–95.

"Changing conditions lead more physicians to union." 1986. *Austin American Statesman,* April 15, p. 4.

Chen, May Ying. 1993. "Reaching for their rights: Asian workers in New York City." Pp. 133–150 in Glenn Adler and Doris Suarez (editors), *Union Voices.* Albany, N.Y.: State University of New York Press.

Cherlin, Andrew J., Frank F. Furstenberg, Jr., P. Lindsay Chase-Lansdale, Kathleen E. Kiernan, Philip K. Robins, Donna Ruane Morrison, and Julien O. Teitler. 1991. "Longitudinal studies on effects of divorce on children in Great Britain and the United States." *Science* 252,5011 (June 7): 1386–1390.

Cherry, Mike. 1974. *On High Steel: The Education of an Ironworker.* New York: Quadrangle.

Claes, Dag Harald. 2000. *The Politics of Oil-Producer Cooperation.* Boulder, Colo.: Westview.

Childe, Gordon V. 1964. *What Happened in History.* Baltimore: Penguin.

Chomsky, Noam. 1994. "Time bombs." *In These Times* (February 21): 14–17.

Cipolla, Carlo. 1980. *Before the Industrial Revolution,* 2nd ed. New York: W. W. Norton.

Clayton, R., and Jonas Pontusson. 1998. "Entitlement cuts, public sector restructuring, and inegalitarian trends in advanced capitalist societies." *World Politics* 51,1 (October): 67–79.

Cleveland, Harlan, and W.T. Anderson. 1999. "Transnational: A word whose time has come." *Futures* 31,9–10 (November): 879–885.

Clogg, Clifford C. 1979. *Measuring Underemployment*. New York: Academic Press.

Clogg, Clifford C., and Teresa A. Sullivan. 1983. "Labor force composition and underemployment trends, 1969–1980." *Social Indicators Research* 12,2 (February): 117–152.

Cockburn, Cynthia. 1983. *Brothers: Male Dominance and Technological Change*. London: Pluto.

Cohn, Samuel Ross. 1985a. "Clerical labor intensity and the feminization of clerical labor in Great Britain, 1857–1937." *Social Forces* 63,4 (June): 1060–1068.

Cohn, Samuel. 1985b. *The Process of Occupational Sex-Typing: The Feminization of Clerical Labor in Great Britain*. Philadelphia: Temple University Press.

Colclough, Glenna, and Charles M. Tolbert, II. 1992. *Work in the Fast Lane*. Albany, N.Y.: State University of New York Press.

Cole, Robert E. 1989. *Strategies for Learning: Small-Group Activities in American, Japanese, and Swedish Industry*. Berkeley: University of California Press.

Coleman, James W. 1998. *The Criminal Elite: The Sociology of White-Collar Crime*, 4th ed. New York: St. Martin's.

Collins, Sharon M. 1997. *Black Corporate Executives: The Making and Breaking of the Black Middle Class*. Philadelphia: Temple University Press.

Collinson, David L. 1992. *Managing the Shopfloor*. Berlin, Germany: Walter de Gruyter.

Company, Sarah O., and Martin E. Personick. 1992. "Profiles in safety and health: Retail grocery stores." *Monthly Labor Review* 115,9 (September): 9–25.

Cook, David T. 1983. "High tech versus U.S. labor unions." *Christian Science Monitor*, September 9, p. 1.

Cooke, William N. 1990. "Factors Influencing the Effect of Joint Union–Management Programs on Employee–Supervisor Relations." *Industrial and Labor Relations Review* 43,5 (July): 587–603.

Cooley, Mike. 1982. "New Technologies—some trade union concerns and possible solutions." Pp. 193–206 in Niels Bjorn-Anderson (editor), *Information Society: For Richer, for Poorer*. Amsterdam: North Holland.

Cooley, Mike. 1986. "Socially useful design: A form of anticipatory democracy." *Economic and Industrial Democracy* 7,4 (November): 553–559.

Corn, Jacqueline Karnell. 1996. "Historical perspectives on mining." Pp. 3–12 in Philip Harber, Marc B. Schenker, and John R. Balmes (editors), *Occupational and Environmental Respiratory Disease*. St. Louis: Mosby.

Cornfield, Daniel B. 1985. "Economic segmentation and expression of labor unrest: Striking versus quitting in the manufacturing sector." *Social Science Quarterly* 66, 2 (June): 247–265.

Cornfield, Daniel B. (editor). 1987. *Workers, Managers, and Technological Change: Emerging Patterns of Labor Relations*. New York: Plenum.

Cornfield, Daniel B. 1989. *Becoming a Mighty Voice: Conflict and Change in the United Furniture Workers of America*. New York: Russell Sage Foundation.

Cornfield, Daniel B., and Teresa A. Sullivan. 1983. "Fieldwork in the oligopoly: Protecting the corporate subject." *Human Organization* 42,3 (Fall): 258–263.

Cornforth, Chris. 1992. "Co-operatives." Pp. 186–192 in Gyorgy Szell (editor), *Concise Encyclopedia of Participation and Co-Management*. Berlin: Walter de Gruyter.

Costello, Cynthia. 1984. "Women's work in the office." *Social Science Journal* 21,4 (October): 116–121.

Costello, Cynthia. 1991. *We're Worth It! Women and Collective Action in the Insurance Workplace*. Urbana, Ill.: University of Illinois Press.

Coughlin, Ellen K. 1989. "Making a business of belief: Sociologist examines the direct-selling industry in America." *The Chronicle of Higher Education* 35,45 (July 19): 4–6.

Cowan, Ruth Schwartz. 1983. *More Work for Mother*. New York: Basic Books.

Craypo, Charles. 1986. "The deindustrialization of a factory town." Pp. 105–113 in Gerald P. Glyde and Donald Kennedy (editors), *Contemporary Readings in Labor Issues*. Dubuque, Iowa: Kendall-Hunt.

Creighton, Sean, and Randy Hodson. 1995. "Whose side are they on? Technical workers and management ideology." In Steven Barley (editor), *Technical Workers and Technical Work*. Ithaca, N.Y.: Cornell University Press.

Cressey, Peter, and Robin Williams. 1990. *Participation in Change*. Shankill, Ireland: European Foundation for the Improvement of Living and Working Conditions.

Crompton, Rosemary, and Gareth Stedman Jones. 1984. *White-Collar Proletariat: Deskilling and Gender in Clerical Work*. Philadelphia: Temple University Press.

Crosby, Faye J. 1982. *Relative Deprivation and Working Women*. New York: Oxford University Press.

Cross, Michael. 1985. *Towards the Flexible Craftsman*. London: Technical Change Center.

Crozier, Michel. 1964. *The Bureaucratic Phenomenon*. Chicago: University of Chicago Press.

Csikszentmihalyi, Mihaly, and Barbara Schneider. 2000. *Becoming Adult: How Teenagers Prepare for the World of Work*. New York: Basic.

Cullen, John B. 1985. "Professional differentiation and occupational earnings." *Work and Occupations* 12,3 (August): 351–372.

Daday, Gerhard, and Beverly Burris. 2001. "Technocratic teamwork." Pp. 241–262 in Steven Vallas (editor), *The Transformation of Work*, Volume 10 of *Research in the Sociology of Work*. Greenwich, Conn.: Elsevier.

Dahl, Jonathan. 1986. "Ripping out asbestos endangers more lives as laws are ignored." *Wall Street Journal* (March 5): 1.

Dana, Bramel, and Ronald Friend. 1981. "Hawthorne, the myth of the docile worker, and class bias in psychology." *American Psychologist* 36,8 (August): 867–878.

Daniels, Arlene Kaplan. 1988. *Invisible Careers: Women Civic Leaders from the Volunteer World*. Chicago: University of Chicago Press.

Danziger, James N., K. L. Kraemer, D. E. Dunkle, and J. L. King. 1993. "Enhancing the Quality of Computing Service." *Public Administration Review* 53,2 (March): 161–169.

Davey, Caroline L., and Marilyn J. Davidson. 2000. "The right of passage? The experiences of female pilots in commercial aviation." *Feminism and Psychology* 10,2 (May): 195–225.

Davidson, Kenneth M. 1985. *Megamergers: Corporate America's Billion-Dollar Takeovers*. Cambridge, Mass.: Ballinger.

Davies, Margery W. 1982. *Woman's Place Is at the Typewriter: Office Work and Office Workers, 1870–1930*. Philadelphia: Temple University Press.

Davis, Devra Lee, Kenneth Bridbord, and Marvin Schneiderman. 1981. "Estimating cancer causes." Pp. 285–316 in R. Peto and M. Schneiderman (editors), Banbury Report #9 *Quantification of Occupational Cancer*. Cold Spring Harbor, N.Y.: Cold Spring Harbor Laboratory.

Davis, Gerald F., and Mark S. Mizruchi. 1999. "The money center cannot hold." *Administrative Science Quarterly* 44,2 (June): 215–239.

Deal, Terrence E., and Allan A. Kennedy. 1982. *Corporate Cultures: The Rites and Rituals of Corporate Life*. Reading, Mass.: Addison-Wesley.

DeAnda, Roberto M. 1996. "Falling back: Mexican-origin men and women in the U.S. economy." Pp. 41–50 in Roberto M. DeAnda (editor), *Chicanas and Chicanos in Contemporary Society*. Boston: Allyn and Bacon.

Deaux, Kay, and Joseph C. Ullman. 1983. *Women of Steel*. New York: Praeger.

Deckard, Barbara Sinclair. 1983. *The Women's Movement*, 2nd ed. New York: Harper and Row.

DeFreitas, Gregory. 1993. "Unionization among racial and ethnic minorities." *Industrial and Labor Relations Review* 46,2 (January): 284–301.

DeGrazia, Raffaele. 1980. "Clandestine employment: A problem of our times." *International Labour Review* 119,5 (September–October): 549–563.

Delbridge, Rick. 1998. *Life on the Line in Contemporary Manufacturing*. Oxford: Oxford University Press.

Deming, W. Edwards. 1982. *Quality, Productivity, and Competitive Position*. Cambridge, Mass.: MIT Pewaa.

Denby, Charles. 1978. *Indignant Heart*. Boston: South End Press.

Derber, Charles (editor). 1982. *Professionals as Workers: Mental Labor in Advanced Capitalism*. Boston: G. K. Hall.

Desmarez, Pierre. 2002. "The sociology of work: From work to the firm and beyond." In Daniel B. Cornfield and Randy Hodson (editors), *Worlds of Work*. New York: Plenum.

Desmond, Kevin. 1986. *Inventions Innovations Discoveries*. London: Constable.

de Soto, Hernando. 2000. *The Mystery of Capital: Why Capitalism Triumphs in the West and Fails Everywhere Else*. New York: Bantam.

Dicken, Peter. 1998. *Global Shift*, 3rd ed. New York: Guilford.

Diesenhouse, Susan. 1993. "In a shaky economy, even professionals are 'temps.' " *New York Times* (May 16).

DiMartino, Vittorio, and Linda Wirth. 1990. "Telework: A new way of working and living." *International Labour Review* 129, 5: 529–554.

Dinnerstein, Myra. 1992. *Women between Two Worlds: Midlife Reflections on Work and Family*. Philadelphia: Temple University Press.

DiTomaso, Nancy, and R. Hooijberg. 1996. "Diversity and the demands of leadership." *Leadership Quarterly* 7,2 (Summer): 163–187.

Dixon, Roland B. 1977. "The Northern Maidu." Pp. 262–291 in Carleton S. Coon (editor), *A Reader in Cultural Anthropology*. Huntington, N.Y.: Krieger.

Doeringer, Peter B., Philip I. Moss, and David G. Terkla. 1986. "Capitalism and kinship." *Industrial and Labor Relations Review* 40,1 (October): 48–60.

Dohm, Arlene. 2000. "Gauging the labor force effects of retiring baby boomers." *Monthly Labor Review* 123,1 (July): 17–25.

Dohse, Knuth, Ulrich Jurgens, and Thomas Malsch. 1985. "From 'Fordism' to 'Toyotism'? The social organization of the labor process in the Japanese automobile factory." *Politics and Society* 14, 2: 115–145.

Domhoff, G. William. 1998. *Who Rules America?*, 3rd ed. Mountain View, Calif.: Mayfield.

Doner, Richard F. 1991. *Driving a Bargain: Automobile Industrialization and Japanese Firms in Southeast Asia*. Berkeley: University of California Press.

Dong, Ok Lee. 1992. "Commodification of ethnicity: The sociospacial reproduction of immigrant entrepreneurs." *Urban Affairs Quarterly* 28,2 (December): 258–276.

Dore, Ronald P., and Mari Sako. 1998. *How the Japanese Learn to Work*. London: Routledge.

Dore, Ronald. 1986. *Flexible Rigidities: Industrial Policy and Structural Adjustment in the Japanese Economy, 1970–80*. Stanford, Calif.: Stanford University Press.

Douglass, David, and Joel Krieger. 1983. *A Miner's Life*. London: Routledge and Kegan Paul.

Dowie, Mark. 1977. "Pinto madness." *Mother Jones* 2,8 (September–October).

Dreeben, Robert. 1968. *On What Is Learned in School*. Reading, Mass.: Addison-Wesley.

Dreiling, Michael C. 2000. "Corporate political action leadership in the defense of NAFTA." *Social Problems* 47,1 (February): 21–48.

"Dual careers." 1987. *Wall Street Journal* (June 2): 33.

Dubose, Louis. 1987. "Letter from northern Mexico." *Texas Observer*, March 20, p. 23.

Duchin, Faye. 1998. *Structural Economics: Measuring Change in Technology, Lifestyles, and the Environment*. Washington, D.C.: Island.

Dulles, Foster Rhea, and Melvyn Dubofsky. 1984. *Labor in America*, 4th ed. Arlington Heights, Ill.: Harlan Davidson.

Duncan, O. D. 1961. "A socioeconomic index for all occupations." Pp. 109–138 in A. J. Reiss et al. (editors), *Occupations and Social Status.* New York: Free Press.

Dun and Bradstreet, Inc. 1994. *Million Dollar Directory.* Bethlehem, Penn.: Dun and Bradstreet, Inc.

Duneier, Mitchell. 1999. *Sidewalk.* New York: Farrar, Straus and Giroux.

Dunn, L. F. 1985. "Nonpecuniary job preferences and welfare losses among migrant agricultural workers." *American Journal of Agricultural Economics* 67,2 (May): 257–265.

Durkheim, Emile. 1966 [1897]. *Suicide.* (Translated by John A. Spaulding and George Simpson.) New York: Free Press.

Durkheim, Emile. 1984 [1933]. *The Division of Labor in Society* (Translated by W. D. Halls). New York: Free Press.

Dyer-Witheford, Nick. 1999. *Cyber-Marx: Cycles and Circuits of Struggle in High-Technology Capitalism.* Urbana, Ill.: University of Illinois Press.

Ebaugh, Helen Rose Fuchs. 1988. *Becoming an Ex: The Process of Role Exit.* Chicago: University of Chicago Press.

Edelman, Lauren B. 1990. "Legal environments and organizational governance: The expansion of due process in the American workplace." *American Journal of Sociology* 95,6 (May): 1401–1440.

Edgar, R. W. 2000. "Jubilee 2000: Paying our debts." *Nation,* April 24, pp. 20–21.

Edin, Kathryn, and Laura Lein. 1997. *Making Ends Meet: How Single Mothers Survive Welfare and Low-Wage Work.* New York: Russell Sage Foundation.

Edwards, Richard C. 1979. *Contested Terrain.* New York: Basic Books.

Edwards, Richard. 1993. *Rights at Work: Employment Relation in the Post-Union Era.* Washington, D.C.: Brookings Institute.

Egan, Timothy. 1988. "Workers to assume control of a bankrupt shipbuilder." *New York Times* (August 23): 5.

Eisenscher, Michael. 1984. *Silicon Valley: A Digest of Electronics Data.* San Jose, Calif.

Elliott, Philip. 1972. *The Sociology of the Professions.* London: Macmillan.

Ellul, Jacques. 1964. *The Technological Society.* New York: Vintage.

Emmanuel, Arghiri. 1984. *Profit and Crises.* London: Verso.

Endo, Koshi. 1994. "*Satei* (Personal Assessment) and Interworker Competition in Japanese Firms." *Industrial Relations* 33,1 (January): 70–82.

England, Paula. 1992. *Comparable Worth.* New York: Aldine de Gruyter.

England, Paula, and N. Folbre. 1999. "Who should pay for the kids?" *Annals of the American Academy of Political and Social Science* 563 (May): 194–207.

Epstein, Cynthia Fuchs. 1970. *Woman's Place: Options and Limits in Professional Careers.* Berkeley: University of California Press.

Epstein, Cynthia Fuchs. 1982. "Ambiguity as social control: Woman in professional elites." Pp. 61–72 in Phyllis L. Stewart and Muriel G. Cantor (editors), *Varieties of Work.* Beverly Hills, Calif.: Sage.

Epstein, Edythe. 1984. "Negotiating over technological change in banking and insurance." *International Labor Review* 123,4 (July–August): 405–422.

Erickson, Bonnie H., Patricia Albanese, and Slobodan Drakulic. 2000. "Gender on a jagged edge: The security industry, its clients, and the reproduction and revision of gender." *Work and Occupations* 27,3 (August): 294–318.

Erikson, Erik H. 1963. *Childhood and Society,* 2nd ed. New York: Norton.

Ermann, M. David, and Richard J. Lundman. 2000. *Corporate and Governmental Deviance,* 6th ed. New York: Oxford University Press.

Essed, Philomena. 1991. *Understanding Everyday Racism.* Newbury Park, Calif.: Sage.

Estey, Marten. 1981. *The Unions,* 3rd ed. New York: Harcourt Brace Jovanovich.

Everett, Melissa. 1995. *Making a Living While Making a Difference: A Guide to Creating Careers with a Conscience.* New York: Bantam.

Ezzamel, Mahmoud, and Hugh Willmott. 1998. "Accounting for teamwork: A critical study of group-based systems of organizational control." *Administrative Science Quarterly* 43,2 (June): 358–396.

Fantasia, Rick. 1988. *Cultures of Solidarity.* Berkeley: University of California Press.

Faunce, William A. 1981. *Problems of an Industrial Society.* New York: McGraw-Hill.

Feagin, Joe R. 1991. "The continuing significance of race: Antiblack discrimination in public places." *American Sociological Review* 56,1 (February): 101–116.

Feagin, Joe R., Anthony M. Orum, and Gideon Sjoberg. 1991. *A Case for the Case Study.* Chapel Hill, N.C.: University of North Carolina Press.

Fein, Mitchell. 1976. "Motivation for work." Pp. 465–530 in Robert Dubin (editor), *Handbook of Work, Organization, and Society.* Chicago: Rand McNally.

Ferber, Marianne A., and Brigid O'Farrell (editors). 1991. *Work and Family: Policies for a Changing Work Force.* Washington, D.C.: National Academy Press.

Ferman, Louis A., Michele Hoyman, Joel Cutcher-Gershenfeld, and Ernest J. Savoie. 1990a. "Editors' introduction." Pp. 1–23 in *New Developments in Worker Training.* Madison, Wisc.: University of Wisconsin, Industrial Relations Research Association.

Ferman, Louis A., Michele Hoyman, and Joel Cutcher-Gershenfeld. 1990b. "Joint union–management training programs: A synthesis in the evolution of jointism in training." Pp. 157–190 in L. Ferman, M. Hoyman, J. Cutcher-Gershenfeld, and E. Savoie (editors), *New Developments in Worker Training: A Legacy for the 1990s.* Madison, Wisc.: Industrial Relations Research Association.

Fernandez, John P., and Julie Davis. 1999. *Race, Gender, and Rhetoric: The True State of Race and Gender Relations in Corporate America.* New York: McGraw-Hill.

Fernandez, R. M., E. J. Castilla, and P. Moore. 2000. "Social capital at work: Networks and employment in a phone center." *American Journal of Sociology.* 105,5: 1288–1356.

Ferree, Myra Marx. 1984. "Class, housework, and happiness: Women's work and life satisfaction." *Sex Roles* 11,11/12 (December): 1057–1074.

Fine, Gary Alan. 1984. "Negotiated orders and organizational cultures." Pp. 239–262 in Ralph H. Turner and James F. Short, Jr. (editors), *Annual Review of Sociology,* Vol. 10. Palo Alto, Calif.: Annual Reviews.

Fink, Deborah. 1998. *Cutting into the Meatpacking Line.* Chapel Hill, N.C.: University of North Carolina Press.

Fisher, C. D. 1980. "On the dubious wisdom of expecting job satisfaction to correlate with performance." *Academy of Management Review* 5: 607–612.

Fishman, Paula M. 1978. "Interaction: The Work Women Do." *Social Problems* 75: 397–406.

Fleischer, Arthur, Jr., Geoffrey C. Hazard, Jr., and Miriam Z. Klipper. 1988. *Board Games: The Changing Shape of Corporate Power.* Boston, Mass.: Little Brown.

Fleischer, Mitchell, and Jonathan A. Morell. 1985. "The organizational and managerial consequences of computer technology." *Computers in Human Behavior* 1,1: 83–93.

Fletcher, Joyce K. 1999. *Disappearing Acts: Gender, Power and Relational Practice at Work.* Cambridge, Mass.: MIT Press.

Fligstein, Neil. 1996. "Markets as politics." *American Sociological Review* 61,4 (August): 656–673.

Florida, Richard, and Martin Kenney. 1991. "Transplanted organizations: The transfer of Japanese industrial organization to the U.S." *American Sociological Review* 56,3 (June): 381–398.

Foner, Philip S. 1982. *Organized Labor and the Black Worker, 1619–1981.* New York: International Publishers.

Forbes. 1993. (April 26): 92, 204–238.

Forester, Tom. 1981. *The Microelectronics Revolution.* Cambridge, Mass.: MIT Press.

Form, William. 1985. *Divided We Stand.* Urbana, Ill.: University of Illinois Press.

Form, William, and Claudine Hanson. 1985. "The consistency of stratal ideologies of economic justice." Pp. 239–270 in Robert V. Robinson (editor), *Research in Social Stratification and Mobility.* Vol. 4. Greenwich, Conn.: JAI Press.

Form, William, and David Byron McMillen. 1983. "Women, men, and machines." *Work and Occupations* 10,2: 147–178.

Forsyth, Patrick B., and Thomas J. Danisiewicz. 1985. "Toward a theory of professionalization." *Work and Occupations* 12,1 (February): 59–76.

Fossum, E. 1983. *Computerization of Working Life.* New York: Wiley.

Fox, Bonnie, and Pamela Sugiman. 1999. "Flexible work, flexible workers: The restructuring of clerical work in a large telecommunications company." *Studies in Political Economy* 60 (Autumn): 59–84.

Fox, Steve. 1991. *Toxic Work.* Philadelphia: Temple University Press.

Francis, Arthur. 1986. *New Technology at Work.* Oxford: Oxford University Press.

Francis, Arthur, Mandy Snell, Paul Willman, and Graham Winch. 1981. "The impact of information technology at work: The case of CAD/CAM and MIS in engineering plants." Pp. 182–193 in *Information Technology: Impact on the Way of Life.* Dublin, Ireland: Tycooly International Publishers.

Frank, Andre Gunder, and Barry K. Gills. 1993. *The World System.* London: Routledge.

Franklin, James C. 1993. "Industry output and employment." *Monthly Labor Review* 116,11 (November): 41–57.

Freeman, Richard B. (editor). 1994. *Working under Different Rules.* New York: Russell Sage.

Freeman, Richard B., and James L. Medoff. 1984. *What Do Unions Do?* New York: Basic Books.

Freeman, Richard B., and Joel Rogers. 1999. *What Workers Want.* Ithaca, N.Y.: Industrial and Labor Relations Press.

Freidson, Eliot. 1970. *The Profession of Medicine: A Study in the Sociology of Applied Knowledge.* New York: Harper and Row.

Freidson, Eliot. 1986. *Professional Powers.* Chicago: University of Chicago Press.

Frenkel, Stephen J. (editor). 1993. *Organized Labor in the Asia-Pacific Region.* Ithaca, N.Y.: Cornell University School of Industrial and Labor Relations Press.

Frenkel, Stephen J., Marek Korczynski, Karen A. Shire, and May Tam. 1999. *On the Front Line: Organization of Work in the Information Society.* Ithaca, N.Y.: Cornell University Press.

Friedl, Ernestine. 1975. *Women and Men: An Anthropologist's View.* New York: Holt, Rinehart and Winston.

Friedman, Samuel R. 1982. *Teamster Rank and File.* New York: Columbia University Press.

Fromm, Erich. 1968. *The Revolution of Hope.* New York: Harper and Row.

Fuchs, Victor. 1968. *The Service Economy.* New York: National Bureau of Economic Research.

Fuller, Linda. 1992. *Work and Democracy in Socialist Cuba.* Philadelphia: Temple University Press.

Fuller, Linda, and Vicki Smith. 1996. "Consumers' reports: Management by customers in a changing economy." Pp. 74–90 in Cameron Lynn Macdonald and Carmen Sirianni (editors), *Working in the Service Society.* Philadelphia: Temple University Press.

Furchtgott-Roth, Diana, and Christine Stolba. 1999. *Women's Figures: An Illustrated Guide to the Economic Progress of Women in America.* Washington, D.C.: American Enterprise Institute.

Fusselman, Kay. 1985. "Secretaries wanted." *The Secretary* 45, 1 (January): 3–5.

Galenson, David W. 1984. "The rise and fall of indentured servitude in the Americas." *Journal of Economic History* 71, 1 (March): 1–26.

Galle, Omer R., Candace Hinson Wiswell, and Jeffrey A. Burr. 1985. "Racial mix and industrial productivity." *American Sociological Review* 50, 1 (February): 20–33.

Gallop-Goodman, Gerda. 2000. "Retire with style." *American Demographics* (August): 13–15.

Gamst, Frederick C. (editor). 1995. *Meanings of Work.* Albany, N.Y.: State University of New York Press.

Gannon, Martin J. 1984. "Preferences of temporary workers: Time, variety, and flexibility." *Monthly Labor Review* 107, 8 (August): 26–28.

Gardner, John. 1993. "Community organizing: Seeds of justice." *In These Times,* May 17: 18–19.

Garson, Barbara. 1994. *All the Livelong Day,* 2nd ed. New York: Penguin.

Garson, Barbara. 1988. *The Electronic Sweatshop.* London: Penguin.

Gaventa, John. 1980. *Power and Powerlessness: Quiescence and Rebellion in an Appalachian Valley.* Urbana, Ill.: University of Illinois Press.

Gecas, Victor. 1979. "The influence of social class on socialization." In W. R. Burr et al. (editors), *Contemporary Theories about the Family.* Vol. 1. New York: Free Press.

Geis, Gilbert, Robert F. Meier, and Lawrence M. Salinger (editors). 1995. *White-Collar Crime.* New York: Free Press.

Geison, Gerald L. (editor). 1983. *Professions and Professional Ideologies in America.* Chapel Hill, N.C.: University of North Carolina Press.

Gereffi, Gary. 1999. "International trade and industrial upgrading in the apparel commodity chain." *Journal of International Economics* 48, 1 (June): 37–70.

Gerson, Kathleen. 1985. *Hard Choices: How Women Decide about Work, Career, and Motherhood.* Berkeley: University of California Press.

Gerstel, N. R. 1977. "The feasibility of commuter marriage." Pp. 357–367 in P. J. Stein, J. Richman, and N. Hannon (editors), *The Family: Functions and Conflicts and Symbols.* Reading, Mass.: Addison-Wesley.

Gevitz, Norman. 1982. *The D.O.'s: Osteopathic Medicine in America.* Baltimore: Johns Hopkins University Press.

Gifford, Court (editor). 1999. *Directory of U.S. Labor Organizations.* Washington, D.C.: Bureau of National Affairs.

Giles, Anthony, and Jacques Belanger. 2002. "The sociology of work in Canada." In Daniel B. Cornfield and Randy Hodson (editors), *Worlds of Work.* New York: Plenum.

Gillespie, Richard. 1991. *Manufacturing Knowledge: A History of the Hawthorne Experiments.* New York: Cambridge University Press.

Gittleman, Maury, Michael Horrigan, and Mary Joyce. 1998. " 'Flexible' workplace practices: Evidence from a nationally representative survey." *Industrial and Labor Relations Review* 52, 1 (October): 99–115.

Glass, Jennifer. 2000. "Envisioning the integration of family and work: Toward a kinder, gentler workplace." *Contemporary Sociology* 29, 1 (January): 129–143.

Glass, Jennifer, and Valerie Camarigg. 1992. "Gender, parenthood, and job-family compatibility." *American Journal of Sociology* 98, 1 (July): 131–151.

Glenn, Evelyn Nakano, and Roslyn L. Feldberg. 1977. "Degraded and deskilled: The proletarianization of clerical work." *Social Problems* 25, 1 (October): 52–64.

Glenn, Norval D., and Charles N. Weaver. 1982. "Enjoyment of work by full-time workers in the U.S., 1955 and 1980." *Public Opinion Quarterly* 46: 459–470.

Glenn, Norval D., and Charles N. Weaver. 1985. "Age, cohort, and reported job satisfaction in the United States." *Current Perspectives on Aging and the Life Cycle* 1: 89–109.

Godish, Thad. 1997. *Air Quality.* Boca Raton, Fl.: Lewis.

Goffman, Erving. 1955. "On face-work: An analysis of ritual elements in social interaction." *Psychiatry* 18: 213–231.

Goldthorpe, John H., David Lockwood, Frank Bechhofer, and Jennifer Platt. 1969. *The Affluent Worker in the Class Structure.* London: Cambridge University Press.

Gordon, David M. 1996. *Fat and Mean: The Corporate Squeeze of Working Americans and the Myth of Managerial Downsizing.* New York: Free Press.

Gore, Albert. 1993. *Creating a Government that Works Better and Costs Less.* New York: Plume.

Gouldner, Alvin W. 1964. *Patterns of Industrial Bureaucracy.* New York: Free Press.

Gouldner, Alvin W. 1979. *The Future of the Intellectuals and the Rise of the New Class.* London: Macmillan Publishing.

Graham, Laurie. 1995. *On the Line at Subaru-Isuzu.* Ithaca, N.Y.: Industrial and Labor Relations Press.

Granovetter, Mark. 1995. *Getting A Job: A Study of Contacts and Careers,* 2nd ed. Chicago: University of Chicago Press.

Granovetter, Mark, and Richard Swedberg. 1992. *The Sociology of Economic Life.* Boulder, Colo.: Westview.

Gray, Charles D. 1987. "Putting human rights into trade policy." *New York Times* (May 31): G1.

Green, Gareth M., and Frank Baker. 1991. *Work, Health, and Productivity.* New York: Oxford University Press.

Green, Gloria, Khoan tan Dinh, John A. Priebe, and Ronald R. Tucker. 1983. "Revisions in the current population survey beginning in January 1983." *Employment and Earnings* 30, 2 (February): 7–15.

Green, Hardy. 1990. *On Strike at Hormel.* Philadelphia: Temple University Press.

Green, James R. 1980. *The World of the Worker.* New York: Hill and Wang.

Greenberg, Edward S. 1980. "Participation in industrial decision making and work satisfaction: The case of producer cooperatives." *Social Science Quarterly* 60, 4 (March): 551–569.

Greenberg, Edward S. 1986. *Workplace Democracy.* Ithaca, N.Y.: Cornell University Press.

Greenwood, R., and R. Lachman. 1996. "Change as an underlying theme in professional service organizations." *Organization Studies* 17, 4: 563–572.

Grenier, Guillermo J. 1988. *Inhuman Relations: Quality Circles and Anti-Unionism in American Industry.* Philadelphia: Temple University Press.

Griffin, Larry J., Philip J. O'Connell, and Holly J. McCammon. 1989. "National variation in the context of struggle." *Canadian Review of Sociology and Anthropology* 26, 1 (February): 37–68.

Griffith, David. 1993. *Jones's Minimal: Low-Wage Labor in the United States.* Albany, N.Y.: State University of New York Press.

Grint, Keith. 1991. *The Sociology of Work.* Cambridge, England: Basil Blackwell.

Gross, Harriet E. 1980. "Dual-career couples who live apart: Two types." *Journal of Marriage and the Family* 42, 3 (August): 567–576.

Gruber, James E., and Lars Bjorn. 1982. "Blue-collar blues: The sexual harassment of women autoworkers." *Work and Occupations* 9 (August): 271–298.

Gruber, James E., and Lars Bjorn. 1986. "Women's responses to sexual harassment." *Social Science Quarterly* 67, 4 (December): 814–826.

Gruenberg, Barry. 1980. "The happy worker: An analysis of educational and occupational differences in determinants of job satisfaction." *American Journal of Sociology* 86, 2 (September): 247–271.

Grunberg, Leon, R. Anderson-Connolly, and Edward S. Greenberg. 2000. "Surviving layoffs." *Work and Occupations* 27: 7–31.

Guest, Robert H. 1987. "Industrial sociology: The competitive edge." *American Sociological Association Footnotes* 15, 1 (January): 5.

Gullickson, Gay L. 1983. "Agriculture and cottage industry: Redefining the causes of proto-industrialization." *Journal of Economic History* 43, 4 (December): 831–850.

Gunderson, Morley, and Allen Ponak. 2000. *Union-Management Relations in Canada,* 4th ed. Reading, Mass.: Addison-Wesley.

Gutek, Barbara A. 1985. *Sex and the Workplace.* San Francisco, Calif: Jossey-Bass.

Guthrie, Douglas, and L.M. Roth. 1999. "The states, courts, and equal opportunities for female CEOs." *Social Forces* 78: 511–542.

Guy, Mary E. 1985. *Professionals in Organizations: Debunking a Myth.* New York: Praeger Press.

Gyllenhammar, Pehr G. 1977. *People at Work.* Reading, Mass.: Addison-Wesley.

Haas, Ain. 1983. "The aftermath of Sweden's codetermination law." Pp. 19–46 in *Economic and Industrial Democracy.* London: Sage.

Haas, Jack. 1977. "Learning real feelings: A study of high steel ironworkers' reactions to fear and danger." *Sociology of Work and Occupations* 4, 2 (May): 147–170.

Haber, Samuel. 1991. *The Quest for Authority and Honor in the American Professions, 1750–1900.* Chicago: University of Chicago Press.

Hackman, J. Richard. 1990. *Groups That Work (and Those That Don't).* San Francisco: Jossey-Bass.

Hall, Richard H. 1994. *Sociology of Work.* Thousand Oaks, Calif.: Pine Forge.

Hall, Richard H. 1996. *Organizations,* 6th ed. New York: Prentice Hall.

Halle, David. 1984. *America's Working Man.* Chicago: University of Chicago Press.

Haller, Max, Wolfgang Konig, Peter Krause, and Karin Kurz. 1985. "Patterns of career mobility and structural positions in advanced capitalist societies: A comparison of men in Austria, France, and the United States." *American Sociological Review* 50, 5 (October): 579–603.

Halliday, Terence C. 1987. *Beyond Monopoly: Lawyers, State Crises, and Professional Empowerment.* Chicago: University of Chicago Press.

Halliday, Terence C., and Lucien Karpik. 1997. "Postscript: Lawyers, political liberalism, and globalization." Pp. 350–370 in Terence C. Halliday and Lucien Karpik (editors), *Lawyers and the Rise of Western Political Liberalism.* New York: Oxford University Press.

Halliday, Terence C., Michael J. Powell, and Mark Granfors. 1993. "After minimalism: Transformations of state bar associations, 1918–1950." *American Sociological Review* 58, 4 (August): 515–535.

Halpern, Sydney A. 1992. "Dynamics of professional control: Internal coalitions and crossprofessional boundaries." *American Journal of Sociology* 97, 4 (January): 994–1022.

Hamilton, Richard F., and James D. Wright. 1986. *The State of the Masses.* New York: Aldine.

Hansmann, Henry. 1996. *The Ownership of Enterprise.* Cambridge, Mass.: Harvard University Press.

Haraszti, Miklos. 1978. *A Worker in a Worker's State.* New York: Universe.

Hareven, Tamara. 1982. *Family Time and Industrial Time: The Relationship between the Family and Work in a New England Industrial Community.* Cambridge, England: Cambridge University Press.

Harper, Douglas. 1987. *Working Knowledge.* Chicago: University of Chicago Press.

Harriman, Ann. 1985. *Women/Men/Management.* New York: Praeger.

Harris, Candee S. 1987. "Magnitude of job loss." Pp. 89–100 in Paul D. Staudohar and Holly E. Brown (editors), *Deindustrialization and Plant Closure.* Lexington, Mass.: D. C. Heath.

Harris, Kathleen Mullan. 1993. "Work and welfare among single mothers in poverty." *American Journal of Sociology* 99, 2 (September): 317–352.

Harrison, Bennett. 1994. *Lean and Mean: The Changing Landscape of Corporate Power in the Age of Flexibility.* New York: Basic Books.

Hartmann, Heidi, Robert E. Kraut, and Louise A. Tilly (editors). 1986. *Computer Chips and Paper Clips: Technology and Women's Employment.* Washington, D.C.: National Academy Press.

Hartmann, Heidi, Patricia A. Roos, and Donald J. Treiman. 1985. "An agenda of basic research on comparable worth." Pp. 3–33 in Heidi I. Hartmann (editor), *Comparable Worth.* Washington, D.C.: National Academy Press.

Hauser, Philip M. 1964. "Labor force." Pp. 161–191 in Robert E. L. Faris (editor), *Handbook of Modern Sociology.* Chicago: Rand McNally.

Hauser, Robert M., and David L. Featherman. 1977. *The Process of Stratification: Trends and Analyses.* New York: Academic Press.

Hayano, David M. 1982. *Poker Faces: The Life and Work of Professional Card Players.* Berkeley: University of California Press.

Hayes, Robert H., and William J. Abernathy. 1980. "Managing our way to economic decline." *Harvard Business Review* 58, 4 (July–August): 67–77.

Healey, James R. 1992. "U.S. Steel learns from experience." *U.S. Today,* April 10, pp. B1–B2.

Heckscher, Charles C. 1988. *The New Unionism.* New York: Basic Books.

Heinz, John P., and Edward O. Laumann. 1982. *Chicago Lawyers: The Social Structure of the Bar.* New York and Chicago: Russell Sage Foundation and American Bar Association.

Helmick, Sandra A., and Judith D. Zimmerman. 1984. "Trends in the distribution of children among households and families." *Child Welfare* 62, 5 (September–October): 401–409.

Henson, Kevin D. 1996. *Just a Temp.* Philadelphia: Temple University Press.

Herzberg, Frederick. 1966. *Work and the Nature of Man.* New York: Harcourt Brace Jovanovich.

Hill, Reuben. 1986. "Life cycle stages for types of single parent families: Of family development theory." *Family Relations* 35 (January): 19–29.

Hill, Stephen. 1981. *Competition and Control at Work.* Cambridge, Mass.: MIT Press.

Hills, Stuart L. 1987. *Corporate Violence: Injury and Death for Profit.* Totowa, N.J.: Rowman and Littlefield.

Hirsch, Paul M. 1986. "From ambushes to golden parachutes: Corporate takeovers as an instance of cultural framing and institutional integration." *American Journal of Sociology* 91, 4 (January): 800–837.

Hirschhorn, Larry. 1997. *Reworking Authority.* Cambridge, Mass.: MIT Press.

Hobsbawm, Eric J. 1969. *Industry and Empire.* Middlesex, England: Penguin.

Hobsbawm, Eric J. 1996. *The Age of Extremes.* New York: Vintage.

Hochschild, Arlie. 1983. *The Managed Heart: Commercialization of Human Feeling.* Berkeley: University of California Press.

Hochschild, Arlie. 1989. *The Second Shift: Working Parents and the Revolution at Home.* New York: Viking.

Hochschild, Arlie R. 1997a. *The Time Bind: When Work Becomes Home and Home Becomes Work.* New York: Metropolitan Books.

Hochschild, Arlie R. 1997b. "Work: The great escape." *New York Times Magazine* (April 20): 51–55+, 81, 84.

Hochstedler, Ellen (editor). 1984. *Corporations as Criminals.* Beverly Hills, Calif.: Sage.

Hodge, Robert W., Paul M. Siegel, and Peter M. Rossi. 1964. "Occupational prestige in the United States, 1925–1963." *American Journal of Sociology* 70, 3 (November): 286–302.

Hodson, Randy. 1983. *Workers' Earnings and Corporate Economic Structure.* New York: Academic Press.

Hodson, Randy. 1984. "Corporate structure and job satisfaction: A focus on employer characteristics." *Sociology and Social Research* 69, 1 (October): 22–49.

Hodson, Randy. 1985. "Workers' comparisons and job satisfaction." *Social Science Quarterly* 66, 2 (June): 266–280.

Hodson, Randy. 1989. "Gender differences in job satisfaction: Why aren't women more dissatisfied?" *Sociological Quarterly* 30, 3 (September): 385–399.

Hodson, Randy. 1991. "Good soldiers, smooth operators, and saboteurs: A model of workplace behaviors." *Work and Occupations* 18, 3 (August): 271–290.

Hodson, Randy. 1999. "Management citizenship behavior: A new concept and an empirical test." *Social Problems* 46, 3: 460–478.

Hodson, Randy. 2001. *Dignity at Work.* New York: Cambridge University Press.

Hodson, Randy, and Robert L. Kaufman. 1982. "Economic dualism: A critical review." *American Sociological Review* 47, 6 (December): 727–739.

Hodson, Randy, Sandy Welsh, Sabine Rieble, Cheryl Sorenson Jamison, and Sean Creighton. 1993. "Is worker solidarity undermined by autonomy and participation?" *American Sociological Review* 58, 3 (June): 398–416.

Hoffman, Lily M. 1989. *The Politics of Knowledge: Activist Movements in Medicine and Planning.* Albany, N.Y.: State University of New York Press.

Hogan, Dennis P. 1981. *Transitions and Social Change: The Early Lives of American Men.* New York: Academic Press.

Hollinger, Richard C., and John P. Clark. 1983. "Deterrence in the workplace: Perceived certainty, perceived severity, and employee theft." *Social Forces* 62, 2 (December): 398–418.

Honig, Marjorie, and Giora Hanoch. 1985. "Partial retirement as a separate mode of retirement behavior." *Journal of Human Resources* 20, 1 (Winter): 21–46.

Hood, Jane C. 1988. "From night to day: Timing and the management of custodial work." *Journal of Contemporary Ethnography* 17, 1 (April): 96–116.

Hooks, Gregory. 1987. "Comparison of the United States, Sweden and France." Pp. 245–258 in Paul D. Staudohar and Holly E. Brown (editors), *Deindustrialization and Plant Closure.* Lexington, Mass.: D. C. Heath.

House, James S. 1981. *Work Stress and Social Support.* Reading, Mass.: Addison-Wesley.

Howard, Robert. 1981. "Microshock in the information society." *In These Times* (January 21–27): 11–13, 22.

Howard, Robert. 1985. *Brave New Workplace.* New York: Viking.

Howe, Louise Kapp. 1977. *Pink Collar Workers: Inside the World of Women's Work.* New York: G.P. Putnam's Sons Publishing.

Howell, Jon P., and Peter W. Dorfman. 1986. "Leadership and substitutes for leadership among professional and nonprofessional workers." *Journal of Applied Behavioral Science* 22, 1: 29–46.

Howitt, A. W. 1904. *The Native Tribes of Southeast Australia.* London: Macmillan.

Hsiung, Ping-Chun. 1996. *Living Rooms as Factories: Class, Gender and the Satellite Factory System in Taiwan.* Philadelphia: Temple University Press.

Hughes, Everett Cherrington. 1958. *Men and Their Work.* Glencoe, Ill.: Free Press.

Hughes, Everett Cherrington. 1965. "Professions." Pp. 1–14 in Kenneth S. Lynn and the editors of Daedalus (editors), *The Professions in America.* Boston: Houghton Mifflin.

Human Development Report. 1998. New York: Oxford University Press.

Hunnicutt, Benjamin Kline. 1996. *Kellogg's Six-Hour Day.* Philadelphia: Temple University Press.

Hunt, Morton. 1985. *Profiles of Social Research.* New York: Russell Sage Foundation.

Iams, Howard. 1993. "Working wives and social security: Many still rely on husband's benefit." *Population Today* 21, 11 (November): 6–7.

Illich, Ivan. 1977. *Disabling Professions.* London: Marion Boyars.

Industrial Workers of the World. 1980. *Songs of the Workers.* Chicago: Industrial Workers of the World.

Itzkowitz, Gary. 1996. *Contingency Theory.* Lanham, Md.: University Press of America.

Jackall, Robert. 1988. *Moral Mazes: The World of Corporate Managers.* New York: Oxford University Press.

Jacoby, Neil H. 1973. *Corporate Power and Social Responsibility.* New York: Macmillan.

Jacoby, Sanford M. 1997. *Modern Manors: Welfare Capitalism since the New Deal.* Princeton, N.J.: Princeton University.

Jaffee, David. 1998. *Levels of Socio-Economic Development Theory.* Bridgeport, Conn.: Praeger.

James, David R., and Michael Soref. 1981. "Managerial theory: Unmaking the corporation president." *American Sociological Review* 46, 1 (February): 1–18.

Jasso, Guellermina, Douglas S. Massey, Mark R. Rosenzweig, and James P. Smith. 2000. "The new immigrant survey pilot." *Demography* 37, 1 (February): 127–138.

Jaynes, Gerald David. 1986. *Branches without Roots: Genesis of the Black Working Class in the American South, 1862–1882.* New York: Oxford University Press.

Jencks, Christopher. 1985. "Affirmative action for blacks." *American Behavioral Scientist* 28, 6 (July–August): 731–760.

Johnson, William G., and James Lambrinos. 1985. "Wage discrimination against handicapped men and women." *Journal of Human Resources* 20, 2 (Spring): 264–277.

Jonas, Norman. 1986. "The hollow corporation." *Business Week* (March 3): 57–59.

Jones, Stephen R. G. 1992. "Was there a Hawthorne effect?" *American Journal of Sociology* 98, 3 (November): 451–468.

Judis, John B. 1987. "Making a mockery of 'Made in the U.S.A.'" *In These Times* (April 15–21): 7.

Juravich, Tom. 1985. *Chaos on the Shop Floor: A Worker's View of Quality, Productivity, and Management.* Philadelphia: Temple University Press.

Jureidini, R., and K. White. 2000. "Life insurance, the medical examination, cultural values." *Journal of Historical Sociology* 13, 2 (June): 190–214.

Jurgens, Ulrich, Larissa Klinzing, and Lowell Turner. 1993. "The transformation of industrial relations in Eastern Germany." *Industrial and Labor Relations Review* 46, 2 (January): 229–244.

Kahn, Robert L. 1972. "The meaning of work: Interpretation and proposals for measurement." Pp. 159–203 in Angus S. Campbell and Philip E. Converse (editors), *The Human Meaning of Social Change.* New York: Russell Sage Foundation.

Kalleberg, Arne L. 1977. "Work values and job rewards: A theory of job satisfaction." *American Sociological Review* 42, 1 (February): 124–143.

Kalleberg, Arne L., and Ivar Berg. 1987. *Work and Industry.* New York: Plenum.

Kalleberg, Arne L., David Knoke, Peter V. Marsden, and Joe L. Spaeth. 1996. *Organizations in America.* Thousand Oaks, Calif.: Sage.

Kalleberg, Arne L., Barbara F. Reskin, and Ken Hudson. 2000. "Bad jobs in America: Standard and nonstandard employment relations and job quality in the United States." *American Sociological Review* 65, 2 (April): 256–278.

Kamerman, Sheila B. 1986. "Maternity, paternity, and patenting policies: How does the United States compare?" Pp. 53–66 in Sylvia Ann Hewlett, Alice S. Ilchman, and John J. Sweeney (editors), *Family and Work: Bridging the Gap.* Cambridge, Mass.: Ballinger.

Kanter, Rosabeth Moss. 1977. *Men and Women of the Corporation.* New York: Basic Books.

Kanter, Rosabeth Moss. 1983. *The Change Masters: Innovation for Productivity in the American Corporation.* New York: Simon and Schuster.

Kaplan, Robert E., Michael M. Lombardo, and Mignon S. Mazique. 1985. "A mirror for managers: Using simulation to develop management teams." *Journal of Applied Behavioral Science* 21, 3: 241–253.

Kapur, Kanicka. 1998. "The impact of health on job mobility." *Industrial and Labor Relations Review* 51, 2 (January): 282–297.

Katz, Daniel, and Robert L. Kahn. 1978. *The Social Psychology of Organizations,* 2nd ed. New York: Wiley.

Katzman, David M. 1978a. "Domestic service: Women's work." Pp. 377–391 in Ann H. Stromberg and Shirley Harkness (editors), *Women Working: Theories and Facts in Perspective.* Palo Alto, Calif.: Mayfield.

Kauffman, Scott. 1979. "Plywood strikers back fired woman." *In These Times* (December 5–11): 5.

Keef, Stephen P. 1998. "The causal association between employee share ownership and attitudes." *British Journal of Industrial Relations.* 36,1 (March): 73–82.

Keefe, Jeffrey H., and Adrienne E. Eaton (editors). 1999. *Employment Dispute Resolution and Worker Rights in the Changing Workplace.* Champaign, Ill.: Industrial Relations Research Association.

Keister, Lisa A. 2000. *Chinese Business Groups.* New York: Oxford University Press.

Kelley, Maryellen R. 1990. "New process technology, job design and work organization." *American Sociological Review* 55,2 (April): 191–208.

Kelley, Robert E. 1985. *The Gold Collar Worker: Harnessing the Brain Power of the New Work Force.* Reading, Mass.: Addison-Wesley.

Kellner, Hansfried, and Frank W. Heuberger (editors). 1992. *Hidden Technocrats: The New Class and New Capitalism.* New Brunswick, N.J.: Transaction.

Kenyon, R. 2000. "Changes in unemployment insurance legislation." *Monthly Labor Review* 123: 27–35.

Kessler, J. A. 1999. "NAFTA, emerging apparel production networks and industrial upgrading." *Review of International Political Economy* 6,4 (Winter): 565–608.

Kessler-Harris, Alice. 1982. *Out to Work: A History of Wage-Earning Women in the United States.* New York: Oxford University Press.

Kidder, Tracy. 1981. *The Soul of a New Machine.* New York: Avon Books.

Kim, Kwang Chung, Won Moo Hurh, and Marilyn Fernandez. 1989. "Intra-group differences in business participation: Three Asian immigrant groups." *International Migration Review* 23,1 (Spring): 73–96.

Kluegel, James R., and Lawrence Bobo. 1993. "Opposition to race-targeting." *American Sociological Review* 58,4 (August): 443–464.

Kohn, Melvin L. 1976. "Social class and parental values: Another confirmation of the relationship." *American Sociological Review* 41,3 (June): 538–545.

Kohn, Melvin L. 1990. *Social Structure and Self-Direction.* Cambridge, Mass: Blackwell.

Kohn, Melvin L., Atsushi Naoi, Carrie Schoenbach, Carmi Schooler, and Kazimierz M. Slomczynski. 1990. "Position in the class structure and psychological functioning in the United States, Japan, and Poland." *American Journal of Sociology* 95,4 (January): 964–1008.

Kondo, Dorinne K. 1990. *Crafting Selves: Power, Gender and Discourses of Identify in a Japanese Workplace.* Chicago: University of Chicago Press.

Kono, Toyohiro. 1992. *Long-Range Planning of Japanese Corporations.* Berlin: Walter de Gruyter.

Kotter, John P. 1982. *The General Managers.* New York: Free Press.

Krahn, Harvey J., and Graham S. Lowe. 1998. *Work, Industry, and Canadian Society.* Toronto: Nelson.

Kranzberg, Melvin, and Joseph Gies. 1986. "Medieval work: Guilds and the putting-out system." Pp. 2–6 in Lauri Perman (editor), *Work in Modern Society.* Dubuque, Iowa: Kendall-Hunt.

Kritzer, Herbert M. 1999. "The professions are dead, long live the professions: Legal practice in a postprofessional world." *Law and Society Review* 33,3 (October): 713–759.

Kunda, Gideon. 1992. *Engineering Culture.* Philadelphia: Temple University Press.

Kusterer, Ken C. 1978. *Know-How on the Job: The Important Working Knowledge of "Unskilled" Workers.* Boulder, Colo.: Westview.

Kuttner, Robert. 1996. *Everything for Sale: The Virtues and Limits of Markets.* New York: Alfred A. Knopf.

Lamphere, Louise (editor). 1992. *Structuring Diversity: Ethnographic Perspectives on the New Immigration.* Chicago: University of Chicago Press.

Langfred, C. W. 2000. "The paradox of self-management: Individual and group autonomy in work groups." *Journal of Organizational Behavior* 21,5 (August): 563–585.

Lapan, R. T., Albert Adams, S. Turner, and J. M. Hinkelman. 2000. "Seventh graders' vocational expectations." *Journal of Career Development* 26,3 (Spring): 215–229.

Larson, Magali Sarfatti. 1977. *The Rise of Professionalism: A Sociological Analysis.* Berkeley: University of California Press.

Lasley, Paul, F. Larry Leistritz, Linda M. Labao, and Katherine Meyer. 1995. *Beyond Amber Waves of Grain: Social and Economic Restructuring in the Heartland.* Boulder, Colo.: Westview.

Lee, Ching Kwan. 1998. *Gender and the South China Miracle: Two Worlds of Factory Women.* Berkeley: University of California Press.

Lee, Richard B. 1981. "Politics, sexual and nonsexual, in an equalitarian society: The !Kung San." Pp. 83–102 in General D. Berreman (editor), *Social Inequality: Comparative and Developmental Approaches.* New York: Academic Press.

Leicht, Kevin. 1989a. "Unions, plants, jobs and workers: An analysis of union satisfaction and participation." *Sociological Quarterly* 30,2 (Summer): 331–362.

Leicht, Kevin. 1989b. "On the estimation of union threat effects." *American Sociological Review* 54: 1035–1047.

Leidner, Robin. 1993. *Fast Food, Fast Talk: Service Work and the Routinization of Everyday Life.* Berkeley: University of California Press.

Leigh, Duane E., and Kirk D. Gifford. 1999. "Workplace transformation and worker upskilling." *Industrial Relations* 38,2 (April): 174–191.

Leigh, J. Paul. 1986. "Correlates of absence from work due to illness." *Human Relations* 39,1 (January): 81–100.

Leiter, Jeffrey. 1986. "Reactions to subordination: Attitudes of Southern textile workers." *Social Forces* 64,4 (June): 948–974.

Lengyel, Gyorgy, and Laszlo Neumann. 2002. "The sociology of work in Hungary." In Daniel B. Cornfield and Randy Hodson (editors), *Worlds of Work.* New York: Plenum.

Lenski, Gerhard. 1966. *Power and Privilege: A Theory of Social Stratification.* New York: McGraw-Hill.

Leventman, Paula Goldman. 1981. *Professionals out of Work.* New York: Free Press.

Levine, Adeline Gordon. 1982. *Love Canal.* Lexington, Mass.: D. C. Heath.

Levine, David I. 1995. *Reinventing the Workplace: How Business and Employees Can Both Win.* Washington, D.C.: Brookings Institute.

Levitan, Sar A., and Isaac Shapiro. 1987. *Working but Poor: America's Contradiction.* Baltimore: Johns Hopkins University Press.

Levitan, Sar A., and Robert Taggart. 1977. *Jobs for the Disabled.* Baltimore: Johns Hopkins University Press.

Levitt, Martin Jay. 1993. *Confessions of a Union Buster.* New York: Crown.

Levy, Frank. 1987. *Dollars and Dreams: The Changing American Income Distribution.* New York: Russell Sage Foundation.

Lewin, Tamar. 1984. "Atari layoff case highlights question of workers' rights." *Austin American Statesman,* July 25, p. G6.

Li, J. H., and Joachim Singelmann. 1998. "Gender differences in class mobility: A comparative study of the United States, Sweden, and West Germany." *Acta Sociologica* 41,4: 315–333.

Liebow, Elliot. 1967. *Tally's Corner.* Boston: Little, Brown.

Light, Ivan H. 1972. *Ethnic Enterprise in America: Business and Welfare Among Chinese, Japanese, and Blacks.* Berkeley: University of California Press.

Liker, Jeffrey K., W. Mark Fruin, and Paul S. Adler. 1999. *Remade in America: Transplanting and Transforming Japanese Management Systems.* New York: Oxford University Press.

Lillrank, Paul, and Noriaki Kano. 1989. *Continuous Improvement: Quality Circles in Japanese Industry.* Ann Arbor: University of Michigan Center for Japanese Studies.

Lincoln, James R., and Arne L. Kalleberg. 1990. *Culture, Control, and Commitment.* New York: Cambridge University Press.

Lincoln, James R., and Arne L. Kalleberg. 1996. "Commitment, Quits, and Work Organization in Japanese and U.S. Plants." *Industrial and Labor Relations Review* 50: 39–59.

Link, Bruce G., Bruce P. Dohrenwend, and Andrew E. Skodol. 1986. "Socioeconomic status and schizophrenia: Noisome occupational characteristics as a risk factor." *American Sociological Review* 51,2 (April): 242–258.

Lipset, Seymour Martin, and Gary Marks. 2000. *It Didn't Happen Here: Why Socialism Failed in the U.S.* New York: W. W. Norton.

Lipset, Seymour Martin, Martin A. Trow, and James S. Coleman. 1956. *Union Democracy.* Glencoe, Ill.: Free Press.

Little, Malcolm [Malcolm X] as told to Alex Haley. 1965. *The Autobiography of Malcolm X.* New York: Grove Press.

Litwack, Leon. 1962. *The American Labor Movement.* Englewood Cliffs, N.J.: Prentice Hall.

Lo, Jeannie. 1990. *Office Ladies Factory Women: Life and Work at a Japanese Company.* Armonk, N.Y.: M. E. Sharpe.

Logue, J., and J.S. Yates. 1999. "Worker ownership American style." *Economic and Industrial Democracy* 20,2 (May): 225–253.

Long, Richard J. 1984. "Microelectronics and quality of working life in the office: A Canadian perspective." Pp. 273–293 in Malcolm Warner (editor), *Microprocessors, Manpower and Society.* New York: St. Martin's.

Lorber, Judith. 1984. *Women Physicians: Careers, Status, and Power.* New York: Tavistock.

Lorenz, Edward H. 1992. "Trust and the flexible firm: International comparisons." *Industrial Relations* 31,3 (Fall): 455–472.

"The lowdown on literacy." 1994. *American Demographics* 16,6 (June): 6.

Lowe, Graham S. 1987. *Women in the Administrative Revolution: The Feminization of Clerical Work.* Toronto: University of Toronto Press.

Lowe, N., and M. Kenney. 1999. "Foreign investment and the global geography of production." *World Development* 27,8 (August): 1427–1443.

Lueck, Thomas J. 1988. "Thousands of executives seeking Wall Street jobs in bleak market." *New York Times* (January 17): 1, 32.

Luger, Michael Ian. 1996. "Quality-of-life differences and urban and regional outcomes." *Housing Policy Debate* 7,4: 749–771.

Lundman, Richard J. 2002. *In the Company of Cops.* New York: Oxford University Press.

Lustig, R. Jeffrey. 1985. "The politics of shutdown." *Journal of Economic Issues* 19,1 (March): 123–152.

Luthans, Fred, and D. L. Lockwood. 1984. "Toward an observation system for measuring leadership in natural settings." Pp. 117–141 in James G. Hunt, D. Hosking, C. Schriesheim, and R. Stewart (editors), *Leaders and Managers: International Perspectives on Managerial Behavior and Leadership.* New York: Pergamon.

Luthans, Fred, Stuart A. Rosenkrantz, and Harry W. Hennessey. 1985. "What do successful managers really do? An observation study of management activities." *Journal of Applied Behavioral Science* 21,3: 255–270.

MacCorquodale, Patricia M., and Gary Jensen. 1993. "Women in the law: Partners or tokens?" *Gender and Society* 7,4 (December): 582–593.

Majka, Linda C., and Theo J. Majka. 1982. *Farm Workers, Agribusiness, and the State.* Philadelphia: Temple University Press.

Maklan, D. M. 1977a. *The Four-Day Workweek: Blue Collar Adjustment to a Nonconventional Arrangement of Work and Leisure Time.* New York: Praeger.

Maklan, D. M. 1977b. "How blue collar workers on 4-day work weeks use their time." *Monthly Labor Review* 100, 8: 18–26.

Makower, Joel. 1981. *Office Hazards: How Your Job Can Make You Sick.* Washington, D.C.: Tilden.

Malambre, Alfred L., Jr., and Lindley H. Clark, Jr. 1992. "Productivity statistics for the service sector may understate gains." *Wall Street Journal* (August 12): 1.

Malecki, Edward J. 1984. "High technology and local economic development." *Journal of the American Planning Association* 50, 3 (Summer): 262–269.

Mangum, Garth, Donald Mayall, and Kristen Nelson. 1985. "The temporary help industry: A response to the dual internal labor market." *Industrial and Labor Relations Review* 38, 4 (July): 599–611.

Marchak, Patricia. 1983. *Green Gold.* Vancouver: University of British Columbia Press.

Marchak, M. Patricia. 1991. *The Integrated Circus: The New Right and the Restructuring of Global Markets.* Montreal: McGill-Queen's University Press.

Margolis, Diane Rothbard. 1979. *The Managers: Corporate Life in America.* New York: Morrow.

Marks, Abigail, Patricia Findlay, James Hine, Alan McKinlay, and Paul Thompson. 1998. "The politics of partnership? Innovation in employment relations in the Scottish spirits industry." *British Journal of Industrial Relations* 36, 2 (June): 209–226.

Mars, Gerald. 1982. *Cheats at Work.* London: Unwin.

Marsh, Robert M. 1992. "The difference between participation and power in Japanese factories." *Industrial and Labor Relations Review* 45, 2 (January): 250–257.

Marshall, F. Ray. 1987. *Unheard Voices: Labor and Economic Policy in a Competitive World.* New York: Basic Books.

Marshall, F. Ray, and Vernon M. Briggs, Jr. 1989. *Labor Economics,* 6th ed. Homewood, Ill.: Irwin.

Marshall, F. Ray, and Marc Tucker. 1992. *Thinking for a Living: Work, Skills, and the Future of the American Economy.* New York: Basic.

Martin, Susan Ehrlich. 1980. *Breaking and Entering: Policewomen on Patrol.* Berkeley: University of California Press.

Martinez, Sue, and Alan Ramo. 1980. "In the valley of the shadow of death." *In These Times* (October 8): 12–13.

Marx, Gary T. 1999. "Measuring everything that moves: The new surveillance at work." Pp. 165–189 in Ida Harper Simpson and Richard L. Simpson (editors), *Research in the Sociology of Work,* Volume 8: *Deviance in the Workplace.* Greenwich, Conn.: JAI Press.

Marx, Karl. 1959 [1844]. *The Economic and Philosophic Manuscripts of 1844.* Translated by Martin Milligan. Moscow: International Publishers.

Marx, Karl. 1967 [1887]. *Capital.* Vol. 1. New York: International Publishers.

Maslow, Abraham H. 1954. *Motivation and Personality.* New York: Harper and Brothers.

Massey, Douglas S. 1996. "Inequality in the age of extremes." *Demography* 33: 1–20.

Massey, Douglas S., and Nancy A. Denton. 1993. *American Apartheid.* Cambridge, Mass.: Harvard University Press.

Matthei, Julia A. 1982. *An Economic History of Women in America.* Brighton, England: Harvester.

Maume, David J., Jr. 1999. "Glass ceilings and glass escalators." *Work and Occupations* 26: 483–509.

McCarthy, Michael J. 1988. "Managers face dilemma with 'temps.' " *Wall Street Journal* (April 5): 27.

McDonald, Charles. 1992. "U.S. union membership in future decades." *Industrial Relations* 31, 1 (Winter): 13–30.

McElroy, Michael P., and Michael B. Hazzard. 1994. "Occupational staffing patterns at the four-digit SIC level." *Monthly Labor Review* 117, 2 (February): 30–33.

McGee, Leo, and Robert Boone, 1979. *The Black Rural Landowner.* Westport, Conn.: Greenwood Press.

McGregor, Douglas Murray. 1967. *The Professional Manager.* New York: McGraw-Hill.

McHugh, P.P., Joel Cutcher-Gershenfeld, and M. Polzin. 1999. "Employee stock ownership plans." *Economic and Industrial Democracy* 20, 4 (November): 535–560.

McMahon, Patrick J., and John H. Tschetter. 1986. "The declining middle class: A further analysis." *Monthly Labor Review* 109, 9 (September): 22–27.

McManus, Jim. 1987. "Boston lobstermen co-op has archdiocese backing." *National Catholic Reporter* August 28: 11.

Meiksins, Peter F., and Chris Smith. 1999. *Engineering Labour.* London: Verso.

Mendoza, Martha. 1997. "Federal work rules see many changes." *Columbus Dispatch* (December 14).

Mers, Gilbert. 1988. *Working the Waterfront: The Ups and Downs of a Rebel Longshoreman.* Austin: University of Texas Press.

Merton, Robert K. 1968. *Social Theory and Social Structure.* New York: Free Press.

Metcalfe, L. 2000. "European integration and globalization." *International Review of Administrative Sciences* 66, 1 (March): 119–124.

Miceli, Marcia P., M. Rehg, Janet P. Near, and K. C. Ryan. 1999. "Can laws protect whistle-blowers?" *Work and Occupations* 26, 1: 129–151.

Michel, Lawrence, and Paula B. Voos (editors). 1992. *Unions and Economic Competitiveness.* Armonk, N.Y.: M.E. Sharpe.

Milkman, Ruth (editor). 1985. *Women, Work and Protest: A Century of U.S. Women's Labor History.* Boston: Routledge and Kegan Paul.

Milkman, Ruth. 1991. *Japan's California Factories.* Los Angeles, Calif.: University of California Institute of Industrial Relations.

Milkman, Ruth. 1997. *Farewell to the Factory: Auto Workers in the Late Twentieth Century.* Berkeley: University of California Press.

Milkman, Ruth, and Cydney Pullman. 1991. "Technological change in an auto assembly plant." *Work and Occupations* 18, 2 (May): 123–147.

Miller, Ann R., Donald J. Treiman, Pamela S. Cain, and Patricia A. Roos (editors). 1980. *Work, Jobs, and Occupations: A Critical Review of the Dictionary of Occupational Titles.* Washington, D.C.: National Academy Press.

Miller, Delbert C., and William H. Form. 1980. *Industrial Sociology,* 3rd ed. New York: Harper and Row.

Miller, Gale. 1978. *Odd Jobs: The World of Deviant Work.* Englewood Cliffs, N.J.: Prentice Hall.

Miller, Gale. 1981. *It's a Living.* New York: St. Martin's.

Miller, Gale. 1991. *Enforcing the Work Ethic.* Albany, N.Y.: State University of New York Press.

Miller, Joanne, and Howard H. Garrison. 1982. "Sex roles: The division of labor at home and in the workplace." *Annual Review of Sociology* 8: 237–262.

Mills, C. Wright. 1951. *White Collar: The American Middle Classes.* New York: Oxford University Press.

Mills, C. Wright. 1959. *The Sociological Imagination.* London: Oxford University Press.

Mintzberg, Henry. 1973. *The Nature of Managerial Work.* New York: Harper and Row.

Mishel, Lawrence, Jared Bernstein, and John Schmitt. 2001. *The State of Working America.* Ithaca, N.Y.: Cornell University Press.

Mishel, Lawrence, and Paula B. Voos (editors). 1992. *Unions and Economic Competitiveness.* Armonk, N.Y.: Sharpe.

Mizruchi, Mark S. 1992. *The Structure of Corporate Power.* Cambridge, Mass.: Harvard University Press.

Moberg, David. 1986. "Steel contracts: Forging new alloys." *In These Times* (April 23): 2.

Moberg, David. 1993. "Striking back without striking." *In These Times* (April 5): 25–27.

Moberg, David. 1994. "Buy the friendly skies." *In These Times* (February 21): 26.

Moberg, David. 1998. "Child labor indicts the global marketplace." *In These Times* (June 14): 22–23.

Moberg, David. 1993. "Beyond borders: Some in U.S. labor are seeking solidarity with Mexican unions." *In These Times* (May 17): 8.

Molstad, Clark. 1986. "Choosing and coping with boring work," *Urban Life* 15, 2 (July): 215–236.

Moody's Industrial Manual. 1999. New York: Mergent.

Moore, Richard, and Elizabeth Marsis. 1984. "Telecommuting: Sweatshop at home sweet home?" *In These Times* (April 18–24): 12–13.

Morgan, Jerry. 1983. "Working in the automated office." *Newsday* (August 15).

Mogensen, Vernon L. 1996. "Office politics: Computers, labor, and the fight for safety and health." New Brunswick, N.J.: Rutgers University Press.

Mort, Jo-Ann. 2000. *Not Your Father's Union Movement.* New York: Verso.

Mortimer, Jeylan T., Jon Lorence, and Donald S. Kumka. 1986. *Work, Family, and Personality: Transition to Adulthood.* Norwood, N.J.: Ablex.

Mortimer, Jeylan T., and Michael D. Finch. 1996. *Adolescents, Work and Family.* Thousand Oaks, Calif.: Sage.

Mowshowitz, Abbe. 1997. "On the theory of virtual organization." *Systems Research and Behavioral Science* 14, 6 (November): 373–384.

Moyers, Gene. 1985. "The comparable worth debate." *Public Management* 67, 8 (August): 2–4.

Mueller, Charles W., Ashley Finley, Roderick D. Iverson, and James D. Price. 1999. "The effects of group racial composition on job satisfaction, organizational commitment, and career commitment." *Work and Occupations* 26: 187–219.

Muller-Jentsch, Walther. 2002. "The sociology of work in Germany." In Daniel B. Cornfield and Randy Hodson (editors), *Worlds of Work.* New York: Plenum.

Murphree, Mary C. 1981. *Rationalization and Satisfaction in Clerical Work: A Case Study of Wall Street Legal Secretaries.* Ph.D. dissertation, Department of Sociology, Columbia University.

Murphree, Mary C. 1987. "New technology and office tradition: The not-so-changing world of the secretary." Pp. 98–135 in Heidi I. Hartmann (editor), *Computer Chips and Paper Clips: Technology and Women's Employment,* Vol. 2. Washington, D.C.: National Academy Press.

Murphy, Michelle. 2000. "Toxicity in the details: The history of the women's office worker movement and occupational health in the late-capitalist office." *Labor History* 41, 2 (May): 189–213.

Murray, Michael A., and Tom Atkinson. 1981. "Gender differences in correlates of job satisfaction." *Canadian Journal of Behavioral Science* 13, 1 (June): 44–52.

Nahapiet, Janine, and Sumantra Ghoshal. 1998. "Social capital, intellectual capital, and the organizational advantage." *Academy of Management Review,* 23, 2 (April): 242–266.

Naisbitt, John. 1999. *High Tech/High Touch.* New York: Broadway.

Nasar, Sylvia. 1992. "The 1980s: A very good time for the very rich." *New York Times* (March 5): A1.

Nash, June. 1984. "We eat the mines and the mines eat us." Pp. 138–149 in Herbert Applebaum (editor), *Work in Market and Industrial Societies.* Albany, N.Y.: State University of New York Press.

Nash, June C. 1989. *From Tank Town to High Tech.* Albany, N.Y.: State University of New York Press.

National Commission for Employment Policy. 1979. *Final Report of the American Assembly on Youth Employment.* Harriman, N.Y.: Arden House.

National Commission on Employment and Unemployment Statistics. 1979. *Counting the Labor Force.* Washington, D.C.: U.S. Government Printing Office.

National Research Council. 1985. *Office Workstations in the Home.* Washington, D.C.: National Academy Press.

Naughton, Keith, and Mark Hossenball. 2000. "Ford vs. Firestone." *Newsweek,* September 11. www.msnbc.com/news/457861.

Neff, Walter S. 1985. *Work and Human Behavior,* 3rd ed. New York: Aldine.

Nelson, Daniel. 1975. *Managers and Workers: Origins of the New Factory System in the United States, 1880–1920.* Madison, Wisc.: University of Wisconsin Press.

Nelson, Harry. 1984. "VDTs questioned in pregnancy problems." *Austin American Statesman,* February 19.

Nelson, Robert L., and William P. Bridges. 1999. *Legalizing Gender Inequality: Courts, Markets and Unequal Pay for Women in the United States.* Cambridge, England: Cambridge University Press.

Neuberger, Albert. 1977. "The technical skills of the Romans." Pp. 516–539 in Carleton S. Coon (editor), *A Reader in Cultural Anthropology.* Huntington, N.Y.: Krieger.

"The New Salesperson." 1984. *Management Review* 73, 5 (May): 54.

Newman, Katherine S. 1999. *Falling from Grace: Downward Mobility in the Age of Affluence.* New York: Vintage.

Newman, Katherine S. 2000. *No Shame in My Game.* New York: Vintage.

Nippert-Eng, Christena E. 1996. *Home and Work: Negotiating Boundaries through Everyday Life.* Chicago: University of Chicago Press.

Noble, David F. 1997. *The Religion of Technology: The Divinity of Man and the Spirit of Invention.* New York: Knopf.

Noble, Kenneth B. 1986. "Unions push retraining plans." *New York Times* (March 23, Employment Outlook Supplement): 10–11.

Nordheimer, Jon. 1988. "Older migrants' years of toil in sun end in cold twilight." *New York Times* (May 29): 1, 14.

Norr, James L., and Kathleen L. Norr. 1978. "Work organization in modern fishing." *Human Organization* 37,2 (Summer): 163–171.

North, C. C., and P. K. Hatt. 1947. "Jobs and occupations: A popular evaluation." *Public Opinion News* 9,3 (September 1).

Norwood, Stephen H. 1990. *Labor's Flaming Youth: Telephone Operators and Worker Militancy.* Urbana, Ill.: University of Illinois Press.

Oakes, Guy. 1990. *The Soul of the Salesman: The Moral Ethos of Personal Sales.* London: Humanities Press International.

Oakley, Ann. 1975. *The Sociology of Housework.* New York: Pantheon.

O'Connell, Martin, and David E. Bloom. 1987. "Juggling jobs and babies: America's child care challenge." *Population Trends and Public Policy* 12 (February).

O'Connor, James. 1998. *Natural Causes.* New York: Gilford.

Ong, Aihwa. 1983. "Global industries and Malay peasants in Peninsular Malaysia." Pp. 426–441 in June Nash and Maria Patricia Fernandez-Kelly (editors), *Women, Men, and the International Division of Labor.* Albany, N.Y.: State University of New York Press.

Oppenheimer, Andres. 1986. "Haitians regret cashing in pigs for 'rich' U.S. hogs." *Austin American Statesman* (June 8): K7.

Oppenheimer, Valerie Kincade. 1970. *The Female Labor Force in the United States.* Population Monograph Series, No. 5. Berkeley, Calif.: Institute of International Studies, University of California.

Oppenheimer, Valerie Kincade. 1974. "The life cycle squeeze." *Demography* 11,2 (May): 227–246.

O'Reilly, James T. 1998. *Keeping Buildings Healthy.* New York: Wiley.

Orbach, Michael K. 1977. *Hunters, Seamen, and Entrepreneurs: The Tuna Seinermen of San Diego.* Berkeley: University of California Press.

O'Relley, Z. Edward. 1986. "The changing status of collectivized and private agriculture under central planning." *American Journal of Economics and Sociology* 45,1 (January): 9–16.

Ornstein, Michael D. 1976. *Entry into the American Labor Force.* New York: Academic Press.

Orr, Julian E. 1996. *Talking about Machines: An Ethnography of a Modern Job.* Ithaca, N.Y.: Cornell University Press.

Ortiz, Luis. 1998. "Union Response to Teamwork: The Case of Opel Spain." *Industrial Relations Journal* 29: 42–57.

Orwell, George. 1933. *Down and Out in London and Paris.* New York: Harcourt Brace Jovanovich.

Orwell, George. 1958. *The Road to Wigan Pier.* New York: Harcourt Brace Jovanovich.

Ospina, Sonia. 1996. *Illusions of Opportunity.* Ithaca, New York: Cornell University Press.

Osterman, Paul. 1999. *Securing Prosperity: The American Labor Market: How It Has Changed and What to Do about It.* Princeton: Princeton University Press.

Osterman, Paul. 1986. "The impact of computers on the employment of clerks and managers." *Industrial and Labor Relations Review* 39,2 (January): 175–186.

Ouchi, William G. 1981. *Theory Z: How American Business Can Meet the Japanese Challenge.* Reading, Mass.: Addison-Wesley.

Ozaki, Muneto (editor). 1999. *Negotiating Flexibility.* Geneva, Switzerland: International Labour Office.

Pacheco, Ferdie. 1997. *Pacheco's Art of Ybor City.* Gainesville, Fl.: University of Florida Press.

Palmer, Bryan D. 1983. *Working-Class Experience: The Rise of Canadian Labor, 1800–1980.* Toronto: Butterworth.

Papanek, Hanna. 1973. "Men, women, and work: Reflections on the two-person career." *American Journal of Sociology* 78,4 (January): 852–872.

Parcel, Toby (editor). 1999. *Work and Family. Research in the Sociology of Work,* Volume 7. Greenwich, Conn.: JAI Press.

Parker, Mike, and Jane Slaughter. 1994. *Working Smart: A Union Guide to Participation Programs and Reengineering.* Detroit, Mich.: Labor Notes.

Parker, Robert E. 1994. *Flesh Peddlers and Warm Bodies: The Temporary Help Industry and Its Workers.* New Brunswick, N.J.: Rutgers University Press.

Parsons, Talcott. 1968. "Professions." In David L. Sills (editor), *International Encyclopedia of the Social Sciences.* New York: Macmillan.

Pascarella, Perry. 1984. *The New Achievers: Creating a Modern Work Ethic.* New York: Macmillan.

Paules, Greta Foff. 1991. *Dishing It Out: Power and Resistance among Waitresses in a New Jersey Restaurant.* Philadelphia: Temple University Press.

Pavalko, Eliza K., and Glen H. Elder, Jr. 1993. "Women behind the men: Variations in wives' support of husbands' careers." *Gender and Society* 7,4 (December): 548–567.

Pear, Robert. 1987. "Women reduce lag in earnings but disparities with men remain." *New York Times* (September 4): 1, 7.

Peirce, Neal R., and Carol Steinbach. 1981. "But will the small businesses survive?" Pp. 4–9 in Robert Friedman and William Schweke (editors), *Expanding the Opportunity to Produce: Revitalizing the American Economy through New Enterprise Development.* Washington, D.C.: Corporation for Enterprise Development.

Pelled, Lisa Hope, Kathleen M. Eisenhardt, and Katherine R. Xin. 1999. "Exploring the black box: An analysis of work group diversity, conflict, and performance." *Administrative Science Quarterly,* 44,1 (March): 1–28.

Peltz, Michael, and Marc A. Weiss. 1984. "State and local government roles in industrial innovation." *Journal of the American Planning Association* 50,3 (Summer): 270–279.

Pendleton, Andrew, Nicholas Wilson, and Mike Wright. 1998. "The perception and effects of share ownership: Empirical evidence from employee buy-outs." *British Journal of Industrial Relations* 36,1 (March): 99–123.

Penn, Roger, and Hilda Scattergood. 1985. "Deskilling or enskilling? An empirical investigation of recent theories of the labour process." *The British Journal of Sociology* 36,4 (December): 611–630.

Perlin, Michael L. 2000. *The Hidden Prejudice: Mental Disability on Trial.* Washington, D.C.: American Psychological Association.

Perrow, Charles. 1986. *Complex Organizations: A Critical Essay,* 3rd ed. Glenview, Ill.: Scott, Foresman.

Perrow, Charles. 1984. *Normal Accidents: Living with High-Risk Technologies.* New York: Basic.

Perrucci, Robert. 1994. *Japanese Auto Transplants in the Heartland.* Hawthorne, N.Y.: Aldine de Gruyter.

Perrucci, Robert, and Earl Wysong. 1999. *The New Class Society.* Lanham, Md.: Rowman and Littlefield.

Perry, Stewart E. 1978. *San Francisco Scavengers: Dirty Work and the Pride of Ownership.* Berkeley, Calif.: University of California Press.

Personick, Valerie A. 1987. "Industry output and employment through the end of the century." *Monthly Labor Review* 110,9 (September): 30–45.

Peters, Thomas J. 1997. *The Circle of Innovation.* New York: Knopf.

Peterson, Richard B., Thomas W. Lee, and Barbara Finnegan. 1992. "Strategies and tactics in union organizing campaigns." *Industrial Relations* 31,2 (Spring): 370–381.

Peterson, Trond. 1992. "Payment systems and the structure of inequality: Conceptual issues and an analysis of salespersons in department stores." *American Journal of Sociology* 98,1 (July): 67–104.

Pfeffer, Jeffrey. 1998. *The Human Equation: Building Profits by Putting People First.* Boston: Harvard Business School Press.

Phelps, Edmund S. 1997. *Rewarding Work: How to Restore Participation and Self-Support to Free Enterprise.* Cambridge, Mass.: Harvard University Press.

Picou, J. Steven, and Duane A. Gill. 1993. "Disaster and long-term stress: Impacts of the *Exxon Valdez* oil spill on commercial fishermen." Paper presented at the Southwestern Sociological Association meetings, New Orleans.

Piel, Gerard. 1994. "Tech-eyed optimist." *Nation* 258,18 (May 9): 614.

"Pink slips for white collars." 1986. *U.S. News and World Report* (March 17): 46–47.

Pisato, M. 1997. "Mobility regimes and generative mechanisms: A comparative analysis of Italy and the United States." *European Journal of Sociology* 13,2 (September): 179–198.

Pleck, Joseph H. 1977. "The work-family role system." *Social Problems* 24,4 (April): 417–427.

Polanyi, Karl. 1957. *The Great Transformation.* Boston: Beacon.

Polivka, Anne E., and Jennifer M. Rothgeb. 1993. "Redesigning the CPS Questionnaire." *Monthly Labor Review* 116,9 (September): 10–28.

Polsky, Andrew J. 1991. *The Rise of the Therapeutic State.* Princeton, N.J.: Princeton University Press.

Portes, Alejandro, Carlos Dore-Cabral, and Patricia Landolt. 1997. *The Urban Caribbean.* Baltimore, Md.: Johns Hopkins University Press.

Portes, Alejandro, and Leif Jensen. 1989. "The enclave and the entrants: Patterns of ethnic enterprise in Miami before and after Mariel." *American Sociological Review* 54,6 (December): 929–950.

Portes, Alejandro, and Saskia Sassen-Koob. 1987. "Making it underground: Comparative material on the informal sector in western market economies." *American Journal of Sociology* 93,1 (July): 30–61.

Poulantzas, Nicos. 1975. *Classes in Contemporary Capitalism.* Translated by David Fernbach. London: New Left Books.

Powell, Gary N. (editor). 1999. *Handbook of Gender and Work.* Thousand Oaks, Calif.: Sage.

Powell, Michael J. 1989. *From Patrician to Professional Elite: The Transformation of the New York City Bar Association.* New York: Russell Sage Foundation.

Presser, Harriet B. 1988. "Shift work and child care among young dual-earner American parents." *Journal of Marriage and the Family* 50,1 (February): 133–149.

Presser, Harriet B. 2000. "Nonstandard work schedules and marital instability." *Journal of Marriage and the Family.* 62: 93–110.

Pringle, Rosemary. 1988. *Secretaries Talk: Sexuality, Power, and Work.* Sydney: Allen and Unwin.

Punch, Maurice (editor). 1996. *Dirty Business: Exploring Corporate Misconduct.* Thousand Oaks, Calif.: Sage.

Rabinow, Paul. 1999. *French DNA: Trouble in Purgatory.* Chicago: University of Chicago Press.

Radforth, Ian. 1986. "Logging pulpwood in Northern Ontario." Pp. 245–280 in Craig Heron and Robert Storey (editors), *On the Job: Confronting the Labour Process in Canada.* Kingston: McGill-Queen's University Press.

Raeburn, Nicole C. 2002. *Inside Out: The Struggle for Lesbian, Gay, and Bisexual Rights in the Workplace.* Minneapolis: University of Minnesota Press.

Rajan, Amin. 1985. "Office technology and clerical skills." *Futures: The Journal of Forecasting and Planning* 17, 4 (August): 410–413.

Randall, Adrian. 1991. *Before the Luddites.* New York: Cambridge University Press.

Rankin, Tom. 1990. *New Forms of Work Organization.* Toronto, Canada: University of Toronto Press.

Rees, Albert. 1989. *The Economics of Trade Unions,* 3rd ed. Chicago: University of Chicago Press.

Reeves, Joy B., and Ray Darville. 1986. "Female clerical workers in academic settings: An empirical test of the gender model." *Sociological Inquiry* 56, 1 (Winter): 105–124.

Reich, Robert B. 1992. *The Work of Nations: Preparing Ourselves for the 21st Century.* New York: Knopf.

Reid, Samuel Richardson. 1976. *The New Industrial Order: Concentration, Regulation, and Public Policy.* New York: McGraw-Hill.

Reiter, Ester. 1991. *Making Fast Food: From the Frying Pan into the Fryer.* Montreal: McGill-Queen's University Press.

Reskin, Barbara F. 1998. *The Realities of Affirmative Action in Employment.* Washington, D.C.: American Sociological Association.

Reskin, Barbara F., and Debra Branch McBrier. 2000. "Why not ascription? Organizations' employment of male and female managers." *American Sociological Review* 65, 2 (April): 210–233.

Reskin, Barbara F., and Irene Padavic. 1994. *Women and Men at Work.* Thousand Oaks, Calif.: Pine Forge.

Riemer, Jeffrey. 1977. "Becoming a journeyman electrician." *Work and Occupations* 4: 87–98.

Rifkin, Jeremy. 1995. *The End of Work.* New York: Putnam.

Rinehart, James, Christopher Huxley, and David Robertson. 1997. *Just Another Car Factory? Lean Production and Its Discontents.* Ithaca, N.Y.: Cornell University Press.

Risen, James. 1984. "Lacking skills, displaced auto workers feel crunch." *Los Angeles Times* (December 30): 119.

Ritzer, George. 2000. *The McDonaldization of Society.* Thousand Oaks, Calif.: Pine Forge.

Ritzer, George A., and David Walczak. 1988. "Rationalization and the deprofessionalization of physicians." *Social Forces* 67, 1 (September): 1–22.

Roberson, James E. 1998. *Japanese Working Class Lives: An Ethnographic Study of Factory Workers.* London: Routledge.

Roberts, Glenda S. 1994. *Staying on the Line: Blue-Collar Women in Contemporary Japan.* Honolulu: University of Hawaii Press.

Roethlisberger, F. J., and William J. Dickson. 1939. *Management and the Worker.* Cambridge, Mass.: Harvard University Press.

Rohlen, Thomas P. 1974. *For Harmony and Strength.* Berkeley: University of California Press.

Rollins, Judith. 1985. *Between Women: Domestics and Their Employers.* Philadelphia: Temple University Press.

Romero, Mary. 1992. *Maid in the USA.* New York: Routledge.

Roscigno, Vincent J., and Martha Crowley. 2001. "Rurality, institutional disadvantage, and achievement." *Rural Sociology* (June).

Roscigno, Vincent J., and William F. Danaher. 2001. "Media and mobilization: The case of radio and Southern textile worker insurgency, 1929–1934." *American Sociological Review* 66, 1 (February): 21–48.

Rosenfeld, Rachel Ann, M. E. Van Buren, and Arne L. Kalleberg. 1998. "Gender differences in supervisory authority: Variations among industrialized democracies." *Social Science Research* 27, 1 (March): 23–49.

Rosenthal, Neal H. 1985. "The shrinking middle class: Myth or reality?" *Monthly Labor Review* 108, 3 (March): 3–10.

Rosner, David, and Gerald Markowitz. 1991. *Deadly Dust.* Princeton, N.J.: Princeton University Press.

Ross, Bob. 1993. "Bottom line for topless dancers." *USA Today* (June 4): 2.

Ross, Catherine, and Marylyn P. Wright. 1998. "Women's work, men's work, and the sense of control." *Work and Occupations* 25: 333–355.

Ross, R., and K. Trachte. 1983. "Global cities and global classes: The peripheralisation of labour in New York City." *Review* 6: 393–431.

Rossi, Peter M. 1989. *Down and Out in America: The Origins of Homelessness in the United States.* Chicago: University of Chicago Press.

Rostow, Walt W. 1998. *The Great Population Spike and After.* New York: Oxford University Press.

Rothman, Robert A. 1984. "Deprofessionalization: The case of law in America." *Work and Occupations* 11, 2 (May): 183–206.

Rothman, Robert A. 1998. *Working: Sociological Perspectives,* 2nd ed. Englewood Cliffs, N.J.: Prentice Hall.

Rothschild, Joyce, and Terance D. Miethe. 1999. "Whistle-blower disclosures and management retaliation." *Work and Occupations* 26, 1 (February): 107–128.

Rothschild, Joyce. 2000. "Creating a just and democratic workplace." *Contemporary Sociology* 29, 1 (January): 195–213.

Rothschild, Joyce, and Raymond Russell. 1986. "Alternatives to bureaucracy: Democratic participation in the economy." Pp. 307–328 in Ralph H. Turner and James F. Short, Jr. (editors), *The Annual Review of Sociology,* Vol. 12. Palo Alto, Calif.: Annual Reviews.

Roy, Donald. 1952. "Quota restriction and goldbricking in a machine shop." *American Sociological Review* 57, 5 (March): 427–442.

Rueschemeyer, Dietrich. 1983. "Professional autonomy and the social control of expertise." Pp. 38–58 in Robert Dingwall and Philip Lewis (editors), *The Sociology of the Professions.* New York: St. Martin's.

Rumberger, Russell W. 1981. "The changing skill requirements of jobs in the U.S. economy." *Industrial and Labor Relations Review* 34, 4: 578–589.

Russell, Raymond, and Veljko Rus (editors). 1991. *International Handbook of Participation in Organizations,* Volume 2, *Ownership and Participation.* Oxford: Oxford University Press.

Rutten, Rosanne. 1982. *Women Workers of Hacienda Milagros.* Amsterdam: Universiteit van Amsterdam Press.

Sabel, Charles F., and Michael J. Piore. 1984. *The Second Industrial Divide.* New York: Basic Books.

Sabel, Charles F., and Jonathan Zeitlin. 1997. *Worlds of Possibilities: Flexibility and Mass Production in Western Industrialization.* New York: Cambridge University Press.

Salpukas, Agis. 1984. "Taking the pulse of trucking." *New York Times* (September 8).

Sampson, Anthony. 1995. *Company Man: The Rise and Fall of Corporate Life.* New York: Random House.

Sanchez, Laura. 1993. "Women's power and the gendered division of domestic labor in the Third World." *Gender and Society* 7, 3 (September): 434–489.

Sanday, Peggy Reeves. 1981. *Female Power and Male Dominance.* Cambridge, England: Cambridge University Press.

Santa Clara County v. Southern Pacific Railroad Co. 1886. 118 U.S. 394, 396.

Santiago, Jack. 1989. *Miles of Smiles, Years of Struggle: Stories of Black Pullman Porters.* Urbana, Ill.: University of Illinois Press.

Saunders, Conrad. 1981. *Social Stigma of Occupations.* Westmead, England: Gower.

Scherer, Frederick M. 1999. *New Perspectives on Economic Growth and Technological Innovation.* Washington, D.C.: Brookings Institute.

Schlesinger, Robert J. 1984. "Industrial robots, work and industry: Past, present and future." Pp. 11–30 in Malcolm Warner (editor), *Microprocessors, Manpower and Society.* New York: St. Martin's.

Schlossberg, Stephen I., and Steven M. Fetter. 1986. *U.S. Labor Law and the Future of Labor–Management Cooperation.* Washington, D.C.: U.S. Department of Labor.

Schneider, Keith. 1987. "Oklahoma suicides show crisis on farm has not yet passed." *New York Times* (August 17): 1.

Schor, Juliet B. 1992. *The Overworked American.* New York: Basic Books.

Schroedel, Jean Reith. 1985. *Alone in a Crowd: Women in the Trades Tell Their Stories.* Philadelphia: Temple University Press.

Schudson, Michael. 1984. *Advertising: The Uneasy Persuasion.* New York: Basic Books.

Schwartzman, Helen B. 1993. *Ethnography in Organizations.* Qualitative Research Methods Series, #27. Newbury Park, Calif.: Sage.

Scott, W. Richard. 1998. *Organizations: Rational, Natural, and Open Systems,* 4th ed. Englewood Cliffs, N.J.: Prentice Hall.

Sears, David O., Jim Sidanius, and Lawrence Bobo (editors). 2000. *Racialized Politics.* Chicago: University of Chicago Press.

Sears, Robert R., Eleanor E. Maccoby, and Harry Levin. 1957. *Patterns of Child Rearing.* New York: Harper and Row.

Seeman, Melvin. 1959. "On the meaning of alienation." *American Sociological Review* 24, 6 (December): 783–791.

Seifert, Hartmut. 1991. "Employment effects of working time reduction in the former Federal Republic of Germany." *International Labour Review* 130, 4: 495–510.

Senker, Peter, and Mark Beasley. 1986. "Computerized production and inventory control systems: Some skill and employment implications." *Industrial Relations Journal* 16, 3 (Special Issue): 52–57.

Sennett, Richard, and Jonathan Cobb. 1972. *The Hidden Injuries of Class.* New York: Random House.

Shaevitz, Marjorie Hansen, and Morton H. Shaevitz. 1980. *Making It Together as a Two-Couple.* Boston: Houghton Mifflin.

Shaiken, Harley. 1984. *Work Transformed: Automation and Labor in the Computer Age.* New York: Holt, Rinehart and Winston.

Shaiken, Harley, S. Lopez, and I. Mankita. 1997. "Two routes to team production: Saturn and Chrysler compared." *Industrial Relations* 31, 1 (January): 17–45.

Shalom, Stephen Rosskamm. 1993. *Imperial Alibis: Rationalizing U.S. Intervention after the Cold War.* Boston: South End.

Shenhav, Yehouda A. 1988. "Abandoning the research bench: Individual, organizational, and environmental accounts." *Work and Occupations* 15, 1 (February): 5–23.

Shorrock, Tim. 1983. "Fifteen years of neglect have left the steel industry in crisis." *Multinational Monitor* 4, 6 (June): 12–13.

Shostak, Arthur B. 1999. *CyberUnion: Empowering Labor through Computer Technology.* Armonk, N.Y.: M.E. Sharpe.

Shrivastava, Paul. 1987. *Bhopal: Anatomy of a Crisis.* Cambridge, Mass.: Ballinger.

Siegel, Lenny, and John Markoff. 1985. *The High Cost Of High Tech: The Dark Side of the Chip.* New York: Harper and Row.

Sikora, Martin. 1993. "Megadeal response to market pressures." *Mergers and Acquisitions* 28, 2 (September/October): 6–8.

Silverman, Bertram, and Murray Yanowitch. 2000. *New Rich, New Poor, New Russia.* Armonk, N.Y.: M.E. Sharpe.

Simpson, Richard L. 1985. "Social control of occupations and work." *Annual Review of Sociology* 11: 415–436.

Skocpol, Theda. 2000. *The Missing Middle: Working Families and the Future of American Social Policy.* New York: W.W. Norton.

Slessarev, Helene. 1997. *The Betrayal of the Urban Poor.* Philadelphia: Temple University Press.

Smith, Adam. 1937 [1776]. *The Wealth of Nations.* New York: Random House.

Smith, John P. 1987. "The social and economic correlates of bankruptcy during the farm fiscal crisis 1970–1987." *Mid-American Review of Sociology* 12 (Spring): 35–53.

Smith, Margo L. 1989. "Where is Maria now? Former domestic workers in Peru." Pp. 127–142 in Elsa M. Chaney and Mary Garcia Castro (editors), *Muchachas No More: Household Workers in Latin America and the Caribbean.* Philadelphia: Temple University Press.

Smith, Roy C., and Ingo Walter. 1997. *Global Banking.* New York: Oxford University Press.

Smith, Timothy K. 1993. "Chiropractors seeking to expand practices take aim at children." *Wall Street Journal* (March 18).

Smith, Vicki. 1990. *Managing the Corporate Interest: Control and Resistance in an American Bank.* Berkeley: University of California Press.

Smith, Vicki. 1996. "Employee involvement, involved employees: Participative work arrangements in a white-collar service occupation." *Social Problems* 43,2 (May): 166–179.

Smith, Vicki. 1998. "The fractured world of the temporary worker: Power, participation, and fragmentation in the contemporary workplace." *Social Problems* 45,4 (November): 411–430.

Snow, David A., and Leon Anderson. 1993. *Down on their Luck: A Study of Homeless People.* Berkeley, Calif.: University of California Press.

Sorenson, Olav, and Pino G. Audia. 2000. "The social structure of entrepreneurial activity: Geographic concentration of footwear production in the United States, 1940–1989." *American Journal of Sociology* 106,2 (September): 424–462.

South, Scott J., Charles M. Bonjean, Judy Corder, and William T. Markham. 1982. "Sex and power in the federal bureaucracy." *Work and Occupations* 9,2 (May): 233–254.

Spain, Daphne, and Suzanne M. Bianchi. 1996. *Balancing Act: Motherhood, Marriage and Employment Among American Women.* New York: Russell Sage.

Spanier, Graham G., and Paul C. Glick. 1980. "The life cycle of American families: An expanded analysis." *Journal of Family History,* 5,1 (Spring): 97–111.

Spenner, Kenneth I. 1985. "The upgrading and downgrading of occupations: Issues, evidence, and implications for education." *Review of Educational Research* 55,2 (Summer): 125–154.

Spenner, Kenneth I. 1990. "Skill: Meanings, methods, and measures." *Work and Occupations* 17,4 (November): 339–421.

Spradley, James P., and Brenda J. Mann. 1975. *The Cocktail Waitress.* New York: Wiley.

Sprouse, Martin. 1992. *Sabotage in the American Workplace.* San Francisco, Calif.: Pressure Drop.

Spruill, Charles R. 1982. *Conglomerates and the Evolution of Capitalism.* Carbondale: Southern Illinois University Press.

Staines, Graham L., and Joseph H. Pleck. 1983. *The Impact of Work Schedules on the Family.* Ann Arbor: Survey Research Center, Institute for Social Research, University of Michigan.

Standing, Guy. 1996. *Russian Unemployment and Enterprise Restructuring.* New York: St. Martin's.

Standing, Guy, and Victor Tokman (editors). 1991. *Towards Social Adjustment: Labour Market Issues in Structural Adjustment.* Geneva: International Labour Office.

Stark, David C., and Laszlo Bruszt. 1998. *Postsocialist Pathways.* Cambridge, England: Cambridge University Press.

Starr, Paul. 1982. *The Social Transformation of American Medicine.* New York: Basic Books.

Stearns, Linda Brewster, and Kenneth D. Allan. 1996. "Economic behavior in institutional environments: The corporate merger wage of the 1980s." *American Sociological Review* 61,4 (August): 699–718.

Steele, Paul D., and Robert L. Hubbard. 1985. "Management styles, perceptions of substance abuse, and employee assistance programs in organizations." *Journal of Applied Behavioral Science* 21,3: 271–286.

Steinbeck, John. 1939. *The Grapes of Wrath.* New York: Viking.

Steinman, David, and Samuel S. Epstein. 1995. *The Safe Shopper's Bible.* New York: Macmillan.

Stevens, Fred, Frans van der Horst, Frans Nijhuis, and Silvia Bours. 2000. "The division of labour in vision care: Professional competence in a system of professions." *Sociology of Health and Illness* 22,4 (July): 431–452.

Stevens, Gillian, and Joo Hyun Cho. 1985. "Socioeconomic indexes and the new 1980 Census occupational classification scheme." *Social Science Research* 14,2 (June): 142–168.

Steward, Barbara. 2000. "Changing times: The meaning, measurement, and use of time in teleworking." *Time and Society* 9,1 (March): 57–74.

Stone, Katherine. 1974. "The origins of job structures in the steel industry." *Review of Radical Political Economy* 6: 113–173.

Strauss, A. 1978. *Negotiations.* San Francisco: Jossey Bass.

Streeck, Wolfgang. 1996. *German Capitalism.* Notre Dame, Ind.: The Helen Kellogg Institute.

Strober, Myra H., and Agnes Miling Kaneko Chan. 1999. *The Road Winds Uphill all the Way: Gender, Work, and Family in the United States and Japan.* Cambridge, Mass.: MIT Press.

Strober, Myra H., and Carolyn L. Arnold. 1987. "Integrated circuits/segregated labor: Women in computer-related occupations and high-tech industries." Pp. 136–182 in Heidi I. Hartmann (editor), *Computer Chips and Paper Clips: Technology and Women's Employment.* Vol. 2. Washington, D.C.: National Academy Press.

Strom, Sharon Hartman. 1987. " 'Machines instead of clerks': Technology and the feminization of bookkeeping, 1910–1950." Pp. 63–97 in Heidi I. Hartmann (editor), *Computer Chips and Paper Clips: Technology and Women's Employment.* Vol. 2. Washington, D.C.: National Academy Press.

Strom, Sharon Hartman. 1992. *Beyond the Typewriter: Gender, Class, and the Origins of Modern American Office Work, 1900–1930.* Urbana, Ill.: University of Illinois Press.

Sullivan, Kathleen. 1985. "Technology keeps track of workers." *Austin American Statesman* (December 12): HI.

Sullivan, Teresa A. 1978. *Marginal Workers, Marginal Jobs: The Underutilization of American Workers.* Austin: University of Texas Press.

Sullivan, Teresa A. 1989. "Women and minority workers in the new economy: Optimistic, pessimistic, and mixed scenarios." *Work and Occupations* 16,4 (November): 395–415.

Sullivan, Teresa A. 1990. "The decline of occupation." Pp. 1–31 in M. Hallinan et al. (editors), *Change in Societal Institutions.* New York: Plenum.

Sullivan, Teresa A. 1992. "Women and minority workers in the new economy: Optimistic, pessimistic, and mixed scenarios." *Work and Occupations* 16,4 (November): 393–415.

Sullivan, Teresa A. 1996. "The cashier complex and the transformation of occupations." Pp. 127–141 in Dennis L. Peck and J. Selwyn Hollingsworth (editors), *Demographic and Structural Change: The Effects of the 1980s on American Society.* Westport, Conn.: Greenwood.

Sullivan, Teresa A., and Jan E. Mutchler. 1985. "Equal pay or equal work? The implications of underemployment for comparable worth studies." Pp. 175–192 in R. L. Simpson and I. H. Simpson (editors), *Research in the Sociology of Work.* Greenwich, Conn.: JAI Press.

Sullivan, Teresa A., Elizabeth Warren, and Jay L. Westbrook. 2000. *The Fragile Middle Class: Americans in Debt.* New Haven, Conn.: Yale University Press.

Sussman, Marvin B., Suzanne K. Steinmetz, and Gary W. Peterson. 1999. *Handbook of Marriage and the Family.* New York: Plenum.

Szell, Gyorgy (editor). 1992. *Concise Encyclopedia of Participation and Co-Management.* Berlin: Walter de Gruyter.

Taggart, James H., and Michael C. McDermott. 1993. *The Essence of International Business.* New York: Prentice Hall.

Tausky, Carl. 1984. *Work and Society.* Itasca, Ill.: F. E. Peacock.

Tausky, Curt. 1991. "*Perestroika* in the USSR and China." *Work and Occupations* 18,1 (February): 94–108.

Taylor, Frederick W. 1911. *The Principles of Scientific Management.* New York: Harper and Row.

Terkel, Studs. 1974. *Working.* New York: Avon.

Thackray, John. 1982. "The American takeover war." *Management Today* (September): 82–86.

Thomas, Robert J. 1990. "Blue-collar careers." Pp. 354–379 in M. B. Arthur, D. T. Hall, and B. S. Lawrence (editors), *Handbook of Career Theory.* New York: Cambridge University Press.

Thompson, Paul. 1983. *The Nature of Work.* London: Macmillan.

Thurow, Lester C. 1975. *Generating Inequality.* New York: Basic Books.

Tilly, Chris. 1996. *Half a Job: Bad and Good Part-Time Jobs in a Changing Labor Market.* Philadelphia: Temple University Press.

Tilly, Chris, and Charles Tilly. 1998. *Work Under Capitalism.* Boulder, Colo.: Westview.

Tilly, Louise A., and Joan W. Scott. 1978. *Women, Work and Family.* New York: Holt, Rinehart and Winston.

Tivendell, J., and C. Bourbonnais. 2000. "Job insecurity in a sample of Canadian civil servants as a function of personality and perceived job characteristics." *Psychological Reports* 87,1 (August): 55–60.

Tolbert, Charles M., Thomas A. Lyson, and Michael D. Irwin. 1998. "Local capitalism, civic engagement, and socioeconomic well-being." *Social Forces* 77,2 (December): 401–427.

Tolbert, Pamela S., and Phyllis Moen. 1998. "Men's and women's definitions of 'good' jobs." *Work and Occupations* 25: 168–194.

Toledo, Enrique de la Garza. 2002. "The sociology of work in Mexico." In Daniel B. Cornfield and Randy Hodson (editors), *Worlds of Work.* New York: Plenum.

Tomaskovic-Devey, Donald. 1993. *Gender and Racial Inequality at Work.* Ithaca, N.Y.: Industrial Relations Review Press.

Tonelson, Alan. 2000. *The Race to the Bottom: Why a Worldwide Worker Surplus and Uncontrolled Free Trade are Sinking American Living Standards.* Boulder, Colo.: Westview.

Torstendahl, Rolf, and Michael Burrage (editors). 1990. *The Formation of Professions: Knowledge, State, and Strategy.* Newbury Park, Calif.: Sage.

Treiman, Donald J. 1977. *Occupational Prestige in Comparative Perspective.* New York: Academic Press.

Trice, Harrison M. 1993. *Occupational Subcultures in the Workplace.* Ithaca, N.Y.: Industrial and Labor Relations.

Trice, Harrison H., and Janice M. Beyer. 1992. *The Cultures of Work Organizations.* Englewood Cliffs, N.J.: Prentice Hall.

Troyer, Lisa, Charles W. Mueller, and Pavel I. Osinsky. 2000. "Who's the boss? A role-theoretic analysis of customer work." *Work and Occupations* 27,3 (August): 406–427.

Tucker, James. 1999. *The Therapeutic Corporation.* New York: Oxford University Press.

Tulin, Roger C. 1984. *A Machinist's Semi-Automated Life.* San Pedro, Calif.: Singlejack.

Turner, Lowell. 1998. *Fighting for Partnership: Labor and Politics in Unified Germany.* Ithaca, N.Y.: Cornell University Press.

Uchitelle, Louis. 1994. "Job losses don't let up even as hard times ease." *New York Times* (March 22).

United Nations, Industrial Development Organization. 2000. *International Yearbook of Industrial Statistics.* Vienna: U.N. Industrial Development Organization.

U.S. Bureau of the Census. 1943. *U.S. Census of Population, 1940: Comparative Occupational Statistics, 1870–1940.* Authored by Alba M. Edwards. Washington, D.C.: U.S. Government Printing Office.

U.S. Bureau of the Census. 1952. *Statistical Abstract of the United States,* 72nd ed. Washington, D.C.: U.S. Government Printing Office.

U.S. Bureau of the Census. 1975. *Historical Statistics of the United States, Colonial Times to 1970.* Bicentennial edition, part 1. Washington, D.C.: U.S. Government Printing Office.

U.S. Bureau of the Census. 1981. *Statistical Abstract of the United States,* 101st ed. Washington, D.C.: U.S. Government Printing Office.

U.S. Bureau of the Census. 1982. *Statistical Abstract of the United States,* 102nd ed. Washington, D.C.: U.S. Government Printing Office.

U.S. Bureau of the Census. 1984. *Census of Population, 1980: Detailed Occupation of the Experienced Civilian Labor Force by Sex for the United States and Regions, 1980 and 1970.* Supplemental Report 80-S1-15. Washington, D.C.: U.S. Government Printing Office.

U.S. Bureau of the Census. 1986. *After-School Care of School-Age Children.* Current Population Reports, Series P. 23, No. 149. Washington, D.C.: U.S. Government Printing Office.

U.S. Bureau of the Census. 1987a. *Census of Manufacturers.* Washington, D.C.: U.S. Government Printing Office.

U.S. Bureau of the Census. 1987b. *Census of Retail Trade.* Washington, D.C.: U.S. Government Printing Office.

U.S. Bureau of the Census. 1987c. *Census of Service Industries.* Washington, D.C.: U.S. Government Printing Office.

U.S. Bureau of the Census. 1987d. "Who's minding the kids?" *Statistical Brief* SB-2-87. Washington, D.C.: U.S. Government Printing Office.

U.S. Bureau of the Census. 1991. *Statistical Abstract of the United States,* 111th ed. Washington, D.C.: U.S. Government Printing Office.

U.S. Bureau of the Census. 1991a. *Money Income of Households, Families, and Persons in the U.S., 1990.* Washington, D.C.: U.S. Government Printing Office.

U.S. Bureau of the Census. 1991b. "Poverty in the United States: 1990." *Current Population Reports.* Series P-60, no. 16. Washington, D.C.: U.S. Government Printing Office.

U.S. Bureau of the Census. 1991c. "Who's supporting the kids?" *Statistical Brief* SB-91-18, October.

U.S. Bureau of the Census. 1992a. *1987 Census of Manufacturing: Concentration Ratios in Manufacturing.* MC87-S-6, February. Washington, D.C.: U.S. Government Printing Office.

U.S. Bureau of the Census. 1992b. "Where the jobs were: Job creation in the late 1980s." *Statistical Brief* SB-92-3: 2.

U.S. Bureau of the Census. 1993. *Statistical Abstract of the United States,* 113th ed. Washington, D.C.: U.S. Government Printing Office.

U.S. Bureau of the Census. 1993a. "Education: The ticket to higher earnings." *Statistical Brief* SB-93-7 (April).

U.S. Bureau of the Census. 1993b. "Preparing for retirement: Who had pension coverage in 1991?" *Statistical Brief* SB-93-6 (April).

U.S. Bureau of the Census. 1993c. "Public use microdata samples of the 1990 census, 5% sample." Machine-readable microdata file.

U.S. Bureau of the Census. 1993d. "Retirement plans: How many of us are covered?" *Census and You* 28, 7 (July): 1.

U.S. Bureau of the Census. 1994a. "Almost fifty million have a disability." *Census and You* 29, 2 (February):1, 12.

U.S. Bureau of the Census. 1994b. "Two-thirds of U.S. households with children were maintained by married couples." *Census and You* 29, 3 (March): 8.

U.S. Bureau of the Census. 1994c. "Who's minding the kids?" *Current Population Reports* P70-36 (May).

U.S. Bureau of the Census. 1999. *Statistical Abstract of the United States,* 119th ed. Washington, D.C.: U.S. Government Printing Office.

U.S. Bureau of the Census. 1999a. *Quarterly Financial Report for Manufacturing, Mining and Trade.* Washington, D.C.: U.S. Government Printing Office.

U.S. Bureau of the Census. 2000. *Statistical Abstract of the United States,* 120th ed. Washington, D.C.: U.S. Government Printing Office.

U.S. Congress. 1901. *Report of the Industrial Commission on the Relations of Capital and Labor,* 56th Congress, House Document 495. Washington, D.C.: U. S. Government Printing Office.

U.S. Congress, Office of Technology Assessment. 1985. *Automation of America's Offices, 1985–2000.* OTA-CIT-287. December. Washington, D.C.: U.S. Government Printing Office.

U.S. Congressional Research Service. 1985. *The Computer Revolution and the U.S. Labor Force.* Washington, D.C.: U.S. Government Printing Office.

U.S. Department of Agriculture. 1982. "Agricultural labor in the 1980s." Berkeley, Calif.: Division of Agricultural Sciences.

U.S. Department of Commerce, International Trade Administration. 1987. *U.S. Foreign Trade Highlights,* C61.2812: 987. Washington D.C.: U.S. Government Printing Office.

U.S. Department of Health, Education, and Welfare. 1973. *Work in America.* Cambridge, Mass.: MIT.

U.S. Department of Labor, Bureau of Labor Statistics. 1984. *Employment Projections for 1995.* Bulletin 2197 (March).

U.S. Department of Labor, Bureau of Labor Statistics. 1985. *Employment and Earnings* 32, 1 (January).

U.S. Department of Labor, Bureau of Labor Statistics. 1986a. "BLS survey reports on work patterns and preferences of American workers." Press release (August 7).

U.S. Department of Labor, Bureau of Labor Statistics. 1986b. *Occupational Outlook Handbook,* 1986/87 ed. Washington, D.C.: U.S. Government Printing Office.

U.S. Department of Labor, Bureau of Labor Statistics. 1987a. *Employment and Earnings* 34, 1 (January).

U.S. Department of Labor, Bureau of Labor Statistics. 1987b. "Employment in perspective: Women in the labor force." Report 740 (First Quarter).

U.S. Department of Labor, Bureau of Labor Statistics. 1989a. *Employment and Earnings* 36, 1 (January).

U.S. Department of Labor, Bureau of Labor Statistics. 1989b. *Handbook of Labor Statistics.* Washington, D.C.: U.S. Government Printing Office.

U.S. Department of Labor, Bureau of Labor Statistics. 1991. *How Workers Get Their Training.* Washington, D.C.: U.S. Government Printing Office.

U.S. Department of Labor, Bureau of Labor Statistics. 1992a. "Employee tenure and occupational mobility in the early 1990s." Press Release 92-386 (June 26).

U.S. Department of Labor, Bureau of Labor Statistics. 1992b. "The unemployed: Who they are and how they are counted." Summary 92-4. Washington, D.C.: U.S. Government Printing Office.

U.S. Department of Labor, Bureau of Labor Statistics. 1992c. "Workers on flexible and shift schedules." Press Release 92-491 (August 14).

U.S. Department of Labor, Bureau of Labor Statistics. 1993a. "Proportion of 1992 high school graduates enrolled in college." Press Release 93-226 (June 22).

U.S. Department of Labor, Bureau of Labor Statistics. 1993b. "Work and family: Changes in wages and benefits among young adults." Report 849, July.

U.S. Department of Labor, Bureau of Labor Statistics. 1994a. *Employment and Earnings* 41, 3 (March).

U.S. Department of Labor, Bureau of Labor Statistics. 1994b. "The employment situation, January 1994." Press Release 94-57 (February 4).

U.S. Department of Labor, Bureau of Labor Statistics. 1998a. "Displaced workers summary." Press Release 98-347 (August 19).

U.S. Department of Labor, Bureau of Labor Statistics. 1998b. "Employment characteristics of families in 1997." USDL 98-217 (May 21).

U.S. Department of Labor, Bureau of Labor Statistics. 1998c. "Workers on flexible schedules in 1997." USDL 98-119 (March 26).

U.S. Department of Labor, Bureau of Labor Statistics. 1999. "Contingent and alternative employment arrangements, February 1999." USDL 99-362 (December 21).

U.S. Department of Labor, Bureau of Labor Statistics. 2000a. "Employees on Nonfarm Payrolls." Available at *http://www.bls.gov.*

U.S. Department of Labor, Bureau of Labor Statistics. 2000b. "Worker displacement during the late 1990s." USDL 00-223 (August 9).

U.S. Department of Labor, Bureau of Labor Statistics. 2001a. *Employment and Earnings* 48, 1 (January).

U.S. Department of Labor, Bureau of Labor Statistics. 2001b. "The employment situation, January 2001." USDL 01-35 (February 2).

U.S. Department of Labor, Bureau of Labor Statistics. 2001c. "Union members in 2000." USDL 01-21 (January 18).

U.S. Department of Labor, Bureau of Labor Statistics, CPS Questionnaire Evaluation Work Group. 1993. "Composite questionnaire for CATI/CAPI overlap (CCO) test." Photoduplication, March 4.

U.S. Department of Labor, Employment and Training Administration. 1977. *Dictionary of Occupational Titles,* 4th ed. Washington, D.C.: U.S. Government Printing Office.

U.S. Department of Labor, Employment and Training Administration. 1993a. *Selected Characteristics of Occupations Defined in the Revised Dictionary of Occupational Titles.* Washington, D.C.: Government Printing Office.

U.S. Department of Labor, Employment and Training Administration, Advisory Panel for the Dictionary of Occupational Titles. 1993b. "The new DOT: A database of occupational titles for the twenty-first century." Final Report, May. Washington, D.C.: U.S. Government Printing Office.

U.S. General Accounting Office. 1986. *School Dropouts: The Extent and Nature of the Problem.* Washington, D.C.: U.S. Government Printing Office.

U.S. Senate. 1986. Hearings before the Subcommittee on Innovation and Technology. Oversight on the Small Business and Innovation and Research Program. 99th Congress. Washington, D.C.: U.S. Government Printing Office.

Useem, Michael. 1996. *Investor Capitalism: How Money Managers are Changing the Face of Corporate America.* New York: Basic Books.

Useem, Michael, and Jerome Karabel. 1986. "Pathways to top corporate management." *American Sociological Review* 51, 2 (April): 184–200.

Valian, Virginia. 1999. *Why So Slow? The Advancement of Women.* Cambridge, Mass.: MIT Press.

Vallas, Steven Peter. 1990. "The concept of skill." *Work and Occupations* 17,4 (November): 379–398.

Vallas, Steven Peter. 1993. *Power in the Workplace: The Politics of Production at AT&T.* Albany: State University of New York Press.

Vallas, Steven Peter. 1999. "Rethinking post-Fordism: The meaning of workplace flexibility." *Sociological Theory* 17, 1 (March): 68–101.

Vallas, Steven Peter. 2001. *The Transformation of Work,* Volume 10 of *Research in the Sociology of Work.* Greenwich, Conn.: Elsevier.

Vanek, Joann. 1988. "Housewives as workers." Pp. 392–414 in Ann H. Stromberg and Shirley Harkess (editors), *Working Women,* 2nd ed. Mountain View, Calif.: Mayfield.

Vaughan, Diane. 1996. *The Challenger Launch Decision: Risky Technology, Culture, and Deviance at NASA.* Chicago: University of Chicago Press.

Vaughan, Diane. 1999. "The dark side of organizations: mistake, misconduct, and disaster." Pp. 271–305 in John Hagan and Carol S. Cook (editors), *Annual Review of Sociology,* Vol. 12. Palo Alto, Calif.: Annual Reviews.

Veblen, Thorstein. 1921. *The Engineers and the Price System.* New York: Viking.

Vecsey, George. 1974. *One Sunset a Week: The Story of a Coal Miner.* New York: E. P. Dutton.

Verna, Rohit, and Kenneth K. Boyer. 2000. "Service classification and management challenges." *Journal of Business Strategies* 17, 1 (Spring): 5–24.

Vernon, Raymond. 1998. *In the Hurricane's Eye: The Troubled Prospects of Multinational Enterprises.* Cambridge, Mass.: Harvard University Press.

Vickers, Margaret H. 2000. "The 'invisibly' chronically ill as unexamined organizational fringe-dwellers." Pp. 3–21 in Randy Hodson (editor), *Research in the Sociology of Work,* Volume 9: *Marginal Employment.* Greenwich, Conn.: JAI Press.

Vogel, Ezra F. 1991. *The Four Little Dragons: The Spread of Industrialization in East Asia.* Cambridge, Mass.: Harvard University Press.

Vogel, Lisa. 1993. *Mothers on the Job.* New Brunswick, N. J.: Rutgers University Press.

Vonnegut, Kurt, Jr. 1952. *Player Piano.* New York: Dell.

Von Otter, Casten. 2002. "The sociology of work in Sweden." In Daniel B. Cornfield and Randy Hodson (editors), *Worlds of Work.* New York: Plenum.

Vreeland, Eleanor P. 1985. "Trouble in the office." *Small Business Report* 10, 4 (April): 8.

Wacjman, Judy. 1998. *Managing Like a Man: Women and Men in Corporate Management.* Cambridge, England: Polity.

Waddoups, C. Jeffrey. 1999. "Union wage effects in Nevada's hotel and casino industry." *Industrial Relations* 38, 4: 577–583.

Waldfogel, Jane. 1997. "Working mothers then and now." Pp. 82–126 in Francine Blau and Ronald Ehrenberg (editors), *Gender and Family Issues in the Workplace.* New York: Russell Sage.

Waldinger, Roger David. 1996. *Still the Promised City?* Cambridge, Mass.: Harvard University Press.

Walker, Charles R., and Robert H. Guest. 1952. *The Man on the Assembly Line.* Cambridge, Mass.: Harvard University Press.

Wallace, Michael, and Arne L. Kalleberg. 1982. "Industrial transformation and the decline of craft: The decomposition of skill in the printing industry, 1931–1978." *American Sociological Review* 47, 3 (June): 307–324.

Wallace, Michael, Kevin T. Leicht, and Lawrence E. Raffalovich. 1999. "Unions, strikes and labor's share of income." *Social Science Research* 28: 265–88.

Wallace, Phyllis A. (editor). 1982. *Women in the Workplace.* Boston: Auburn.

Wallerstein, Immanuel M. 1999. *The End of the World as We Know It.* Minneapolis: University of Minnesota Press.

Walsh, Joan. 1985a. "Comparable worth advocates optimistic." *In These Times* (October 9): 5.

Walsh, Joan. 1985b. "Cultural exchange on the production line." *In These Times* (April 17): 8–9.

Walsh, Joan. 1985c. "Union threatens GM boycott." *In These Times* (November 27): 5.

Walsh, Joan. 1986. "Layoff settlement zaps Atari." *In These Times* (June 25): 7.

Walsh, John P. 1994. *Supermarkets Transformed: Understanding Organizational and Technological Innovations.* New Brunswick, N.J.: Rutgers University Press.

Walsh, John P., and Shu-fen Tseng. 1998. "The Effects of Job Characteristics on Active Effort at Work." *Work and Occupations* 25,1 (February): 74–96.

Wanner, R. A., and B. C. Hayes. 1996. "Intergenerational occupational mobility among men in Canada and Australia." *Canadian Journal of Sociology/Cahiers Canadiens de Sociologie* 21, 1 (Winter): 43–76.

Ward, Kathryn (editor). 1990. *Women Workers and Global Restructuring.* Ithaca, N.Y.: Cornell University School of Industrial and Labor Relations Press.

Watson, Tony J. 1980. *Sociology, Work and Industry.* London: Routledge and Kegan Paul.

Weber, Max. 1946. "Bureaucracy." In H. H. Gerth and C. Wright Mills (editors), *From Max Weber.* New York: Oxford University Press.

Weber, Max. 1958 [1904]. *The Protestant Ethic and the Spirit of Capitalism.* New York: Scribner's.

Webster, Eddie. 2002. "The sociology of work in South Africa." In Daniel B. Cornfield and Randy Hodson (editors), *Worlds of Work.* New York: Plenum.

Webster, Juliet. 2000. "Today's second sex and tomorrow's first? Women and work in the European information society." Pp. 119–140 in Ken Ducatel, Juliet Webster, and Werner Herrmann (editors), *The Information Society in Europe.* Lanham, Md.: Rowman and Litlefield.

Weil, David. 1991. "Enforcing OSHA: The role of labor unions." *Industrial Relations* 30, 1 (Winter): 20–36.

Weinfeld, Morton. 1999. "The challenge of ethnic match: Minority origin professionals in health and social services." Pp. 117–141 in Harold Troper and Morton Weinfeld (editors), *Ethnicity, Politics, and Public Policy: Case Studies in Canadian Diversity.* Toronto: University of Toronto Press.

Weiss, Leonard W. 1971. *Case Studies in American Industry.* 2nd ed. New York: Wiley.

Wellman, David. 1995. *The Union Makes Us Strong: Radical Unionism on the San Francisco Waterfront.* Cambridge, England: Cambridge University Press.

Wells, Miriam J. 1996. *Strawberry Fields: Politics, Class, and Work in California Agriculture.* Ithaca, N.Y.: Cornell University Press.

Werth, James L. 1999. *Contemporary Perspectives on Suicide.* Philadelphia: Brunner and Mazel.

Wertheimer, Barbara Mayer. 1977. *We Were There: The Story of Working Women in America.* New York: Pantheon.

Westwood, Sallie. 1984. *All Day, Every Day.* London: Pluto.

Wharton, Amy S. 1993. "The affective consequences of service work." *Work and Occupations* 20, 2 (May): 205–232.

Wilensky, Harold, and Charles N. Lebeaux. 1986. "The early impact of industrialization on society." Pp. 28–39 in Lauri Perman (editor), *Work in Modern Society.* Dubuque, Iowa: Kendall-Hunt.

Williams, Bruce B. 1987. *Black Workers in an Industrial Suburb.* New Brunswick, N.J.: Rutgers University Press.

Williams, Christine L. 1995. *Still a Man's World.* Berkeley, Calif.: University of California Press.

Williams, Christine L. 1992. "The glass escalator: Hidden advantages for men in the 'female' professions." *Social Problems* 39, 3 (August): 253–268.

Williams, Christine L., Patti A. Giuffre, and Kirsten Dellinger. 1999. "Sexuality in the workplace: Organizational control, sexual harassment, and the pursuit of pleasure." *Annual Review of Sociology* 25 (August): 73–93.

Williams, Mike. 1993. " 'Sick-building' epidemic silently sweeps the workplace." *Houston Chronicle* (February 21).

Wilson, F. 2000. "The social construction of sexual harassment and assault of university students." *Journal of Gender Studies* 9, 2: 171–187.

Wilson, William Julius. 1987. *The Truly Disadvantaged: The Inner City, the Underclass and Public Policy.* Chicago: University of Chicago Press.

Wilson, William Julius. 1997. *When Work Disappears: The World of the New Urban Poor.* (Reprinted edition.) New York: Vintage.

Winslow, Ron. 1993. "Shortage of primary-care doctors spurs schools, HMOs to try to redress balance." *Wall Street Journal* (February 24).

Winter, Michael P. 1988. *The Culture and Control of Expertise: Toward a Sociological Understanding of Librarianship.* New York: Greenwood.

Wolf, Harald. 1995. "Introduction." *International Journal of Political Economy* (special issue on *Democracy at Work?*) 25, 3 (Fall): 3–19.

Wolff, Edward N. 2000. "How persistent is industry specialization over time in industrialized countries?" *International Journal of Technology Management* 19, 1: 194–205.

Woodard, Michael D. 1997. *Black Entrepreneurs in America: Stories of Struggle and Success.* New Brunswick, N.J.: Rutgers University Press.

Woodworth, Warner P., and Christopher B. Meek. 1995. *Creating Labor Management Partnerships.* Reading, Mass.: Addison-Wesley.

"Workers active." 1985. *Global Electronics* 59 (November): 1.

"World's biggest trade deal approved." 1993. *Herald-Times* (December 16): A3.

Wright, Erik Olin. 2000. "Working-class power, capitalist-class interests, and class compromise." *American Journal of Sociology* 105, 4 (January): 957–1002.

Wright, John W. 2000. *The American Almanac of Jobs and Salaries.* New York: Avon.

Wysocki, Bernard, Jr. 1998. "Oil megadeal signals a world of low inflation." *Wall Street Journal* (December 2): 112.

Yoon, In_Jin. 1991. "The changing significance of ethnic and class resources in immigrant businesses: The case of Korean immigrant businesses in Chicago." *International Migration Review* 25, 2 (Summer): 303–333.

Yount, Kristen R. 1991. "Ladies, flirts, and tomboys: Strategies for managing sexual harassment in an underground coal mine." *Journal of Contemporary Ethnography* 19, 4 (January): 396–422.

Zachary, G. Pascal. 1993. "Like factory workers, professionals face loss of jobs to foreigners." *Wall Street Journal* (March 17): 1.

Zachary, G. Pascal, and Bob Ortega. 1993. "Workplace revolution boosts productivity at cost of job security." *Wall Street Journal* (March 10): A1, A8.

Zaucher, Sabine, Christian Korunka, A. Weiss, A. Kafka-Lutzow. 2000. "Gender-related effects of information technology implementation." *Gender, Work and Organization* 7, 2 (April): 119–132.

Zelizer, Viviana A. Rotman. 1979. *Morals and Markets: The Development of Life Insurance in the United States.* New York: Columbia University Press.

Zey, Mary. 1998. *Rational Choice Theory and Organizational Theory.* Thousand Oaks, Calif.: Sage.

Zhou, Xueguang. 1993. "Occupational power, state capacities, and the diffusion of licensing in the American states, 1890 to 1950." *American Sociological Review* 58, 4 (August): 536–552.

Zisman, Michael. 1978. "Office automation: Revolution or evolution?" *Sloan Management Review* 19 (Spring): 1–16.

Zuboff, Shoshana. 1988. *In the Age of the Smart Machine.* New York: Basic.

Zunz, Olivier. 1998. *Why the American Century?* Chicago: University of Chicago Press.

Index

A

Absenteeism, 104
Abstract, specialized knowledge, 283–285
Accountability, 285
Achieved demographic characteristics, 41
Acquisitions, 190, 310
Acuna, Roberto, 200
Administrators. *See* Managers
Advanced industrial society, 199–200
Aerospace industry, 126
Aetna Life & Casualty Company, 364
Affirmative action, 68, 112–114, 313
AFL. *See* American Federation of Labor
AFL-CIO, 153, 154, 223, 427
  growth of, 160, 162
  publicity campaigns, 167
  and telecommuting, 241
  worker retraining, 250
African Americans
  and affirmative action, 313
  appendix table, 455–460
  as clerical workers, 334
  as farmland owners, 201
  and job satisfaction, 100
  and marginal work, 371
  median income for families by race, ethnicity, and gender, 59, 60
  in occupations, 113
  as private household workers, 361
  in professions, 292
  in sales, 349
  sectoral shift of labor force, 258–259
  in temporary work, 365
  unemployment rates, 124
  and unions, 150, 153, 165
  wages of college graduates, 114
  and white-collar work, 333
  See also Minorities
AFSCME. *See* American Federation of State, County, and Municipal Employees
AFT. *See* American Federation of Teachers

Age
  division of labor by, 181
  and job satisfaction, 100
  and marginal jobs, 369
  See also Older people
Agricultural industry, 200–204
  federal price supports, 202–203
  rising costs and falling prices, 201–202
  rising productivity, 200–201
  technological advances, 202
  and technology, 171–172
  workers in, 130, 203–204
Agricultural Labor Relations Act of 1975, 155–156
Agricultural societies, 10–13
Airbus, 126
Airline industry, 262
Airlines reservations clerks, 256
Akzo (Belgian-based company), 429
Alger, Horatio, 191
Alienation and Freedom (Blauner), 96–97
Alienation from work, 6
Alienation theories, 90–92, 94, 104, 108
Alternative medicine, 304
Altruism, 287–288
America Online, 190, 387
American Airlines, 240
American Arbitration Association, 158
American Association of University Professionals (AAUP), 155
American Can, 385
American Civil Liberties Union (ACLU), 249
American College of Obstetricians and Gynecologists, 289
American Electronics Association, 250
American Express Company, 240
American Federation of Labor (AFL), 25, 149–150
American Federation of State, County, and Municipal Employees (AFSCME), 162
American Federation of Teachers (AFT), 154–155, 162

American Medical Association, 155, 289
American Motors, 396
American Railway Union (ARU), 145
American Revolution, 406
American Telephone and Telegraph Company (AT&T), 111, 387
American Tobacco Company, 26, 385
Americans with Disabilities Act of 1990, 137
Analysis of labor, 182
Analysis techniques, 34–40
  case studies, 37–39
  ethnographies, 35–37
  multiple regression, 52–53
  sample surveys, 39–40
Analysis units, 40–54
  industry as, 46–47
  occupation as, 57–53
  workers and labor force as, 40–46
  workplaces as, 54
Andaman Islanders, 10
Anderson, Jonathan, 249
Anomie, 92
Antitrust laws, 381, 383, 386
Apartheid, 68, 418
Appalachia, 207
Appendix table, 455–460
Apple Computer, 227
Apprenticeship programs, 181, 210, 443
Aquaculture, 206
Argyris, Chris, 92, 93
Artisans, 16–17
As You Like It (Shakespeare), 61
Asbestos, 131–132
Ascribed demographic characteristics, 40–41
Ash, Mary Kay, 346
Asian Americans and unions, 153
Assembly jobs, 236–237
Assembly lines, 27–28, 94, 96
AT&T. *See* American Telephone and Telegraph Company
Atari, 128
ATMs, 338
Audio recording equipment, 337

Authority and the professions, 285–287
Automated messages, 256
Automation, 32, 94, 179, 434
 and clerical work, 341
 continuous-process, 229
 office, 238
Automatix, 248
Automobile industry, 96, 161,
 215–217, 382
 complaints against bureaucratic
  procedures, 187
 displaced workers in, 222
 and Ford, 398
Autonomous work groups, 419–421
Autonomy, 177
Average hourly earnings of workers, 266

B

"Back-office" jobs, 333
Backward linkage, 381
Bank tellers, 337–338
Barath, Bill, 442
Barbers, 264–265
Barnard, Chester, 28
Barriers to entering the labor force, 69
Bartering goods, 55
BASF, 410
"Battle of the Running Bulls," 151
Bednorz, J. Georg, 299
Behavioral approach to managerial
 performance, 321–322
Behavioral responses to work,
 103–107
Bell, Daniel, *The Coming of Post-
 Industrial Society,* 179
Belongingness needs, 92, 95
Bench workers, 211–212
Benefits. *See* Fringe benefits
Berle, Adolf, 383
Bhopal, 136
Big Kahuna, The (film), 351
Billing-machine operators, 331
Birch, David, 398
Birth rates, 72, 82
Blacklists, 149
Blacks. *See* African Americans;
 Minorities
Blauner, Robert, *Alienation and Freedom,*
 96–97
Block scheduling, 86
Blue Cross-Blue Shield, 239, 291
Blue-collar workers, 200, 227
 changes in, 278
 and high technology, 241, 246
Boeing Aircraft, 154
Bookkeepers, 331, 332
Border plants, 414–415
Boredom in work, 212
Boston Tea Party, 406
Boundaries of home and work, 79, 80
Bowden, Doris, 374

Boy Scouts, 167
Breakfast of Champions (film), 327
Bridges, Beau, 169
Bridgestone/Firestone, 192
British Empire, 406
Brooking, Michelle, 247
Brotherhood of Sleeping Car Porters,
 150
Brown, A. Radcliffe, 10
Buck, Pearl, *The Good Earth,* 33
Buick, 217
Burawoy, Michael, 189
Bureaucracy, 5, 185–193
 and control, 186–188
 and corporate accountability, 191
 and creativity, 191
 customizing of, 188
 defining, 185–186
 efficiencies in, 186
 informal work cultures, 188–190
 internal labor markets, 187–188
 job ladders in, 64
 limitations of, 190–193
 making out on the shop floor, 189
 rules in, 285
 and workers, 26–27
Burger King, 36, 54
Burnout, 272–273
Bus driver in Los Angeles, 263
Business education, 332
Business proprietors, 312
Business services, 261

C

Cable Guy, 275
Cable News Network (CNN), 190
CAD. *See* Computer-assisted design
Cafeteria approach to benefits, 87, 443
California Department of Labor, 132
California Federation of Teachers, 167
Call-waiting, 256
Calvin, John, 20
Canada, 417
 fishing industry in, 205
 labor law in, 152
 Labour Canada Task Force, 248
 unions in, 146, 156, 160
 Winnipeg general strike (1919), 148
Cancer-causing chemicals in the
 workplace, 131
Capital and labor conflict, 19, 147
Capital mobility, 220
Capital-intensive economy, 179
Capitalism, 6, 91, 107–108
Captive markets, 407
Carayol, Rene, 294
Career in life cycle, 63–64
Career mobility, 214
Career shifts, 73–74
Caribbean Data Services (CDS), 240
Carnegie Steel Corporation, 26, 385

Carpal tunnel syndrome, 129
Carradine, John, 374
Carrey, Jim, 275
Cartels, 408
Case studies, 37–39
Cash cow, 386
Cash payments, 55
Cashiers, 341
Casual hiring, 365
Catalog companies, 344–345
Catholic Church, 15
Central planning, 425
Centralization of control, 177,
 190, 247
Certification, 295–296
Chains of stores, 344
Changing job content, 234–241
Chapmin, Ralph, "Solidarity
 Forever," 141
Charlemagne's empire, 14
Chase Manhattan Bank, 393
Chavez, Cesar, 203, 204, 359
Cheap labor, 333
Chelberg, Bruce, 393
Chemical industry, 96, 229
Chemical sensitization, 132
Chemicals, 130–131
Chernobyl, 136
Child abuse, 104
Child care, 154, 263
 on-site and off-site, 83
 and unions, 167
Child labor, 147, 224
Child labor laws, 369
Child rearing, 62
 practices in, 66
Children, 82–83
 and Industrial Revolution, 23–24
China, 424, 425
Chronic stress injuries, 129–130
Chrysler Corporation, 382, 387,
 396, 402
Cigar factories, 146, 147
Cigar Makers' Union, 149
Cigarette sales, 26
CIO. *See* Congress of Industrial
 Organizations
Circle of Innovation, The (Peters), 243
Citibank, 238
Civil Rights Act of 1964, Title VII,
 112, 118, 137, 153, 436
Civil rights movement, 288
Civil War, 330, 331, 332
Class. *See* Social classes
Classical civilization, 15
Clayton Act of 1914, 149
Clean Air Act, 193
Cleaning service workers, 359
Cleanliness taboos, 352
Clergy members, 288
Clerical operatives, 334

Clerical workers, 94, 97, 162, 328–341
and the Communication Workers of America (CWA), 154, 155
demand for, 330–331
feminization of, 330, 331–333, 334
future of, 341
and high technology, 238, 341
history of, 329–335
job content, 338–339
job displacement, 241
number of jobs by year, 329
office technology, 335–338
rationalization of clerical work, 338–339, 340
relations among workers, 339–340
supply of, 331–335
and telecommuting, 240
transforming the clerical occupations, 335–341
trends in, 334–335
variety of occupations in (Appendix Table 1), 331
work reorganization, 338–341
Clerics, 329
Client firms, 362, 364
Climate surveys, 323
CLUW. See Coalition of Labor Union Women
CNC. See Computer-aided numeric control
Coal, 22, 24
Coalition of Labor Union Women (CLUW), 154
Coca-Cola Kid (film), 401
Codes of ethics, 296
Codetermination, 418–419, 440–441
Codetermination Law of 1976, 419
Collective bargaining, 149–150, 157–158, 418
Collective power, 277
Collective responses to work, 140–169
See also Unions
Collins, Sharon, 313
Colorado Fuel and Iron Company, 148
Columbus, Chris, 88
Combined Insurance, 268, 269
Combined and uneven development in the global economy, 413–415
Coming of Post-Industrial Society, The (Bell), 179
Commercial subcontracting, 395
Commission for Racial Equality (CRE), 294
Commissions in sales, 342, 344
Commitment to jobs, 103
Commodification of agriculture, 414
Commodity chains, 414
Commonality bonds, 7
Communications Workers of America (CWA), 229, 234
and clerical workers, 154, 155

Communism, 407, 425
Community colleges, 180–181
Community involvement, 104
Commuter marriages, 79
Commuting, 61, 79
Company resistance to unions, 161–162
Company unions, 424
Comparable worth debate, 120–122
Comparison groups, 100
Compensation in services, 265–266
Competing organizational forms, 426–427
Competition in the world economy, 434
Complexity of skills, 179
Comprehensive Employment Training Act, 449
CompuServe, 190
Computer games, 235, 236
Computer science, 334
Computer-aided numeric control (CNC), 237, 241, 246
Computer-assisted design (CAD), 237, 244
Computers, 434–435
on the job, 181
laptop, 40, 42–43
Computervision, 248
Concentration ratio, 381
Conglomerates, 54, 190, 381, 385
Congress of Industrial Organizations (CIO), 150–152, 156
Conoco Oil, 386, 387
Construction industry, 208–209
Consultants, 355
Consumer goods, 172
Contingency theory, 188
Contingent work, 355–356
Continuing education, 233
Continuous improvement, 421
Continuous learning, 437
Continuous-process automation in the chemical industry, 229
Continuous-process technology, 96, 338
Control
and bureaucracy, 186–188
centralization of, 190, 247
Cooperatives, 98–99, 194
Copyists, 331, 337
Core jobs, 362, 365
Core nations, 405
Corning Glass, 317
Corporate accountability and bureaucracy, 191
Corporate campaign, 166
Corporate medicine, 291
Corporate portfolio management, 325
Corporations, 377–401
effects of increasing size, 390–392
and employees, 391–392
largest multinational corporations, 410
largest in U.S., 380

legal status of, 382–383, 386
linkages between, 392–396
merger jargon, 388–389
mergers, 384–392
monopoly power, 380–381
oligopoly power, 381–382
owners versus managers, 383
power of, 378–384
public concerns about, 378–379
role of banks, 393–394
size as power, 382
small-firm sector, 396–399
and society, 390–391
Corruption, 186
Corvee labor, 15
Cost-of-living raises, 84
Cottage industry, 18
Cowboys, 358
Craft labor, 16–17, 27
Craft settings, 96
Craft technology, 178
Craft unionism, 143, 149–150, 208
Craft workers, 210–211, 312, 341
Creativity and bureaucracy, 191
Criminal careers, 353
Cross-sectional surveys, 39
Crusades, 17
Crystal cube of the future workplace, 438–439
Cultural division of labor, 68, 111
Current Population Survey (CPS), 39, 41, 42–43
Customer standards in services, 267
Customers, exploitation of, 379
Cyclical unemployment, 125

D

Daimler-Benz Corporation, 382, 387
Dangerous conditions, 129
Darwell, Jane, 374
Darwin, Charles, 25
Data entry, 334
Data General, 248
Data stewards, 251
Day laborers, 365
Dead-end jobs, 69–71
Death by overwork (*karoshi*), 423
Debs, Eugene, 145
Debt crisis in the global economy, 415
Deep-sea fishing, 206
Demographic characteristics, 40–41
Department stores, 342
Dependency theory of industrial development, 404–405
Dependent development, 416
Deprofessionalization, 297–300
Deskilling thesis, 179, 230–231, 233, 435
Determinism, technological and organizational, 195
Developing nations, 416–417
Deviant occupations, 353

DeVito, Danny, 351
Dictionary of Occupational Titles (DOT), 50–53
Dignity and respect in work, 6, 98
Dillard, Hub, 208
Direct personal control, 182–183
Direct sales, 345
Direct worker participation, 193–195
Dirty work, 352, 356, 372
Disabilities, 136–137
    and marginal jobs, 368
    social support, 137
    training and, 137
Discouraged workers, 44–45, 126
Discrimination, 370
    against disabled people, 137
    gender, 119–122
    in hiring, 111–118
    in pay and promotions, 118–122
Dispersion of information, 247
"Displaced homemaker," 84
Displacement of workers, 435
Distributive services, 262
Diversification, 385, 388–390
Diversity
    in the workforce, 31
    working-class, 215
Diversity of skills, 179
Diversity of tasks, 93
Division of labor, 4
    advance of, 22–23
    by age, 181
    by gender, 9, 181
    cultural, 68, 111
    international, 245
    manufacturing, 182
    social, 182
Divisions of the larger corporation, 390
Divorce, 65, 104
Domestic industry, 18
Domestic servants, 173
Domestic service in Lima, Peru, 260
Doubling up, 213
Downsizing, 222, 313, 379
Downward mobility, 73
Drugs, 55
Dryfus, Richard, 109
Dual labor markets, 114–115, 363, 372
Duchin, Faye, 228
Duplicating machines, 337
Durkheim, Emile, 17, 92
Dust Bowl, 374

E

E-mail, 40, 338
Early national unions, 144–146
Early retirement, 75
Eastern Europe and Russia, 425–426
Economic revitalization, 398
Economically active population, 40
Economies of scale, 7, 380–381

Education, 233–234, 262, 449
    and life cycles, 76
    and marginal jobs, 368
Educational services, 172
EEC. *See* European Economic Community
EEOC. *See* Equal Employment Opportunity Commission
E.I. DuPont, 26, 386, 387
Eight-Hour League, 145
Elderly people. *See* Older people
"Electronic cottage," 226
Electronic revolution, 241–242
Electronic surveillance, 248, 249, 435
Electronic technology, 227, 434
Electronic workstations, 337
Electronics assembly in Malaysia, 29
Electronics workers, 132
Emotional work, 270
Empire Airlines, 387
Employed people, 41
Employee stock ownership plans (ESOPs), 441–442
Employer commitment, 446
Employer standards in services, 266–267
Employers' role in service industries, 267–271
Employment change in selected industries, 48, 172–173
Employment contracts and marginal jobs, 365–366
Employment and Training Administration, United States Department of Labor, 50
Empowering the consumer, 297
"Empty nest," 65, 85
"Enclosures," 20
Engineering, 235–236
Engineering in the future, 30
Entering the labor force, 69–72
Enthusiasm for work, 104
Entry-port jobs, 71–72
Environmental damage, 379
Environmental degradation, 135–136
Equal Opportunity Employment Commission (EEOC), 112, 122
Equal Pay Act of 1964, 120
Equal Rights Amendment, 122
Equal rights legislation, 112–114
Ergonomics, 133
Erikson, Erik H., 62, 63
Erin Brockovich (film), 57
ESOPs. *See* Employee stock ownership plans
Establishment surveys, 39–40
Establishments, 54, 397
Esteem needs, 92
Ethics and managers, 315
Ethnicity, 27
    and marginal jobs, 370

median income for families by race, ethnicity, and gender, 59, 60
    and occupational steering, 68
    and small-business owners, 310–311
Ethnographies, 35–37
Etienne, Franklin, 249
European Economic Community (EEC), 413
Evans, Darell, 201
Exclusionary practices, 210–211
Executives. *See* Managers
Expanded leisure, 450–451
Experimental bias, 37
Exploitation of customers, 379
Export economies, 417
Exporting jobs, 363
Externalization in manufacturing, 221–222
Externalizing costs, 192
Extractive industries, 199
Extrinsic rewards, 103
Exxon Valdez (ship), 192

F

F International, 240
Face-work, 268, 272
Factory system, 21–26
Fair Labor Standards Act of 1938, 360
Fair trade, 223, 413
Fair Trade and Tariff Act, 223
Families, 61–88
    boundaries of home and work, 79, 80
    children, 82–83
    combining family and work, 85–87
    "empty nest," 85
    homemaking as a career, 83–84
    impact on work, 85
    income squeeze, 84–85
    life-cycles, 64–65
    median income by race, ethnicity, and gender, 59, 60
    role conflict and overload, 76–79
    work arrangements among couples, 79
    working women around the world, 79, 81
    See also Single-parent families
Family leaves, 86
Family life cycle, 64–65
Family and Medical Leave Act of 1993, 86
Farm workers, 155–156, 163
    changes in, 278
    and marginal jobs, 358–359
Fast food industry, 36, 269
    Burger King, 54
    hiring older people, 75, 370
    McDonald's, 268, 269
Featherbedding, 158
Federal Office of Contract Compliance, 112
Federal Trade Commission (FTC), 383, 386

Feminization of clerical workers, 330, 331–333, 334
Feudal society, 15–17, 186
Fields, Sally, 169
Financial control of managers, 324
Financial shell games, 221
Finney, Albert, 327
Firestone, 192
Firms, 54
    free agent, 396
    and marginal jobs, 364–365
First-line supervisors, 316
Fish farming, 206
Fisher Body Plant, 151
Fisher, Robert, 166
Fishing industry, 205–206
    aquaculture, 206
    technology and organization in, 206
Flattening, 313
Flexible schedules, 83
Flextime, 85–86
Flight attendants, 270
Flint sit-down strike, 151
Fonda, Henry, 374
Food service workers, 359
Food surpluses, 11–12
Ford Escort, 408
Ford Explorer, 192
Ford, Henry, 398
Ford Motor Company, 26, 111, 164, 414
    profit sharing plan, 443
    and U.S. Steel, 442
Fordism, 398
Foreign competition, 363, 373
Foreman's control, 183
Forestry, 204–205
Fortune (magazine), 397
Forward linkage, 381
"Four tigers," 424–425
France, 417–418
Franchises, 311
Free agent firms, 396
Free labor, 13–14, 17
Free trade, 223, 412
Frictional unemployment, 124
Fringe benefits, 70, 84, 345, 450–451
    cafeteria approach to, 87
    family related, 86–87
    in marginal jobs, 361, 444–446
    and subcontracting, 395
Frontier Airlines, 387
Functional illiteracy, 368
Future of work, 433–453
F. W. Woolworth Company, 26

G

Gambling, 55, 353
Garbage collectors, 102, 262, 265
Gardeners, 355

Gatekeeper functions in the professions, 296
GATT. *See* General Agreement on Tariffs and Trade
Gays. *See* Homosexuals
Gdansk, Poland shipyards, 146
Gender
    case studies in, 38, 39
    comparison of income by, 120
    discrimination, 119–122
    division of labor by, 9, 181
    and household workers, 362
    job evaluation points, 121
    and job satisfaction, 100
    and marginal jobs, 370
    median income for families by race, ethnicity, and gender, 59, 60
    and occupational steering, 68
    See also Women
Gender-typing of occupations, 111, 116–118, 270
General Agreement on Tariffs and Trade (GATT), 412
General Electric, 26, 111, 190, 221, 385
General Motors, 126, 151, 164, 216, 396
    largest U.S. corporation, 379, 380
    in Mexico, 402
    Roger and Me (film), 139
    and U.S. Steel, 442
General strikes, 144
General unions, 146, 146–149
    See also Unions
Geographic isolation and marginal jobs, 366–368
German codetermination, 418–419
Germanic tribes, 15
Gesundheit Institute, 306
Gheg people of Albania, 12–13
Glasnost (openness), 425
Glass ceiling, 119
Global competition in manufacturing, 220–224
Global economy, 402–432
    combined and uneven development, 413–415
    competition in, 220–224, 434, 435
    contemporary global economy, 406–408
    debt crisis, 415
    development of, 403–408
    fair trade, 413
    free trade, 412
    multinational corporations (MNCs), 408–410
    protectionism, 412
    trading blocks, 413
    United States economic dominance, 411–412
    See also Work practices around the globe
Global Products, Inc., 313

Globalization, 6
GNP. *See* Gross national product
Golden parachutes, 310, 388, 391
Gompers, Samuel, 149
Good Earth, The (film), 33
Gorkom, Van, 393
Government agencies
    as analysis units, 54
    regulation by, 298
Government employees, 262
Government work regulations, 254
Graft, 186
Grapes of Wrath, The (film), 374
Graphic Communication Union, 157
Great Depression, 36, 150, 151, 160, 333, 412
Grievance procedures, 158
Grievances, 141
Grocery stores, 347, 348
Gross national product (GNP), 41
Gross national product (GNP) around the world, 403, 404
Group practice, 290
Group solidarity, 143
Guatemala bordello, 353, 354
Guest, Robert, 437
Guilds, 14, 16–17
    resistance to merchant capitalism, 19–20

H

Harman Industries, 164
Hawthorne (effect) studies, 35, 37, 184–185
Haymarket Square, 145
Haywood, Bill, 148
Hazardous work, 128–136
    asbestos, 131–132
    chemicals, 130–131
    environmental degradation, 135–136
    high-technology, 132
    industrial accidents, 129–130
    occupational diseases, 130
    safety and health in the workplace, 134, 137
    "sick buildings," 133–134
    statistics on, 128
    stressful jobs, 134–135
    and unions, 164
    video display terminals (VDTs), 132–133, 164
Health maintenance organizations (HMOs), 291, 298, 303
Heat risks, 129
Height and weight requirements, 112
Herzberg, Frederick, 92–93
Hidden curriculum, 67
Hierarchical need satisfaction (Maslow), 92–93
Hierarchical organizations, 64, 71
Hierarchy, steepness of, 309–310

High school graduates, 368
High-technology life cycle, 248–250
High-technology workplace, 226–253
    assembly jobs, 236–237
    changing job content, 234–241
    clerical work, 238
    competing views, 227–228
    definition, 227
    deskilling thesis, 230–231
    engineering, 235–236
    hazardous work in, 132
    jargon in, 237
    job creation, 243–244
    job displacement, 241–243
    machine work, 237–238
    and the meaning of work, 246
    middle management, 238–239
    mixed-effects position, 231–233
    and organizational dynamics, 246–250
    public policy and employment,
        245–246
    segmentation, 244–245
    skill requirements, 228–234
    skill-upgrading thesis, 229–230
    skilled maintenance work, 238
    technical workers, 239
    telecommuting, 239–241
    training for changing skill
        requirements, 233–234
    and unions, 250–251
    See also Technology
Higher disposable incomes, 257
Hill, Joe, 149
Hired hands, 358
Hiring discrimination, 111–118
Hirsch, Paul, 388
Hispanics, 114
    appendix table, 455–460
    as clerical workers, 334
    median income for families by race,
        ethnicity, and gender, 59, 60
    in occupations, 113
    as private household workers, 361
    in professions, 292
    in sales, 349
    in temporary work, 365
    unemployment rates, 124
    in unions, 165
    See also Minorities
History of work, 8–32
HMOs. See Health maintenance
    organizations
Hoffa, Jimmy, 153
Holiday rush season, 365
Hollow corporations, 221
Home offices, 61
Home production, 83
Home Shopping Network, 345
Homemakers, 44, 173
    as a career, 83–84
    unpaid, 62

versus home workers, 83
    See also Women
Homosexuals, rights of, 118
Hooker Chemical Company, 136
Horizontal differentiation, 185
Horizontal dimensions of
    bureaucracy, 316
Horizontal mobility, 73
Hostile takeovers, 385
Hotel management, 242
Hotel and Restaurant Employees
    Union, 166
Hotels, turnover of workers, 358
Hourly wages around the globe, 417
Hours of operation, 344
House party sales, 345
Hughes, Everett C., 35
Human capital, 187–188
Human relations, 28, 323
Human relations school, 184
Hunters, 282
Hunting and gathering societies, 8–10
"Hygiene factors" in job satisfaction, 93

I

IBM, 299, 386, 408
IC Industries, Inc., 393
ICFTU. See International Confederation
    of Free Trade Unions
Ideologies, 6
Ideologies about work, 103
Illegal aliens, 55, 359–360
Illegal goods and services, 55
Illegal or morally suspect
    occupations, 353
Illegal work practices, 129
Illich, Ivan, 303
ILO. See International Labour
    Organisation
Immigrant farm workers, 358
Immigrants, 5, 233, 333
    from China, 405
    from Europe, 27, 405
    illegal aliens, 55, 72
    and marginal jobs, 370
Immigration Reform and Control Act
    of 1986, 360
Imperial societies, 13–15
Import substitution, 417
Inadequate work, 356, 359, 360
Income squeeze, 84–85
Income tax avoidance, 55
Indefinite layoffs, 125–126
Indentured laborers, 23
Independent practice, 290
Indirect effects of high technology in
    the workplace, 244
Individual differences in job satisfaction,
    99–100
Individual life cycle, 62–63
Indoor air pollution, 133–134

Indsco, 116
Industrial accidents, 129–130
Industrial capitalists, 25, 26, 426
Industrial chemicals, 130–131
Industrial cities, 24–25
Industrial conglomerates, 386
Industrial development theories,
    403–406
Industrial Revolution, 20–21, 182, 407
Industrial society, 17
Industrial subcontracting, 395
Industrial Supply Corporation
    (Indsco), 38
Industrial unionism, 150–152, 156
Industrial Workers of the World
    (Wobblies), 141, 147–149
Industries
    as analysis units, 46–47
    definition of, 171
    and marginal jobs, 363–364
    and occupations, 200
Inequality, 5
    in the Middle Ages, 15
    and work, 8
Inflation and retirement, 76
Inflexibility of large corporations, 379
Informal economy, 355
Informal sectors, 416
Informal work cultures, 188–190
Information
    dispersion of, 247
    in a post-industrial society, 254
Information flow, 230
Information retrieval, 247
Information technology (IT)
    industry, 294
Injunctions, 149
Innovation, 398–399, 448–450
Innovative employment sector,
    436–444
Inner-city areas, 367
Insatiable demand for services, 256
Inside game, 166
"Insider" situations, 190
Instability of jobs, 355
Institutions of higher learning, 379
Instrumental orientation, 103
Insurance companies, 298
Insurance underwriters, 231, 269
Integrated economic systems, 405
Intellectual obsolescence, 303
Intergenerational mobility, 72
Interlocking directorates, 54, 392–393
International Typographical
    Union, 167
Internal democracy of unions,
    159–160, 167
Internal disciplinary procedures, 296
Internal dynamics, 403
Internal labor markets, 187–188, 373
International agreements, 251

International Brotherhood of Teamsters, 142, 153
   Teamsters for a Democratic Union (TDU), 167
International Confederation of Free Trade Unions (ICFTU), 427
International division of labor, 245
International Federation of Chemical Workers, 429
International Harvester, 385
International labor solidarity, 427–429
International Labour Office, United Nations, 50
International Labour Organisation (ILO), 223, 427
International Ladies' Garment Workers Union (ILGWU), 150
International Longshoremen's Association, 113
International Monetary Fund, 159, 224
International Standard Classification of Occupations (ISCO), 50
International Telephone and Telegraph (ITT), 409
International Woodworkers of America, 154
Interpersonal relations, 104
Interviewing, 70
Intragenerational mobility, 72
Intrinsic rewards, 102
Inventions, 22, 399
Inverted U-curve of technology and alienation, 96, 97
Involuntary part-time work, 356, 365, 370
IPC Electric, 294
Ippanshoku, 311
Iron Molders Union, 144
ISCO. *See* International Standard Classification of Occupations
Isolation, 92, 104
ITT. *See* International Telephone and Telegraph

J

Janitorial work, 265
Japan, 438–440
   automobile industry, 382
   lifetime employment, 193–195
   macroplanning, 421–424
Jim Crow era, 68
Job autonomy, 93
Job commitment, 55, 103
Job content
   changing, 234–241
   clerical workers, 338–339
Job counselors, 50
Job creation in the high-technology workplace, 243–244
Job descriptions, 71

Job displacement, 368–369, 435
   clerical workers, 241
   high-technology workplace, 241–243
Job enlargement, 340
Job enrichment, 340
Job hopping, 71
Job ladders, 84
   in bureaucracies, 64
   in entry-port jobs, 71, 72
Job redesign, 437, 443–444
Job repackaging, 85–86
Job satisfaction, 89–109
   across occupations, 96
   alienation theories, 90–92, 94, 104, 108
   determinants of, 93–101
   and expectations, 101
   future of, 107–108
   individual differences in, 99–100
   measuring, 101–107
   self-actualization theories, 90, 92–93, 94, 104, 108
Job security, 84, 442–443
Job shedding, 310
Job stress, 134–135
Job Training Partnership Act, 223, 449
Job transfers, 79
Job-worker fit, 101
Johnson, Harold, 126
Johnson, Lyndon B., 112
Johnson, Thomas, 442
Joint union-management programs, 440–441
Jonas, Norman, 221
Jones, Mother, 149
Journeyman, 16
Just-in-time delivery systems, 395
"Justice for Janitors," 162

K

Kafka, Franz, *The Trial,* 197
Kanter, Rosabeth Moss, *Men and Women of the Corporation,* 38, 39
Karoshi (death by overwork), 423
Keefe, Jeffrey, 232
Kerr-McGee nuclear fuel plant, 136
Keypunchers, 334
Keys, Julie, 356
Kidder, Tracy, *Soul of a New Machine, The,* 235
King, Martin Luther, Jr., 111, 153, 288
Kmart, 344
Knights of Labor, 25, 146
Knowledge, 176–177
   abstract and specialized, 283–285
   monopolizing, 288–289
Knowledge workers, 323
Knowledge-intensive economy, 179
Kondo, Dorinne, 67
Kruizenga, Jack, 393

L

Labor control and organizational structure, 182–184
Labor Day, 145
Labor force
   as analysis unit, 40–46
   diversity in, 31
   entering, 69–72
   participation rate, 41
Labor unions. *See* Unions
Labor-saving technologies, 173
Labour Canada Task Force, 248
Laissez-faire, 25, 417
Language skills, 333, 370
Laptop computers, 40, 42–43
Latifundia, 406, 415
Law clerks, 330
Layoffs, 125–127
   of older people, 75
   replacement jobs, 127
   seniority-based protection, 153
   social consequences of, 126–127
   See also Unemployment
Least developed nations, 415–416
Lectors, 146, 147
Lee, Pat, 226
Legal secretaries, 340
Legislative controls for unemployment, 127–128
Leisure expansion, 450–451
Leisure and life cycles, 76
Lesbians. *See* Homosexuals
Less productive industries, 364
Lewis, John L., 150
Liberal arts education, 233–234
Librarians, 300
Libraries, 242
Licensing, 289, 295
Liddle, Donald, 165
Life cycle of high-technology, 248–250
Life cycles, 61
Life managers, 309, 318
Lifetime employment, 193–195
Limited liability, 382
Line Maintenance Operating System (LMOS), 247
Line positions, 186
Lipnak, Jessica, 226
Literacy, 329, 330
Little, Malcolm, 68
Lobbying by unions, 159
Local impact of high-technology development, 245–246
Long-distance merchandising, 344
Long-term effects of high technology in the workplace, 244
Long-term planning, 421
"Losing it," 272
Love Canal, 136
Low occupational power, 357–358

Low productivity and services, 256
Low-level radiation, 129
"Low-tech" occupations, 244
Low-wage workers and unions, 165
Lowell factory women, 24
Lower-class, 68
Loyal opposition firms, 396
Luddite movement, 105
Luther, Martin, 20

M

McCarthyism, 152–153
McCaw Cellular Communications, 387
McCormick Harvester plant, 145
"McDoctor Clinics," 290
McDonald's Hamburger University,
    268, 269
McGregor, Douglas, 92, 93
Machine operators and assemblers,
    211–213
Machine work, 237–238
Machinists, 180
McLough Steel, 398
Macroplanning, 421–424
Macworld (magazine), 249
Maids, 355
Maidu tribe, 11
Mail-order houses, 342
Making out on the shop floor, 189
Malcolm X, 68
Management
    from the rear, 248
    top-heavy, 190
Management information system (MIS),
    237, 238–239
Management practices, 364
Management-controlled firms, 383
Managerial control in professions,
    298–300
Managers, 307–327
    administrators, 309
    checklist for, 320–321
    definition, 308–309
    demand for, 309–310
    environmental dimensions,
        316–317
    and ethics, 315
    executives, 308, 384
    failures of, 324–325
    future of, 324–325
    the managerial career, 313–315
    middle, 238–239, 312, 318
    performance tracking, 320–324
    and scale dimensions, 316
    self-employed workers, 310–311
    specialization, 317
    staff and line managers, 309, 318
    supply of, 311–313
    and technology, 319–320
    women and minorities as, 311–312
Mandatory retirement, 74

Manufacturing, 173, 199, 210–224
    automobile industry, 215–217
    craft workers, 210–211
    declining middle class, 223
    division of labor, 182
    downsizing and flexibility, 222
    externalization, 221–222
    global competition, 220–224
    machine operators and assemblers,
        211–213
    plant closings, 221, 222
    steel industry, 217–218
    textile industry, 218–220
    unexplored alternatives, 223–224
    unskilled labor, 213–214
    working-class culture, 215
Maquiladoras, 414–415
Marcos, Ferdinand, 416
Marginal jobs, 352–374, 444–448
    age, 369
    by employment contract, 365–366
    by firm, 364–365
    by industry, 363–364
    contingent work, 355–356
    definition, 353
    disabling conditions, 368
    and dual labor markets, 114–115,
        363, 372
    educational level, 368
    farm laborers, 358–359
    and foreign competition, 363, 373
    future of, 372–373
    gender, 370
    geographic isolation, 366–368
    illegal or morally suspect
        occupations, 353
    internal labor markets, 373
    job displacement, 368–369
    low occupational power, 357–358
    pay and benefits, 444–447
    private household workers, 359,
        360–361
    race and ethnicity, 370
    reduction of, 450
    service workers, 359–360
    small workplaces, 357
    and social classes, 371–372
    and technology, 356, 373
    unattractive work, 357
    underemployment, 356
    and unions, 359, 361
    unregulated work, 353–355
    See also Temporary work
Margolis, Diane, 313–314
Market-driven behavior, 325
Markets, 7
Married women, 116, 368
Married-couple families median
    income, 59, 60
Mars, Gerald, 106
Marshall Plan, 407

Marx, Karl, 17, 20, 215
    alienation of workers, 90–92, 93, 107
Mary Kay Cosmetics, 345, 346
Maslow, Abraham, 92–93
Mass media, 68
Mass production, 27, 96
Mass psychogenic illness, 132
Mass strikes, 144, 146
Mass-production, 150
Mass-production technology, 178
Massachusetts Computer, 248
Master's degree in business
    administration (MBA), 312, 318
Materials, 176
Maternity leaves, 86
Matrix organization, 186
May Day (1886), 144–145
Mayo, Elton, 28, 184
MBA. *See* Master's degree in business
    administration
Meaney, George, 153
Meaning of work, 246
Meaningful work. *See* Job satisfaction
Means, Gardiner, 383
Median income for families by race and
    ethnicity (1999), 59, 60
Medicaid, 290
Medical doctors, 296
Medicare, 290, 298
Medicine man, 10
Megamergers, 386–388
Memorex, 227
Men
    barriers in corporate careers, 314
    and computer science, 334
Men and Women of the Corporation
    (Kanter), 38, 39
Mentors, 314
Mercantilism, 23
Mercedes-Benz, 419
Merchant capitalism, 17–20
Merchants, 16, 18, 341
Mergers, 190, 310
Merit systems, 114
Meritocracies, 292–294
Mi Familia (film), 432
Microchip technology, 178–179
Microcomputers, 337
Microprocessor technologies, 228–234
Middle Ages, 14–21
Middle class, 223, 371, 372
Middle East oil cartel, 387
Middle management, 238–239, 312,
    316, 318, 444
Middle-aged workers, 74, 84, 233
Migrant farm workers, 358
Mill workers, 219
Mills, C. Wright, *Sociological Imagination,
    The,* 110–111
Mini-mills, 218
Minimum wage, 70, 71, 347

Mining industry, 206–208
  occupational solidarity, 207–208
  strip mining, 208
Minorities, 5
  Appendix table, 279, 455–460
  and deviant occupations, 353
  disadvantages of, 446–448
  discrimination in hiring, 111–118
  discrimination in pay and promotions, 118–119
  in management, 311–312
  and occupational steering, 68
  in paid labor, 435–436
  in the paraprofessions, 302
  in professions, 292, 293
  in the semiprofessions, 300, 301
  in service industries, 272
  in unskilled jobs, 214
  women as clerical workers, 333–334
MIS. *See* Management information system
Mitchell, John, 409
Mixed-effects position in high-technology workplace, 231–233
Mobil Oil, 385, 387
Mobility. *See* Occupational mobility
Modern versus traditional societies, 8
Modernization theory of industrial development, 403
Mommy track, 311
Mongolian tribes, 15
Monopolies, 146, 385
Monopolizing knowledge, 288–289
Monopoly capitalism, 26–28
Montgomery Ward, 26, 333, 385
"Moonlighting," 84
Moore, Michael, 139
Morally suspect occupations, 353
Morrill, Calvin, 315
Motivation to work, 10, 30–31
Mr. Holland's Opus (film), 109
Mrs. Doubtfire (film), 88
Mueller, K. Alex, 299
Multidivisional firms, 190
Multinational corporations (MNCs), 54, 363, 408–410, 415
Multiplant companies, 397
Multiple life roles, 77
Multiple regression, 52–53
Muse Air, 387
Muslims, 294
Mystification and the professions, 289

N

NAFTA. *See* North American Free Trade Agreement
National Association of Working Women, 133
National Commission for Employment, 125
National Commission on Employment and Unemployment Statistics, 41

National Education Association (NEA), 142, 154
National health care plan, 444–445
National Labor Relations Act of 1935, 151–152, 158
National Labor Relations Board, 392
National Labor Union (NLU), 144
National Longitudinal Surveys (NLS), United States Bureau of Labor Statistics, 39
National Opinion Research Center (NORC) scale, 50
National Organizations Survey, 40
National Science Foundation, 398
NC. *See* Numeric control
NEA. *See* National Education Association
Neo-imperialism, 407–408
Netherlands, 406
Netscape, 190
New jobs, 398
New manufactured products, 257
"New middle class," 303
New York Stock Exchange, 238
NILF (not in the labor force), 41, 45
Nippon Steel, 217
Nissan, 216
Nobel Prize for physics, 299
Noblesse oblige, 287
Noise, 129, 130
Nonparticipant observations, 35–37
NORC scale. *See* National Opinion Research Center (NORC) scale
Norma Rae (film), 169
Normlessness, 92
Norms, 35–36, 62
Norris-LaGuardia Act of 1932, 149
North American Free Trade Agreement (NAFTA), 413
North American labor history, 142–156
  American Federation of Labor (AFL), 149–150
  Congress of Industrial Organizations (CIO), 150–152, 156
  craft unionism, 149–150
  early national unions, 144–146
  farm workers, 155–156
  general unions, 146–149
  industrial unionism, 150–152, 156
  Industrial Workers of the World (Wobblies), 147–149
  Knights of Labor, 146
  legislative gains, 151–152
  local craft unions, 143
  McCarthyism and right-wing attacks, 152–153
  May Day (1886), 144–145
  postwar period, 152–153
  professional workers, 154–155
  public-sector unions, 154
  Pullman Company strike, 145–146
  racial equality, 153

  railroads, 144
  Taft-Hartley amendments (1947), 152, 161
  women in unions, 154
  workers' political parties, 143–144
  See also Unions
Northwest Airlines, 387
Not in the labor force (NILF), 41, 45
Nuclear families, 65
Numeric control (NC), 237, 241, 246
Nurses, 300
Nursing assistants, 357

O

O★Net, United States Department of Labor, 50
Occupational communities, 207–208
Occupational differences in unemployment, 125
Occupational diseases, 130
Occupational inheritance, 66, 371–372
Occupational mobility, 72–76
  alternative cycles, 76
  career shifts, 73–74
  downward mobility, 73
  opportunity structure, 72–73
  upward mobility, 73
  See also Retirement
Occupational power, 357–358
Occupational prestige, 50, 52
Occupational Safety and Health Administration (OSHA), 130, 134
Occupational segregation, 112
Occupational steering, 67, 68
Occupational transformation in the United States, 277–278
Occupations
  as analysis units, 52–53
  appendix table, 455–460
  fastest growing and rapidly declining, 48
  and industries, 200
  major occupational groups, 49
Office automation, 238
Office boys, 330
Office equipment, 331
Office Space (film), 453
Office technology, 335–338
"Offshore" telecommuting, 240–241
"Old boy" networks, 190
Older people
  hiring of, 75, 370
  lay-offs of, 75
  See also Age; Retirement
Oligopolies, 26, 363
Olmos, Edward James, 432
On-the-job training, 71, 181
OPEC. *See* Organization of Petroleum Exporting Countries
Operations technology, 176
Opportunity structure, 72–73
Orderlies, 357

Orderly career, 64
Organization of Petroleum Exporting Countries (OPEC), 408, 414
Organization of work, 4–6, 8
time line, 18
Organizational changes, 310
Organizational climate, 323
Organizational culture approach to managerial performance, 322–323
Organizational dynamics in the high-technology workplace, 246–250
Organizational size, 7
Organizational structure, 177–178
and determinism, 195
influencing work, 181–185
in job satisfaction, 97–98
as labor control, 182–184
Orientation, 71
Orwell, George, 3
OSHA. *See* Occupational Safety and Health Administration
Osteopathic doctors, 296
Outsourcing, 394–396
Overflow work, 70
Overseas Private Investment Corporation, 427
Ozark Airlines, 387

P

Pacific Rim, 424
Paid production of services, 257
Pan American Airways, 387
Paper entrepreneurialism, 220
Paper mills, 229
Paralegals, 302
Paramedics, 302
Paraprofessions, 302
Parent companies, 54
Parents, elderly, 84, 85
Parker, Robert, 365, 366
Parliamentary systems, 144
Parsons, Albert, 145
Part-time work, 70
involuntary, 356
and pensions, 74
for retirees, 76
in sales, 344, 345
for women, 83
Participant observation, 35
Participation continuum, 445
Participation and job satisfaction, 98–99
Patch Adams (film), 306
Paternity leaves, 86
Paul, Jemoschick, 250
Pay, 97
discrimination in, 118–122
in marginal jobs, 444–446
Payment in kind, 360
Peddlers, 342
Peer support and solidarity, 95
People's Express, 387
Perestroika (restructuring), 425

Perez, Lupita, 263
Perfect Storm, The (film), 225
Performance tracking of managers, 320–324
Peripheral jobs, 362
Peripheral nations, 405
Peripheral support functions, 395
Perks (perquisites), 70
Personal services, 172, 264–265, 311
Personal shopping, 347
Personal service workers, 359
Peters, Tom, *Circle of Innovation, The,* 243
Petrochemicals, 26
Peugeot-Citroen, 418
Pharmacists, 300
Pheanis, Robert, 442
Philippines, 416
Photocopying, 331, 337, 340
Physiological needs, 92
Piece-rate basis, 240
Piece-rate earnings, 22–23
Piedmont Airlines, 387
Piel, Gerald, 228
Pillsbury, 54
Pink-collar jobs, 47, 49
Pinkerton detective agency, 149
Pirate capitalism, 426
Plant closings, 221, 222
Player Piano (Vonnegut), 241, 242
Plumbers, Electricians, and Sheet Metal Workers, 153
Plutonium contamination, 136
Plywood industry, 99, 194
Police brutality against strikers, 147
Police work, 264
Policewomen, sexual harassment of, 123
Political parties, 143–144
Political power of corporations, 379
Pollution, 135–136, 206
Population in sampling theory, 39
Postbureaucratic, 437
Postindustrial society, 28–32, 199–200, 254
Postprofessionalism, 297
Poverty in single-parent families, 78
Powderly, Terence, 146
Power
and the professions, 288–290
without boundaries, 409–410
of workers, 434, 448–449
Powerlessness, 92
Practical information, 284
"Praise-addiction," 38, 39
Precision Robotics, 248
Predatory pricing, 381
Pregnancy, 116
Pregnancy leave, 120
Preindustrial society, 282, 283
Prevailing wage, 121
Pride in work, 95, 104
Priests, 329
Primitive accumulation, 406

Princeton Packing, Inc., 157
Printing industry, 96, 242
Privacy for Consumers and Workers Act, 249
Private household workers, 359, 360–361
Privatization, 426
Pro bono publico work, 287
Producer cooperatives, 98–99
Producer services, 261–262
Product design, 247
Product marketing, 342–345
Productivity, 76
Professional associations, 40, 284, 295
as analysis units, 54
and disseminating information, 297
Professional culture, 284–285
Professional managers, 318
Professional schools, 288–289
Professional services, 172, 261, 283
Professional workers, 154–155
Professionalization, 282, 295–297, 334
Professions, 6, 30, 281–306
abstract, specialized knowledge, 283–285
altruism, 287–288
authority, 285–287
autonomy, 285
changes in degrees of, 295–300
changes in, 290–292
and clerical workers, 329–330
corporate medicine, 291
definition, 282
deprofessionalization, 297–300
future of, 303–304
licensing, 289, 295
meritocracies, 292–294
monopolizing knowledge, 288–289
mystification, 289
paraprofessions, 302
power of, 288–290
professionalization, 282, 295–297
recognizing, 282–288
self-employed, 312
semiprofessions, 300–302
and Theory Z, 323
Profit distribution, 443
Prohibition, 353
Promotion policies, 98
Promotions, 71
discrimination in, 118–122
Prostitution, 55, 353
in Guatemala, 353, 354
Protectionism in the global economy, 412
Protestant work ethic, 20
Pruning period, 314
Public assistance, 359
Public education, 331
Public employment, 264
Public goods expansion, 451
Public policy and employment, 245–246
Public utilities, 243
Public-sector unions, 154, 163

Pullman Company, 145, 150
Pullman Company strike, 145–146
Pullman, Standard Inc., 393
Pulpwood industry, 204–205
Putting-out industry, 18–19

Q

Quality circles, 164–165, 323
Quality Control Circles, 421, 426, 438–440
Quality of Employment Surveys (QES), United States Bureau of Labor Statistics, 39
Quality of work life, 437, 450–451
Questionnaires by laptop computer, 42–43
Quitting jobs, 104–105

R

Race
 and household workers, 362
 and job satisfaction, 100
 and marginal jobs, 370
 median income for families by race, ethnicity, and gender, 59, 60
 and occupational steering, 68
 in professional workplace, 294
 and unions, 153
Railroads, 144, 241–242, 262
Ratio of supervisors to production workers, 177
Rationalization of clerical work, 338–339, 340
Raw materials, 200
RCA, 418
Reading a table, 44–45
Reagan, Ronald, 386, 393
Reliability, 34
Rechanneling managers, 314
Recruiting, 70, 71
Regional specialization, 17
Regulation of professions, 297–298
Reiter, Ester, 36
Release time for union business, 158
Renault, 396, 418
Reorganization of work, 331
Reorganizations, 310
Repetitive work, 211–213
Replacement jobs, 127
Republic Airlines, 387
Resistance in work, 105
Response error, 40
Restaurants, 256
 turnover of workers, 358
Retail trade, 172, 242
Retirement, 74–76, 85
 early, 75
 mandatory, 74, 75
 pension plans, 74
 preparation for, 75, 76
 and women, 75
 See also Older people

Retraining, 449
Reuther, Walter, 153
Riddell, Paul, 294
Right-to-know legislation, 134
Right-wing attacks, 152–153
Rights of property, 149
Ritzer, George, 291
Roberts, Julia, 57
Robot, 236
Robotics, 233, 434
Rockefeller family, 26, 392
Rockefeller, John D., 148, 394
Rodriguez, Victor, 192
Rodriquez, Mark Anthony, 192
Roger and Me (film), 139
Role conflict, 76–79
Role models, 68
Role overload, 76–79
Role specialization, 177
Roles, 62
Roman Empire, 14, 15
Roosevelt, Franklin D., 159
Roosevelt, Theodore, 207
Routinization of professional judgment, 303
Royal Northwest Mounted Police, 148
Ruling elite, 190–191
Rural areas, 367
Russia, 425–426
Russian Federation, 425–426

S

Sabbaticals, 76
Sabel, Charles, 398
Sabotage, 105
Safety and health in the workplace, 224
 and unions, 164, 167
Safety needs, 92
Sales workers
 demand for, 342–348
 direct sales, 345, 346
 future of, 349–350
 history of, 341–342
 knowledge base, 347
 number of jobs by year, 329
 product marketing, 342–345
 supply of, 348–349
 and technology, 349
 type of firm, 345–347
Sample of people, 39
Sample surveys, 39–40
Santa Clara County v. Southern Pacific Railroad Co., 382
Santa Cruz Electronic Export Processing Zone, 250–251
Satellite firms, 396
Scabs, 149
Scandinavian autonomous work groups, 419–421
School dropouts, 371
School-to-Work website, 69

Schools
 primary source of child care, 83
 as socializing agents, 66
Scientific American (magazine), 228
Scientific management, 27, 183, 184, 323
Seagram, 387
Sears, Roebuck, and Company, 26, 344
Seasonal work, 70, 76, 355
Secondary boycotts, 145, 163
Secondary labor markets, 372, 373
Secretaries. *See* Clerical workers
Sectoral transformation of the labor force, 258–259
Seeman, Melvin, 92, 93
Segmentation in the high-technology workplace, 244–245
Segregation, 113
Selected Characteristics of Occupations Defined in the Dictionary of Occupational Titles, 50
Self-actualization needs, 92
Self-actualization theories, 90, 92–93, 94, 104, 108
Self-direction, 93–95
Self-employed workers, 310–311
Self-estrangement, 92
Self-fulfilling prophecies, 371
Self-interest, 7
Self-service stores, 344
Semiconductor companies, 132
Semiprofessions, 300–302
Semiskilled jobs, 27, 211
Semiskilled machine operators, 210
Seniority, 75, 158, 233, 438
Seniority-based protection against layoffs, 153
Sequential life plan, 62, 76
Serfs, 15
Service industries, 29–30, 199, 254–275
 bus driver in Los Angeles, 263
 business services, 261
 characteristics of, 255–257
 compensation in services, 265–266
 definition, 255
 distributive services, 262
 domestic service in Lima, Peru, 260
 employers' role, 267–271
 fast food, 269
 flight attendants, 270
 future of, 273
 and high technology, 242–243
 increase in, 108
 insurance, 269
 and marginal jobs, 359–360
 personal services, 264–265
 producer services, 261–262
 professional services, 261
 sectoral transformation of the labor force, 258–259
 service interaction, 266–273
 social services, 262–264
 sources of demand for, 257–258

standards, 266–267
tertiarization, 259, 261
and theft, 105–107
workers' perspective, 271–273
Service interaction, 266–273
Service Workers International Union, 160, 165
SES scores. *See* Socioeconomic status (SES) scores
Severance pay, 127
Sexual blackmail, 122
Sexual harassment, 122–123, 218
Sexual orientation, 118
Shadow economy, 355
Shakespeare, William, *As You Like It,* 61
Shaking-out process, 201
Shaman, 10, 282
Sharecroppers, 358
Sheraton, 249
Sherman Antitrust Act of 1890, 149
Shop stewards, 158, 247
Shop-floor actions, 146
Short-term contracts, 355
Shorthand, 337
Shrinking middle class, 371, 372
Shultz, George, 393
SIC codes. *See* Standard Industrial Classification (SIC) codes
"Sick buildings," 133–134
Silicon Technology, 250
Silicon Valley, 226
Silkwood, Karen, 136
Simon, Paul, 249
Simple tool technology, 178
Simulations, 322
Single-parent families
by race, 77
in family life cycle, 65
mother-child versus father-child, 78
See also Families
Sit-down strikes, 150–151
Size of companies, 97
Skill requirements in the high-technology workplace, 228–234
Skill-upgrading thesis, 229–230
Skilled craft workers, 210
Skilled maintenance work, 238
Skilled trades, 25
Skills, 179–181
acquiring new skills, 180–181
Slave labor, 13–14
in the Roman mines, 14
in United States, 23
Small workplaces and marginal jobs, 357
Small-business owners and ethnicity, 310–311
Smith, Adam, 22, 182, 378
Smith, Margo, 260
Smuggling goods, 55
Social classes, 10
deeply divided, 451
and household workers, 362

and managers, 312
and marginal jobs, 371–372
structure of, 28, 30
Social control, 271
Social Darwinism, 25
Social division of labor, 182
Social isolation, 104
Social mobility, 6
Social organization of work. *See* Organization of work
Social programs, 220–221
Social relations of production, 5, 175–176
Social scientists, 34
Social Security, 75
avoidance in paying taxes for, 55, 361
and farm workers, 359
and unregulated work, 353, 355
Social services, 263
Social stratification, 8, 63
Social structure, 177
Social welfare, 262
Social workers, 300
Socialism, 107–108
Socialization
formal, 66–67
informal, 65–66
in the workplace, 67–69
Socializing agents, 65
Socioeconomic status (SES) scores, 52, 53
Sociological Imagination, The (Mills), 110–111
Soderberg, Steven, 57
Soldiering, 183
Solidarity, 143
"Solidarity Forever" (Chaplin), 141
Songs of the Workers, 141
Soul of a New Machine, The (Kidder), 235
South Africa, 418
Southern Christian Leadership Conference, 153
Southwest Airlines, 387
Southwest Railroad System, 146
Soviet Union, 425, 426
Spacey, Kevin, 351
Span of control, 71, 309
Spatial differentiation, 185
Specialty steel products, 218
Specialization, 91, 303
in clerical work, 330–331, 334
in less developed nations, 406–407
managers, 317
Speed versus quality, 212–213
Spenner, Kenneth, 232
Spies, 149, 161–162
Spinner, Francis Elias, 332
Spouse abuse, 104
Staff managers, 309, 318
Staff positions, 186
Stages of life (Erikson), 62, 63
Standard Industrial Classification (SIC) codes, 46–47, 48

Standard Oil, 26, 385, 392, 394
Standardization, 177, 295
Standards in service industries, 266–267
Star Trek, 253
State-regulated capitalism, 417–418
Statistical discrimination, 115–116
Steel, 22
Steel industry, 26, 217–218, 398, 399
Steepness of hierarchy, 309–310
Steinbeck, John, *The Grapes of Wrath,* 374
Stereotypes, 371
Stigma in jobs, 265
Stiller, Ben, 275
Stockholders, 382
Stone, Peter, 128
Stressful jobs, 134–135
Strikebreakers, 149
Strikes, 149–150, 158–159
Strip mining, 208
Stripping, 353
Structural ambiguities, 7
Structural unemployment, 125
Students in temporary work, 365
Subcontracting, 240, 364, 394–396
Subjective screening criteria, 115
Subjective theory of alienation, 92, 93
Submerged economy, 355
Subminimum wage, 369
Subsidiaries, 54, 390
Substance abuse, 104
Substitute teachers, 365
Summer jobs, 70
Supermarkets, 347, 348
Supplemental unemployment benefits, 127
Surplus of food, 11–12
Surveillance, electronic, 248, 249
Survey of Income and Program Participation (SIPP), United States Bureau of the Census, 39
Sutherland, Duchess of, 20
Sweatshops, 18, 19, 129, 219–220, 361, 446
Sweden's "safety stewards," 134
Switchboard operators, 331, 339
Sylvis, Bill, 144
Sympathy strikes, 163

T

Tables, how to read, 44–45
Taft-Hartley amendments (1947), 152, 161
Tandon Magnetics, 250, 251
Tax avoidance, 55, 361
Taxi drivers, 262
Taylor, Frederick, 27, 183, 184, 193
Teachers, 296, 300, 365
Teamsters for a Democratic Union (TDU), International Brotherhood of Teamsters, 167
Teamwork systems, 421, 437–440
Technical control, 183

Technical workers, 239
Technique, 284
Technological displacement, 241
Technology, 5, 8
  and clerical workers, 238, 341
  defining, 176–178
  and determinism, 195
  and industries, 171–177
  influencing work, 178–181
  and job satisfaction, 96–97
  and managers, 319–320
  and marginal jobs, 356, 373
  office, 335–338
  See also High-technology workplace
Teenage workers, 85
Telecommunications networks, 173, 433
Telecommuting, 239–241, 240, 340
Teledyne, 248
Telemarketing, 342, 345
Telephone industry, 241
Telephone operators, 135
Telephones, 330, 335
Temporary work, 245, 279, 344
  and marginal jobs, 355, 365, 366–367
Tendonitis, 129
Tenure and job satisfaction, 100
Tenured positions, 355
Tertiarization, 259, 261
Texas Air, 387
Texas Department of Agriculture, 136
Textile industry, 218–220, 364
Textile strike in Lawrence,
  Massachusetts, 147
Textiles and the Industrial Revolution, 22
Theft, 105–107
Theoretical knowledge, 283
Theory X and Theory Y, 93, 193,
  322–323
Theory Z, 193, 322–323
Third parties, 290
Third political parties, 144
Third World nations, 220
Thompson Electronics, 418
"Tie-ins," 54
"Time poverty," 433
Time-boundedness of services, 255–256
Time-Warner, 190, 387
Toffler, Alvin, 226
Tokenism, 119
Top management, 316
Top-heavy management, 190
Topless dancers, 166
Totems, 10
Toxic work environment, 135
Toyota, 164, 216
Trade Adjustment Act, 223
Trade associations, 40
  as analysis units, 54
Trade unions. *See* Unions
Trading blocks in the global economy, 413
Traditional versus modern societies, 8

Training, 443, 449
  for changing skill requirements,
    233–234
  for jobs, 67
  on-the-job, 71
  in services, 268–271
Trans Union, 393
Transfer prices, 414
Transportation, 173
Trial, The (film), 197
Triangle fire, 147
Truck drivers, 262
Tuna fishing, 205
Turnover, 70, 104–105
TWA, 387
Two-career couples, 368
Typewriters, 331, 332, 335, 337, 339

U

UAW. *See* United Auto Workers
Ubhey, Sarabjit, 294
Unattractive work, 357
Underclass, 371, 446
Underemployment, 55, 356
Unemployed people, 41
Unemployment, 124–128
  in clerical work, 331
  coping with, 127–128
  differences between social groups, 125
  legislative controls, 127–128
  occupational differences in, 125
  in service industries, 257
  See also Layoffs
Unemployment compensation, 127
Unemployment rates, 41–42, 310
Union of American Physicians and
  Dentists, 155
Union Carbide Corporation, 136
Union Carbide plant in Bhopal,
  India, 410
Union shop contracts, 152
Unions, 449
  as analysis units, 54
  collective bargaining, 157–158
  company resistance, 161–162
  current union roles, 157–160
  growing and declining unions,
    160–163
  growth areas, 162
  and high-technology workplace,
    250–251
  image of unions, 166–167
  industrial shifts, 161
  Industrial Workers of the World,
    147–149
  innovative programs for the 2000s,
    163–167
  internal democracy of, 159–160, 167
  international comparisons, 162–163
  international competition, 161
  international labor solidarity, 427–429

  in Japan, 424
  and job satisfaction, 98
  Knights of Labor, 146
  largest in the U.S., 143
  lobbying, 159
  low-wage workers, 165
  and marginal jobs, 359, 361
  membership concerns, 159–160
  membership in, 142
  membership trends, 160
  national legislative agenda, 163–164
  need for, 141–142
  origins of, 25–26, 28
  racially separate, 113
  strikes, 158–159
  and subcontracting, 395
  in textile industry, 219
  for topless dancers, 166
  See also North American labor history
United Airlines, 387
United Auto Workers (UAW), 151,
  160, 164, 217, 428
  and Ford, 443
  New Directions Movement, 167
United Farm Workers, 156, 203,
  204, 359
United Food and Commercial
  Workers, 162
United Fruit, 385
United Mine Workers, 150
United Nations, International Labour
  Office, 50
United States
  economic dominance, 411–412
  in the global economy, 407–408
  occupational transformation in,
    277–278
  political system of, 144
United States Bureau of the Census
  Current Population Survey (CPS),
    39, 41, 42–43
  major occupational groups, 49
  Survey of Income and Program
    Participation (SIPP), 39
  unemployed people, 46
United States Bureau of Labor Statistics
  Current Population Survey (CPS),
    39, 41, 42–43
  on discouraged workers, 45
  earnings in industries, 46
  National Longitudinal Surveys
    (NLS), 39
  Quality of Employment Surveys
    (QES), 39
United States Department of
  Agriculture, 202, 204
United States Department of Labor
  Dictionary of Occupational Titles
    (DOT), 50–53
  Employment and Training
    Administration, 50

O★Net, 50
*Selected Characteristics of Occupations
Defined in the Dictionary of
Occupational Titles,* 50
United States Steel (now USX),
217, 385
United Steelworkers of America
(USWA), 151
United Students Against Sweatshops, 220
United Way, 167
Universal Mechanical and Electrical
Components, 115
Universal public education, 331
University professors, 355
Unpaid housework, 361
Unregulated work and marginal jobs,
353–355
Unskilled labor, 213–214
Unstable work, 360
Upper class, 190–191
Upward mobility, 73, 333
Urban artisans, 17
U.S. Steel Gary Works plant, 441, 442
USWA. *See* United Steelworkers of
America
USX (formerly United States Steel),
217, 385
Utilities, 173

V

Vacations, 76
Vagabondage, 20–21
Validity, 34
Varian Associates, 132
Vaughan, Diane, 136
VDT. *See* Video display terminal
Vertical differentiation, 185
Vertical dimensions of bureaucracy, 316
Vertical integration, 381, 385
Vertical mobility. *See* Occupational
mobility
Vesting, 74
Vibrating tools, 129
Video display terminals (VDTs), 237
hazardous work, 132–133, 164
Video game production, 235, 236
Virginia Company, 23
Vocational education, 180–181, 449
Voice recognition systems, 338
Volunteer work, 44, 85
Volvo, 99, 420
Vonnegut, Kurt, Jr.
Breakfast of Champions, 327
Player Piano, 241, 242

W

Wage labor, 18
Wage scale of unions, 157
Waitpersons, 272
Wal-Mart, 244, 379, 380, 397
Walczak, David, 291

Wallace, Glenn K., 202
Walters, Bob, 129
Warner Brothers, 190
Washington State comparable worth
study, 121
WCL. *See* World Confederation of
Labour
Weapons, 55
Web-based surveys, 40
Weber, Max, 17, 185–186
Welfare, 127
Welfare reform, 372
Wells, Orson, 197
Western Union, 26
Westinghouse Electric Company, 26,
184–185, 385
WFTU. *See* World Federation of Trade
Unions
Whistle-blowing, 193
White backlash, 114
White-collar workers, 27, 84, 227
African Americans as, 333
changes in, 278
clerical and sales, 328
and high technology, 241,
242–243, 246
strikes by, 154
White-collar-blue-collar division, 47,
49, 53
Whites
median income for families by race,
ethnicity, and gender, 59, 60
Protestant men, 311
unemployment rates, 124
Wholesale trade, 172
Williams, Robin, 88, 306
Willis, Bruce, 327
Winnipeg Trades and Labour
Council, 148
Wobblies. *See* Industrial Workers of the
World
Women
appendix table, 279, 455–460
barriers in corporate careers, 314
and child rearing, 62
comparable worth, 120–122
in construction work, 209
disadvantages of, 446–448
educational attainment of, 118
federal employment by Civil Service
grade, 117
feminization of clerical workers, 330,
331–333, 334
in the guild system, 20
home duties, 119–120
home job ghetto, 240
and Industrial Revolution, 23–24
in involuntary part-time work, 370
labor force participation, 5, 41, 82
liberation of, 31
in the Lowell factories, 24

in management, 311–312
married, 116, 368
middle-class, 333
minorities as clerical workers,
333–334
in occupations, 113
in paid labor, 435–436
in the paraprofessions, 302
poverty rate, 119
pregnancy leave, 120
in professions, 293
and retirement, 75
in sales, 348
sectoral shift of labor force, 258–259
in the semiprofessions, 300, 301
in steel industry, 218
stereotypical jobs for, 67, 111, 113
in temporary work, 365
in top management positions, 116
and unions, 146, 147, 154
in unskilled jobs, 214
See also Gender; Homemakers
Women-supported families median
income, 59, 60
Woolnough, Roisin, 294
Word processors, 331, 334, 337, 339, 340
Work
behavioral responses, 103–107
changes in, 4–8
consequences for individuals, 6–7
consequences for society, 7–8
definition of, 3–4
future of, 31–32
history of, 8–32
ideologies, 103
meaning and dignity in, 6, 246
problems in studying, 54–55
trends in, 434–436, 438–439
Work arrangements among couples, 79
Work ethic, 6, 20
Work flow, 70
Work groups, 419–421, 437–438
Work groups at the Kalmar Volvo
plant, 420
Work practice innovations, 421–422
Work practices around the globe, 415–429
Canada, 417
China, 424
competing organizational forms,
426–427
developing nations, 416–417
Eastern Europe and Russia, 425–426
"four tigers," 424–425
France, 417–418
German codetermination, 418–419
hourly wages, 417
international labor solidarity, 427–429
Japan's macroplanning, 421–424
least developed nations, 415–416
Scandinavian autonomous work
groups, 419–421

South Africa, 418
state-regulated capitalism, 417–418
See also Global economy
Work regulations, 254
Work sharing, 86
Worker education, 437
Worker ownership, 441–442
Worker participation, 437
Workers
  as analysis units, 40–46
  average hourly earnings of, 266
  direct participation by, 193–195
  discouraged, 44–45, 126
  managing diversity, 124
  participation of, 445
  power of, 434, 448–449
  rediscovering, 184–185

resistance of, 183–184
in service industries, 271–273
standards in services, 267
training for, 443
Workers' political parties, 143–144
Working ahead on the line, 213
Working at home, 61–62
Working years, 69–76
Working-class culture, 215
Working-class ideology, 103
Workplace as analysis unit, 54
Workplace experimentation, 449–450
Works councils, 418
World Bank, 224
World Confederation of Labour
  (WCL), 427
World economy. *See* Global economy

World Federation of Trade Unions
  (WFTU), 427
World labor confederations, 427–429
World systems theory of industrial
  development, 405–406
World Trade Organization (WTO), 159,
  224, 412

Y
Yellow-dog contracts, 149
Young, William, 136

Z
Zeitlin, Jonathan, 398
Zeitlin, Maurice, 383
Zuboff, Shoshana, 230